Advances in Intelligent Systems and Computing

Volume 1119

The series "Advances in Intelligent Systems and Computing" contains publications on theory, applications, and design methods of Intelligent Systems and Intelligent Computing. Virtually all disciplines such as engineering, natural sciences, computer and information science, ICT, economics, business, e-commerce, environment, healthcare, life science are covered. The list of topics spans all the areas of modern intelligent systems and computing such as: computational intelligence, soft computing including neural networks, fuzzy systems, evolutionary computing and the fusion of these paradigms, social intelligence, ambient intelligence, computational neuroscience, artificial life, virtual worlds and society, cognitive science and systems, Perception and Vision, DNA and immune based systems, self-organizing and adaptive systems, e-Learning and teaching, human-centered and human-centric computing, recommender systems, intelligent control, robotics and mechatronics including human-machine teaming, knowledge-based paradigms, learning paradigms, machine ethics, intelligent data analysis, knowledge management, intelligent agents, intelligent decision making and support, intelligent network security, trust management, interactive entertainment, Web intelligence and multimedia.

The publications within "Advances in Intelligent Systems and Computing" are primarily proceedings of important conferences, symposia and congresses. They cover significant recent developments in the field, both of a foundational and applicable character. An important characteristic feature of the series is the short publication time and world-wide distribution. This permits a rapid and broad dissemination of research results.

**** Indexing: The books of this series are submitted to ISI Proceedings, EI-Compendex, DBLP, SCOPUS, Google Scholar and Springerlink ****

More information about this series at http://www.springer.com/series/11156

Himansu Das · Prasant Kumar Pattnaik ·
Siddharth Swarup Rautaray · Kuan-Ching Li
Editors

Progress in Computing, Analytics and Networking

Proceedings of ICCAN 2019

Springer

Editors
Himansu Das
School of Computer Engineering
KIIT Deemed to be University
Bhubaneswar, Odisha, India

Prasant Kumar Pattnaik
School of Computer Engineering
KIIT Deemed to be University
Bhubaneswar, Odisha, India

Siddharth Swarup Rautaray
School of Computer Engineering
KIIT Deemed to be University
Bhubaneswar, Odisha, India

Kuan-Ching Li
Department of Computer Science
and Information Engineering (CSIE)
Providence University
Taichung, Taiwan

ISSN 2194-5357 ISSN 2194-5365 (electronic)
Advances in Intelligent Systems and Computing
ISBN 978-981-15-2413-4 ISBN 978-981-15-2414-1 (eBook)
https://doi.org/10.1007/978-981-15-2414-1

This Springer imprint is published by the registered company Springer Nature Singapore Pte Ltd.
The registered company address is: 152 Beach Road, #21-01/04 Gateway East, Singapore 189721, Singapore

Committees

Advisory Committee

Rajkumar Buyya, FIEEE, The University of Melbourne, Australia
Prasant Mohapatra, FIEEE, University of California, USA
Mohammad S. Obaidat, FIEEE, Fordham University, USA
Vincenzo Piuri, FIEEE, University of Milano, Italy
Carlos A. Coello Coello, FIEEE, CINVESTAV, Mexico
Shekhar Verma, IIIT Allahabad, India
Valentina Emilia Balas, Aurel Vlaicu University of Arad, Romania
Tirthankar Gayen, Jawaharlal Nehru University, India
Dilip Kumar Partihary, IIT Kharagpur, India
Sasmita Samanta, KIIT University, India
Joao Manuel R. S. Tavares, Universidade do Porto, Portugal
Abhaprakash Praharaj, Tata Consultancy Services Limited, Germany
Nilanjan Dey, Techno India College of Technology, India
Sanjeevikumar Padmanaban, University of Johannesburg, South Africa
Ganapati Panda, IIT Bhubaneswar, India
Kenji Suzuki, Illinois Institute of Technology, USA
Pabitra Pal Choudhury, ISI Kolkata, India
Andrey V. Savchenko, National Research University Higher School of Economics, Russia
Rajib Mall, IIT Kharagpur, India
Kuan-Ching Li, Providence University, Taiwan
Bidyut B. Chaudhuri, LFIEEE, Indian Statistical Institute Kolkata, India
Goutam Chakraborty, Iwate Prefectural University, Japan
Pramod Kumar Meher, NTU, Singapore
Ishwar K. Sethi, Oakland University, Rochester, MI
Raj Jain, LFIEEE, Washington University, USA
P. N. Suganthan, NTU, Singapore
Sagar Naik, University of Waterloo, Canada

Ujjwal Maulik, FNAE, Jadavpur University, India
Dhiya AL-Jumeily, Liverpool John Moores University, UK
Anupam Agrawal, IIIT Allahabad, India
Hugo Pedro Proenca, University of Beira Interior, Portugal
L. M. Patnaik, INSA Senior Scientist, IISc Bangalore, India
Hrushikesha Mohanty, KIIT University, India
Tandra Pal, NIT Durgapur, India
Xiao-Zhi Gao, Aalto University, Finland
Prabhat Kumar Mahanti, University of New Brunswick, Canada
Swapan Bhattacharya, Jadavpur University, India
Prashant Pillai, University of Bradford, UK
Arun K. Somani, Iowa State University, Ames, IA
R. V. Rajakumar, Director, IIT Bhubaneswar, India
Ashok Deshpande, University of California, Berkeley, USA
Amitava Chatterjee, Jadavpur University, India
Malay K. Kundu, ISI Kolkata, India
Sushil K. Prasad, Georgia State University, USA
Nabanita Das, ISI Kolkata, India
B. K. Panigrahi, IIT Delhi, India

Program Committee

Wenwu Wang, University of Surrey, UK
Padmalochan Bera, IIT Bhubaneswar, India
Ashutosh Bhatia, BITS Pilani, India
Pierluigi Siano, Universita degli Studi di Salerno, Italy
D. Janakiram, IIT Madras, India
Roberto Caldelli, Universita degli Studi Firenze, Italy
Satchidananda Dehuri, Fakir Mohan University, India
C. W. Chow, National Chiao Tung University, Taiwan
Mohd Helmy Abd Wahab, Universiti Tun Hussein Onn, Malaysia
Kuntal Ghosh, ISI Kolkata, India
G. C. Deka, Ministry of Skill Development and Entrepreneurship, India
Chinmaya Mahapatra, University of British Columbia, Canada
M. S. Kakkasageri, Basaveshwar Engineering College (Autonomous), India
Y-Chuang Chen, Minghsin University of Science and Technology, Taiwan
Bighnaraj Naik, VSSUT, India
Prasant Kumar Pattnaik, KIIT University, India
Aniket Mahanti, University of Auckland, New Zealand
Dilip Ranjan Mohanty, Wellington Management, Massachusetts, USA
S. S. Rautaray, KIIT University, India
Qurban A. Memon, UAE University
Hajjam El Hassani, Universite de Bourgogne Franche Comte, France

Anjali Mathur, Jaipur National University, India
B. K. Rout, FIE, Birla Institute of Technology and Science, Pilani, India
Somanath Tripathy, IIT Patna, India
Veena Goswami, KIIT University, India
Gautam K. Das, IIT Guwahati, India
Hemant Kumar Rath, TCS Research and Innovation, Singapore
Ruggero Donida Labati, Universita' degli Studi di Milano, Italy
Gayadhar Panda, NIT Meghalaya, India
Ala' Aburumman, University of South Australia, Australia
Himansu Das, KIIT Deemed to be University, India
Gunter Fahrnberger, University of Hagen, Germany
Venkata Rao Kasukurthi, Andhra University College of Engineering, India
Pratyay Kuila, NIT Sikkim, India
Babita Majhi, Central University Bilaspur, India
Durga Prasad Mohapatra, NIT Rourkela, India
Chandal Nahak, IIT Kharagpur, India
Gyanendra Kr. Verma, NIT Kurukshetra, India
Pradheepkumar Singaravelu, NEC India Pvt Ltd, India
Marco Mussetta, Politecnico Di Milano, Italy
Indrajit Saha, NITTTR Kolkata, India
Vasanth Iyer, Florida International University, Warrensburg, MO
Felix Albu, Valahia University of Targoviste, Romania
Mohamed Aminc Ferrag, Guelma University, Algeria
Neha Sharma, ZIBACAR, India
Monish Chatterjee, Asansol Engineering College, India
Angelo Genovese, Universita degli Studi di Milano, Italy
Pankaj Kumar Sa, NIT Rourkela, India
Carlo Vallati, University of Pisa, Italy
Radu-Emil Precup, SMIEEE, Politehnica University of Timisoara, Romania
Bernd E. Wolfinger, University of Hamburg, Germany
Meenakshi D'Souza, IIIT Bangalore, India
P. G. Sapna, Coimbatore Institute of Technology, India
Vinod Kumar Singh, SRGI, Jhansi (U.P.), India
Sudhakar Pandey, NIT Raipur
Neeraj Kumar, Thapar University, Patiala, India
Natarajan Meghanathan, Jackson State University, USA
J. Mohanty, KIIT Deemed to be University, India
Malka Halgamuge, The University of Melbourne, Australia
Sarangapani Jagannathan, Missouri University of Science and Technology, USA
Alok kumar Jagadev, KIIT University, India
Subhankar Dhar, San Jose State University, USA
Santi Maity, IIEST, India
Srinivasan R, M.S. Ramaiah Institute of Technology, India

Annappa B., National Institute of Technology Karnataka Surathkal, India
Ranjit Rajak, Dr. HariSingh Gour Central University, India
K. Sridhar Patnaik, BIT, Mesra, Ranchi, India
Ali Mirza Mahmood, DMS SVH College of Engineering, India
G. Sahoo, BIT, Mesra, Ranchi, India
Sachin D. Ruikar, Walchand College of Engineering, Sangli, Maharashtra, India
Chandan Jyoti Kumar, Don Bosco College of Engineering and Technology, India
Sourodeep Bhattacharjee, CSIR-Central Mechanical Engineering Research Institute, India
K. M. Azharul Hasan, Khulna University of Engineering and Technology (KUET), Bangladesh
Kamaljit Lakhtaria, Sir Padampat Singhania University, India
Ujwala Baruah, NIT, Silchar, India
Samarjeet Borah, Sikkim Manipal Institute of Technology, India
Rabindra K. Barik, KIIT Deemed to be University, India
Subrat Kumar Pattanayak, National Institute of Technology, Raipur, India
Dibya Jyoti Bora, The Assam Kaziranga University, Jorhat, Assam
Jagannath Singh, KIIT Deemed to be University, India
Shamim H. Ripon, East West University, Dhaka, Bangladesh
Subhansu Bandyopadhyay, University of Calcutta, India
Virender Kadyan, Chitkara University, India
Deepak Ranjan Nayak, IIITDM Kanchipuram, India
Asis Tripathy, VIT Vellore, India
Abinash Tripathy, Raghu Engineering College, Visakhapatnam, India
Jyotsna Kumar Mandal, Kalyani University, India
Monalisa Biswal, National Institute of Technology, Raipur, India
Sanjeet Nayak, Manipal University Jaipur, India
Suvendu Nayak, C. V. Raman College of Engineering, Bhubanneswar, India
Ranjita Kumari Dash, PES University, Bengaluru, India

Chief Patron

Achyuta Samanta, KIIT Deemed to be University, India

Patron

Hrushikesha Mohanty, KIIT Deemed to be University, India
Sasmita Samanta, KIIT Deemed to be University, India
Jnyana Ranjan Mohanty, KIIT Deemed to be University, India

Honorary General Chairs

Rajkumar Buyya, The University of Melbourne, Australia
Xiao-Zhi Gao, Aalto University, Finland

General Chairs

Valentina Emilia Balas, Aurel Vlaicu University of Arad, Romania
Samaresh Mishra, KIIT Deemed to be University, India
Kuan-Ching Li, Providence University, Taiwan
Prasant Kumar Pattnaik, KIIT Deemed to be University, India

Program Chairs

Bighnaraj Naik, VSSUT, India
Jitendra Kumar Rout, KIIT Deemed to be University, India

Organizing Chair

Himansu Das, KIIT Deemed to be University, India
Siddharth S. Rautaray, KIIT Deemed to be University, India

Session Management Chair

Manjusha Pandey, KIIT Deemed to be University, India
Ajay Kumar Jena, KIIT Deemed to be University, India
Suneeta Mohanty, KIIT Deemed to be University, India
Minakhi Rout, KIIT Deemed to be University, India
Suchismita Das, KIIT Deemed to be University, India

Publication Chairs

Suresh Chandra Moharana, KIIT Deemed to be University, India
Ajaya Kumar Parida, KIIT Deemed to be University, India

Chittaranjan Pradhan, KIIT Deemed to be University, India
Mahendra Kumar Gourisaria, KIIT Deemed to be University, India

Publicity Chairs

Raghvendra Kumar Agrawal, Lakshmi Narain College of Technology, Bhopal, India
Krishna Chakravarty, KIIT Deemed to be University, India
Souvik Pal, Brainware University, India
Kunal Anand, KIIT Deemed to be University, India

Registration Chairs

Sital Dash, KIIT Deemed to be University, India
R. N. Ramakant Parida, KIIT Deemed to be University, India
Abhaya Kumar Sahoo, KIIT Deemed to be University, India
Roshni Pradhan, KIIT Deemed to be University, India

Preface

The 2nd International Conference on Computing Analytics and Networking (ICCAN 2019) took place in Bhubaneswar, India, during December 14–15, 2019. It was hosted by the School of Computer Engineering, Kalinga Institute of Industrial Technology (KIIT), Deemed to be University.

The ICCAN is a premier international open forum for scientists, researchers and technocrats in academia as well as in industries from different parts of the world to present, interact and exchange the state of art of concepts, prototypes, innovative research ideas in several diversified fields. The primary focus of the conference is to foster new and original research ideas and results in the three board tracks: computing, analytics and networking with its prospective applications in the various interdisciplinary domains of engineering. This is an exciting and emerging interdisciplinary area in which a wide range of theory and methodologies are being investigated and developed to tackle complex and challenging real-world problems. The conference includes invited keynote talks and oral paper presentations from both academia and industry to initiate and ignite our young minds in the meadow of momentous research and thereby enrich their existing knowledge.

ICCAN 2019 received more than 200 submissions. Each submission was reviewed by at least three Program Committee members. The committee decided to accept only 66 full papers. Papers were accepted on the basis of technical merit, presentation and relevance to the conference. ICCAN 2019 was enriched by the lectures and insights given by the following five distinguished invited speakers: Laxmi N. Bhuyan, Distinguished Professor, Department of Computer Science and Engineering, University of California, Riverside (UCR), California, USA; Rajkumar Buyya, Director, Cloud Computing and Distributed Systems (CLOUDS) Laboratory, The University of Melbourne, Australia; Kaushal Kumar Shukla, Professor, Department of Computer Science and Engineering, Indian Institute of Technology (BHU), Varanasi, India; Shekhar Verma, Professor, Indian Institute of Information Technology, Allahabad, U.P., India; and Deepak Puthal, School of Computing, Newcastle University, Newcastle upon Tyne, UK. We thank the invited speakers for sharing the enthusiasm for research and accepting our invitation to share their expertise as well as contributing papers for inclusion in the proceedings.

ICCAN 2019 has been able to maintain standards in terms of the quality of papers due to the contribution made by many stakeholders.

We are thankful to the General Chairs Valentina Emilia Balas, Aurel Vlaicu University of Arad, Romania, Samaresh Mishra, KIIT Deemed to be University, India, Kuan-Ching Li, Providence University, Taiwan, Prasant Kumar Pattnaik, KIIT Deemed to be University, India. We further thank the Program Chair Bighnaraj Naik, VSSUT, India, and Jitendra Kumar Rout, KIIT Deemed to be University, India, for their guidance and valuable inputs. We are grateful to Achyuta Samanta (Founder of KIIT and KISS) for his constant support for the ICCAN. We are thankful to higher authorities of KIIT specially Hrushikesha Mohanty, Vice Chancellor, Sasmita Samanta, Pro-Vice-Chancellor, Jnyana Ranjan Mohanty, Registrar, D. N. Dwivedy, Mentor, for providing the infrastructure and resources to organize the conference.

Thanks are due to the Advisory and Program Committee members for their guidance related to the conference. We would also like to thank the Session Management Chairs, Publications Chairs, Publicity Chairs, Registration Chairs and Web Management Chair who have made an invaluable contribution to the conference. We acknowledge the contribution of Easy Chair in enabling an efficient and effective way in the management of paper submissions, reviews and preparation of proceedings. Finally, we thank all the authors and participants for their enthusiastic support. We sincerely hope that you find the book to be of value in the pursuit of academic and professional excellence.

Bhubaneswar, India

Himansu Das
Siddharth Swarup Rautaray

Organizing Chair, ICCAN 2019

Acknowledgements

The making of this edited volume was like a journey that we had undertaken for several months. We wish to express our heartfelt gratitude to our families, friends, colleagues and well-wishers for their constant support throughout this journey. We express our gratitude to all the paper contributors, who allowed us to quote their remarks and work in this volume. In particular, we would like to acknowledge the hard work of authors and their cooperation during the revisions of their papers. We would also like to acknowledge the valuable comments of the reviewers which have enabled us to select these papers out of so many papers we received and also improve the quality of the papers. We wish to acknowledge and appreciate Springer team for their continuous support throughout the entire process of publication. Our gratitude is extended to the readers, who gave us their trust, and we hope this work guides and inspires them.

About Conference

The International Conference on Computing, Analytics and Networking (ICCAN 2019) is a premier international open forum for scientists, researchers and technocrats in academia as well as in industries from different parts of the world to present, interact and exchange the state of art of concepts, prototypes, innovative research ideas in several diversified fields. The primary focus of the conference is to foster new and original research ideas and results in the three board tracks: computing, analytics and networking with its prospective applications in the various interdisciplinary domains of engineering. This is an exciting and emerging interdisciplinary area in which a wide range of theory and methodologies are being investigated and developed to tackle complex and challenging real-world problems. The conference will include invited keynote talks and oral paper presentations from both academia and industry to initiate and ignite our young minds in the field of momentous research and thereby enrich their existing knowledge. We wish the participants will enrich their knowledge by new perspectives and views on current research topics from leading scientists, researchers and academicians around the globe.

Contents

About the Editors

Himansu Das is an as Assistant Professor at the School of Computer Engineering, KIIT University, Bhubaneswar, India. He holds B. Tech and M. Tech degrees from Biju Pattnaik University of Technology (BPUT), Odisha, India. He has published several research papers in various international journals and conferences and has also edited several international books. He is a member of the editorial and review boards of a number of journals and conferences and has served as the organizing chair, publicity chair and as a member of the program committees of several national and international conferences. He is also associated with various educational and research societies like IACSIT, ISTE, UACEE, CSI, IET, IAENG, and ISCA. His research interests include grid computing, cloud computing, and machine learning.

Prasant Kumar Pattnaik, PhD (Computer Science) fellow of the IETE, and senior member of the IEEE, is a Professor at the School of Computer Engineering, KIIT University, Bhubaneswar. He has more than a decade of teaching and research experience, and has published several papers in peer-reviewed international journals and conferences. His areas of specialization are mobile computing, cloud computing, brain computer interface and privacy preservation. He has also edited various volumes for Springer and IGI Global.

Siddharth Swarup Rautaray, PhD (Information Technology) and member of the IET, IEEE, ACM is an Assistant Professor at the School of Computer Engineering, KIIT University, Bhubaneswar. His research interests include big data analytics, human computer interaction and image processing. He has published research papers in various international journals and conferences. He has been an organizing chair/committee member for a number of international conferences, workshops and symposiums, and a reviewer for various journals and conferences.

Kuan-Ching Li is currently a Distinguished Professor at Providence University, Taiwan. He is a recipient of awards and funding support from several agencies and industrial companies, as he also received distinguished chair professorships from universities in China and other countries. Besides publication in refereed journals and top conference papers, he is co-author/co-editor of more than 25 technical professional books published by CRC Press/Taylor & Francis, Springer, and McGraw-Hill. Dr. Li's research interests include GPU/manycore computing, Big Data, and cloud. He is a senior member of the IEEE and a fellow of the IET.

Implementation of Session-to-Mobility Ratio Based Mobility Management Scheme for Wireless Mesh Network to Handle Internet and Intranet Packets

Abhishek Majumder and Sudipta Roy

Abstract IEEE 802.11s was designed to implement Wireless Mesh Network (WMN). Hybrid Wireless Mesh Protocol (HWMP) is the routing protocol used in IEEE 802.11s. But external station's (STA) mobility is not considered by IEEE 802.11s. Therefore, for handling both Internet and intranet packets a Session-to-Mobility Ratio (SMR)-based technique is implemented in this paper. The scheme efficiently handles mobility of external STA by reducing propagation of route management packets. Performance analysis and comparison of SMR-based scheme and IEEE 802.11s have been carried out. It has been observed that SMR-based scheme outperforms IEEE 802.11s.

Keywords Wireless mesh network · Session-to-mobility ratio · Mobility management · Portal

1 Introduction

For enabling the use of IEEE 802.11 in Wireless Mesh network (WMN) [1–4] IEEE 802.11s [5–9] has been introduced. WMN consists of portal, mesh station (STA), and external STA. Portal is a node that connects the network to the Internet. Mesh STAs are the nodes which are responsible to route the packets from source to the destination. The external STAs communicate with the WMN through the mesh gate associated with access point (APMG). It is desirable from mobility management technique to have less handoff cost for ensuring seamless handoff. This will enhance the performance of the network. Hybrid Wireless Mesh Protocol (HWMP) [8, 10–12] is the routing protocol used in IEEE 802.11s. But external STA's mobility issue has not been solved by IEEE 802.11s [13]. An IEEE 802.11s-based mobility management

A. Majumder (✉)
Department of Computer Science & Engineering, Tripura University, Suryamaninagar, India
e-mail: abhi2012@gmail.com

S. Roy
Department of Computer Science & Engineering, Assam University, Silchar, India

H. Das et al. (eds.), *Progress in Computing, Analytics and Networking*,
Advances in Intelligent Systems and Computing 1119,
https://doi.org/10.1007/978-981-15-2414-1_1

technique named adaptive mobility management scheme [14, 15] has been developed for improving performance of the network considering external STA's mobility. But it cannot handle the intranet traffic. Therefore, for handling both intranet and Internet traffic, in this paper an IEEE 802.11s-based mobility management scheme is implemented.

Section 2 discusses the existing technique. Section 3 presents the session-to-mobility ratio-based technique. Simulation results are discussed in Sect. 4 and it also discusses the comparison. In Sect. 5, the paper has been concluded.

2 Related Work

Many schemes have been proposed for mobility management of the clients [16–21] such as Mesh Mobility Management (M^3) [22], iMesh [23], MEMO [24], WMM [25] and IEEE 802.11s uses HWMP as routing protocol. The external STA sends upstream Internet packets to its host APMG. The host APMG then sends these packets toward portal. Portal receives downstream Internet packet and forwards it toward host APMG. External STA sends upstream intranet packets to its host APMG. The source external STA's host APMG then sends packets to host APMG of destination external STA. If route to the destination is not known for setting up the route, the host APMG of source external STA broadcasts route request message (PREQ). The destination sends back route reply (PREP) message. Proxy Update Confirmation (PXUC) and Proxy Update (PXU) are used to update proxy table. But external STA's mobility is not considered by IEEE 802.11s. An IEEE 802.11s-based mobility management scheme named adaptive mobility management scheme [14, 15] has been proposed and implemented for efficiently managing the mobility of external STA. But the adaptive mobility management scheme is not capable of handling the Internet packets.

3 Session-to-Mobility Ratio Based Mobility Management Scheme

This section presents detailed working of Session-to-Mobility Ration (SMR) based mobility management technique [26] and different tables and frames used for its implementation. It has mainly three parts: mobility management, route setup, and routing of data packets.

Three new fields have been added in proxy table, namely, proxy_ESTA_IP, proxy_part_of_forward_chain, and proxy_serving_APMG. Proxy_ESTA_IP stores the IP address of the external STA. Proxy_part_of_forward_chain is set to 1 if the current APMG is not serving APMG of corresponding external STA. Otherwise, it is set to 0. Proxy_serving_APMG stores the address of external STA's serving APMG. Two new fields have been added in the routing table, namely, is_mobile_station and part_of_forward_chain. Is_mobile_station field is equal to 1 if node is an external STA. Otherwise, it is set to 0. Part_of_forward_chain will be set to 1 if APMG is forward chain's part for corresponding external STA. A new table named corresponding node table has been introduced. The table has seven fields, namely, corresponding_serving_APMG_MAC, corresponding_ESTA_MAC, corresponding_ESTA_seq_num, corresponding_ESTA_IP, source_ESTA_MAC, source_ESTA_IP, and corresponding_ESTA_status. The fields corresponding_serving_APMG_MAC, Corresponding_ESTA_MAC, Corresponding_ESTA_seq_num, and Corresponding_ESTA_IP store the MAC address of the serving APMG, MAC address, sequence number, and IP address of the corresponding external STA, respectively. On the other hand, the fields source_ESTA_MAC and source_ESTA_IP store MAC address and IP address of the source external STA, respectively. Corresponding_ESTA_status field stores the status of the corresponding external STA. In the MAC header of re-association request frame two new fields, namely, SMR_field and ESTA_IP have been added. SMR_field is a 1-byte field. It stores the session-to-mobility ratio of external STA which has sent the re-association request frame to the APMG. ESTA_IP is a 4-byte field. It stores the IP address of the external STA which is the originator of re-association request frame. In the PREQ frame format, three new fields have been added, namely, Src_ESTA_IP_address, Dest_ESTA_IP_address, and ESTA_forward_chain. Src_ESTA_IP_address is a 4-byte field. It stores the IP address of originator external STA. Dest_ESTA_IP_address is a 4-byte field. It stores the IP address of the destination external STA. ESTA_forward_chain is a two-bit field. The field is set to 00 if the PREQ is originated by a mesh STA for setting up the path. It is set to 01 if PREQ is originated for a mobile STA. It is set to 10 if PREQ is generated for extending the forward chain. The field is set to 11 if PREQ is generated for terminating the forward chain. In PREP, five new fields have been added, namely, Dest_ESTA_IP_address, Src_ESTA_IP_address, Src_ESTA_MAC_address, Src_APMG_MAC_address, and Src_sequence_num. Dest_ESTA_IP_address is a 4-byte field. It stores the IP address of the destination external STA to whom the PREP is sent. Src_ESTA_IP_address is a 4-byte field. It stores the IP address of the external STA for whom PREP is generated. Src_ESTA_MAC_address is a 6-byte field. It stores the MAC address of the external STA for whom PREP is generated. Src_APMG_MAC_address is a 6-byte field. It stores the MAC address of the serving APMG of external STA for whom PREP is generated. Src_sequence_num is a 4-byte field. It stores the content of sequence number field corresponding to the proxy table entry of external STA for which the PREP is generated.

In PXU message four new fields, namely, PXU_Src_ESTA_MAC, PXU_dest_ESTA_MAC, PXU_cancel_flag, and PXU_part_of_forward_chain. PXU_Src_ESTA_MAC is a 6-byte field. It stores the MAC address of the source external STA for whom PXU message is generated. PXU_dest_ESTA_MAC is a 6-byte field. It stores the MAC address of the destination external STA. PXU_cancel_flag field is a 1-bit field. It is set to 1 if PXU message is sent to old APMG to delete the proxy table entry of PXU_Src_ESTA_MAC.

Otherwise, it is set to 0. PXU_part_of_forward_chain is a 1-bit field. It is set to 1 if source APMG is the part of the forward chain. Otherwise, it is set to 0. Two new subfields have been added in the proxy information field, namely, ESTA_IP_addr and proxy_information_serving_APMG. ESTA_IP_addr carries the content of Proxy_ESTA_IP field of the sender's proxy table. Proxy_information_serving_APMG carries the content of proxy_serving_APMG of the sender's proxy table. In PXUC message, three new fields have been added, namely, PXUC_serving_APMG, PXUC_cancel_flag, and PXUC_corresponding_node_table. PXUC_serving_APMG is a 6-byte field. If PXU_part_of_forward_chain field of the received PXU is 1, PXUC_serving_APMG will be the address of serving APMG of external STA. Otherwise, the field will be null. PXUC_cancel_flag is a 1-bit field. If the PXU_cancel_flag field of received PXU is 1, the PXUC_cancel_flag field will also be set to 1. Otherwise, it will be 0. The PXUC_corresponding_node_table field stores all the entries of corresponding node table which correspond to the PXU_Src_ESTA_MAC of PXU message received by originator of PXUC. Otherwise, it will be null.

3.1 Mobility Management

When handoff takes place external STA calculates its SMR value. While re-associating with new APMG external STA sends its SMR value and IP address. On receiving the message, APMG creates external STA's new entry in proxy table. The new APMG then checks whether the value of SMR_field is greater than threshold (δ_{th}) or not.

If no, the new APMG sends PXU message to the old APMG. On receiving the PXU message, the old APMG sends back PXUC message to the new APMG. The old APMG deletes external STA's proxy table entry. The PXUC will carry all the entries of external STA in the corresponding node table. The old APMG deletes all entries of external STA from the corresponding node table. When the PXUC message is received by the new APMG, it updates corresponding node table. Then it updates proxy table. The new APMG transmits PREQ message toward old APMG. On receiving PREQ message, the entry of external STA will be created or updated in routing table of old APMG.

If yes, new APMG sends PXU message to old APMG. When the old APMG receives the PXU message, it removes external STA's entry from the proxy table. The old APMG sends back PXUC message to the new APMG. In the field PXUC_corresponding_node_table, the PUC message will carry all the entries of corresponding node table which is having source_ESTA_MAC equal to PXU_Src_ESTA_MAC. Then all such entries will be deleted from corresponding node table. When the new APMG receives the PXUC message, it updates its corresponding node table. Through these fields it will also update its proxy table. New APMG then sends PXU message to the portal and all the corresponding nodes. Since portal transmits RANN message periodically, each APMG has a path toward portal. Using this path PXU message will be sent to the portal. Corresponding_serving_APMG_MAC and corresponding_ESTA_MAC fields of corresponding node table are used by current APMG of the external STA to send PXU message to the corresponding nodes. On receiving PXU message, the corresponding APMG checks whether there is an entry in proxy table where station address is same as content of the PXU_dest_ESTA_MAC field. If there is an entry, the corresponding APMG and portal updates its proxy. The corresponding node table will also be updated. If there is no entry, the PXU message is forwarded to next hop using routing table entry of the PXU_dest_ESTA_MAC. This process will continue till PXU reaches the current APMG of the destination external STA. When the PXU message is received by the current APMG, it updates the proxy table. It also updates the corresponding node table entry of the source external STA.

When handoff takes place, old APMG finds entries in its corresponding node table having source_ESTA_MAC same as PXU_Src_ESTA_MAC. It takes the contents of corresponding_ESTA_MAC field corresponding to those entries. Then it counts the number of entries of corresponding_ESTA_MAC in corresponding node table. If the count is 1, the entries of the proxy table where station address is same as the address of corresponding_ESTA_MAC will be deleted from the proxy table. This is done to remove the possibility of stale cache from proxy table.

3.2 Route Setup

When the external STA wants to send packets to another external STA, the current APMG of the external STA looks for destination external STA entry in the proxy table. If entry is found, a new entry will be created in the corresponding node table. If no entry is found, new APMG sends PREQ to the portal. When portal receives PREQ, it searches entry of dest_ESTA_IP_address in its proxy table. If it finds, it sends PREP to the originator APMG of the PREQ. When the originator of the PREQ receives PREP it updates its proxy table. The new APMG then sends PXU message to the serving APMG of the destination external STA. When the PXU message is received by the destination APMG, it searches the entry of PXU_dest_ESTA_MAC in its proxy table. If no entry is found in the proxy table, it searches the entry of PXU_dest_ESTA_MAC in the routing table. The PXU is then forwarded to the next hop. The PXU finally reaches the external STA's current APMG. The current APMG searches entry of PXU_dest_ESTA_MAC and finds it in the proxy table. Then current APMG updates its proxy table. The corresponding node table of current APMG of external STA is also updated. Thus, path is established between source and destination APMG.

3.3 Routing of Data Packets

External STA sends upstream Internet packets toward its current APMG. Using the routing table the packet is forwarded toward next hop node for reaching portal.

When the portal receives the downstream Internet packet destined to the external STA, external STA's entry is searched in proxy table using IP address of the destination external STA. MAC address of serving APMG of the external STA is found from that entry. The address is stored in the address 3 field of the MAC header. The MAC address of the external STA which corresponds to that entry will be stored in mesh address extension subfield of the mesh control field. The packet is then forwarded to the next hop mesh STA to reach the serving APMG of the destination. When an intermediate mesh STA receives the packet, it searches the entry of serving APMG's address in its routing table. Then packet is forwarded to next hop mesh STA. This process continues till the packet reaches external STA's serving APMG. When serving APMG receives that packet it searches destination external STA's address in its proxy table. If the entry is not found the address of destination external STA is searched in the routing table. The packet is then forwarded to the next hop mesh STA corresponding to that entry. This process will continue till the packet reaches destination external STA's current APMG. When packet reaches current APMG, address of destination external STA is searched in the proxy table. The packet is then delivered to the destination external STA. On the other hand, if external STA's entry is present in serving APMG's proxy table, packet is delivered to external APMG.

When external STA wants to send packet, it is first checked if destination external STA's entry is present in the proxy table. If there is no entry a path will be established between the source external STA and destination external STA. Then destination external STA's corresponding node table entry is created. Address of the serving APMG of the destination is stored in address 3 field of MAC header. The address of the destination external STA is stored in the mesh address extension subfield of the mesh control field. The packet is then forwarded to the next hop to reach the serving APMG of the destination external STA using routing table entry. When it reaches the serving APMG of the destination external STA, the entry of external STA is searched in the proxy table. If no entry is found, the address of destination external STA is searched in the routing table. The packet is then forwarded to the next hop corresponding to that entry. The process continues till the packet reaches external STA's current APMG. In the current APMG's proxy table, entry of destination external STA will be found. The packet is then delivered to the destination external STA.

4 Simulation Result and Comparison

NS-2 [27] is used to carry out simulation. The simulation topology has been shown in Fig. 1. Node number 1 has been set as the gateway. IEEE 802.11 has been used in MAC layer. The external STAs follow random waypoint mobility model pause

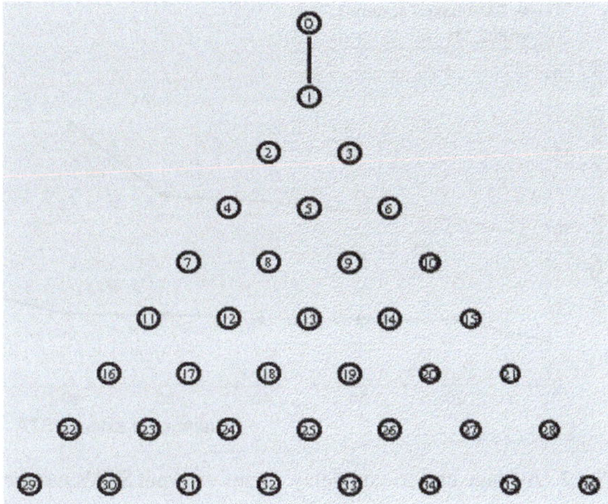

Fig. 1 Topology used for simulation

time 0 s. Number of mesh STA, traffic type, radio range, simulation time, maximum number of connections, packet size are set to 36, CBR, 41.04 m, 100 s, 10 and 258 bytes, respectively. Distance between two neighboring mesh STA is 40 m. Threshold SMR value of the proposed scheme is set to 0.5.

Figure 2 shows that throughput of IEEE 802.11s is less than SMR-based scheme. This is because in IEEE 802.11s PREQ is broadcasted during every handoff which causes congestion. In IEEE 802.11s as external STA's number increase, number of PREQ broadcasted also increases. It causes congestion. In case of the proposed scheme, with increase in number of external STA, number of PREQ sent also increases. Therefore, throughput of IEEE 802.11s and the proposed scheme decreases as external STA's number increases.

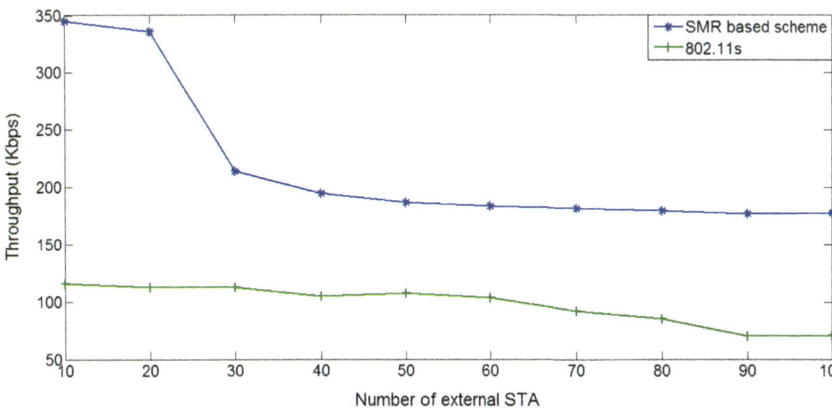

Fig. 2 Throughput versus external STA's number

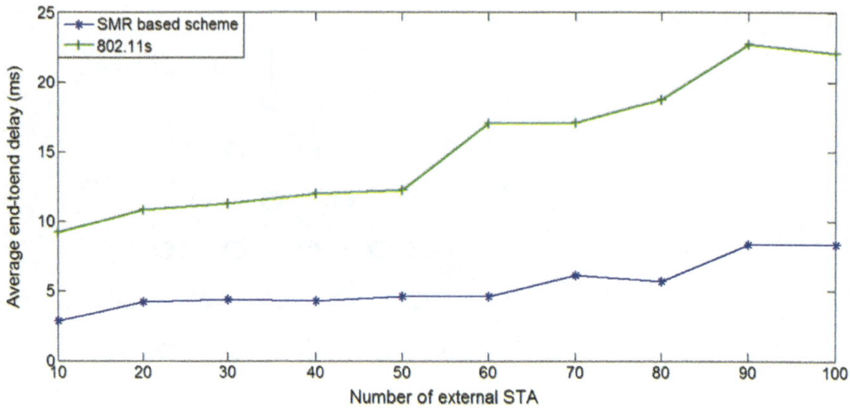

Fig. 3 Average end-to-end delay versus external STA's number

Figure 3 shows that end-to-end delay of IEEE 802.11s is higher than that of the proposed scheme because of congestion caused by broadcast of PREQ. When external STA's number is increased, number of PREQ transmission also gets increased. As external STA number is increased, there is limited increase in transmitted PREQ's number. Therefore, with increase in number of external STA, SMR-based scheme's average end-to-end delay increases, but it remains less compared to IEEE 802.11s. Figure 4 shows that SMR-based scheme's normalized routing overhead is less than IEEE 802.11s. In SMR-based scheme, propagation of route management packet becomes limited. So, normalized routing overhead is less than IEEE 802.11s. As node's number increases propagation of route management packet in SMR-based scheme also increases but the propagation is of type unicast. Therefore, normalized routing overhead increases as external STA's number increases but increase is very

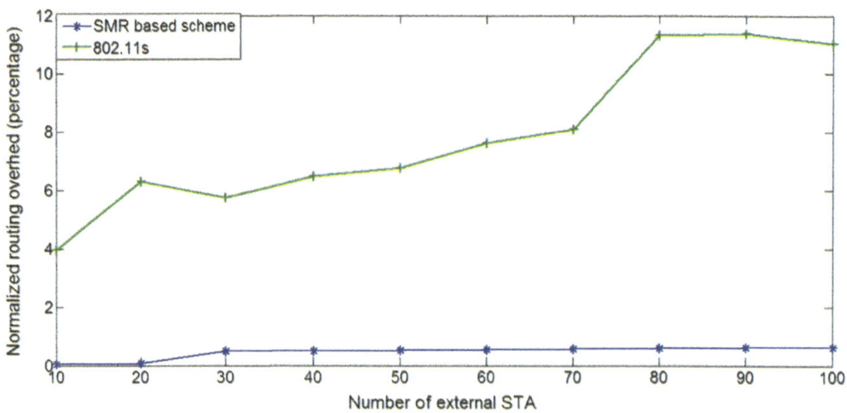

Fig. 4 Normalized routing overhead versus number of external STA

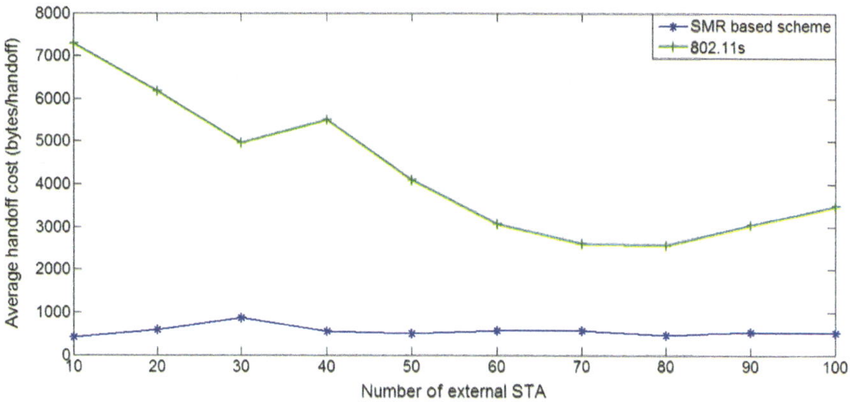

Fig. 5 Average handoff cost versus number of external STA

limited. Figure 5 shows that SMR-based scheme's handoff cost is less than that of IEEE 802.11s because of broadcast of PREQ for route setup in IEEE 802.11s.

5 Conclusion and Future Work

An SMR-based mobility management scheme is implemented in this paper. It is integrated with HWMP. In the scheme, location update messages are sent to the portal and corresponding external STA only if SMR value of external STA is greater than the threshold. By this way number of location update propagation is reduced. Simulation of both IEEE 802.11s and SMR-based scheme has been carried out. It has been found that SMR-based scheme outperforms IEEE 802.11s. Extending the SMR-based scheme to work with multiple portals remains as future work.

References

1. Akyildiz, F.: A survey on wireless mesh networks. IEEE Commun. Mag. **43**(9), S23–S30 (2005)
2. Akyildiz, I.F., Wang, X., Wang, W.: Wireless mesh networks: a survey. Comput. Netw. **47**(4), 445–487 (2005)
3. Benyamina, D., Hafid, A.S., Gendreau, M.: Wireless mesh networks design—a survey. IEEE Commun. Surv. Tutor. **14**(2), 299–310 (2012)
4. Zhang, Y., Luo, J., Hu, H.: Wireless mesh networking architectures. Protocols and standards. CRC Press, Boca Raton (2006)
5. Hiertz, G.R., Denteneer, D., Max, S., Taori, R., Cardona, J., Berlemann, L., et al.: IEEE 802.11s: the WLAN mesh standard. IEEE Wirel. Commun. **17**(1), 104–111 (2010)

6. IEEE Std 802.11-2012.: Information technology—telecommunications and information exchange between systems—local and metropolitan area networks—specific requirements—part 11: wireless LAN medium access control (MAC) and physical layer (PHY) specifications (2010)
7. IEEE P802.11s/D1.07-2007.: Information technology—telecommunications and information exchange between systems—local and metropolitan area networks—specific requirements—part 11: wireless medium access control (MAC) and physical layer (PHY) specifications (2007)
8. Camp, J.D., Knightly, E.W.: The IEEE 802.11s extended service set mesh networking standard. IEEE Commun. Mag. **46**(8), 120–126 (2008)
9. Wang, X., Lim, A.O.: IEEE 802.11s wireless mesh networks: framework and challenges. Ad Hoc Netw. **6**(6), 970–984 (2008)
10. Sooriyaarachchi, S.J., Fernando, W.A.C., Gamage, C.D.: Evaluation of scalability of hybrid wireless mesh protocol in IEEE 802.11. In: Fifteenth international conference on advances in ICT for emerging regions (ICTer), pp. 152–159 (2015)
11. Bari, S.M.S., Anwar, F., Masud, M.H.: Performance study of hybrid wireless mesh protocol (HWMP) for IEEE 802.11s WLAN mesh networks. In: International conference on computer and communication engineering, pp. 712–716 (2012)
12. Bahr, M.: Proposed routing for IEEE 802.11s WLAN mesh networks. In: 2nd annual international workshop on wireless internet, p. 5 (2006)
13. Sampaio, S., Soutp, P., Vasques, F.: Are view of scalability and topological stability issues in IEEE 802.11s wireless mesh networks deploymants. Int. J. Commun. Syst. **29**(4), 671–693 (2015)
14. Majumder, A., Roy, S., Dhar, K.K.: Design and analysis of an adaptive mobility management scheme for handling internet traffic in wireless mesh network. In: Annual International Conference on Emerging Research Areas and International Conference on Microelectronics, Communications and Renewable Energy, pp. 1–6 (2013)
15. Majumder, A., Roy, S.: Implementation of adaptive mobility management technique for wireless mesh network to handle internet packets. In: 2nd international conference on communication, devices and computing (2019) (In Press)
16. Mase, K.: Layer 3 wireless mesh networks: mobility management issues. IEEE Commun. Mag. **49**(7), 156–163 (2011)
17. Xie, J., Wang, X.: A survey of mobility management in hybrid wireless mesh networks. IEEE Netw. **22**(6), 34–40 (2008)
18. Mishra, B.B., Dehuri, S., Panigrahi, B. K., Nayak, A.K., Mishra, B.S.P., Das, H. (eds.).: Computational intelligence in sensor networks. Springer (2019)
19. Sarkar, J.L., Panigrahi, C.R., Pati, B., Das, H.: A novel approach for real-time data management in wireless sensor networks. In: Proceedings of 3rd international conference on advanced computing, networking and informatics, pp. 599–607. Springer, New Delhi (2016)
20. Das, H., Naik, B., Pati, B., Panigrahi, C.R.: A survey on virtual sensor networks framework. Int. J. Grid Distrib. Comput. **7**(5), 121–130 (2014)
21. Panigrahi, C.R., Sarkar, J.L., Pati, B., Das, H.: S2S: a novel approach for source to sink node communication in wireless sensor networks. In: International conference on mining intelligence and knowledge exploration, pp. 406–414. Springer, Cham (2015, December)
22. Huang, R., Zhang, C., Fang, Y.: A mobility management scheme for wireless mesh networks. In: Global telecommunication conference, pp. 5092–5096 (2007)
23. Navda, V., Kashyap, A., Das, S.R.: Design and evaluation of iMesh: an infrastructure-mode wireless mesh network. In: 6th IEEE international symphosia on a world wireless mobile multimedia networks, pp. 164–170 (2005)
24. Maoshen, R., Chao, L., Huizhou, Z., Tong, Z., Wei, Y.: MEMO: an applied wireless mesh network with client support and mobility management. In: IEEE global telecommunication conference, pp. 5075–5079 (2007)

25. Huang, D.W., Lin, P., Gan, C.H.: Design and performance study for a mobility management mechanism (WMM) using location cache for wireless mesh networks. IEEE Trans. Mob. Comput. **7**(5), 546–556 (2008)
26. Majumder, A., Roy, S.: Design and analysis of a dynamic mobility management scheme for wireless mesh network. Sci World J, Article ID 656259, 16 (2013)
27. Fall, K., Varadhan, K.: The network simulator (ns-2). http://www.isi.edu/nsnam/ns (2007)

An Ensemble Approach for Classification of Thyroid Using Machine Learning

Bhavna Dharamkar, Praneet Saurabh, Ritu Prasad and Pradeep Mewada

Abstract Medical diagnosis and extracting patterns that can be converted into useful knowledge is a challenging work for researchers. Medical records are based on real-time data that have high dimensionality which makes task of extracting patterns even more complex. Prediction of various diseases often suffers due to large dimensionality that includes thyroid, diabetes, cancer, etc. Data mining is a process of finding out information and constructing a knowledge base from a huge amount of data, which remains useful. Data mining achieves these tasks involving classification, clustering, regression, and prediction. Reduction of dimensionality with the motive of knowledge discovery is an essential aspect of data mining as it helps in prediction/determination for various data analytics filed like medical diagnosis, business modeling, and government data analysis. This paper presents cooperative method for classification of thyroid using machine learning fusing C4.5 and random forest classification technique (CCTML). Experimental results are compared with other conventional techniques and the results obtained through CCTML reported better classification accuracy and outperform conventional methods.

Keywords Thyroid · Classification · Feature selection · C4.5 · Random forest · Multilayer perceptron · Bayesian net

1 Introduction

In human body, thyroid is a very important organ and thyroid hormones act throughout the body, influencing metabolism, growth and development, and body temperature. Thyroid is a butterfly-shaped organ, which is orchestrated at the base of the throat responsible for making dynamic thyroid hormones, T3 and T4, that sway some breaking point of the body [1]. The thyroid organ is one of the best endocrine organs which are found in the neck [2]. Triodothyronine (T3) and thyroxine [3] which can

---

B. Dharamkar · R. Prasad (✉) · P. Mewada
Technocrats Institute of Technology Advance, Bhopal 462021, M.P., India

P. Saurabh (✉)
Mody University of Science and Technology, Lakshmangarh, Rajasthan, India

H. Das et al. (eds.), *Progress in Computing, Analytics and Networking*,
Advances in Intelligent Systems and Computing 1119,
https://doi.org/10.1007/978-981-15-2414-1_2

once in a while be suggested as tetraiodothyronine (T4) are hormones which control the rate of digestion and affect the progression and rate of cutoff of different particular frameworks in the body [3]. During infancy and childhood, adequate thyroid hormone is crucial for brain development. Thyroid hormones are also responsible for controlling the ingestion. These hormones sway all tissues of the body and augments cell movement. Thyroid is inclined to several particular issues, for instance, goiter, thyroid affliction, hyperthyroidism, and hypothyroidism [4]. Thyroid ailments are very common in individuals and late detection leads to complex diagnosis and effects are of severe magnitude.

Data mining is the process of analyzing hidden patterns of data using different techniques for categorization to gather and acquire useful information. These patterns are collected and assembled in common areas to create data warehouses. Now it becomes relevant to data discovery and knowledge discovery. Data mining remains one of the different logical instruments for researching data and mining patterns that have the potential to yield some solution-oriented patterns [5]. In reality, data mining is the path toward finding associations or precedents among numerous fields even in social databases. There are diverse AI frameworks, which can be used to perform following data mining functionalities that include clustering, classification, etc.

In recent times, various works have used conventional methods to find trends in disease-related datasets but due to dataset's complex organization and structure with high dimensionality many a times fail to give optimum results. Keeping this in mind this paper introduces a cooperative method for classification of thyroid using machine learning after fusing C4.5 and random forest classification technique (CCTML). Experimental results are compared with other conventional techniques and the results obtained through CCTML reported better classification accuracy and outperform conventional methods. This paper is sub-divided into various subsections to improve readability. Section 2 puts forward the related work of the domain while Sect. 3 outlines the proposed methodology. Section 4 explains the experimental results and Sect. 5 presents the conclusion.

2 Related Work

Data mining strives to find information from the repository and then constructs a knowledge base from the available data. It achieves these tasks involving classification, clustering, regression, and prediction. Reduction of dimensionality is an important aspect of data mining and knowledge discovery since it helps in determining various trends of various application areas like medical diagnosis, business modeling, and government data analysis. Medical diagnosis and extracting patterns that can be converted into useful knowledge is a challenging work for researchers [6]. Medical records are based on real-time data that have high dimensionality which makes task of extracting patterns even more complex. In recent years, there has been significant growth in popularity of different machine learning techniques to solve complex computational problems.

In an interesting work, Jagdeeshkannan et al. [6] developed an estimation mechanism to measure thyroid lesions for any patient and automated the detection and classification of thyroid cancer cells. Chandel et al. [7] did an extensive and comparative survey on thyroid disease detection using K-nearest neighbor and Naive Bayes classification techniques. Amina and Parkavi [8] also surveyed all the thyroid detection methods using data mining with their novelty and limitations. Dogantekin et al. [9] explored the possibilities of automatic diagnosis system based for thyroid in their work. Cetişli [10] proposed an expert system for developing an adaptive neuro-fuzzy classifier with using linguistic hedges. Azar et al. [11] developed a linguistic hedges neural-fuzzy classifier that selected features in its quest of detection and subsequent diagnosis of thyroid diseases. The performance evaluation of this system is estimated by using classification accuracy and k-fold cross-validation. Omiotek et al. [12] presented a novel technique for classification of ultrasound thyroid images. This method was based on the theory of good and bad thyroid images. This work used artificial neural networks which have been used to build classification models and then made a starting point to build a support system in the process of medical diagnosis. These explore show that thyroid diagnosis and detection are not only limited to conventional methods but machine learning can be also deployed to get better results. Later on, various methods have been introduced [13–15] but still there existed some room for improving the efficiency as conventional methods are not designed for self-configuring itself. Also, limitations for conventional methods in detection include choosing the right technique, given the high dimensionality, poor interpretation, and varying class prediction. Next section presents cooperative method for classification of thyroid using machine learning fusing C4.5 and random forest classification technique (CCTML).

3 Proposed Methodology

This paper presents cooperative method for classification of thyroid using machine learning (CCTML) that fuses C4.5 and random forest classification technique. Conventional methods for detection and prediction suffered from problem of large dimensionality and adaptability. CCTML works with the motive of reduction of dimensionality for knowledge discovery. CCTML determines the characterization of thyroid ailments using machine learning features since CCTML classifies the different instances of dataset for thyroid ailment through various classifiers. Therefore, now it becomes natural to measure the performance of classifier in terms of error rate.

3.1 Algorithm for Classification of Thyroid Disease

Step 1: Collected dataset from UCI repository.

Step2: Apply datasets on decision tree techniques and calculate the performance of individual model.

Step 3: Ensemble the C4.5 and random forest model using voting scheme and calculate the performance.

Step 4: Comparing measures of ensemble model with individual model and ensemble model gives better performance.

Step 5: Apply the feature selection on best ensemble model (C4.5+RF) with different feature subsets.

Step 6: Achieved high classification accuracy with reduced five-feature subset using together C4.5 and random forest with thyroid dataset.

Step 7: Recommended together C4.5 and random forest model for thyroid dataset.

Step 8: End.

3.2 *Flowchart for Detection and Classification of Thyroid*

The above algorithm and Fig. 1 represent the working of the proposed CCTML. Decision tree technique calculates the performance of thyroid detection model. Thereafter, C4.5 and random forest models calculate the performance using voting scheme. After that a comparison has been made to identify the better performance. The classifier predicts the classes of each instance, if it is correct it is counted as success otherwise fail. The error rate is the number of instances that has been predicted incorrectly over the whole dataset and measures the performance of classifier.

4 Experimental Results

This section presents the experimental finding for the experiments carried on proposed cooperative method for classification of thyroid using machine learning after fusing C4.5 and random forest classification technique (CCTML) and then a comparison has been drawn for CCTML with the current state of art. CCTML is implemented using Orange canvas and investigation purpose. Dataset is taken from UCI repository. This dataset contains the instances of thyroid and non-thyroid sicknesses. Thyroid informational index contains 7547 records in which 776 have a place with thyroid and 6771 have a place with non-thyroid information. Thyroid information comprises both hypothyroid and hyperthyroid information. Informational index had 29 highlights and 1 class. Table 1 highlights the arrangement of dataset classes to either thyroid or non-thyroid, and it also contains feature_id and attribute names.

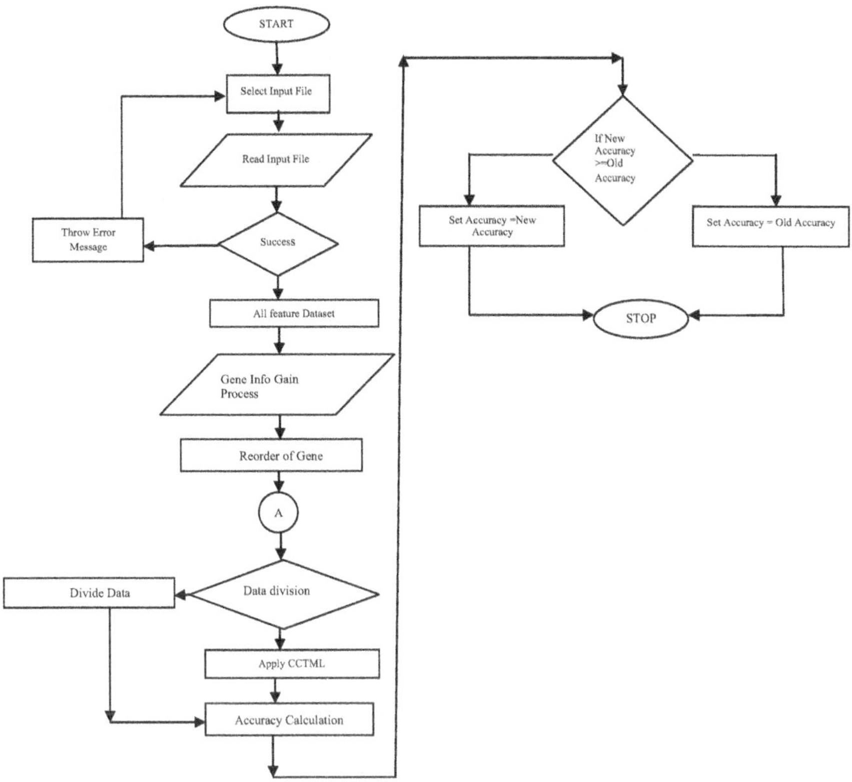

Fig. 1 Flowchart of CCTML

Table 1 Dataset features

Id	Attribute name	Id	Attribute name	Id	Attribute name
1	Age	11	Query hyperthyroid	21	TT4 measured
2	Sex	12	Lithium	22	TT4
3	On thyroxine	13	Goiter	23	T4U measured
4	Query on thyroxine	14	Tumor	24	T4U
5	On antithyroid medication	15	Hypopituitary	25	FTI measured
6	Sick	16	Psych	26	FTI
7	Pregnant	17	TSH measured	27	TBG measured
8	Thyroid surgery	18	TSH	28	TBG
9	I131 treatment	19	T3 measured	29	Referral source
10	Query hypothyroid	20	T3	Class	Thyroid or non-thyroid

4.1 Performance Determining Parameters for Classification

Performance of any classifier can be tested through the occurrence of true positives (TP) which is the amount of positive points for positive while true negatives (TN) denote an outcome where the model correctly predicts the negative class. In the same way, a false positive (FP) is an outcome where the model incorrectly predicts the positive class while false negative (FN) is a test result that indicates that a condition does not hold, while in fact it does.

Precision: It is the measure of true positive with true negative in total number of instances out of 100. Higher the precision the better the method is, represented in Eq. (1).

$$\text{Precision} = (\text{TP} + \text{TN})/\text{N} * 100 \quad (1)$$

where N is total number of instances.

Affectability: It measures the extent of the genuine positives over the sum of true positive and false negative given in Eq. (2).

$$\text{Affectability} = \text{TP}/\ (\text{TP} + \text{FN}) * 100 \quad (2)$$

Explicitness: It measures the genuine negatives over sum of true negative and false positive, lower value is desired and the formula is presented in Eq. (3).

$$\text{Particularity} = \text{TN}/(\text{TN} + \text{FP}) * 100 \quad (3)$$

4.2 Experimentation and Result Analysis

This subsection presents the experimental results carried over different classifiers and the proposed CCTML to measure their respective performance under the parameters of classification accuracy, affectability, and explicitness. Experimental results presented through Table 2 and Fig. 2 indicate the comparative performance of classification accuracy with different classifiers including proposed CCTML. Experiments are performed after diving the UCI dataset into three different categories like Dataset I having testing data of 70% and training data of 30% while Dataset II having testing data of 80% and training data of 20%, and Dataset III having testing data of 90% and training data of 10%. Experiments are done for three different test conditions, test condition I employed dataset I while test condition II used dataset II and test condition III worked on dataset III. From the results, it is evident that as the training sample grows from 10 to 30% under all the classifiers the results get better. This indicates that with growth in training sample size the accuracy of classification increases. It is

Table 2 Classification accuracy of different classifiers

Methods	Test condition I	Test condition II	Test condition III
	Accuracy (%)	Accuracy (%)	Accuracy (%)
C4.5	95	94	92
Random forest	91	92	91
Decision stump	91	91	92
Multilayer perceptron	93	94	92
Bayes net	93	93	93
Decision table	93	91	95
CCTML	96	95	97

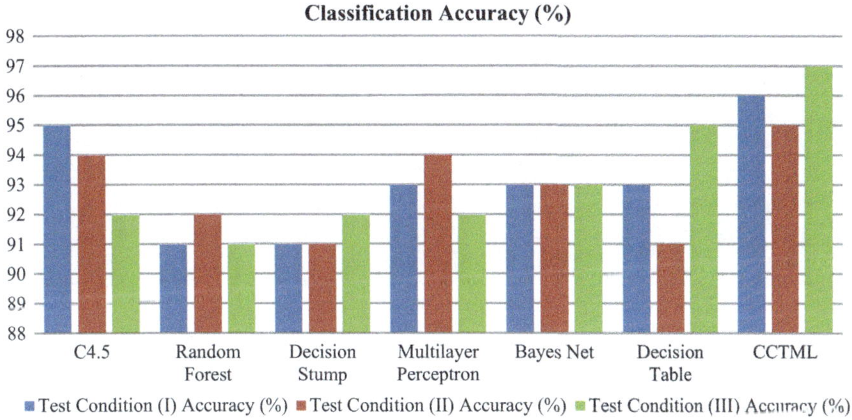

Fig. 2 Classification accuracy (%)

also worth mentioning that out of all classifiers the proposed CCTML yields highest classification accuracy.

Experimental results presented through Table 3 and Fig. 3 present the comparative performance of classification affectability with different classifiers including proposed CCTML. Experiments are performed for all the three different test conditions with varying percentage of training and test set. Experimental results demonstrate that the proposed CCTML outperforms other classifiers and yields better results.

Table 4 and Fig. 4 illustrate comparative performance of classification explicitness with current state of art. CCTML reports lower explicitness for three different test conditions, test condition I employed dataset I while test condition II used dataset II and test condition III worked on dataset III. Experimental results very clearly highlighted the need for larger training set data size as explicitness of classifier reduced with increase in training set dataset size. CCTML under all the test conditions reported lowest explicitness which is desired.

Table 3 Classification affectability different classifiers

Methods	Test condition I	Test condition II	Test condition III
	Affectability (%)	Affectability (%)	Affectability (%)
C4.5	91	91	90
Random forest	90	86	88
Decision stump	88	86	82
Multilayer perceptron	84	85	88
Bayes net	91	89	86
Decision table	87	84	85
CCTML	92	91	90

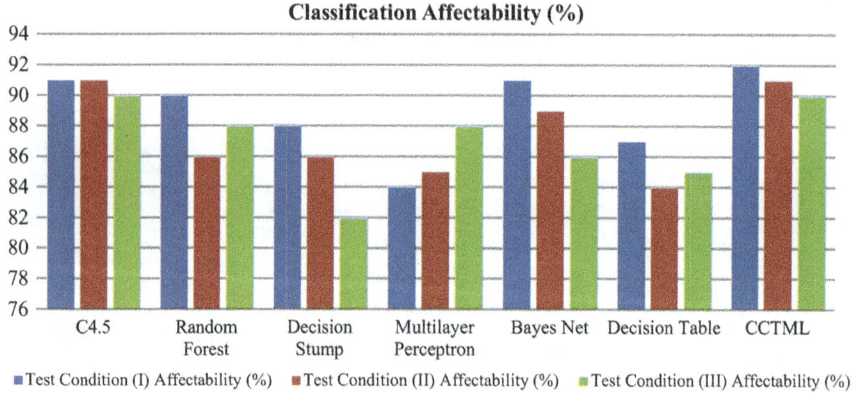

Fig. 3 Classification affectability (%)

Table 4 Classification explicitness of different classifiers

Methods	Test condition I	Test condition II	Test condition III
	Explicitness (%)	Explicitness (%)	Explicitness (%)
C4.5	52	56	59
Random forest	56	54	57
Decision stump	53	55	58
Multilayer perceptron	58	57	59
Bayes net	53	59	65
Decision table	55	62	61
CCTML	51	54	56

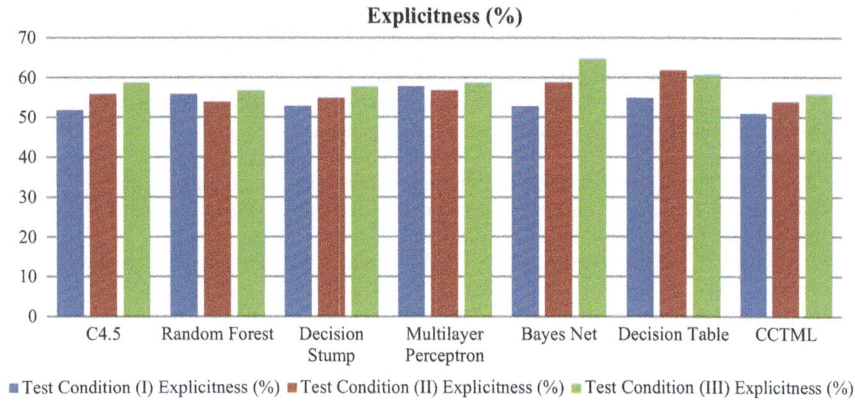

Fig. 4 Classification explicitness (%)

5 Conclusion

Pattern extraction from real-time medical records is a challenging task and complex arrangement makes it even more complex. Prediction of various diseases often suffers due to large dimensionality that includes thyroid, diabetes, cancer, etc. This paper presents cooperative method for classification of thyroid using machine learning fusing C4.5 and random forest classification technique (CCTML). Experimental results measure classification accuracy, affectability, and explicitness for CCTML and current state of art. Experimental results demonstrated that performance of all the classifiers including proposed CCTML improved with increase in training sample size from 10 to 30%. Also, CCTML reported highest classifier accuracy and affectability with lowest explicitness among its peers for all test conditions.

References

1. Zadeh, H.G., Janianpour, S., Haddadnia, J.: Recognition and classification of the cancer cells by using image processing and lab view. Int. J. Comput. Theor. Eng. (2009)
2. Leung, C.C., Chan, F.H.Y., Lam, K.Y., Kwok, P.C.K., Chen, W.F.: Thyroid cancer cells boundary location by a fuzzy edge detection method. In: Proceedings of the 15th international conference on pattern recognition (2000)
3. Hudli, S.A., Hudli, A.V., Hudli, A.A.: Application of data mining to candidate screening. In: 2012 IEEE international conference on advanced communication control and computing technologies (ICACCCT), pp. 287–290. IEEE (2012)
4. Munla, N., Khalil, M., Shahin, A., Mourad, A.: Driver stress level detection using HRV analysis. In: 2015 international conference on advances in biomedical engineering (ICABME), pp. 61–64. IEEE (2015)
5. Shariati, S., Haghighi, M.M.: Comparison of Anfis neural network with several other ANNs and support vector machine for diagnosing hepatitis and thyroid diseases. In: Proceedings of IEEE IACSIT 2010, pp. 596–599 (2010)

6. Jagdeeshkannan, R., Aarthi, G., Akshaya, L., Ravy, K.: An approach for automated detection and classification of thyroid cancer cells. In: ICT and critical infrastructure, proceedings of the 48th annual convention of computer society of India, 2, pp. 379–389 (2014)
7. Chandel, K., Kunwar, V., Sabitha, S.: A comparative study on thyroid disease detection using K-nearest neighbor and Naive Bayes classification techniques. CSI Trans. ICT **4**(2–4), 313–319 (2016)
8. Amina, B., Parkavi, A.: Prediction of thyroid disease using data mining techniques. In: 5th international conference on advanced computing and communication systems (ICACCS), pp. 342–345 (2019)
9. Dogantekin, E., Dogantekin, A., Avci, D.: An automatic diagnosis system based on thyroid gland: ADSTG. Expert Syst. Appl. **37**(9), 6368–6372 (2010)
10. Cetişli, B.: Development of an adaptive neuro-fuzzy classifier using linguistic hedges: Part 1. Expert Syst. Appl. **37**(8), 6093–6101 (2010)
11. Azar, A.T., Hassanien, A.E., Kim, T.: Expert system based on neural-fuzzy rules for thyroid diseases diagnosis, pp. 94–105. Springer (2012)
12. Omiotek, Z., Burda, A., Wójcik, W.: Application of selected classification methods for detection of Hashimoto's thyroiditis on the basis of ultrasound images, computational intelligence, medicine and biology, pp. 23–37. Springer (2015)
13. Sahu, S., Saurabh, P., Rai, S.: An enhancement in clustering for sequential pattern mining through neural algorithm using web logs. In: International conference on computational intelligence and communication networks, pp. 758–764 (2015)
14. Saxena, M., Saurabh, P., Verma, B.: A new hashing scheme to overcome the problem of overloading of articles in Usenet, pp. 967–975. Springer, AISC (2012)
15. Mishra, B.K., Saurabh, P., Verma, B.: A novel approach to classify high dimensional datasets using supervised manifold learning, pp. 22–30. Springer, CCIS (2012)

Text Classification with K-Nearest Neighbors Algorithm Using Gain Ratio

Manjari Singh Rathore, Praneet Saurabh, Ritu Prasad and Pradeep Mewada

Abstract Content classification is the errand of naturally arranging a lot of records into classifications from a predefined set. This implies that it allocates predefined classifications to free-content archives. This paper introduces a unique two-phase determination strategy for content classification using data gain (CCDG) that will guide and examine the hereditary calculation of the given dataset. In the first phase of CCDG, each term inside the archive is positioned depending on its significance for grouping and data gain. In the second stage, hereditary calculation through GA and main segment investigation through PCA determines and highlights the relevant extraction of the trend of the given stream of bits in decreasing impact. In this manner, all the content that has lesser significance can be overlooked while only impactful content remains for providing details. Experiments show encouraging and better results for proposed CCDG as compared to conventional methods under all the dataset and test conditions.

Keywords Text · Classification · Feature selection · PCA

1 Introduction

Content order is one of the recent research explores due to the need to arrange and sort developing number of electronic archives around the world. Up till now, content characterization has been effectively connected to different areas, for example, theme discovery [1], spam email sifting [2], SMS spam separating [3], creator identification [4], website page order [5], and opinion examination. A normal content order structure comprises preprocessing, include extraction, highlighting the choice, and characterization stages. It takes care of the issue of allocating content substance to predefined classifications [6]. Content classification has fundamental significance in applications utilized in reality. For instance, news stories are normally sorted out

M. S. Rathore · R. Prasad (✉) · P. Mewada
Technocrats Institute of Technology Advance, Bhopal 462021, M.P., India

P. Saurabh (✉)
Mody University of Science and Technology, Lakshmangarh, Rajasthan, India

H. Das et al. (eds.), *Progress in Computing, Analytics and Networking*,
Advances in Intelligent Systems and Computing 1119,
https://doi.org/10.1007/978-981-15-2414-1_3

by subject classifications (themes) or geological codes; scholarly papers are regularly grouped by specialized areas and sub-spaces; quiet reports in human services associations are frequently recorded from numerous perspectives, utilizing scientific classifications of ailment classifications, kinds of surgeries, protection repayment codes, etc. Different explores have been introduced to achieve this task but somehow they fail to address the complex input from the users [7].

This paper presents a unique two-phase determination strategy for content classification by using data gain (CCDG) that will guide and examine the hereditary calculation. In the first phase of CCDG, each term inside the archive is positioned depending on its significance for grouping and data gain. In the second stage, hereditary calculation through GA and main segment investigation through PCA will determine and highlight the relevant extraction. This paper is organized as follows: Sect. 2 provides the related literature. Section 3 represents proposed methodology, Sect. 4 provides the result and analysis while Sect. 5 provides conclusion.

2 Related Work

Content order is characterized as doling out pre-characterized classifications to a given arrangement of archives dependent on the grouping designs. Although numerous data recovery applications exist, for example, sifting and hunting down relevant data help content order examining. These processes often suffer from a noteworthy issue of content classification which is the high dimensionality of the component space. Some element extractions have been effectively utilized in content arrangement, for example, chief segment examination, dormant semantic ordering, grouping techniques, and so on. Among the numerous strategies that are utilized for highlight extraction, PCA performance is exemplary [8]. PCA is a method for decreasing dimensionality of the given dataset or input stream. Content order in this regard is the method of arranging a report into predefined classes dependent on the content of the record [9]. As of now, an ever-increasing number of techniques have been associated to the content classification assignments depending on measurable speculations and AI, for example, KNN, Naive Bayes, Rocchio, decision tree, and support vector machine. There also exist a few systems that have been proposed for the content arrangement, for example, multi-mark content arrangement dependent on another direct classifier learning technique and a classification-sensitive refinement strategy [10].

Another methodology for managing multi-name content classification is dependent on another direct classifier learning philosophy. It utilizes a substitution weighted characterization procedure to build a multi-name direct classifier [11]. It, in general, utilizes the degrees of comparability between classes to manage the association of voluminous classes with pertinence to a testing archive. The testing archives are regularly legitimately characterized into numerous classes by utilizing a predefined edge worth. Content element determination utilizes insect state streamlining [12]. Highlight determination and highlight extraction are the most essential in grouping frameworks. Highlight choice is normally used to reduce the spatiality of datasets with tens or a large number of highlights which may be impractical to any process [13]. One of the issues in which highlight choice is vital is content order.

At present, there are a few different ways to influence content element decision. To assist content classification, Roweis [13] introduced a novel dimensionality reduction algorithm. These properties streamline the calculation and determine better results. Over the years, various methods have been introduced [14–17] but none of them yielded desired results. Therefore, next section presents a unique two-phase determination strategy for content classification by utilizing data gain (CCDG).

3 Proposed Methodology

This section introduces the proposed content classification by utilizing data gain (CCDG) that will guide and examine the hereditary calculation. In the first phase of CCDG, each term inside the archive is positioned depending on its significance for grouping and data gain. In the second stage, hereditary calculation through GA and main segment investigation through PCA will determine and highlight the relevant extraction. In this manner, all the content that has lesser significance can be overlooked while only impactful content remains for providing details and reducing computational complexities.

In the primary stage, each term inside the content is positioned on the basis of their significance for grouping in decreasing order. In this way, terms with higher significance are allocated to the top positions and terms of less significance are assigned to end resulting positions. In second stage of the proposed CCDG, PCA highlights the choice and GA is used to choose and highlight the already extractions that are independent. Along these lines, amid content classification, terms of less significance are overlooked. The terms which are most significant are extraction and highlighted. Hence, the computational time and unpredictability of the class are reduced. CCDG reduces the dimensionality of datasets, this arrangement can now be utilized for informational index for further investigation and reference purpose.

Figure 1 describes the flow diagram of CCDG for text classification. This diagram very clearly puts forward the five different steps employed. KNN is majorly used for this purpose since it is very simple classification algorithm. Therefore, it is extensively used in data mining. Major KNN steps are given as follows:

KNN Algorithm

Step 1: *Assume k is a number of training datasets and l can be any random state.*

Step 2: *Assign an array arr[ ]approaching a set of tuple (a,b) and set training dataset.*

Step 3: *for i = 0 to k:*

Step 4: *Calculate Euclidean distance d(arr[i], l).*

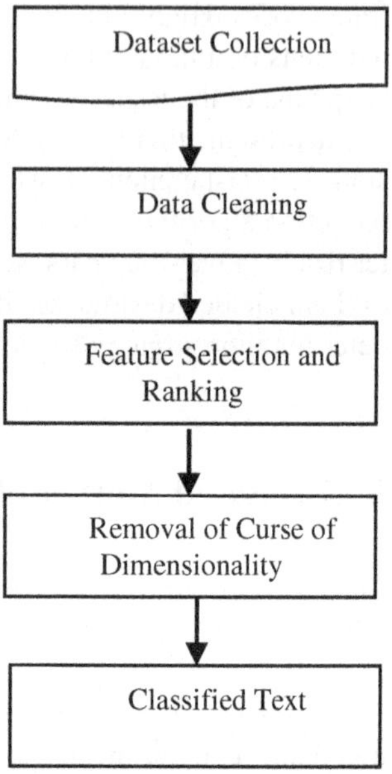

Fig. 1 Flow diagram of CCDG

Step 5: *Make set j of x shortest distances calculated. For every distance related to test.*
Step 6: *Return and apply for j up to highest order.*

Iterative Dichotomiser 3 algorithm is a type of classification method where selection of value is very important for making a choice of selection or ranking. Entropy is used for making a calculation in this algorithm. In any dataset, represented quantity of bits is expected to define properties and its entropy values. A log capacity with base 2 is utilized, in light of the fact that the data is encoded in bits. Given "A" is a chance to be a dataset and B_i is the Nth class. Let B (i, A) be the arrangement of tuples of class B (i) in "A". Let A and B (i, A) mean the quantity of tuples in A and B (i, A), individually and "n" signifies the quantity of classes. Entropy (A) is only the normal measure of data expected to distinguish the class name of a tuple in A given in Eq. (1).

$$\mathrm{Info}(A) = \sum_{i=1}^{n} \left(p_i * \log_2 p_i\right) \quad (1)$$

where p_i is the likelihood that a subjective tuple in "A" has a place with class B (i) and is evaluated by B (i, A)/A. If entropy is maximum it is difficult to select the class based on this information. The dataset has progressive number of bits that speaks of itself. CCDG underlines the impact of entropy in the dataset. It also helps to find a choice tree for that one root needs to be picked that guarantees the quality of tree. It should also not overfit the information by going till the most extreme end. Therefore, entropy of each property is determined using Eq. (2).

$$\mathrm{Info_A(A)} = \sum_{J=1}^{v} (|A_j|/|A|) * \mathrm{Info}(A_j) \tag{2}$$

where $\mathrm{Info_A(A)}$ is the entropy or extra data required to make an order for a characteristic (|Aj|/|A|), which is also the volume sum of the entropies of every one of the v parcels of the characteristic "D". High volume of data can be classified with incremental entropy as the underlying entropy diminishes gradually. This is the reason for calculation of information gain (IG) of an attribute stated in Eq. (3).

$$\mathrm{Gain(D)} = \mathrm{Info(A)} - \mathrm{Info_A(A)} \tag{3}$$

More the value of Eq. (3) more will be the impact of information gain. In any case, iterative dichotomiser3 check has few apprehensions. C4.5 broadens the probability of iterative dichotomiser3 and later discharges its impact. C4.5 purges the dataset by eliminating split data. This process standardizes the estimation of the data gain across all dataset. Flowchart (Fig. 2) and algorithm for content classification by utilizing data gain (CCDG) are given.

Algorithm for Content classification by utilizing data gain (CCDG)

Step 1 *Find k-clusters to form.*
Step 2 *Create random values centroid equivalent to k-clusters.*
Step 3 *Apply Euclidean distance formula and calculate nearest distance between centroid and input value up to closest distance given in Eq.* (4)

$$d_{ij} = \sqrt{\sum_{k=1}^{p} (X_{ik} - X_{jk})^2 + Y_{ik} - Y_{jk})^2} \tag{4}$$

where d_{ij} *is the Euclidean distance between the cluster* X_{ik} *with the cluster center* X_{jk}.
Step 4 *Reorder distance results based on its nearest distance to the centroid.*
Step 5 *Reset centroid value generated from the step 4 and satisfy following Eq.* (5):

$$C_k = \frac{1}{n_k} \sum \mathrm{d_i} \tag{5}$$

C_k *= cluster value,* n_k *= total data in the cluster,* d_i *= sum of incoming cluster distances.*
Step 6 *Repeat steps 2 to 5 until no value changed.*

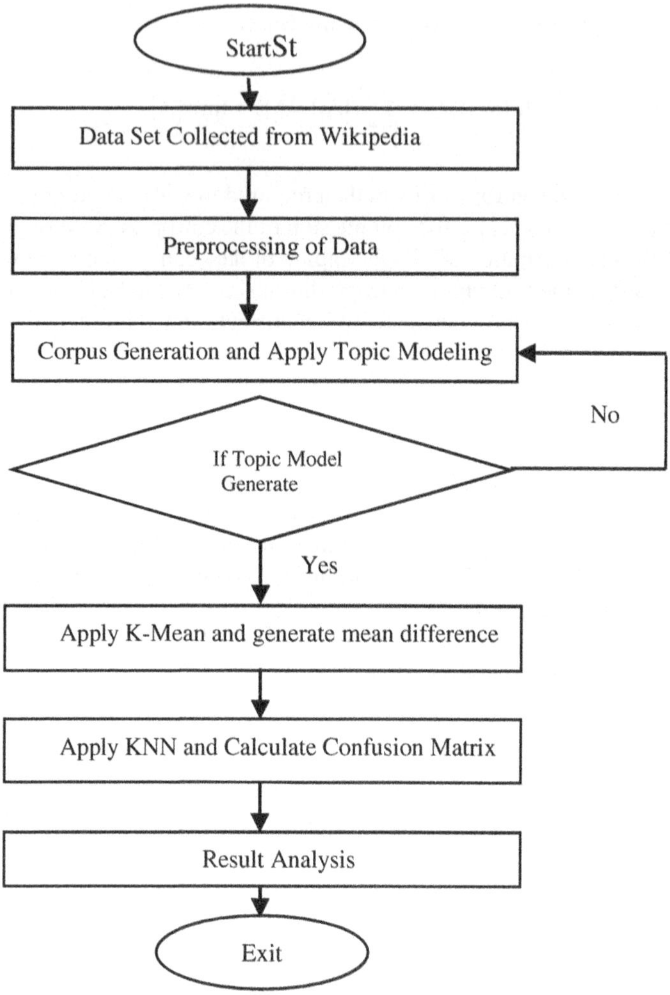

Fig. 2 Flow graph of content classification by utilizing data gain (CCDG)

Step 7 *On the off chance that stage 6 has been satisfied, at that point the mean estimation of the bunch focus (μ_j) in the last cycle will be utilized as a parameter to decide the order of the information.*

The aim of this system is that when the partition of the dataset is resolved, the point where the last detachment is done is picked as the neighbor (k times). Division can decide the total of squares and a detachment involving individual properties. This is the reason if division of dataset occurs that partition will be given priority that has more number of base properties. Therefore, when a partition having elements with more base properties is there then chances of estimation of its inclusion in data gain increase. Next section puts forward experimental result of the proposed CCDG and state of art.

4 Result and Analysis

This section presents subsequent implementation of the proposed content classification by utilizing data gain (CCDG). Experiments are performed to calculate the results with proposed CCDG and current state of art (KNN). CCDG is implemented in Python and tested with five different datasets (Scale, Weight and Distance Car estimate Mushroom Breast Cancer Wisconsin and Thyroid datasets). Table 1 gives a view of the same. Principle estimation of K is calculated by general approach and is represented through square base of the quantity of the dataset. Experiments are performed to analyze CCDG first and then a comparison has been made for CCDG and current state of the art (KNN) under different test conditions for different datasets. Experimental result and its subsequent analysis are given in Tables 1 and 2.

Experimental results in Table 2 and then through Figs. 3 and 4 show the comparison graph between KNN and the proposed CCDG. These results very clearly illustrate that proposed CCDG works well as compared to the existing KNN method. New incorporations in CCDG enable it to yield better results for all the datasets for both number of rows and features.

Table 1 Dataset optimization using CCDG

S. no.	Dataset	No. of rows	No. of features	Type
1	Scale, weight and distance	625	4	Unqualified
2	Car estimate	1728	6	Unqualified
3	Mushroom	2500	21	Mixed
4	Breast cancer Wisconsin	730	10	Mixed
5	Thyroid	1225	21	Mixed

Table 2 Dataset accuracy in percentage with respect to K-NN and CCDG

S. no.	Dataset	No. of rows		No. of features	
		KNN%	CCDG%	KNN%	CCDG%
1	Scale, weight and distance	22	20	79	64
2	Car estimate	47	41	70	69
3	Mushroom	50	42	91	88
4	Breast cancer Wisconsin	27	21	88	80
5	Thyroid	35	29	94	92

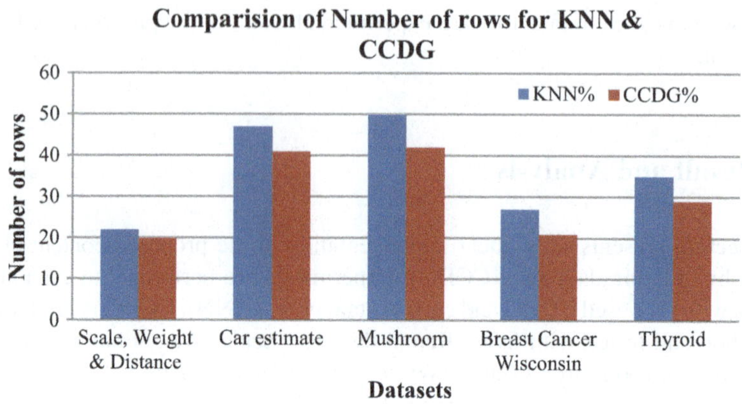

Fig. 3 Result comparison graph of number of rows for KNN and CCDG

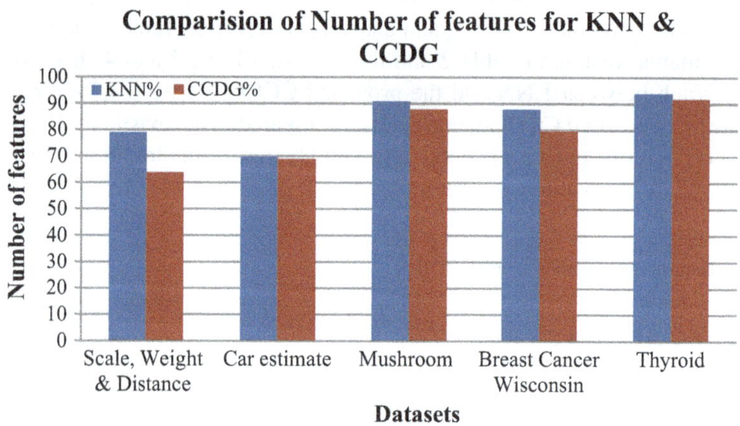

Fig. 4 Result comparison graph of number of features for KNN and CCDG

5 Conclusion

Content classification task is naturally arranging a lot of records into classifications from a predefined set using different methods for data representation and trend analysis. Over the years, it has shown significant significance in real-world applications. This paper introduces two-phase determination strategy for content classification (CCDG) by utilizing data gain. First phase of CCDG involved each term inside the archive is positioned according to their significance for grouping while second phase puts forward the hereditary calculation (GA) and PCA investigation. This included determination and highlighting extraction techniques independently to the terms which are positioned in decreasing request of significance. Experimental results of CCDG with new incorporations yield better results for all the datasets (Scale, Weight and Distance Car estimate Mushroom Breast Cancer Wisconsin and Thyroid datasets) under all the parameters for all the test conditions as compared to the current state of art.

References

1. Ertugrul, Ö.F., Tagluk, M.E.: A novel version of k nearest neighbor: dependent nearest neighbor. Appl. Soft Comput. **55**, 480–490 (2017)
2. Singh, A., Deep, K., Grover, P.: A novel approach to accelerate calibration process of a k-nearest neighbor classifier using GPU. J. Parallel Distrib. Comput. **104**, 114–129 (2017)
3. Parvin, H., Alizadeh, H., Minati, B.: A modification on k-nearest neighbor classifier. Glob. J. Comput. Sci. Technol. (2010)
4. Keller, J.M., Gray, M.R., Givens, J.A.: A fuzzy k-nearest neighbor algorithm. IEEE Trans. Syst. Man. Cybern. **4**, 580–585 (1985)
5. Faziludeen, S., Sankaran, P.: ECG beat classification using evidential K-nearest neighbours. Proc. Comput. Sci. **89**, 499–505 (2016)
6. Song, Y., Liang, J., Lu, J., Zhao, X.: An efficient instance selection algorithm for k nearest neighbor regression. Neurocomputing **251**, 26–34 (2017)
7. Bian W.: Fuzzy-rough nearest-neighbor classification method: an integrated framework. In: Proceedings of the IASTED internet. conference on applied informatics, pp. 160–164, Austria (2002)
8. Nguyen, B., Morell, C., De Baets, B.: Large-scale distance metric learning for k-nearest neighbors regression. Neurocomputing **214**, 805–814 (2016)
9. Lin, Y., Li, J., Lin, M., Chen, J.: A new nearest neighbor classifier via fusing neighborhood information. Neurocomputing **143**, 164–169 (2014)
10. Manocha, S., Girolami, M.A.: An empirical analysis of the probabilistic k-nearest neighbour classifier. Pattern Recogn. Lett. **28**(13), 1818–1824 (2007)
11. Sarkar, M.: Fuzzy-rough nearest neighbor algorithms in classification. Fuzzy Sets Syst. **158**(19), 2134–2152 (2007)
12. Timofte, R., Van Gool, L.: Iterative nearest neighbors. Pattern Recogn. **48**(1), 60–72 (2015)
13. Roweis S.T., Saul L.K.: Nonlinear dimensionality reduction by locally linear embedding. Science **290**(5500), 2323–2326 (2000)
14. Aharon, M., Elad, M., Bruckstein, A.: SVD: an algorithm for designing over complete dictionaries for sparse representation. IEEE Trans. Signal Process. **54**(11), 4311–4322 (2006)
15. Sahu S., Saurabh P., Rai S.: An enhancement in clustering for sequential pattern mining through neural algorithm using web logs. In: International conference on computational intelligence and communication networks, pp. 758–764 (2015)
16. Saxena, M., Saurabh, P., Verma, B.: A new hashing scheme to overcome the problem of overloading of articles in Usenet, pp. 967–975. Springer, AISC (2012)
17. Mishra, B.K., Saurabh, P., Verma, B.: A novel approach to classify high dimensional datasets using supervised manifold learning, pp. 22–30. Springer, CCIS (2012)

References

An Effective Framework for Skyline Queries Using Principal Component Analysis

Meenakshi Karsh, Sandeep Rai, Rajesh Boghey and Praneet Saurabh

Abstract Skyline operators are fascinating concepts that let the users extend and evolve a database system. SKY-MR+ algorithm is an efficient framework implemented for skyline operators and queries which uses quad-tree-based histogram, but faces serious limitations and provides inconsistent execution time especially for High datasets. In such cases, it also reports higher processing time with the increase of number of machines in the system. In this paper, an effective framework for skyline queries using principal component analysis (EFSQ-PCA) is proposed and developed which reduces the execution time for High datasets even in the cases of increase in number of machines in the system. The proposed mechanism finds the "Points in Region" using principal component analysis and this forms the base to increase the processing capabilities of skyline queries on various synthetic datasets. Experimental results show improvement in execution time of the proposed EFSQ-PCA as compared to current state of the art under different numbers of dimensions for dataset.

Keywords Skyline · Dimensions · Execution time · PCA

M. Karsh (✉) · S. Rai
Technocrats Institute of Technology Advance, Bhopal 462021, M.P, India
e-mail: meenakshikarsh11@gmail.com

S. Rai
e-mail: sandtec@gmail.com

R. Boghey
Technocrats Institute of Technology, Bhopal 462021, M.P, India
e-mail: rajeshboghey@gmail.com

P. Saurabh (✉)
Mody University of Science and Technology, Lakshmangarh, Rajasthan, India
e-mail: praneetsaurabh@gmail.com

H. Das et al. (eds.), *Progress in Computing, Analytics and Networking*,
Advances in Intelligent Systems and Computing 1119,
https://doi.org/10.1007/978-981-15-2414-1_4

1 Introduction

The skyline operators are enthralling and very important as it returns objects that cannot be able to determine by any other objects [1]. Past decades have witnessed various skyline queries that were very relevant as it involved multi-criteria decision support. The skyline queries are very useful for finding tuple from large dataset. Nowadays, the size of dataset is increasing constantly as a result skyline queries are implemented in dynamic spaces, multidimensional spaces, in subspace in metric spaces, streaming environments, and in time-series data [2, 3]. There are many types of algorithms that were implemented for skyline query handling like window-based, dynamic, conveyed, geometric-based, record-based, separate and vanquish, and dynamic programming algorithms. Also, several variations like k-predominant skyline, top-k overwhelming inquiries, spatial skyline queries, and others were proposed to understand application-explicit problems [4]. Skyline operator identifies balance with possible multiple attributes in the time of exploring unknown data that are not familiar. The query retrieves the number of interesting hotel in our motivating example which determines useful or more interesting spatial queries [5]. Spatial skyline queries (SSQ) introduced for that given dataset of points P, query set of points Q with a "d"-dimensional space, find points "p" that are not determined by any other SSQ, and for every point of data attributes are its distances from query point. In crisis management domain, the residential buildings that have got to be exhausted first within the event of many explosion/fires area unit those that area unit within the abstraction skyline with relation to the unearth locations. The explanation is that these places of area units either most likely treed inside fireplaces or set at the edges of the expanding fire. Since presenting the skyline operator, many economical algorithms were projected for the final skyline question. These algorithms use methods, for example, separate and vanquish, closest neighbor search [5], arranging [3], and file structures [7, 8] to answer the general skyline questions. Many examinations have an additional target on the skyline query preparing during drawback settings like knowledge streams [6] and information living on cell phones [4].

In this paper, an effective framework for skyline queries using principal component analysis (EFSQ-PCA) is proposed and developed which reduces the execution time for High datasets even in the cases of increase in number of machines in the system. The paper is organized as follows: Section 2 provides the related literature. Section 3 represents proposed methodology, Sect. 4 provides the experimental setup and result analysis while Sect. 5 provides conclusion.

2 Related Work

This section reviews the various advances that have been made in skyline queries, Zou et al. [9] implemented efficient skyline queries for the processing using MapReduce algorithm. The skyline queries are useful for finding tuple from a large dataset.

Nowadays, the size of dataset is increasing constantly switching from single-node environment to nonconventional paradigm like MapReduce. Chomicki et al. [10] planned and enforced a presorting formula for skyline supported theory and optimizations. There has been intriguing as of late in skyline queries, conjointly known as Pareto queries, on relative databases. In another work, [11] have investigated what the skyline recommends and why skyline questions are helpful, remarkably for communicating inclination. Deng et al. [12] in their work enforced a brand new dynamic skyline queries to perform execution of giant graphs. Given a collection of question points, a dynamic skyline question reports all knowledge points that do not seem to be dominated by alternative knowledge points consistent with the distances between knowledge points and question points. Various other works contemplated dynamic skyline in mathematical space [13], street network [14], and mathematical space [15] in their work. Hieu et al. [16] enforced a MapReduce be a piece of method for key worth storage. Here, Map reduce is used to reduce and analyze giant database for significant trends. Wang et al. [17] planned to associate energy economical Skycube question process algorithms in wireless device networks. Bartolini et al. [18] planned to associate economical kind based analysis of skyline operators. Kossmann et al. [19] enforced a web formula for skyline queries supported that the standards of Shooting Stars within the Sky. Skyline queries request a lot of fascinating focuses from a possibly substantial arrangement of information focuses. Over the years, various methods have been introduced [20, 21] but none of them yielded desired results. This paper puts forward an effective framework for skyline queries using principal component analysis (EFSQ-PCA) which is proposed and developed and which reduces the execution time for High datasets even in the cases of increase in number of machines in the system. The proposed mechanism finds the "Points in Region" using principal component analysis and this forms the base to increase in the processing capabilities of skyline queries on various synthetic datasets.

3 Proposed Methodology

This section presents an effective framework for skyline queries using principal component analysis (EFSQ-PCA) which is proposed and developed and which reduces the execution time for High datasets even in the cases of increase in number of machines in the system. Here is a quad-tree-based histogram for space partitioning that has been designed by the basis of the leaf hubs based on the benefit of the part as far as the evaluated execution time. Dominance control separating strategy to successfully prune non-skyline point in advance is applied and then data based on the regions divided by the quad-tree too, resulting in the skyline applicant point representing a skyline.

3.1 Principle Component Analysis

The feature space is calculated in the following way. Given a set of centered observations $\sum_{i=1}^{M} \mathrm{Xi} = 0$, Xk, where k = 1,..., M, the traditional way of formulating the covariance matrix using PCA is given in Eq. (1) as follows:

$$C = \frac{1}{M \sum\limits_{j=1}^{\mathrm{M}} Xj\ \mathrm{Xj}^{\mathrm{t}}} \tag{1}$$

Now the nonlinear feature space F must be defined. F is related to the input space by a possibly nonlinear map in (2)

$$\emptyset : R^{N} \rightarrow F \tag{2}$$

The covariance matrix in F can now be defined as given in Eq. (3) as

$$C' = 1/M \sum\nolimits_{j=1}^{\mathrm{M}} \Phi(Xj)\Phi(\mathrm{Xj}^{\mathrm{t}}). \tag{3}$$

We then determine each eigenvalue λ and corresponding eigenvectors V and C' which satisfy as stated in Eq. (4):

$$\lambda \mathrm{V} = \mathrm{C'V} \tag{4}$$

All situations of V with 0 lie in the span ($\mathrm{X_1}$) and ($\mathrm{X_M}$), and there also exists coefficient αi such that stated in (5):

$$V = \sum_{i=1}^{\mathrm{M}} \alpha\mathrm{i}\Phi(Xi) \tag{5}$$

The M * M Kernel matrix Kij, where j = 1.........M, is defined as stated in Eq. (6):

$$\mathrm{Kij} = (\Phi(\mathrm{Xi}).\Phi(\mathrm{Xj})) \tag{6}$$

where (*) denotes dot product. Thus, the KPCA problem is determining k to satisfy Eq. (7):

$$\mathrm{M}\lambda\mathrm{K}\alpha = \mathrm{K.K}\alpha \tag{7}$$

where a denotes the column vector with entries α1, α2.....m αM to find the solutions of previous equation, one solves $\mathrm{M}\lambda\alpha = \mathrm{K}\alpha$.

3.2 Skewness Algorithm for the Avoidance of Overload

It applies the skewness calculation to maintain a strategic distance from the overload by applying the unevenness in the allocations of the resource. The "N" number of resource allotment with the "R" number of resources can apply to their individual servers P based on Eq. 8:

$$skewness\,P = \sqrt{\sum_{i=1}^{n}\left(\frac{ri}{\bar{r}}\right)^2} \tag{8}$$

where is the normal usage of the considerable number of resources for the "N" number of resources for servers P, the mitigation of resources should be done by looking at the limit.

3.3 EFSQ-PCA Algorithm

1. *Input (D,σ, m, ∂)*
2. *Given a 'D' dataset of 'd' dimensional dataset of sample size 'σ'.*
3. *'m' denotes as number of machines, '∂' a size of threshold value.*
4. *The algorithm initiates with the statistical analysis of dataset based on the size of samples in dataset.*

   S=statistical analysis (σ,D)

5. *The sampled dataset is then passed as an input to the Sky-QTree+ based on the number of machines*

   Q=SKY-QTree+(S,m)

6. *The sampled dataset is then passed as an input to the Sky-QTree+ based on the number of machines.*

   Q=SKY-QTree+(S,m)

7. *The SKY-Tree+ algorithm is then load balanced using local load balancing and number of machines and Sampled data including to apply skewness algorithm to minimize the chances of overloading.*
   A_L*=LocalBalance (Q,S,m)*

8. Broadcast Q and $\boldsymbol{A_L}$

9. Applying Principal Component Analysis on the input Dataset and the broadcast Q and (LocalSL, Vmax, FILTER, COUNT)=RunPCA (boradcast (Q,$\boldsymbol{A_L}$*), D)*

10. If LocalSL.totalSize< threshold (∂)
11. SL= ∀(LocalSL,VMax,FILTER)
12. Else
13. $\boldsymbol{A_G}$*= GlobalBalance (Q, COUNT,m)*
14. Broadcast Q, VMax, FILTER and Ag
15. SL=RunPCA(∀+,LocalSl)
16. Return SL

Consider a d-dimensional dataset denoted as "D" which consists of points {p1, p2, p3…pn}, the points in the "D"-dimensional dataset are represented by 26 <*pi*1, 2, …, *pid*> where *pi*(*k*) denotes the k-th coordinate of pi. It also contains "m" as number of machines and "∂" as size of threshold value. The algorithm initiates with the sampling of dataset based on the size of samples in dataset. Sampling function takes σ (sample size of data) of D-dimensional as an input and it draws the arbitrary example S from the dataset "D", and it at that point parts it into two sections S1 and S2 containing |S1|, |S2| focuses individually and processes their skyline cardinalities *SKYS*1, |*SKYS*2|. Since S1 and S2 have a similar dissemination as D, their skyline cardinalities are relied upon to follow a similar model as that of D. Hence, $SKYS1 = AlogBS1$ and $SKYS2 = AlogBS2|$.

The sampled dataset is then passed as an input to the Sky-QTree+ based on the number of machines which takes the sampled data and number of machines available to process data. Sky-QTREE+ algorithm find constants "A" and "B" from the sampled Data "S" using

$$A = \frac{|SKYs1|}{\log B|S1|},\ B = \frac{log|SKYs2| - log|SKYs1|}{\log(\log|S2| - \log(log|S2|)}$$

Sky-QTREE+ is constructed and will return "Q" using [1].

After the tree is constructed, local balance function is used for the balancing of tree with sampled data and machines available. To adjust the remaining tasks at hand of machines in the worldwide skyline stage, the principal capacity gathers the quantity of the local skyline focuses on each of the unpruned leaf node n and computes up (R (n)). Finally, the generated Q-tree and local balanced Al are broadcasted and principal component analysis will take L-SKY-MR+ and dataset D to return local cardinality point. The covariance matrix is used for the repetitive process of the local point which is then compared with the threshold value. If the value of local cardinality point is less than threshold value then G-SKY function is called with local point and Vmax and filters value as an input and returns Sl. Otherwise, global balancing function is called with Q and count and available machines which will return AG. These Q and Vmax and filter and Ag are again broadcasted and PCA is computed again based on G-SKY-MR+ and local cardinality point to give SL which is then returned again and again till the query is fetched.

3.4 Flowchart

This subsection presents the flowchart of the proposed effective framework for skyline queries using principal component analysis (EFSQ-PCA) in Fig. 1.

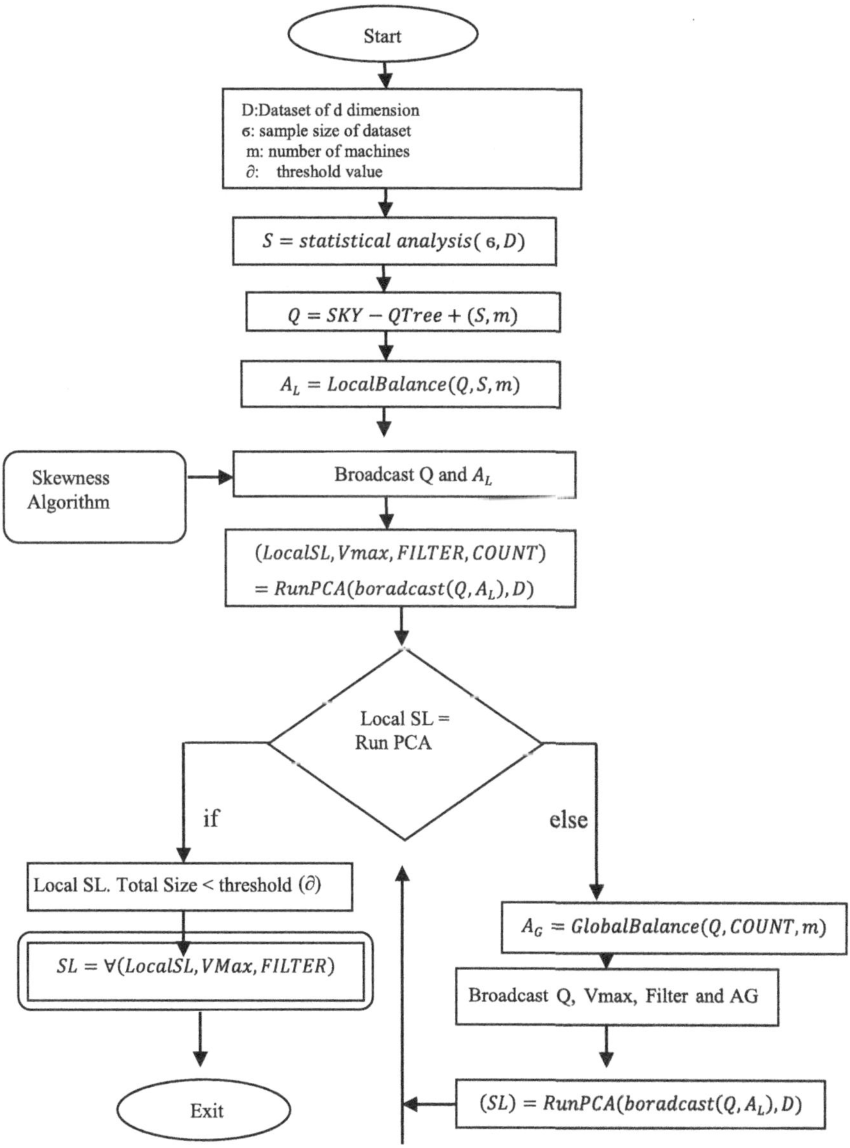

Fig. 1 Flowchart of the EFSQ-PCA mechanism

4 Experimental Setup and Result Analysis

This section measures the performance of skyline queries SKY-MR+ and EFSQ-PCA and then a comparison can be drawn under same test condition using Intel Core i3 processor and 4 GB RAM. The operators on the simulation environment are based on java JDK 1.8 environment and Net Beans 7.4.

4.1 Result Analysis

Experimental results in Table 1 show the analysis of execution time in seconds for k = 10, on various sampled points from 100 to 10000 and for both the existing SKY-MR+ and EFSQ-PCA.

Experimental results show that under the sampled points EFSQ-PCA outperforms the current SKY-MR+ algorithm. EFSQ-PCA consumes less execution time in giving result and becomes more efficient due to new integrations. Figure 2 shows the analysis of execution time for various sampled points from 100 to 10000 for SKY-MR+ and

Table 1 Analysis of execution time with k = 10

Execution time (Sec) with k = 10		
# No. of sampled points	SKY-MR+	EFSQ-PCA
100	76	63
200	75	65
400	78	69
1000	77	67
2000	82	72
4000	84	74
10000	92	78

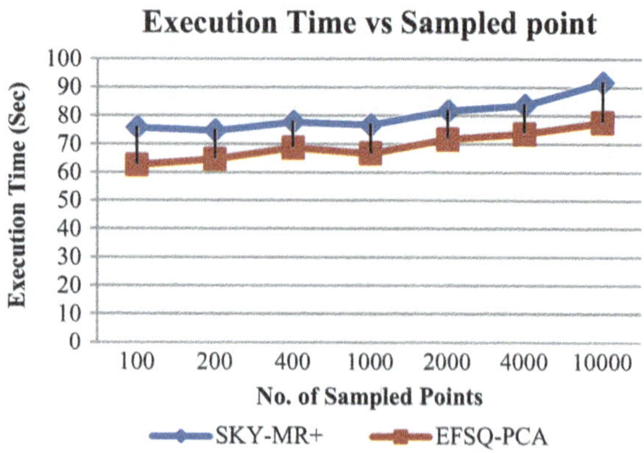

Fig. 2 Comparison of execution time on k = 10

Table 2 Analysis of execution time on ANTI time

Execution Time (Sec) on ANTI		
# No. of sampled points	SKY-MR+	EFSQ-PCA
2	100	80
4	250	160
6	380	300
8	550	450
10	1000	700
12	5000	3500

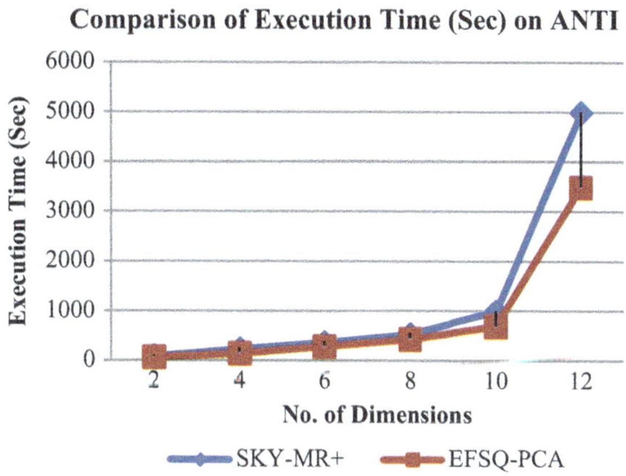

Fig. 3 Comparison of execution time on ANTI

EFSQ-PCA. Under all the sampled points, EFSQ-PCA algorithm provides efficient and less execution time in comparison with existing SKY-MR+.

Table 2 and Fig. 3 present experimental results of SKY-MR+ and EFSQ-PCA under different dimensions from 2 to 12 for ANTI dataset. These results very clearly show that the proposed EFSQ-PCA yields better performance as compared to the current state-of-the-art SKY-MR+. In some cases, EFSQ-PCA reported even 20% better performance. These results also indicate that the newly introduced mechanism is helping EFSQ-PCA to outperform SKY-MR+.

Table 3 and Fig. 4 represent the experimental results and subsequent analysis of relative speed of execution on various numbers of machines from 10 to 40 for SKY-MR+ and EFSQ-PCA. EFSQ-PCA demonstrates its superiority and provides efficient and high relative speedup in comparison with existing Sky-MR+ algorithm.

Experimental results presented in both Table 3 and Fig. 4 show the analysis of relative speed on various numbers of machines from 10 to 40 for both SKY-MR+ and EFSQ- PCA enabled the proposed EFSQ- PCA to give better speed as compared to SKY-MR+.

Table 3 Analysis of relative speed

Execution time (Sec) with k = 10		
# No. of sampled points	SKY-MR+	EFSQ-PCA
10	1	1.3
15	1.5	1.8
20	2	2.2
25	2.5	2.9
30	3	3.5
35	3.5	4.1
40	4	4.6

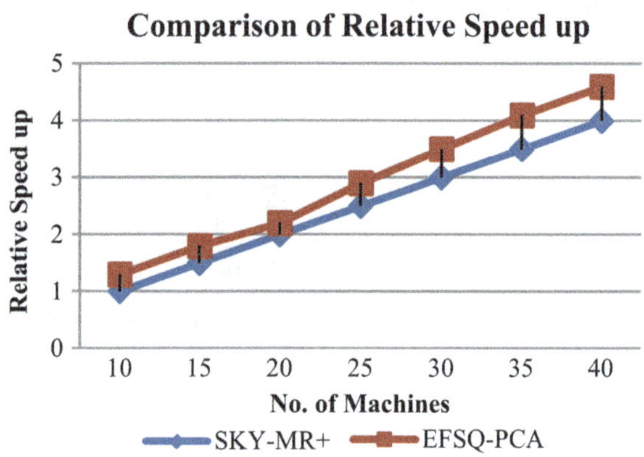

Fig. 4 Comparison of relative speedup

5 Conclusion

The skyline operator could also be actualized straightforwardly in SQL utilizing the current SQL builds. Other algorithms use lists, MapReduce, and general processing on graphics cards. This paper introduced an effective framework for skyline queries using principal component analysis (EFSQ-PCA) to reduce the execution time for High datasets even in the cases of increase in number of machines in the system. The proposed EFSQ-PCA finds the "Points in Region" using principal component analysis and this forms the base to increase in the processing capabilities of skyline queries on various synthetic datasets. All the experimental results show improvement in execution time and relative speed for EFSQ-PCA as compared to the current state of the art.

References

1. Borzsonyi, S., Kossman, D., Creature K.: The skyline operator. In: Proceedings of the ICDE, pp. 421–430 (2001)
2. Chomicki, J., Godfrey, P., Gryz, J., Liang, D.: Skyline with presorting. In: Proceedings of ICDE, pp. 717–816. IEEE Laptop Society (2003)
3. Huang Z., Jensen, C.S., Lu, H., Ooi, B.C.: Skyline queries against mobile light-weight devices in MANETs. In: Proceedings of ICDE'06. IEEE Laptop Society (2006)
4. Kossmann, D., Ramsak, F., Rost, S.: Shooting stars among the sky: a web algorithmic program for skyline queries. In: Proceedings of VLDB'02, pp. 275–286 (2002)
5. Lin, X., Yuan, Y., Wang, W., Lu, H.: Stabbing the sky: economical skyline computation over slippery windows. In: Proceedings of ICDE'05, pp. 502–513 (2005)
6. Papadias, D., Tao, Y., Fu, G., Seeger, B.: Progressive skyline computation in information systems. ACM Trans. Info Syst. **30**(1), 41–82 (2005)
7. Tan, K.-L., Eng, P.-K., Ooi, B.C.: Economical progressive skyline computation. In: Proceedings of VLDB'01, pp. 301–310 (2001)
8. Park, Y., Min, J., Wedge, K.: Economical method of skyline queries exploitation map reduce. IEEE Trans. Inf. Knowl. Eng. (2016)
9. Zou, L., Chen, L., Zsu, M.T.O., Zhao, D.: Dynamic skyline queries in large graphs. In: DASFAA, pp. 62–78 (2010)
10. Chomicki, J., Godfrey, P., Gryz, J., Liang, D.: Skyline with presorting: theory and optimizations. In: ICDE, pp. 717–719 (2003)
11. Sharifzadeh, M., Shahabi, C.: The spatial skyline queries. In: VLDB, pp. 751–762 (2006)
12. Deng, K., Zhou, X., Shen, H.T.: Multi-source skyline query processing in road networks. In: Proceedings of ICDE conference, pp. 796–805 (2007)
13. Chen, L., Lian, X.: Dynamic skyline queries in metric areas. In: EDBT, pp. 333–343 (2008)
14. Gao, Y., Qin, X., Zheng, B., Chen, G.: Efficient reverse top-k boolean spatial keyword queries on road networks. IEEE Trans. Knowl. Data Eng. **27**(5), 1205–1218 (2015)
15. Ganesan, P., Yang, B., Garcia-Molina, H.: One torus to rule them all: multidimensional queries in p2p systems. In: Proceedings of WebDB workshop, pp. 19–24 (2004)
16. Lin, X., Xu, J., Hu, H.: Range-based skyline queries in mobile environments. IEEE Trans. Knowl. Data Eng. **25**(4), 835–849 (2013)
17. Wang, Z., Jin, S., Gong, K.: Energy-efficient skycube question method in wireless detector networks. Telkomnika **11**(10), 6240–6249 (2013)
18. Bartolini, I., Ciaccia, P., Patella, M.: Efficient sort-based skyline analysis. ACM Trans. Info. Syst. **33**(4), 1–49 (2008)
19. Kossmann, D., Ramsak, F., Rost, S.: Shooting stars among the sky: a web algorithmic program for skyline queries. In: Proceedings of the twenty eighth VLDB conference, Hong Kong, China (2002)
20. Saxena, M., Saurabh, P., Verma, B.: A new hashing scheme to overcome the problem of overloading of articles in Usenet, pp. 967–975. Springer, AISC (2012)
21. Mishra, B.K., Saurabh, P., Verma, B.: A novel approach to classify high dimensional datasets using supervised manifold learning, pp. 22–30. Springer, CCIS (2012)

Disease Classification Using Linguistic Neuro-Fuzzy Model

Himansu Das, Bighnaraj Naik and H. S. Behera

Abstract In recent years, due to advancement in medical technologies and its devices, a large volume of medical data is generated continuously from different sources at every moment. Analyzing these large volumes of medical data and correctly diagnosing the diseases are challenging tasks. Generally, these medical data contain uncertain, imprecise, and incomplete information that affects the performance of the classification model. In this paper, Linguistic Neuro-Fuzzy (LNF) model is used for the classification of diseases. First, this model uses a linguistic fuzzification process that computes the membership values of each feature to overcome the uncertainty issues. Second, these membership values are passed to the ANN-based model to predict the disease. The objective of this research work lies in the applications of this LNF model that predicts the diseases. The effectiveness of this model is tested and validated using six benchmark medical datasets. The performance of the LNF model is compared with ANN and observed that the LNF model outperforms than ANN to handle the uncertainty problem.

Keywords Classification · Machine learning · Neuro-fuzzy · ANN · Linguistic variable

---

H. Das (✉) · H. S. Behera
Department of Information Technology, Veer Surendra Sai University of Technology, Burla, Sambalpur 768018, Odisha, India
e-mail: das.himansu2007@gmail.com

H. S. Behera
e-mail: hsbehera_india@yahoo.com

B. Naik
Department of Computer Application, Veer Surendra Sai University of Technology, Burla, Sambalpur 768018, Odisha, India
e-mail: mailtobnaik@gmail.com

H. Das et al. (eds.), *Progress in Computing, Analytics and Networking*,
Advances in Intelligent Systems and Computing 1119,
https://doi.org/10.1007/978-981-15-2414-1_5

1 Introduction

Nowadays disease analysis with machine learning [1] techniques is one of the emerging researches among the academic and industry research community due to the increase in the use of computationally intensive methods. This medical data analysis is a sensitive issue that always needs to be a correct prediction, detection, and analysis of the diseases. So, it is essential to develop and use the appropriate machine learning classification algorithms that are correctly detecting and diagnosing the diseases in a more accurate and efficient manner. But still, there are some clinical issues such as accuracy and reliability need to be addressed that accurately diagnose the diseases. Generally, medical datasets contain noisy, irrelevant, redundant, and incomplete information that affects the performance of the classification model. It is essential to provide the correct and accurate data to the classification model. However, the usefulness of these diagnostic techniques has also some limitations and sometimes overfits or underfits. So, it essential to use the classifiers potentially to analyze the sensitive medical data is correct and accurate way to predict and diagnose the diseases.

Classification [2–4] is the process of constructing a model that accurately predicts the target class from the data into different class levels. The most frequently used classification algorithm called Artificial Neural Network (ANN) [5] is having some pitfalls such as slow convergence rate, a trap into local minima, black box in nature, and low flexible for complex problems. It is a massive parallel model that has the self-adaptive and self-learning capability but it takes more time to predict the result due to its massive parallel in structure. ANN can process numerical data more accurately but it is not suitable to process linguistic variables. It is also not suitable to handle uncertain or imprecise data. Fuzzy Logic (FL) [6] is suitable to handle such data by transforming each feature into its corresponding linguistic variables such as *low*, *medium*, and *high*. The membership value of each feature is expanded to three times the original features which make the network more complex. Every individual technique such as FL and ANN is having its own advantages and disadvantages. To overcome this issue, these individual techniques (FL and ANN) are hybridized to form a hybrid model called Neuro-Fuzzy (NF) [7, 8] model that takes the advantages of both the individual techniques. Apart from this hybrid model, there are some other hybrid models like the Adaptive Neuro-Fuzzy Inference System (ANFIS) [9] that inherits the properties of both the NN and FL. ANN learns itself from data and self-adaptable accordingly but it cannot able to interpret the knowledge from data. The major drawback of the ANFIS model is that it is governed by the rule-based system. It is also essential to generate the correct set of rule to predict the output. In the case of large datasets, the ANFIS model is also not suitable for data analysis due to its large set of rule set generation which makes the network more complex. To address aforesaid issues, Linguistic Neuro-Fuzzy (LNF) [10] model is hybridized that took the advantages of both ANN and FL to address the uncertainty and imprecise input information issues. Pal et al. [10] developed a model in which each feature is fuzzified with the membership values based on the linguistic properties. First, this LNF model transforms the input feature information into its corresponding membership values of the linguistic variables in the fuzzification process. Second, this fuzzified matrix is passed to the NN to predict the disease. The major drawback of the LNF model is that it increases the feature space into three times that makes the network more complex. Apart from this, it does not rely on class label attribute. Some other variants of NF [11–13] models are used for medical disease analysis and the applications of the NF model are described in [14]. The objective of this research work lies in the applications of this LNF model that predicts the diseases more accurately and efficiently. This paper is organized as follows: the detailed architecture of the LNF model is described in Sect. 2. In Sect. 3, the LNF model is implemented using several medical datasets for disease classification. This model also compares the result (performance) with the ANN-based model, and finally Sect. 4 concludes the paper.

2 Linguistic Neuro-Fuzzy Classification Model

In this paper, the LNF model is used to predict the disease for medical data. This model works on the two phases: (1) fuzzification process and (2) the ANN process. In the fuzzification process, each feature of the input pattern is transformed into three membership values with respect to the linguistic three linguistic membership functions such as low, medium, and high. Here, the Π-type membership function is used to represent each feature of any pattern into its corresponding membership value. Due to this fuzzy expansion, the number of features is expanded into its corresponding linguistic membership values (three times) as compared to the original input features to address the uncertainty issue. The detailed fuzzification process is described as follows.

Let us have dataset D, the ith pattern of all the features for the dataset D is presented in Eq. 1.

$$P_i = [F_{i,1}, F_{i,2,} \ldots, F_{i,n,}] \tag{1}$$

The membership value of jth feature of the ith pattern is presented in Eq. 2 by using Π-type membership function. The Π-type membership function used in fuzzification process is described in Eq. 3. Figure 1 presents the linguistic properties of Π-type membership function in which any feature value is transformed into its corresponding fuzzified value based on the linguistic properties. Similarly, all the features of ith pattern contain 3 × n number of fuzzified features, where n is the number of input features available in the dataset which is described in Eq. 4.

$$F_{i,j} = [\mu_{low}(f_{i,j}), \mu_{medium}(f_{i,j}), \mu_{high}(f_{i,j})] \tag{2}$$

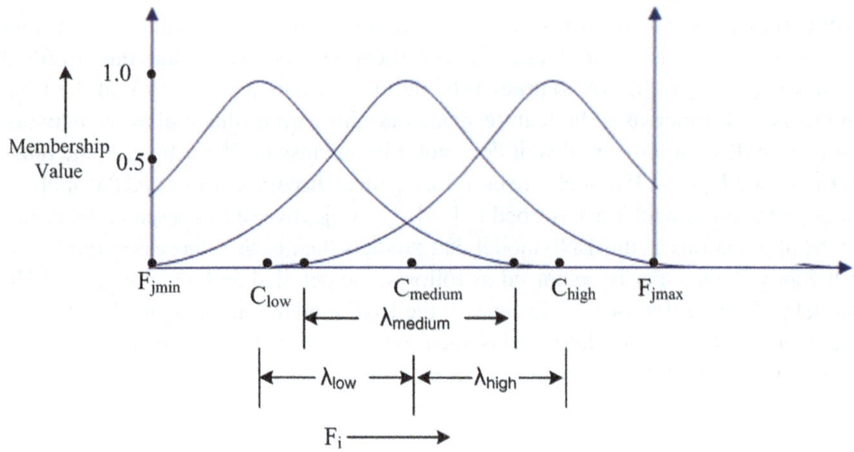

Fig. 1 Linguistic properties representation of Π-type membership function

$$\pi(f_{i,j}; C, \lambda) = \begin{cases} 1 - 2\left(\frac{\|f_{i,j}-C\|}{\lambda}\right)^2 & for\, 0 \leqslant \|f_{i,j} - C\| \leqslant \frac{\lambda}{2} \\ 2\left(1 - \frac{\|f_{i,j}-C\|}{\lambda}\right)^2 & \text{for } \frac{\lambda}{2} \leqslant \|f_{i,j} - C\| \leqslant \lambda \\ 0 & Otherwise \end{cases} \tag{3}$$

where λ and C represent the radius and centroid of the membership function, respectively.

$$F_{j,i} = [\mu_{low}(f_{1,i}), \mu_{medium}(f_{1,i}), \mu_{high}(f_{1,i}), \mu_{low}(f_{2,i}), \mu_{medium}(f_{2,i}), \mu_{high}(f_{2,i}), \ldots, \mu_{low}(f_{n,i}), \mu_{medium}(f_{n,i}), \mu_{high}(f_{n,i})] \tag{4}$$

The maximum and minimum values of the jth feature is computed and presented as $F_{j(\max)}$ and $F_{j(\min)}$, respectively. The radius and central point of the medium linguistic membership function are described in Eq. 5 and Eq. 6, respectively.

$$\lambda_{medium}(F_j) = \frac{1}{2}(F_{j\max} - F_{j\min}) \tag{5}$$

$$C_{medium}(F_j) = F_{j(\min)} + \lambda_{medium}(F_j) \tag{6}$$

The radius and central point of the low linguistic membership function are presented in Eq. 7 and Eq. 8, respectively, where *cp* is the controlling parameter of overlapping.

$$\lambda_{low}(F_j) = \frac{1}{cp}(C_{medium}(F_j) - F_{j(\min)}) \tag{7}$$

$$C_{low}(F_j) = C_{medium}(F_j) - 0.5 \times \lambda_{low}(F_j) \tag{8}$$

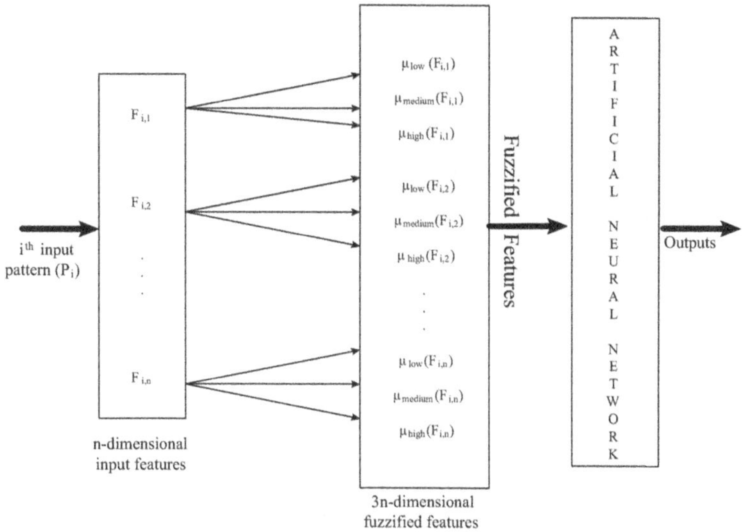

Fig. 2 LNF model

Similarly, the radius and central point of the high linguistic membership function is presented in Eq. 9 and Eq. 10, respectively.

$$\lambda_{high}(F_j) = \frac{1}{cp}(F_{j(\max)} - C_{medium}(F_j) \tag{9}$$

$$C_{high}(F_j) = C_{medium}(F_j) + 0.5 \times \lambda_{high}(F_j) \tag{10}$$

In this LNF model, each feature of input pattern is transformed into its corresponding fuzzified value based on the three linguistic fuzzy variable called *Low*, *Medium*, and *High*. These expanded fuzzified values are determined from its corresponding linguistic membership values of the Π-type membership function.

In the second step, these fuzzified membership values pass to the ANN-based model to predict the diseases. The detailed working model of LNF model is presented in Fig. 2. This model can be successfully applied in several diversified applications [15] such as biomedical signal processing [16], healthcare [17–19], big data analytics [20], forecasting [21], nature inspired computing [22] and machine translation [23].

3 Result Analysis

This section investigates the performance of two classification algorithms such as ANN and LNF by using six medical datasets which are observed. These medical datasets are collected from the UCI machine learning repository [24]. The detailed information (total number of instances, the total number of attributes, number of classes, class label distributions) about these medical datasets are presented in Table 1. The implementation of these classification algorithms is performed by using Python 3.6.5 and the classification accuracy, precession, f-measure, and recall are presented in Table 2. It is noticed from the performance measures that the overall performance of the LNF model is superior to that of the ANN model with these six medical datasets. The error plot of these six datasets with two models (ANN and LNF) is presented in Fig. 3. The results presented here are based on experimentation and observation. These two models are executed ten times on each dataset and the average result is presented in Table 2.

Table 1 Dataset information

Name	Total no. of instances	Total no. of attributes	No. of classes	Class distribution	
				Class 1	Class 2
Mammographic mass	830	5	2	403	427
Breast cancer	699(683)	9	2	444	239
Pima Indian diabetes	768	8	2	268	500
Heart Statlog	270	13	2	120	150
Liver	345	6	2	145	200
Blood transfusion service center	748	5	2	178	570

Table 2 Performance measure matrix

Datasets/models	Accuracy		Precession		F-measure		Recall	
	ANN	LNF	ANN	LNF	ANN	LNF	ANN	LNF
Mammographic mass	79.21	74.615	0.733	0.74058	0.61	0.67795	0.523	0.62697
Breast cancer	91.07	92.34	0.971	0.97231	0.6	0.87327	0.658	0.7934
Pima Indian diabetes	72.96	78.47	0.793	0.7947	0.646	0.7378	0.631	0.6406
Heart Statlog	66.5	77.18	0.648	0.72984	0.617	0.72524	0.644	0.72059
Liver	67.26	67.32	0.703	0.71507	0.63	0.65557	0.54	0.59979
Blood transfusion service	75.86	79.36	0.701	0.7993	0.57	0.7101	0.583	0.64229

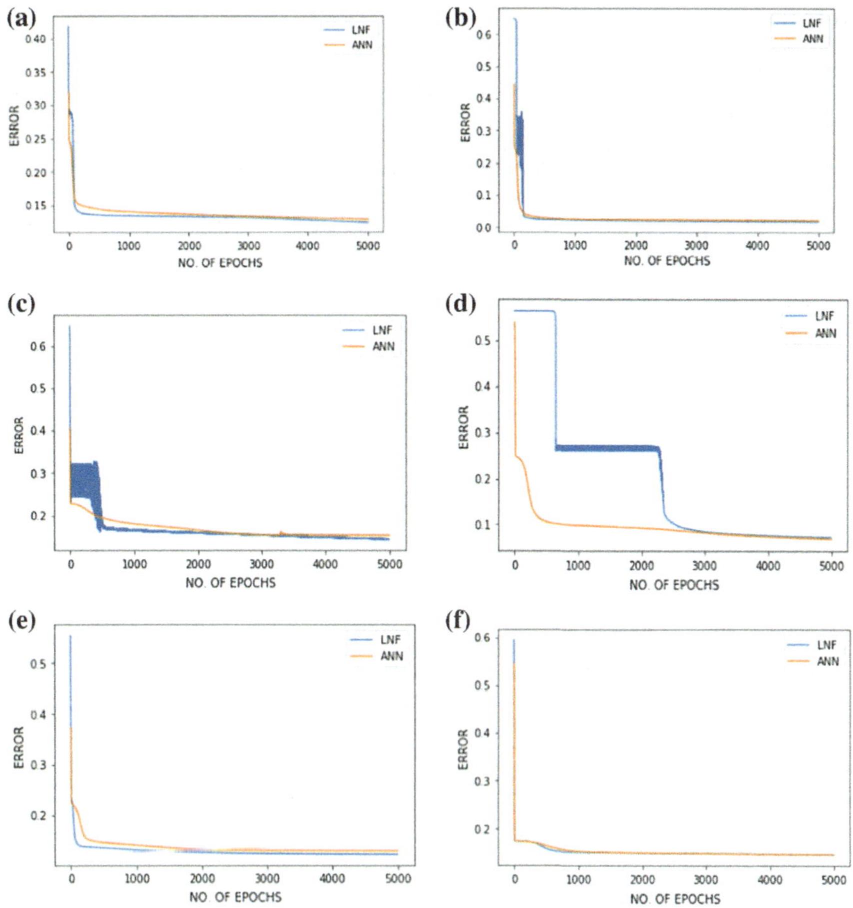

Fig. 3 Error plot of datasets **a** Mammographic mass, **B** Breast cancer, **c** Pima Indian diabetes, **d** Heart Statlog, **e** Liver, **f** Blood transfusion service

4 Conclusion

This work presents the LNF model for medical disease classification for medical data. In this model, input features are fuzzified using a linguistic Π-type membership function to handle the imprecise and uncertain data. These linguistic membership values are passed to the ANN-based model for the classification of the diseases. Based on the experimental analysis, it is observed that the LNF model predicts the diseases more accurately than the ANN model. But the limitation of this LNF model is that the complexity of the model increases due to fuzzy expansion of the input features for which it takes more time to train and test the model. To address this issue, feature extraction and feature selection algorithms will be addressed in our future work.

References

1. Michie, D., Spiegelhalter, D.J., Taylor, C.C.: Machine learning, neural and statistical classification, vol. 13 (1994)
2. Das, H., Jena, A.K., Nayak, J., Naik, B., Behera, H.S.: A novel PSO based back propagation learning-MLP (PSO-BP-MLP) for classification. In: Computational intelligence in data mining-volume 2, pp. 461–471. Springer, New Delhi (2015)
3. Sahani, R., Rout, C., Badajena, J.C., Jena, A.K., Das, H.: Classification of intrusion detection using data mining techniques. In: Progress in computing, analytics and networking, pp. 753–764. Springer, Singapore (2018)
4. Das, H., Naik, B., Behera, H.S.: Classification of diabetes mellitus disease (DMD): a data mining (DM) approach. In: Progress in computing, analytics and networking, pp. 539–549. Springer, Singapore (2018)
5. Haykin, S.: Neural networks, vol. 2. Prentice Hall, New York (1994)
6. Zadeh, L. A.: Fuzzy sets, Inform. Control. **8**, 338–353 (1965)
7. Meher, S.K.: Efficient pattern classification model with neuro-fuzzy networks. Soft. Comput. **21**(12), 3317–3334 (2017)
8. Meher, S.K.: Explicit rough–fuzzy pattern classification model. Pattern Recogn. Lett. **36**, 54–61 (2014)
9. Ubeyli, E.D.: Adaptive neuro-fuzzy inference system for classification of ECG signals using Lyapunov exponents. Comput. Method. Programs Biomed. **93**(3), 313–321 (2009)
10. Pal, S.K., Mitra, S.: Multilayer perceptron, fuzzy sets, and classification. IEEE Trans. Neural Netw. **3**(5), 683–697 (1992)
11. Azar, A.T.: Neuro-fuzzy feature selection approach based on linguistic hedges for medical diagnosis. Int. J. Model. Ident. Control **22**(3), 195–206 (2014)
12. Nilashi, M., Ahmadi, H., Shahmoradi, L., Ibrahim, O., Akbari, E.: A predictive method for hepatitis disease diagnosis using ensembles of neuro-fuzzy technique. J. Infect. Public Health (2018)
13. Shihabudheen, K.V., Pillai, G.N.: Recent advances in neuro-fuzzy system: a survey. Knowl.-Based Syst. **152**, 136–162 (2018)
14. Kar, S., Das, S., Ghosh, P.K.: Applications of neuro fuzzy systems: a brief review and future outline. Appl. Soft Comput. **15**, 243–259 (2014)
15. Pattnaik, P.K., Rautaray, S.S., Das, H., Nayak, J.: Progress in computing, analytics and networking. In: Proceedings of ICCAN, 710 (2017)
16. Pradhan, C., Das, H., Naik, B., Dey, N.: Handbook of Research on Information Security in Biomedical Signal Processing, pp. 1–414. IGI Global, Hershey, PA (2018)
17. Dey, N., Ashour, A.S., Kalia, H., Goswami, R., Das, H.: Histopathological Image Analysis in Medical Decision Making, pp. 1–340. IGI Global, Hershey, PA. https://doi.org/10.4018/978-1-5225-6316-7 (2019)
18. Sahoo, A.K., Mallik, S., Pradhan, C., Mishra, B.S.P., Barik, R.K., Das, H.: Intelligence-based health recommendation system using big data analytics. In: Big Data Analytics for Intelligent Healthcare Management, pp. 227–246. Academic Press (2019)
19. Sahoo, A.K., Pradhan, C., Das, H.: Performance evaluation of different machine learning methods and deep-learning based convolutional neural network for health decision making. In: Nature Inspired Computing for Data Science, pp. 201–212. Springer, Cham (2020)
20. Dey, N., Das, H., Naik, B., Behera, H.S. (eds.): Big Data Analytics for Intelligent Healthcare Management. Academic Press (2019)
21. Rout, M., Jena, A.K., Rout, J.K., Das, H.: Teaching–learning optimization based cascaded low-complexity neural network model for exchange rates forecasting. In: Smart Intelligent Computing and Applications, pp. 635–645. Springer, Singapore (2020)
22. Nayak, J., Naik, B., Jena, A.K., Barik, R.K., Das, H.: Nature inspired optimizations in cloud computing: applications and challenges. In: Cloud Computing for Optimization: Foundations, Applications, and Challenges, pp. 1–26. Springer, Cham (2018)

23. Das, A.K., Pradhan, M., Dash, A.K., Pradhan, C., Das, H.: A constructive machine translation system for English to Odia translation. In: 2018 International Conference on Communication and Signal Processing (ICCSP), pp. 0854–0857. IEEE (2018, April)
24. Bache, K., Lichman, M.: UCI machine learning repository [http://archive.ics.uci.edu/ml], University of California, School of Information and Computer Science, Irvine, CA (2013)

Ant Colony Optimization (ACO-Min) Algorithm for Test Suite Minimization

Subhasish Mohanty, Sudhir Kumar Mohapatra and Sultan Feisso Meko

Abstract This paper presents a test suite (TS) minimization algorithm based on ant colony optimization. The algorithm represents all the test cases in the test suite as nodes of a complete graph. Each test case execution time and corresponding test requirement are stored in the form of a matrix. Ants start from the nodes of the complete graph. The selection of neighbor nodes depends on maximizing requirements and minimizing execution time, for which the ant takes the help of the matrix. The algorithm finds a representative set (RS) of test case which satisfies all the requirements and takes least execution time. The derived representative set satisfies (i) $|RS| \sqsubseteq |TS|$ (ii) $\tau_{RS} \leq \tau_{TS}$. The representative set also preserves the same fault detection capability, like that of the original test suite, which ensures zero compromise on the overall effectiveness of the program.

Keywords Regression testing · Test case reduction · Test case minimization · Ant colony optimization · Representative set · Test suite · Test case

1 Introduction

In any regression testing, software testing and retesting is performed regularly. The growth of software in regression testing involves adding a new test case, to test new functionality that is added to the existing software. The evolutions of software, which are denoted as versions, create redundant test cases within the test suite. These test cases become redundant, as more test cases satisfying the same requirement or covering the same code. The availability of limited resources and time demands the

S. Mohanty
G.I.E.T University, Gunupur, Odisha 765022, India
e-mail: subhasish2307@gmail.com

S. K. Mohapatra (✉) · S. F. Meko
Addis Ababa Science & Technology University, 16417 Addis Ababa, Ethiopia
e-mail: sudhir.kumar@aastu.edu.et; sudhirmohapatra@hotmail.com

S. F. Meko
e-mail: sultan.feisso@aastu.edu.et

H. Das et al. (eds.), *Progress in Computing, Analytics and Networking*,
Advances in Intelligent Systems and Computing 1119,
https://doi.org/10.1007/978-981-15-2414-1_6

detection of those redundant test cases, which exercise the same requirement. The process of removing these redundant test cases is called as test suit minimization. The reduced test case which is derived after the removal of redundant test cases is called as representative set [1].

Problem Statement: Given test case set $T = \{t_1, t_2, t_3, \ldots, t_n\}$ *and testing requirements set* $R = \{r_1, r_2, \ldots, r_m\}$*; the testing requirements must be achieved for the entire coverage of the program. The test case minimization problem is to find a representative set* $T' \sqsubseteq T$*, so that* T' *achieves the same requirements as that of* T *[2, 3].*

The problem of finding $T' \sqsubseteq T$, which covers the entire set of requirements, is NP-complete [4, 5]. Due to the problem's inherent hardness, heuristics-based methods for solving this problem become indispensable [6, 7]. We propose an ant colony based heuristic for solving test case minimization problem, which we call (Ant Colony Optimization) ACO-Min. We perform experiments using five subject programs, out of which, three are open-source programs from Software-artifact Infrastructure Repository (SIR[1]). As per the outcome of our experiments, ACO-Min produces better results by generating a representative set of smaller size. The fault exposing potential (FEP) of the test case is better, as compared to other heuristics. The major findings of the paper are as follows:

1. Guarantee of producing a representative set of the same or reduced size.
2. The minimized test suite takes less time.

This paper discusses related work in Sect. 2. Section 3 contains the algorithm using ant colony optimization. Some experimental results are presented in Sect. 4. In the Conclusion section, the findings of the paper are summarized.

2 Review of Related Work

As it is an NP-complete problem [5, 8], research on test suite minimization has been primarily focused on heuristics that approximate solutions. In [9, 10], the researchers used the Greedy algorithm to detect redundant test cases. Chen and Lau [11] also used the Greedy algorithm, but first, they identified important test cases, which should be present for testing. Then form the remaining test cases, they chose the representative set by using the Greedy algorithm. Jeffrey and Gupta [5] used a searching methodology called as selective redundancy. Jones and Harrold [12] used modified condition/decision coverage (MC/DC) for their minimization technique. Harrold et al. [1] consider representative test case for an individual test case and consider it for the final representative set. Kapfhammer et al. [13] proposed a modified algorithm using irreplaceability. Ma et al. [14], Zhang et al. [15], and Hao et al. [16] used a genetic algorithm (GA) to minimize the test case based on requirement achievement. Mohapatra et al. [17, 18] use genetic algorithm to derived representative set

[1]SIR repository: website, http://sir.unl.edu/portal/index.php/ (Accessed September 15, 2018).

and prioritization of test cases [19, 20]. These research works are based on the fault detection capability of the representative test case. In software testing, time is also a major factor, which we have to consider apart from coverage of requirements.

Mansour and El-Fakih [21] and Black et al. [22] proposed a heuristic-based approach. Yoo and Harman [23] proposed a hybrid algorithm based on evolutionary algorithm and greedy technique for test case minimization. They considered it as a multi-objective optimization problem. On-demand test suite minimization was proposed by Hao et al. [24] where the authors set a limit for test detection capability. Interaction-based technique was proposed by Blue et al. [25]. Zhang et al. [26] proposed a technique, which speeded up mutation technique.

3 Proposed Test Suite Minimization Technique

This paper uses $T = \{t_1, t_2, t_3, \ldots, t_m\}$ as the set of test cases. The requirement set is $R = \{r_1, r_2, r_3, \ldots, r_n\}$. Each test case will satisfy some requirements, i.e., if t_i satisfies r_i, where $r_i \sqsubseteq R$. The test case set is used as the base for construction of a complete graph. Each vertex of the graph is represented by the test cases. The number of ants is at least equal to the number of test cases. The solution will start with the ant in all the vertex. To get the best path, each ant adds new edges to its existing path. Adding of new path is done by the help of find_next() function. This function finds adjacent edges with the highest pheromone deposit. In the case of a tie between edges, it selects edge randomly. This adding of edge stops when no other edge left for the path. The algorithm and its notations are given below.

```
  Input: Complete Graph(Representing Test Case, Matrix Containing test
requirements and execution time).
  Output: A representative test case
  ACO-Min (Complete Graph, Number of ants n , Maximum Iteration  MaxIt)
  {
  it=1
  While  it< MaxIt  do
                        For i = 1 to n
                               Pathi =ϕ
                        End
                        Req = {}
                     For i = 1 to n
                              Pathi = Pathi U {Ti }
                             V=Ti
                            Tmi =Time to execute (Ti)
                           While Req≠R do
                                     t = Find_Next (Ai, V)
                                    Pathi = Pathi U {t}
                                   Tmi = Tmi + Time to execute (t)
                                  V=t
                                  Req = Req U Requirement Staisfied by t
                         End
                    End
```

```
Tmin = Min{Path1,Path2,Path3,....,Pathn} according to { Tm1, Tm2, Tm3 , ........,
Tmn }

Update_Path_Pheromone(Tmin )
it=it+1
End
}
Find_Next (Ai, V)
{
Ev={Aadjacent edges of V)
if  (All Ev with equal pheromone)
Choose Random edge
Else
Choose edge with max pheromone
}
Update_Path_Pheromone(Tmin )
{
Add 1 to the Tmin.
Evaporate 10% of pheromone in the all edge associated with Tmin .
}
```

4 Experimental Results

We take five subject programs for our experiments. Out of the five subject programs, power equiliser and binary search program are written by me. The rest of the three programs along with their test cases are downloaded from SIR repository. The execution of the programs is carried out using Eclipse IDE[2]. Eclipse is configured with JUnit [3], Ant[4], and EclEmma[5] for execution of test cases and their requirement generation. Jumble[6] is used to seed fault into the programs.

The whole setup of the experiment starts by executing the initial version of the program with the test case. Next, their coverage information is collected from EclEmma and Ant report. The initial requirement-test case information is prepared based on the data collected from the coverage information. The next version of the program with the added test case is executed on the IDE. Their execution information is also collected, which was later used for the creation of details presented in Table 1. For our experimental purpose, we take statement coverage, because statement coverage can detect faults. Table 1 is used by our ant colony based algorithm for finding a representative test case. The whole process of experimentation is presented in Fig. 1.

For the comparison of our results, we considered four existing techniques. The first algorithm is Harrold et al.'s Heuristic [1], the second technique is that of Chen and Lau in [11], the third one is Mansour and El-Fakin's approach [21], and the last technique

[2]The Eclipse Foundation website, http://www.eclipse.org/ (Accessed September 15, 2018).

[3]JUnit's official website, http://www.junit.org/ (Accessed September 15, 2018).

[4]http://www.eclipse.org/eclipse/ant/ (Accessed September 15, 2018).

[5]EclEmma's official website, http://www.eclemma.org/ (Accessed September 15, 2018).

[6]Jumble home page, http://jumble.sourceforge.net (Accessed September 15, 2018).

Table 1 Summary of programs used in the experiment

Program	BST	PEQ	JMeter	org.jacoco.report	XML-Security
Source file (KLOC)	1.86	1.46	84.26	2.6	24.54
Test suite pools (T × R)	169 × 98	156 × 124	556 × 167	235 × 23	462 × 187

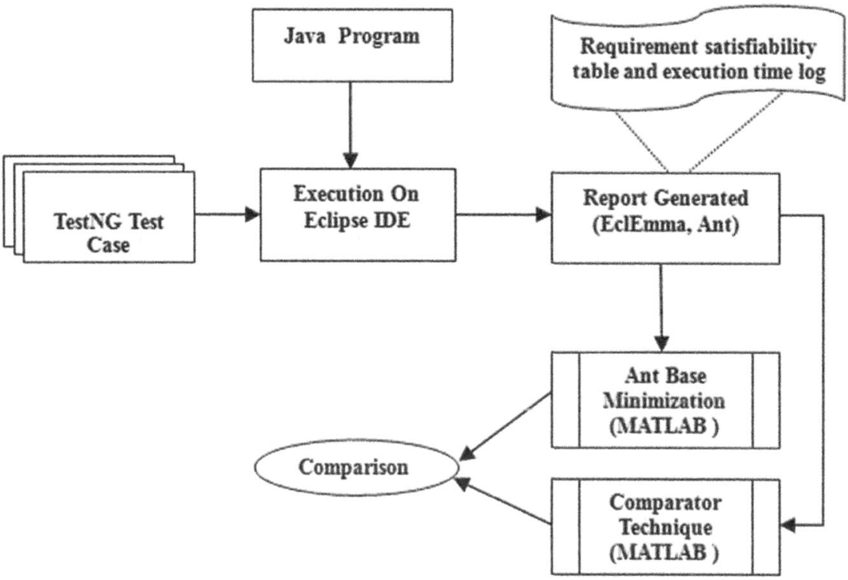

Fig. 1 Procedure of experiment

is of Black et al.'s approach [22]. All these techniques are programmed in MATLAB. After the creation of the table representing test case, satisfying requirement and execution time, our approach and the rest four techniques are executed. The factors like time complexity, representative set size, and number of requirements satisfied are recorded in a log file. This log file is used for result and performance analysis.

4.1 *Experimental Subjects*

The experimental evaluation is done on a PC with Linux operating system, Intel Pentium 2.5 GHz CPU and 8 GB RAM. The feature of programs is presented in Table 1. Five Java programs were considered for our experimental study, ranging

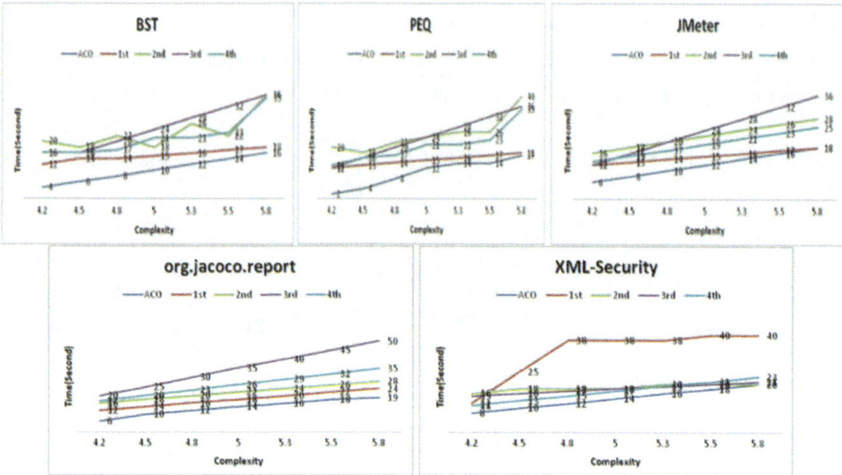

Fig. 2 Scalability of ACO-Min, first, second, third, fourth for five programs (**first**: Harrold et al.'s Heuristic [1], **second**: Chen and Lau [11] **third**: Mansour and El-Fakin [21], **fourth**: Black et al. [22])

from 1.46 to 84.26 KLOC. The test cases are simple unit test cases written using JUnit. Jumble is used to seed fault into the subject program.

4.2 Comparison Criteria

Scalability and representative set size of the subject programs are analyzed using Zhong et al.'s method [27].

4.2.1 Experiment for Scalability and Representative Set Size

The scalability of the test suite is measured using its complexity. The complexity of the test suite is represented by the following equation:

$$Complexity\ (t) = \log_{10}(m \times n) \tag{1}$$

Scalability

The scalability analysis of the subject program is represented in Fig. 2. It is evident from the plots in Fig. 2 that the ant colony based minimization technique is using the minimum time to search for an optimal representative set. The observation on

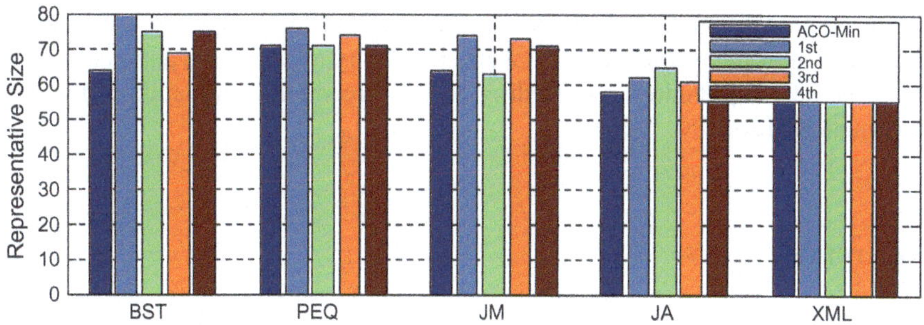

Fig. 3 Sizes of representative sets of ACO-Min, first, second, third, fourth for five programs (**first**: Harrold et al.'s Heuristic [1], **second**: Chen and Lau [11] **third**: Mansour and El-Fakin [21], **fourth**: Black et al. [22])

execution time is τACO-Min ≤ {τ1st, τ2nd, τ3rd, τ4th}. The complexities of the programs are calculated by Eq. (1).

Representative Set Size

The representative set size generated by the proposed algorithm (ACO-Min) and other techniques (First: Harrold et al.'s Heuristic [1], Second: Chen and Lau [11], Third: Mansour and El-Fakin [21], Fourth: Black et al. [22]) are compared and presented in Fig. 3. The x-axis represents our subject program along with different techniques we use for comparison. The y-axis shows the percentage of the representative set to the original test suite size. In all the experiments, we found that the ACO-Min algorithm gives better test suite minimization than the other techniques.

5 Conclusion

In this paper, we proposed an ant colony based test case minimization algorithm ACO-Min. This algorithm generates a representative set which is the minimum cardinality set of the test suite. We compared the proposed algorithm, with four existing heuristic-based algorithms. Our experiments are based on two own programs and three benchmark programs from the SIR repository. The comparison of the results shows that the proposed algorithm greatly reduces the test case size. The reduced test suite's RF factor is zero for all the subject programs studied. This indicates that the fault detection probability of the representative set is the same as that of the original test suite. The derived representative test case contains fewer test cases as compared to the original test suite and it covers all the requirements that are covered by the original test suite. The scalability analysis reveals better time complexity for the ACO-Min algorithm.

One major limitation of the algorithm is that it takes the complete graph as input. The creation of this complete graph adds complexity to the whole automation process. Our next explorations will be directed to overcome this limitation and will be communicated soon.

References

1. Harrold, M.J., Gupta, R., Soffa, M.L.: A methodology for controlling the size of a test suite. ACM Trans. Softw. Eng. Methodol **2**(3), 270–285 (1993)
2. Cormen, T.H., Leiserson, C.E., Rivest, R.L., Stein, C.: Introduction to Algorithms, 2nd edn. MIT Press, Sept (2001)
3. Gupta, R.: A reconfigurable LIW architecture and its compiler. Tech. Rep. 87-3, Dept. Computer Science, Univ. Pittsburgh, Pittsburgh, Pa. (1987)
4. Gupta, R., Soffa, M.L. Compile-time techniques for improving scalar access performance in parallel memories. IEEE Trans. Parallel Distrib. Syst. 2(2), 138–148 (1991)
5. Jeffrey, D., Gupta, N.: Improving fault detection capability by selectively retaining test cases during test suite reduction. IEEE Trans. Softw. Eng. **33**(2), 108–123 (2007)
6. Lin, J.W., Huang, C.Y.: Analysis of test suite reduction with enhanced tie-breaking techniques. Inf. Softw. Technol. **51**(4), 679–690 (2009)
7. Garey, M.R., Johnson, D.S.: Computers and Intractability: A Guide to the Theory of NP-Completeness. Freeman and Company (1979)
8. Karp, R.M.: Reducibility among combinatorial problems. In: Complexity of Computer Computations, pp. 85–103. Plenum Press (1972)
9. Chvatal, V.: A greedy heuristic for the set-covering problem. Math. Oper. Res. **4**(3), 233–235 (1979)
10. Yo, S., Harman, M.: Regression testing minimization, selection and prioritization: a survey. Softw. Test. Verif. Reliab. 22(2) (2012)
11. Chen, T.Y., Lau, M.F.: A new heuristic for test suite reduction. Inf. Softw. Technol. **40**(5–6), 347–354 (1998)
12. Jones, J.A., Harrold, M.J.: Test-suite reduction and prioritization for modified condition/decision coverage. IEEE Trans. Softw. Eng. **29**(3), 195–209 (2003)
13. Lin, C.-T., Tang, K.-W., Chen, C.-D., Kapfhammer, G.M.: Reducing the cost of regression testing by identifying irreplaceable test cases. In: Proceedings of the 6th ICGEC '12
14. Ma, X.Y., He, Z.F., Sheng, B.K., Ye, C.Q.: A genetic algorithm for test-suite reduction. In: Proceedings of the International Conference on Systems, Man and Cybernetics, pp. 133–139, Oct 2005
15. Zhang, Y., Liu, J., Cui, Y., Hei, X.: An improved quantum genetic algorithm for test suite reduction. In: IEEE International Conference on Computer Science and Automation Engineering (CSAE) (2011)
16. Hao, D., Xie, T., Zhang, L., Wang, X., Sun, J., Mei, H.: Test input reduction for result inspection to facilitate fault localization. Autom. Softw. Eng. 17(1), 5–31(2010) (Springer)
17. Mohapatra, S.K., Prasad, S., Minimizing test cases to reduce the cost of regression testing. In: IEEE Proceedings of the 8th INDIACom; INDIACom (2014)
18. Mohapatra, S.K., Prasad, S, Kar, B.P.: Test suit reduction by finding cost optimal representative set. Int. J. Adv. Technol. Eng. Res. (IJATER) 4(3) (2014)
19. Mohapatra, S.K., Prasad, S.: Evolutionary search algorithm for test case prioritization. In: IEEE International Conference on Machine Intelligence Research and Advancement. 978-0-7695-5013-8/13 (2013)
20. Chen, T.Y., Lau, M.: A simulation study on some heuristics for test suite reduction. Inf. Softw. Technol. **40**(13), 777–787 (1998)

21. Mansour, N., El-Fakih, K.: Simulated annealing and genetic algorithms for optimal regression testing. J. Softw. Maint. **11**(1), 19–34 (1999)
22. Black, J., Melachrinoudis, E., Kaeli, D.: Bi-criteria models for all-uses test suite reduction. In: Proceedings of 26th International Conference on Software Engineering, pp. 106–115. IEEE Computer Society, Washington, DC, USA (2004)
23. Yoo, S., Harman, M.: Using hybrid algorithm for pareto efficient multi-objective test suite minimisation. JSS **83**(4), 689–701 (2010)
24. Hao, D., Zhang, L., Wu, X., Mei, H., Rothermel, G.: On-demand test suite reduction. In: ICSE, pp. 738–748 (2012)
25. Blue, D., Segall, I., Tzoref-Brill, R., Zlotnick, A.: Interaction-based test-suite minimization. In: ICSE, pp. 182–191 (2013)
26. Zhang, L., Marinov, D., Khurshid, S.: Faster mutation testing inspired by test prioritization and reduction. In: ISSTA, pp. 235–245
27. Zhong, H., Zhang, L., Mei, H.: An experimental study of four typical test suite reduction techniques. Inf. Softw. Technol. **50**(6), 534–546 (2008)

Peer Analysis of "Sanguj" with Other Sanskrit Morphological Analyzers

Jatinderkumar R. Saini and Jaideepsinh K. Raulji

Abstract In linguistics, morphology is a study regarding word, word formation, its analysis, and generation. A morphological analyzer is a tool to understand grammatical characteristics and constituent's part-of-speech information. A morphological analyzer is a useful tool in many NLP implementations such as syntactic parser, spell checker, information retrieval, and machine translation. Here, 328 Sanskrit words are tested through four morphological analyzers namely—*Samsaadhanii*, morphological analyzers by JNU and TDIL, both of which are available online and locally developed and installed *Sanguj* morphological analyzer. There is a negligible divergence in the reflected results.

Keywords Indeclinable · Inflection · Lemmatization · Morphology · Sanskrit

1 Introduction

The Indo-European families of languages are rich in morphological characteristics. Sanskrit, a language from the Indo-Aryan branch of Indo-European language family, is considered highly inflected in the family but endows regular inflectional patterns for its grammatical constituents. Sanskrit is believed to be one of the oldest existing languages and the mother of most Indian languages. Hence, the same tradition of Sanskrit is also inherited and imbibed in many languages derived from Sanskrit. It

J. R. Saini (✉)
Symbiosis Institute of Computer Studies and Research (SICSR), Symbiosis International (Deemed University) (SIU), Pune, India
e-mail: saini_expert@yahoo.com

J. R. Saini · J. K. Raulji
Dr. Babasaheb Ambedkar Open University, Ahmedabad, Gujarat, India

J. K. Raulji
Ahmedabad University, Ahmedabad, Gujarat, India
e-mail: jaideepraulji@gmail.com

H. Das et al. (eds.), *Progress in Computing, Analytics and Networking*, Advances in Intelligent Systems and Computing 1119,
https://doi.org/10.1007/978-981-15-2414-1_7

is spoken mostly in South Asia and Southeast Asia. The word is formed by a single unit of sound, which collectively might form a root word. These root words undergo added inflections forming word-form in a language. The study of formation of words from such units is known as morphology. The root word undergoes several added inflections through prefix, infix, or suffix to form a final word-form. The decisive factors are gender, number, person, tense, mood, and aspect on which the inflection depends. Hence, an in-depth study of morphological characteristics proves boon in many Natural Language Processing (NLP) tasks, especially for Indian languages which are highly inflected.

The written tradition of Sanskrit language based on archeological study dates back around 500 BCE, while spoken tradition is believed to be around 2500 BCE. Sanskrit is written through many scripts, but Devanagari is widely used due to its rich coverage to encode every gradation of sound using consonants and vowel markers. Additionally, it also includes a rich set of diacritic marks for vowels.

Morphological analysis is vital for building any basic NLP application. For an inflectionally rich language like Sanskrit, it provides ample information of word with its syntactic and semantic role played in a sentence. As Sanskrit is rich in its morphology, the morphological study helps to develop morphological parser which tags each word based on its formation rules for parts of speech. The syntactic parser tags the word with valid parts of speech but is based on the arrangement of words in a sentence. Many Indian languages are like free word order and hence for parts-of-speech tagging, morphological parser plays an important role than the syntactic parser. This also happens to be one of the important motivating factors for this research work. Morphological parsers can be developed using hand-coded rules of formation of a word for a particular language. If tagged corpus is available in size appropriate for training an algorithm, then machine learning techniques are also suitable. The literature study concludes on the fact that most available morphological parsers for Indian languages, particularly Sanskrit, are rule based due to the unavailability of tagged corpus.

The usage of noun, pronoun, and adjective in Sanskrit has inflection based on gender, number, and case. Indeclinable (*Avyaya*) is a word that is independent of inflections. The plural form of "indeclinable" is called "indeclinables". The verb decline is based on tense, aspect, modality, number, and person. The grammatical meta-information regarding Sanskrit is listed in Table 1.

For the remainder of the paper, this section on introduction is followed by a review of literature through related works. This is followed by a section each on the study of existing morphological analyzers and results. This is followed by the last section on the concluding remarks and future work.

Table 1 Grammatical features of Sanskrit language

Sr. no.	Language feature	Description of Sanskrit language
1	Consonants	33
2	Vowels	12
3	Gender	3 [masculine, feminine, neuter]
4	Number	3 (singular, dual, plural)
5	Case markers	8 Cases (*Vibhakti*) [nominative, accusative, instrumental, dative, ablative, genitive, locative, vocative]
6	Persons	3 [first person (*Uttam*), second person (*Madhya*), third person (*Pratham*)]
7	Tense	6 (present, aorist, imperfect past, perfect past, first future, second future)
8	Mood	4 [imperative, potential, conditional, benedictive]
9	Word order	Free

2 Related Works

A bilingual dictionary [13], a rule-based architecture [15] and a constituency mapper [17] were developed by the researchers for the efficient machine translation from Sanskrit language to the Gujarati language. Working for the development of a Morphological Analyzer (MA) is like developing a nonconventional classifier to classify tokens of a given text corpus into classes like pronouns, nouns, adjectives, and verbs. A lot of research has been dedicated to the conventional tasks of classification of text and dataset [18, 19]. Also, though the machine learning algorithms [20, 22, 25], semantic discretization [21], and data mining approaches [23, 24] have been deployed for the classification task, the area of MA remains mostly untouched by the researchers. It is additionally noteworthy to mention that the area of MA for Sanskrit is further more unexplored by the researchers.

Sanskrit being free word order language, its syntacto-semantic relations exclusively depend on word inflections. Jha et al. [1] developed Sanskrit MA which recognizes noun and verb forms. The *subanta* forms are identified using example and rule database while verbs through verb roots using reverse morphology *paninian* techniques. A list of stop words of Sanskrit language [14] and their analysis through MA [16] were presented by researchers. Sanskrit morphological analyzer is also built by Bharati et al. [2]. Their *subanta* module analyzes nouns, pronouns, and adjectives. It uses grammatical rules along with lexicon from Monier William's dictionary. Indeclinables are identified using separate lexicon list and verbs are analyzed by storing generated forms using verb formation rules in database.

Raulji and Saini [3] developed a morphological analyzer, here referred to as *Sanguj*, through Sanskrit lemmatizer. Therein the lemmatizer extracts lemma to match with Sanskrit lemma database containing approximately 30000 lemmas with surface Part-Of-Speech (POS) information. Using morphological rules of inflections for verbs and nouns, detailed POS information is retrieved. Due to complex inflectional nature of pronoun and immutable indeclinables, they are directly added to database.

Srivastava et al. [4] developed unsupervised deep learning based Sanskrit POS tagger due to scarcely available annotated corpus and did sequence modeling using bidirectional Long Short Term Memory (LSTM) auto-encoder. A deep learning based POS tagger by embedding character level features using Recurrent Neural Network (RNN), LSTM, and Gated Recurrent Units (GRU) was presented and implemented by Soman and Poornachandran [5]. A dependency parser is built for Sanskrit using deterministic finite automata for morphological analysis by Goyal et al. [6]. They developed nominal, verb, and particle database as part of implementation.

A Sanskrit stochastic tagger is built by Hellwig [7] using the Markov Model after tokenizing the text and a manually annotated corpus was created containing 1.5 million words. A rule-based POS tagger is built by Tapaswi and Jain [8]. The rules are stored in database as suffix tables, lexical rules tables and context sensitive rules by them. An analysis of indeclinables (*Avyaya*) was carried out by *Avyaya* analyzer using finite-state transducer by Murali et al. [9]. They have also presented the discussion on the various types and roles of the Sanskrit indeclinables.

3 Study of Morphological Analyzers

The following four morphological analyzers were studied and tested against 328 Sanskrit words.

a. *Samsaadhanii*—A Sanskrit Computational Toolkit (Morphological Analyzer) [10].
b. Sanskrit Morphological Analyzer by School of Sanskrit and Indic Studies, Jawaharlal Nehru University (JNU) [11].
c. Morph Analyzer developed by Technology for Development of Indian Languages (TDIL) [12].
d. *Sanguj*/*SanGuj* Morphological Analyzer [3].

To analyze inflectionally rich languages, morphological analyzer with a wide spectrum of grammatical rules coverage is required. And the above-listed analyzers proved to be the best tools looking at results and quality of lucidness in its output.

In *Samsaadhanii*, *subanta*, *tinanta*, and *kridanta* are handled by separate modules also integrated with Monier William's Sanskrit dictionary. The *subanta* analyzer analyzes nouns, pronouns, adjectives, and indeclinables. Indeclinables are added separately as they are independent of change in grammatical functions [10].

A morphological analyzer developed by JNU analyzes inflected nouns, verbs, and indeclinables in sandhi-free text. The indeclinables are identified and analyzed using *avyaya* database. A list of common verb roots is stored in database and analyzed using reverse morphology based on *Paninian* techniques. Nouns are identified based on the example and rule database by this analyzer [11].

The analyzer developed by TDIL at the Department of Information Technology at Ministry of Communications and Information Technology (DoIT at MCIT) (New Delhi) is supported by consortium of seven institutes namely Indian Institute of Information Technology (IIIT) (Hyderabad), Jawaharlal Nehru University (JNU) (New Delhi), University of Hyderabad (Hyderabad), Sanskrit Academy (Hyderabad), *Rashtriya Sanskrit Vidyapeeth* (Tirupati), *Poornaprajna Vidyapeeth* (Bangalore), and JRR Sanskrit University (Jaipur) [12].

The *Sanguj* morphological analyzer [3] is built based on lemmatizer, exclusively developed Sanskrit dictionary and affix rules. As indeclinables do not decline, they are added directly into the indeclinable database for its identification. Pronouns in Sanskrit undergo complex inflectional affixes and hence they are added in database with complete grammatical information.

For analyzing nouns and verbs, the lemma and surface POS information is retrieved from the Sanskrit lemmatizer by the researchers. The detailed grammatical POS information is retrieved by pattern matching of inflectional ending rules of verbs and nouns. The lemmatizer is built by a suffix stripping word and simultaneously matched with the Sanskrit dictionary. The implementation is through open-source products namely, MySQL database and Java. The Java programming language has been used, in addition to being open source, it provides good string handling functions. The algorithmic flowchart for *Sanguj* analyzer is depicted diagrammatically Fig. 1.

4 Results

The Sanskrit morphological analyzers namely the one developed by TDIL, *Samsaadhanii,* and Analyzer by JNU are freely available online while *Sanguj* analyzer is in finalization phase. All the four analyzers were tested with the same 328 Sanskrit words and the results are listed in Table 2. The analyzers mostly categorized output in four main POS categories namely indeclinables, pronouns, verbs, and nouns. The results are fairly similar with few exceptions.

Sanguj analyzer shows a few more indeclinables than the other three analyzers. It included "*Asi*" (असि), "*Asti*" (अस्ति), "*Aam*" (आम) also as indeclinables. There is not much divergence in the results of pronouns. Also, divergence is noted in verbs between MA by TDIL, MA by JNU, and *Samsaadhanii* versus *Sanguj*. The *Sanguj* analyzer shows more verbs, however multiple POS tags in the form of nouns and verbs are also analyzed by the rest of the three. The JNU analyzer shows more nouns. The number of unavailable words is more in TDIL MA and *Samsaadhanii*, fewer in JNU and none in *Sanguj*. The reason is *Sanguj* purely works on affix matching of words after lemma comparison from lemmatizer. Even if the word is not available in the dictionary, still it analyzes using affix pattern matching for verbs and nouns making it more robust than other counterparts. The sample output of the morphological analyzers is listed in Table 3.

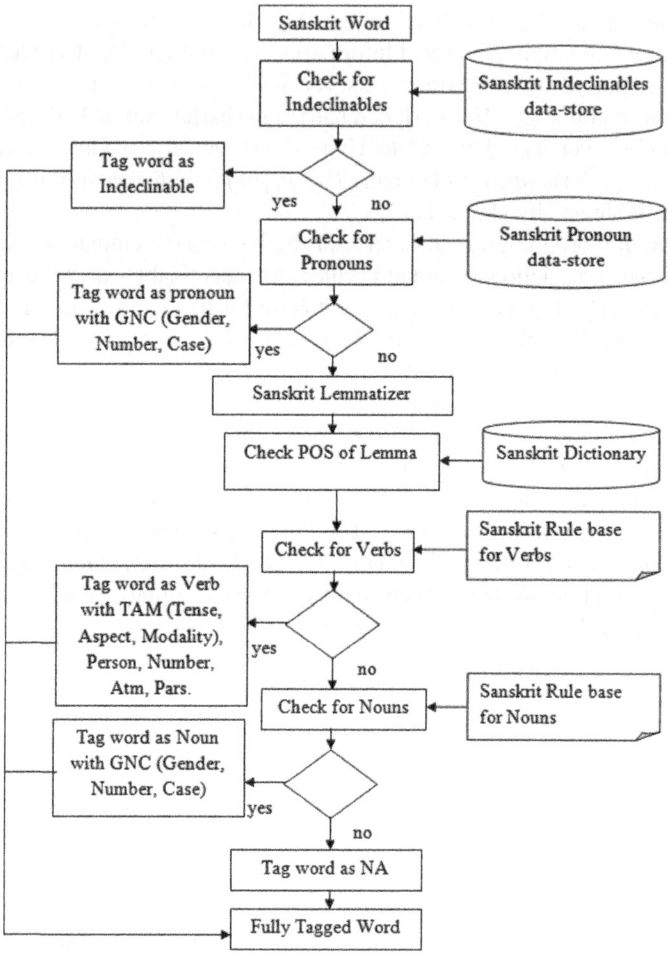

Fig. 1 The flowchart for "*Sanguj*" morphological analyzer

Table 2 POS category-wise results for four morphological analyzers

Sr. no.	Sanskrit morphological analyzers	Indeclinable	Pronoun	Verb	Noun	Unavailable	Total
1	Morphological Analyzer by TDIL	13	32	116	141	26	328
2	*Samsaadhanii*	12	29	106	151	30	
3	*Sanguj*	16	33	130	144	5	
4	Morphological Analyzer by JNU	12	31	109	166	10	

Table 3 Sample output of morphological analyzers, [Nom—nominative, Gen—genitive, Mas—masculine, Neu—neuter, Sin—singular, Im—imperfect, Pars—*Parasmaipada*, Atm—*Aatmanepada*, FP—first person, SP—second person, TP—third person]

Sr. No.	Input Word	MA by TDIL	Output *Samsaadhanii*	*Sanguj*	MA by JNU
1	नगरस्य	नगरस्य=नगर{नपुं}{6;एक}	नगर नपुं 6 एक	नगरस्य,Noun,Gen,Neu,Sin	नगरस्य (नगर _ ङस्, षष्ठी, एकवचन)
2	अटति	अटति=अटत्{पुं}{7;एक} अटत्{नपुं}{7;एक} अट्1{कर्तरि;लट्;प्र;एक;परस्मैपदी अटँ;भ्वादिः	अटत् पुं 7 एक{कृदन्त}(अट्1 शतृ_लट् धातुः अटँ}{गणः भ्वादिः}) अटत् नपुं 7 एक{कृदन्त}(अट्1 शतृ_लट्{धातुः अटँ}{गणः भ्वादिः}) अट् कर्तरि लट् प्र एक परस्मैपदी{धातुः अटँ}{गणः भ्वादिः	अटति,Verb,1,Present,Pars,TP,Sin	अटति ([P_laT_1.1:90])
3	इयं	इयम्=इदम्{स्त्री}{1;एक}	इदम् सर्वनाम स्त्री 1 एक	इयं,Pro,Nom,Fem,Sin	इयं ([SND_f_1.1])
4	अत्र	अत्र=अत्र{अव्य}	अत्र अव्य	अत्र,Indec,null,null,null	अत्र ([AVKVD])
5	ग्रिष्मस्य	Unavailable	Unavailable	ग्रिष्मस्य,Noun,Gen,Neu,Sin	ग्रिष्मस्य (ग्रिष्म _ ङस्, षष्ठी, एकवचन)

5 Conclusion

The importance of morphological analyzers cannot be ruled out for analysis and processing of morphologically rich natural languages. They play a pivotal role in the development of syntactic parsers, POS taggers, information retrieval, and rationale machine translation approach. The conceptual and logical design of morphological analyzer remains the same but output can be mutated to cater needs of underlying NLP application. Here, the study of all the morphological analyzers for Sanskrit language and little divergence in results proves the fact that Sanskrit is most precise and regular in its grammar formulations. But still for a rule-based system, fine-tuning is difficult due to exceptions instilled in natural languages. Additionally, maintaining rule-based system is a herculean task due to the large and complex nature of language characteristics. Hence, an ample amount of tagged corpus, if available, can solve the purpose with better accuracy by employing machine or deep learning techniques. It is concluded that the *Sanguj* morphological analyzer performs better than the other three existing counterparts. Generation of POS for incorrect Sanskrit words, like कोम्पुटरमहे(*Computermahe*) which does not make any semantic sense, is a limitation of the "*Sanguj*" MA. In the future, we plan to further fine-tune the Sanguj morphological analyzer for Sanskrit language, including with more usage of data for its training and testing.

References

1. Jha, G.N., Agrawal, M., Mishra, S.K., Mani, D., Mishra, D., Bhadra, M., Singh, S.K.: Inflectional morphology analyzer for Sanskrit. In: Sanskrit Computational Linguistics, pp. 219–238. Springer (2007)
2. Bharati, A., Kulkarni, A.P., Sheeba, V.: Building a wide coverage Sanskrit morphological analyser: a practical approach. In: The First National Symposium on Modelling and Shallow Parsing of Indian Languages, IIT-Bombay
3. Raulji, J.K., Saini, J.R.: Sanskrit lemmatizer for improvisation of morphological analyzer. J. Stat. Manag. Syst. (Taylor and Francis) **22**(4), 613–625 (2019)
4. Srivastava, P., Chauhan, K., Aggarwal, D., Shukla, A., Dhar, J., Jain V.P.: Deep learning based unsupervised POS tagging for Sanskrit. In: Algorithms, Computing and Artificial Intelligence, p. 56. ACM (2018)
5. Soman, K.P., Poornachandran, P.: A deep learning based part-of-speech (POS) tagger for Sanskrit language by embedding character level features. In: Forum for Information Retrieval Evaluation, pp. 56–60. ACM (2018)
6. Goyal, P., Arora, V., Behera, L.: Analysis of Sanskrit text: parsing and semantic relations. In Sanskrit Computational Linguistics, pp. 200–218. Springer (2007)
7. Hellwig, O.: Sanskrit tagger: a stochastic lexical and POS tagger for Sanskrit. In Sanskrit Computational Linguistics, pp. 266–277. Springer (2007)
8. Tapaswi, N., Jain, S.: Treebank based deep grammar acquisition and part-of-speech tagging for Sanskrit sentences. In CSI Sixth International Conference on Software Engineering (CONSEG), pp. 1–4. IEEE (2012)
9. Murali, N., Ramasree, R.J., Acharyulu, K.V.R.K.: Avyaya analyzer: analysis of indeclinables using finite state transducers. Int. J. Comput. Appl. **38**(6), 7–11 (2012)
10. Samsaadhanii—A Sanskrit Computational Toolkit (Morphological Analyzer). http://scl.Samsaadhanii.in/scl/. Accessed May 2019
11. Sanskrit Morphological Analyzer, Computational Linguistics R&D, School of Sanskrit and Indic studies, Jawaharlal Nehru University. http://Sanskrit.jnu.ac.in/morph/analyze.jsp. Accessed May 2019
12. Morph–Analyzer, Technology for Development of Indian Languages (TDIL). http://tdil-dc.in/san/morph/index_dit.html. Accessed May 2019
13. Raulji, J.K., Saini, J.R.: Bilingual dictionary for Sanskrit–Gujarati MT Implementation. In: ICT for Sustainable Development. Springer (in press)
14. Raulji, J.K., Saini, J.R.: Generating stopword list for Sanskrit language. In: Advance Computing, pp. 799–802. IEEE (2017)
15. Raulji, J.K., Saini, J.R.: A rule based architecture for Sanskrit to Gujarati machine translation system. In: Emerging Trends in Engineering, Science and Technology. IEEE (2018) (in press)
16. Raulji, J.K., Saini, J.R.: Sanskrit stopword analysis through morphological analyzer and its Gujarati equivalent for MT system. In: ICT for Sustainable Development. Springer (2019) (in press)

17. Raulji, J.K., Saini, J.R.: Sanskrit-Gujarati constituency mapper for machine translation system. In: IEEE Bombay Section Signature Conference (IBSSC-2019). IEEE (2019) (in press)
18. Saini, J.R., Desai, A.A.: Analysis of classifications of unsolicited bulk emails. Int. J. Comput Inf Eng. **4**(2), 91–95 (2010)
19. Chandrakar, O.S., Saini, J.R.: Empirical study to suggest optimal classification techniques for given dataset. In: Computational Intelligence and Communication Technology, pp. 30–35. IEEE Computer Society (2015)
20. Kaur, J., Saini, J.R. Punjabi poetry classification: the test of 10 machine learning algorithms. In: Machine Learning and Computing, pp. 01–05. ACM (2017)
21. Chandrakar, O.S., Saini, J.R., Bhatti, D.G.: Novel semantic discretization technique for type-2 diabetes classification model. In: Innovations in Computer Science and Engineering. Lecture Notes in Networks and Systems, vol. 74, pp. 135–141. Springer (2019)
22. Rakholia, R.M., Saini, J.R.: Classification of Gujarati documents using Naïve Bayes classifier. Indian J. Sci. Technol. **10**(5), 1–9 (2017)
23. Das, H., Naik, B., Behera, H.S.: Classification of diabetes mellitus disease (DMD): a data mining (DM) approach. In: Progress in Computing, Analytics and Networking, pp. 539–549. Springer (2018)
24. Sahani, R., Rout, C., Badajena, J.C., Jena, A.K., Das, H.: Classification of intrusion detection using data mining techniques. In: Progress in Computing, Analytics and Networking, pp. 753–764. Springer (2018)
25. Das, H., Jena, A.K., Nayak, J., Naik, B., Behera, H.S.: A novel PSO based back propagation learning-MLP (PSO-BP-MLP) for classification. In: Computational Intelligence in Data Mining, vol. 2, pp. 461–471. Springer (2015)

Color Image Encryption Technique Using 4D Logistic Map

Shubham Kumar and Chittaranjan Pradhan

Abstract Due to the increased use of data transmission over the public channel, data security is highly required. Chaotic maps are performing better in speed and robustness than the cryptographic techniques for data security. The 2D logistic and 3D logistic maps are the commonly used chaotic maps. But, in case of sensitive information, the extension of 3D logistic map is used. Here, a 4D logistic map has been proposed for the data encryption. It is performing better than the 3D logistic map; which is being verified by the NPCR, UACI parameters as well as the encryption and decryption timings.

Keywords 2D logistic map · 3D logistic map · 4D logistic map · Chaotic map · Decryption · Encryption

1 Introduction

Digitization has made human life much easier. Nowadays, multimedia content is shared everywhere, every time. This uprise of digitization leads to the rise of data stealing and unauthorized information gathering. In order to keep our data secure and protected, we can perform some information security practices such as steganography and cryptography. Steganography is the practice of concealing a digital signal inside another digital signal. Cryptography deals with the encryption/decryption of the message [1].

The advantage of steganography over cryptography is that the digital signal does not look suspicious and attract attention. Digital signals that are encrypted look like white noise and simply do not make sense which makes them look suspicious and arouse interest. But, only steganography is not enough to protect data from

S. Kumar (✉) · C. Pradhan
Kalinga Institute of Industrial Technology, Bhubaneswar, India
e-mail: shubham.kumar71185@gmail.com

C. Pradhan
e-mail: chitaprakash@gmail.com

H. Das et al. (eds.), *Progress in Computing, Analytics and Networking*, Advances in Intelligent Systems and Computing 1119,
https://doi.org/10.1007/978-981-15-2414-1_8

sophisticated attackers. The data should be first encrypted and then embedded to the cover. This technique will not only protect our data, but also avoid any suspicion.

Image encryption technique plays with the digital image pixels, which involves keys. The encrypted image is transmitted over the public channel. With the help of the keys used for encryption, one can decrypt the encrypted image. Every color image can be divided into three planes which are red, green, and blue planes [2]. The pixels of these color planes will be encrypted/decrypted separately.

2 Related Work

The Logistic map is a chaotic map which arises from a nonlinear dynamical equation. Yin Dai et al. proposed one encryption scheme using logistic map and Chebyshev map for the medical images for the performance improvement of grayscale images [3]. Samar M. Ismail et al. proposed one technique for medical image encryption using double-humped logistic map [4].

Ajita Sahay et al. proposed one multidimensional encryption technique of the color image using the combination of Gauss iterated map and logistic map [5]. Md. Billal Hossain et al. proposed an enhanced encryption approach for grayscale image using 3D Chaotic map [6].

Motivated by the extensive work done by Bidyut Jyoti Saha et al. [7] from 3D to 4D, an extension of 3D logistic map has been done to the 4D for color image encryption.

The basis of 3D and 4D logistic maps is the 2D logistic map, which is the advancement of 1D logistic map. As we add more parameters, the digital signal will be more secure and it will be tougher to predict the secret content for the pernicious parties [8]. The color image is first divided into three planes (red, green, and blue). For each plane, chaotic sequences with chaotic behavior are generated which is defined as

$$F(X,Y) = \begin{cases} X_{i+1} = \mu_1 * X_i(1 - X_i) + \beta_1 (Y_i)^2 \\ Y_{i+1} = \mu_2 * Y_i(1 - Y_i) + \beta_2\big((X_i)^2 + X_i * Y_i\big) \end{cases} \tag{1}$$

where $i = 0, 1, 2 \ldots . \mu_1, \beta_1, \mu_2, \beta_2$ are the secret parameters. X_0 *and* Y_0 are the initial parameters. The control parameters range from $2.75 < \mu_1 < 3.4, 0.15 < \beta_1 < 0.21, 2.70 < \mu_2 < 3.45, 0.13 < \beta_2 < 0.15, 0 < X_i, Y_i < 1$. The transposed vector Y is multiplied with the other vector X to generate a 2D matrix K of size equivalent to the size of the inputted image. The quantization of all the elements of the matrix is done by the following equation:

$$P(x) = \begin{cases} 0, \ if \ 0 < K(i,j) \le 0.5 \\ 1, \ if \ K(i,j) > 0.5 \end{cases} \tag{2}$$

Then, XOR operation is performed between the pixel bit of the color plane and the quantized matrix element.

3 Proposed Work

The 3D logistic map is introduced to provide more protection of the data. It has one more parameter than 2D logistic map that makes it more secured. The 3D map is generated by the following equation:

$$F(X,Y,Z) = \begin{cases} X_{i+1} = \mu_1 * X_i(1 - X_i) + \beta_1(Y_i + Z_i)^2 \\ Y_{i+1} = \mu_2 * Y_i(1 - Y_i) + \beta_2\big((X_i)^2 + X_i * (Y_i + Z_i)\big) \\ Z_{i+1} = \mu_3 * Y_i(1 - Y_i) + \beta_3\big((X_i)^2 + (Y_i)^2 + X_i * Y_i * Z_i\big) \end{cases} \quad (3)$$

where $2.75 < \mu_1 < 3.4,\ 0.15 < \beta_1 < 0.21,\ 2.70 < \mu_2 < 3.45,\ 0.13 < \beta_2 < 0.15,\ 2.65 < \mu_3 < 3.5,\ 0.13 < \beta_3 < 0.20,\ 0 < X_i,\ Y_i,\ Z_i < 1.$

This will create three vectors X, Y, and Z. The transposed vector Y is multiplied with the other vector X to generate a 2D matrix. Further, this matrix is again multiplied with the transposed of vector Z to create a 3D matrix. Every element of this matrix is quantized by using Eq. (2).

For the demonstration purpose, we have taken "selena.bmp" color image of size 64 × 64. By taking $\mu_1 = 2.75,\ \mu_2 = 2.7,\ \mu_3 = 2.65,\ \beta_1 = 0.15,\ \beta_2 = 0.13,\ \beta_3 = 0.13,\ X_0 = 0.5,\ Y_0 = 0.5\ and\ Z_0 = 0.5$, the encryption of the image is shown in Fig. 1, which represents the color planes followed by the combination of all planes together.

Using the appropriate keys and the correct initial values, the decryption is performed to recover the image from the concealed or encrypted image. It is shown in Fig. 2.

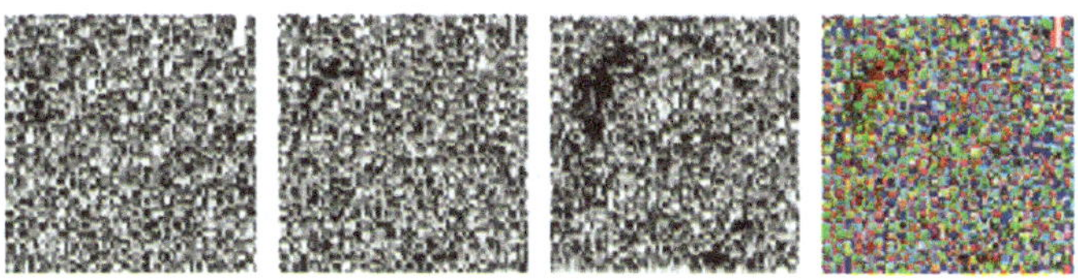

Fig. 1 3D logistic map encryption

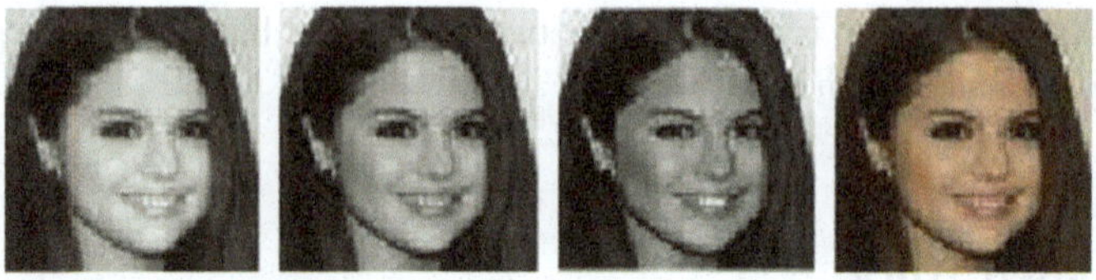

Fig. 2 3D logistic map decryption

3.1 *4D Logistic Map*

To achieve better security over 3D logistic map, a 4D version of it has been proposed. The extended dimension will help in achieving better robustness and better authenticity [7]. This extension is focusing on the distribution of color planes. The three planes (red, green, and blue) are divided into three parts which are shuffled prior to embedding to the load. This technique will confuse the attackers.

3.1.1 Encryption

1. The color planes are being extracted from the original input color image.
2. A matrix of the same size as the input image is created according to the 3D logistic map Eq. (3).
3. Then XOR operation is performed between each pixel and the matrix value having same X and Y coordinates. The value at Z coordinate at any coordinate X and Y is XORed with that layer of bit.
4. Each color plane is divided into three horizontal parts. So, nine parts of the original image will be generated. The nine parts are shuffled in order to confuse the attacker. The image is shuffled in the following manner as shown in Fig. 3.

3.1.2 Decryption

To extract the image in the original form, the encrypted image is decrypted by using the encryption keys and the original initial parameters. The steps are

1. The color planes are being extracted from the encrypted color image.
2. Each color plane is again divided into three horizontal parts.
3. After the reconstruction of each plane, the decryption process happens by taking the same chaotic sequences.

For the illustration purpose, "selena.bmp" color image with height and width of 64 pixels has been taken. By taking $\mu_1 = 2.75,\ \mu_2 = 2.7,\ \mu_3 = 2.65,\ \beta_1 = 0.15,\ \beta_2 = 0.13,\ \beta_3 = 0.13,\ X_0 = 0.5,\ Y_0 = 0.5\ and\ Z_0 = 0.5$, the encryption of the image is shown in Fig. 4, which represents the color planes followed by the combination of all planes together.

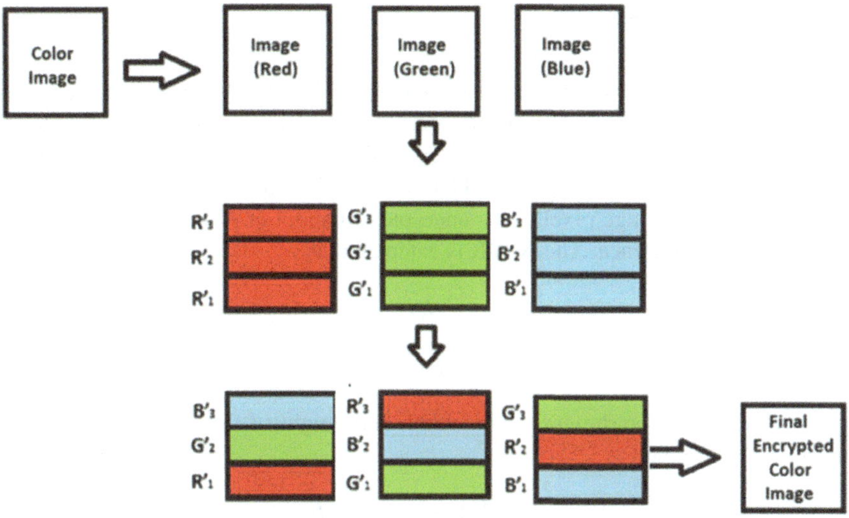

Fig. 3 4D logistic map encryption process

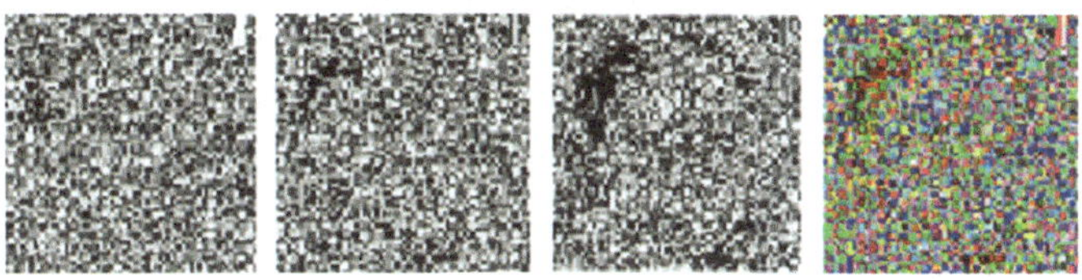

Fig. 4 4D logistic map encryption result

Figure 5 shows the result of decrypting the encrypted image using the same parameter list. It shows the extraction of the original image. The process of extraction is similar to encryption. The 3D logistic map is generated according to the input given by the user. Each pixel of the encrypted image is XORed with the values of this map. Then each plane of the encrypted image is divided into three horizontal parts and rearranged to get the planes of the original image. The planes are concatenated and we obtain the original image.

Fig. 5 4D logistic map decryption result

4 Result Analysis

The security level of the proposed 4D logistic map is checked by using the parameters like UACI (Unified Average Change Rate) and NPCR (Number of Pixel Change Rate) [9, 10]. The relationship between the encrypted image and original image is shown by the differential attack analysis. NPCR parameter is used to verify the effect of changing original image pixel in the encrypted image. Let A and B be two images with one pixel difference. The NPCR is calculated as

$$NPCR = \frac{\sum_{i,j} K(i,j)}{w * h} * 100\% \tag{4}$$

where w and h be the dimensions or width and height of the images A and B. K matrix can be defined as

$$K(i,j) = \begin{cases} 0, & ifA(i,j) = B(i,j) \\ 1, & ifA(i,j) \neq B(i,j) \end{cases} \tag{5}$$

The percentage of pixel differences in the images A and B can be calculated by the parameter UACI as

$$UACI = \frac{\sum_{i,j} \frac{|A(i,j)-B(i,j)|}{255}}{w * h} * 100\% \tag{6}$$

Table 1 shows the NPCR and UACI values by considering different images of different sizes. Similarly, the elapsed time (both encryption and decryption) of three different sizes of images are shown in Table 2.

5 Conclusion

It has been found that 4D logistic map performs similar to the 3D logistic map with respect to NPCR and UACI calculation. Also, it has been observed that the elapsed time for encryption and decryption of 4D logistic map is more than the 3D logistic map. So, it can be concluded that the increased parameters of the 4D logistic map make it more secure and robust as compared to 3D logistic map.

Thus, the 4D logistic map can be used for the sensitive data during the communication. Further investigation can be made on the application of it in the different scientific applications such as digital watermarking.

Table 1 NPCR and UACI values comparison

Input image	Image size	3D logistic map		4D logistic map	
		NPCR	UACI	NPCR	UACI
	32 × 32	91.82%	30.53%	92.02%	30.15%
	64 × 64	91.41%	33.17%	91.65%	33.05%
	128 × 128	93.55%	29.04%	93.84%	28.87%

Table 2 Encryption and Decryption Time

Image	Image size	3D logistic map		4D logistic map	
		Encryption time (s)	Decryption time (s)	Encryption time (s)	Decryption time (s)
	32 × 32	0.0625	0.0625	0.1406	0.1406
	64 × 64	0.2968	0.2812	0.5156	0.5312
	128 × 128	1.5937	1.5937	2.6875	2.5781

References

1. Pradhan, C., Saxena, V., Bisoi, A. K.: Non blind digital watermarking technique using DCT and cross chaos map. In: International conference on communications, devices and intelligent systems, pp. 282–285. IEEE (2012)
2. Cheng, J., Guo, J.: A new chaotic key-based design for image encryption and decryption. In: International Symposium on Circuits and Systems, vol. 4, pp. 49–52, IEEE (2000)
3. Dai, Y., Wang, X.: Medical image encryption based on a composition of logistic maps and Chebyshev maps. In: International Conference on Information and Automation, pp. 210–214, IEEE (2012)
4. Ismail, S.M., Said, L.A., Radwan, A.G., Madian, A.H., Abu-Elyzeed, M.F.: Generalized doubled-humped logistic map-based medical image encryption. J. Adv. Res. Elsevier **10**, 85–98 (2018)
5. Sahay, A., Pradhan, C.: Multidimensional comparative analysis of image encryption using gauss iterated and logistic maps. In: International Conference on Communication and Signal Processing, pp. 1347–1351, IEEE (2017)
6. Hossain, M.B., Rahman, M.T., Rahman, A.B.M.S., Islam, S.: A new approach of image encryption using 3D chaotic map to enhance security of multimedia component. In: International Conference on Informatics, Electronic and Vision, pp. 1–6, IEEE (2014)
7. Saha, B. J., Kabi, K. K., Pradhan, C.: A new approach on color image encryption using arnold 4D cat map. In: International Conference on Computational Intelligence in Data Mining, Springer, vol. 1, pp. 131–138 (2016)
8. Wang, X.Y., Zhang, Y.Q., Zhao, Y.Y.: A novel image encryption scheme based on 2-D logistic map and DNA sequence operations. Nonlinear Dynamics, Springer, vol. 82, no. 3, pp. 1269–1280 (2015)
9. Wu, Y., Noonam, J. P., Agaian, S.: NPCR and UACI randomness tests for image encryption. Cyber J. Multidiscip. J. Sci. Technol., J. Selected Area Telecommun., 31–38 (2011)
10. Khanzadi, H., Eshghi, M., Borujeni, S.E.: Image encryption using random bit sequence based on chaotic maps. Arab J. Sci. Eng. Springer **39**(2), 1039–1047 (2014)

Deep Learning Models for Analysis of Traffic and Crowd Management from Surveillance Videos

S. Seema, Suhas Goutham, Smaranita Vasudev and Rakshith R. Putane

Abstract Deep learning models have been used in the field of object detection and object counting. The problem statement dealt with in this paper aims to achieve the objectives of traffic and crowd management. The Single Shot MultiBox Detector (SSD) model is used in conjunction with a line of counting approach to count the objects of interest in a video captured using surveillance cameras. The proposed model has been used for analyzing traffic surveillance videos to make intelligent traffic decisions to prioritize traffic signals based on the traffic densities. As a sub case of traffic management, a Tesseract OCR model is used to capture the license plate of vehicles violating any traffic regulations. For crowd management, surveillance videos are analyzed to obtain the crowd statistics to handle crowd management in cases of emergencies and huge public gatherings for safety and security.

Keywords Deep learning · Neural networks · Single Shot MultiBox Detector · Line of interest counting · Tesseract · Optical character recognition · Surveillance videos

1 Introduction

With the advent of technology in this digital era, surveillance video cameras have been installed in abundance in public places for monitoring purposes. Analyzing the surveillance video data using deep learning models can help in making informed

S. Seema (✉) · S. Goutham · S. Vasudev · R. R. Putane
Department of Computer Science and Engineering,
M.S. Ramaiah Institute of Technology, Bangaluru, Karnataka, India
e-mail: seemas@msrit.edu

S. Goutham
e-mail: suhas201227@gmail.com

S. Vasudev
e-mail: smaranitav@gmail.com

R. R. Putane
e-mail: rputane4316@gmail.com

H. Das et al. (eds.), *Progress in Computing, Analytics and Networking*,
Advances in Intelligent Systems and Computing 1119,
https://doi.org/10.1007/978-981-15-2414-1_9

decisions, with an added advantage of accuracy in prediction. In the past decade, urban traffic congestion has been a major concern for commuters leading to delays in commute time and a hurdle for the traffic officials for managing traffic. An effective way to solve the problem of traffic congestion is to use the vehicle count from traffic surveillance videos as input data for traffic management.

As a part of traffic management, to keep a track of the vehicles which have violated the traffic rules and regulations in terms of signal bypassing, Tesseract OCR (Optical Character Recognition) is used to extract the license plate number of vehicles from the surveillance videos.

The right use of these crowd counting techniques can help in taking care of unprecedented events such as strikes or stampedes. In case of emergencies and casualties, people counting can provide a rough estimate of the people in the area and this could lead to easier emergency management decisions. The SSD model with the line-of-interest counting can achieve this objective of finding the crowd count from a surveillance video.

2 Literature Survey

The existing research works aligning with the work have been outlined below. Application of the Fast-RCNN framework for detection of vehicle type in a traffic scene proposed by Suhao et al. [1] has been proved to achieve a higher target detection accuracy after experimenting by applying the framework for an actual traffic scenario. To deal with challenging road scenarios and external factors, Li et al. [2] discuss about the vehicle detection from a traffic video using the Darknet framework, considering properties such as color changing and various road scenarios. The object detection problem statement is transformed to a binary classification problem and solved using the YOLO-vocRV network.

Object detection frameworks based on deep learning to handle problems such as clutter, occlusion, and low resolution based on different degrees of variations on R-CNN have been outlined in the work of Zhao et al. [3]. Wang et al. [4] propose a new multiple object detection framework in a traffic scenario, namely, AP-SSD (Single Shot Multi Box Detector), an improvement to the object detection framework SSD, by designing a feature extraction convolution kernel library composed of multi-shape Gabor and color Gabor.

Ren et al. [5] talk about the object detection system YOLO by proposing a new model called YOLO-PC (YOLO-based People Counting) to count the number of people by processing real-time videos. Sun et al. [6] propose an efficient method for counting people in real-world cluttered scenes which computes a point cloud from the depth video frame and re-projects it onto a ground plane to normalize the depth information which is later analyzed for identifying potential human heads.

A three-step methodology which uses an Expectation Maximization (EM)-based method to handle identification of people in a lower resolution is introduced by Hou et al. [7] to deal with the people counting problem in video surveillances. Badr et al. [8] introduce an Automatic Number Plate Recognition System (ANPR) using morphological operations, histogram manipulation, and edge detection techniques for plate localization and characters segmentation.

3 System Architecture

The system architecture is illustrated in Fig. 1. The Android application will be targeted for two kinds of users, namely, the traffic and police authorities. This application provides an easy-to-use interface with options to select the operation to be carried out. The choice of operations include traffic management for traffic signal prioritization, license plate detection (a subcase of traffic management), crowd management and lastly, sending reports such as the vehicle or people count witnessed over a period of time.

Based on the user's input of the choice, a request is sent to a Flask server. The Flask server processes this incoming request and runs a test script. Based on the options chosen, surveillance videos are chosen accordingly. For instance, if a traffic management option is chosen, relevant surveillance traffic videos are selected. The video links are sent as input to the deep learning SSD model or to the script running the Tesseract's OCR algorithm based on the operation to be performed.

The SSD model is used for handling two use cases, namely, traffic and crowd management. The optical character recognition used by Tesseract is used for license plate detection of vehicles from traffic surveillance videos for monitoring traffic violation, a subcase of traffic management.

Traffic Management For cases of vehicle count, the SSD model with Mobilenet [9] is used, which runs on the videos and generates output videos which show the count of the number of vehicles detected in a particular surveillance video. For instance, consider a particular traffic signal which handles three different lanes, in a round-robin fashion. In this scenario, three videos are sent to the SSD model and the model predicts the counts of vehicles detected in the traffic videos using the line of counting approach.

The predicted counts of detected vehicles in all the videos are sent as an email using the SMTP protocol to the registered email addresses of the traffic authorities. The mail contains details of the lane number with the associated number of vehicles detected on that lane. Based on this information, the traffic authorities can take decisions to prioritize signals based on the lane which shows a periodic higher vehicle count. This is how traffic management is dealt with.

Traffic Violation Management In cases of license plate detection, Tesseract optical character recognition extracts the license plate number of the moving vehicles in a surveillance video. The extracted license plate number can be sent as an email to the traffic officials in cases of violation of traffic rules, thereby automating the process of monitoring violations on roads, such as signal jumping or bypassing. The method employed in the paper is similar to License Plate Recognition (LPR) system

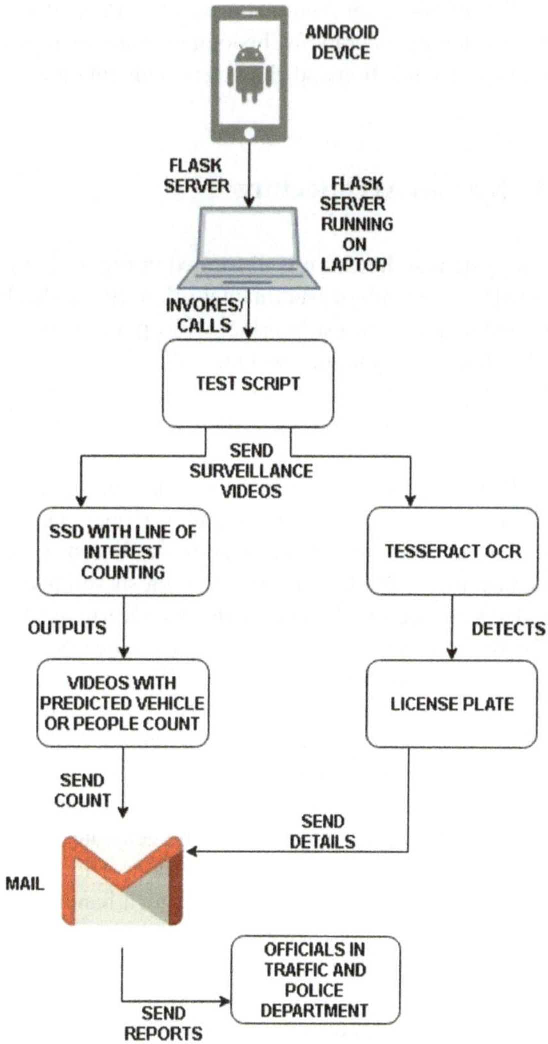

Fig. 1 System architecture depicting the working flow of sending the surveillance videos to a model running on a Flask server

proposed by Prabhakar et al. [10] with the fundamental steps such as detection of number plate, segmentation of characters, and recognition of each characters.

Crowd Management In the previous works related to crowd management, Zhang et al. [11] propose a deep convolutional neural network for crowd counting, where the network is trained for achieving crowd density and crowd count. Using the same basis, this paper shows the usage of the SSD model with Mobilenet. This model uses the line of counting approach on the input videos to generate output videos which show the count of people detected in a surveillance video. Based on the predicted counts of the people in the videos sent as an email to the concerned officials, any crowd management decisions are dealt with.

4 Working

The application has been designed as a client–server architecture. The clients would be the users of the Android application. As this application is intended for usage by the traffic and police authorities, for easier monitoring, a mobile application build on an Android platform is developed. The choice of an Android application includes easy accessibility, wide usage along with the advantage of an easy-to-use interface. This Android application establishes a connection with a Flask server on an available port. The prerequisite for the choice of port is that no other incoming traffic must be directed to this port on which the Flask server is running.

The Flask server invokes a test script which sends surveillance videos as input to the SSD model [12] as per the request sent via the Flask server from an Android client. The model's outputs are sent via email using the Simple Mail Transfer Protocol (SMTP) to the email address registered on the Android client. This summarizes the overall working of sending requests through a server and handling requests by a test script followed by sending the predictions to the user.

4.1 Dataset

In our work, the SSD model is used with Mobilenet, which has been pre-trained on the COCO dataset, consisting of 80 different classes such as people, bicycles, vehicles and so on. The test dataset for traffic and crowd management includes over 80 surveillance videos collected from YouTube and real-time videos captured from surveillance video footage. For license plate recognition, the Tesseract model has been tested on 35 surveillance videos collected from YouTube and videos captured in real time using a surveillance video footage.

4.2 Single Shot MultiBox Detector

The deep learning model used in our work is SSD, which leverages deep convolutional neural networks to perform the tasks of classification of objects as well as locating the objects. Kanatov et al. [13] have demonstrated in their work the advantages of using object recognition using deep convolutional neural networks and the usage of rectangle filters for object detection. Hsieh et al. [14] establish and highlight the drawbacks of regression methods to detect target objects. Hence, SSD model is preferred over regression methods.

The SSD model is used in conjunction with the line of interest counting approach. Line Of Interest (LOI) [15] Counting is used to count the number of objects crossing the line. It differs from the Region of interest (ROI) which monitors all locations of a given existing region whereas LOI only needs to monitor the entrances or the exits of a given area of space. By counting the number of people or vehicles across the line of interest, the total number of people or vehicles can be easily determined.

The SSD model uses OpenCV's libraries to read the surveillance video to analyze the various components of the video. The model is tuned to count objects of interest only using the line of interest counting approach. The objects detected by the SSD model are counted as they pass the line-of-interest, set horizontally or vertically at a particular coordinate for the video. The line of interest is set horizontally (to a y-coordinate) for vehicle counting and is set vertically (to a x-coordinate) for crowd counting. This line serves as a reference for counting the objects of interest in a surveillance video. The model operates on each frame individually and after processing all frames, a new output video is generated denoting the bounding boxes with the respective class probabilities.

4.3 Optical Character Recognition

A deep learning-based text recognition called Optical Character Recognition using Tesseract is used for license plate detection of cars in surveillance videos. Our work makes use of the Tesseract which supports deep learning-based optical character recognition, hence increasing accuracy of character recognition. In simple terms, Tesseract OCR [16] makes use of a recurrent neural network called Long Short-Term Memory (LSTM) network.

The working involves using the OpenCV's libraries to read the input video of vehicles in a scene. The code is written to extract the license plate of vehicles by using the bounding box concept. Each of these detected license plates is passed to the deep learning text recognition algorithm of Tesseract's LSTM. The algorithm's output is the license plate number of the vehicle in the surveillance video. Using optical character recognition which recognizes a license plate by extracting one character at a time tends to provide more accurate results.

5 Use Cases

The use cases under different categories which are handled by our proposed model have been enlisted.

5.1 Traffic Management

Consider a scenario with a traffic signal periodically handling two different lanes, switching signals at every 2-minute intervals. In this case, two videos are sent as input to the SSD model with the line of interest counting. Three output videos with

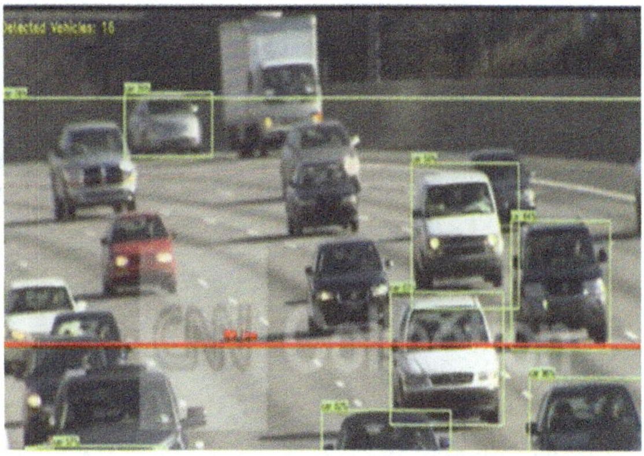

Fig. 2 Lane 1 with 16 detected vehicles

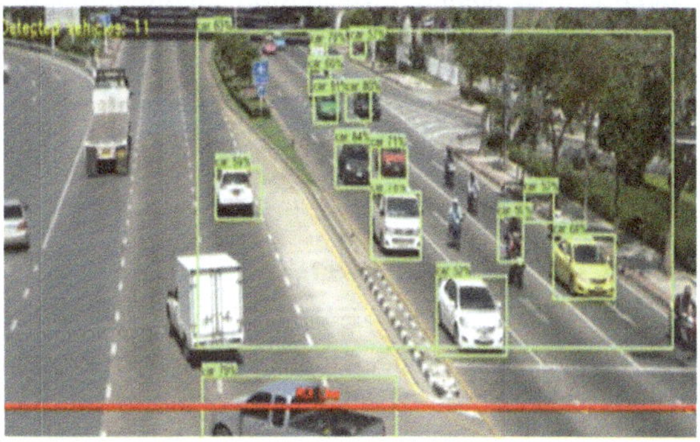

Fig. 3 Lane 2 with 11 detected vehicles

the respective count of detected vehicles are generated. Based on the higher vehicle count among these three lanes, the traffic signals can be prioritized to reduce the traffic congestion on the heavily congested road. Keeping a track of the number of vehicles which has crossed the line helps in estimating the number of vehicles approaching closer to the traffic signal, which directly implies traffic congestion on a road. Over a period of time, if the traffic official sees a continued congestion on Lane 1 over Lane 2, traffic prioritization of signal for Lane 1 can be considered to reduce the traffic congestion. As shown in Figs. 2 and 3, the detected vehicles imply the number of vehicles which has crossed the region of interest line (shown by red in the snapshots).

5.2 *Traffic Violation Management*

Consider a scenario where a vehicle violates the traffic rules and jumps a traffic signal when it is red. To keep a track of these vehicles for violating traffic regulations, the traffic officials can use the Tesseract optical character recognition to extract

Fig. 4 A snapshot of one of the surveillance video frames showing a car

Fig. 5 The license plate extracted by the model from the video

the license plate of the vehicle of interest. This would facilitate an efficient traffic violation management instead of manually noting the number plates of vehicles by seeing surveillance videos. Figure 4 shows an image of a car which has violated the traffic signal rules. On applying the model with OCR, the extracted license plate is shown in Fig. 5.

5.3 Crowd Management

Analysis of a surveillance video at the entrance of a building can help in emergency management in case of any casualty. Taking an example of any fire in a building, the surveillance video camera installed at the entrance can be considered as an input video and running the SSD model on this will provide a rough estimate of the people inside the building premises. Based on this count detected, a mail can be sent to the concerned authorities to operate effectively in case of casualties. To deal with crowd management for security reasons, consider a scenario of a public event with multiple gate entries. Consider three different surveillance videos capturing the people entering the event. These videos can be sent as input for the model and the predictions can help in determining the gate which has the most number of people entering, directly implying the need for more security to ensure smooth operation of the event.

From Figs. 6 and 7, one can infer that inside the building a total of 11 people exist at time t1. This is how crowd counting can help in emergency management to reduce casualties in a building. The crowd count can prove beneficial to have an approximate of the people inside a building during any event or in calculating a people toll inside building premises.

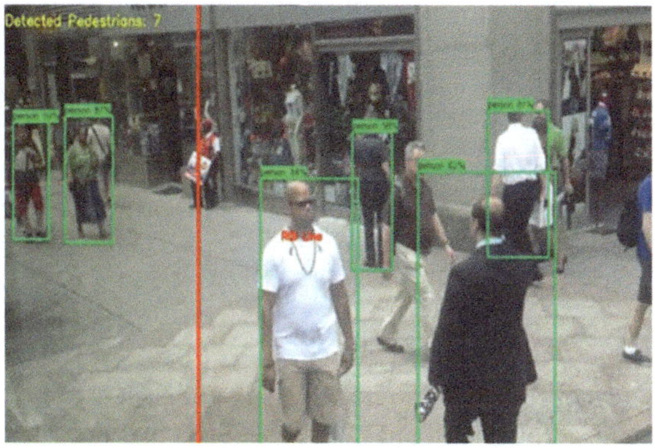

Fig. 6 Building A's entrance "1" with a crowd count of 7 people (as shown on the top left-hand corner)

Fig. 7 Building A's entrance "2" with a crowd count of 4 people (as shown on the top left-hand corner)

Table 1 Performance of the SSD model for vehicle counting

Traffic management performance			
Sl. no.	True vehicle count in video	Detected vehicle count in video	Accuracy (%)
1	9	7	77.77
2	14	11	78.57
3	28	22	78.57
4	46	36	78.26

6 Performance Evaluation

The performance of the model can be evaluated using metrics. For object counting, the metric of accuracy used is the number of objects detected to the number of objects actually present in the surveillance video. The results for accuracy of vehicle count for traffic management is shown in Table 1 and the results for crowd count in Table 2.

Table 2 Performance of the SSD model for crowd counting

Crowd management performance			
Sl. no.	True people count in video	Detected people count in video	Accuracy (%)
1	25	22	88
2	38	32	84.21
3	45	39	86.66
4	47	40	85.10

After running on over 40 videos for each use case, the accuracy of the predicted vehicle count is found to be 78.302% and the accuracy of the predicted vehicle count is found to be 85.68%. The accuracy for crowd counting is found to be higher compared to the vehicle count due to heterogenous factors such as different sizes, shapes of vehicles whereas there is homogeneity in the people feature.

For measuring the performance of license plate detection using the deep learning-based optical character recognition, the metric used is number of mismatched characters in the license plate. Testing on a majority of Indian license plates, results showed that out of the nine characters visible on the license plate, in 96% of the cases, all the nine characters are predicted correctly. If any noise or disturbance exists in the video, the accuracy drops with a mismatch or error in two of the characters on the license plate.

7 Conclusion and Future Work

In conclusion, comparing the performance of the model with respect to other deep learning models, the SSD model performs relatively better. In comparison to YOLO model, SSD's performance tends to be better for larger videos given as input to the model. The object localization and detection techniques used by SSD can be characterized and attributed to its good performance. A few observations made regarding the performance of the SSD model is the accuracy in object detection. The line of counting approach with SSD provided a good estimate of count of objects of interest. In cases of license plate recognition, this proposed model works well as license plates are analyzed on a character basis which tends to improve accuracy in prediction. In the near future, we are aiming to extend this work to cover cases of surveillance video analysis by including object tracking.

References

1. Suhao, L., Jinzhao, L., Guoquan, L., Tong, B., Huiqian W., Yu, P.: Vehicle type detection based on deep learning in traffic scene. In: 8th International Congress of Information and Communication Technology. China (2018)
2. Li, X., Liu, Y., Zhao, Z., Zhang, Y., He, L.: A deep learning approach of vehicle multitarget detection from traffic video. J. Adv. Transp. **2018** (2018)
3. Zhao, Z.-Q., Zheng, P., Xu, S.-Ta., Wu, X.: Object detection with deep learning: a review. J. LATEX Cl. Files **14**(8) (2018)
4. Wang, X., Hua, X., Xiao, F., Li, Y., Hu, X., Sun, P.: Multi-Object Detection in Traffic Scene Based on Improved SSD. Machine Learning and Embedded Computing in Advanced Driver Assistance Systems (ADAS) (2018)
5. Ren, P., Fang, W., Djahel, S.: A novel YOLO-based real-time people counting approach. In: Third IEEE Annual International Smart Cities Conference (2017)
6. Sun, S., Akhtar, N., Song, H., Zhang, C., Li, J., Mian, A.: Benchmark data and method for real-time people counting in cluttered scenes using depth sensors. J. LATEX **14** (2018)
7. Hou, Y.-L., Pang, G.K.H.: People counting and human detection in a challenging situation. IEEE Trans. Syst. Man Cybern. Part A Syst. Hum. **41**(1) (2011)
8. Badr, A., Abdelwahab, M.M., Thabet, A.M., Abdelsadek, A.M.: Automatic number plate recognition system. Ann. Univ. Craiova **38**(1), 62–71 (2011)
9. Howard, A.G., Zhu, M., Chen, B., Kalenichenko, D., Wang, W., Weyand, T., Andreetto, M., Adam, H.: MobileNets: Efficient Convolutional Neural Networks for Mobile Vision Applications (2017). arXiv:1704.04861v1 [cs.CV]
10. Prabhakar, P., Anupama, P., Resmi, S.R.: Automatic vehicle number plate detection and recognition. In: International Conference on Control. Instrumentation, Communication and Computational Technologies (ICCICCT) (2014)
11. Zhang, C., Li, H., Wang, X., Yang, X.: Cross-scene crowd counting via deep convolutional neural networks. In: 2015 IEEE Conference on Computer Vision and Pattern Recognition (CVPR) (2015)
12. Single Shot MultiBox Detector: https://ai.google/research/pubs/pub44872
13. Kanatov, M., Atymtayeva, L.: Deep convolutional neural network based person detection and people counting system. Adv. Eng. Technol. Appl. **7**(3), 21–25 (2018)
14. Hsieh, M.-R., Lin, Y.-L., Hsu, W.H.: Drone-based object counting by spatially regularized regional proposal network. In: IEEE International Conference on Computer Vision (ICCV) (2017)
15. Kowcika, A., Sridhar, S.: A literature survey on crowd (people) counting with the help of surveillance videos. Int. J. Inf. Technol. Decis. Mak. **3**(4), 2353–2361 (2015)
16. Tesseract Optical Character Recognition: https://www.learnopencv.com/deep-learning-based-text-recognition-ocr-using-tesseract-and-opencv/

An Innovative Image-Based Tabular Data Extraction Parallel Algorithm

Varun Nagesh Jolly Behera, Ashish Ranjan and Motahar Reza

Abstract We propose a novel algorithm for extracting data from images of tabular documents having a specific structure. Our proposed method is able to maintain the original table format and structure, and offers better efficiency over existing methodologies due to its high scalability and parallel architecture. The findings of this paper will increase efficiency in the data extraction process from image-based tabular data and help ease the digitization process of tabular records.

Keywords Parallel algorithm · Tabular data · Computer vision

1 Introduction

A major challenge being faced by today's world that is driven by data analysis is the dawdling rate at which data is sourced. One of the noteworthy reasons for this is the unavailability of older data in digital format, instead, it being in a written or physical format [1–4], requiring manual labor to digitize it. Data like this is generally found as older government records, or as analog records kept by small businesses. A lot of this data is tabular in nature, and in this paper, we propose a solution for extracting this specifically structured data from images with the help of computer vision, and try to increase the rate of data extraction by utilizing parallel processing.

In this paper, we use an image dataset containing multiple images of tabular data, both in handwritten and typed out manner. We separate the individual table cells with the help of contours, and then identify the cell data using optical character recognition

V. N. J. Behera (✉) · A. Ranjan · M. Reza
High Performance Computing Lab, National Institute of Science and Technology, Berhampur 761008, India
e-mail: varunbeheralego@gmail.com

A. Ranjan
e-mail: ashishranjan6282@gmail.com

M. Reza
e-mail: reza@nist.edu

H. Das et al. (eds.), *Progress in Computing, Analytics and Networking*, Advances in Intelligent Systems and Computing 1119,
https://doi.org/10.1007/978-981-15-2414-1_10

[5]. We use the multiple cores on a CPU to make the data extraction process more efficient by utilizing its parallel structure. The algorithm consists of three phases. First, using image processing techniques, the image of the document or record is segmented and analyzed to determine potential table locations. In the second phase, within the determined locations all contours are found and are compared with the expected cell area to identify the different cells within the table, and, the text within the cell is identified using an Optical Character Recognition Engine [5]. The second phase occurs in parallel for each individual cell to increase the overall efficiency of the algorithm. Finally, for each location a new table is generated based on the identified cells and the text within them.

2 Goals

Although there exist a number of different research efforts focusing on analyzing tables in documents, they all focus on various parts of the same problem [6–8]. Some mobile applications allow the user to capture and perform analysis on the fly, but this approach cannot handle large datasets with many images within reasonable time. Our primary goal is to extract handwritten tabular data assisted by an OCR software [5] while preserving the table structure. Due to our generalized approach, this method can be applied to non-handwritten data or images containing only the table. This multipurpose methodology allows for usage in many different scenarios, being able to identify and extract tables from any image data. We further extend our goal by adding the ability of storing the identified tables in some higher level format which includes spreadsheets and relational databases.

3 Methodology

As mentioned previously, our method has three phases. This can be observed in Fig. 1. A detailed description of each of these phases follows. The first phase is purely based on obtaining the location and area of the tables. The second phase involves finding the individual cells and identifying its contents with the help of optical character recognition. The third and final phase involves generating the table in its correct structure, and storing it in a higher level format.

3.1 Table Location Identification

The analysis that occurs in this phase involves identifying and isolating potential table areas on the image. Due to variety of possible types of images of documents, identifying a possible table with the help of loose constraints is necessary. This is

Fig. 1 Algorithm flow chart

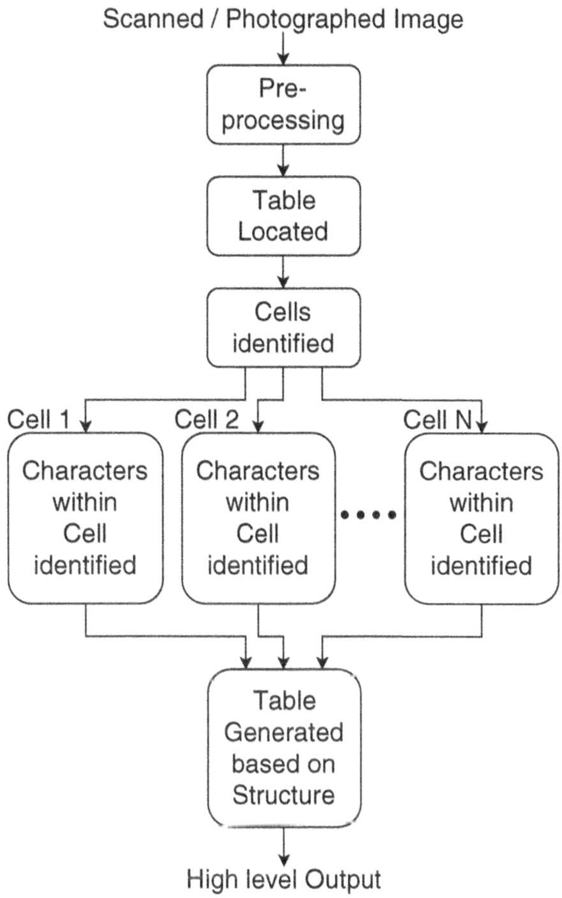

done by marking rectangular zones with the help of white space analysis and keyword analysis. Prior to this, scaling and rotation operations may be required, so that the table itself is correctly aligned. The input image for this process may not be perfect, often containing either noise, or more commonly, being skewed. This imperfection occurs due to the very nature of obtaining the image via scanning, fax or a photograph. Since the OCR Engine [5] expects the text to be exactly horizontal, the skewness needs to be fixed. We use techniques outlined in [9–11] to determine the skew angle of the image.

3.2 Cell Analysis and Character Recognition

Once the table areas have been identified, we can proceed with identifying the contents of said tables. We do this by first finding all contours present within the image segment and comparing it to the expected cell area value, if such a region is found, then with the help of an OCR engine [5] cell data is obtained. Contours [12, 13] can be explained simply as a curve joining all the continuous points (along the boundary), having same color or intensity. These contours are a useful tool for shape analysis and object detection and recognition. Here, we use it to detect the borders of the cells. Since the boundary is made up of continuous points having same color intensities, they form the contours. These contours are equivalent to the cells in the table. Within each cell, characters are identified using an OCR engine [5].

Algorithm 1: Proposed Algorithm

```
Input: Image segment containing tabular data
Result: The Generated Table

READ Input Image
APPLY Preprocessing and de-skewing algorithm
FIND all contours in image
SET number of parallel processes
foreach contour in image do
    DISTRIBUTE contours among workers
    if area of contour = expected cell area then
        USE OCR engine to detect text within cell
        APPEND text, yValue to list L
    end
end
prevY ← ∞
p ← 0
foreach i in 0 to length of L do
    if prevY < L[i].yValue then
        APPEND L[i-p:i] to list T
        p ← 0
    end
    p ← p + 1
    prevY ← L[i].yValue
end
```

In order to detect handwritten data, proper preprocessing is required before the above algorithm is used to extract the data from it. Scanned images and photographs of documents are highly susceptible to noise and orientation issues. The noise from the image is removed using Adaptive Gaussian Thresholding [14, 15] and Otsu's Binarization [16]. These preprocessing steps are essential for successful text recognition. Finally, the resultant image is ready for use in our algorithm to extract the text as mentioned before. To decrease the time complexity of this process we use CPU parallelism. The processing for each cell occurs in parallel on seperate cores of the CPU.

3.3 Table Generation

The final phase, in which the table structure is determined which includes identifying its headers and footers. The table is then saved in the required high-level format.

4 Results

The algorithm is implemented in Python 3.7 programming language and it utilizes the open source computer vision library, OpenCV and the Tesseract OCR engine [5]. For our experimental evaluation, we used a system with an Intel Core i7 7700HQ processor, with 4 cores and 8 threads.

Our algorithm was tested with several input data. One of such input data is shown in Fig. 2 which is from a sample excel spreadsheet provided by Microsoft. It contains 65 rows and 11 columns.

Each of the cells is then allowed to be processed in parallel. The region of image (ROI) containing the cells are separated out to carry out the operation. Next, the Tesseract OCR is run for each cell. The text may be rotated in the input image. This can be solved by rotating the image at different angles. Both scaling and rotation operations need to be done in every combination to consider every possible orientation.

Figure 3 shows the contents of the final list of characters segmented cell wise and row wise. The implementation of Optical Character Recognition is working correctly as proper output is produced. This needs manual verification to ensure that all of the data retrieved is correct because this project aims to develop a system which can convert visual data to textual data which can easily produce wrong output because of the presence of noise and visual artifacts in the input image.

We used multiple test cases to test the handwritten images. One of those inputs is shown in the Fig. 4.

The algorithm was tested using many such test inputs and the output was found to be accurate in about 90% of the cases with typed text data and 80% of the cases with handwritten data. This reduction in accuracy can be attributed to variations in handwritings of people and tendency of certain characters to look similar. This is a limitation of the OCR engine we use [5], and not of our algorithm. This means that the system is very reliable and if clean and systematic data is supplied, proper extraction of data in the tabular form can be done from any source, given that the data can be supplied in the form of an image.

Government	Canada	Carretera	None	1618.5	$ 3.00	$ 20.00	$ 32,370.00	$ -	$ 32,370.00	$ 16,185.00
Government	Germany	Carretera	None	1321	$ 3.00	$ 20.00	$ 26,420.00	$ -	$ 26,420.00	$ 13,210.00
Midmarket	France	Carretera	None	2178	$ 3.00	$ 15.00	$ 32,670.00	$ -	$ 32,670.00	$ 21,780.00
Midmarket	Germany	Carretera	None	888	$ 3.00	$ 15.00	$ 13,320.00	$ -	$ 13,320.00	$ 8,880.00
Midmarket	Mexico	Carretera	None	2470	$ 3.00	$ 15.00	$ 37,050.00	$ -	$ 37,050.00	$ 24,700.00
Government	Germany	Carretera	None	1513	$ 3.00	$ 350.00	$ 5,29,550.00	$ -	$ 5,29,550.00	#########
Midmarket	Germany	Montana	None	921	$ 5.00	$ 15.00	$ 13,815.00	$ -	$ 13,815.00	$ 9,210.00
Channel Partners	Canada	Montana	None	2518	$ 5.00	$ 12.00	$ 30,216.00	$ -	$ 30,216.00	$ 7,554.00
Government	France	Montana	None	1899	$ 5.00	$ 20.00	$ 37,980.00	$ -	$ 37,980.00	$ 18,990.00
Channel Partners	Germany	Montana	None	1545	$ 5.00	$ 12.00	$ 18,540.00	$ -	$ 18,540.00	$ 4,635.00
Midmarket	Mexico	Montana	None	2470	$ 5.00	$ 15.00	$ 37,050.00	$ -	$ 37,050.00	$ 24,700.00
Enterprise	Canada	Montana	None	2665.5	$ 5.00	$ 125.00	$ 3,33,187.50	$ -	$ 3,33,187.50	#########
Small Business	Mexico	Montana	None	958	$ 5.00	$ 300.00	$ 2,87,400.00	$ -	$ 2,87,400.00	#########
Government	Germany	Montana	None	2146	$ 5.00	$ 7.00	$ 15,022.00	$ -	$ 15,022.00	$ 10,730.00
Enterprise	Canada	Montana	None	345	$ 5.00	$ 125.00	$ 43,125.00	$ -	$ 43,125.00	$ 41,400.00
Midmarket	United States of America	Montana	None	615	$ 5.00	$ 15.00	$ 9,225.00	$ -	$ 9,225.00	$ 6,150.00
Government	Canada	Paseo	None	292	$ 10.00	$ 20.00	$ 5,840.00	$ -	$ 5,840.00	$ 2,920.00
Midmarket	Mexico	Paseo	None	974	$ 10.00	$ 15.00	$ 14,610.00	$ -	$ 14,610.00	$ 9,740.00
Channel Partners	Canada	Paseo	None	2518	$ 10.00	$ 12.00	$ 30,216.00	$ -	$ 30,216.00	$ 7,554.00
Government	Germany	Paseo	None	1006	$ 10.00	$ 350.00	$ 3,52,100.00	$ -	$ 3,52,100.00	#########
Channel Partners	Germany	Paseo	None	367	$ 10.00	$ 12.00	$ 4,404.00	$ -	$ 4,404.00	$ 1,101.00
Government	Mexico	Paseo	None	883	$ 10.00	$ 7.00	$ 6,181.00	$ -	$ 6,181.00	$ 4,415.00
Midmarket	France	Paseo	None	549	$ 10.00	$ 15.00	$ 8,235.00	$ -	$ 8,235.00	$ 5,490.00
Small Business	Mexico	Paseo	None	788	$ 10.00	$ 300.00	$ 2,36,400.00	$ -	$ 2,36,400.00	#########
Midmarket	Mexico	Paseo	None	2472	$ 10.00	$ 15.00	$ 37,080.00	$ -	$ 37,080.00	$ 24,720.00
Government	United States of America	Paseo	None	1143	$ 10.00	$ 7.00	$ 8,001.00	$ -	$ 8,001.00	$ 5,715.00
Government	Canada	Paseo	None	1725	$ 10.00	$ 350.00	$ 6,03,750.00	$ -	$ 6,03,750.00	#########
Channel Partners	United States of America	Paseo	None	912	$ 10.00	$ 12.00	$ 10,944.00	$ -	$ 10,944.00	$ 2,736.00
Midmarket	Canada	Paseo	None	2152	$ 10.00	$ 15.00	$ 32,280.00	$ -	$ 32,280.00	$ 21,520.00
Government	Canada	Paseo	None	1817	$ 10.00	$ 20.00	$ 36,340.00	$ -	$ 36,340.00	$ 18,170.00
Government	Germany	Paseo	None	1513	$ 10.00	$ 350.00	$ 5,29,550.00	$ -	$ 5,29,550.00	#########
Government	Mexico	Velo	None	1493	$ 120.00	$ 7.00	$ 10,451.00	$ -	$ 10,451.00	$ 7,465.00
Enterprise	France	Velo	None	1804	$ 120.00	$ 125.00	$ 2,25,500.00	$ -	$ 2,25,500.00	#########
Channel Partners	Germany	Velo	None	2161	$ 120.00	$ 12.00	$ 25,932.00	$ -	$ 25,932.00	$ 6,483.00
Government	Germany	Velo	None	1006	$ 120.00	$ 350.00	$ 3,52,100.00	$ -	$ 3,52,100.00	#########
Channel Partners	Germany	Velo	None	1545	$ 120.00	$ 12.00	$ 18,540.00	$ -	$ 18,540.00	$ 4,635.00
Enterprise	United States of America	Velo	None	2821	$ 120.00	$ 125.00	$ 3,52,625.00	$ -	$ 3,52,625.00	#########
Enterprise	Canada	Velo	None	345	$ 120.00	$ 125.00	$ 43,125.00	$ -	$ 43,125.00	$ 41,400.00
Small Business	Canada	VTT	None	2001	$ 250.00	$ 300.00	$ 6,00,300.00	$ -	$ 6,00,300.00	#########
Channel Partners	Germany	VTT	None	2838	$ 250.00	$ 12.00	$ 34,056.00	$ -	$ 34,056.00	$ 8,514.00
Midmarket	France	VTT	None	2178	$ 250.00	$ 15.00	$ 32,670.00	$ -	$ 32,670.00	$ 21,780.00
Midmarket	Germany	VTT	None	888	$ 250.00	$ 15.00	$ 13,320.00	$ -	$ 13,320.00	$ 8,880.00
Government	France	VTT	None	1527	$ 250.00	$ 350.00	$ 5,34,450.00	$ -	$ 5,34,450.00	#########
Small Business	France	VTT	None	2151	$ 250.00	$ 300.00	$ 6,45,300.00	$ -	$ 6,45,300.00	#########
Government	Canada	VTT	None	1817	$ 250.00	$ 20.00	$ 36,340.00	$ -	$ 36,340.00	$ 18,170.00
Government	France	Amarilla	None	2750	$ 260.00	$ 350.00	$ 9,62,500.00	$ -	$ 9,62,500.00	#########
Channel Partners	United States of America	Amarilla	None	1953	$ 260.00	$ 12.00	$ 23,436.00	$ -	$ 23,436.00	$ 5,859.00
Enterprise	Germany	Amarilla	None	4219.5	$ 260.00	$ 125.00	$ 5,27,437.50	$ -	$ 5,27,437.50	#########
Government	France	Amarilla	None	1899	$ 260.00	$ 20.00	$ 37,980.00	$ -	$ 37,980.00	$ 18,990.00
Government	Germany	Amarilla	None	1686	$ 260.00	$ 7.00	$ 11,802.00	$ -	$ 11,802.00	$ 8,430.00
Channel Partners	United States of America	Amarilla	None	2141	$ 260.00	$ 12.00	$ 25,692.00	$ -	$ 25,692.00	$ 6,423.00
Government	United States of America	Amarilla	None	1143	$ 260.00	$ 7.00	$ 8,001.00	$ -	$ 8,001.00	$ 5,715.00
Midmarket	United States of America	Amarilla	None	615	$ 260.00	$ 15.00	$ 9,225.00	$ -	$ 9,225.00	$ 6,150.00
Government	France	Paseo	Low	3945	$ 10.00	$ 7.00	$ 27,615.00	$ 276.15	$ 27,338.85	$ 19,725.00
Midmarket	France	Paseo	Low	2296	$ 10.00	$ 15.00	$ 34,440.00	$ 344.40	$ 34,095.60	$ 22,960.00
Government	France	Paseo	Low	1030	$ 10.00	$ 7.00	$ 7,210.00	$ 72.10	$ 7,137.90	$ 5,150.00
Government	France	Velo	Low	639	$ 120.00	$ 7.00	$ 4,473.00	$ 44.73	$ 4,428.27	$ 3,195.00
Government	Canada	VTT	Low	1326	$ 250.00	$ 7.00	$ 9,282.00	$ 92.82	$ 9,189.18	$ 6,630.00
Channel Partners	United States of America	Carretera	Low	1858	$ 3.00	$ 12.00	$ 22,296.00	$ 222.96	$ 22,073.04	$ 5,574.00
Government	Mexico	Carretera	Low	1210	$ 3.00	$ 350.00	$ 4,23,500.00	$ 4,235.00	$ 4,19,265.00	#########
Government	United States of America	Carretera	Low	2529	$ 3.00	$ 7.00	$ 17,703.00	$ 177.03	$ 17,525.97	$ 12,645.00
Channel Partners	Canada	Carretera	Low	1445	$ 3.00	$ 12.00	$ 17,340.00	$ 173.40	$ 17,166.60	$ 4,335.00
Enterprise	United States of America	Carretera	Low	330	$ 3.00	$ 125.00	$ 41,250.00	$ 412.50	$ 40,837.50	$ 39,600.00
Channel Partners	France	Carretera	Low	2671	$ 3.00	$ 12.00	$ 32,052.00	$ 320.52	$ 31,731.48	$ 8,013.00

Fig. 2 Sample input data

This parallel approach was found to be nearly four times faster than serial approach if the number of cells in the image was huge. The performance metrics can be seen in the Fig. 5. As can be seen, our algorithm scales very well over large numbers of cells within a table, and this high scalability improves efficiency of the data extraction process.

5 Conclusion

A robust solution was created for the defined problem statement. We were able to model a solution that could identify, analyze, and extrapolate tabular data from images of tables. Due to our usage of the CPU parallelism the algorithm works more efficiently, without sacrificing accuracy. Our model can be used for various

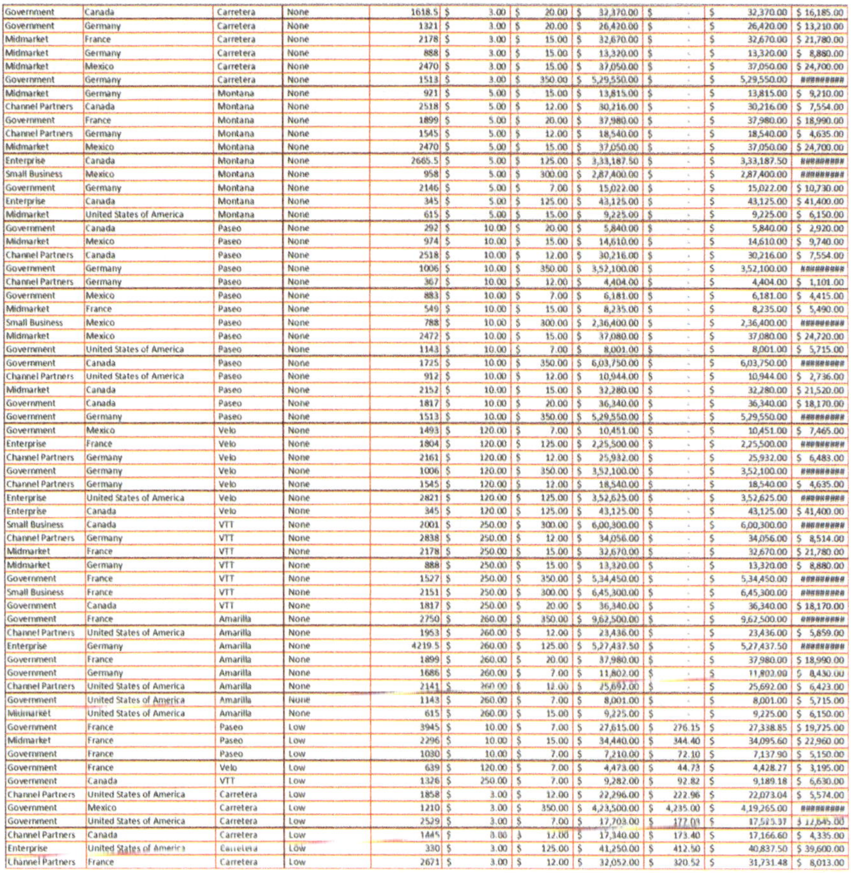

Government	Canada	Carretera	None	1618.5	$ 3.00	$ 20.00	$ 32,370.00	$ -	$ 32,370.00	$ 16,185.00
Government	Germany	Carretera	None	1321	$ 3.00	$ 20.00	$ 26,420.00	$ -	$ 26,420.00	$ 13,210.00
Midmarket	France	Carretera	None	2178	$ 3.00	$ 15.00	$ 32,670.00	$ -	$ 32,670.00	$ 21,780.00
Midmarket	Germany	Carretera	None	888	$ 3.00	$ 15.00	$ 13,320.00	$ -	$ 13,320.00	$ 8,880.00
Midmarket	Mexico	Carretera	None	2470	$ 3.00	$ 15.00	$ 37,050.00	$ -	$ 37,050.00	$ 24,700.00
Government	Germany	Carretera	None	1513	$ 3.00	$ 350.00	$ 5,29,550.00	$ -	$ 5,29,550.00	#########
Midmarket	Germany	Montana	None	921	$ 5.00	$ 15.00	$ 13,815.00	$ -	$ 13,815.00	$ 9,210.00
Channel Partners	Canada	Montana	None	2518	$ 5.00	$ 12.00	$ 30,216.00	$ -	$ 30,216.00	$ 7,554.00
Government	France	Montana	None	1899	$ 5.00	$ 20.00	$ 37,980.00	$ -	$ 37,980.00	$ 18,990.00
Channel Partners	Germany	Montana	None	1545	$ 5.00	$ 12.00	$ 18,540.00	$ -	$ 18,540.00	$ 4,635.00
Midmarket	Mexico	Montana	None	2470	$ 5.00	$ 15.00	$ 37,050.00	$ -	$ 37,050.00	$ 24,700.00
Enterprise	Canada	Montana	None	2665.5	$ 5.00	$ 125.00	$ 3,33,187.50	$ -	$ 3,33,187.50	#########
Small Business	Mexico	Montana	None	958	$ 5.00	$ 300.00	$ 2,87,400.00	$ -	$ 2,87,400.00	#########
Government	Germany	Montana	None	2146	$ 5.00	$ 7.00	$ 15,022.00	$ -	$ 15,022.00	$ 10,730.00
Enterprise	Canada	Montana	None	345	$ 5.00	$ 125.00	$ 43,125.00	$ -	$ 43,125.00	$ 41,400.00
Midmarket	United States of America	Montana	None	615	$ 5.00	$ 15.00	$ 9,225.00	$ -	$ 9,225.00	$ 6,150.00
Government	Canada	Paseo	None	292	$ 10.00	$ 20.00	$ 5,840.00	$ -	$ 5,840.00	$ 2,920.00
Midmarket	Mexico	Paseo	None	974	$ 10.00	$ 15.00	$ 14,610.00	$ -	$ 14,610.00	$ 9,740.00
Channel Partners	Canada	Paseo	None	2518	$ 10.00	$ 12.00	$ 30,216.00	$ -	$ 30,216.00	$ 7,554.00
Government	Germany	Paseo	None	1006	$ 10.00	$ 350.00	$ 3,52,100.00	$ -	$ 3,52,100.00	#########
Channel Partners	Germany	Paseo	None	367	$ 10.00	$ 12.00	$ 4,404.00	$ -	$ 4,404.00	$ 1,101.00
Government	Mexico	Paseo	None	883	$ 10.00	$ 7.00	$ 6,181.00	$ -	$ 6,181.00	$ 4,415.00
Midmarket	France	Paseo	None	549	$ 10.00	$ 15.00	$ 8,235.00	$ -	$ 8,235.00	$ 5,490.00
Small Business	Mexico	Paseo	None	788	$ 10.00	$ 300.00	$ 2,36,400.00	$ -	$ 2,36,400.00	#########
Midmarket	Mexico	Paseo	None	2472	$ 10.00	$ 15.00	$ 37,080.00	$ -	$ 37,080.00	$ 24,720.00
Government	United States of America	Paseo	None	1143	$ 10.00	$ 7.00	$ 8,001.00	$ -	$ 8,001.00	$ 5,715.00
Government	Canada	Paseo	None	1725	$ 10.00	$ 350.00	$ 6,03,750.00	$ -	$ 6,03,750.00	#########
Channel Partners	United States of America	Paseo	None	912	$ 10.00	$ 12.00	$ 10,944.00	$ -	$ 10,944.00	$ 2,736.00
Midmarket	Canada	Paseo	None	2152	$ 10.00	$ 15.00	$ 32,280.00	$ -	$ 32,280.00	$ 21,520.00
Government	Canada	Paseo	None	1817	$ 10.00	$ 20.00	$ 36,340.00	$ -	$ 36,340.00	$ 18,170.00
Government	Germany	Paseo	None	1513	$ 10.00	$ 350.00	$ 5,29,550.00	$ -	$ 5,29,550.00	#########
Government	Mexico	Velo	None	1493	$ 120.00	$ 7.00	$ 10,451.00	$ -	$ 10,451.00	$ 7,465.00
Enterprise	France	Velo	None	1804	$ 120.00	$ 125.00	$ 2,25,500.00	$ -	$ 2,25,500.00	#########
Channel Partners	Germany	Velo	None	2161	$ 120.00	$ 12.00	$ 25,932.00	$ -	$ 25,932.00	$ 6,483.00
Government	Germany	Velo	None	1006	$ 120.00	$ 350.00	$ 3,52,100.00	$ -	$ 3,52,100.00	#########
Channel Partners	Germany	Velo	None	1545	$ 120.00	$ 12.00	$ 18,540.00	$ -	$ 18,540.00	$ 4,635.00
Enterprise	United States of America	Velo	None	2821	$ 120.00	$ 125.00	$ 3,52,625.00	$ -	$ 3,52,625.00	#########
Enterprise	Canada	Velo	None	345	$ 120.00	$ 125.00	$ 43,125.00	$ -	$ 43,125.00	$ 41,400.00
Small Business	Canada	VTT	None	2001	$ 250.00	$ 300.00	$ 6,00,300.00	$ -	$ 6,00,300.00	#########
Channel Partners	Germany	VTT	None	2838	$ 250.00	$ 12.00	$ 34,056.00	$ -	$ 34,056.00	$ 8,514.00
Midmarket	France	VTT	None	2178	$ 250.00	$ 15.00	$ 32,670.00	$ -	$ 32,670.00	$ 21,780.00
Midmarket	Germany	VTT	None	888	$ 250.00	$ 15.00	$ 13,320.00	$ -	$ 13,320.00	$ 8,880.00
Government	France	VTT	None	1527	$ 250.00	$ 350.00	$ 5,34,450.00	$ -	$ 5,34,450.00	#########
Small Business	France	VTT	None	2151	$ 250.00	$ 300.00	$ 6,45,300.00	$ -	$ 6,45,300.00	#########
Government	Canada	VTT	None	1817	$ 250.00	$ 20.00	$ 36,340.00	$ -	$ 36,340.00	$ 18,170.00
Government	France	Amarilla	None	2750	$ 260.00	$ 350.00	$ 9,62,500.00	$ -	$ 9,62,500.00	#########
Channel Partners	United States of America	Amarilla	None	1953	$ 260.00	$ 12.00	$ 23,436.00	$ -	$ 23,436.00	$ 5,859.00
Enterprise	Germany	Amarilla	None	4219.5	$ 260.00	$ 125.00	$ 5,27,437.50	$ -	$ 5,27,437.50	#########
Government	France	Amarilla	None	1899	$ 260.00	$ 20.00	$ 37,980.00	$ -	$ 37,980.00	$ 18,990.00
Government	Germany	Amarilla	None	1686	$ 260.00	$ 7.00	$ 11,802.00	$ -	$ 11,802.00	$ 8,430.00
Channel Partners	United States of America	Amarilla	None	2141	$ 260.00	$ 12.00	$ 25,692.00	$ -	$ 25,692.00	$ 6,423.00
Government	United States of America	Amarilla	None	1143	$ 260.00	$ 7.00	$ 8,001.00	$ -	$ 8,001.00	$ 5,715.00
Midmarket	United States of America	Amarilla	None	615	$ 260.00	$ 15.00	$ 9,225.00	$ -	$ 9,225.00	$ 6,150.00
Government	France	Paseo	Low	3945	$ 10.00	$ 7.00	$ 27,615.00	$ 276.15	$ 27,338.85	$ 19,725.00
Midmarket	France	Paseo	Low	2296	$ 10.00	$ 15.00	$ 34,440.00	$ 344.40	$ 34,095.60	$ 22,960.00
Government	France	Paseo	Low	1030	$ 10.00	$ 7.00	$ 7,210.00	$ 72.10	$ 7,137.90	$ 5,150.00
Government	France	Velo	Low	639	$ 120.00	$ 7.00	$ 4,473.00	$ 44.73	$ 4,428.27	$ 3,195.00
Government	Canada	VTT	Low	1326	$ 250.00	$ 7.00	$ 9,282.00	$ 92.82	$ 9,189.18	$ 6,630.00
Channel Partners	United States of America	Carretera	Low	1858	$ 3.00	$ 12.00	$ 22,296.00	$ 222.96	$ 22,073.04	$ 5,574.00
Government	Mexico	Carretera	Low	1210	$ 3.00	$ 350.00	$ 4,23,500.00	$ 4,235.00	$ 4,19,265.00	#########
Government	United States of America	Carretera	Low	2529	$ 3.00	$ 7.00	$ 17,703.00	$ 177.03	$ 17,525.37	$ 12,645.00
Channel Partners	Canada	Carretera	Low	1445	$ 3.00	$ 12.00	$ 17,340.00	$ 173.40	$ 17,166.60	$ 4,335.00
Enterprise	United States of America	Carretera	Low	330	$ 3.00	$ 125.00	$ 41,250.00	$ 412.50	$ 40,837.50	$ 39,600.00
Channel Partners	France	Carretera	Low	2671	$ 3.00	$ 12.00	$ 32,052.00	$ 320.52	$ 31,731.48	$ 8,013.00

Fig. 3 Data after cell identification

applications, either specific or multipurpose. Primarily, it can be used for knowledge discovery, making it easier to source the data required for data analysis, when the data is only available in written format, for example, like those used in surveys. Another important use-case would be the digitization of old records currently kept either in written or scanned format, both of which are inconvenient. Written data will, sans this system, require manual labor to digitize, and scanned documents require larger amount of storage. We believe that advancements made in this area are beneficial, and will find many applications in the digitization of non-digital datasets.

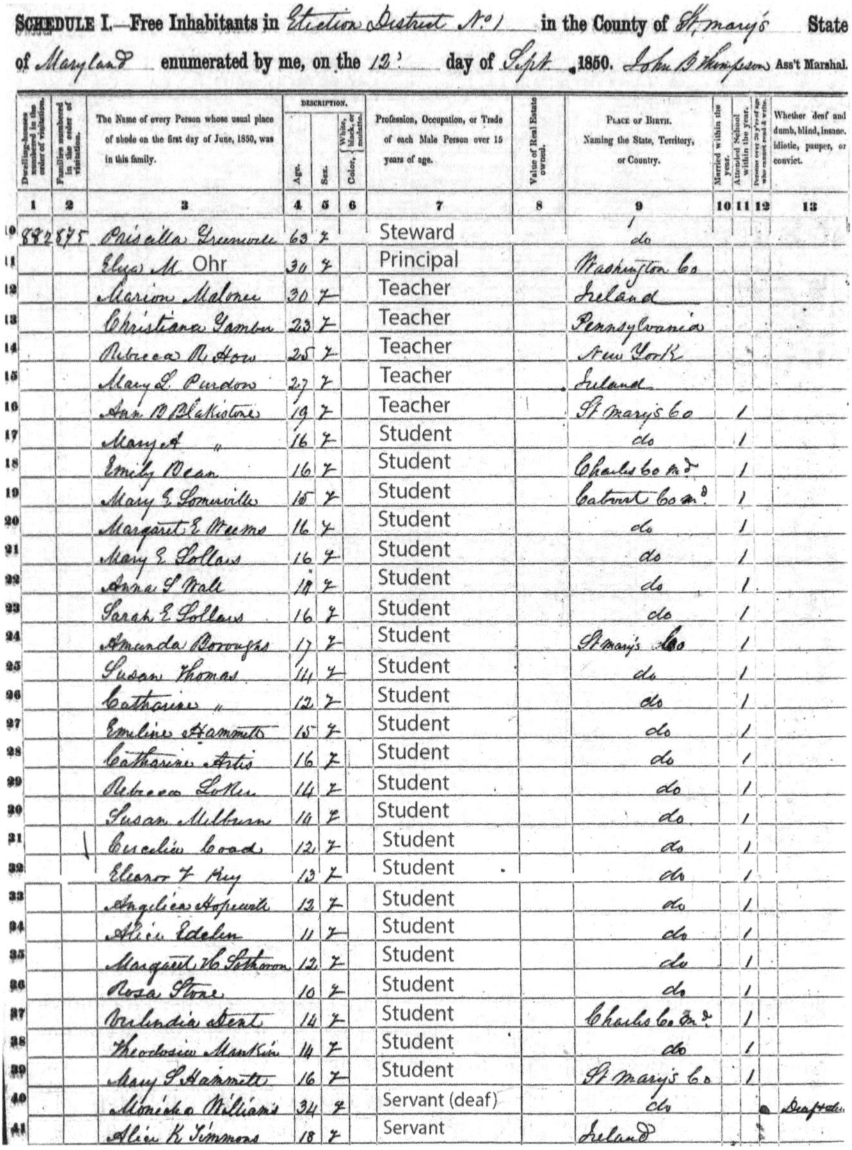

SCHEDULE I.—Free Inhabitants in Election District No. 1 in the County of St. Mary's State of Maryland enumerated by me, on the 12th day of Sept. 1850. John B Thompson Ass't Marshal.

	Dwelling-houses numbered in the order of visitation.	Families numbered in the order of visitation.	The Name of every Person whose usual place of abode on the first day of June, 1850, was in this family.	Age.	Sex.	Color, White, black, or mulatto.	Profession, Occupation, or Trade of each Male Person over 15 years of age.	Value of Real Estate owned.	Place of Birth. Naming the State, Territory, or Country.	Married within the year.	Attended School within the year.	Persons over 20 y'rs of age who cannot read & write.	Whether deaf and dumb, blind, insane, idiotic, pauper, or convict.
	1	2	3	4	5	6	7	8	9	10	11	12	13
10	882	875	Priscilla Greenwell	63	F		Steward		do				
11			Eliza M. Ohr	30	F		Principal		Washington Co				
12			Marion Maloney	30	F		Teacher		Ireland				
13			Christiana Gamber	23	F		Teacher		Pennsylvania				
14			Rebecca R. How	25	F		Teacher		New York				
15			Mary L. Purdon	27	F		Teacher		Ireland				
16			Ann B Blakistone	19	F		Teacher		St Mary's Co		1		
17			Mary A "	16	F		Student		do		1		
18			Emily Bean	16	F		Student		Charles Co Md		1		
19			Mary E Somerville	15	F		Student		Calvert Co Md		1		
20			Margaret E Weems	16	F		Student		do		1		
21			Mary E Sollars	16	F		Student		do		1		
22			Anna S Wall	19	F		Student		do		1		
23			Sarah E Sollars	16	F		Student		do		1		
24			Amanda Boroughs	17	F		Student		St Mary's Co		1		
25			Susan Thomas	14	F		Student		do		1		
26			Catharine "	12	F		Student		do		1		
27			Emeline Hammett	15	F		Student		do		1		
28			Catharine Artis	16	F		Student		do		1		
29			Rebecca Lotten	14	F		Student		do		1		
30			Susan Milburn	14	F		Student		do		1		
31			Cecelia Coad	12	F		Student		do		1		
32			Eleanor F Key	13	F		Student		do		1		
33			Angelica Hopewell	12	F		Student		do		1		
34			Alice Edelen	11	F		Student		do		1		
35			Margaret H Sothoron	12	F		Student		do		1		
36			Rosa Stone	10	F		Student		do		1		
37			Verlinda Dent	14	F		Student		Charles Co Md		1		
38			Theodosia Mankin	14	F		Student		do		1		
39			Mary S Hammett	16	F		Student		St Mary's Co		1		
40			Monica Williams	34	F		Servant (deaf)		do				Deaf & du[illegible]
41			Alice K Timmons	18	F		Servant		Ireland				

Fig. 4 Handwritten data

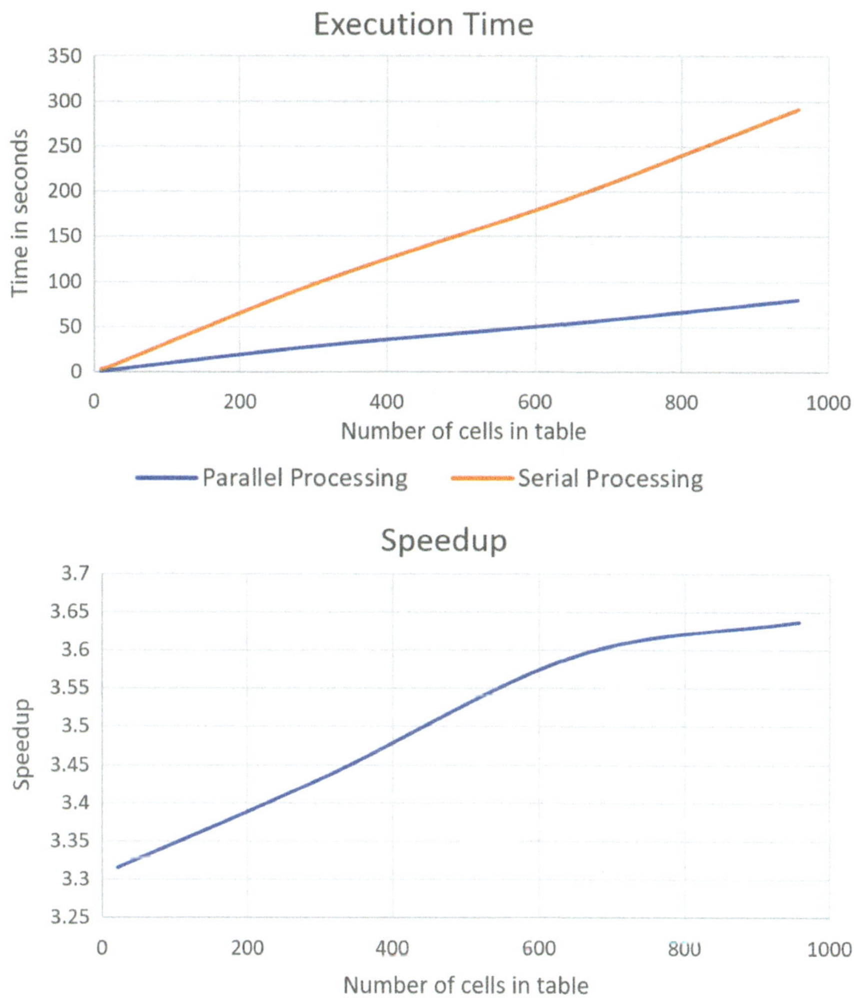

Fig. 5 Performance metrics

References

1. Henry, C., Smith, K.: Ghostlier demarcations: large-scale text digitization projects and their utility for contemporary humanities scholarship. In: The Idea of Order: Transforming Research Collections for 21st Century Scholarship, no. 147, pp. 106 (2010)
2. Balk, Hildelies, Ploeger, Lieke: Impact: working together to address the challenges involving mass digitization of historical printed text. OCLC Syst. Serv. Int. Dig. Library Perspect. **25**(4), 233–248 (2009)
3. Pal, U., Chaudhuri, B.B.: Machine-printed and hand-written text lines identification. Pattern Recognit. Lett. **22**(3–4), 431–441 (2001)
4. Borghi, M., Karapapa, S.: Copyright and Mass Digitization: A Cross-Jurisdictional Perspective. Oxford University Press (2013)

5. Smith, R.: An overview of the tesseract OCR engine. In: Ninth international conference on document analysis and recognition (ICDAR 2007), vol. 2, pp. 629–633. IEEE (2007)
6. Pinto, D, McCallum, A, Wei, W., Croft, W.B.: Table extraction using conditional random fields. In: Proceedings of the 26th Annual International ACM SIGIR Conference on Research and Development in Information Retrieval, pp. 235–242. ACM (2003)
7. Shafait, F., Smith, R.: Table detection in heterogeneous documents. In: Proceedings of the 9th IAPR international workshop on document analysis systems, pp. 65–72. ACM (2010)
8. Wang, S-Y., Yagasaki, T.: Feature extraction system for identifying text within a table image, December 8 1998. US Patent 5,848,186
9. Srihari, S.N., Govindaraju, V.: Analysis of textual images using the Hough transform. Mach. Vis. Appl. **2**(3), 141–153 (1989)
10. Jia, C.C., Woodson, A-M., Sansom-Wai, C., Tretter, D.R.: Image processing system with automatic image cropping and skew correction, August 6 2002. US Patent 6,430,320
11. Sansom-Wai, C.Y., Williams, I.H., Tretter, D.R.: Image processing system with image cropping and skew correction, October 30 2001. US Patent 6,310,984
12. Arbelaez, Pablo, Maire, Michael, Fowlkes, Charless, Malik, Jitendra: Contour detection and hierarchical image segmentation. IEEE Trans. Pattern Anal. Mach. Intell. **33**(5), 898–916 (2010)
13. Malik, Jitendra, Belongie, Serge, Leung, Thomas, Shi, Jianbo: Contour and texture analysis for image segmentation. Int. J. Comput. Vis. **43**(1), 7–27 (2001)
14. Deng, G., Cahill, L.W.: An adaptive Gaussian filter for noise reduction and edge detection. In: 1993 IEEE conference record nuclear science symposium and medical imaging conference, pp. 1615–1619. IEEE (1993)
15. Chang, S.G., Yu, B., Vetterli, M.: Adaptive wavelet thresholding for image denoising and compression. IEEE Trans. Image Process. **9**(9), 1532–1546 (2000)
16. Otsu, Nobuyuki: A threshold selection method from gray-level histograms. IEEE Trans. Syst. Man Cybern. **9**(1), 62–66 (1979)

Brain Tumor Segmentation from MRI Images Using Deep Learning Framework

Suchismita Das

Abstract Brain tumors are the most common and aggressive diseases which lead to a very short life expectancy in their highest grade. Thus, automatic brain tumor detection is necessary at early stage to reduce death frequency due to this. Magnetic Resonance Imaging (MRI) is a widely used imaging technique to assess but it is not practically possible to perform manual segmentation on the large amount of data produced by MRI in due time. This paper focuses on tumor detection from magnetic resonance images having dataset of brain images to classify the tumor and non-tumor symptoms using deep learning fully convolution neural network. In particular, it has been developed using image enhancement, segmentation, and classification techniques. The U-Net architecture is used for segmenting the tumor followed by fully CNN to classify the extracted portion to improve the performance. The outcomes are recorded and compared with existing techniques and found the proposed framework gives better performance by 10%. This method can be extended in 3D images using standard brain tumor database.

Keywords Medical image processing · Deep learning · U-Net architecture · Brain tumor segmentation · Fully convolutional neural networks

1 Introduction

As the world is advancing in technology and health, the field of diseases seems slower in their progress. Apart from the ailments caused by the microbes, the world now seems to be more terrified of the fatal blows of cancer on mortality, which starts its journey from being a tumor. The abnormal growth of cells in any part of the body is generally treated as a tumor. If it is not taken care, it may lead to one's death. So its detection at early stages is vital to save lives. Tumor has a tendency to occur in any part of the body, including brain which may lead to entire system collapse in future. The treatment of the brain tumor differs from person to person based on the size and

S. Das (✉)
KIIT Deemed to be University, Bhubaneswar 751024, Odisha, India
e-mail: suchismita.dasfcs@kiit.ac.in

H. Das et al. (eds.), *Progress in Computing, Analytics and Networking*,
Advances in Intelligent Systems and Computing 1119,
https://doi.org/10.1007/978-981-15-2414-1_11

shape of the tumor. It is estimated that brain tumor has about 120 types. The primary tumors stay within brain, and they do not spread to other parts of the body but the metastatic tumors spread to other parts from its place of origin and occur majorly in adults. It is found statistically that primary tumor occurs more in older adults and children.

Generally, tumors are of two types, one is fast growing and fatal cancer in nature and also spreads to other parts of the brain and spine known as malignant type and other one is slow growing and non-fatal known as benign type. The World Health Organization (WHO) has graded the brain tumors into four distinct categories as pilocytic astrocytoma: least aggressive and slow growth, low-grade astrocytoma: reproduces and affects the nearby tissue, anaplastic astrocytoma: malignant, reproduces and affects tissues and glioblastoma: Most malignant reproduces rapidly and affects tissues.

According to National Brain Tumor Foundation (NBFT), in developed countries, the rate of death of people due to brain tumor is increasing every year [1]. According to American Brain Tumor Association, in 2015, nearly 78,000 cases of primary brain tumor were submitted out of which 25,000 were malignant. Due to the complex structure of brain it is too difficult to detect tumor at an early stage. There are different tissues present in the normal brain image [2] like WM, GM, and CSF. Its structure is so intricate that the manual segmentation of the tumor involves huge processing time and ends with inaccurate results. Tumor detection from Magnetic Resonance (MR) images are nowadays the main focus of the radiologists or clinical experts as the segmenting and extracting the infected tumor area from MRI is a time-consuming task whose result depends only on their expertise and experience in respective field. To overcome this limitation and to get better performance, the computer-aided technology should be implemented by considering the knowledge of the experts. So to help its diagnosis, the research in automatic segmentation techniques is gaining more importance.

Automatic segmentation of tumor is still challenging due to tumor's complex shape and varied position and size. Multimodal imaging techniques (MRI), computed tomography (CT), and other imaging techniques contribute toward more accurate brain tumor segmentation. However, further development in the fields can lead the modern medical science to a higher success rate of the tumor treatment. CNN can prove to be one of such developments and this paper focuses on this very aspect, i.e., brain tumor segmentation using complex neural network algorithms.

There are various effective algorithms present for brain tumor segmentation and classification, but they have different limitations in general image classification problem [3]. Overall on averaging performance of the CNNs model is independent of individual CNN performance as for each specific input, only one CNN can perform well, but other CNNs may harm the overall performance. When test images and training images are of the same scale, traditional frameworks can only perform well because they cannot generalize well beyond training data. The cost of training is high due to the training of several CNNs separately.

2 Related Works

In previous years, many research works have been done using fully connected CNNs to segment brain tumors as well as for other medical data segmentations. From the previous extensive experiment and research, it is well known that 3D CNNs [4] produce significantly better results than 2D CNNs on segmentation problems. However, 3D CNNs are expensive to design in terms of hardware cost and time. Therefore, a U-Net architecture was designed to give higher quality results than traditional CNNs.

In CNN convolution, operation is applied to the input by convolution layer and the result is passed to the next layer. Each convolution neuron processes data only for its receptive field. Even if fully connected feedforward neural networks are used for learning the features as well as for classifying the data in various applications. But practically, it is difficult to apply this architecture for image segmentation. As the size of the input image is large, even in shallow (opposite of deep) architecture a very high number of neurons would be necessary. For instance, a fully connected layer for a (small) image of size 100×100 has 10000 weights for each neuron in the second layer. The convolution operation allows the network to be deeper with fewer parameters by reducing the number of free parameters [5].

In 2018, the author H. Abdalla describes a method for segmenting the brain tumor from MRI images using feedforward backpropagation neural network with supervised learning [6], which was applied for automatic classification of the images under a controlled experiment. The network performances were evaluated and reported best results with accuracy of 99% and sensitivity of 97.9%. The similar type of work was also experimented by the author S. V. Telrandhe in 2015 where he proposed a method [7] based on backpropagation neural network technique for the segmentation of tumor from MRI image. It combines the morphological operations and thresholding processes for segmenting the region from MRI images and used ANN for training and testing the network based on suitable extracted features.

Previously there are different methodologies proposed by researcher using K-means clustering, watershed algorithm, and combination of any two techniques. The method combining the morphological operation of erosions [8] and watershed algorithm was applied to detect the tumor in 2014. K-means clustering and Gray-Level Co-occurrence Matrix (GLCM) is depicted in [9] as a unique framework for neoplasm classification from the brain magnetic resonance imaging pictures. In this paper, author uses the support vector machine to improve the performance.

3 Proposed Framework

The proposed framework adapts the U-Net architecture which is used in segmenting the tumor from the input images both training and testing. It enables our algorithm to focus on different features of the MRI images for segmentation and for classification. The output of the segmentation is fed to the sequential CNN for the classification to achieve more accuracy. The deep learning U-Net is used for segmenting the tumor from the masked images and through the sequential CNN we performed the classification of the tumor for further medical diagnosis.

The methodology consists of following steps: image preprocessing, segmentation using U-Net architecture, and prediction of tumor using CNN as depicted in Fig. 1. First stage is preprocessing and converting to grayscale images of collected MRI images to enhance it and make it more suitable for analysis. Then in second stage the processed images are fed to U-Net for automatic segmentation of brain tumor by training the dataset. In last stage, the sequential CNN was designed to refine the result of the previous stage and minimizing the prediction error.

Our U-Net network architecture consists of a contraction (downsampling) path and an expansion (upsampling) path [10] as shown in Fig. 2. For contraction path, every block has two times of 3 × 3 Conv filter layer and for the downsampling, 2 × 2 max-pooling is used to extract more advanced features and reducing the size of feature maps. Ultimately in downsampling the size of the feature maps decrease from 240 × 240 to 15 × 15. Expansion path of consecutive of 2 × 2 Up-conv and two times of 3 × 3 Conv is done to recover the size of segmentation map with loss of the localization information. Finally, the number of feature maps is reduced, by using 1 × 1 convolutional layer, to segment the whole images into foreground (tumor) and background, respectively.

The generalized model of U-Net is used having different input image sizes. In our experiment, there are a total of 20 normal convoluted layers. There are four pooling layers and four up-convoluted layers. However, the filter size of each layer is shown in Table 1.

The segmentation is done by training the network through different activation functions. The Rectified Linear Unit (RelU) Activation function is used for all hidden layers to compute and train the network. The RelU is a very simple dynamic activation function introduced by [11], having biological and mathematical underpinning. I. It computes the output by thresholding values at 0, i.e., $f(x) = \max(0, x)$ for each input value x. Similarly, the output function produces the result based on the input fed from

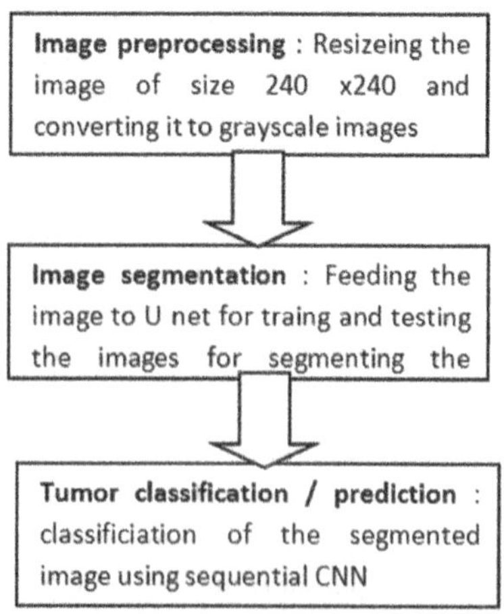

Fig. 1 Steps involved in the proposed methodology

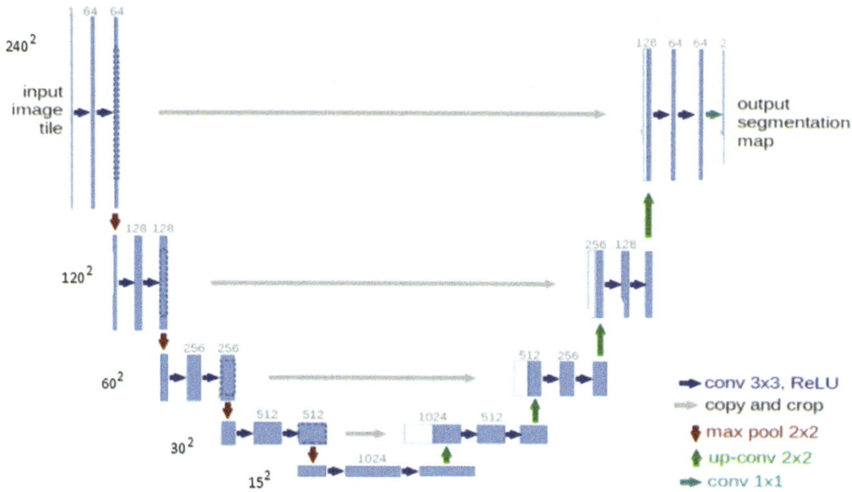

Fig. 2 U-Net architecture

Table 1 Filter size in each layer

Layer 1–2	64 Kernel Size:3 × 3
Layer 3–4	128 Kernel Size:3 × 3
Layer 5–6	256 Kernel Size:3 × 3
Layer 7–8	512 Kernel Size:3 × 3
Layer 9–10	1024 Kernel Size:3 × 3
Layer 11–12	512 Kernel Size:3 × 3
Layer 13–14	256 Kernel Size:3 × 3
Layer 15–16	128 Kernel Size:3 × 3
Layer 17–18	64 Kernel Size:3 × 3
Layer 19	2 Kernel Size:3 × 3
Layer 20	1 Kernel Size:1 × 1

the output of previous hidden layer. The output layer uses the sigmoid function for all the input y fed to it and produces the result as a part of tumor or non-tumor. The sigmoid function can be defined as

$$f(x) = \frac{1}{1 + e^{-(x)}} \tag{1}$$

Here a U-Net, i.e., a fully connected convolution neural network is implemented for segmenting the brain tumor from MRI images. After computing the input brain volume through a series of downsampling max-pooling convolutional layers, data is fed through a series of upsampling transpose convolutional layers. The final output of the network is the same shape as the input, but each pixel of the output contains the predicted the probability of the corresponding pixel in the input belongs to the tumor rather than containing visual information. The predicted segmentation is generated by binarizing the predicted probabilities [12] that the pixels that round up to 1 are labeled as tumorous and pixels that round down to 0 are labeled as non-tumorous. Here, binary cross-entropy function is used as loss function.

4 Results

The proposed framework is implemented on pilot dataset consisting of 200 brain MRI images. These images were acquired from the Kalinga Institute of Medical Science, Odisha. The labels associated with these volumes are manual approved by medical specialist. Actual collected images of different sizes have been preprocessed and converted to size 240×240 as input image. Further image size changes continuously in each layer based on filter size.

The training set was divided into training images and labeled (masked) images as shown in Fig. 3. The training images would be the input to the CNN, whereas the labeled images would be the predicted output. The CNN performs backpropagation if the predicted output does not match to the label, and hence readjusts the weights of the CNN.

First, the training of the U-Net has been done. For training the U-Net, CNN takes two parameters as input. The first parameter is the brain MRI image itself. The second parameter is the label of the image, i.e., its mask. The label or mask provides the CNN with the correct segmentation output that it needs to produce on any given test image. The model trains itself on a mere 160 images for 7 epochs reaching an accuracy of 73% on the training images itself. The accuracy can be achieved more by training the model with higher number of datasets. A set of training images and their corresponding labeled or masked set are represented in Figs. 4 and 5.

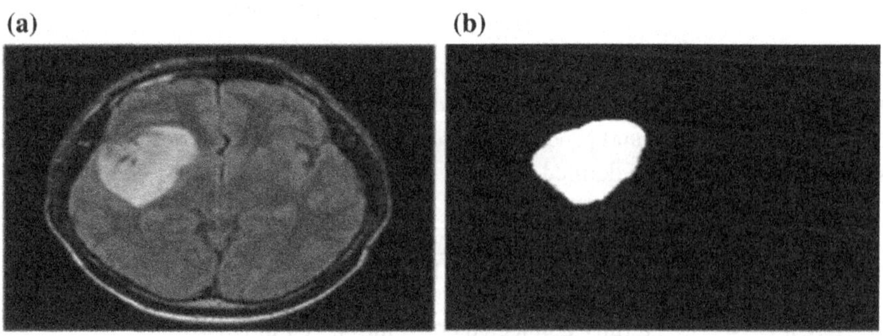

Fig. 3 **a** Original image, **b** masked image

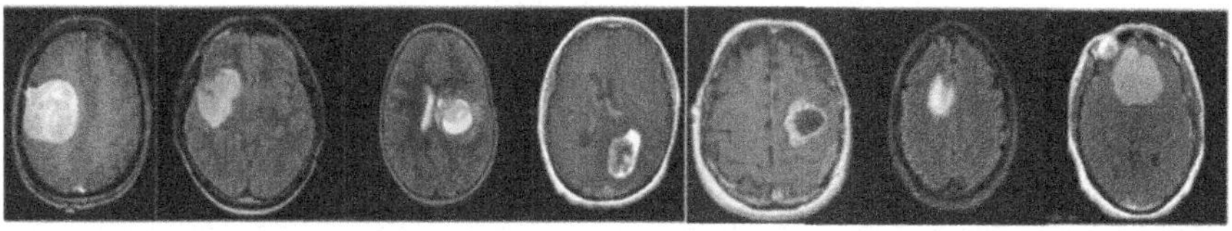

Fig. 4 Training set

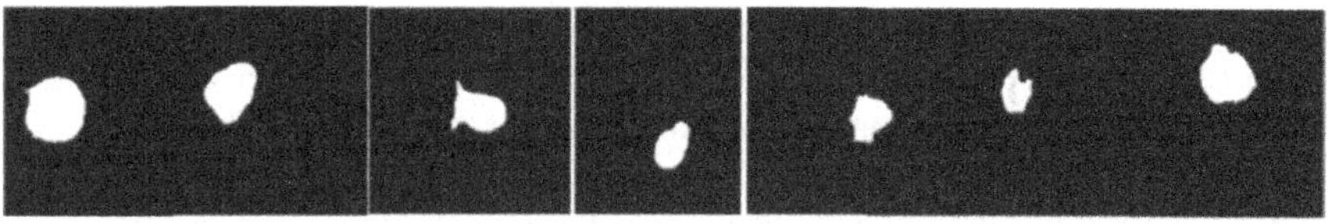

Fig. 5 Labeled set

The U-Net CNN takes the input data and predicts the pixel intensities as a probability of being a tumor. The prediction is then compared with the label set and the weights are adjusted by backpropagation.

During the test phase, the U-Net CNN takes a test image as input and produces an output as shown in Figs. 6 and 7. The pixels of the output are in grayscale and it represents the probability of a pixel being a tumor. The lower is the intensity in pixel, the more is the probability that pixel belongs to a tumor. This output, consisting of

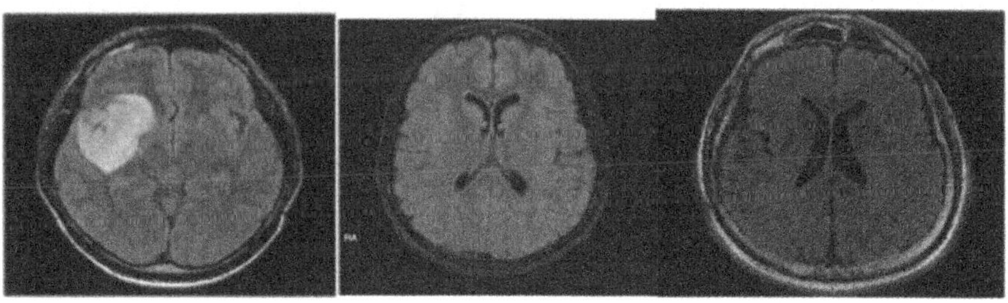

Fig. 6 Test images

Fig. 7 Segmented images

raw predicted tumorous pixels, has been threshold to a probability of 0.822 [13]. Any pixel having a probability higher than 0.822 has been changed to white indicating the segmented tumorous region.

The U-Net architecture merely performs a segmentation of a region it predicts to be tumorous. As the size of the training dataset for the U-Net architecture was very small, many features could have been wrongly taken [14]. Therefore, to eliminate certain wrongly extracted features, the result of the U-Net was furthered into a sequential CNN of three layers. The output of this CNN predicts whether the brain MRI image has tumors.

The CNN has been trained over some of the results of the U-Net CNN dividing them into two classes: tumorous and non-tumorous shown in Figs. 8 and 9. The model has been trained with 300 epochs/step.

A binary classification is done. It predicts the probability of a brain MRI image to be non-tumorous. The probability of class division is threshold at 0.5.

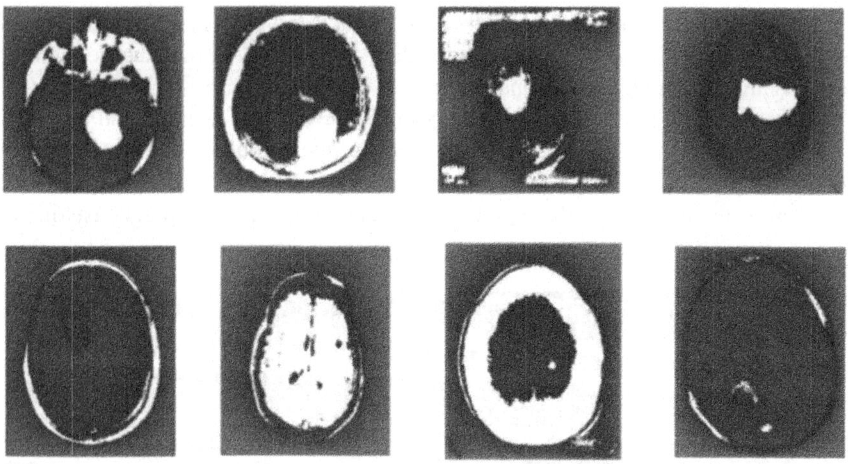

Fig. 8 Non-tumor class

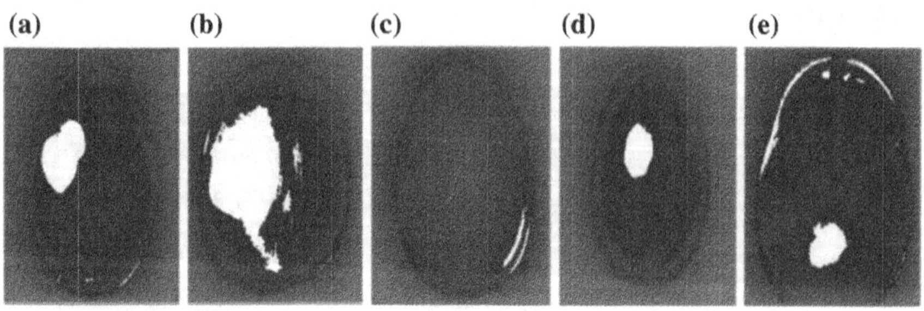

Fig. 9 Predictions: **a** tumor, **b** tumor, **c** non-tumor, **d** tumor, **e** tumor

A very accurate training requires a huge dataset and powerful hardware. Due to limited resources, the CNN would be trained for 20 epochs with a batch size of 32 over 160 images verified over 40 images.

5 Conclusion

A major challenge for radiologist in treatment of brain tumor is to identify the tumor extent from images collected through the noninvasive Magnetic Resonance Imaging (MRI) technique. In this paper, we modeled a U-Net architectural framework for brain tumor segmentation which helps the radiologist for early detection and diagnosis of tumor. As this fully connected CNN combines localization with contextual information in downsampling and upsampling, respectively, it results in a good prediction of segmentation map. This one can be used for different sizes of input images because this model does not have dense layers. The U-Net architecture allows prediction of accurate results on very shallow neural network trained over a very small dataset. The features extracted by the U-Net might not be accurate all time, and hence the results are required to be passed through another sequential neural network [18]. In future the performance of the network can be analyzed by considering the number of convolution layers and different upsampling techniques used for the U-Net architecture.

References

1. El-Dahshan, E.A., Mohsen, H.M., Revett, K., Salem, A.M.: Computer-aided diagnosis of human brain tumor through MRI: a survey and a new algorithm. Expert Syst. Appl. **41**, 5526–5545 (2014)
2. Angulakshmi, M., LaxmiPriya, G.: Automated brain tumour segmentation techniques—a review. Int. J. Imaging Syst. Technol. **27**, 66–77 (2017)
3. Jiang, Y., Chi1, Z: A scale-invariant framework for image classification with deep learning. In: IEEE International Conference on Systems, Man, and Cybernetics (SMC), pp. 1019–1024. Banff Center, Canada, 5–8 October (2017)
4. Bernal, J., Kushibar, K., Cabezas, M., Valverde, S., Oliver, A., Llad, X.: Quantitative analysis of patch-based fully convolutional neural networks for tissue segmentation on brain magnetic resonance imaging (2018). arXiv:1801.06457
5. Aghdam, H.H., Heravi, E.J.: Guide to Convolutional Neural Networks, pp. 247–258 (2017)
6. Elnoor, H., Abdalla, M. Esmail, M.: Brain tumor detection by using artificial neural network. In: International Conference on Computer, Control, Electrical, and Electronics Engineering (2018)
7. Telrandhe, S.V., Chikate, D., Banode, P.: Automated brain tumor detection using back propagation neural network. Int, J, Soft Comput. Artif. Intell. **3** (2015)
8. Zin, S., Khaing, A.S.: Brain tumor detection and segmentation using watershed segmentation and morphological operation. Int. J. Res. Eng. Technol. **3**, 367–374 (2014)
9. Wankhade, A., Malviya, A.V.: Brain tumor detection using K-mean clustering and SVM. Int. Res. J. Eng. Technol **5**, 3186–3194 (2018)

10. Dong, H., Yang, G., Liu, F., Mo1, Y., Guo, Y.: Automatic brain tumor detection and segmentation using U-Net based fully convolutional networks. In: Conference on Medical Image Understanding and Analysis (2017)
11. Agarap, A.: Deep learning using rectified linear units (ReLU) (2018). https://arxiv.org/pdf/1803.08375.pdf
12. Bakas, S., Akbari, H., Sotiras, A., Bilello, M., Rozyscki, M., Kirby, S., Davatzikos, C.: Segmentation labels and radiomic features for the pre-operative scans from the TCGA-GBM collection. The Cancer Imaging Archive (2017)
13. Otsu, N.: A threshold selection method from gray-level histograms. IEEE Trans. Syst. Man Cybern. **9**, 62–66 (2007)
14. Noh, H., Hong, S., Han, B.: Learning deconvolution network for semantic segmentation. In: IEEE International Conference on Computer Vision, pp. 1520–1528. Santiago, Chile (2015)

Satellite Image Enhancement Using Hybrid Denoising Method for Fusion Application

Anju Asokan and J. Anitha

Abstract Image fusion involves combining useful details from input images into a single image and the image can convey the complete particulars. Image fusion finds wide application in remote sensing, change detection, and medical imaging. The presence of noise in the input images limits the accuracy of fusion. To overcome this limitation, a hybrid filtering technique using gradient and guided filter is proposed to fuse satellite data. Source images are denoised using a hybrid filtering framework comprising of a gradient filter followed by an edge-preserving guided filter. The denoised images are fused using the traditional discrete wavelet transform. The results are compared against the fused outputs for traditional filters like median filter, Wiener filter, and guided filter by computing performance metrics such as entropy, Peak Signal-to-Noise Ratio(PSNR), Structural Similarity Index (SSIM), Feature Similarity Index (FSIM), gradient-based quality index ($Q^{AB/F}$), and CPU time. The results show that the hybrid filtering based fusion outperforms other filtering-based fusion techniques.

Keywords Image fusion · Multitemporal · Remote sensing · Guided filter · Gradient filter · Wiener filter

1 Introduction

Satellite images nowadays are widely used in the area of image processing. There are different kinds of satellite images which give different information content on agricultural lands, water bodies, disaster-affected areas, and so on. Two such images are the panchromatic and multispectral images. While panchromatic images contain different bands of the spectrum, they have a very poor spatial resolution. Remote

A. Asokan (✉) · J. Anitha
Department of Electronics and Communication Engineering, Karunya Institute of Technology and Sciences, Coimbatore, India
e-mail: anju.asok@yahoo.com

J. Anitha
e-mail: anithaj@karunya.edu

H. Das et al. (eds.), *Progress in Computing, Analytics and Networking*, Advances in Intelligent Systems and Computing 1119,
https://doi.org/10.1007/978-981-15-2414-1_12

sensing application for detecting features on the earth is possible only if the image has a very good spatial resolution. The satellite image is affected by various atmospheric interferences due to its acquisition from long distance. This affects the overall image quality and makes the analysis difficult. Noise removal is a very relevant step in the image processing applications. Hence, prior to carrying out the analysis, the images need to be treated in order to remove the noise components.

Image fusion technique combines the source images and aims at reducing the redundancy in source images. The fused image contains more information than the individual source images. Image fusion is very important for computer vision and robotics where the outcome is useful for further processing steps. They are applicable in many areas such as biomedical, satellite sensing, surveillance, and consumer applications.

The remainder of the paper is arranged as follows: Sect. 2 describes the related works. Section 3 gives the proposed hybrid denoising method. Section 4 gives the results and discussion and Sect. 5 presents the conclusion.

2 Related Works

Transform-based fusion techniques such as Shearlet and Contoulet find wide application. But these transforms do not consider the geometric structure in the satellite images. To overcome this drawback, Shearlet transform which incorporates shift invariance is used [1]. This method combines the entropy principle analysis along with the shift invariance to extract the high-frequency bands which contain the maximum information of the satellite images and perform the image fusion. Another method which focuses on sparse coding is proposed in [2].

A spider optimization based algorithm is proposed in [3]. The cartoon components are fused by a spatial fusion method. The textural components are combined using sparse representation. The main advantage of Non-subsampled Contourlet Transform (NSCT) is proper representation of edges in the output image [4]. The multidimensional expansion of fused image enhances the image for analysis. In order to obtain a fused image which includes all the details from the source images, it is necessary that the correlation between the images is analyzed. The spatial and temporal fusion method is done after extracting the correlation coefficient between time-series images pixel by pixel [5]. A prominent denoising technique is the transform-based denoising technique which has the limitation that it is computationally expensive [6].

A filtering method to remove the impulse noise based on a joint bilateral filter is proposed, which has the advantage of preserving the edge details in the image [7]. It is an improved form of the simple bilateral filter which calculates the color distance between pixels using a reference image which is characterized by sharp edges and less noise content. Hyperspectral pansharpening is a commonly used method when trying to fuse together panchromatic and hyperspectral images of different resolutions. Pansharpening using guided filter and Gaussian filter is used in [8]. A Gaussian noise removal method using fast guided filter and noise thresholding is used in the application of medical imaging [9]. An infrared and visible image fusion technique using redundant lifting non-separable wavelet-based shearlet transform and guided filter is presented in [10]. An image fusion using guided filter based difference image is presented in [11]. The highlight of using a guided filter compared to bilateral filter is that it is another edge-preserving filter which finds the filtered output using a guidance image which is the image or a reference. Compared to bilateral filter, which is a simple edge-preserving filter, this filter has the advantage that it can replicate the features of the guidance image onto the image to be filtered.

Classification algorithm for early diabetes detection is proposed in [12]. Decision tree based intrusion detection system is presented in [13]. Data mining classification using evolutionary algorithm is described in [14]. Image processing in biomedical, computing, and networking applications is proposed in [15–17]. Optimization algorithms for cloud computing are presented in [18].

Proposed work focuses on a DWT-based fusion method which is done on denoised satellite images and compared using performance metrics such as PSNR, entropy, FSIM, SSIM, $Q^{AB/F}$, and CPU time. Here a hybrid denoising method using gradient and guided filter is proposed. The results are compared against fused outputs for traditional filters like median filter, Wiener filter, and guided filter. Source images used are LANDSAT 7 multitemporal images of 648 × 1462.

3 Proposed Method

The work is executed in Matlab 2018a on an Intel® Core™ i3-4005U CPU @1.70 GHz system by applying on different sets of satellite images of the same location but recorded at different times. The images are from global land survey. Out of 100 available images, we have used 5 sets for analysis.

In the proposed framework, the satellite images of an area at different times or the multitemporal images are acquired. LANDSAT images of same region are used as the source images. The schematic diagram for the DWT-based image fusion scheme is depicted in Fig. 1.

The mutitemporal source images are individually subjected to hybrid filtering. Hybrid filter is composed of gradient and a guided filter. The preprocessed images

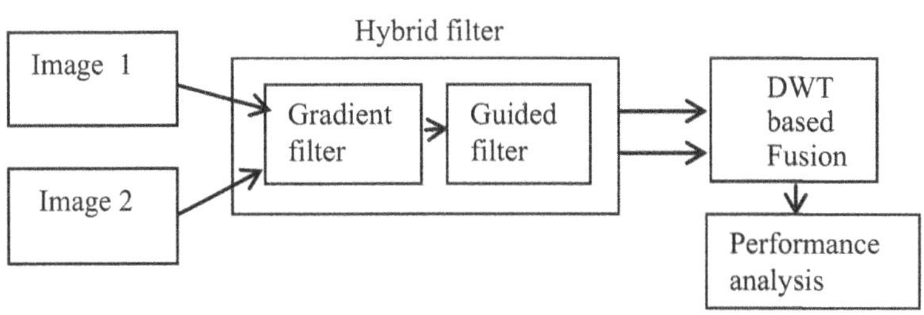

Fig. 1 Block diagram of the proposed framework

are fused using the DWT-based method and the performance is analyzed. Then, the two input images are filtered using the median filter, Wiener filter, and guided filter. Then the filtered images are fused using DWT. The relative performance comparison is done to analyze the different fused results. The gradient filter is a smoothening filter and the guided filter is an edge-preserving filter which can capture the smoothened edge details and preserve them. The median filter sets the output pixel depending on the median of the neighborhood pixels. Wiener filter removes the noise and counters the blurring simultaneously.

3.1 Hybrid Filtering Based Denoising

The impulse noise affected satellite images are passed through the hybrid filter. Hybrid filter comprises the gradient filter and the guided filter. Gradient filter can detect the low-frequency regions and does the filtering operation by smoothing the noise components. It does not alter the high-frequency regions. The image thus obtained is a blurred image in which the sharp transitions such as the corners and edges are also blurred. The gradient filter is designed using Eq. (1) and magnitude of the gradient is given in Eq. (2):

$$\text{grad}(f) = \begin{bmatrix} g_x \\ g_y \end{bmatrix} = \begin{bmatrix} df/dx \\ df/dy \end{bmatrix} \quad (1)$$

$$M(x, y) = \text{mag}(\text{grad}(f)) = \sqrt{gx^2 + gy^2} \quad (2)$$

where $M(x, y)$ gives the magnitude of the gradient, and g_x and g_y denote the gradient in x- and y-directions, respectively.

Image is smoothed using the gradient filter and passed through a guided filter. This filter can detect the edges in the image and is a fast edge-preserving method. It also eliminates the noise in the remaining parts of the image. It is based on a linear transformable filtering in which the guidance image I is modified as per the objective and it can be same as input image x. Linear relation ensures that the output has identical edge details as the guidance image. The output value for pixel I is calculated using Eq. (3) as

$$y_i = \sum_j W_{ij}(I)x_j \quad (3)$$

where i and j are the pixel indices and W_{ij} denotes the kernel function of the filter.

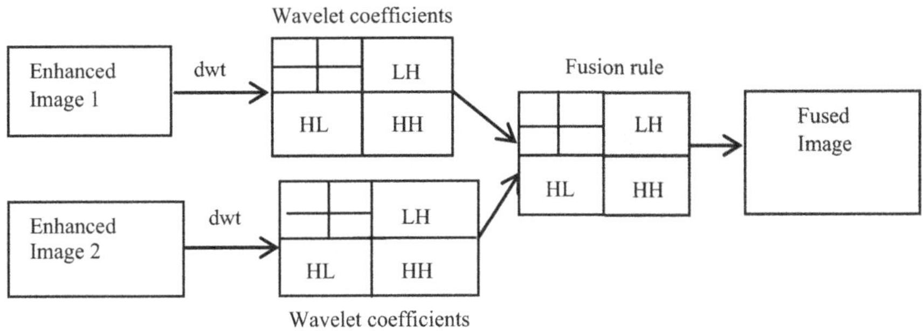

Fig. 2 DWT-based image fusion

3.2 DWT-Based Fusion

The main assumption done before the fusion performed is that since the images of the same location are taken using the same sensor, the images are registered pixel to pixel, and hence no correction procedures are to be carried out. Figure 2 represents the DWT-based image fusion framework.

Each of the registered satellite images is passed individually through two digital filters H0 and H1. H0 is considered as a high-pass filter and H1 is a low-pass filter. The outputs of H0 and H1 are subsampled by a factor of two. This gives low-frequency (L) and high-frequency components (H). These outputs are then given to two digital filters. Once the two stages are completed the output frequency components are HH-High High, HL-High Low, LH-Low High, and LL-Low Low. The fusion of the HH, HL, LH, and LL frequency components of the first and second images is done using the averaging rule to get final image.

4 Results and Discussion

The performance of the fusion method is evaluated using quantitative metrics such as PSNR, entropy, FSIM, SSIM, $Q^{AB/F}$, and CPU time (in sec).

PSNR measures the accuracy of the output image and is controlled by intensity values of the image. Table 1 shows the quantitative results for image fusion with various filters.

From the table, it is observed that the fusion of hybrid filtered images gives better results in comparison to traditional filter based images. Smoothing leads to the overall blurring of the image and the sharp transitions such as edges and feature borders are blurred. It lowers the noise components in the image and improves the PSNR. Guided filter is a smoothing filter. Even though it smoothens the overall image, it detects the edges and does not oversmoothen them, thereby improving the PSNR values when compared to a median filter and Wiener filter. In the hybrid filter, gradient filter

Table 1 Quantitative results for image fusion with various filters

Technique	Database	PSNR	H	FSIM	SSIM	$Q^{AB/F}$	Time
With median filtering	Sample 1	28.922	4.982	0.864	0.850	0.822	0.1766
	Sample 2	26.414	5.174	0.844	0.833	0.810	0.1822
	Sample 3	27.117	5.254	0.818	0.801	0.789	0.1843
	Sample 4	28.397	5.529	0.833	0.821	0.796	0.1793
	Sample 5	26.794	5.524	0.846	0.832	0.784	0.1802
With Wiener filtering	Sample 1	35.990	4.998	0.894	0.877	0.829	0.2543
	Sample 2	34.786	4.922	0.870	0.873	0.831	0.2622
	Sample 3	33.675	5.133	0.855	0.866	0.833	0.2813
	Sample 4	32.887	5.322	0.876	0.871	0.849	0.2476
	Sample 5	35.675	5.312	0.869	0.881	0.840	0.2544
With guided filtering	Sample 1	40.988	5.334	0.889	0.876	0.865	0.3192
	Sample 2	37.909	5.412	0.891	0.887	0.873	0.3213
	Sample 3	42.112	5.499	0.897	0.897	0.887	0.3235
	Sample 4	45.655	5.356	0.894	0.901	0.877	0.3042
	Sample 5	44.909	5.411	0.900	0.903	0.871	0.3322
With hybrid filtering	Sample 1	56.219	5.826	0.922	0.946	0.902	0.3656
	Sample 2	54.120	5.806	0.916	0.954	0.899	0.3612
	Sample 3	55.484	5.742	0.900	0.934	0.900	0.3632
	Sample 4	56.496	5.821	0.938	0.941	0.887	0.3508
	Sample 5	55.991	5.826	0.946	0.947	0.878	0.3532

smoothens the image but the guided filter can detect the presence of edges and preserve them thereby improving the PSNR values. The entropy H is defined as the information content in the image.

Due to the over blurring of the sharp transitions in the image after median filtering and fusing the images, the entropy value is less when compared to the Weiner filter where the edges are less blurred. Guided filter can preserve the edge information and gives improved entropy results.

FSIM measures the feature equivalence of input and output image.

$Q^{AB/F}$ represents a gradient-based quality index which measures edge details of the input images in comparison to the output image. It is given using Eq. (4) as

$$Q^{AB/F} = \frac{\sum_{i,j}\left(Q^{AF}(i,j)w^{A}(i,j) + Q^{BF}(i,j)w^{B}(i,j)\right)}{\sum_{i,j}\left(w^{A}(i,j) + w^{B}(i,j)\right)} \tag{4}$$

where $Q^{AF} = Q_g^{AF} Q_0^{AF}$ which is the product of edge strength and orientation preservation values at (i, j). Q^{BF} is also calculated in the same way, $w_A(i, j)$ and $w_B(i, j)$ are the weights corresponding to Q^{AF} and Q^{BF}.

SSIM is the structural equivalence of input and output image.

From table, features in the image after median filter and Wiener filter are distorted to a greater extent when compared to those after guided filtered fusion. Hence, the SSIM, FSIM, and $Q^{AB/F}$ values are improved for guided filtered image fusion when compared to traditional filters. The hybrid filtering consumes more CPU time over traditional filters because of its two-stage filtering framework. The presence of gradient filter in hybrid filter adds an additional level of image denoising, thus improving SSIM, FSIM, and $Q^{AB/F}$ values. Figure 3 shows outputs for all the techniques.

From the table, it is concluded that the hybrid filtered image fusion produces improved outcome when compared to the median filter, Wiener filter, and guided filter results.

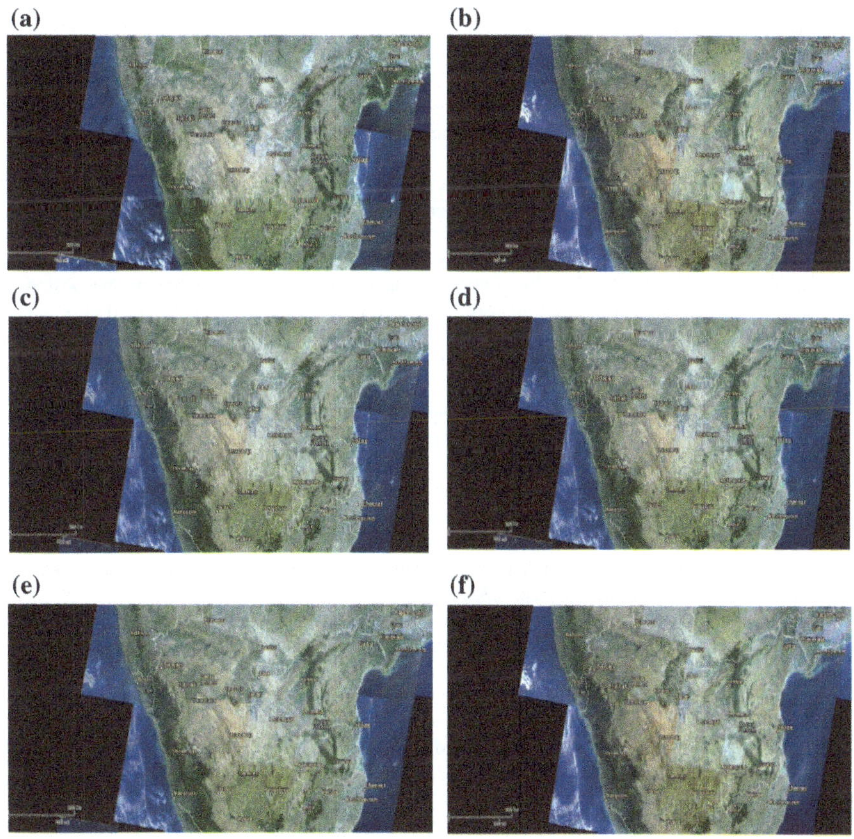

Fig. 3 **a** Image 1, **b** image 2, **c** median filter based fusion, **d** wiener filter based fusion, **e** guided filter based fusion and **f** proposed method

5 Conclusion

Fusion techniques find wide usage in various remote sensing applications. In this paper, a hybrid filter based image fusion using DWT is presented. The hybrid filter is composed of gradient and guided filter. The results are compared against traditional filters. Major filtering procedures used are median, Wiener, and guided filter. The main observation is the improvement in the image quality on fusing the images after combined gradient and guided filtered images are used and this hybrid filter gives better performance than traditional filters but at the cost of high CPU time. As a future work, modification on the method could be done for generalizing fusion under varying conditions. Multi-scale detail enhancement can be explored to improve the entropy values in satellite images. Optimized filtering techniques could also be used for greater efficiency.

References

1. Luoa, X., Zhang, Z., Wua, X.: A novel algorithm of remote sensing image fusion based on shift-invariant Shearlet transform and regional selection. Intl. J. Electr. Commun. **70**, 186–197 (2015)
2. Zhu, Z., Yin, H., Chai, Y., Li, Y., Qi, G.: A novel multi-modality image fusion method based on image decomposition and sparse representation. Inf. Sci. **432**, 516–529 (2017)
3. Mauryaa, L., Mahapatraa, P.K., Kumara, A.: A social spider optimized image fusion approach for contrast enhancement and brightness preservation. Appl. Soft Comput. **52**, 575–592 (2016)
4. Anandhi, D., Valli, S.: An algorithm for multi-sensor image fusion using maximum a posteriori and nonsubsampled contourlet transform. Comput. Electr. Eng. **65**, 139–152 (2017)
5. Wu, M., Huang, W., Niu, Z., Wang, C., Li, W., Yu, B.: Validation of synthetic daily Landsat NDVI time series data generated by the improved spatial and temporal data fusion approach. Inform. Fusion. **40**, 34–44 (2017)
6. Bhandari, A.K., Kumar, D., Kumar, A., Singh, G.K.: Optimal sub-band adaptive thresholding based edge preserved satellite image denoising using adaptive differential evolution algorithm. Neurocomputing. **174**, 698–721 (2015)
7. Cui, B., Ma, X., Xie, X., Ren, G., Ma, Y.: Classification of visible and infrared hyperspectral images based on image segmentation and edge-preserving filtering. Infrared Phys. Technol. **81**, 79–88 (2016)
8. Dong, W., Xiao, S., Li, Y.: Hyperspectral pansharpening based on guided filter and gaussian filter. J. Vis. Commun. Image Repr. **53**, 171–179 (2018)
9. Majeeth, S.S., Babu, C.N.K.: Gaussian noise removal in an image using fast guided filter and its method noise thresholding in medical healthcare application. Image Signal Process. **43**, 1–9 (2019)
10. Liu, L., Song, M., Peng, Y., Li, J.: A novel fusion framework of infrared and visible images based on RLNSST and guided filter. Infrared Phys. Technol. **100**, 99–108 (2019)
11. Yan, X., Qin, H., Li, J., Zhou, H., Yang, T.: Multi-focus image fusion using a guided-filter-based difference image. Appl. Opt. **55**, 2230–2239 (2016)
12. Das, H., Naik, B., Behera, H. S.: Classification of diabetes mellitus disease (DMD): a data mining (DM) approach. In: Progress in Computing, Analytics and Networking, pp. 539–549. Springer, Singapore (2018)
13. Sahani, R., Rout, C., Badajena, J. C., Jena, A. K., Das, H.: Classification of intrusion detection using data mining techniques. In: Progress in Computing, Analytics and Networking, pp. 753–764. Springer, Singapore (2018)

14. Das, H., Jena, A.K., Nayak, J., Naik, B., Behera, H.S.: A novel PSO based back propagation learning-MLP (PSO-BP-MLP) for classification. In: Computational Intelligence in Data Mining, vol. 2, pp. 461–471. Springer, New Delhi (2015)
15. Pradhan, C., Das, H., Naik, B., Dey, N.: Handbook of Research on Information Security in Biomedical Signal Processing, pp. 1–414. IGI Global, Hershey, PA (2018)
16. Sahoo, A.K., Mallik, S., Pradhan, C., Mishra, B.S.P., Barik, R.K., Das, H.: Intelligence-based health recommendation system using big data analytics. In: Big Data Analytics for Intelligent Healthcare Management, pp. 227–246. Academic Press (2019)
17. Pattnaik, P.K., Rautaray, S.S., Das, H., Nayak, J.: Progress in computing, analytics and networking. In: Proceedings of ICCAN, p. 710 (2017)
18. Nayak, J., Naik, B., Jena, A.K., Barik, R.K., Das, H.: Nature inspired optimizations in cloud computing: applications and challenges. In: Cloud Computing for Optimization: Foundations, Applications, and Challenges, pp. 1–26. Springer, Cham (2018)

Classification of Amino Acid Using Micro-electrical Model

Tanusree Roy and Pranabesh Bhattacharjee

Abstract Amino acids play a vital role in the field of genomics. In this paper, amino acids are designed and classified using an equivalent micro-electrical model. The main objective of this paper is to design simplified electrical models for individual amino acid by passive and active components based on their physicochemical properties. The network responses are analyzed to characterize their property and establish a correspondence between biological attributes and electrical properties.

Keywords Amino acid · Electrical model · MOS · Simulation

1 Introduction

Amino acids are the basic building blocks of protein [1] and perform important roles in growth, damaged cell repair, metabolism, and many other biological activities. The chemical properties of the amino acid are very important in the field of genomics to determine the nature of the protein. Several researchers developed various models to analyze the biological behavior of amino acid, DNA, RNA, gene sequence, protein structure, etc. considering their physical and chemical properties. For example, codon-based electrical models are developed to characterize 20 essential amino acids [2, 3]. RLC (M) circuit is modeled to analyze protein structure [4]. The electrical gene network model is developed to study the behavior of cancerous gene [5]. The architecture of a VLSI chip is proposed to study the behavior of amino acid in protein sequence [6]. This kind of mimic design is not only beneficial for disease diagnosis but also equally important to process or recognize/sense any equivalent biomedical signal for electronic information/data security purposes nowadays [7].

T. Roy (✉) · P. Bhattacharjee
Department of Electrical and Electronics Engineering, University of Engineering and Management, Kolkata 700156, India
e-mail: tanusree.rinki@gmail.com

P. Bhattacharjee
e-mail: pranabesh.dream@gmail.com

H. Das et al. (eds.), *Progress in Computing, Analytics and Networking*, Advances in Intelligent Systems and Computing 1119,
https://doi.org/10.1007/978-981-15-2414-1_13

Realizing the importance of the electrical network model in the present society, the authors in this paper design passive (resistor, capacitor based) as well as active (PMOS, NMOS) electrical model for 20 individual amino acids based on their hydropathy scale [8] and investigate the electrical responses to correlate the biological attributes of amino acids. Here, the physicochemical properties [9–12] (Table 1) of the amino acids (hydropathy index value, pK value, physical property, and polarity of side chain) are considered for electrical simulation. Designing circuit models for individual codon (a combination of three nucleotides) [2] in primary gene structure is very complex regarding computational load. To overcome this problem, authors realize amino acid circuit model which is simpler and sensitive in different applications, especially in predicting genetic abnormalities.

The main objective of the paper is twofold:

- Classifying 20 amino acids (Table 1) based on their physicochemical property using proposed amino acid circuit models.
- Correlating the electrical model responses with the biological properties of amino acid.

Table 1 Amino acid properties [7, 9, 10]

List of 20 essential amino acids	Abbreviations		Hydropathy nature (with index values)	Chemical properties
Alanine	Ala	A	Hydrophobic(1.8)	Aliphatic
Arginine	Arg	R	Hydrophilic(−3.8)	Basic
Asparagine	Asn	N	Hydrophilic(−3.5)	Amide
Aspartic acid	Asp	D	Hydrophilic(−3.5)	Acidic
Cysteine	Cys	C	Hydrophobic(2.5)	Sulfur
Glutamine	Gln	Q	Hydrophilic(−3.5)	Amide
Glutamic acid	Glu	E	Hydrophilic(−3.5)	Acidic
Glycine	Gly	G	Hydrophilic(−0.4)	Aliphatic
Histidine	His	H	Hydrophilic(−3.2)	Basic/Aromatic
Isoleucine	Ile	I	Hydrophobic(4.5)	Aliphatic
Leucine	Leu	L	Hydrophobic(3.8)	Aliphatic
Lysine	Lys	K	Hydrophilic(−3.9)	Basic
Methionine	Met	M	Hydrophobic(1.9)	Sulfur
Phenylalanine	Phe	F	Hydrophobic(2.8)	Aromatic
Proline	Pro	P	Hydrophilic(−1.6)	Aliphatic
Serine	Ser	S	Hydrophilic(−0.8)	Hydroxyl
Threonine	Thr	T	Hydrophilic(−0.7)	Hydroxyl
Tryptophan	Trp	W	Hydrophilic(−0.9)	Aromatic
Tyrosine	Tyr	Y	Hydrophilic(−1.3)	Aromatic
Valine	Val	V	Hydrophobic(4.2)	Aliphatic

2 Micro-electrical Modeling of Amino Acid

The proposed equivalent electrical model of amino acid is developed using Mentor Graphics HEP1 EDA tool with 0.13 micrometer (μm) MOS technology. The electrical modeling technique of individual amino acid is described as following:

- First, the chemical structures of individual amino acids are modeled using passive (resistor and capacitor) and active (MOSFET) electrical components based on their hydropathy index value [8].
- The simulated responses of amino acid models are analyzed to characterize 20 amino acids based on their physicochemical property.

There are 20 essential amino acids which have fixed combination of the amino group (NH2) and carboxyl group (COOH), known as backbone, and a variable side-chain group (–r) that gives unique amino acid structure [9–11]. Both groups are attached to a central alpha-carbon (–CH) (Fig. 1). The backbone part of the proposed electrical model is represented by a 7 ohm resistor because the total number of atoms in the backbone group is 7 [13]. Unlike the previous impedance-based model [4, 5, 13] the side-chain part is represented only by a capacitor with the respective hydropathy index value for all amino acids. The use of inductor is replaced by a capacitor in this proposed model. According to the previous work [13], the hydrophilic circuit structure reflects the nature of low-pass filter and hydrophobic circuit structure reflects high-pass filter. For this reason, the shape of the RC circuit model is considered as "L" model [14]. The connection of R and C totally depends on the HI (hydropathy index) sign, if the HI value is negative then the capacitor placed as load to form a low-pass filter (Fig. 2a), and if the HI value positive then the capacitor connected at the input side to form a high-pass filter (Fig. 2b). Positive phase values of the hydrophobic amino acids resemble positive hydropathy index and negative phase values of hydrophilic amino acids resemble negative hydropathy index [8]. Depending on this established concept [13], the proposed hydrophobic amino acid model is designed as a high-pass filter; on the other hand, the hydrophilic amino acid model is designed as low-pass filter. In this paper, both passive and active (MOS based) models (Fig. 3) are designed for 20 amino acids individually in a similar way

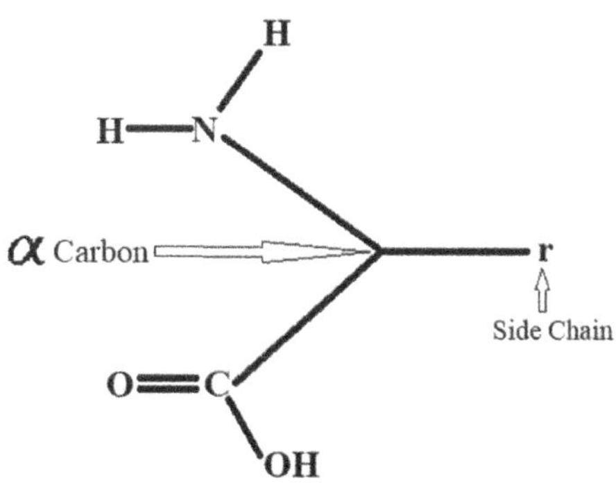

Fig. 1 The basic structure of amino acid

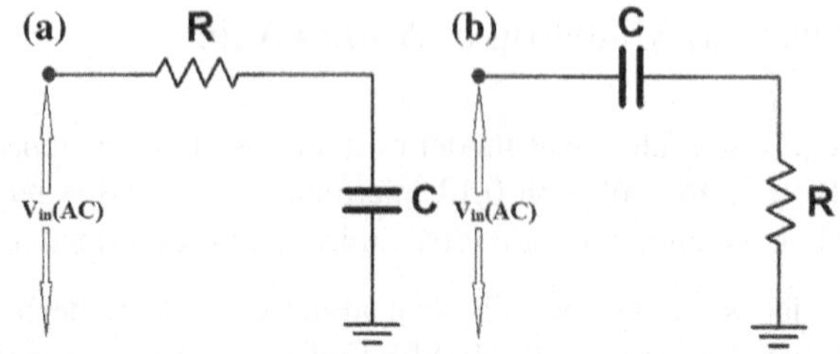

Fig. 2 Passive model for **a** hydrophilic and **b** hydrophobic amino acid

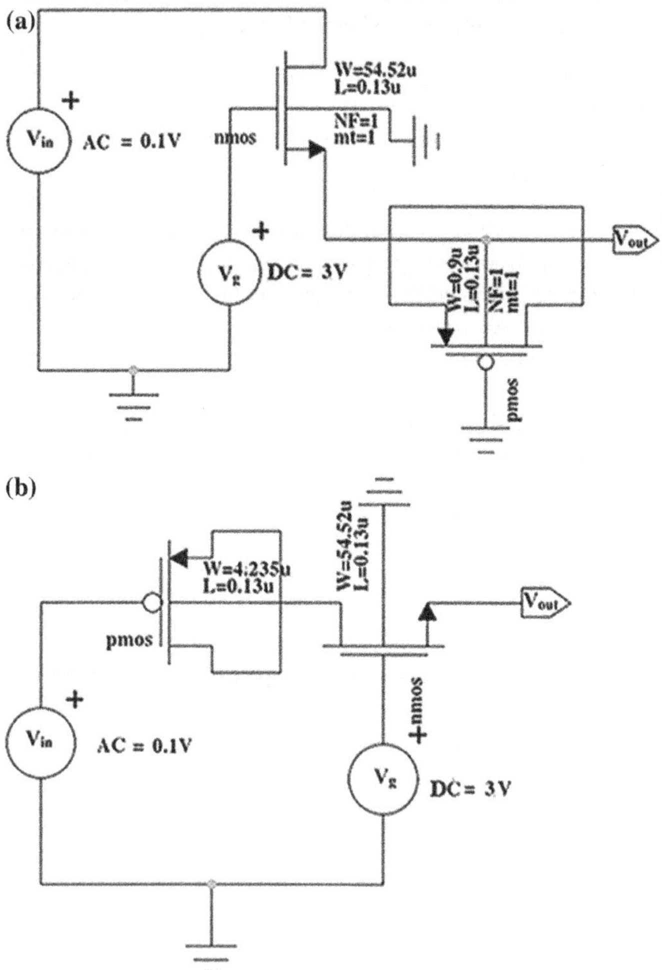

Fig. 3 Active model for **a** hydrophilic and **b** hydrophobic amino acid

as shown in Figs. 1 and 2. In the active model, the 7 O resistance is modeled using a single NMOS transistor by keeping the gate-to-source voltage V_{gs} larger than the threshold voltage V_t and drain-to-source voltage V_{ds} smaller than $V_{gs}-V_t$ in triode region [15].

$$R_{ds} = \frac{\Delta V_{ds}}{\Delta I_d} \tag{1}$$

where R_{ds} is MOSFET drain-to-source resistance, V_{ds} is drain-to-source voltage, and I_d is drain current. The channel W/L ratio of the MOSFET is 54.52u/0.13u. The above equation is for small V_{ds} in triode operation of MOSFET, and for this experiment, a 3 V dc supply is taken for V_{gs}. Equation (1) shows every point of V/I slope is giving a constant value of R_{ds}. For the capacitor part, the side chain is designed using PMOS gate oxide capacitance C_g (in linear region) [14–16]. The capacitor value is changed with different regions of operation, like for cutoff region total gate capacitance is [14].

$$C_g = A \cdot \frac{\varepsilon_{ox}}{\tau_{ox}} + 2C_{ov} \tag{2}$$

where ε_{ox} is the permittivity of the gate oxide, τ_{ox} is the thickness of the gate oxide, C_{ox} is the gate oxide capacitance per unit area which is the ratio of permittivity to thickness, and A is the total gate area ($W*L$). But, for linear operation [14] the equation is different as following (when $V_D = V_S$):

$$C_{gs} = C_{gd} = \frac{1}{2} C_{ox} WL + C_{ov} \tag{3}$$

Here, C_{gs} is a gate-to-source capacitance, C_{ds} is a gate-to-drain capacitance, and C_{ov} is gate overlap capacitance. There is also some overlap capacitance between the gate and source and the gate and drain, which is minimized in such a way that it becomes negligible. Equation (3) is used here to evaluate the required side-chain capacitance, because of the linear operation. To compute the value of MOS gate capacitance C_g according to the respective HI value, the parameters considered are $\varepsilon_{ox} = 34.53*10^{-12}$ F/m, $\tau_{ox} = 2.5*10^{-9}$ m, which are taken from mentor 130 nm technology development kit (TDK) file, and the value of "W" is varied with the corresponding HI values to match the gate capacitance value with respective HI value.

3 Results and Discussions

After modeling 20 individual amino acids, each circuit model is analyzed (for both passive and active) using a small 100 mV ac input source. The corresponding magnitude and phase responses are observed within the frequency range of 1 Hz to

100 GHz to classify individual properties, i.e., hydrophobic/hydrophilic acidic/basic, aromatic/aliphatic (Figs. 4 and 5).

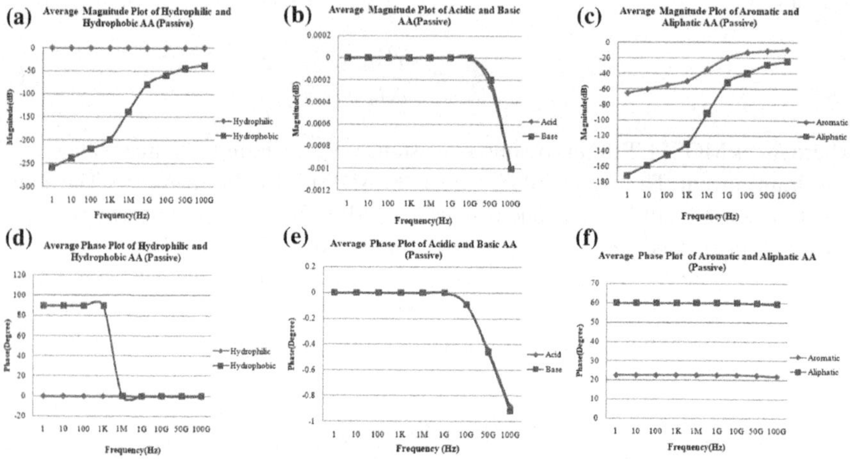

Fig. 4 Passive amino acid model average magnitude response: **a** hydrophilic/hydrophobic, **b** acidic/basic, **c** aromatic/aliphatic; and average phase plot for (**d**) hydrophilic/hydrophobic, **e** acidic/basic, **f** aromatic/aliphatic

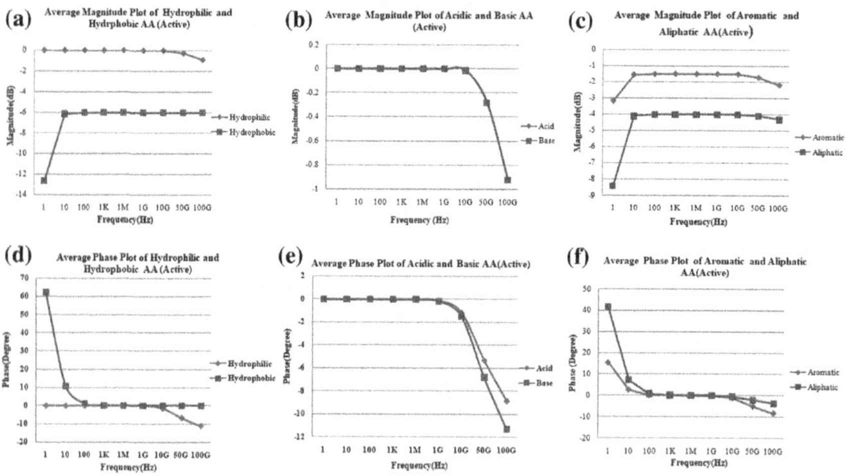

Fig. 5 Active amino acid model average magnitude response: **a** hydrophilic/hydrophobic, **b** acidic/basic, **c** aromatic/aliphatic; and average phase plot for (**d**) hydrophilic/hydrophobic, **e** acidic/basic, **f** aromatic/aliphatic

3.1 Magnitude Response

The magnitude response for each amino acid with positive hydropathy index shows rising characteristics, whereas the amino acids having negative hydropathy index show falling characteristics (Figs. 4a and 5a). Both active and passive circuits give a similar response, but the active response is much better and precise than a passive one. These responses reflect the biological properties [5]. Amino acids are also characterized as acidic/basic and aromatic/aliphatic as shown in Figs. 4 and 5. For acidic/basic property, both responses are decreasing in nature as shown in Figs. 4b and 5b, and no such changes can be noticed. Similarly, the plots in Figs. 4c and 5c show rising nature in lower frequency for both aromatic and aliphatic amino acid and falling nature in higher frequency.

3.2 Phase Response

In the case of phase response, both hydrophilic and hydrophobic are falling in nature, but the hydrophilic curve shows negative phase, whereas the hydrophobic shows positive phase as shown in Figs. 4d and 5d. These responses are also similar to the previously referred biological responses [5]. For acidic/basic property, both responses are decreasing in nature and negative in phase as shown in Figs. 4e and 5e, same as magnitude response. For aromatic and aliphatic, both responses are falling in phase but aromatic one is lower than aliphatic one as shown in Figs. 4f and 5f, which correlates that aliphatic amino acids having larger HI value than aromatic one.

4 Conclusion

Amino acid structures and properties are important for determining protein structures and properties. In this context, the realization of micro-electrical models for 20 essential amino acids and the investigation of individual model responses for their classification play a significant role. The electrical responses reflect a high correspondence with functional biology. This network model concept may help in structure-based drug design and also be useful in developing electronic and bio-sensing devices.

5 Future Work

The proposed design is useful to characterize the properties of amino acids electrically and could be helpful in detecting any kind of genetic abnormality or disease by forming an equivalent gene network model. Further, the gene model response may be used to identify the abnormality of the gene sequence. This concept can be applied by taking healthy and abnormal gene sequence as soft databases from the national website like National Center for Biotechnology Information (NCBI).

Acknowledgements The research work is funded by the DST, Science and Engineering Research Board, Govt. of India (Grant No. EEQ/2017/000293). The working facility is provided by University of Engineering and Management, Kolkata-700156.

References

1. McKee, T., McKee, J.R.: Biochemistry: The Molecular Basis of Life, vol. 3. McGraw-Hill, New York (2003)
2. Dutta, M., Barman, S.: Codon characterization based on electrical response. Int. J. Comput. Appl. **975** (2015)
3. Marshall, R.: Modeling DNA/RNA strings using resistor-capacitor (rc) ladder networks. Comput. J. **53**(6), 644–660 (2009)
4. Sampath, G.: RLC (M) circuit models of protein structure: analysis, visualization, shape synthesis, and pattern matching. In: 40th Annual Conference on Information Sciences and Systems, pp. 1623–1628. IEEE (2006)
5. Roy, T., Das, S., Barman, S.: Electrical Network Modeling of Amino Acid String and Its Application in Cancer Cell prediction, Advances in Intelligent Systems and Computing, vol. 243, pp. 293–301. Springer, India (2014)
6. Mitra, S., Das, S., Chaudhuri, P.P, Nandi, S.: Architecture of a VLSI chip for modeling amino acid sequence in proteins. In: IEEE Proceedings of Ninth International Conference on VLSI Design, pp. 316–317. IEEE, India (1996)
7. Pradhan, C., Das, H., Naik, B., Dey, N.: Handbook of Research on Information Security in Biomedical Signal Processing. IGI Global, Hershey, PA (2018)
8. Kyte, J., Doolittle, R.F.: A simple method for displaying the hydropathic character of a protein. J. Mol. Biol. **157**(1), 105–132 (1982)
9. Timberlake, K.C.: Chemistry, 5th edn. Haper-Collins Publishers, New York (1992)
10. Devlin, T.M.: The Textbook of Biochemistry, 3rd edn. Wiley, New York (1992)
11. Nelson, D.L., Lehninger, A.L., Cox, M.M.: Lehninger Principles of Biochemistry. Macmillan (2008)
12. Lesk, A.M.: Introduction to Protein Architecture. Oxford University Press, New York (2000)
13. Roy, T., Barman, S.: A behavioral study of healthy and cancer genes by modeling electrical network. Gene **550**(1), 81–92 (2014). Elsevier
14. Temes, G.C., LaPatra, J.W.: Introduction to Circuit Synthesis and Design. McGraw-Hill, New York (1977)
15. Razavi, B.: Design of Analog CMOS Integrated Circuits. McGraw-Hill, India (2018)
16. Allen, P.E., Holberg, D.R.: CMOS Analog Circuit Design. Oxford University Press, Oxford (2002)

Personality Prediction of Social Network Users Using Ensemble and XGBoost

Aditi Kunte and Suja Panicker

Abstract Machine learning has gained tremendous attention from researchers recently. It has wide applications in tasks such as prediction and classification. Current work focuses on the effective detection of the personality of social network users. Personality is a combination of one's thinking and behavior. Having knowledge about personality of a person has many applications in real world such as varied recommendation systems or HR departments. Personality of a person can be better understood by interacting with him/her. Predicting personality using social media is a new approach where direct interaction with people can be eliminated and accurate predictions can be built. Although different machine learning methods have been used by researchers recently for the task of prediction, the use of Ensembles has not been explored. Current work focuses on advanced classifiers such as XGBoost and Ensemble for prediction. Experimentation on the real-time Twitter dataset indicates high accuracy of 82.59% with an Ensemble. These results are encouraging for future research.

Keywords Personality prediction · Social media · Machine learning · Psychological tests

1 Introduction

Predicting personality is a new domain of machine learning applications. Machine learning is an exciting technology that has gained a lot of attention from researchers. It deals with training machines to be able to cognize like humans. Machine learning is broadly defined as a branch of artificial intelligence that learns from previous data

A. Kunte · S. Panicker (✉)
School of Computer Science and Engineering, Dr. Vishwanath Karad MIT World Peace University, Pune, India
e-mail: suja.panickar@mitwpu.edu.in

A. Kunte
e-mail: aditikunte95@gmail.com

H. Das et al. (eds.), *Progress in Computing, Analytics and Networking*, Advances in Intelligent Systems and Computing 1119,
https://doi.org/10.1007/978-981-15-2414-1_14

and detects patterns from data to make decisions which does not require human intervention.

Personality is an important paradigm of human life. Personality of a person is reflected through the way he thinks, behaves, and deals with situations. Knowing the personality of a person has many applications in the real world. Personality prediction is useful in organizations to hire persons, motivate employees; it can decrease conflicts and improve collaboration in the working environment. Having knowledge about personality is also useful in recommendation systems or HR departments. Traditional methods of personality prediction used questionnaires wherein participants had to answer some questions, and the result was used to determine the individual's personality traits. However, this method is time-consuming and laborious hence social media serves as a promising platform for this purpose.

Considering the previous research, we observed that social networking sites such as Twitter and Facebook have been used popularly to gather information followed by performing prediction. On surveying literature, we noted that the following psychological tests have been prevalently used to examine the type of personality—Big five personality test [1], Myers–Briggs type indicator [2, 3], and DISC inventory [4]. One significant limitation of past research is the unavailability of a sufficient benchmarked dataset. Also, several datasets used previously are not publicly available.

Use of social media is increasing rapidly. According to the survey in January 2019, among a worldwide population of 7.7 billion, Internet users have reached 4.2 billion. 3.397 billion among them are active social media users. Approximately 116 min per day has been spent daily by users on social media. Thus social media is the most promising platform in order to gather information from user as it is the most convenient platform for a person to express himself/herself easily [5].

Existing work [6, 7] used Twitter API to generate dataset and classifying personality. Datasets used earlier are not publicly available. In the current work, we have employed Twitter streaming API to generate a real-time dataset. This dataset is collected on the basis of big five personality traits. Also, we have used advanced machine learning algorithms such as boosting to improve accuracy and have obtained significant results using these algorithms.

This paper is organized as follows: Sect. 2 presents an overview of literature survey, Sect. 3 covers system architecture, Sect. 4 covers experimentation and results, Sect. 5 presents conclusion and future scope.

2 Literature Survey

Following is an overview of the recent survey.

Stochastic Gradient Descent, Gradient Boosting, and stacking were used [8] for personality recognition on Twitter in Indonesia where SGD and super learner outperform over XGBoost. Support vector machine and linear regression were used [9] for finding user demographics from information of listening to music by using LFM-1b dataset. Personality recognition is performed using decision tree and support vector machine [10] using datasets of Twitter, facebook and YouTube where decision tree outperforms over SVM. Personalities on the basis of selfies using correlation analysis over Sina Weibo Chinese microblogging site was performed in [11]. Deep neural network and linear regression were used for finding user profile using deep multimodal fusion [12]. Here, age and gender were also predicted along with personality.

Digital footprints over social media were used to determine personality in [13] which is employed in recommending services to the user according to his interest. Sentiments are analyzed from marijuana-related posts from Facebook in [14] using NLP processing. In [15], trolled tweets are detected from Twitter using sequential mining optimization, Naive Bayes, and random forest algorithms. In contrast to major works cited so far, [16] finds the dark side of personality using status updates from Facebook with the help of machine learning algorithms such as Ordinary Least Squares regression and Least Absolute Shrinkage and Selection Operator regression. Predicting emotions from users' data using techniques as Zero-R, Random forest, and RBF [17] was done in Instagram dataset.

As mentioned earlier, researchers have also used psychological tests for determining the type of personality. From the survey, we observed that big five is most commonly used by researchers. Also, different machine learning algorithms such as kNN, SVM and Naive Bayes have been used by researchers for the task of prediction.

3 System Architecture

System architecture of current work is illustrated in Fig. 1 [18]. Dataset is collected manually using Twitter streaming API. After preprocessing, data is visualized using visualization techniques in Python. After this, machine learning algorithms such as Ensemble and XGBoost are applied to yield one of the five categories of personality.

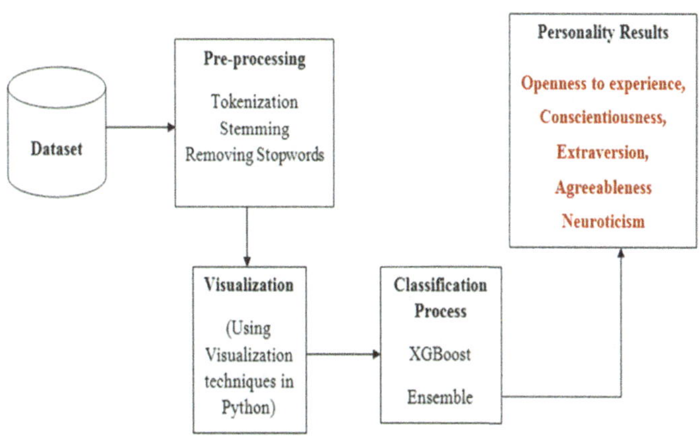

Fig. 1 System architecture

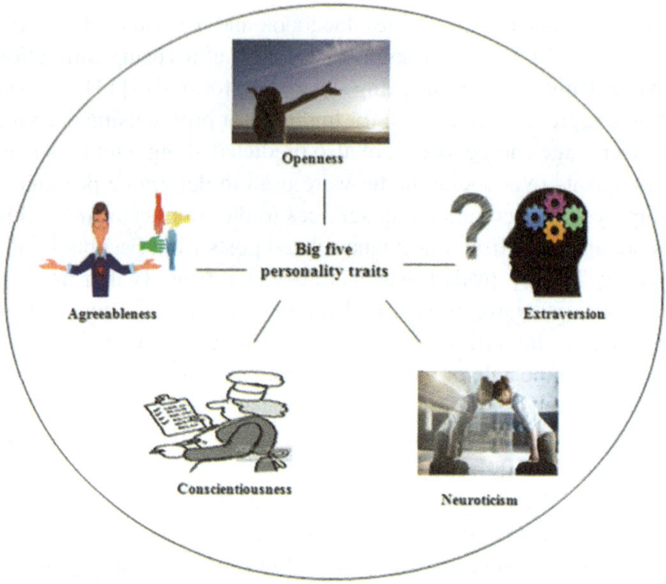

Fig. 2 Big five personality traits

Characteristics of the big five traits are as shown in Fig. 2:

We have developed a simple diagram in Fig. 2 for a pictorial representation of the five personality traits under the big five assessment. As illustrated in Fig. 2, people having "openness" category are open-minded; people with "Agreeableness" characteristic are kind and considerate; "Conscientious" people are disciplined and they focus on work; people with "Neuroticism" category have characteristics like anxiety, envy, depressed mood; and "Extraversion" people are socially connected and full of energy.

4 Experimentation and Results

4.1 Dataset Collection

For the current work, we have fetched real-time Twitter data using Twitter streaming API [19]. The extracted dataset contains 9918 rows and six columns. Labels of the datasets are classified into five categories automatically with Twitter API. Features of this dataset have been divided using big five personality traits hence all the five columns from the dataset represent class labels and the column named "status" contains posts fetched from Twitter.

Steps for fetching data using Twitter API are as given below:

Fig. 3 Output of TF-IDF

```
print(tf[1, tfidf.vocabulary_['sleepy']])
print(tf[1, tfidf.vocabulary_['funny']])
print(tf[1, tfidf.vocabulary_['song']])

0.4087316918523252
0.37365784710020533
0.0

song word tfidf is minimun, that mean this word is very rarer
```

i. Login to https://apps.twitter.com/ using your Twitter username and password.
ii. Click on "Create new app" and "Create your Twitter application" by clicking on agree terms and conditions.
iii. After the successful app creation, you will get "API key" and "API secret"
iv. Using Python library called "tweepy" and Twitter credentials extract the tweets.
v. Store the extracted tweets in ".csv" file.

4.2 Preprocessing

Raw data is converted into an understandable format during the data preprocessing. This step involves steps such as tokenization, stemming and removing stop words. For the current dataset, we have also applied TF-IDF weighting which is used to determine how important a word is with respect to a document. It is useful in order to determine the frequency of word in a document.

As a demonstration of TF-IDF, Fig. 3 demonstrates output of TF-IDF where words "sleepy" and "funny" have 40 and 37% of occurrence in the whole document whereas the word "song" has not occurred anywhere in the document.

4.3 Experimental Results

The dataset was tested using XGBoost and Ensemble algorithms. XGBoost is a highly flexible, portable, and efficient algorithm. Gradient boosting framework is adopted by XGBoost to work on machine learning algorithms. XGBoost is a solution for various data science problems as it gives fast and accurate results [20]. On the other hand, Ensemble method combines multiple machine learning algorithms to solve a particular problem. It improves accuracy over single machine learning algorithm. For the current work, we have employed logistic regression, decision tree, and SVM classifiers in the Ensemble model for prediction purpose. Experimentation results are mentioned in the following subsections.

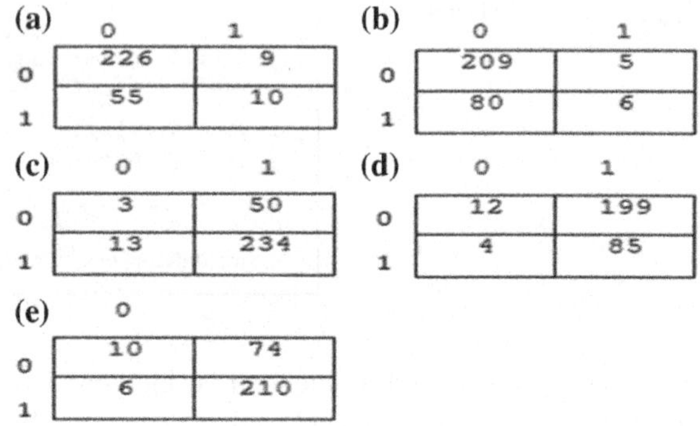

Fig. 4 Confusion matrices for XGBoost

Table 1 Accuracy of XGBoost model

Feature	AGR	CON	OPN	EXT	NEU
Accuracy (%)	78.66	71.66	79	32.33	73.33

Table 2 Accuracy of Ensemble model

Feature	AGR	CON	OPN	EXT	NEU
Accuracy (%)	76.09	71.99	82.59	63.69	68.59

4.3.1 Confusion Matrices for Features of Twitter Dataset

Confusion matrix is used to describe the performance of classification model [21]. Confusion matrices for each of the features agreeableness, conscientiousness, openness, extraversion, and neuroticism are represented in Fig. 4a, b, c, d, and e, respectively.

Based on Fig. 4, the accuracies of XGBoost and Ensemble for the five class labels are presented in Tables 1 and 2. XGBoost has the highest accuracy of 79% for openness while Ensemble has 82.59% for openness.

4.3.2 Findings from Experimentation

On comparing the two methods for prediction, it is observed that Ensemble has higher accuracy over XGBoost. These results suggest a strong predictive power of Ensemble classifiers.

Figure 5 demonstrates the graphical representation of accuracies for two methods where the Ensemble method has the highest accuracy of 82.59% for openness category.

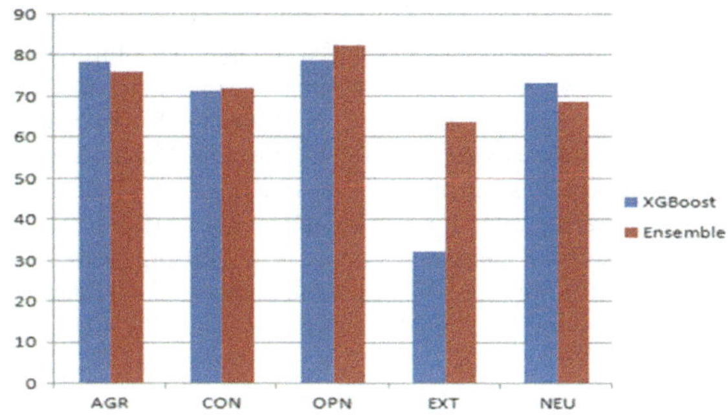

Fig. 5 Comparison of accuracies of Ensemble and XGBoost

5 Conclusion and Future Scope

Personality is an important aspect of human life. Personality prediction has roots in psychology and different psychological tests have been developed for predicting the same. The use of machine learning algorithms in this domain is encouraging. It is beneficial if there are automated systems that can predict personality without much human intervention. Current work focuses on real-time Twitter dataset and uses various machine learning algorithms for predicting personality. Experimental results indicate that Ensemble and XGBoost gave accurate results over the current dataset. Ensemble slightly outperforms XGBoost with 82.59% accuracy. Future work would be to extend the current system with more features such as users' images or audiovisual data for prediction purpose. Also, the following points can be useful for further research.

(1) In recent works, researchers have been focusing on traditional machine learning algorithms such as Naive Bayes, SVM and kNN for the task of personality prediction. The use of machine learning approaches such as bagging and boosting outperforms traditional predictive models, thus there is a scope to work on these models elaborately.
(2) In future, we shall consider more modalities of user input such as images and other visual data for further improving the accuracy.

References

1. Dandannavar, P.S., Mangalwede, S.R., Kulkarni, P.M.: Social media text—a source for personality prediction. International Conference on Computational Techniques, Electronics and Mechanism Systems (CTEMS) (2018)
2. Lima, A.C.E.S., de Castro, L.N.: Predicting temperament from Twitter data. In: International Congress on Advanced Applied Informatics (2016)
3. Lukito, L.C., Erwin, A., Purnama, J., Danoekoesoemo, W.: Social media user personality classification using computational linguistic. In: 8th International Conference on Information Technology and Electrical Engineering (ICITEE) (2016)

4. Ahmad, N., Siddique, J.: Personality assessment using Twitter Tweets. In: 21st International Conference on Knowledge Based and Intelligent Information and Engineering Systems (2017)
5. https://www.brandwatch.com/blog/amazing-social-media-statistics-and-facts/
6. Al-Samarraie, H., Eldenfria, A., Dawoud, H.: The impact of personality traits on users' information-seeking behavior. Inf. Process. Manag. **53**(1), 237–247 (2017)
7. Vioulès, MJ, Moulahi, B., Azè, J., Bringay, S.: Detection of suicide-related posts in Twitter data streams. IBM J. Res. Dev. **62**(1), 7–1 (2018)
8. Maheswari, K., Packia, P., Priya, A.: Predicting customer behaviour in online shopping using SVM classifier. In: 2017 IEEE International Conference on Intelligent Techniques in Control, Optimization and Signal Processing (INCOS) (2017)
9. Adi, G.Y.N., Tandio, M.H., Ong, V., Suhartono, D.: Optimization for automatic personality recognition on Twitter in Bahasa Indonesia. In: 3rd International Conference on Computer Science and Computational Intelligence, pp. 473–480 (2018)
10. Krismayer, T., Schedl, M., Knees, P., Rabiser, R.: Predicting user demographics from music listening information. Multimed. Tools Appl. **78**(3), 2897–2920 (2019)
11. Farnadi, G., Sitaraman, G., Sushmita, S., Celli, F., Kosinski, M., Stillwell, D., Davalos, S., Moens, M.-F., De Cock, M.: Computational personality recognition in social media. User Model. User Adap. Inter. 109–142 (2016)
12. Guntuku, S.C., Qiu, L., Roy, S., Lin, W., Jakhetiya, V.: Do others perceive you as you want them to? Modeling personality based on selfies. In: Proceedings of the 1st international workshop on Affect and sentiment in multimedia, pp. 21–26. Brisbane, Australia (2015)
13. Farnadi, G., Tang, J., De Cock, M., Moens, M.-F.: User profiling through deep multimodal fusion 5–9 (2018)
14. Azucar, D., Marengo, D., Settanni, M.: Predicting the Big 5 personality traits from digital footprints on social media: a meta-analysis. Personal. Individ. Differ. Comput. Hum. Behav. 150–159 (2018)
15. Tran, T., Nguyeny, D., Nguyeny, A., Golenz, E.: Sentiment analysis of Emoji-based reactions on Marijuana-related topical posts on Facebook. In: IEEE International Conference on Communications (ICC) (2018)
16. Fornacciari, P., Mordonini, M., Poggi, A., Sani, L., Tomaiuolo, M.: A holistic system for troll detection on Twitter. Comput. Hum. Behav. (2018)
17. Akhtara, R., Winsboroughb, D., Ortd, U., Johnsonb, A., Chamorro-Premuzic, T.: Detecting the dark side of personality using social media status updates. Personal. Individ. Differ. 90–97 (2018)
18. Tato, A., Nkambou, R., Frasson, C.: Predicting emotions from multimodal users' data (2018)
19. Panicker, S., Kunte, A.: Personality prediction using social media. In: 5th International Conference for Convergence in Technology (I2CT) (2019)
20. Tandera, T., Suhartono, D., Wongso, R., Prasetio, Y.L.: Personality prediction system from Facebook users. In: 2nd International Conference on Computer Science and Computational Intelligence (2017)
21. https://xgboost.readthedocs.io/en/latest/index.html

Optimizing Performance of Text Searching Using CPU and GPUs

M. Musthafa Baig, S. Sivakumar and Soumya Ranjan Nayak

Abstract In this work, we are solving the major problem of reducing the time complexity of searching a string in huge corpus by using GPU as our computational environment (utilizing GPGPU and CUDA as programming platform) and Knuth–Morris–Pratt (KMP) and BMH (Boyer–Moore–Horspool) as string matching algorithms. String matching is a widely used technique in current research interest of various application areas such as bioinformatics, network intrusion detection, and computer virus scan. Although data are memorized in various ways, text remains the main form to exchange information. This is particularly evident in literature or linguistics where data are composed of huge corpus and dictionaries. These analytics are required in computer science where a large amount of data is stored in linear files. To search a particular string from these huge corpus takes more time in traditional CPU's and this is a major problem.

Keywords KMP algorithm · BMH algorithm · Heterogeneous computing · GPGPU · CUDA · Performance factors

1 Introduction

Text mining is a creation of new information that is not obvious in a collection of documents. New information could be pattern, trend, Relationship and documents refers unit of text (web page, email, plain text file, etc.). Recently Heterogeneous High-Performance Computing (HPC) architecture based on general-purpose multicore microprocessors and Graphical Processing Units (GPUs) is becoming most popular in a wide range of applications. These architecture innovations have led

M. M. Baig · S. Sivakumar
Department of Computer Science and Engineering, Koneru Lakshamiah Education Foundation, Vaddeswaram, Guntur, Andhra Pradesh, India
e-mail: sivaiit79@gmail.com

S. R. Nayak (✉)
Chitkara University Institute of Engineering and Technology, Chitkara University, Punjab, India
e-mail: nayak.soumya17@gmail.com

H. Das et al. (eds.), *Progress in Computing, Analytics and Networking*, Advances in Intelligent Systems and Computing 1119,
https://doi.org/10.1007/978-981-15-2414-1_15

to drastic performance improvements to surpass several Teraflops for the double-precision arithmetic per chip. In addition, user-friendly programming environment such as CUDA from Nvidia have been developed for the General-Purpose Computing User GPUs (GPGPU).

String matching algorithms are commonly used in various applications such as the keyword matching for the given input text, the intrusion detection in the network system, the human genome, and sequence matching [1, 2]. The Knuth–Morris–Pratt (KMP) algorithm, Boyer–Moore–Horspool (BMH) is superior to the other similar pattern matching algorithms for its fast execution time when it is applied to large-sized input texts and reference patterns. Existing system implements the String matching algorithms in CPU environment which is very limited in terms of number of cores as the most number of cores provided by a CPU till date is "72" which is provided by the manufacturer "intel" and is called as "Xeon-phi-supercomputing chip" which is very less compared to that of the modern GPU which has a maximum cores count of "4352" provided by the manufacturer "NVidia" named "RTX 2080TI" that is "60.44X" times more than that of the CPU, hence by utilizing all the cores provided by the GPU, we can highly reduce the running time of the string matching algorithms. In context to running time, various many works were reported in the literature in the field of fractal geometry in terms of computational running time [3–7].

In this work, we propose the idea of implementing the string matching algorithms namely KMP and BMH in GPU environment thereby reducing the running time of these algorithms. The use of GPUs as processing elements was very limited in terms of execution time. The concept of General-Purpose computing on Graphics Processing Units (GPGPU) was introduced to exploit the processing power and the memory bandwidth of the GPUs with the use of APIs that hide the GPU hardware from programmers, one such API is the Compute Unified Device Architecture (CUDA).

The rest of the paper is organized as follows: Sect. 2 describes existing methodologies of KMP algorithm and BMH algorithm on CPU, Sect. 3 explains our proposed techniques for parallelizing and optimizing the KMP and BMH algorithms on GPU, Sect. 4 shows the experimental results with analyses. Section 5 concludes the paper with future research directions.

2 Literature Survey

2.1 Improvements to String Matching in GPU Using CUDA

In 2007, NVIDIA introduced a parallel computing platform and programming model named CUDA. It allows a substantial increase in computing performance by utilizing the Graphics Processing Unit (GPU) to accelerate various kinds of applications. We review the different work that has been done using string matching algorithms in GPU.

Zha and Sahni [8] developed in 2011 an adaption of the Aho–Corasick and Boyer–Moore algorithms and implemented them in GPU using a NVIDIA Tesla GT200 GPU and a Xeon 2.8 GHz quad-core CPU. Their work showed a speedup for the AC and BM algorithms in single-threaded implementations by a factor of 3.1–9.5. However, the BM algorithm showed to run 7–10% slower when compared to a multi-threaded implementation in the quad-core CPU.

In 2013, Xu [9] implemented in GPU the string matching algorithm MASM and extension of the BPR algorithm for multiple patterns. Using a NVIDIA GeForce 310 M GPU and a Core i3 2.27 GHz CPU, they achieved a speedup by a factor of 28 relative to a single-thread CPU implementation. The same year, Bellekens [2] implemented the Knuth–Morris–Pratt algorithm in a NVIDIA Tesla K20M GPU and compared it versus Xeon E5-2620. The results showed a 29-fold increase in speed where the GPU was used instead of the CPUs.

Nagaveni [10] in 2014 presented string matching algorithms to find DNA Sequences to detect breast cancer. Using a NVIDIA Tesla C2070 GPU and a Core i7 CPU, his experiment showed the GPU implementation 30 times more efficient than the serial implementation in the CPU.

In 2015, Kouzinopoulos [11] did an experiment using a GTX 280 GPU and a Xeon 2.4 GHz CPU. They tested the Aho–Corasick, Set Horspool, Set Backward Oracle Matching, Wu-Manber, and SOG multiple pattern matching algorithms showing the basic implementation of these algorithms in GPU were between 2.5 and 10.9 faster than the CPU implementation. The Set Horspool and the Set Backward Oracle Matching algorithms showed the most gains when using a GPU. Another work by Sharma [12] using a NVIDIA GeForce GT 635 M, implemented the Rabin–Karp pattern matching algorithm used for Deep Packet Inspection in NIDS. The GPU implementation achieved a speedup by a factor of 14 when compared against an Intel quad-core processor.

A recent work by Ashkiani [13] in 2016, using a NVIDIA Tesla K40c GPU, implemented a string matching algorithm based on the Rabin–Karp algorithm showing a 4.81 speedup against a CPU implementation.

2.2 Heterogeneous Architecture

A typical heterogeneous compute node nowadays consists of two multicore CPU sockets and two or more many-core GPUs. A GPU is currently not a stand-alone platform but a coprocessor to a CPU. Therefore, GPUs must operate in conjunction with a CPU-based host through a PCI-Express bus, as shown in Fig. 1. That is why, in GPU computing terms, the CPU is called the host and the GPU is called the device.

A heterogeneous application consists of two parts:

- Host code
- Device code.

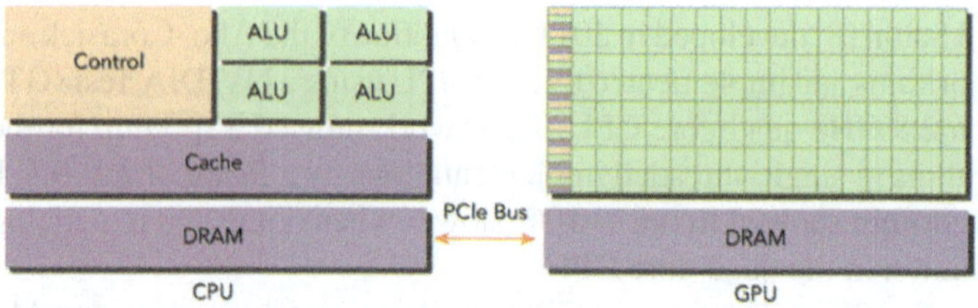

Fig. 1 Heterogeneous architecture

Host code runs on CPUs and device code runs on GPUs. An application executing on a heterogeneous platform is typically initialized by the CPU. The CPU code is responsible for managing the environment, code, and data for the device before loading compute-intensive tasks on the device.

2.3 Cuda Programming Structure

The CUDA programming model enables you to execute applications on heterogeneous computing systems by simply annotating code with a small set of extensions to the C programming language. A heterogeneous environment consists of CPUs complemented by GPUs, each with its own memory separated by a PCI-Express bus. Therefore, you should note the following distinction:

- **Host**: the CPU and its memory (host memory)
- **Device**: the GPU and its memory (device memory).

In context to this, NVIDIA introduced a programming model called Unified Memory, which bridges the divide between host and device memory spaces. This improvement allows you to access both the CPU and GPU memory using a single pointer, while the system automatically migrates the data between the host and device. This programmer-managed control of memory and data gives you the power to optimize your application and maximize hardware utilization.

A key component of the CUDA programming model is the kernel, the code that runs on the GPU device. As the developer, you can express a kernel as a sequential program. Behind the scenes, CUDA manages scheduling programmer-written kernels on GPU threads. From the host, you define how your algorithm is mapped to the device based on application data and GPU device capability. The host can operate independently of the device for most operations. When a kernel has been launched, control is returned immediately to the host, freeing the CPU to perform additional tasks complemented by data parallel code running on the device. A typical CUDA program consists of serial code complemented by parallel code. The serial code (as well as task parallel code) is executed on the host, while the parallel code is executed on the GPU device. The host code is written in ANSI C, and the device code is written using CUDA C.

Table 1 Host and device memory functions

Standard C functions	CUDA C functions
Malloc	cudaMalloc
Memcpy	cudaMemcpy
Memset	cudaMemset
Free	cudaFree

A typical processing flow of a CUDA program follows this pattern:

1. Copy data from CPU memory to GPU memory.
2. Invoke kernels to operate on the data stored in GPU memory.
3. Copy data back from GPU memory to CPU memory (Table 1).

The function used to perform GPU memory allocation is cudaMalloc, and its function signature is

cudaError_t cudaMalloc (void** devPtr, size_t size).

This function allocates a linear range of device memory with the specified size in bytes.

The function used to transfer data between the host and device is cudaMemcpy, and its function signature is

cudaError_t cudaMemcpy (void* dst, const void* src, size_t count, udaMemcpyKind kind). The main procedure is as follows:

main ()

{

1. Execute the preprocessing phase of the algorithms;
2. Where relevant represent tries using arrays;
3. Allocate one-dimensional and two-dimensional arrays in the global memory of the device using the cudaMalloc () and cudaMallocPitch () functions, respectively;
4. Copy the data from the host arrays to the device arrays using the cudaMemcpy () And cudaMemcpy2D() functions;
5. Launch CUDA kernel;
6. Copy the results array back to host memory;
7. Calculate the results;

}

3 BMH and KMP Algorithms

Given a text *txt*[0…*n*−1] and a pattern *pat*[0…*m*−1], write a function *search(char pat[], char txt[])* that prints all occurrences of *pat[]* in *txt[]*. You may assume that $n > m$.

Main Features

Performs the comparisons from left to right.
Preprocessing phase in O(m) space and time complexity.
Searching phase in O(m + n) time complexity (independent from the alphabet size).
Performs at most 2n−1 text character comparisons during the searching phase.

Algorithm 1
Knuth–Morris–Pratt Implemented in CPU

Step 1: Initialize the input variables:
n = Length of the Text
m = Length of the Pattern
u = Prefix −function of pattern (p)
q = Number of characters matched
Step 2: Define the variable: q=0, the beginning of the match
Step 3: Compare the first character of the pattern with first character of text. If match is not found, substitute the value of u [q] to q. If match is found, then increment the value of q by 1
Step 4: Check whether all the pattern elements are matched with the text elements. If not, repeat the search process. If yes, print the number of shifts taken by the pattern
Step 5: look for the next match

Pseudocode for prefix function:
n ← length[S]
m ← length[P]
a ← Compute Prefix Function
q ← 0
for i ← 1 to n do
while q > 0 and p[q+1] != S[i] do q ← a[q]

if p[q+1] = S[i] then q ← q+1
end if
if q == m then q ← a[q]
endif
end **while**
end **for**
end while

Algorithm 2
Knuth–Morris–Pratt Implemented in GPU

Declare and define an index x
n ← length[S]
m ← length[P]
preKmp← compute prefix function
result←stores the index of the occurrence of the pattern
x←blockDim.x*blockIdx.x+threadIdx.x, **if** x < n **then** start ← x * m * chunk
stop ← (x + 1) * m * chunk + m (-)1
if stop > n **then** stop ← n
endif
i ← 0, j ← start
whilej < stop
while i >(-)l and P[i] != T[j]
i ← preKmp[i] **end while**
i++ and j++
if i >= m **then** result[j] ← j – i ;
i ← preKmp[i]
end if, end while, end if

Algorithm 3
Boyer–Moore–Horspool Implemented in CPU

The Boyer–Moore–Horspool is a string matching algorithm that compares characters from the end of the pattern to it is beginning. When characters do not match, searching jumps to the next matching position in the pattern. In other words, the Boyer–Moore–Horspool Algorithm is an algorithm for finding substrings into strings.

Steps for pre-processing
Step 1: Initialize pattern length m ← |P|
Step 2: Initialize text length n ← |S|
Step 3: Initialize skip table K ← m for all symbols
Step 4: Initialize pattern index j ← 1
Step 5: for jth character p[j] in the pattern set p[j] ← m(-)j
Step 6: increment pattern index j ← j+1
Step 7: if j < m(-)1 then go to step 4

Pseudocode for pre-processing

```
P ← pattern, m ← pattern length
ASIZE ← 256
preBmh←compute bad character function, i ← 0
while i < ASIZE
    preBmh[i] ← m end while
i ← 0
while i <m(-)1
    preBmh[P[i]] ←m(-)i(-)1
end while
```

Algorithm 4
Boyer–Moore–Horspool Implemented in GPU

Pseudocode

```
T ← text,
Pre compute shift (preComp)
    i ← 0
while i <= n(-)m
i ← i + preBmh[T[i+m(-)1]]
        preBmh[i] ← i
end while
x ← current index
x←blockDim.x* blockIdx.x+threadIdx.x
ifx <=n(-)m &&preComp[x] == x
found ← true
    c ← T[x+m(-)1], i ← 0
while i < m(-)1
  if P [m(-)1] != c or P[i] != T [x +i]
found←false then break
end if, end while
if found == true
        result[x] ← true
      end if
```

4 Result and Discussion

The sample text data used in this work ranges from the sizes of 2, 4 to452 MB and the word count ranges from 1000, 100000 to 6000000 (>60 million); the text data resides in the host machine's external memory. For this work, the maximum threads supported per block are used to increase the overall efficiency of the system, the threads are locked to 1024 per each block used and also grids and blocks are variables as they adjust themselves depending on the algorithmic needs.

In Table 2, the KMP algorithms were implemented in CPU and GPU and the number of words in the text file was increased above 60 million and the pattern length was of the range 10–100 characters, the result shows that the difference in the running time between the CPU and GPU

In Table 3, the BMH algorithms were implemented on CPU and GPU and the number of words in the text file was same as they were in CPU environment, that is

Table 2 Processing time for KMP algorithm implementing on CPU and GPU

Size of text file(MB)	No. of words in the text file	Length of pattern (m)	Running time (ms) KMP (CPU)	Running time (ms) KMP (GPU)
2.4	306076	<10	14	0.004717
4.8	652151	<10	30	0.009381
77.29	10434401	<10	538	0.149518
452	62606401	<10	2.95 s	0.877666

Table 3 Processing time for BMH algorithm implementing on CPU and GPU

Length of the pattern (m)	Running time (sec) BMH (CPU)	Running time (milli sec) BMH (GPU)
>10	6.54	13.48
>50	8.94	23.66
>80	10.52	34.21
>100	13.64	41.47

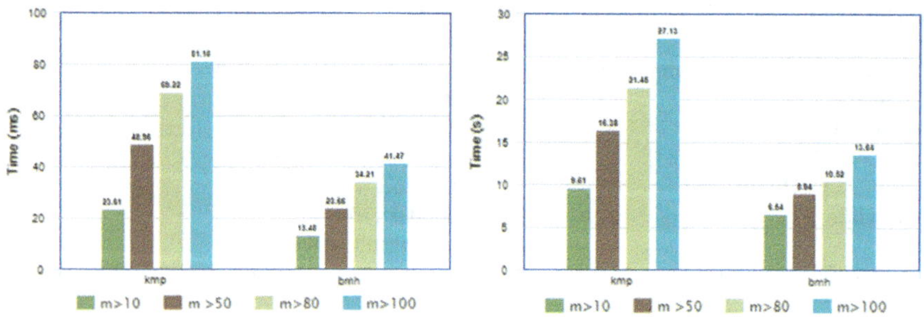

Fig. 2 KMP and BMH algorithms computed in GPU

above 60 million and the pattern length was of the range 10–100 characters, the result shows that the difference in the running time between the two algorithms. From this observation, we have seen that kmp in GPU is 18 times faster than the kmp in CPU from taking the average of all the pattern sizes and keeping the thread count in each block to its maximum available which is 1024 as shown in Fig. 2.

5 Conclusion

In this paper, parallel implementations were presented of the Knuth–Morris–Pratt and Boyer–Moore–Horspool exact string matching algorithms using the CUDA toolkit. Both the serial and the parallel implementations were compared in terms of running

Table 4 Characteristics of GeForce 950 M card

Property	Value
CUDA core	640
Graphics clock	914 MHz
Process clock	2500 MHz
Memory clock	1000 or 2500 MHz
Memory interface	128 bit
Total available graphics	4156 MB
Dedicated video memory	DDR3
Shared system memory	4028 MB

time for different reference sequences, pattern sizes while keeping the thread size to maximum. It was shown that the parallel implementation of the algorithms was up to 24× faster than the serial implementation, especially when larger text and smaller pattern sizes were used. In addition, it was discussed that in order to achieve peak performance on a GPU, the hardware must be as utilized as possible and the thread size per block should be kept at maximum levels that is maximum supported per block. The detailed summarization of GPU characteristic with the use of NVIDIA GT950M card is demonstrated in Table 4.

References

1. Park, S., Kim, D., Park, N.: High performance parallel KMP algorithm on heterogeneous architecture. In: IEEE 3rd International Journal of Foundation and Application (2018)
2. Rajesh, S., Prathima, S., Reddy, D.: Unusual pattern detection in DNA database using KMP algorithm. Int. J. Comput. Appl. **1**, 1–7 (2010)
3. Nayak, S.R., Mishra, J., Palai, G.: Analysing roughness of surface through fractal dimension: a review. Image Vis. Comput. **89**, 21–34 (2019)
4. Nayak, S.R., Khandual, A., Mishra, J.: Ground truth study of fractal dimension of color images of similar textures. J. Text. Inst. **109**, 1159–1167 (2018)
5. Nayak, S.R., Mishra, J., Khandual, A., Palai, G.: Fractal dimension of RGB color images. Int. J. Light Electron Opt. **162**, 196–205 (2018)
6. Nayak, S.R., Mishra, J., Palai, G.: An extended DBC approach by using maximum Euclidian distance for fractal dimension of color images. Int. J. Light Electron Opt. **166**, 110–115 (2018)
7. Nayak, S.R., Mishra, J.: A modified triangle box-counting with precision in error fit. J. Inform. Optim. Sci. **39**(1), 113–128 (2018)
8. Zha, X., Sahni, S.: GPU-to-GPU and Host-to-Host multipattern string matching on a GPU. IEEE Trans. Comput. **62**, 1156–1169 (2013)
9. Xu, K., Cui, W., Hu, Y., Guo, L.: Bit-parallel multiple approximate string matching based on GPU. Procedia Comput. Sci. **17**, 523–529 (2013)
10. Nagaveni, V., Raju, G.: Various string matching algorithms for DNA sequences to detect breast cancer using CUDA processors. Int. J. Eng. Technol. (2014)
11. Kouzinopoulos, C.S., Michailidis, P.D., Margaritis, K.G.: Scalable Comput. Pract. Exp. (2015)

12. Sharma, J., Singh, M.: CUDA based Rabin-Karp patterns matching for deep packet inspection on a multicore GPU. Int. J. Comput. Netw. Inform. Secur. **7** (2015)
13. Ashkiani, S., Amenta, N., Owens, J.D.: Parallel approaches to the string matching problem on the GPU. In: 28th ACM Symposium on Parallelism in Algorithms and Architectures (2016)

Automatic Parking Service Through VANET: A Convenience Application

Biswa Ranjan Senapati and Pabitra Mohan Khilar

Abstract Determination of available parking slots with optimum time at public places like shopping malls, cinema halls, railway station, airport, etc., is the major concern for all vehicle users, particularly for four-wheelers. Searching for the available parking slot randomly increases the delay, the wastage of fuel, etc. Also, monitoring of the parking slot, collection of the parking fee is done manually. An increase in the number of vehicles gradually puts overhead on the parking services. Vehicular ad hoc network (VANET) is the emerging research area which basically focusses on the safety applications. Due to the presence of the communication unit, the application area of VANET is widely increasing day-by-day. In this paper, an automated determination of available parking slots and an automatic parking fee collection is proposed through VANET. The proposed method is compared with the random searching method in terms of average waiting time and total service time. The proposed work is simulated using Network Simulator NS-2. For the maintenance of the parking slot, automatic collection of parking fees is simulated using an open-source database called MongoDB and Node-RED.

Keywords Ad hoc network · MANET · VANET · Total service time

1 Introduction

Entertainment and luxury in life are now becoming an essential part of all human beings. Due to this reason frequent visits to different public places like shopping malls, cinema halls, parks, etc. are increasing day by day. An increase in the population and increase in the number of vehicles (particularly four-wheelers) put an

B. R. Senapati (✉) · P. M. Khilar
Department of Computer Science and Engineering, National Institute of Technology, Rourkela, India
e-mail: biswa.rnjn@gmail.com

P. M. Khilar
e-mail: pmkhilar@nitrkl.ac.in

H. Das et al. (eds.), *Progress in Computing, Analytics and Networking*, Advances in Intelligent Systems and Computing 1119,
https://doi.org/10.1007/978-981-15-2414-1_16

overhead on determining the available parking slot. Determination of available parking slot is a challenging task in the public places during holidays. Also, for the maintenance of parking places, parking fees are collected manually. Manual collection at the entry point also increases the delay, creates traffic congestion, etc. The above real-life problem demands the automatic determination of an available parking slot and automatic collection of parking fees. VANET can play the role of the automation of the above parking problems.

VANET is a subclass of Mobile Ad hoc NETwork (MANET) [1]. Due to technological advancement and demand of the increasing population, number of vehicles manufactured per year is increasing day by day. Figure 1 shows the number of vehicles manufactured per year.

Nowadays vehicles are not considered as a traditional carrier. Due to the technical advancements in the electro-mechanical-material-science department, various sensors for various applications are developed at an affordable cost [2–8]. The availability of various sensors and the presence of communication unit in the vehicle makes the vehicle smart and intelligent. Figure 2 shows various components of modern vehicles.

Vehicular network mainly consists of two types of nodes. The first one is the vehicle which acts as the mobile nodes. The second one is the road side unit (RSU) which

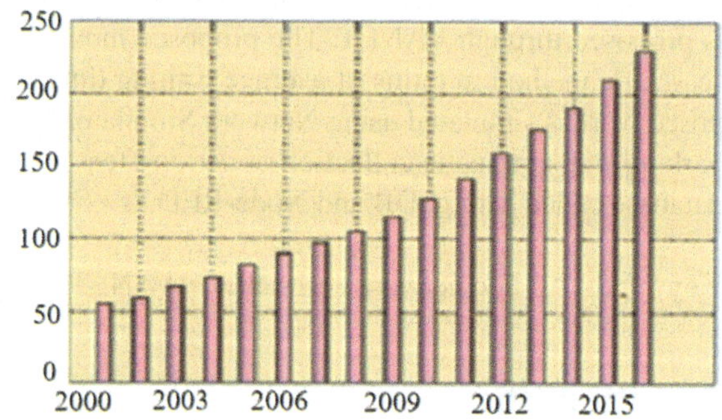

Fig. 1 Vehicles manufactured per year

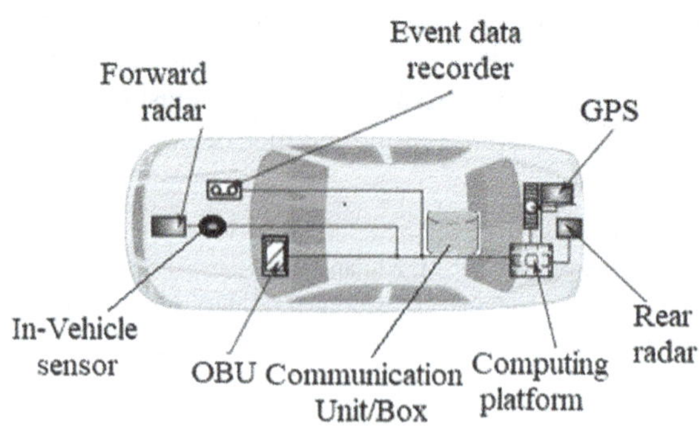

Fig. 2 Smart and intelligent vehicles

acts as the static node and have a fixed infrastructure. Based on these components the communication in VANET is classified into two categories [9].

1 Vehicle to vehicle (V2V): Communication between one vehicle with other vehicles present within the communication range is called V2V communication.
2 Vehicle to infrastructure (V2I or I2V): Communication between a vehicle with RSU or vice versa present within the communication range is called V2I or I2V communication.

For the effective use of VANET for various applications, routing protocol plays an important role [10]. Figure 3 shows the types of communication and types of routing protocol used in VANET.

Due to random motion of the vehicles, variable speed of the vehicles, and frequent link disconnection between the vehicles; VANET prefers position-based routing protocols. Some of the examples of position-based routing protocols are geographic source routing (GSR), greedy perimeter stateless routing (GPSR), anchor-based street and Traffic-Aware Routing (A.STAR) protocol, Greedy Traffic-Aware Routing (GyTAR), etc.

The major contribution of this paper is as follows.

(a) Computation of total service time for the overall parking service using M/M/1 queueing model.
(b) Determination of availability of free parking slot using a non-sql-based database MongoDB and performing the required operation of searching using Node-RED.
(c) Determination of the type of vehicle and automatic deduction of parking fees using MongoDB.
(d) Comparison of average searching time for the proposed search and random search using the network simulator NS-2.

The rest of the paper is organized as follows. Section 2 presents the motivation for this paper. Section 3 discusses the literature survey. Section 4 presents the proposed work. Simulation set up and simulation result is shown in Sect. 5. Finally, the conclusion and future scope are mentioned in Sect. 6.

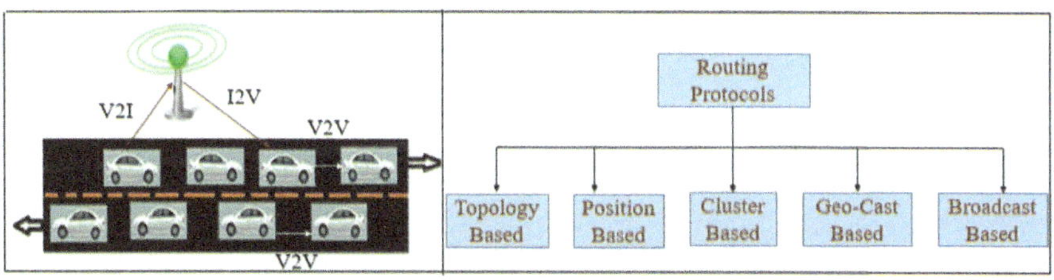

Fig. 3 Types of communication and routing in VANET

2 Motivation

1 Random search for the determination of available parking slots incurs an increase in delay as well as an increase in overall service time.
2 Manual collection of the parking fee at the entry or exit point also increases the delay.
3 Many vehicles are going through the parking slot every day. Providing a hard copy of receipt through the paper is costly which is valid only for 3–4 h.
4 Also, searching for the available parking slot, drivers may neglect the presence of other drivers, vehicles, obstacles, etc., which causes accidents in the parking lot.

Thus, presence of communication unit in the vehicle, to reduce the average searching time for the determination of available parking slot, to decrease the wastage of paper receipt, to minimize the overall waiting time and service time, to avoid accidents in the parking lot motivates for the development of automatic parking service through VANET.

3 Literature Survey

Due to the presence of communication unit and the development of various sensors at affordable cost, the vehicular network is used for the wide range of applications. The application of VANET is classified into four categories which is shown in Fig. 4.

Combined with intelligent transportation system (ITS) VANET provides safety applications like intelligent traffic systems [11], reduction in traffic congestion [12]. Convenience application of VANET includes the toll tax collection [13], movement of emergency vehicles [14]. Commercial application of VANET includes marketing

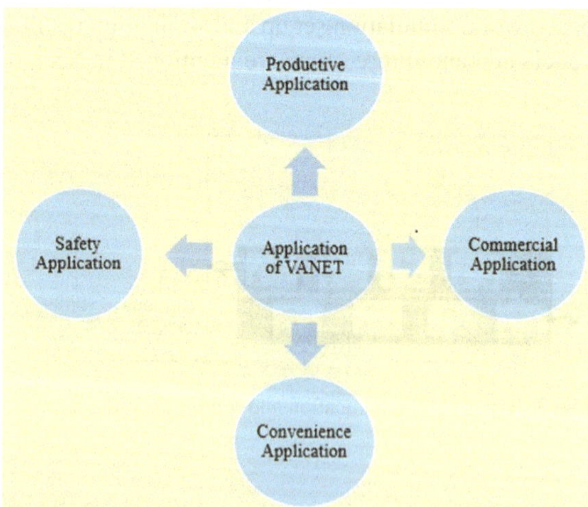

Fig. 4 Classification of different application of VANET

on wheels [15], sharing of Wi-Fi through VANET [16]. The productive application of VANET includes environmental monitoring through VANET [17].

Various approaches are also proposed for the determination of an available parking slot. Sensor-based determination for the available parking slot using the probabilistic approach was proposed by Suhr et al. [18]. But the probabilistic approach may not always predict the accurate result as the demand for the searching of the available parking slot varies with respect to time. Sharma et al. proposed an IoT-based approach for the determination of an available parking slot by displaying the available parking slot at the entry point through LCD [19]. But during the peak demand for the parking slot, the available parking slot is updated very quickly. So while reaching the predetermined parking slot may not be available after reaching the location of the available parking slot. This paper thus proposed an automatic parking service through VANET which not only reduces the search time, also reduces the waiting time at the entry of the gate for the collection of the parking fee.

4 Proposed Model

The proposed work focuses on two important aspects.

(a) Determination of available parking slot.
(b) Automatic parking fee collection.

4.1 Determination of Available Parking Slot

RSU is assumed to be present at the entry and exit point of the parking slot. The vehicle which wants to enter into the parking slot sends the request consisting of vehicle number to the RSU. A database is maintained at the RSU which consists of the three fields. The first field is the parking lot no. which refers to every single location where the vehicle can be stored. The second one is the parking status which is either available (A) or not available (NA). When the request arrives, then the available parking slot present nearer to entry point is sent as the reply. At the same time, the status of the parking slot is changed from available to not available and the corresponding vehicle number is stored in the database as the third attribute. When the vehicle leaves the parking slot again sends the request to the RSU present at the exit side of the parking slot. At that instant, the same database is updated. The parking status changes from not available to available. The corresponding value in the vehicle no. becomes null.

Fig. 5 Attributes of the database for the automatic parking service

Parking Slot No.	Parking Status	Vehicle No.

(a) Attributes of the database to determine the available parking slot

Vehicle No.	Owner_Name	Account_No

(b) Attributes of the database for the automatic parking fee collection

4.2 Automatic Parking Fee Collection

Another database is maintained which consists of the three fields and are Vehicle No, Owner_Name, and Account_No. When the vehicle sends the request with vehicle number as one of its attributes, then the corresponding vehicle number is searched from the database. If the information is present in the database, then the corresponding parking fee is deducted from the account registered for the corresponding vehicle. If the vehicle is not registered in the database, then the corresponding vehicle information is added to the database for the future automatic parking fee deduction. Figure 5 shows the attributes of the database used in both phases.

The pseudocode for the overall parking service is mentioned in Algorithm 1.

```
Input: RSU with two database at the entry and exit point.
Output: Automatic determination of parking slot and collection of parking fee.
If vehicle wants to enter a parking place and distance(Veh-RSU) < Transmission range
    Send requests to the RSU with vehicle no as the input.
    From the 1st database Search the nearest available parking lot.
    if ∀ parking status ∈ Not available
        Send reply not to enter the parking area.
    else
        Send the nearest available parking lot number.
        Change the status of the parking slot Not available and add the vehicle No.
    end if
    Search the vehicle number in the second database.
    if Vehicle No ∈ 2nd database
        Deduct the parking fee from the registered account number.
        Transmit the acknowledgement of deduction to the registered mail-id and
mobile number.
    else
        Register the information of the vehicle in the 2nd database for the
automatic parking fee service.
    end if
end if
STOP
```

(Algorithm-1: Automatic parking service)

Definition of performance metric:

- **Average search time (AST)**: Search time refers to the time for the identification of available parking slots.
- **Total service time (TST)**: Total service time refers to the sum of searching time, collection of parking fee time, and entry of the vehicle to the predetermined parking slot.

TST is a queueing problem and is evaluated using M/M/1 queueing model.

5 Simulation Setup and Simulation Result

The computation of AST and TST is performed by using a network simulator tool NS-2. For the database operation tools used are MongoDB and Node-RED. The simulation area is set to 2000 m × 2000 m. The communication range between the nodes is set to 200 m. The size of the transmission packet is set to 512 bytes. Priority queue is used having queue length set to 50. IEEE 802.11p is set as the MAC layer protocol. The overall network setup parameters are represented in Table 1.

Figure 6 shows the comparison of the proposed search with a random search with respect to average search time and total search time.

From the above comparison, the proposed work performs better. Searching from the database for the available parking slot takes almost constant time which is independent of the number of vehicles present in the parking slot. Also, due to limited data present in the database, the operation on the database is faster than manual searching of the available parking slot and manual parking fee collection.

6 Conclusion and Future Scope

In this paper, an automated parking service with the help of vehicles and RSU is proposed. The performance of the proposed work performs better as compared to

Table 1 Simulation network setup parameters

Parameters	Parameter value
Simulation area	2000 m × 2000 m
Communication range	200 m
Size of packet	512 bytes
Speed of the vehicles	20–50 km/hr
MAC protocol	IEEE 802.11p
Number of vehicles	5–85
Simulation time	250 s

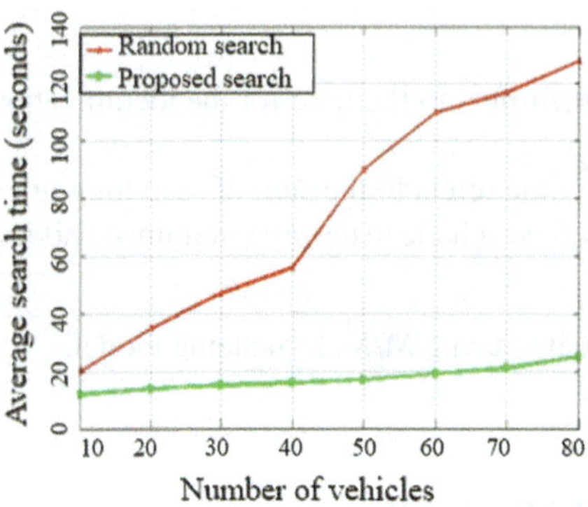

Fig. 6 Comparison of the proposed search with random search

random search in terms of average search time (AST) and total service time (TST). In the future, a real test on VANET could be performed using a trans-receiver module in the vehicles and RSU. Also, some security aspects could be added during the transaction of the parking fee from the registered account number.

References

1. Bhoi, S.K., Khilar, P.M.: Vehicular communication: a survey. IET Netw **3**(3), 204–217 (2013)
2. Das, H., Naik, B., Pati, B., Panigrahi, C.R.: A survey on virtual sensor networks framework. Int. J. Grid Distrib. Comput. **7**(5), 121–130 (2014)
3. Mohapatra, S., Khilar, P.M.: Forest fire monitoring and detection of faulty nodes using wireless sensor network. In: 2016 IEEE Region 10 Conference (TENCON), pp. 3232–3236. IEEE (2016)
4. Mishra, B.B., Dehuri, S., Panigrahi, B.K., Nayak, A.K., Mishra, B.S.P., Das, H.: Computational Intelligence in Sensor Networks, vol. 776. Studies in Computational Intelligence. Springer, Berlin (2018)
5. Mohapatra, S., Khilar, P.M.: Artificial immune system based fault diagnosis in large wireless sensor network topology. In: TENCON 2017-2017 IEEE Region 10 Conference, pp. 2687–2692. IEEE (2017)
6. Sarkar, J.L., Panigrahi, C.R., Pati, B., Das, H.: A novel approach for real-time data management in wireless sensor networks. In: Proceedings of 3rd International Conference on Advanced Computing, Networking and Informatics, pp. 599–607. Springer, New Delhi (2016)
7. Panigrahi, C.R., Sarkar, J.L., Pati, B., Das, H.: S2S: a novel approach for source to sink node communication in wireless sensor networks. In: International Conference on Mining Intelligence and Knowledge Exploration, December 2015, pp. 406–414. Springer, Cham (2015)
8. Mohapatra, S., Khilar, P.M., Swain, R.R.: Fault diagnosis in wireless sensor network using clonal selection principle and probabilistic neural network approach. Int. J. Commun. Syst. e4138 (2019)
9. Schoch, E., Kargl, F., Weber, M., Leinmuller, T.: Communication patterns in VANETs. IEEE Commun. Mag. **46**(11), 119–125 (2008)
10. Singh, S., Agrawal, S.: VANET routing protocols: issues and challenges. In: 2014 Recent Advances in Engineering and Computational Sciences (RAECS), pp. 1–5. IEEE (2014)

11. Khekare, G.S., Sakhare, A.V.: A smart city framework for intelligent traffic system using VANET. In: 2013 International Mutli-conference on Automation, Computing, Communication, Control and Compressed Sensing (iMac4s), pp. 302–305. IEEE (2013)
12. Knorr, F., Baselt, D., Schreckenberg, M., Mauve, M.: Reducing traffic jams via VANETs. IEEE Trans. Veh. Technol. **61**(8), 3490–3498 (2012)
13. Senapati, B.R., Khilar, P.M., Sabat, N.K.: An automated toll gate system using VANET (2019)
14. Buchenscheit, A., Schaub, F., Kargl, F., Weber, M.: A VANET-based emergency vehicle warning system. In: 2009 IEEE Vehicular Networking Conference (VNC), pp. 1–8. IEEE (2009)
15. Bhoi, S.K., Puthal, D., Khilar, P.M., Rodrigues, J.J.P.C., Panda, S.K., Yang, L.T.: Adaptive routing protocol for urban vehicular networks to support sellers and buyers on wheels. Comput. Netw. **142**, 168–178 (2018)
16. Tufail, A., Fraser, M., Hammad, A., Hyung, K.K., Yoo, S.-W.: An empirical study to analyze the feasibility of WIFI for VANETs. In: 2008 12th International Conference on Computer Supported Cooperative Work in Design, pp. 553–558. IEEE (2008)
17. Senapati, B.R., Swain, R.R., Khilar, P.M.: Environmental monitoring under uncertainty using smart vehicular ad hoc network. In: Smart Intelligent Computing and Applications, pp. 229–238. Springer, Singapore (2020)
18. Suhr, J.K., Jung, H.G.: Sensor fusion-based vacant parking slot detection and tracking. IEEE Trans. Intell. Transp. Syst. **15**(1), 21–36 (2013)
19. Sharma, A., Bansal, S., Mishra, S., Kumar, P., Choudhury, T.: Automated Car Parking with Empty Slot Detection Using IoT. SSRN 3403921 (2019)

Fault Diagnosis in Wireless Sensor Network Using Self/Non-self Discrimination Principle

Santoshinee Mohapatra and Pabitra Mohan Khilar

Abstract For the detection of faulty sensor nodes in the wireless sensor network (WSN), a self/non-self discrimination principle was discussed. The performance parameters such as detection accuracy, false alarm rate, and false positive rate are used to evaluate the proposed method. The outcome of the simulation shows that the proposed method gives better results in terms of performance metrics compared to the existing algorithms.

Keywords Fault diagnosis · Wireless sensor network · Self/non-self discrimination principle · Artificial immune system

1 Introduction

Wireless sensor network (WSN) is a set of small and cheap sensor nodes, which can sense the data from the environment and send it to other sensor nodes or base station. The sensor nodes communicate with each other through a wireless medium. Due to the characteristics of a sensor node, the WSN can be used in various application areas like agriculture, monitoring forest fire, military application, environmental monitoring [1–3], and so on. Due to the deployment scenario, sensor nodes are subjected to different types of faults.

Due to technological advancement of sensor network, sensors are used in vehicles [4]. Sensor node may fail due to various issues such as security, data communication, limited amount of energy, and so on [5]. To handle this various, data management techniques [6], data communication techniques [7], computational intelligence techniques [8] have been developed. Researchers also developed virtual sensor network which can provide multiple services in a single WSN infrastructure [9].

S. Mohapatra (✉) · P. M. Khilar
Department of CSE, NIT Rourkela, Rourkela 769008, Odisha, India
e-mail: santoshinee88@gmail.com

P. M. Khilar
e-mail: pmkhilar@nitrkl.ac.in

H. Das et al. (eds.), *Progress in Computing, Analytics and Networking*,
Advances in Intelligent Systems and Computing 1119,
https://doi.org/10.1007/978-981-15-2414-1_17

The behavioral fault of WSN is of two types such as hard and soft fault. Hard faults are nothing but, permanent faults in which sensor nodes are unable to perform an operation because of power failure or environmental condition. In soft fault, the sensor node transmits the data which are erroneous [10]. To detect the faulty nodes, it is necessary to develop an algorithm which can detect the faulty nodes accurately.

Different types of algorithms are there to detect the faulty nodes such as neural network, statistical method, neighboring coordination, comparison. In comparison-based approaches, due to comparison, message exchange is too high which leads to consumption of more energy. The statistical method and neighboring coordination based approaches depend on the number of nodes. For less nodes, the estimation of the result will not be accurate. Neural network approaches mainly depends on datasets. It will give better results if the dataset is good. From the principle and strategies of human immune system (HIS), the operation of the immune system of human being can be related to the diagnosis process of WSN.

There was a greater interest seen among scientists and researchers in developing biologically inspired algorithms in the past. Artificial immune system (AIS) is considered as one of the most popular approaches due to its principle. AIS is influenced by the principle of HIS, which can expertly saved our bodies from bacteria and viruses. In this paper, we have discussed about self/non-self discrimination principle of AIS. It is used to distinguish between the self and non-self. The self refers to the body's own cells and the non-self refers to foreign cells. Initially, Forrest et al. [12] proposed the self/non-self discrimination principle and then it is widely used in anomaly detection, pattern recognition, computer security, and fault detection. The contribution in this paper is outlined as follows.

- A self/non-self discrimination principle has been used to distinguish between faulty and fault-free sensor nodes.
- To validate the proposed method, an extensive simulation and comparison has been done using network simulator.
- Using different performance parameters, the proposed work is evaluated.

Arrangement of different sections of this paper is given below. The study of literature is given in Sect. 2. The principle of self/non-self discrimination is mentioned in Sect. 3. Section 4 discusses about the proposed work. The experimental results are discussed in Sect. 5. This paper concludes in Sect. 6.

2 Literature Survey

The self/non-self discrimination principle (alternatively called as negative selection algorithm) is widely used in literature for various types of applications.

The authors in [13] proposed a multi-operational algorithm using the negative selection principle. For comparison with other algorithms, they have used the fault model of DC motor as a benchmark. In [14], the authors have proposed two novel negative selection algorithms. Usually, the detectors are generated randomly, but in this work the detectors are generated in nonrandom ways, which eliminate the training time of detectors. The performances of both the experiments performed on the iris dataset, ball bearing fault data set and two-dimensional synthetic datasets were examined. The result shows that in most cases they give better results than the others.

In [15], the authors have proposed one type of data flow attack called a Sybil attack. They have implemented an improved version of NSA with learning capability, and they have also used r-contiguous bit matching rule. They have compared their work with other works by taking three performance parameters such as false positive, false negative, and detection rate and their work shows better results than others. In [16] the authors have proposed a method for classification of real noise in speech sentences based on NSA. To validate the proposed method they have taken six types of real noise. This method shows better results than the classical classifiers in terms of accuracy.

In [17], a fault diagnosis algorithm is proposed by combining both the clonal selection principle and negative selection algorithm, which determines the fault types properly. By optimizing the mutation operator, the convergence rate of antibody generation in the detector set was also improved. The fault diagnosis model was tested by experiments. The vibrating signals were collected and transferred by a WSN. The data were analyzed and diagnosed based on the fault diagnosis model. In [18] using NSA the authors have proposed a motor fault diagnosis scheme. The motor faults can be encountered using a hierarchical structure. This structure efficiently detects the incipient motor faults as well as the fault types. In the simulation, the authors have examined the fault diagnosis method using two real-world problems.

In [19] inspired by an immune system, fault detection and isolation of a wind turbine system has been proposed. To detect and isolate both individually and simultaneously occurring faults, the authors have designed an NSA which is hierarchical in nature. To evaluate the proposed work, a nonparametric statistical comparison test has been put under various fault circumstances. The simulation result shows that the NSA and SVM give the same performances while in certain fault circumstances NSA gives a better result than SVM.

3 Self/Non-self discrimination Principle

A human body's immune system has a self-defense mechanism that can protect the body from being attacked by various organisms, such as bacteria or viruses. The immune system's main goal is to find out all the cells and molecules in the body and classify them into self and non-self (fault-free). Self generally refers to appropriate or fault-free data, and non-self refers to the data being calculated. There are two steps which are shown in Figs. 1 and 2. In Fig. 1, self-strings and the strings which are generated randomly are matched. The matched string gets rejected and the detector set holds the strings that do not match. In Fig. 2, protected strings are matched with those in the detector set. If there is a match the string will be identified as non-self.

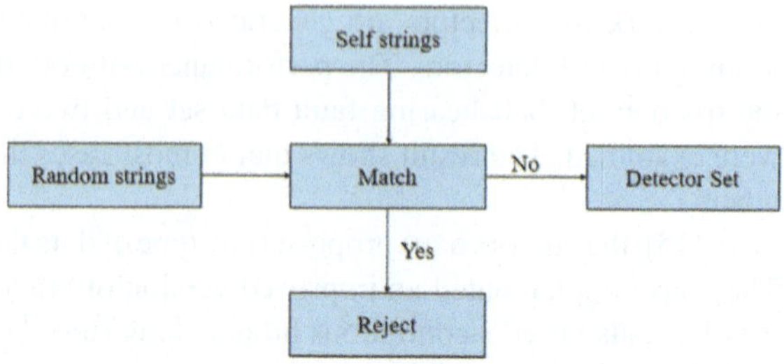

Fig. 1 Censoring

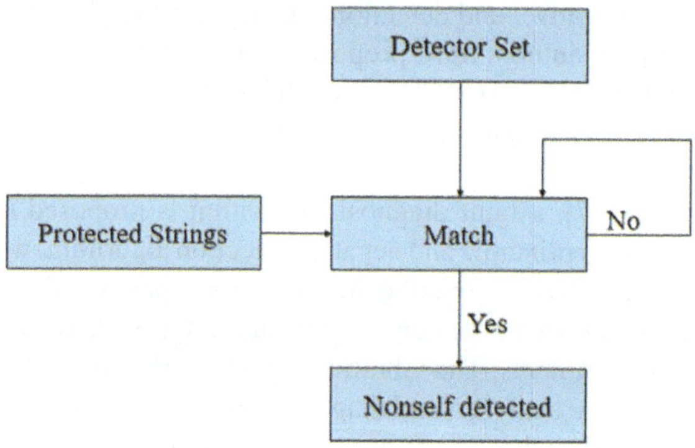

Fig. 2 Detection

4 Proposed Work

The proposed method has two phases to identify the faulty nodes. (I.) detector generation phase and (II.) matching phase.

I. **Detector generation phase**
   In this phase, a detector set consisting of a set of binary strings is generated randomly which is used to detect faulty nodes. It is necessary to ensure that the detector set does not match with the self set. New patterns can then be collected from the monitoring scheme. These patterns are converted into and compared to the detector set in the suitable binary form. A pattern can be expected to be non-self if it matches the detector set and then the necessary action can be taken accordingly.

II. **Matching phase**
   Different matching rules are there such as binary matching, r-contiguous matching, r-chunk matching, hamming distance matching, and Euclidean distance matching [20]. Here we have used r-contiguous matching rule. If the matching rule requires a precise match, the detector set should contain all possible strings that might happen which is impractical and increase the computational overhead. For a match to take place, only r adjacent bits are needed to be

identical. For example, the strings "00111010" and "01111011" match when the r = 4 where r is matching threshold. The non-self can be identified by using the following self-/non-self discrimination (SNSD) Algorithm 1.

Algorithm 1 SNSD Algorithm

```
Input: Self data;
Output: Detection of non-self;
Initialize: Detector set = ϕ ;
while (¬ Stop condition()) do
  Detectors ← Generate Random Detectors();
  for (Detector_i ∈ Detector set) do
    if (¬ Matches (Detector_i, Self data)) then
      Detector set ← Detector_i;
    end if
  end for
end while
Return (Detector set);
Match the protected string with the Detector set;
if ( Matches (protected string, Detector set)) then
  Non-self detected;
end if
```

5 Results and Discussions

In this work, the NS-2 simulator [21] has been used to measure the performance of the proposed algorithm and it is compared with existing algorithms such as WAIS [22] and DSFD [11]. Different performance parameters are used to evaluate the performance such as detection accuracy (DA), false alarm rate (FAR), and false positive rate (FPR). Deployment of 500 sensor nodes are done in an area of $500 \times 500\text{m}^2$. The parameters for simulation are set in Table 1.

The plot between DA versus the probability of fault is shown in Fig. 3, which shows that the proposed SNSD algorithm gives better accuracy than other existing approaches. When the probability of fault increases detection accuracy decreases. The proposed SNSD algorithm gives an average of 98.18% accuracy, whereas the existing WAIS and DSFD algorithms give 97.8% and 95.93% accuracy, respectively.

FAR versus probability of fault is plotted in Fig. 4 which indicates that when the probability of faults increases false alarm rate increases. The proposed SNSD algorithm gives an average of 3.56% FAR, whereas the existing WAIS and DSFD algorithms gives 3.85% and 4.15% of FAR, respectively.

Figure 5 shows the FPR versus probability of fault. Increase in probability of fault also increases FPR. The proposed SNSD algorithm gives an average of 1.81% FPR, whereas the existing WAIS and DSFD algorithms gives 2.2% and 4.06% of FPR, respectively.

Table 1 Simulation Parameters

Parameter	Value
Total nodes	500
Simulation time	200 s
Transmission range	200 m
Channel rate	250 kbps
MAC protocol	IEEE 802.15.4
Propagation model	TwoRayGround
Initial Energy	10J
Packet size	512 bytes
Packet rate	1 pkt/s
Antenna type	Omni antenna
Grid size	$500 \times 500\,m^2$
Network type	Arbitrary network

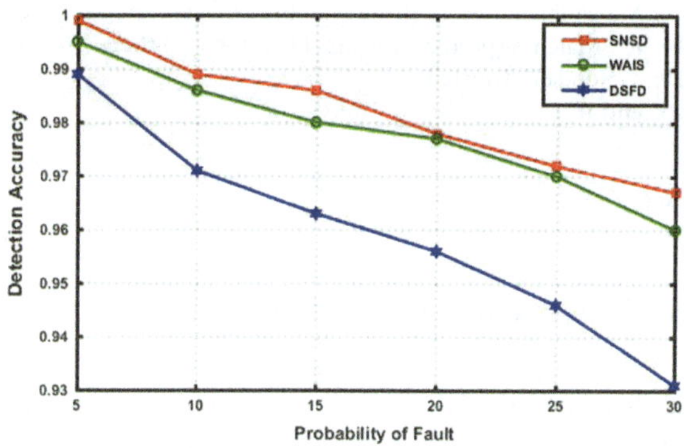

Fig. 3 Detection accuracy versus probability of fault

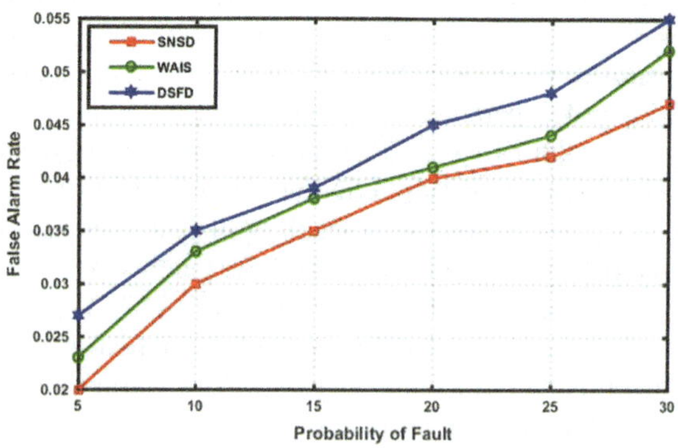

Fig. 4 False alarm rate versus probability of fault

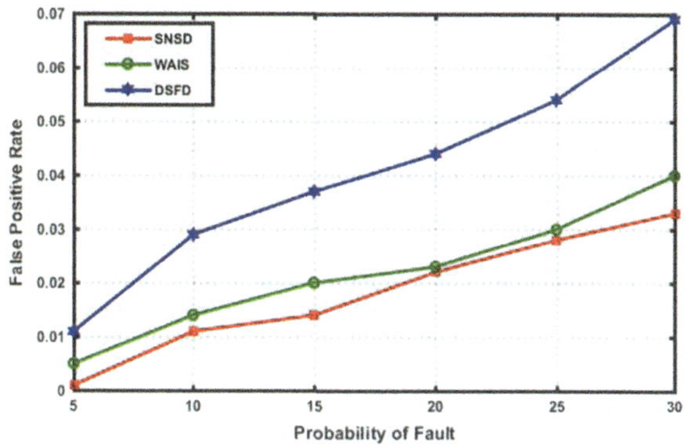

Fig. 5 False positive rate versus probability of fault

6 Conclusion

In this paper, a self/non-self discrimination principle of AIS has been used to detect faulty nodes. Various performance metrics have been taken into consideration for measuring the performances. The proposed SNSD algorithm is compared with WAIS and DSFD algorithm. The proposed algorithm gives more accuracy and less FAR, FPR as compared to WAIS and DSFD as shown in the simulation. In the future, for the diagnosis of various faults in WSNs, different immune inspired techniques can be used.

References

1. Yick, J., Mukherjee, B., Ghosal, D.: Wireless sensor network survey. Comput. Netw. **52**(12), 2292–2330 (2008)
2. Mohapatra, S., Khilar, P.M.. Forest fire monitoring and detection of faulty nodes using wireless sensor network. In: 2016 IEEE Region 10 Conference (TENCON). IEEE (2016)
3. Senapati, B.R., Swain, R.R., Khilar, P.M.: Environmental monitoring under uncertainty using smart vehicular ad hoc network. In: Smart Intelligent Computing and Applications, pp. 229–238. Springer, Singapore (2020)
4. Senapati, B.R., Khilar, P.M., Sabat, N.K.: An automated toll gate system using VANET (2019)
5. Pattnaik, P.K., et al. (eds.): Progress in Computing, Analytics and Networking: Proceedings of ICCAN 2017, vol. 710. Springer, Singapore (2018)
6. Sarkar, J.L., et al.: A novel approach for real-time data management in wireless sensor networks. In: Proceedings of 3rd International Conference on Advanced Computing, Networking and Informatics. Springer, New Delhi (2016)
7. Panigrahi, C.R., et al.: S2S: a novel approach for source to sink node communication in wireless sensor networks. In: International Conference on Mining Intelligence and Knowledge Exploration. Springer, Cham (2015)
8. Mishra, B.B., et al. (eds.): Computational Intelligence in Sensor Networks. Springer, Berlin (2019)
9. Das, H., et al.: A survey on virtual sensor networks framework. Int. J. Grid Distrib. Comput. **7**(5), 121–130 (2014)

10. Mohapatra, S., Khilar, P.M., Swain, R.R.: Fault diagnosis in wireless sensor network using clonal selection principle and probabilistic neural network approach. Int. J. Commun. Syst. e4138 (2019)
11. Panda, M., Khilar, P.M.: Distributed self fault diagnosis algorithm for large scale wireless sensor networks using modified three sigma edit test. Ad Hoc Netw. **25**, 170–184 (2015)
12. Forrest, S., et al.: Self-nonself discrimination in a computer. In: 1994 IEEE Computer Society Symposium on Research in Security and Privacy. Proceedings. IEEE (1994)
13. Laurentys, C.A., et al.: Design of an artificial immune system for fault detection: a negative selection approach. Expert Syst. Appl. **37**(7), 5507–5513 (2010)
14. Li, D., Liu, S., Zhang, H.: Negative selection algorithm with constant detectors for anomaly detection. Appl. Soft Comput. **36**, 618–632 (2015)
15. Zeeshan, M., et al.: An immunology inspired flow control attack detection using negative selection with R-contiguous bit matching for wireless sensor networks. Int. J. Distrib. Sens. Netw. **11**(11), 169654 (2015)
16. de Abreu, C.C.E., Duarte, M.A.Q., Villarreal, F.: An immunological approach based on the negative selection algorithm for real noise classification in speech signals. AEU-Int. J. Electron. Commun. **72**, 125–133 (2017)
17. Chen, G., Zhang, L., Bao, J.: An improved negative selection algorithm and its application in the fault diagnosis of vibrating screen by wireless sensor networks. J. Comput. Theoret. Nanosci. **10**(10), 2418–2426 (2013)
18. Gao, X.Z., Wang, X., Zenger, K.: Motor fault diagnosis using negative selection algorithm. Neural Comput. Appl. **25**(1), 55–65 (2014)
19. Alizadeh, E., Meskin, N., Khorasani, K.: A negative selection immune system inspired methodology for fault diagnosis of wind turbines. IEEE Trans. Cybern. **47**(11), 3799–3813 (2017)
20. González, F., Dasgupta, D., Gómez, J.: The effect of binary matching rules in negative selection. In: Genetic and Evolutionary Computation Conference. Springer, Berlin (2003)
21. The Network Simulator NS-2. http://www.isi.edu/nsnam/ns/ (2010)
22. Mohapatra, S., Khilar, P.M.: Artificial immune system based fault diagnosis in large wireless sensor network topology. In: TENCON 2017-2017 IEEE Region 10 Conference, pp. 2687–2692. IEEE (2017)

Efficient Data Structure for the Top-*k* Variation of Sports Auction

Biswajit Sanyal, Subhashis Majumder and Priya Ranjan Sinha Mahapatra

Abstract In any sports auction, bidders may have fund constraints and purchasing the best team may not be possible for all. So, reporting k best teams in non-increasing order of total expenses might be helpful to them. Now, they have multiple options in their hands from which they can choose the best available team according to their fund constraints. In this paper, we deal with the top-k variation of sports auction, where we report k best teams (top-k teams) of fixed size in non-increasing order of total cost, which might help franchises to make decisions on buying players as per available budget. Teams are formed by choosing a predefined number of players from various categories where each player has his own cost. We initially present a basic technique where we construct a *metadata structure* G even before costs of players of various categories and k are known, so that we can later use G to report the top-k purchases efficiently when costs are available. We then extend our work by generating the required portions of G on the fly, so that no preprocessing is needed, which in turn improves the space complexity of the algorithm remarkably.

Keywords One shift · Metadata structure · DAG · Top-k query · Cost array · Max heap

---

B. Sanyal (✉)
Department of Information Technology,
Government College of Engineering and Textile Technology, Serampore,
West Bengal, India
e-mail: biswajit_sanyal@yahoo.co.in

S. Majumder
Department of Computer Science and Engineering, Heritage Institute of Technology,
Kolkata, West Bengal, India
e-mail: subhashis.majumder@heritageit.edu

P. R. Sinha Mahapatra
Department of Computer Science, University of Kalyani, Kalyani,
West Bengal, India
e-mail: priya@klyuniv.ac.in

H. Das et al. (eds.), *Progress in Computing, Analytics and Networking*,
Advances in Intelligent Systems and Computing 1119,
https://doi.org/10.1007/978-981-15-2414-1_18

1 Introduction

Top-k query retrieval is nowadays a very popular choice in many application domains since it reports the most relevant data rather than generating a (potentially huge) list of all the data items that satisfy a certain query. Such an approach is studied in many different domains including multimedia [4], business analytics [2], information retrieval [3], computational geometry [1, 5, 8], databases [6, 10], and so on.

In this paper, we consider the top-k variation of a typical sports auction where we report k best teams (top-k teams) of fixed size in non-increasing order of total cost, which might help franchises to make decisions on buying players as per available budget. Teams are formed by choosing a predefined number of players from various categories. It has a widespread application in different sports auctions like (i) Auction in IPL Cricket League, (ii) Auction in Pro Kabadi League, etc. In any sports auction, bidders may have fund constraints and purchasing the best team may not be possible for all. So, reporting k best teams in non-increasing order of total expenses might be helpful to them.

Other than sports auction, it has other applications also. Building a team for developing software is somewhat a similar problem. The only difference is that here reporting the teams in the reverse order (in nondecreasing order) might make more sense and there also our solution methodology can be applied with a minor change. Suppose a software team is to be built with some people having skillsets in back-end development, front-end development, core engineering etc. If there is a hiring price for each person and a minimum number of people is to be taken for each type of skill, then reporting k best teams according to the total expenses will be helpful to the organization. The organization now can get different options in its hand among which it can choose a team according to its budget constraints.

In our problem, the input consists of (i) an integer value n, where n is the number of categories of players participating in the auction, (ii) a set P of positive integers, $\{p_1, p_2, \ldots, p_n\}$, where p_i is the number of players in category i ($i \in [1 \ldots n]$), (iii) the cost vector $C_i = (C_{i,1}, C_{i,2}, \ldots, C_{i,p_i})$ for each category i, where $C_{i,t}$ is the cost of purchasing the tth ($t \in [1 \ldots p_i]$) player of category i, (iv) a set Np of positive integers, $\{np_1, np_2, \ldots, np_n\}$, where np_i ($np_i \leq p_i$) is the number of players, we want to purchase from some category i. In this current work, we report k best purchases (top-k purchases) of $\sum np_i$ players from n different categories in non-increasing order of total cost. Our solution is based on constructing a metadata structure G as part of preprocessing (without knowing the cost vectors) to speed up the reporting steps. The metadata structure G consists of (i) a DAG (Directed Acyclic Graph) M_i for each category i, $i \in [1 \ldots n]$, and (ii) a DAG M_g that coordinates the information from all the n number of DAGs, M_1 to M_n. We further extend our work by generating the required portions of G on the fly, thereby saving a considerable portion of storage space.

The paper is organized as follows. In Sect. 2, we introduce the DAG M_i for each category i, $i \in [1 \ldots n]$. Clearly, we have in total n such DAGs, M_1 to M_n. In Sect. 3, we define the DAG M_g that coordinates the information from all the M_i ($i \in [1 \ldots n]$).

In Sect. 4, we discuss how we can report our desired result efficiently by using n local max heaps H_1 to H_n and a global max heap H_g, where we use the master DAG M_g and also the n DAGs M_1 to M_n, on demand to reduce the burden of preprocessing step. In this section, we also give an algorithm and analyze its space and run-time complexity and proves its correctness. Finally, in Sect. 5, we conclude our work and mention some open problems.

2 Directed Acyclic Graph M_i ($i \in [1 \ldots n]$)

Using a similar concept of our earlier work [7, 9], we here construct a DAG M_i for each category i of players. Clearly, there are n such DAGs, one for each category. Here also, we assume that the given cost vector C_i, $i \in [1 \ldots n]$ is kept in a list sorted in non-increasing order and $P_i = \{1, 2, \ldots, p_i\}$ be the set of positions of costs in the list. A purchase of np_i players from ith category is now viewed as a sorted list of np_i distinct positions chosen from P_i. We construct the DAG M_i as a layered DAG (V, E), where each node $v \in V$ contains the information of np_i positions out of p_i available positions of a purchase. As in our previous work [7], in DAG M_i, for each node we store the purchase information as a bit vector $B = \{b_1, b_2, \ldots, b_{p_i}\}$. In the bit vector, $b_j = 1$, $j \in [1 \ldots p_i]$ indicates that the jth player is purchased and $b_j = 0$ says that the player is unsold. Clearly, in the bit vectors corresponding to ith category, there are np_i 1s and $(p_i - np_i)$ 0s. Here also, we draw a directed edge from node v_i to node v_j using the concept of "One Shift" as introduced by Majumder et al. [7].

Like our previous work [7], we can use a max heap H_i to report the desired top-k purchases of np_i players for some category i ($i \in [1 \ldots n]$). Note that, like n DAGs M_1 to M_n, we require n max heaps H_1 to H_n (local max heaps) also. In order to avoid the duplicate insertions in the heap, here also, we use the concept of mandatory one shift [7, 9]. Figure 1 gives an example of the DAG M_i for the case $p_i = 6$ and $np_i = 3$.

From the above discussion, it is obvious that the edges in M_i can be drawn without knowing anything about the actual costs in any particular category of players. The DAGs actually exploit the partial order that inherently exists among the different configurations of the lists of flag bits.

3 Directed Acyclic Graph M_g

The master DAG M_g that we build for n categories, can also be represented as a layered DAG (V, E), where each node $v \in V$ contains a list of n digits, $d = (d_1, d_2, \ldots, d_n)$. Note that, $d_i = k$ denotes the kth best purchase for the ith category, in non-increasing order of total costs, is chosen. A directed edge from node V_i to node V_j denotes that the total cost of purchasing $\sum np_i$ players from all n categories

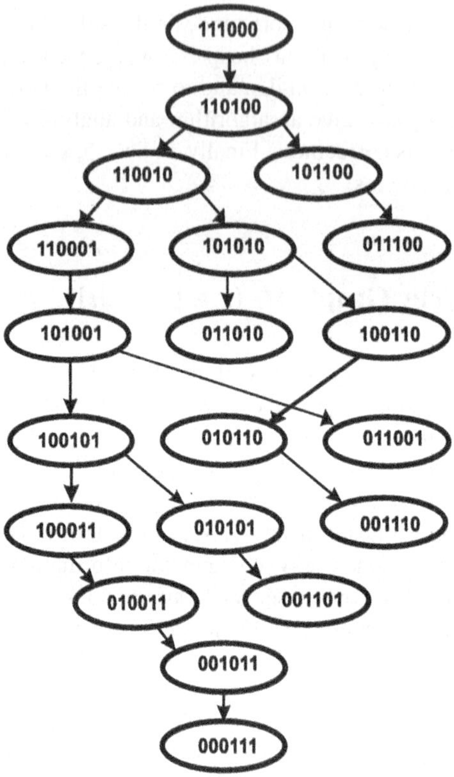

Fig. 1 The DAG M_i for the case $p_i = 6$ and $np_i = 3$

corresponding to node V_i, is better than the total cost of purchasing $\sum np_i$ players from all n categories corresponding to node V_j, and hence V_i will occupy a higher position compared to that of V_j in the *top-k* list of required purchases.

Definition 1 A node V_i is said to be **better than** node V_j, if the total cost of purchasing $\sum np_i$ players from all n categories corresponding to node V_i, is always greater than or equal to the total cost of purchasing $\sum np_i$ players from all n categories of node V_j.

In DAG M_g, the root node has n number of 1 digits, represents the top-1 purchase of $\sum np_i$ players from all n categories. In order to get the rest of the $(k-1)$ best purchases, we need to generate the other nodes of M_g, for which, we consider the relative values of the n number of digits, contained in the nodes.

Definition 2 If an existing node V_i has a certain sequence of n digits, represented by d and if for another node V_j, the sequence of n digits is represented by d', where only one digit of d, gets incremented by 1 and rest of the $(n-1)$ digits are the same, we say that node d' is global one shift of d.

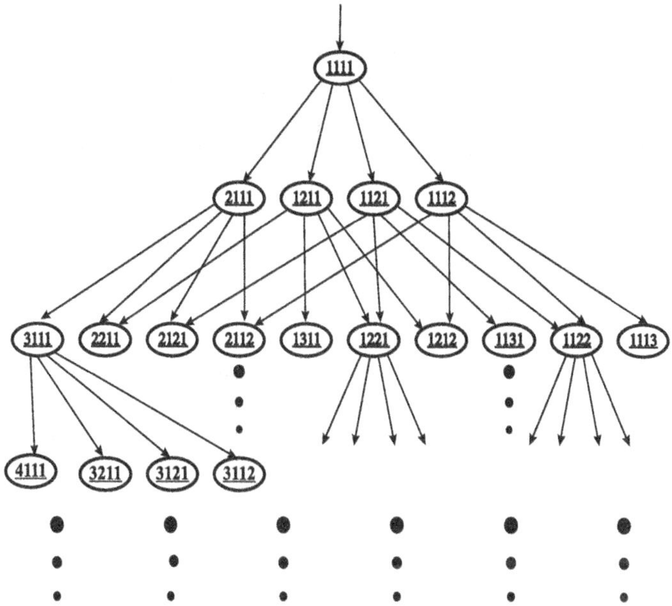

Fig. 2 Directed acyclic graph M_g for $n = 4$

Here also, like the DAG M_i ($i \in [1 \ldots n]$), we draw a directed edge from node V_i to node V_j using the concept of "Global One Shift". See Fig. 2 of DAG M_g for the case $n = 4$.

4 Solution Methodology

In this section, we present an algorithm that describes how we can implicitly construct the n DAGs M_1 to M_n and M_g on the fly (on demand), using n number of min-heaps H_1 to H_n and a global min-heap H_g, to generate the *top-k* purchases from all n categories.

4.1 Implicit Generation of Portions of M_i ($i \in [1 \ldots n]$) and M_g

Initially, for each category i, we construct the root node of all the local DAGs M_i and place the root nodes into the corresponding local heap H_i. We use n number of local heaps in total, one for each category. From the purchase information of the nodes in M_i, we calculate the corresponding costs of the nodes before inserting it in the heap.

The root node of heap H_i now represents the top-1 overall cost of purchasing np_i players from the ith category. Additionally, for each category i we maintain a cost array and we store the top-1 cost of the category i at position 1 of the corresponding cost array, so that in future its top-1 cost can be retrieved from there, if required. Clearly, we have in total n cost arrays. Now, we construct the root node of DAG M_g and insert it into global heap H_g. Note that it has n number of 1s as choice info, indicating the top-1 costs of each category i, $i \in [1 \ldots n]$. Its cost is then calculated as the summation of top-1 costs of all the n categories, retrieved from the n local cost arrays. This cost will be the top-1 cost of purchasing $\sum np_i$ players from all n categories.

Next, we extract the top-1 node from the global heap H_g and insert its child nodes from the DAG M_g into the global heap H_g. Note that M_g has not been explicitly constructed but since we can predict the shifts (global one shifts) very easily, we can easily construct the choice info of each node and also calculate the total costs of each of the child nodes, before insertion into H_g. If before insertion of any child node into H_g, we find the choice info has a digit d for any position p, we inspect the dth position of the cost array of pth category, if we find some cost there we use it, otherwise we calculate the dth best cost for the category p and store it into that cost array. In order to calculate the dth best cost for the category p, we consider the $(d-1)^{\text{th}}$ extraction from the heap H_p and insert its child nodes from the corresponding M_p into the heap H_p after calculating the cost of each child node from its purchase information. Note that here also we do not construct the M_p explicitly but from the knowledge of the shift mechanism, we figure out which child nodes have to be inserted in the corresponding H_p. So basically the whole multi-heap data structure (consisting of n local heaps and 1 global heap) gets created on demand from our implicit understanding of the structures of the DAGs M_i and M_g. The dth extraction from the heap H_p gives us the dth best cost of the category p and it is stored into the dth position of the pth cost array, so that it can be used to calculate the cost of all those global heap nodes that has a digit d at position p. We can now retrieve the present root node of H_g that gives the next best cost of purchase from all the categories. We continue the above steps after the extraction of each node from the global heap H_g till we get the required top-k purchases. After the kth extraction, our algorithm stops.

We repeat here that we construct the nodes of different levels of the DAGs (meta-data structure) M_g and M_is on demand, as and when required, keeping their structure in mind as described in Sect. 2 and Sect. 3. We give the major steps of our algorithm in Algorithm 1.

4.2 Proof of Correctness of Algortihm 1

Theorem 1 *Algortihm 1 reports k teams* $(T_1, T_2, \ldots, T_k)$ *in non-increasing order of total cost.*

Algorithm 1 Top-k-Team-Prob(n: # categories, k: number, P: category wise available players, C: cost vectors, Np: category wise required players)

for each category i of players, sort the cost vector $C_i=\{c_{i,1}, c_{i,2}, \ldots, c_{i,p_i}\}$ in non-increasing order;
//construction of root nodes of n DAGs M_1 to M_n
//store top-1 costs into cost arrays
for $i = 1$ to n **do**
  Construct root node T_i of DAG M_i as bit vector of p_i bits, where first np_i bits are 1s and rests are 0s;
  Calculate cost of T_i;
  Insert T_i into local heap H_i; // T_i holds purchase info as a bit vector and cost of the purchase from a single category
  for $p = 1$ to k initialize cost[i][p] = 0;
  Extract root object R_i from local heap H_i;
  cost[i][1] = R_i.cost;
end for
//construct root node T of DAG M_g
Construct the root node T of M_g, containing n number of 1s;
Calculate $T.cost = \sum_{i}^{n} cost[i][1]$;
Insert T into global heap H_g; //T holds n digits choice info and cost of total purchase from n categories
Extract root object R from global heap H_g;
Print R.cost; //top-1 cost of the total purchase from n categories
for $s = 2$ to k **do**
  Construct listChildGlobal(R); //get child nodes of R from construction method of M_g
  for each node I_g in listChildGlobal(R) **do**
    I_g.cost = 0;
    for $i = 1$ to n **do**
      d=I_g.digit[i]; //I_g.digit[i] is the ith digit of node I_g
      if cost[i][d] = 0 **then**
        Construct listChildLocal(R_i);
        //get child nodes from construction method of M_i, if failed to get next best cost of the category i from cost array
        for each node I_l in listChildLocal(R_i) calculate cost of I_l and then insert I_l into local heap H_i;
        //I_l holds purchase info as a bit vector and cost of the purchase from a single category
        Extract root object R_i from local heap H_i;
        Reheap local heap H_i;
        //store next best cost of the category i into cost array
        cost[i][d]=$R_i.cost$;
      end if
      //calculate cost of the node of DAG M_g
      I_g.cost=I_g.cost + cost[i][d];
    end for
    insert I_g into global heap H_g; //I_g holds n digits choice info and cost of total purchase from n categories
  end for
  Extract root object R from global heap H_g;
  Reheap global heap H_g;
  Print R.cost;//next best cost of the total procurement
end for

Proof Assume for contradiction that for some $p, l \in [2 \ldots k]$ the team T_p is reported before the team T_l but $cost(T_l) > cost(T_p)$.

Let v_l and v_p be the nodes in global DAG M_g that report the teams T_l and T_p, respectively. Now, consider three possible cases for the nodes v_p and v_l:

Case 1, a directed edge from v_l to v_p: From the concept of "Global One Shift" in the DAG M_g, it is obvious that no child node in the DAG can be reported before its parent node. It contradicts our assumption that the team T_p is reported before the team T_l.

Case 2, a directed edge from v_p to v_l: From the concept of "Global One Shift" in the DAG M_g, it is obvious that the node v_p is **better than** the node v_l. It contradicts our assumption that $cost(T_l) > cost(T_p)$.

Case 3, no edge between v_l and v_p: In this case, the reporting order is decided by the global max heap H_g where key is considered as the total cost of corresponding team structure. We assume that the team T_p is reported before the team T_l. During extraction of the node v_p (node for team T_p) from the root of heap H_g, the node v_l (node for team T_l) may be in the heap or may not. It is obvious from the insertion logic into the heap that the node v_p is **better than** than the node v_l. It contradicts our assumption that $cost(T_l) > cost(T_p)$.

Since all possible cases lead to contradictions, our original assumption must be invalid. This proves the theorem.

4.3 Computational Complexity

(a) Space complexity: Between two consecutive extractions from the global heap H_g, we need to insert at most n nodes into the heap H_g from the DAG M_g. Hence, total number of entries into H_g is bounded by $1 + (k - 1)n$, i.e., $O(kn)$ and each node contains a list of n digits.

For each node of global heap, if it contains a digit d for a category i, for which we fail to get the cost from the cost array of i, we need to extend (implicitly though) the corresponding DAG M_i for the category i and we have to insert the child nodes of the previous root node into the local heap H_i from the DAG M_i, where a node in the DAG M_i has at most 2 [9] children. However, at each step of generating the next best purchase, we may have to handle this pathological situation at most n times. From the above discussion, it is clear that at most $2 \times n$ new nodes will be inserted, spreading overall n local heaps. Initially, each local heap H_i holds root node of DAGs M_i, $i \in [1 \ldots n]$. Hence, for the generation of k best procurements, total number of entries into the local heaps will be $O(n + kn)$. Instead of storing the bit vector in each node of local DAG M_i, we can store it as a number also. So, each node of local DAG M_i is represented by $O(1)$ space. Similarly, we can store n digits of each node of global DAG M_g as a number. So, here also, in case of M_g each node is represented by $O(1)$ space. Hence the overall space requirement of the Algorithm 1, is $O(n + kn)$.

(b) Time complexity: For *top*-1 assignment, global heap H_g and each local heap H_i $(i \in [1 \ldots n])$ needs to be constructed with one node, taking $O(1)$ time. Since, we have n local heaps H_1 to H_n, so, overall time requirement for *top*-1 assignment is, $T(1) = O(n)$. At each step of generating the next best purchase, at most n new nodes need to be inserted into the global heap H_g and at most 2 new nodes need to be inserted in each local heap. When, we are considering the pth best purchase, global heap already holds $O(pn)$ nodes, so inserting n new nodes will take $O(n \log pn)$ time. Now each of these n insertions may trigger the insertion of $O(1)$ new nodes into some local heap, when the total number of nodes present in all the local heaps is $O(pn)$. If we consider that the nodes will be more or less evenly distributed over the n local heaps, inserting these new $O(1)$ number of nodes will take $O(\log p)$ time. The worst-case can be however $O(\log pn)$, but a probabilistic analysis will always give a much better bound. So, worst-case time requirement for the pth extraction from global heap will be $O(n \log pn + n \log pn) = O(n \log pn)$.

Overall time requirement for $(k-1)$ best procurements after the 1st procurement is $\sum_{p=1}^{k-1} O(n \log pn) = n \sum_{p=1}^{k-1} O(\log p) + n \log n \sum_{p=1}^{k-1} O(1) = O(nk \log k + nk \log n) = O(nk \log nk)$. So, the overall time requirement of Algorithm 1 is, $T(k) = O(nk \log nk + n) = O(nk \log nk)$.

5 Conclusions and Future Research

In this paper, we propose a novel solution for generating the *top-k* purchases of $\sum np_i$ players out of n available categories in an unconstrained environment. Our technique is based on constructing $n+1$ DAGs. Here, we consider that a player can only be in a single category. Considering a player in more than one category like a player may be both a bowler as well as a batsman, and adding different constraints to the problem and solving them using similar methods remain possible directions for future research.

References

1. Afshani, P., Brodal, G.S., Zeh, N.: Ordered and unordered top-*k* range reporting in large data sets. In: ACM-SIAM SODA, pp. 390–400 (2011)
2. Agarwal, S., Chaudhuri, S., Das, G., Gionis, A.: Automated ranking of database query results. In: Biennial Conference on Innovative Data Systems Research, pp. 1–12 (2003)
3. Buckley, C., Salton, G., Allan, J.: The effect of adding relevance information in a relevance feedback environment. In: ACM SIGIR, pp. 292–300 (1994)
4. Chaudhuri, S., Gravano, L., Marian, A.: Optimizing top-*k* selection queries over multimedia repositories. IEEE TKDE **16**(8), 992–1009 (2004)
5. Karpinski, M., Nekrich, Y.: Top-*k* color queries for document retrieval. In: ACM-SIAM SODA, pp. 401–411 (2011)
6. Marian, A., Bruno, N., Gravano, L.: Evaluating *top-k* queries over web-accessible databases. ACM Trans. Database Syst. **29**(2), 319–362 (2004)

7. Majumder, S., Sanyal, B., Gupta, P., Sinha, S., Pande, S., Hon, W.K.: Top-*K* query retrieval of combinations with sum-of-subsets ranking. In: COCOA, pp. 490–505 (2014)
8. Rahul, S., Gupta, P., Janardan, R., Rajan, K.S.: Efficient top-*k* queries for orthogonal ranges. In: WALCOM, pp. 110–121 (2011)
9. Sanyal, B., Majumder, S., Hon, W.K.: Efficient generation of top-k procurements in a multi-item auction In: WALCOM, pp. 181–193 (2014)
10. Suzuki, T., Takasu, A., Adachi, J.: Top-*k* query processing for combinatorial objects using Euclidean distance. In: IDEAS, pp. 209–213 (2011)

A Deep Convolutional Neural Network Approach to Rice Grain Purity Analysis

Mushahidul Islam Shamim, Biprodip Pal, Anhad Singh Arora and Mahbubul Amin Pial

Abstract Traditional rice grain classification is costly, time-consuming and requires sophisticated human expertise. Besides, computer vision based methods are still based on predefined morphological features that are often not transferable across different types of grains. In this paper, the feasibility of automated feature extraction for rice grain purity analysis has been demonstrated using a Convolutional Neural Network (CNN) based deep learning approach. Due to the lack of benchmark datasets, the paper defines a dataset with technician-verified, labeled images of different types of rice grains with a background of uniform illumination. Moreover, the paper also proposes the architecture of a CNN for automated rice grain feature extraction. The performance of a classifier trained on these features is compared to classifiers trained on morphological features used by modern computer vision approaches. It is found that in this dataset, the proposed method can detect the presence of native and foreign grains in a given sample of rice grains with superior accuracy which is at least 25% better in case of a multiclass classification scenario.

Keywords Convolutional neural network · Rice grain · Morphological features · Deep learning · Computer vision

---

M. I. Shamim · B. Pal (✉)
Rajshahi University of Engineering & Technology, Rajshahi, Bangladesh
e-mail: biprodip.cse@gmail.com

M. I. Shamim
e-mail: mushahidshamim@gmail.com

A. S. Arora
Foods One, Ludhiana, India
e-mail: anhadarora@nyu.edu

M. A. Pial
Bangladesh University of Engineering and Technology, Dhaka, Bangladesh
e-mail: aminpial@linearaxon.com

H. Das et al. (eds.), *Progress in Computing, Analytics and Networking*,
Advances in Intelligent Systems and Computing 1119,
https://doi.org/10.1007/978-981-15-2414-1_19

1 Introduction

The selection of grains for analysis is typically carried out in two steps of the nine-spoon method to pick 9 spoonfuls of rice from different places and Quadrant Method to analyse samples that are divided into different quadrants. The final readings are so sensitive that even variations of ~4% of the same characteristic are considered vastly different. At times, even two technicians with different levels of expertise may generate a different report for the same sample. It is an immensely time-consuming and expensive process. Machine learning has been successfully used in several applications like healthcare or intrusion detections [1, 2]. Machine vision techniques have started to be utilized very often in analysing rice grains by many researchers. Qing et al. [3] developed an automatic inspection and image processing unit to evaluate rice chalkiness and shape using a camera and a lightbox. They calculated the chalkiness of grain using an automatic multi-threshold method based on maximum entropy, and also estimated rice grain shape parameters by finding the Minimum Enclosing Rectangle (MER). Their objective analysis methods were strongly correlated with manual estimations by trained technicians [3]. Devi et al. used Canny Edge Detection to extract morphological features to classify grains on their predicted lengths [4]. Their method was tested on four different samples of rice grains and the accuracy of the predictions was within 2% of manual calculations [4]. In a similar case, the research considered the problem of the touching of grains in image preprocessing by using the Shrinkage algorithm [5]. In an attempt to replace the industry-standard Satake RSQI10A grain scanner software, Ali et al. enumerated similar morphological features of rice grains in their low-cost solution [6]. The results from such projects paved the way for an objective, appearance-based quantifier that would be superior and more reliable than manual inspection. Although several works used other features like shape, texture, colour features to generate optimal morphological features for grains and achieved good accuracy, these features are also depended on rice specific expertise knowledge [7–9] and are not often transferable to different rice categories. Moreover, no publicly available database exists that has technician-verified, classified image samples of different categories of rice grains with uniform illumination background. Because of the intrinsic limitations of classical and previous automated methods, this work proposed a deep learning based system to, demonstrate the potential of Convolutional Neural Networks (CNNs) for automated transferable feature extraction and classification of rice grains. Apart from the methodology, the paper also defines a dataset with technician-verified, labeled images described in the data collection section.

2 Materials and Methods

2.1 Data Collection

This paper defines a rice grain dataset with the help of specialists from rice distributors and retailers in India that is available for research.[1] After obtaining a significant random sample using the *Nine-Spoons* and *Quadrant Methods*, the rice grains were scanned using a generic Flat-Bed Scanner (FBS). The images were all scanned with a black background at 300 ppi as depicted in Fig. 1. The dataset contains the rice varieties most commonly found in North India. The total number of available images for each grain category and the total individual grain samples that were extracted from these images are summarized in Table 1. Initially, the dimension of every image was 3507 × 2550 pixels, with around 100–200 individual rice grains in each image as depicted in Fig. 1. The grains were manually placed far apart from each other so as to avoid the algorithm from considering multiple grains as a single grain. All samples were scanned using a similar arrangement.

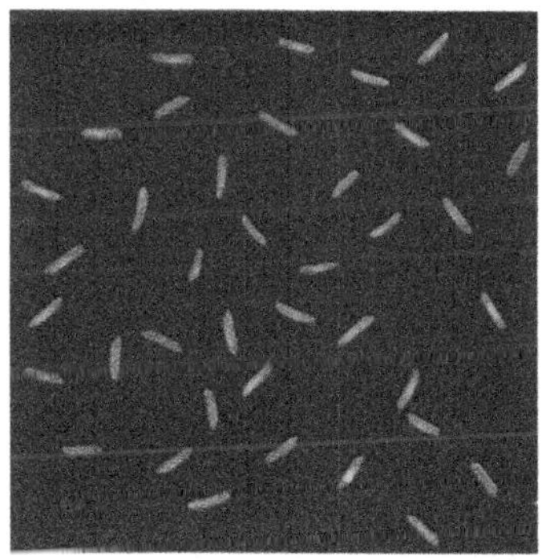

Fig. 1 Scanned image example

Table 1 Summary samples in dataset

Grain type	Total images	Grain samples
PR	192	15,277
1121	155	17,345
Sharbati	51	13,596
1401	52	12,559
1509	105	17,262
Sona Masoori	24	4,024
RH-10	17	3,852
Sugandha	59	11,933

[1]https://github.com/Mushahid2521/Rice-Grain-Purity-Analysis-Using-Deep-Learning.

2.2 Preprocessing

The scanned rice grain image (SI), was first segmented into individual grain images. Canny Edge Detection algorithm was employed to detect the grain boundaries [10]. SI was first converted into a grayscale image. Noises were reduced using *Gaussian Smoothing*. *Sobel kernel* was used to find intensity gradients in the horizontal (G_x) and in the vertical direction (G_y) of the image. Edge gradient (G) and each pixel direction (ϴ) was computed according to Eqs. 1 and 2 followed by *Non-Maximum Suppression* to get the binary image with thin edges.

$$\mathrm{G} = \sqrt{G_x^2 + G_y^2} \tag{1}$$

$$\ominus = \frac{G_y}{G_x} \tag{2}$$

Finally, *Hysteresis Thresholding* performed thresholding twice using an upper threshold (maxVal = 100) and lower threshold (minVal = 100) based on validation.

The next step in the process is to find and grab contours (boundary of a grain) from the output of the previous step. On a white colour mask, contours of black colour were drawn using the tuples that we grabbed in the previous step. This made the image binary with the colours black (0) in place of the grains and white (255) as the background. The mask was then used to extract the grains from the SI. The contour values were also used to generate the Minimum Enclosing Rectangles (MER) around the grains and crop them into individual grain images (CI). The progress of this process is depicted in Fig. 2 using cropped images for visual clarity. When the grains were cropped from the SI, the CI were of different sizes. To feed into a predictive model, these images were resized to the same length and width by padding the CI of the grains with black pixels (0). Considering different height and width of the grain images, we have used a size of 128 × 128 pixels as the Standard Dimensions for individual grain images. Since fewer samples and a lot of model parameters often lead to overfitted deep learning models, data augmentation was performed. Original images were rotated from different angles to obtain around 24,000 individual grains for every class.

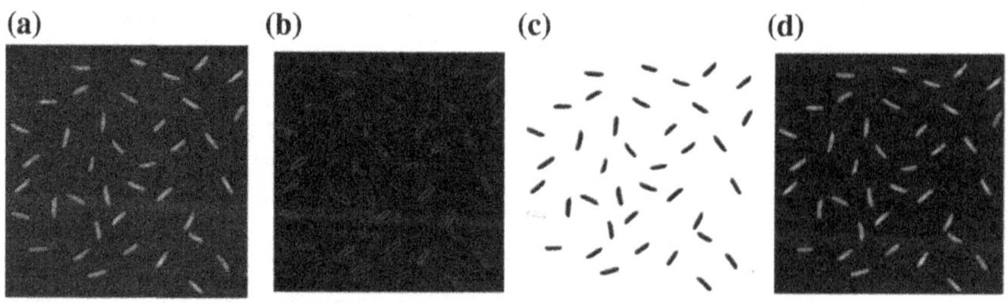

Fig. 2 Edge detection for individual grains. **a** Input scanned image (SI). **b** Canny edge detection. **c** Grain extraction mask. **d** Final output

2.3 Morphological Feature Extraction

Quantifiers that represent size namely: the length (l_r) and width (w_r) of Minimum Enclosing Rectangle (MER), the length of the major axis (l_{a1}) and minor axis (l_{a2}), length of the grain considering it an ellipse (l_e), the perimeter (p), the area of each grain (A), the area of a shape (A_S) and the area of the MER (A_R) were selected. These basic features were combined into six complex features (Eqs. 3–8) that contributed significantly to rice grain classification and were subjected to normalization before use [9].

$$Area\text{ by }Perimeter\text{ ratio} = \frac{A}{p} \tag{3}$$

$$\text{The ratio of }Area\text{ to the total of Area and Perimeter} = \frac{A}{A + p} \tag{4}$$

$$\text{Aspect Ratio: Ratio of }Major\ Axis\text{ length to }Minor\ Axis\text{ length} = \frac{l_{a1}}{l_{a2}} \tag{5}$$

$$\text{Rectangularity} = \frac{A_S}{A_R} \tag{6}$$

$$\text{Equivalent Diameter} = \sqrt{\frac{4A}{\pi}} \tag{7}$$

$$\text{Shape Factor 3} = \frac{A}{l_{a1} \times l_{a2}} \tag{8}$$

2.4 Convolutional Neural Network Architecture

The input of the proposed CNN architecture is a 128 by 128 image patch. The kernel size in each layer was chosen to be minimal. The samples of CI for each type of grain were passed through four pairs of Convolution and Max-Pooling layers. In the first two convolution layers, a 5×5 filter size was used, and in the next two layers, a 3×3 filter size was used. The number of filters in the Convolution layers is, respectively 6, 16, 64 and 128 as depicted in Fig. 3. In the Max-Pooling layers, a 2×2 pool size was used. Due to faster convergence and non-saturating values in gradient descent compared to sigmoid functions, Rectified Linear Unit (ReLU), computing $f(z) = \max(0, z)$, on corresponding weighted sum z, was used as an activation function for all the layers. A softmax function on corresponding input x computing $f(x) = (1 + e^{-x})^{-1}$ was used in the last *dense* layer for classification, computing the class probabilities. A 40% *dropout* was used between the last two dense layers to reduce *overfitting*. The loss function used was Categorical Cross-Entropy (CE) as given in Eq. 9.

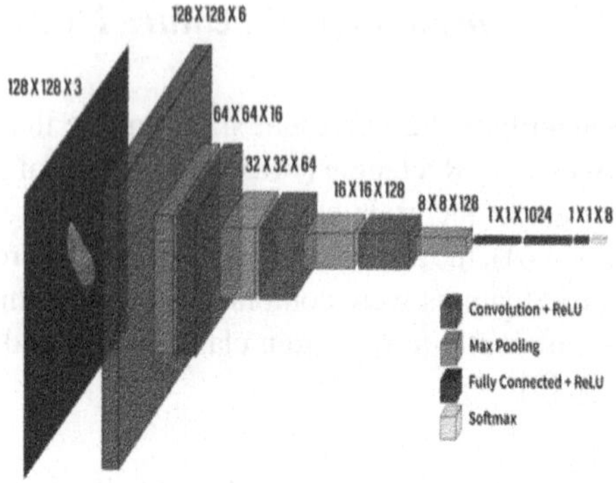

Fig. 3 Convolutional neural network architecture

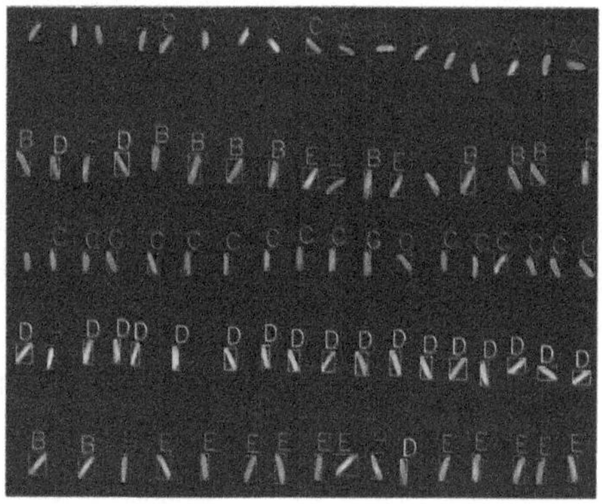

Fig. 4 Program's prediction

$$\mathrm{CE} = -\sum_{i}^{C} t_i \log\log f(s)_i \tag{9}$$

where, for C classes, t_i is the ground truth and *f(s)* is the corresponding activation. The model was trained with *Adam* optimizers because of its faster convergence. The network architecture is depicted in Fig. 3. The hyperparameter selection was performed based on cross-validation as discussed in the result section. Figure 4 shows an example with different types of grains arranged in different rows and corresponding identification as drawn by the program. The purity of the rice sample is calculated as the percentage of native grains in the sample by count.

3 Results and Discussion

To compare the outcome of the proposed approach with a traditional approach, morphological features of samples were classified using classifiers kNN [11], Decision Tree (DT) [11] and Artificial Neural Network (ANN) [12]. For the parameter selection process, 900 samples were selected and used in a threefold cross-validation fashion. The value of k was selected as 3 for kNN based on best average validation accuracy of 39% as depicted in Fig. 5. Similarly, for ANN classification, the number of nodes was selected as 140 with the best average accuracy of 49% as depicted in Fig. 6.

Structural parameter selection in the CNN setup, is enlisted in Table 2 for the selection of total convolutional layers and dropout. With 3 layers the network performs better for training data compared to others (4 or 5 layers) which comes with a cost of lower validation accuracy. It indicates possible overfitting as it has fewer parameters compared to 4 or 5 convolutional layer setup. A 5 convolutional layer setup shows poor performance compared to 4 layer setup in both training and validation data. Thus 4 convolutional layers were used for further experiment. Without dropout in fully connected layers, the selected model performs better for training data since dropout is used to reduce overfitting, but without dropout, the model achieves the same accuracy with lower standard deviation for validation data. Thus 4 convolutional layers with dropout are proposed for this experiment.

In the same experimental setup as the validation environment, with 3000 test samples for each category and rest as train data, average classification accuracy

Fig. 5 kNN parameter (k) selection

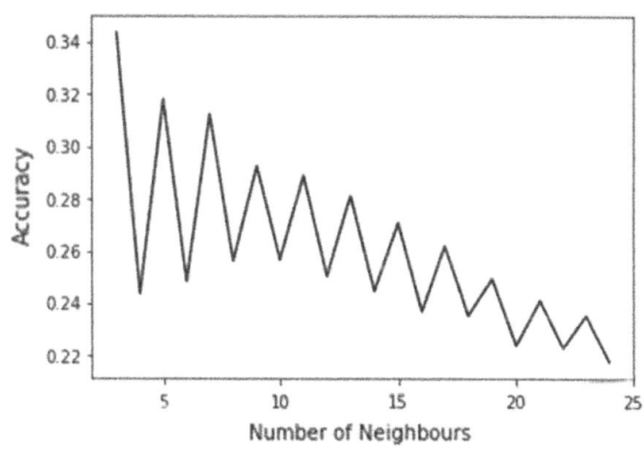

Fig. 6 Hidden layer nodes selection for ANN

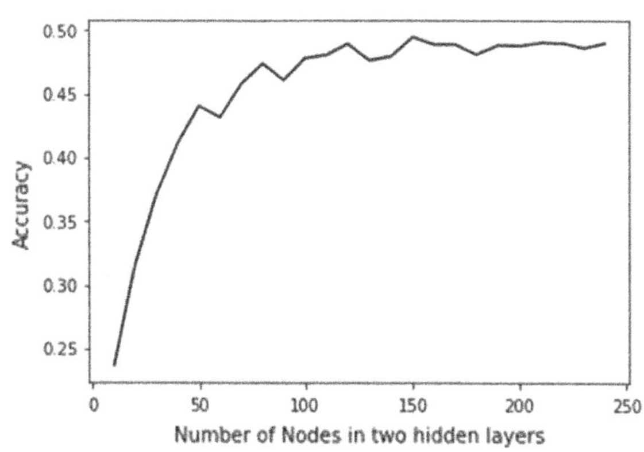

Table 2 Average performance for parameter selection of CNN

Experiments	Training accuracy	Validation accuracy
Without dropout (4 convolution layers)	64 ± 2.2	42 ± 1.6
With dropout (4 convolution layers)	62 ± 6.5	42 ± 0.9
Convolutional layers 5	48 ± 8.4	42 ± 0.4
Convolutional layers 4	62 ± 6.5	42 ± 0.9
Convolutional layers 3	78 ± 2.6	40 ± 0.5

of the kNN, Decision Tree and ANN classifier on manual morphological features are listed in Table 3. These experiments are based on the selected parameters as discussed previously. While the latest machine vision approaches often using ANN with morphological features [8, 9], the accuracy of those features for this dataset has been found to be poor. Besides classifiers like decision tree and kNN also depicts poor performance on those features. On the other hand, proposed convolutional features with ANN classifiers provide better accuracy of 77% on the same test data.

The confusion matrices for test data classification are given in Figs. 7. From the comparison of the confusion matrix, it can be concluded that some grain types could be classified with high accuracy using the morphological features but some would not. In these cases, CNN works considerably well. The further experiment reveals that if the model is formulated into binary-classification, then the accuracy increases to 81% for all grain types. CNN-based extracted features are highly transferable. It extracts specific as well as generic features in the intermediate layers [13]. These features learned from one type of grain samples can be tuned to identify a different category of grain samples. Thus the performance of CNN is better across different categories of rice grains. This is clear from the confusion matrices that CNN classifies at least around 1760 samples of every category correctly while the performance of others may drop significantly up to 430 for particular categories. In Table 4, the

Table 3 Rice grain classification accuracy

Experiments	Average testing accuracy (%)
DT on morphological features	52
kNN on morphological features	48
ANN on morphological features [6, 7]	51
Proposed approach (convolutional feature + ANN)	77

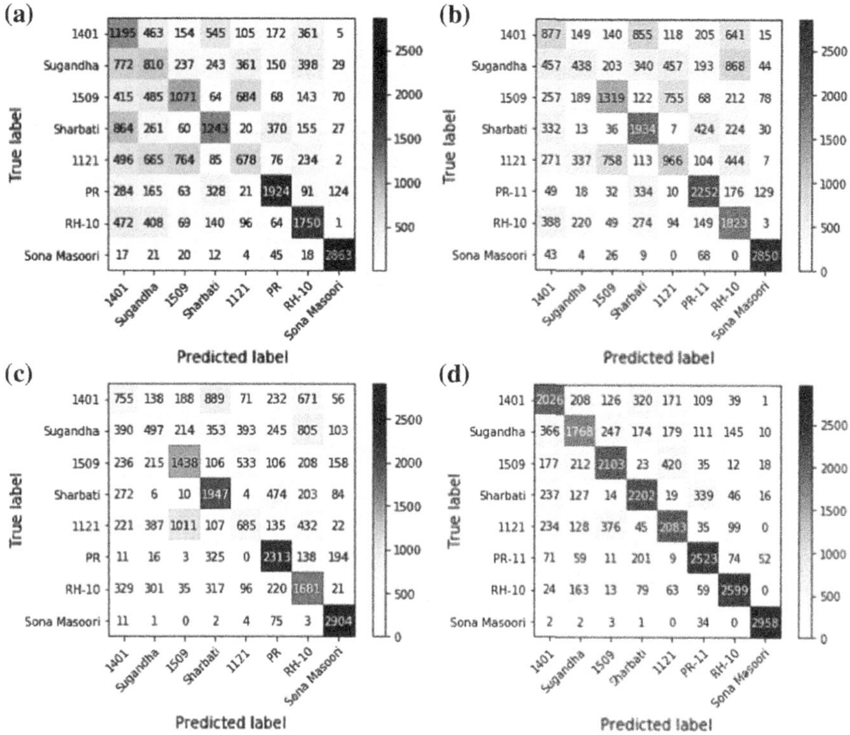

Fig. 7 **a** Confusion matrix for KNN. **b** Confusion matrix for decision tree. **c** Confusion matrix for neural network. **d** Confusion matrix for CNN

precision, recall and F1 scores for different classifiers are enumerated for comparison in multi-class classification scenario which also elucidates the success of CNN.

4 Conclusion

This paper proposed a CNN-based rice grain classification approach that is more accurate compared to morphological features used by earlier researchers to identify foreign grains in a given sample of rice grains. Compared to the manual human expertise based approach as discussed in the introduction, this proposed method is consistent and fast. Moreover, an expert-defined dataset has been introduced for experiment and a neural network based approach was suggested on automated deep convolutional features due to its superior accuracy. In the test dataset, the total classification accuracy was 77% which increases up to 82% in binary classification settings which is better than some modern approaches. Apart from the ad-mixture, which represents the impurity of a rice sample, there are other factors such as the percentage of Pin-Broken grains, Damaged and Discoloured (DDC) grains, immature grains,

Table 4 An objective comparison of different classifiers

Type of grains	KNN			Decision tree			ANN			CNN		
	Precision	Recall	F1	Precision	Recall	F1	Precision	Recall	F1	Precision	Recall	F1
PR	0.67	0.64	0.66	0.65	0.75	0.7	0.61	0.77	0.68	0.79	0.83	0.81
1121	0.34	0.23	0.27	0.40	0.32	0.36	0.38	0.23	0.29	0.67	0.74	0.71
Sharbati	0.47	0.41	0.44	0.49	0.64	0.55	0.48	0.65	0.55	0.73	0.77	0.75
1401	0.26	0.40	0.32	0.33	0.29	0.31	0.34	0.25	0.29	0.67	0.68	0.67
1509	0.44	0.36	0.39	0.52	0.44	0.47	0.50	0.48	0.49	0.79	0.62	0.69
Sona Masoori	0.92	0.95	0.94	0.90	0.95	0.93	0.82	0.97	0.89	0.98	0.99	0.98
RH-10	0.56	0.58	0.57	0.42	0.61	0.49	0.41	0.56	0.47	0.86	0.9	0.88
Sugandha	0.25	0.27	0.26	0.32	0.15	0.20	0.32	0.17	0.22	0.67	0.64	0.66

and chalky grains. These factors together influence the sale and purchase of a particular batch of milled rice. Extended research may consider identifying the other 4 important factors and building low-cost, full-service software that is time efficient and objective.

References

1. Das, H., Naik, B., Behera, H.S.: Classification of diabetes mellitus disease (DMD): a data mining (DM) approach. In: Progress in Computing, Analytics and Networking, pp. 539–549. Springer, Singapore (2018)
2. Sahani, R., Rout, C., Badajena, J.C., Jena, A.K., Das, H.: Classification of intrusion detection using data mining techniques. In: Progress in Computing, Analytics and Networking, pp. 753–764. Springer, Singapore (2018)
3. Qing, Y., Jianhua, C., Zexin, G., Chengxiao, S., Zhiwei, Z.: Inspection of rice appearance quality using machine vision. In: 2010 Second WRI Global Congress on Intelligent Systems, vol. 4, pp. 274–279 (2009). https://doi.org/10.1109/gcis.2009.91
4. Gayathri Devi, T., Neelamegam, P., Sudha, S.: Machine vision based quality analysis of rice grains, pp. 1052–1055 (2017). https://doi.org/10.1109/icpcsi.2017.8391871
5. Mahale, B., Korde, S.: Rice quality analysis using image processing techniques. In: International Conference for Convergence for Technology, pp. 1–5. IEEE (2014). https://doi.org/10.1109/i2ct.2014.7092300
6. Ali, S.F., Jamil, H., Jamil, R., Torij, I., Naz, S.: Low cost solution for rice quality analysis using morphological parameters and its comparison with standard measurements. In: 2017 International Multi-topic Conference (INMIC), pp. 1–6. IEEE, (2017). https://doi.org/10.1109/inmic.2017.8289475
7. Zareiforoush, H., Minaei, S., Alizadeh, M.R., Banakar, A.: Qualitative classification of milled rice grains using computer vision and metaheuristic techniques. J. Food Sci. Technol. **53**(1), 118–131 (2016)
8. Kolkure, V.S., Shaikh, B.N.: Identification and quality testing of rice grains using image processing and neural network. Int. J. Recent Trends Eng. Res. **3**(01), 130–135 (2017)
9. Mousavirad, S.J., Tab, F.A., Mollazade, K.: Design of an expert system for rice kernel identification using optimal morphological features and back propagation neural network. Int. J. Appl. Inf. Syst. **3**(2), 33–37 (2012)
10. Canny, J.: A computational approach to edge detection. IEEE Trans. Pattern Anal. Mach. Intell. **PAMI-8**(6), 679–698 (1986). https://doi.org/10.1109/tpami.1986.4767851
11. Kirkos, E., Charalambos, S., Yannis, M.: Data mining techniques for the detection of fraudulent financial statements. Expert Syst. Appl. **32**(4), 995–1003 (2007)
12. Pal, B., Hasan, M.A.M.: Neural network & genetic algorithm based approach to network intrusion detection & comparative analysis of performance. In: 2012 15th International Conference on Computer and Information Technology (ICCIT), pp. 150–154. IEEE (2012)
13. Pal, B., Ahmed, B.: A deep domain adaption approach for object recognition using Multiple Model Consistency analysis. In : 2016 9th International Conference on Electrical and Computer Engineering (ICECE), pp. 562–565. IEEE (2016)

Design a Microstrip Patch Antenna for Multi-Service Purposes

Afrin Nusrat, Smita Parija, Shubham Kumar, Sobika Singh and Subham Singh

Abstract This paper presents the design, simulation and optimization of different microstrip patch antenna for multiservice purposes. Here we have designed four different types of antennas—E-shaped patch antenna using the substrate Rogers RT/Duroid whose resonating frequency is 2.5 GHz and bandwidth is 500 MHz. Secondly, we have designed a U-shaped patch antenna using the substrate RT/Duroid (2.2) whose resonating frequency is 2.4 GHz and bandwidth is 150 MHz. These two types of antenna are single band antennas and we have an octagonal–rectangular patch antenna using the substrate FR4_epoxy(4.4) which operates at two resonating frequency 8.5 GHz and 15.5 GHz and bandwidth is 550 MHz and 650 MHz, respectively. A dual-band U-shaped patch antenna is designed in succession of single-band U-shaped antenna Then we have compared the parameters like bandwidth, return loss and resonating frequency of 4 types of antenna to find out the best among all and their different applications.

Keywords Return loss · Single band · Optimization cloud computing

A. Nusrat · S. Parija (✉) · S. Kumar · S. Singh · S. Singh
Department of Electronics, CV Raman Engineering College, Bhubaneswar, Odisha, India
e-mail: smita.parija@gmail.com

A. Nusrat
e-mail: afrinnusrat13@gmail.com

S. Kumar
e-mail: shubham1502050@gmail.com

S. Singh
e-mail: sobika124@gmail.com

S. Singh
e-mail: ssubhssingh@gmail.com

H. Das et al. (eds.), *Progress in Computing, Analytics and Networking*,
Advances in Intelligent Systems and Computing 1119,
https://doi.org/10.1007/978-981-15-2414-1_20

1 Introduction

Today, communication plays an important role in this digital world. To transreceive the signal, the antenna is needed. In this nanotech world, microstrip patch antenna has the advantage of low profile, ease of fabrication, low return losses, but suffers a disadvantage of narrow bandwidth. In the hunt for high efficiency and wide bandwidth, many techniques have been proposed in the past three decades. The most common technique used is by cutting slots on the patch in different shapes. The most common shape is E-shaped, U-shaped, and octagonal shape.

This paper proposed four different shapes and parameters have been composed to settle for the better one. The octagonal–rectangular shape patch antenna with 29% enhanced bandwidth is designed to cover both 8.5 and 15.5 GHz frequencies, and have a return loss of −30 dB [1]. The U-shaped patch antenna having one U-slot gives the single resonating frequencies with 10% enhanced bandwidth and by adding another U-slot, one more resonating frequency occurs, with better efficiency. The proposed configurations were designed and simulated used HFSS Software [2].

E-shaped patch antenna configuration for the reduction of antenna size, bandwidth enhancement [3]. This paper compares 4 different types of patch antenna for multiservice purposes and we have compared their bandwidth to find out the best one to be used for communication. E-shaped and U-shaped patch antenna resonating at frequencies of 2.5 and 2.4 GHz for wireless communication, and an orthogonal–rectangular-shaped patch antenna covers a wide band of 2.5–18 GHz which resonates at 2 different frequencies of 8.15 and 15 GHz (dual band). Another rectangular patch antenna resonating at frequency 6.8 GHz for various purposes [4].

2 Antenna Performance

2.1 E-Shaped Patch Antenna

The antenna geometry of E-shaped having wideband and multiband operation. The configuration of antenna size bandwidth enhancement and multiple resonances has proposed by different papers [5]. It is typically used for a high data rate that leads to enhanced bandwidth, according to the requirement of the antenna. The substrate used is RT/Duroid. These designed resonates at a frequency of 2.4 GHz and bandwidth is 500 MHz [6, 7] (Fig. 1 and Table 1).

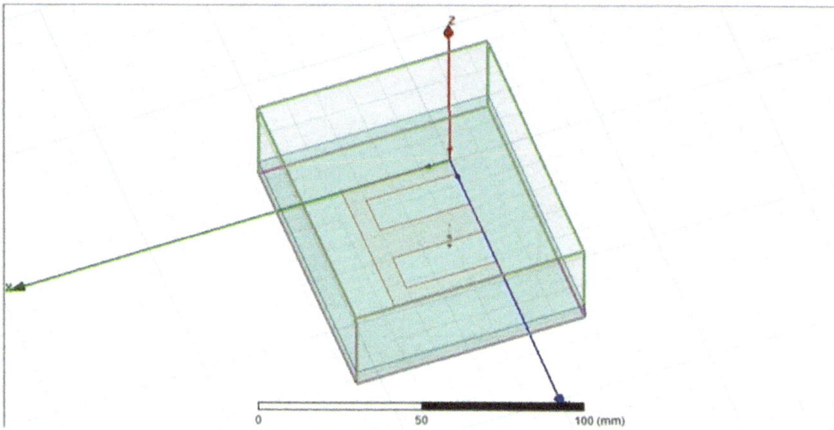

Fig. 1 3D geometry of E-shaped patch antenna

Table 1 Dimensions of E-shaped patch antenna

Width of substrate	Length of substrate	Height of substrate	Dielectric constant ($\in r$)
76 mm	88 mm	6.7 mm	2.2

3 Simulation Result

3.1 U-Shaped Patch Antenna

The U-shaped patch maintains a wide range of impedance bandwidth characterstic. The substrate used is RT/Duroid. The feeding techniques to be used are probe feeding. The impedance bandwidth is 150 MHz, resonating at a frequency of 2.4 GHz [8] (Figs. 2, 3 and Table 2).

3.2 Simulation and Result of Return Loss of U-Shaped Patch Antenna

See Fig. 4.

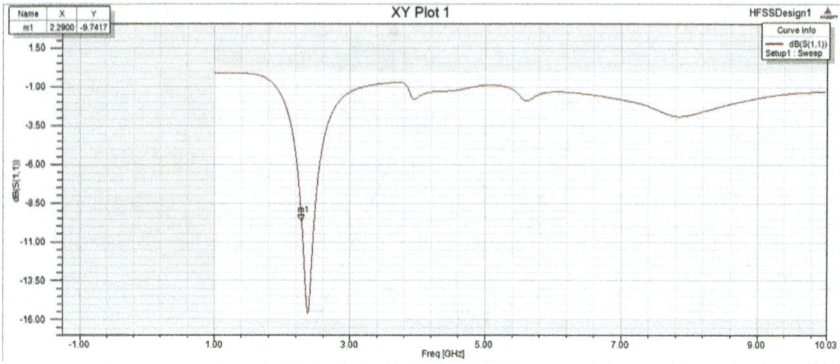

Fig. 2 Plot of return loss E-shaped patch antenna

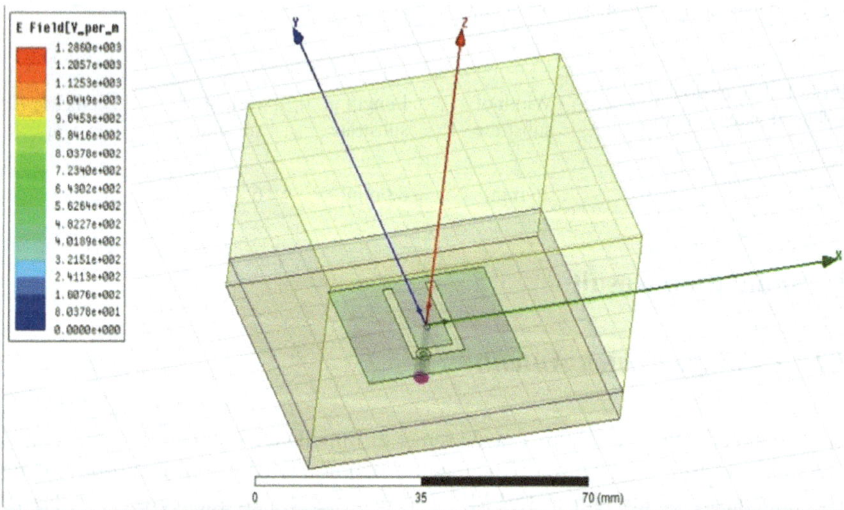

Fig. 3 3D geometry of U-shaped patch antenna

Table 2 Dimension of U-shaped antenna

Width of substrate	Length of substrate	Height of substrate	Dielectric constant ($\in r$)
76 mm	88 mm	6.7 mm	2.2

3.3 Octagonal and Rectangular Patch Antenna

The geometry of octagonal–rectangular patch antenna covers a wide range of frequency (2.5–18 GHz), it is also called a wideband antenna. The substrate used in this

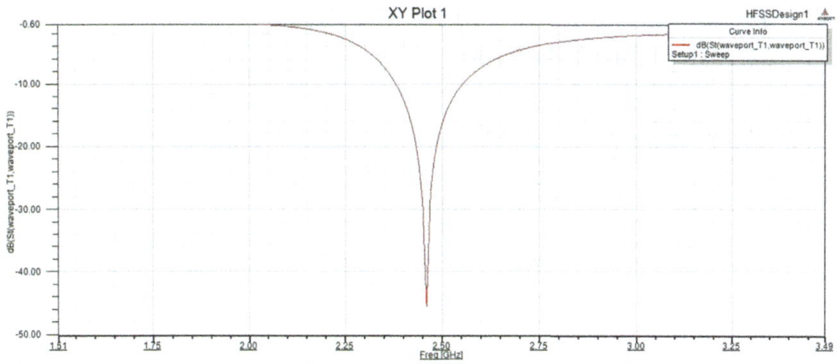

Fig. 4 Plot of return loss U-shaped patch antenna

is FR4_epoxy (4.4). The feeding technique to be used is patch feeding. It operates at two resonating frequencies (dual band). The resonating frequencies are 8.5 GHz and 15.5 GHz and their bandwidths are 550 MHz and 650 MHz, respectively (Fig. 5 and Table 3).

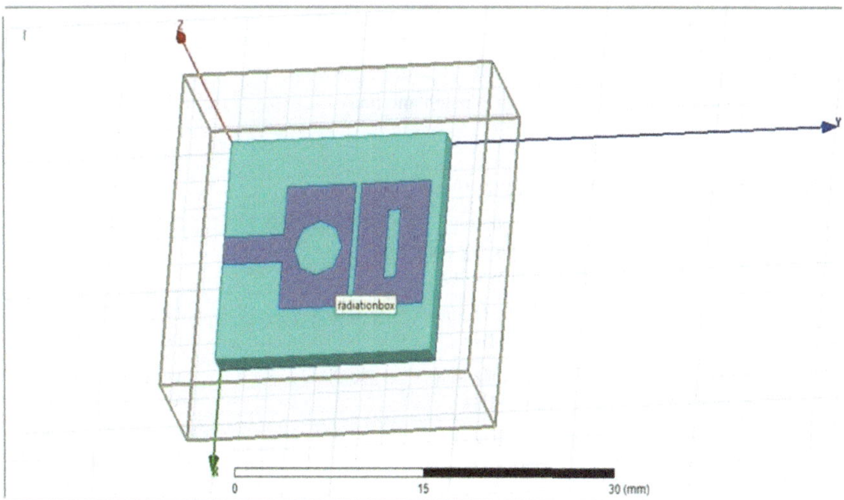

Fig. 5 3D geometry of octagonal–rectangular-shaped patch antenna

Table 3 Dimension of octagonal- and rectangular-shaped antenna

Width of substrate	Length of substrate	Height of substrate	Dielectric constant ($\in r$)
16 mm	18 mm	1.6 mm	4.4

3.4 Simulation and Results of Return Loss of U-Shaped Microstrip Patch Antenna

See Fig. 6 and Table 4.

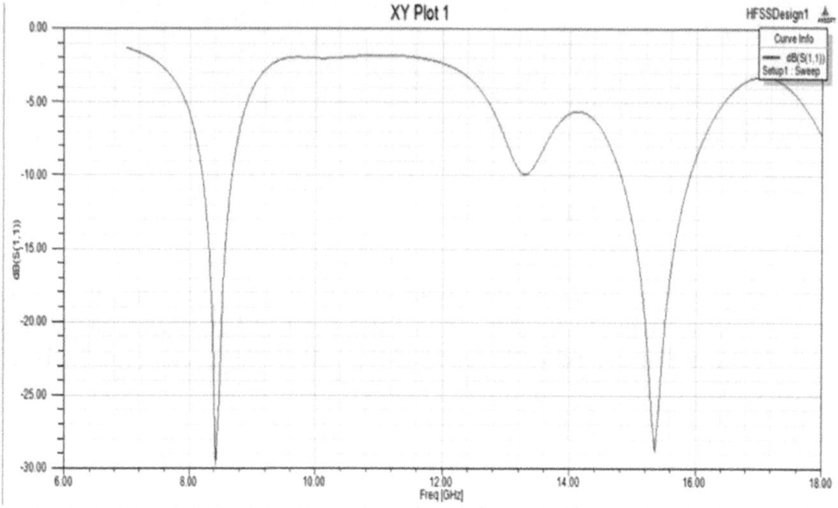

Fig. 6 Return loss of U-shaped microstrip patch antenna

Table 4 Comparative analysis of different shaped antennas

Antenna	Resonating frequency and bandwidth	Return loss	Applications
E-shaped patch antenna	2.5 GHz and 500 MHz	0.025	Used in Wi-Fi (802.11), WiMAX
U-shaped patch antenna	2.4 GHz and 150 MHz	0.0002	Used in ZigBee, WiMAX, WLAN
Octagonal–rectangular-shaped patch antenna (dual band)	8.5 GHz (at 550 MHz) and 15.5 GHz (at 650 MHz)	0.001 (at 8.5 GHz) and 0.0016 (at 15.5 GHz)	Used for military surveillances, GSM 2.5G and 3G
U-shaped patch antenna (dual band)	150 MHz (at 3 GHz) and 210 MHz (at 3.9 GHz)	0.0056 (at 3 GHz) and 0.056 (at 3.9 GHz)	Maritime radio navigation, aeronautical radio navigation

4 Analysis and Comparison

In the above table, we have compared four different types of the patch antenna. Firstly, we designed a E-shaped patch antenna, resonating at a frequency of 2.5 GHz of bandwidth 500 MHz and return loss is 0.025 which is widely used for Wi-Fi, WiMAX purpose. Second, we designed an U-shaped patch antenna resonating at the frequency of 2.4 GHz of bandwidth 150 MHz and return loss is 0.0002 widely used for ZigBee, WLAN, and WiMAX. When we compared the above antennas we found that one can tradeoff between bandwidth and return loss(power) according to their requirements. Third, we have designed an octagonal–rectangular–shaped patch antenna resonating at two different frequencies of 8.5 GHz and 15.5 GHz between the frequency band of 6–18 GHz of bandwidth 550 MHz and 650 MHz whose return losses are 0.001 and 0.0016, respectively. The dual-band antenna is widely used for multiple purposes like military surveillances, GSM 2.5G and 3 IMT band. Lastly, we have designed a dual U-shaped antenna resonating at two different frequencies of 3 and 3.9 Ghz between the frequency band 3–4Ghz of bandwidth 150 and 210 MHz whose return losses are 0.0056 and 0.056. It is used for maritime radio navigation, aeronautical radio navigation. By comparing both dual-band antennas, the octagonal antenna is better as the range of frequency band is 6–18 GHz of bandwidth 550 and 650 MHz.

5 Conclusion

We have concluded that single and dual Microstrip Patch Antenna in different shapes. E-shaped, U-shaped and octagonal–rectangular patch antenna has been analysed for various wireless communication such as Wi-Fi, WiMAX, WLAN, GSM 2.5G AND 3G, military applications, etc. The parameters like return loss, bandwidth, resonating frequency have been obtained for the given antennas. Among all four different types of antennas, octagonal–rectangular-shaped patch antenna is providing good performance in the range 6–18 GHz and also covers a wide range of frequencies.

References

1. Chair, R., Mak, C.-L., Lee, K.-F., Luk, K.-M., Kishk, A.A.: Miniature wide-band half U-Slot and half E-shaped patch antennas. IEEE Trans. Antenna Propag. **53**(8), 2645–2652 (2005)
2. Razzaqi, A.A., Mustaqim, M., Khawaja, B.A.: Wideband E-shaped antenna design for WLAN application. In: International Conference on Emerging Technology, pp. 1–6 (2013)
3. Srisuji, T.(PG Scholar), Nandagopal, C.: Analysis on microstrip patch antenna for wireless communication. In: International Workshop on Antenna Technology, pp. 1–4 (2015)
4. Li, Y., Liu, Q., Chen, Y., Li, C., Mo, Z., Li, F.: A compact triple wideband-notched UWB antenna. In: International Conference on Electronic and Communication System, pp. 538–541 (2018)

5. Joshi, N.K., Upadhye, P.A.: Study of E-shaped patch antenna with two additional rectangular slots. In: International Conference on Micro-Electronic and Telecommunication, pp. 47–51 (2016)
6. Harfoushi, O., Akhorshaideh, A.H., Aqqad, N., Janini, M.A., Obiedat, R.: Factors affecting the intention of adopting cloud computing in Jordanian hospitals. Commun. Netw. **08**(02), 88–101 (2016). https://doi.org/10.4236/cn.2016.82010
7. Balanis, C.A.: Antenna Theory: Analysis and Design, April 1982
8. Dhupkariya, S., Singh, V.K.: Textile antenna for C-band satellite communication application. J. Telecommun. Switch. Syst. Netw. **2**(2), 20–25 (2015). ISSN: 2394-1987

TwoFold Frisky Algorithm (TFFA): A Fast Frequent Itemset Algorithm

Md. Abdul Aziz, Lalam Rajesh, Saikiran Reddi and G. Durga Prasad

Abstract Finding frequent itemsets and generation of association rules by using frequent items plays an important role in the field of data mining. Many algorithms were proposed to get frequent itemsets, but the most popular algorithm is Apriori which is implemented on the horizontal database. In which the method frequently scans the database and returns the flood of candidates which are the significant disadvantages. A novel algorithm based on the vertical database was introduced, which overcomes the disadvantages of Apriori. The proposed algorithm discards the calculation of a few frequent itemsets by taking the next maximum itemset in the process of generating maximal frequent itemset. That means when the length of a frequent itemset with the higher value in powers of two appears then it neglects the lower valued itemsets though it is in powers of two. The simulation results were compared with Apriori and FP-Growth algorithms. It was shown that the novel implementation performed better than Apriori and FP-Growth.

Keywords Frequent itemset · Apriori algorithm · FP growth algorithm · Next maximum

1 Introduction

In prevailing, Data mining (DM) [1] is extracting and analyzing data from a vast amount of data means. Data analysis is done by data mining functionalities like classification [2], association rules [3], clustering [4] etc. The most fundamental and important concern nowadays is research on one of the data mining functionality which is the preparation of association rules. Obtaining association rule may beneficial for unearthing interesting patterns in marketing [5], statistical interpretation [6], decision-making [7] etc. Association rules can be generated by means of frequent itemsets [8].

Md. Abdul Aziz (✉) · L. Rajesh · Saikiran Reddi · G. Durga Prasad
CSE Department, RGUKT-RK Valley, Kadapa, India
e-mail: ajju198821@gmail.com

H. Das et al. (eds.), *Progress in Computing, Analytics and Networking*,
Advances in Intelligent Systems and Computing 1119,
https://doi.org/10.1007/978-981-15-2414-1_21

Many algorithms were aimed to get frequent itemsets, but the most traditional algorithm is Apriori algorithm introduced by Agrawal and Srikant in [9] which is performed on the horizontal database. In which, the algorithm browses the database to attain frequent itemsets level-wise based on the breadth-first strategy [10]. The algorithm is mild and easy to implement and is suitable for mining scattered data sets. In which the method repeatedly scans the database and returns plenty of candidates which are the notable disadvantages. On the other side of the coin, the FP-Growth algorithm is an improvement of the Apriori algorithm which was proposed by Han [12]. The FP-Growth algorithm uses two scans to compress the dataset into the data structure of the frequent pattern tree (FP—Tree). FP—Tree may not fit in memory and it is expensive to build. FP-Growth algorithm is faster than the Apriori algorithm. It is an array-based algorithm and requires only two database scans. The implementation of FP-Growth uses Depth first technique [11].

We had briefly explained the Apriori algorithm and the FP-Growth algorithm which are performed on the horizontal database. We know that the FP-Growth algorithm is faster than the Apriori algorithm. The newly proposed novel algorithm, i.e.,TFFA is performed on the vertical database and it gives better performance when compared with Apriori algorithm and it is also better than the FP-Growth algorithm after some support. The particulars of the TFFA's performance compared with both Apriori and FP-Growth algorithms are shown in the Experimental results and Analysis part.

In this paper, the basic introduction part which is needed for introducing a new algorithm is provided. After the introduction part, the literature survey gives research related to this paper. After that, this paper comes up with the frequent itemsets mining algorithm which includes basic concepts, the Apriori's weakness, and the horizontal and vertical databases. Then, this paper provides a detailed description of the TFFA. At the end, this paper exposes the experimental results and the analysis which shows the algorithm's performance.

2 Literature Review

Frequent itemsets mining has always been an important part of data mining. Many algorithms were introduced for producing frequent itemsets from the given data sets. Apriori algorithm, FP-Growth algorithm, and Eclat algorithm [13] are the few important algorithms for finding frequent itemsets. Apriori and FP-Growth algorithms use horizontal format and Eclat uses the vertical format for representing the data. Apriori uses BFS, FP-Growth and Eclat uses DFS for searching.

FP-Growth and Eclat algorithms are usually faster than Apriori. FP-Growth is the best among the three algorithms and is thus most scalable. Eclat performs poorer than FP-Growth and the Apriori performs the worst. By using the vertical format different algorithms were proposed, i.e., U-Pro Eclat Algorithm [14], E-ACO Algorithm [15] and IApriori [16]. U-Pro Eclat Algorithm and E-ACO Algorithm are based on Eclat Algorithm and also performs better than Apriori and Eclat. The efficiency of

IApriori is much better than the Apriori algorithm. U-Pro Algorithm firstly converts horizontal dataset to vertical dataset. E-ACO Algorithm uses vertical datasets in the Eclat algorithm with ACO Technique.

This paper introduces a different approach which operates on next maximum value and discards calculation of few itemsets based on a condition which is a new technique in finding frequent itemsets and it gives better result than Apriori and FP-Growth algorithms after some support. It will be explained in the algorithm description field in detail.

3 Frequent Itemsets Mining Algorithm

3.1 Basic Concepts

Let I = i_1, i_2, i_3, ..., i_n be a set of items. Let D be the database which is set of database transactions where each transaction T is a set of items such that $T \subseteq I$. Each transaction is correlated with an identifier named TID. Let X be a set of items. A transaction T is said to contain X if and only if $X \subseteq T$. An association rule is a significance which is of the form $X \Rightarrow Y$ where $X \subset I$, $Y \subset I$ and $X \cap Y = \varnothing$. The rule $X \Rightarrow Y$ holds in the transaction set D with support *d*, where *d* is the percentage of transactions in D that contain $X \subset Y$ (i.e., the union of sets X and Y). This is taken as probability, P($X \subset Y$). The rule $X \Rightarrow Y$ has confidence *e* in the transaction set D, where *e* is the percentage of transactions in D comprises of X that also contain Y. This is taken as the conditional probability, P($Y \mid X$). That is..,

$$\text{support}(X \Rightarrow Y) = P(X \subset Y).$$

$$\text{confidence}(X \Rightarrow Y) = P(Y \mid X).$$

Let I = i_1, i_2, i_3,..., i_n be a set of items. Let D is a database having T = t_1, t_2, t_3,..., t_m be a set of transaction ID's for each transaction. Transaction t_p have items $I_a \subseteq I$ where $1 < p < m$.

Vertical Database Let V be the vertical database having $i_a = T_q$, Where $i_a \in I$ and $T_q \subseteq T$ and $1 < a < n$.

Support $\text{support}(I_j \cup I_k) = \text{Length}(T_j \cap T_k)$, $I_j \subseteq I$ and $I_k \subseteq I$, $T_j \subseteq T$ and $T_k \subseteq T$.

3.2 Apriori's Weakness

The Apriori algorithm uses a bottom-up approach for rule mining. There would be a level-wise search where k-frequent itemsets are used to find k+1 itemsets. The property that apriori follows is 'All subsets of a frequent itemset must be frequent' which is known as Apriori property [9].

The drawbacks of the classic Apriori algorithm are as follows:

(1) Very slow and Multiple scans of the database are required during the frequent itemset generation.

(2) The obstacle is the Candidate Generation.

(3) There is a need for several iterations for mining of data.

(4) Large numbers of in frequent itemsets are generated and thus increase the space complexity.

(5) More search space is required and I/O cost will be increased.

3.3 *Horizontal and Vertical Databases*

A Horizontal database (Table 1) [17] is one which places all data *i* row (or record) when a transaction occurs. A database table is represented as a chain of database pages that contain one or more data rows. Horizontal databases are suitable for applications where we need a set of horizontal records.

A vertical database (Table 2) [18] is one in which the physical layout of the data is column-by-column rather than row-by-row. Rather than being arranged as horizontal records, it is processed vertically. Data in a vertical database is processed through fast logical operators and they allow data to be stored in large pages in which the relevant data items can be retrieved in a single read operation.

Table 1 Horizontal database

TID	Items
1	Apple, Grape, Mango
2	Grape, Banana, Papaya
3	Grape, Orange
4	Apple, Grape, Banana
5	Apple, Orange, Mango
6	Apple, Grape, Orange, Papaya
7	Apple, Orange

Table 2 Vertical Database

Item Set	TID Set
Apple	1, 4, 5, 6, 7
Grape	1, 2, 3, 4, 6
Mango	1, 5
Orange	3, 5, 7
Papaya	2, 6

4 TFF Algorithm Description

1. This algorithm uses a vertical database (V) having two columns, items (V.items) and transactions (V.transactions).

2. Initially, we set the minimum support (min_sup) value to some arbitrary value. Say min_sup = 2.

3. In vertical datadase (V),

Item I_i, $1 \leq i \leq n$, where *i* represents the item number and *n* represents the total number of items.

Transaction T_j, $1 \leq j \leq m$, where *j* represents the transaction number and *m* represents the number of transactions.

Every item I_i presents in set of transactions (T_j), i.e., $I_i \rightarrow T_j$.

4. The number of elements in the transaction set is the support of respective item.

For example I_1 item is present in T_1, T_3, T_5, T_6 transactions then the support of I_1 is 4.

5. Initially the number of rows in V is V.length which represents the number of unique items in the vertical database. Here every unique item represents a set. i.e., in V, V.items contains items as I_1, I_2, I_3,..., I_n.

6. nextMax is a variable that represents the length of the next generated itemsets. Initially, the algorithm sets the nextMax value to 2.

7. The algorithm passes the parameters V (vertical database) and nextMax value to JoinItemsets function. The function performs the following steps

Step1) Unions the column V.items based on the nextMax value.

By the Union operation, if the length of V.items is greater than or equal to nextMax value and the Union of items in V.items is not present in V.items column of Vnew (is same as V, updated table based on the min_support).

Step2) It takes the intersection of the column V.transactions of the respective unioned items in step 1.

By the intersection operation, if the length of V.transactions is greater than or equal to Minimum Support (min_sup) then the algorithm will add V.items , V.transactions to Vnew.

For example consider the vertical database V in Table 3.

Step 1: That means if nextMax value is 2, the algorithm generates the 2 length itemsets by taking the union of V.items in vertical database(V). Suppose, V.items have I_1, I_2, I_3 as shown in the Table 3. Then the generated itemsets are (I_1, I_2), (I_1, I_3), (I_2, I_3).

Step 2: Perform the intersection operation (common transactions) of elements in V.transactions of respective elements in generated itemsets in step 1. The output of step1 and step 2 is shown in the Table 4.

Table 3 Vertical Database V

V.items	V.transaction	Support
I_1	T_1, T_3, T_5, T_6	4
I_2	T_1, T_2, T_3, T_4	4
I_3	$T_1, T_2, T_3, T_4, T_5, T_6$	6

Table 4 2 length itemset table

V.items	V.transaction	Support
I_1, I_2	T_1, T_3	2
I_1, I_3	T_1, T_3, T_5, T_6	4
I_2, I_3	T_1, T_2, T_3, T_4	4

Table 5 V_{new}

V.items	V.transaction
I_1, I_2	T_1, T_3
I_1, I_3	T_1, T_3, T_5, T_6
I_2, I_3	T_1, T_2, T_3, T_4

8. Vnew is same as V (Table 3). It contains V.items and V.transactions which satisfy the condition support ≥ min_sup. In the above example, all V.items satisfied the condition. So all the items will present in Vnew (Table 5).

9. After performing the Union and Intersection operations on the Vertical database V, if the algorithm gets the maximum frequent itemset whose length is product of p and 2 (say p is the length of that item set in V), then the algorithm will update the nextMax value to length of the particular item set of V.items in Vnew or after performing the Union and Intersection operations on the Vertical database V, if the algorithm gets the maximum frequent itemset whose length is greater than nextMax, then the algorithm will update the nextMax value to length of the particular item set of V.items in Vnew or After performing the Union and Intersection operations on the Vertical database V, if the algorithm gets the maximum frequent itemset whose length is same as nextMax, then we will update the nextMax = nextMax+1.

10. After updating the nextMax value, the algorithm will prune the rows of Vnew whose itemsets length is lesser than the nextMax value or after updating the nextMax value, the algorithm passes the parameters Vnew and nextMax to the **Prune()** function, then it deletes the rows of Vnew whose itemsets length is lesser than the nextMax value.

11. After the completion of **JoinItemSets()** function then it returns the updated Vnew and nextMax value.

12. Now it assigns the updated Vnew to V.

13. If the Vnew is empty, then the **JoinItemSets()** function returns the previous V and nextMax value as −1. This V contains maximum frequent item sets.

14. The algorithm repeats step 7 to step 12 until all the maximum frequent itemsets were obtained or The algorithm terminates the process when the **JoinItemSets()** function returns the nextMax value as −1.

4.1 TFF Algorithm

The algorithm takes a vertical database as input and gives maximum frequent itemsets as output.

```
Input – Vertical database.
Output – Maximum frequent item sets.
V = VerticalDatabase
nextMax = 2
TFF(V):
while (True) do
    V, nextMax = JoinItemSets(V,nextMax)
    if (nextMax = -1) then
        Exit from loop
    end
    nextMax += 1
end
return V
```

Procedure Prune (Vnew, nextMax) [0] :

This function will be used inside the function called JointItemSets.

```
index=0;
while (index ≤ Vnew.length) do
    if ((length (Vnew[index].items) < nextMax ) then
        delete Vnew[index]
    else
        index=index+1
    end
end
```

Procedure JoinItemSets(V, nextMax) [0] :

```
Vnew = ∅
for (index1=0; index1 ≤ V.length ; index1++) do
    for (index2 = index1+1; index2 ≤ V.length ; index2++) do
        Union = V[index1].items ∪ V[index2].items
        if (length(Union) ≥ nextMax and Union ∉ Vnew) then
            Intersection=
            V[index1].transactions ∩ V[index2].transactions
            if (length(Intersection) ≥ MinSupport) then
                Vnew.insert(Union, Intersection)
            end
            if (length(Union) > nextMax) then
                nextMax = length(Union)
                Prune(Vnew, nextMax)
            end
        end
    end
end
if Vnew is empty then
    return V, -1
end
return Vnew, nextMax
```

4.2 TFF Example

min_sup = 2
confidence = 0.5
H = HorizonalDatabase (Table 6), V = VerticalDatabase (Table 7)
nextMax = 2
JOIN STEP(V, nextMax): (Table 8)
nextMax = 2.
PRUNING STEP(Vnew, nextMax): (Table 9)
Delete rows from Vnew which Item Sets length < nextMax.
JOIN STEP(V, nextMax): (Table 10)
nextMax = 3
PRUNING STEP(Vnew, nextMax): (Table 11)
Delete rows from Vnew which Item Sets length < nextMax.

JOIN STEP(V, nextMax): (Table 12)
nextMax = 5
Vnew is empty it returns nextMax = -1 and V
if (nextMax == -1)
stop process and return V (Maximum frequent item sets)

Table 6 Horizontal Database

TID	Items
T_1	1, 2, 3, 4
T_2	2, 3, 4, 5
T_3	1, 2, 3, 4
T_4	2, 3, 4, 5
T_5	1, 3, 4, 5
T_6	1, 3, 4, 5, 6

Table 7 Vertical Database

Items	TID
1	T_1, T_3, T_5, T_6
2	T_1, T_2, T_3, T_4
3	$T_1, T_2, T_3, T_4, T_5, T_6$
4	$T_1, T_2, T_3, T_4, T_5, T_6$
5	T_2, T_4, T_5, T_6
6	T_6

Table 8 JoinStep->nextMax=2)

Items	TID	Valid
1, 2	T_1, T_3	YES
1, 3	T_1, T_3, T_5, T_6	YES
1, 4	T_1, T_3, T_5, T_6	YES
1, 5	T_1, T_5, T_6	YES
1, 6	T_6	NO
2, 3	T_1, T_2, T_3, T_4	YES
2, 4	T_1, T_2, T_3, T_4	YES
2, 5	T_1, T_4	YES
2, 6	NULL	NO
3, 4	$T_1, T_2, T_3, T_4, T_5, T_6$	YES
3, 5	T_2, T_4, T_5, T_6	YES
3, 6	T_6	NO
4, 5	T_2, T_4, T_5, T_6	YES
4, 6	T_6	NO

Table 9 PruningStep->nextMax=2)

Items	TID
1, 2	T_1, T_3
1, 3	T_1, T_3, T_5, T_6
1, 4	T_1, T_3, T_5, T_6
1, 5	T_1, T_5, T_6
2, 3	T_1, T_2, T_3, T_4
2, 4	T_1, T_2, T_3, T_4
2, 5	T_1, T_4
3, 4	$T_1, T_2, T_3, T_4, T_5, T_6$
3, 5	T_2, T_4, T_5, T_6
4, 5	T_2, T_4, T_5, T_6

4.3 *Example Process*

1. Algorithm initially scans the database 7. Define min_sup = 2 and nextMax = 2.

2. Since the nextMax value is 2, the algorithm generates 2 length itemsets by performing the union of V.items and intersection of V.transactions. The result is shown in Table 8.

3. The itemsets which satisfy the condition support ≥ min_sup were put into the Table 9.

4. Now updates the value of the nextMax based on the conditions given in the algorithm description. So, the nextMax value will be 3.

Table 10 JoinStep->nextMax=3

Items	TID	Valid
1, 2, 3	T_1, T_3	YES
1, 2, 4	T_1, T_3	YES
1, 2, 5	T_1	NO
1, 2, 3	–	NO
1, 2, 4	–	NO
1, 2, 5	–	NO
	nextMax=4	
1, 2, 3, 4	T_1,T_3	YES
1, 2, 3, 5	–	NO
1, 2, 4, 5	–	NO
1, 3, 2, 5	T_1	NO
1, 3, 2, 4	–	NO
1, 3, 4, 5	T_5, T_6	YES
1, 2, 3, 4	–	NO
1, 2, 4, 5	T_1	NO
1, 3 ,4 ,5	–	NO
1, 2, 3, 5	T_1	NO
1, 5, 2, 4	T_1	NO
1, 3, 4, 5	–	NO
2, 3, 4, 5	T_2, T_4	YES
2, 4, 3, 5	–	NO
2, 5, 3, 4	–	NO

Table 11 PruningStep->nextMax=3

Items	TID
1, 2, 3, 4	T_1, T_3
1, 3, 4, 5	T_5, T_6
2, 3, 4, 5	T_2, T_4

Table 12 JoinStep->NextMax=4

Items	TID	Valid
1, 2, 3, 4, 5	NULL	NO
1, 2, 3, 4, 5	NULL	NO
1, 3, 4, 5, 2	NULL	NO

5. Then, generates 3 length itemsets and after some combinations, the nextMax value is updated to 4. Then, generates 4 length itemsets (Table 10) and prunes itemsets which are less than nextMax value.

6. Now takes valid 4 length itemsets into the Table 11. After this, the nextMax value will be 5 and generates 5 length itemsets (Table 12) which are not valid because there is no common TID's.

7. So, the algorithm will be terminated. Therefore, the maximum frequent itemsets obtained are shown in Table 11.

5 Experimental Results and Analysis

In order to examine the execution of the optimization algorithm, tests were executed in the environment of 3.7 GiB Memory, Intel® Core™ i5-3230M CPU @ 2.60GHz × Processor, Intel® Ivybridge Mobile Graphics and Ubuntu 18.04. The algorithms implemented using C++. The test data was obtained from the database of FIMI (http://fimi.ua.ac.be/data/).

The Accidents data set has a volume of 34678 KB, a whole of 340184 transactions, 572 distinct items, and an average of 45 items per transaction. The Mushroom is 557 KB, the total number of items is 119, the number of records is 8123, and each transaction constitutes an average of 23 distinct items (Table 13).

The experimental results of the accidents data set are shown in Fig. 1a and the mushroom data set are shown in Fig. 1b.

From the experimental results, it was shown that the execution time of the algorithm gradually increases with the decrease in minimum support. Also, we can see that the apriori algorithm takes more time than the other two algorithms.

The reason behind getting more time in apriori is repeatedly scanning the database for generating candidate itemsets [19]. According to the results, the FP-Growth algorithm has better performance as it uses FP-Tree to considerably reduce the volume of the database and need not generate any candidate itemsets. The TFFA introduced in this paper achieves good performance essentially because it needs to look at the database once and no word of generating candidate itemsets. Every time the TFFA tries to generate frequent itemsets having the length in powers of 2. By following TFFA, We can leave the calculation of itemsets that are not having the length in powers of 2.

Table 13 Test database description

Transaction database	Distinct items	Transactions count
Accidents	475	340184
Mushroom	23	8123

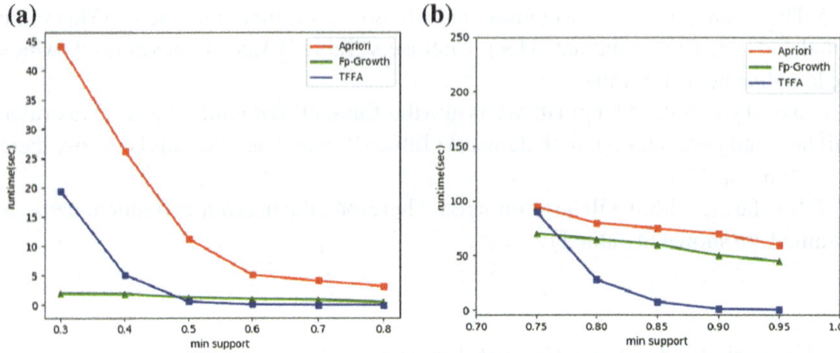

Fig. 1 **a** Comparison of the running time of the three algorithms on the mushroom dataset; **b** Comparison of the running time of the three algorithms on the accidents dataset

6 Conclusion

In this paper, we examined the classical Apriori algorithm and FP-Growth algorithm which are based on the horizontal database. We introduced a novel algorithm named TwoFold Frisky Algorithm based on the vertical database. The Apriori algorithm required to scan the database frequently in order to determine the support degree of the candidate item set, and while generating the candidate item set it also needs to check whether different repeated comparison items exist which affects the efficiency of the algorithm. The proposed novel algorithm discards the calculation of a few frequent itemsets by taking the next maximum itemset in the process of generating maximal frequent itemset. That means when the length of a frequent itemset with the higher value in powers of two appears then it neglects the lower valued itemsets though it is in powers of two. Therefore, the TFFA gives better result than Apriori and FP-Growth algorithms after some support.

References

1. Gopalan, N.P.: Sivaselan book on data mining techniques and trends published by Asoke K. Ghosh. PHI Learning Private Limited
2. Han, J., Kamber, M.: Data Mining—Concepts and Techniques, 2nd edn, Morgan Kaufmann (2006)
3. Road, H., Jose, S.: Fast algorithms for mining association. pp. 487–499
4. Dougherty, J., Kohavi, M., Sahami, M.: Supervised and unsupervised discretization of continuous features. In: International Conference on Machine Learning, pp. 194–202 (1995)
5. Guoxiang, L., Zhiheng, Q.: Data mining applications in marketing strategy. In: International Conference on Intelligent System Design and Engineering Application (ISDEA) (2013)
6. Shan, L., Xuefeng, Z.: The application of data mining in statistics of R&D. In: International Conference on Computer Science and Service System (2012)
7. Adriaans, P., Zantinge, D.: Data Mining. Addison Wesley, New York (1996)

8. Agrawal, R., Imielinski, T., Swami, A.N.: Mining association rules between sets of items in large databases. In: Proceedings of SIGMOD, pp. 207–216 (1993)
9. Agrawal, R., Imielinski, T., Swami, A.: Mining association rules between sets of items in large databases.In: Proceedings of the ACM-SIGMOD Conference on Management of Data, ACM, New York, USA, pp. 207–216 (1993)
10. Kurant, M., Markopoulou, A., Thiran, P.: On the bias of BFS (Breadth First Search). In: 22nd International Teletraffic Congress (ITC 22) (2010)
11. Tarjan, R.E.: Depth-first search and linear graph algorithms. In: 12th Annual Symposium on Switching and Automata Theory (swat) (1971)
12. Han, J., Pei, J., Yin, Y.: Mining frequent patterns without candidate generation. In: Proceedings of 2000 ACM-SIGMOD International Conference on Management of Data (SIGMOD'00), pp. 1–12, Dallas, TX (2000)
13. Zaki, M.J.: Scalable algorithms for association mining. IEEE Trans. Knowl. Data Eng. **12**(3), 372–390 (2000)
14. Yang, J., Zhang, Y., Wei, Y.: An improved vertical algorithm for frequent itemset mining from uncertain database. In: 9th International Conference on Intelligent Human-Machine Systems and Cybernetics (IHMSC) (2017)
15. Sathya, M., Thangadurai, K.: Frequent itemset mining in vertical layout with E-ACO algorithm: in super market. In: World Congress on Computing and Communication Technologies (WCCCT) (2017)
16. Guo, Y.-m., Wang, Z.-j.: A vertical format algorithm for mining frequent item sets. In: 2nd International Conference on Advanced Computer Control (2010)
17. Zaki, M.J., Gouda, K.: Fast vertical mining using diffsets. Technical report 01-1, RPI (2001)
18. Song, W., Liu, Y., Li, J.: Vertical mining for high utility itemsets. In: 2012 IEEE International Conference on Granular Computing
19. Han, J., Pei, J., Yin, Y.: Mining frequent patterns without candidate generation. In: Proceedings of ACM-SIGMOD International Conference on Management of Data, pp. 1–12 (1999)

An Algorithm Based on Next Shortest Path in Large EON Under Dynamic Traffic Grooming

Prasanta Majumdar and Tanmay De

Abstract The elastic light-trail is one of the cutting-edge technologies implemented in *E*lastic *O*ptical *N*etworks (EON). The elasticity in an elastic light-trail or light-path is facilitated by orthogonal frequency division multiplexing. However, in general, to serve a traffic demand generated dynamically in a communication network, a source-destination path is determined by *Dijsktra's* shortest path algorithm. Afterwards, a various spectrum allocation procedures are applied and corresponding data transportation is executed for a certain time duration. In the field of dynamic traffic grooming under EON, $Minimized\ Multi\text{–}hop\ Elastic\ Lightpath\text{–}(m\text{–}MEL)$ and $Multi\text{–}hop\ Elastic\ Lightpath\text{–}(MEL)$ are fundamental algorithms toward the fulfillment of traffic grooming objective. In this study, we investigate the aforesaid algorithms deeply and proposed an innovative algorithm-$Multi\text{–}hop\ Next\ Shortest\ Elastic\ Lightpath\text{–}(MNSEL)$ that utilize $next\ shortest\ path$ to minimize and maximize hop counts and network throughput simultaneously, respectively. The fundamental concept in our research study is the $next\ shortest\ path$ which is determined recursively until the source-destination route setup process is met, when the $shortest\ path$ fails to setup the route depending upon a few criteria. However, here, a thorough and rigorous measurement of the efficiency obtained under the proposed algorithm has been performed.

Keywords Elastic light-trail · Nest shortest path (NSP) · Dynamic traffic grooming · Orthogonal frequency division multiplexing ($OFDM$) · Elastic optical networks (EON)

P. Majumdar (✉)
Dr. B. C. Roy Engineering College, Durgapur, West Bengal, India
e-mail: prasanta.rec.nit.2002@gmail.com

T. De
National Institute of Technology, Durgapur, West Bengal, India
e-mail: tanmayd12@gmail.com

H. Das et al. (eds.), *Progress in Computing, Analytics and Networking*,
Advances in Intelligent Systems and Computing 1119,
https://doi.org/10.1007/978-981-15-2414-1_22

1 Introduction

In this research work, an *E*lastic *O*ptical *N*etworks-EON [1] is investigated from different perspectives under dynamic traffic grooming problem, where, various algorithms are considered and discussed to propose an innovative traffic grooming approaches [1–10] to achieve an enhanced network throughput. However, to setup a source-destination route, traditionally *Dijkstra's shortest path algorithm* is applied in a topology. An elastic lightpath is setup from source node(s) to destination node(d) along a priori determined shortest ($SP(s, d)$) [2] path by allocating same spectrum range common to all optical fiber links ($\in SP(s, d)$) to retain Spectrum Continuity Constraint-SCC [7]. Here, if optical fiber link(s) in $SP(s, d)$ impede SCC constraint, lightpath setup process is canceled as well. This spectrum assigning protocol is followed in $m - MEL$ [2, 4]. Furthermore, if SCC does not hold for an optical fiber link(s) in $SP(s, d)$, an elastic lightpath hoping at node(s) with disparities (in terms of common spectrum) i.e. a multi-hop elastic lightpath is established instead of canceling RSA. The MEL algorithm follows the aforesaid spectrum allocation strategy. However, while RSA [3] process is being executed for a traffic demand $trf(s = source, d = destination, J = required\ spectrum, T = holding\ duration)$, if an established elastic lightpath is found along the $SP(s, d)$, it is expanded in spectrum domain and $s \rightarrow s_1$ and $d_1 \rightarrow d$ both new elastic lightpaths are setup, where, s_1 and d_1 are first and last node, respectively, in the established elastic lightpath.

2 Related Works

A wide variety of research activities in the field of traffic grooming have been performed by several researchers during last decade. However, extension of traditional light-trail concept is incorporated into elastic optical networks is proposed in [2]. Here, network throughput improvement has been observed significantly. The worsening of Quality of Services measured by hop counts met along an elastic light-trail is also considered as another network performance measurement parameter as well. Next, two-stage scheduling algorithm for light-trail under WDM mesh network in association with automatic virtual topology design is proposed in [3]. Static traffic model and related traffic grooming based on integer linear programming is considered for spectrum assignment in [1]. De-multiplexing traffic spectrum and survivable transparent flexible optical network is considered in [4] and [5], respectively. An alternative path to shortest path under dynamic traffic model is proposed in [10]. Unlike the research works discussed so far in this section, next a dynamic traffic grooming under distributed network model is considered and proposed an algorithm in [9]. A survivable traffic grooming with shared protection is discussed in [11]. Next, hybrid and dedicated path protection for traffic grooming problem is proposed and discussed in [12] and [13], respectively. However, in [14] a comprehensive survey for traffic grooming is presented.

The rest part of this paper is organized as follows. Section 3 formulates the proposed dynamic traffic grooming problem. Next, Sect. 4 demonstrates the proposed algorithm. In Sect. 5, numerical results obtained under a few proposed EONs for all concerned algorithms are illustrated and finally Sect. 6 concludes the entire research works briefly.

3 Problem Formulation

The fundamental declarations and definitions for the problem formulation and definition are based on physical and virtual topology which are represented as graphs. Hence, the physical topology is represented as $G_p = (N_p, E_p)$, where $N_p = \{n_1, n_2, \dots n_N\}$ is a set of N network nodes and $E_p = \{e_1, e_2, \dots e_E\}$ is a set of E optical fiber links (edges) such that a $simple\ path(n_i, n_j)_{i \neq j} \in G_p$ for any $(i, j) \in \{1, 2, \dots N\}$. Next, the virtual topology is expressed as $G_v = (N_v, E_v)$, where $N_v \subseteq N_p$ and $E_v = \{\ell_1, \ell_2, \dots \ell_{\overline{E}}\}$ is a set of $\overline{E}$ $(\leq E)$ elastic lightpaths i.e. edges. But, in G_v a $simple\ path(n_i, n_j)_{i \neq j}$ is not necessarily holds for a $(i, j) \in \{1, 2, \dots \overline{N}\}$. Furthermore, a few symbolic representations and assumptions are declared as follows.

1. $trf(s, d, J, T)$ is a traffic demand from source s to destination d with required spectrum J, and holding duration T.
2. $sp_{s,d}(G_p) = \{s \rightarrow n_1 \rightarrow n_2 \rightarrow \dots \rightarrow d\}$ is the source s to destination d $shortest\ path$ in physical topology.
3. $sp_{s,d}(G_v) = \{s \rightarrow v_1 \rightarrow v_2 \rightarrow v_3 \dots \rightarrow d\}$ is the source s to destination d $shortest\ path$ in virtual topology.
4. $\ell^n_{s,d}$ is an yet-to-setup elastic lightpath along $sp_{s,d}(G_p)$.
5. $\ell^e_{s,d}$ is an established elastic lightpath along $sp_{s,d}(G_v)$.
6. $e_i^{x_t}$ denotes ith optical fiber link in the xth lightpath of type t, where $t = n|e$.
7. $F_{s_j}(e_i^{x_t})$ is the jth free spectrum slot in the spectrum domain of $e_i^{x_t}$.
8. $S_{lb}(F_{s_j}(e_i^{x_t}))$ and $S_{ub}(F_{s_j}(e_i^{x_t}))$ represent the lower and upper bound of the corresponding free spectrum slot, respectively.
9. $R_s(F_{s_j}(e_i^{x_t})) = (S_{ub}(F_{s_j}(e_i^{x_t})), S_{lb}(F_{s_j}(e_i^{x_t})))$ is the spectrum range of a free spectrum slot F_{s_j}.
10. $Cap_s(F_{s_j}(e_i^{x_t})) = S_{ub}(F_{s_j}(e_i^{x_t})) - S_{lb}(F_{s_j}(e_i^{x_t}))$ denotes spectrum capacity of a free spectrum slot F_{s_j}.
11. $Cst(S(\ell_i^n)) = L_i \times (J + g)$ is the cost to create an yet-to-setup elastic lightpath ℓ_i^n of length $L_i = L(\ell_i^n)$, where J, g are required spectrum and guard band, respectively.
12. $Cst(Ex(\ell_i^e)) = L_i \times J$ is the cost to expand an established elastic lightpath ℓ_i^e of length $L_i = L(\ell_i^e)$.

3.1 Next Elastic Lightpath (NEL)

To proceed with further discussion under problem definition, in this section, *Next Elastic Lightpath* is considered which is setup along *next shortest path*–*nsp* declared earlier in the previous section. Here, for a source(s)–destination(d) nodes pair $\in G^i_p$ (where, G^i_p is ith instance of the physical topology G_p), a nsp at the corresponding instance is mathematically defined as follows:

$$nsp_{s,d}(G^i_p) = \begin{cases} nsp_{s,d}(G^{i-1}_p - (nsp_{s,d}(G^{i-1}_p) - \{s, d\})), \\ if\, i \geq 1 \\ sp(G^i_p)_{s,d}, \\ otherwise \end{cases} \tag{1}$$

However, the recursive function mentioned in Eq. 1 proceeds as long as the maximum limit (arbitrarily set) is met or no articulation points or bridges are encountered in the subset of a nsp with respect to the instantaneous graph G_p while removing it from this graph G_p. Now, suitable spectrum assignment in all optical fiber links in a nsp produces a *next elastic lightpath*–($NEL_{s,d}$). Here, it is worthwhile to mention that for a (s, d) nodes pair multiple NEL are determined with a various hop counts. The NEL with minimum number of hop counts is termed as victim NEL and expressed as $NEL^{h_m}_{s,d}$ in our research study. Moreover, a $NEL^{h_m}_{s,d}$ converges to $\ell^n_{s,d}$ in the best case.

3.2 Problem Definition

In this sub section, X and Y hypothetical yet-to-setup and established elastic lightpaths (for a s-d nodes pair), respectively, are considered to define the objective function $f(X, Y)$. Here, x and y optical fiber links are used in the underlying routes for X and Y yet-to-setup and established elastic lightpaths, respectively. Now, the objective function $f(X, Y)$ to setup a $mnsel$ is defined as

$$\begin{aligned} f(X, Y) = & \underbrace{\sum_{i=1}^{X} Cst(S(NEL_i)) + \sum_{j=1}^{Y} Cst(Ex(\ell^e_j))}_{A} \\ & + \underbrace{\sum_{i=1}^{X}\sum_{j=1}^{r_i} R_c(F_{s_k}(e^{i_n}_j))}_{B} + \underbrace{\sum_{i=1}^{Y}\sum_{j=1}^{t_i} R_c(F_{s_k}(e^{i_e}_j))}_{C} \end{aligned} \tag{2}$$

where, $R_c(F_{s_k}(e_j^{i_n}))$ denotes the remaining optical spectrum capacity left in $F_{s_k}(e_j^{i_n})$, $x = \sum_{i=1}^{X} r_i = \sum_{i=1}^{X} L(\ell_i)$, $y = \sum_{j=1}^{Y} t_j = \sum_{j=1}^{Y} L(\ell_j)$. Furthermore, A, B and C represent the cost $Cst(R)$ to setup the tentative $mnel$, the total spare optical capacity in the used spectrum slots for yet-to-setup elastic lightpath and the total spare optical capacity in the used spectrum slots for expanded portion in the $mnel$, respectively). Now, the traffic grooming problem in terms of spectrum resources and hops counted (along an elastic lightpath) under the proposed optimization approach is defined as

`Minimize` $f(X, Y)$

`subject to the following constraints`

$$NEL_i = NEL_i^{h_m} \ \forall i \in \{1, 2, \ldots X\} \quad (3)$$

$$\begin{aligned} &((R_s(F_{s_k}(e_1^{i_n})) \cap ..R_s(F_{s_k}(e_{h_1}^{i_n}))) = sr_c > J + g \\ &\wedge (R_s(F_{s_k}(e_{h_1+1}^{i_n})) \cap \ldots R_s(F_{s_k}(e_{h_m}^{i_n}))) = sr_c > J + g \\ &\wedge \ldots \\ &(R_s(F_{s_k}(e_{h_m+1}^{i_n})) \cap \ldots R_s(F_{s_k}(e_{r_i}^{i_n}))) = sr_c > J + g), \ \forall \in \{1, 2, \ldots X\} \end{aligned} \quad (4)$$

where, $h_1, h_2, \ldots h_m$ are hoping points in NEL

$$(R_s(F_{s_k}(e_1^{i_e})) \cap \ldots R_s(F_{s_k}(e_{t_i}^{i_e}))) = sr_c > J, \ \forall \subset \{1, 2, \ldots Y\} \quad (5)$$

$$\begin{aligned} &S_{lb}(F_{s_k}(e_1^{i_e})) = S_{lb}(F_{s_k}(e_2^{i_e})) \ldots = S_{lb}(F_{s_k}(e_{t_i}^{i_e})) \\ &or \\ &S_{ub}(F_{s_k}(e_1^{i_e})) = S_{ub}(F_{s_k}(e_2^{i_e})) \ldots = S_{ub}(F_{s_k}(e_{t_i}^{i_e})) \\ &\forall i \in \{1, 2, \ldots Y\} \end{aligned} \quad (6)$$

Here, minimization of the objective function by restricting an NEL to be along NEL^{h_m} (shown in the Eq. 3) ensures minimum hop counts met along corresponding *multi–hop next shortest elastic lightpath*. Next, an another important constraint is that the free and common spectrum slot for all links in between two successive hoping points must be greater then $J + g$ for all NEL (source and destination are considered to be hoping points in this context only) which is expressed in Eq. 4. Similarly, free and common spectrum slot in a established elastic lightpath must be greater than J (defined in Eq. 5) and adjacent to either end of it (stated in Eq. 6) for all established elastic lightpaths.

4 Algorithm $MNSEL$

In this section, the proposed recursive algorithm $MNSEL$ (shown in Algorithm 1) is discussed briefly. The working principles of the algorithm $MNSEL$ is same as of the algorithm MEL [2]. Here, it is worth mentioning that the proposed algorithm differs in routing mechanism only ($RSA_n()$ is incorporated instead of $RSA()$). Unlike the algorithm MEL, in this case, instead of shortest path (for a source(s) destination(d) nodes pair in a traffic demand $trf(s, d, J, T)$) a nsp (defined in Sec.3.1) is determined under the execution of the module SRA_n followed by spectrum assignment.

Algorithm 1 : $MNSEL(trf(s, d, J, T))$

if $((s, d) \in G_v) \wedge (simple \quad path(s, d) \neq \phi) \wedge (C(Ex(SR_v(s, d))) \leq C(S(SR_p(s, d)))) \wedge (J \leq Q'(SR_v(s, d))) \wedge Ex(SR_v(s, d)$ **then**
  return $true$;
else
  if $(\ell = CandidateLightpath(s, d)) \neq \phi \wedge Ex(\ell)$ **then**
    if $TA_{elastic}(s, s(\ell), B, T)$ **then**
      if $TA_{elastic}(d(\ell), d, B, T)$ **then**
        return $true$;
      else
        return $false$;
      end if
    else
      return $false$;
    end if
  else
    if $RSA_n(s, d, J, G_p)$ **then**
      return $true$;
    else
      return $false$;
    end if
  end if
end if

5 Experimental Results

5.1 Environment and Assumptions for Simulation

In this section, three EONs with network nodes 76, 122 and 172 are considered in which all optical fiber links are equipped with 1000-Gbps OC (Optical capacity) and shown in the Fig. 1a, b, c, respectively. The generated traffic demands belong to the set {OC-100, OC-120, OC-140, OC-160, OC-180} Gbps are dynamic and uniformly distributed over the entire region of an EON. Furthermore, an uni-

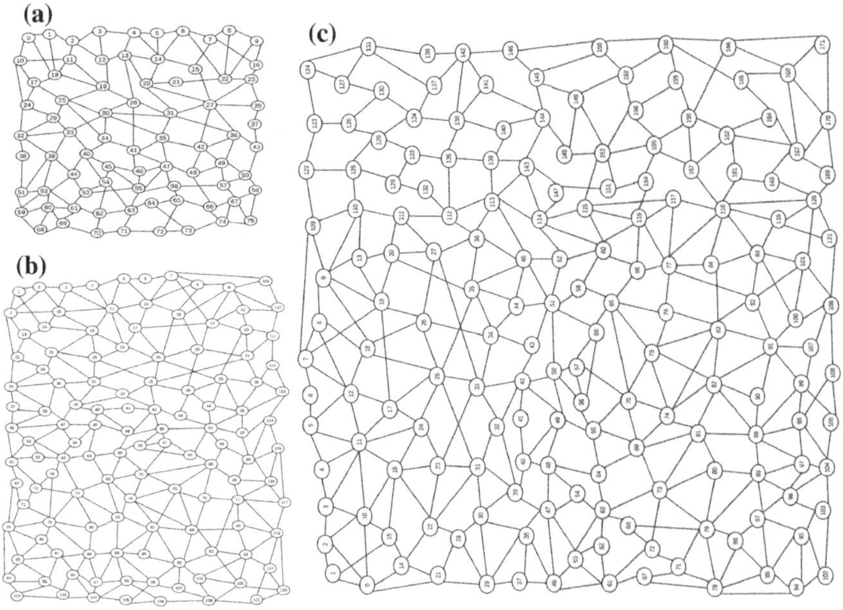

Fig. 1 76 nodes(**a**), 122 nodes(**b**) and 172(**c**) nodes EON

form time distribution on the set {500, 501, 502, ... 600} ms is applied for random generation of delay in between two consecutive traffic demands. However, a simulation session covers the time duration from first traffic demand generation to last traffic demand completion. However, the holding duration of traffic demand is selected dynamically from a time set {30, 40, 50, 60} s based on an uniform distribution. Now, the *network load*–N_L is mathematically defined as

$$N_L = \sum_{i=1}^{\mu} J_i \times T_i \quad (7)$$

in Gb. Next, the most important parameter *network throughput*–N_T to measure the performance obtained under all concerned algorithms is defined as

$$N_T = N_T^e + N_T^n = \sum_{i=1}^{\overline{\mu}} J_i \times T_i \quad (8)$$

where, $\overline{\mu}$ $(\leq \mu)$ is the number of satisfied traffic demands. Moreover, N_T^e and N_T^n are network throughputs incurred by established and yet-to-setup elastic lightpaths, respectively. Next, total number of hops encountered during a simulation is another important parameter to focus and represent the major strength part of the proposed algorithm $MNSEL$, and it is defined as follows:

$$H_t = \sum_{i=1}^{\overline{\mu}} H_i \tag{9}$$

where, H_i represents number of hops met along an elastic lightpath ℓ_i.

5.2 Simulated Numerical Results Based Algorithmic Efficiency Analysis

In this section, at the beginning 76 nodes network is considered to discuss the efficiency of the proposed algorithm $MNSEL$ in comparison to m–MEL and MEL algorithms based on numerical values generated for aforesaid efficiency measurement parameters in various simulation sessions which are shown in Fig. 2a, b. In the Fig. 2a, N_Ts along with N_Ls at various simulation session of different time durations are shown. Moreover, here, N_T segregated into network throughput obtained by established and yet-to-setup elastic lightpaths (N_T^e and N_T^n, respectively) are illustrated in *Gigabyte* and *percentage* calculations which are shown in Fig. 2a, b, respectively. Next, minimization of hop counts (H_t) retaining maximum N_Ts are shown in Fig. 3. In this case, an incremental series of N_Ls {1376980, 2067892, 27785, 3454678, 4078453, 4812342, 5467245, 6178367, 6865452, 7543876, 8253690, 8905180, 9689370} generated in different simulation sessions is considered. To discuss and investigate the efficiency based on the parameters considered so far, network loads 8345670 and 9155080 (belong to the aforesaid set of N_Ls) shown in Fig. 2 are considered. Here, the proposed algorithm $MNSEL$ produces 6494600 and 7073214 N_Ts at these N_Ls, respectively, which are approximately 77.82 (segregated into $N_T^e = 46.23$ and $N_T^e = 31.59$) and 77.26 (segregated into $N_T^e = 51.31$ and $N_T^e = 25.95$), respectively. Next, the N_Ts obtained at the aforesaid N_Ls under the algorithm MEL are 78.02 and 77.57, respectively. Furthermore, a rigorous observation through entire N_Ls list and corresponding N_Ts under both algorithms $MNSEL$ and MEL concludes an equivalent relationship between $N_T(MNSEL)$ and $N_T(MEL)$, which is approximately 2–3% increment in comparison to N_Ts obtained under m–MEL. Now, the total number of hops encountered along elastic lightpaths at the same network loads mentioned earlier are considered and shown in Fig. 2d. Here, the H_t calculated at these N_Ls are 1899 and 2169 under the proposed algorithm. Furthermore, the algorithm MEL produces 2167 and 2356 H_t. In this regard, it is worthwhile to mention that the algorithm $MNSEL$ achieves an enhanced network throughput retaining a minimum number of total hop counts (H_t). Next, approximately same range of network loads generated for 76 nodes EON is also considered for 122 nodes EON and set of corresponding values of N_T (separately shown by N_T^e and N_T^n), H_t are plotted and shown in the Fig. 3a, d, respectively. In this case, it is observed that the algorithm $MNSEL$ produces an enhanced N_T (approximately 4–5% increment in comparison to the previous algorithms) by retaining approximately same hop counts reduction.

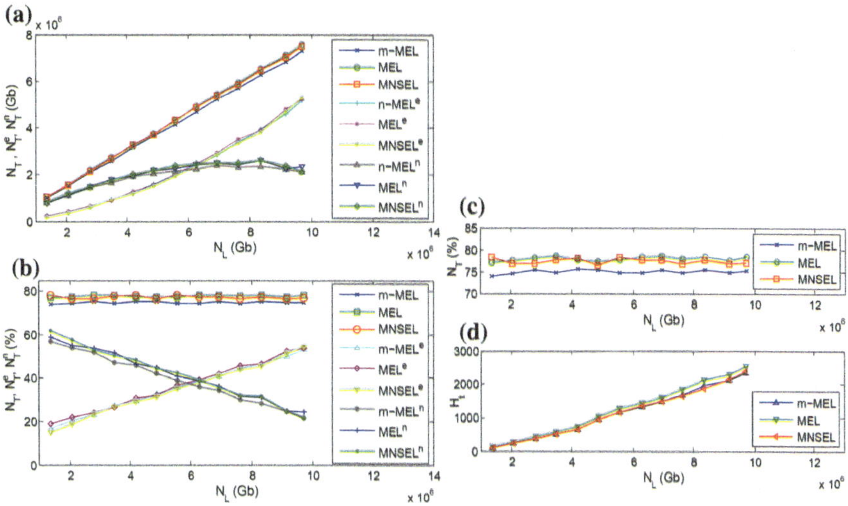

Fig. 2 Network throughputs in Gb (**a**) and % (**b**) calculations, Network throughputs in % (**c**) and H_t (**d**) along with a series of different N_Ls obtained under 76 nodes EON

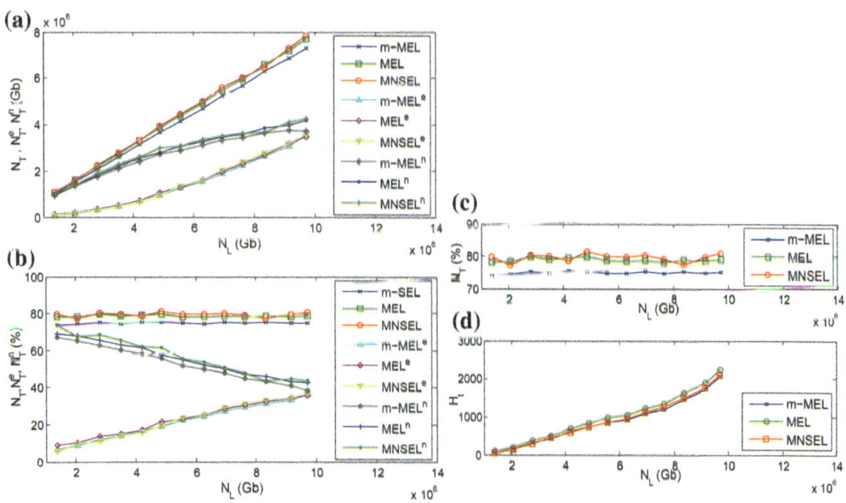

Fig. 3 Network throughputs in Gb (**a**) and % (**b**) calculations, Network throughputs in % (**c**) and H_t (**d**) along with a series of different N_Ls obtained under 122 nodes EON

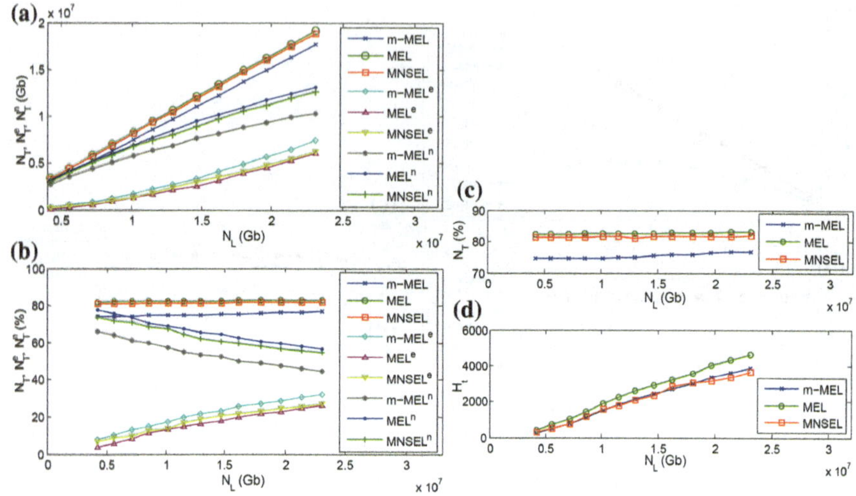

Fig. 4 Network throughputs in Gb (**a**) and % (**b**) calculations, Network throughputs in % (**c**) and H_t (**d**) at different network loads calculated under 122 nodes EON

Next, the most largest proposed EON with 172 nodes is considered for performance measurements and investigations. In this network topology, a different incremental series (unlike the series considered under previous EONs) of N_Ls is considered for practical assessment of the proposed algorithm. Here, the considered incremental series of N_Ls is {4154690, 5567697, 7143892, 8534877, 10156409, 11542672, 12989642, 14467303, 15978456, 17621206, 18983292, 20455172, 21989870}, where, respective simulation sessions are larger (in time domain) in comparison to the earlier series and shown in Fig. 4. Now, network loads 10156409 and 20455172 are considered which are far apart from each other to investigate numerical variations in algorithmic performance parameters in a broad view under all concerned algorithms. The algorithm $MNSEL$ produces 8280421 and 16865289 at these N_Ls (shown in Fig. 4a), respectively, which are approximately 81.52% and 82.44% (shown in Fig. 4b), respectively. Next, the algorithms MEL and m–MEL produce {81.89,82.83} and {74.54,75.21} N_Ts at the concerned N_Ls, respectively. Furthermore, 1584 and 3421 H_ts are obtained under the proposed algorithm at the respective N_Ls, whereas other algorithms MEL and m–MEL produce {1885,4350} and {1533,3538} total number of hops at the mentioned N_Ls, respectively. Here, a significant network throughput improvements (approximately 7–8%) and hop count reductions have been observed simultaneously, which are considerably differ in a positive sense from corresponding values obtained under another concerned algorithms.

6 Conclusion

In this manuscript, a novel concept *next shortest path* has been introduced to setup a *multi-hop next shortest elastic lightpath* to serve a traffic demand under dynamic traffic generation model. The algorithm is applied on three EONs of adequate sizes each in comparison to other existing traditional traffic grooming algorithms to represent the prime objective sharply. Alongside network throughput obtained under various simulation sessions, here, we note the scalability of the algorithm conforming to the recent trend in data communication under large mesh EONs. In this approach for dynamic traffic grooming we note approximately 5–7% network throughput improvement on an average based on applied EON alongside a significant reduction in hop count measurement. However, in this research work, we did not consider degree of an established elastic lightpath i.e., how many times an established elastic lightpath is permissible to expand in frequency domain to accommodate additional intended traffic demands. This aspect can be considered to roll out an another dimension for dynamic traffic grooming which is left as one of the future scopes under the proposed research work.

References

1. Gumaste, A., Das, T., Vaishampayan, R., Wang, J., Somali, A.: Extending light-trails to regional networks: multi-hop light-trails (MLT)–system design and performance. IEEE/OSA J. Opt. Commun. Netw. **4**(12), 1046–1061 (2012). https://doi.org/10.1364/JOCN.4.001046
2. Majumdar, P., Pal, A., De, T.: Extending light-trail into elastic optical networks for dynamic traffic grooming. Opt. Switch. Netw. **20**, 1–15 (2015). https://doi.org/10.1016/j.osn.2015.10.005
3. Gumaste, A., et al.: An autonomic virtual topology design and two-stage scheduling algorithm for light-trail WDM networks. IEEE/OSA J. Opt. Commun. Netw. **3**(4), 372–389 (2011). https://doi.org/10.1364/JOCN.3.000372
4. Majumdar, P., De, T.: De-multiplexing the required spectrum in a traffic demand into multiple non-adjacent granular spectrums for dynamic traffic grooming in EON. Opt. Switch. Netw. **33**, 143–160 (2018). https://doi.org/10.1016/j.osn.2018.04.001
5. Patel, A., Ji, P., Jue, J., Wang, T.: Survivable transparent flexible optical WDM (FWDM) networks. In: Optical Fiber Communication Conference Electronic, Los Angeles, CA, USA (2011). ISBN 978-1-55752-906-0
6. Morea, A., Rival, O.: Advantages of elasticity versus mixed data-rate schemes for restorable optical networks. In: Proceedings of European Conference and Exposition on Optical Communications (2010)
7. Sone, Y., Watanabe, A., Imajuku, W., Tsukishima, Y., Kozicki, B., Takara, H., Jinno, M.: Highly survivable restoration scheme employing optical bandwidth squeezing in spectrum-sliced elastic optical path (SLICE) network. In: Proceedings of Optical Fiber Communication Conference, San Diego, California, United States, (2009). ISBN: 978-1-55752-865-0
8. Dallaglio, M., Giorgetti, A., Sambo, N., Cugini, F., Castoldi, P.: Provisioning and restoration with sliceability in GMPLS-based elastic optical networks [Invited]. IEEE/OSA J. Opt. Commun. Netw. **7**(2), A309–A317 (2015). https://doi.org/10.1364/JOCN.7.00A309
9. De, T., Jain, P., Pal, A.: Distributed dynamic grooming routing and wavelength assignment in WDM optical mesh networks. Photon Netw. Commun. **21**(2), 117–126 (2012). https://doi.org/10.1007/s11107-010-0286-7

10. Majumdar, P., De, T.: An alternative minimum cost route setup algorithm in a large EON for long-haul traffic under dynamic traffic grooming, Article reference: YJNCA2359. J. Netw. Comput. Appl. **140**, 65–86 (2019). https://doi.org/10.1016/j.jnca.2019.04.016
11. Liu, M., Tornatore, M., Mukherjee, B.: Survivable traffic grooming in elastic optical networks shared protection. J. Lightwave Technol. **31**(6), 903–909 (2013). https://doi.org/10.1109/JLT.2012.2231663
12. Georges, A.N., Babri, M, Kora, A.D., Roger, M.F., Aka, B., Lishou, C.: An efficient hybrid protection scheme with shared/dedicated backup paths on elastic optical networks. J. Digital Commun. Netw. **3**(1), 11–18 (2017). https://doi.org/10.1016/j.dcan.2016.05.001. ISSN:2352-8648,
13. DAanko, M., Mikac, B., Furdek, M.: Dedicated path protection for optical networks based on function programmable nodes. J. Opt. Switch. Netw. **27**, 79–87 (2018). https://doi.org/10.1016/j.osn.2017.09.001. ISSN:1573-4277
14. Mata, J., de Miguel, I., Durán, R.J., Merayo, N., Singh, S.K., Jukan, A., Chamania, M.: Artificial intelligence (AI) methods in optical networks: A comprehensive survey. J. Opt. Switch. Netw. **28**, 43–57 (2018). https://doi.org/10.1016/j.osn.2017.12.006. ISSN:1573-4277

A Distance-Based Adaptive Traffic Grooming Algorithm in Large EON Under Dynamic Traffic Model

Prasanta Majumdar and Tanmay De

Abstract The recently emergent technology in optical communication is the elasticity introduced in spectrum domain which provides exact amount of spectrum resources allocation incurred by a data transportation requirement under Elastic Optical Networks (EON). However, in the field of dynamic traffic grooming under EON, to initiate a source to destination data-stream transportation, routing process uses the traditional *Dijsktra's* shortest path algorithm followed by required spectrum allocation in the traditional multi-hop elastic lightpath algorithm(MEL). Here, it is worth mentioning that an established elastic lightpath along a source to destination shortest path is always utilized to setup a $multi\text{–}hop\ elastic\ lightpath$ regardless the distance it covers. In this study, source to destination shortest path length/distance is considered to setup a mel and a $threshold\ path\ length$–(κ) initialized to the diameter of a graph (i.e. the physical topology) is considered as key aspect. In this research work, the prime objective is to achieve a trade-off in between number of hops and cost incurred by a traffic demand. However, toward the fulfillment of objective proposed in this study, a series of simulations has been executed and we note a significant reduction in hop counts in association with a very little reduction in network throughput.

Keywords Elastic light-trail · Threshold shortest path length ($TSPL$) · Dynamic traffic grooming · Orthogonal frequency division multiplexing ($OFDM$) · Elastic optical networks (EON)

P. Majumdar (✉)
Dr. B. C. Roy Engineering College, Durgapur, West Bengal, India
e-mail: prasanta.rec.nit.2002@gmail.com

T. De
National Institute of Technology, Durgapur, West Bengal, India
e-mail: tanmayd12@gmail.com

H. Das et al. (eds.), *Progress in Computing, Analytics and Networking*,
Advances in Intelligent Systems and Computing 1119,
https://doi.org/10.1007/978-981-15-2414-1_23

1 Introduction

In the dynamic traffic grooming under large mesh EON [1, 2], the *Dijkstra's shortest path algorithm* is applied to setup a source to destination shortest path to setup either a lightpath or light-trail [3, 4]. Next, the same spectrum ranges common to all optical fiber links underlying the shortest path are allocated to setup an elastic lightpath that retains Spectrum Continuity Constraint-(SCC) [5]. In this regard, it is worth mentioning that an established elastic lightpath (if found) along that shortest path in the virtual topology [6], is incorporated in the route. In other words, the established elastic lightpath is expanded in spectrum domain to fulfill the purpose of spectrum allocation. Next, the same process is followed for rest part(s) separately until no more established elastic lightpath is found. Now, unlike the traditional MEL [7] algorithm, in this research study under dynamic traffic grooming problem, a restricted variant of MEL based on a source to destination shortest path length is introduced to reduce the value of hop counts encountered along with it. In the proposed algorithm $DAMEL$ (Distance based adaptive multi-hop elastic lightpath), a threshold path length κ (defined in Sect. 4) is considered to determine the nature of route setup process for a traffic demand. Though, to setup a new elastic lightpath in spite of utilizing appropriate established elastic lightpaths the traffic demand incurs more cost, nevertheless, this approach significantly reduces hop counts to provide an enhanced Quality of Services (QoS) [8, 9] standard.

2 Related Works

In the field of traffic grooming under optical networks, a substantial number of research works have been executed during last decade to strengthen potential for enhanced data transportation standard. An advanced variant of traditional light-trail the elastic light-trail technology implemented in $Elastic\ Optical\ Networks$ is proposed in [6]. In this research work a significant network throughput improvement is noted. Furthermore, an another network performance parameter the worsening of $Quality\ of\ Services$ is also considered and measured by number of hops encountered along an elastic light-trail as well. An automatic virtual topology designing algorithm in addition to two-stage scheduling algorithm to setup light-trails in WDM mesh network is proposed and discussed in [3]. In [4], a set of $Integer\ Linear\ Programing$ is considered for multi-hop light-trail establishment under static traffic model. Next, De-multiplexing of an optical spectrum incurred by a traffic demand into multiple granular spectrum slots is elaborated in [7]. The fundamental hardware configuration is also presented to facilitated that de-multiplexing in that research work. Next, an important aspect of traffic grooming in optical network i.e. dynamic traffic grooming in distributed system is considered in [10], in which on network node is assumed to be act as a centralized node in terms of network information. A multi-hop nearest elastic lightpath is proposed in [9]. A survivable traffic grooming algorithm considering shared protection scheme is presented in [11]. In [12] and [13], hybrid and dedicated path protection algorithms are discussed, respectively. A comprehensive survey for traffic grooming is presented in [14].

The rest part of this paper is organized as follows. Section 3 formulates the proposed traffic grooming problem. Next, the main concept $threshold\ path\ length$ is described in Sect. 4 and Sect. 5 discusses the proposed algorithm $DAMEL$. In Sect. 6, numerical results obtained under a few proposed EONs for concerned algorithms are illustrated and finally in Sect. 7 entire research works is concluded briefly.

3 Problem Formulation

To define the proposed problem precisely, physical and virtual topologies are considered and defined as described next. A physical topology is expressed as $G_p = (N_p, E_p)$, where $N_p = \{n_1, n_2, \dots n_N\}$ is a set of N network nodes and $E_p = \{e_1, e_2, \dots e_E\}$ is a set of E optical fiber links (edges) such that a $simple\ path(n_i, n_j)_{i \neq j} \in G_p$ for any $(i, j) \in \{1, 2, \dots N\}$. Next, a virtual topology is represented as $G_v = (N_v, E_v)$, where $N_v \subseteq N_p$ and $E_v = \{\ell_1, \ell_2, \dots \ell_{\overline{E}}\}$ is a set of $\overline{E}$ $(\leq E)$ elastic lightpaths i.e. edges. But, in G_v a $simple\ path(n_i, n_j)_{i \neq j}$ is not necessarily holds for a $(i, j) \in \{1, 2, \dots \overline{N}\}$. Along with the aforesaid fundamental objects, a few additional assumptions and symbols are presented as follows:

1. $trf(s, d, J, T)$ is a traffic demand from source s to destination d with required spectrum J, and holding duration T.
2. $sp_{s,d}G_p = \{s \rightarrow n_1 \rightarrow n_2 \rightarrow \dots \rightarrow d\}$ is the source s to destination d $shortest\ path$ in physical topology.
3. $sp_{s,d}G_v = \{s \rightarrow v_1 \rightarrow v_2 \rightarrow v_3 \dots \rightarrow d\}$ is the source s to destination d $shortest\ path$ in virtual topology.
4. $\ell^n_{s,d}$ is an yet-to-setup elastic lightpath along $sp_{s,d}G_p$
5. $\ell^e_{s,d}$ is an established elastic lightpath along $sp_{s,d}G_v$.
6. A multi-hop elastic elastic lightpath $^m\ell_{s,d}$=$\{\ell^n_1, \ell^n_2, \dots \ell^n_i\} \cup \{\ell^e_1, \ell^e_2, \dots \ell^e_j\}$, where $i \geq 0 \wedge j > 0$
7. $h_c(^m\ell_{s,d}) = i + j - 1$ is the number of hops met along $^m\ell_{s,d}$.
8. $e^{x_t}_i$ denotes i^{th} optical fiber link in the xth lightpath of type t, where $t = n|e$.
9. $F_{s_j}(e^{x_t}_i)$ is the jth free spectrum slot in the spectrum domain of $edg^{x_t}_i$.
10. $S_{lb}(F_{s_j}(e^{x_t}_i))$ and $S_{ub}(F_{s_j}(e^{x_t}_i))$ represent the lower and upper bound of the corresponding free spectrum slot, respectively.
11. $R_s(F_{s_j}(e^{x_t}_i)) = (S_{ub}(F_{s_j}(e^{x_t}_i)), S_{lb}(F_{s_j}(e^{x_t}_i)))$ is the spectrum range of a free spectrum slot F_{s_j}.
12. $Cap_s(F_{s_j}(e^{x_t}_i)) = S_{ub}(F_{s_j}(e^{x_t}_i)) - S_{lb}(F_{s_j}(e^{x_t}_i))$ denotes spectrum capacity of a free spectrum slot F_{s_j}.
13. $Cst(S(\ell^n_i)) = L_i \times (J + g)$ is the cost to create an yet-to-setup elastic lightpath ℓ^n_i of length $L_i = L(\ell^n_i)$, where J, g are required spectrum and guard band, respectively.

14. $Cst(Ex(\ell_i^e)) = L_i \times J$ is the cost to expand an established elastic lightpath ℓ_i^e of length $L_i = L(\ell_i^e)$.

3.1 Problem Definition

In this sub section, x and y hypothetical yet-to-setup and established elastic lightpaths (for a s-d nodes pair), respectively, are considered to define the objective function $f(x, y)$. However, at the beginning, the first sub objective function $f_1(x, y)$ to setup a multi-hop elastic lightpath (mel)–${}^m\ell_{s,d}$ is defined as

$$f_1(x, y) = \sum_{i=1}^{x} Cst(S(\ell_i^n)) + \sum_{j=1}^{y} Cst(Ex(\ell_j^e)) \tag{1}$$

Furthermore, the next sub objective function that counts total number of hops met along the mel is defined as

$$f_2(x, y) = h_c({}^m\ell_{s,d}) \tag{2}$$

Finally, the main objective function is stated as

$$f(x, y) = \frac{f_2(x, y)}{f_1(x, y)} \tag{3}$$

Now, the traffic grooming problem in terms of spectrum resources and hops counted (along an elastic lightpath) under the proposed optimization approach is defined as

```
Minimize f(x, y)
subject to the following constraints
```

$${}^m\ell_{s,d} = \begin{cases} {}^m\ell_{s,i} \rightarrow \ell_{i,j}^e \rightarrow {}^m\ell_{j,d}, & \text{if } L({}^m\ell) > \kappa \\ \ell_{s,d}^n, & \text{otherwise} \end{cases} \tag{4}$$

where, $L({}^m\ell)$ is the length of ${}^m\ell$.

$$\bigwedge_{p \in \{1,2,3,\ldots x\}} \{\bigcap_{i=1}^{e(\ell_p)} (R_s(F_{s_k}(e_i^{p_n}))) = sr_c > J + g\} \tag{5}$$

$$\bigwedge_{p \in \{1,2,3,\ldots y\}} \{\bigcap_{i=1}^{e(\ell_p)} (R_s(F_{s_k}(e_i^{p_e}))) = sr_c > J\} \tag{6}$$

$$
\begin{aligned}
S_{lb}(F_{s_k}(e_1^{i_e})) = S_{lb}(F_{s_k}(e_2^{i_e}))\ldots = S_{lb}(F_{s_k}(e_{e(\ell_i)}^{i_e}))\\
or\\
S_{ub}(F_{s_k}(e_1^{i_e})) = S_{ub}(F_{s_k}(e_2^{i_e}))\ldots = S_{ub}(F_{s_k}(e_{e(\ell_i)}^{i_e}))\\
\forall\, i \in \{1, 2, \ldots y\}
\end{aligned}
\tag{7}
$$

where, κ is the $threshold\ path\ length$ that must be crossed by $L(sp_{s,d})$ for a traffic demand $trf(s, d, J, T)$ to setup a mel or a ${}^m\ell_{s,d}$ converges to $\ell^n(s, d)$, otherwise. Moreover, $e(\ell_p)$ represents total optical fiber links in the pth elastic lightpath, where $p \in \{1, 2, 3, \ldots (x\ or\ y)\}$. However, here, it is worth mentioning that the the ${}^m\ell_{s,d}$ is determined recursively until the length of an instantaneous multi-hop elastic lightpath (i.e. $L({}^m\ell_{s,d})$) is calculated to be under κ. Moreover, minimization of the objective function by retaining the condition provided in Eq. 4 ensures minimum hop counts met along corresponding $multi\text{–}hop\ elastic\ lightpath$. Next, the free and common spectrum slot for all links in all yet-to-setup elastic lightpaths is greater then $J + g$ which is defined by Eq. 5. Similarly, free and common spectrum slot in all links for all established elastic lightpaths must be greater than or equal to J (defined in Eq. 6) which is adjacent to either end of this lightpath (stated in Eq. 7).

4 Threshold Path Length

The traffic grooming problem and corresponding algorithm declared in this manuscript are assumed to be held under dynamic traffic model. In this model, traffic demands are generated dynamically over the execution time of the same i.e. no set of pre defined traffic demands are determined a priori. In this regard, $threshold\ path\ length$ (κ) (the basic concept) is considered which is calculated dynamically each time on entering a traffic demand into the system. However, here, $threshold\ path\ length$ is defined as follows:

$$
\kappa_i = \begin{cases} (\kappa_{i-1} + L(sp^i_{s,d}G_p))/2, & \text{if } i > 0 \\ d_m(G_p), & \text{otherwise} \end{cases} \tag{8}
$$

where, $L(sp^i_{s,d}(G_p))$ is the shortest path length incurred by ith traffic demand, κ_i is the ith $threshold\ path\ length$ and $d_m(G_p)$ = daimeter of the graph i.e. physical topology. Furthermore, κ is initialized to $d_m(G_p)$.

5 Algorithm $DASEL$

A brief description of working steps in the proposed algorithm is presented in this section. Here, it is worthwhile to mention that the algorithm $DAMEL$ (shown in Algorithm 1) differs from (shown at line no. 4) MEL algorithm [6] in one respect only and it is $L(sp_{s,d})$ for a traffic demand $trf(s, d, J, T)$. On entering a traffic demand into the system, if the length of the computed shortest route is greater than a priori computed $threshold\ path\ lentgh–\kappa$ (defined in Sect. 4), algorithm consider an established elastic lightpath along this pre-computed shortest path (if found any) for expansion in spectrum domain, and two more recursion calls are invoked for the rest part.

Algorithm 1 : $DASEL(trf(s, d, J, T))$

if $((s, d) \in G_v) \wedge (simple\quad path(s, d) \neq \phi) \wedge (C(Ex(SR_v(s, d))) \leq C(S(SR_p(s, d)))) \wedge (J \leq Q'(SR_v(s, d))) \wedge Ex(SR_v(s, d)$ **then**
    return $true$;
else
    if $(\ell = CandidateLightpath(s, d)) \neq \phi \wedge Ex(\ell) \wedge L(sp_{s,d}) > \kappa$ **then**
        if $TA_{elastic}(s, s(\ell), B, T)$ **then**
            if $TA_{elastic}(d(\ell), d, B, T)$ **then**
                return $true$;
            else
                return $false$;
            end if
        else
            return $false$;
        end if
    else
        if $RSA(s, d, J, G_p)$ **then**
            return $true$;
        else
            return $false$;
        end if
    end if
end if

6 Experimental Results

6.1 Environment and Assumptions for Simulation

In this section, 76, 122 and 224 nodes mesh EONs (shown in Fig. 1a, b, c, respectively) are considered for practical simulation purpose of the proposed algorithm in comparison to the existing multi-hop elastic lightpath algorithm. In these EONs all optical fiber links assumed to be equipped with 1000-Gbps OC (Optical Capacity). The generated traffic demands are dynamic in nature and assumed to be distributed uniformly over the entire region of an EON. The optical capacity incurred by a traffic demand belongs to the set {OC-100, OC-120, OC-140, OC-160, OC-180} Gbps is selected equally likely. The generation of a traffic demand is detained by a

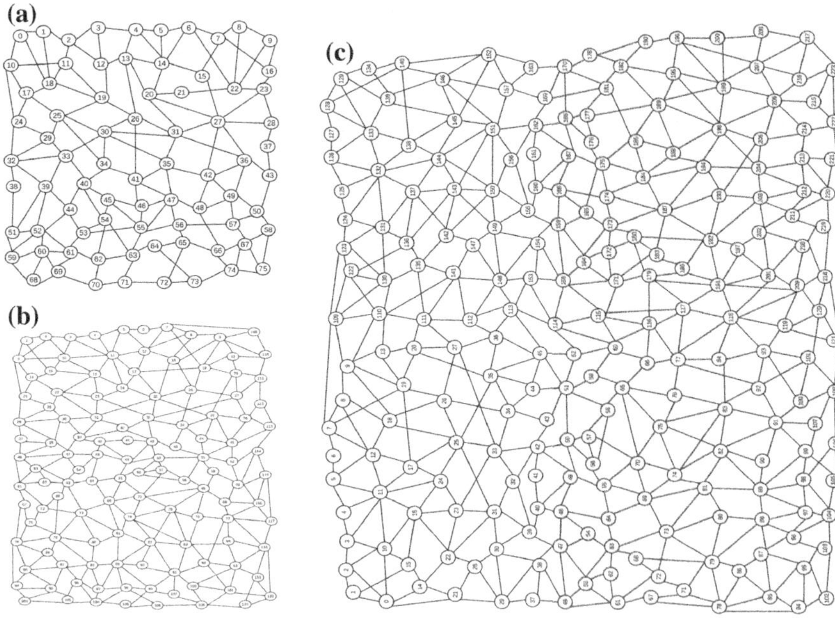

Fig. 1 76, 122 and 224 nodes EON

time delay selected with equal probability from the set {500, 501, 502, ...600} ms from the time of last generated traffic. Moreover, the holding duration of a traffic demand is set to a time slot in {500, 501, 502, ...600} ms randomly. Now, a few parameters for performance measurements are declared and defined as follows:

1. *Network Load*- N_L:

$$N_L = \sum_{i=1}^{\mu} B_i \times T_i \tag{9}$$

   in Gb, where, μ is the maximum traffic demands. Here, a 20 Gb spectrum guard band is embedded in between a pair of adjacent and parallel elastic lightpaths for undistorted transportation of optical signals.

2. *Network Throughput*–H_T:

$$N_T = N_T^e + N_T^n = \sum_{i=1}^{\overline{\mu}} B_i \times T_i \tag{10}$$

   where, $\overline{\mu}$ ($\leq \mu$) is the number of satisfied traffic demands. Moreover, N_T^e and N_T^n are network throughputs incurred by established and yet-to-setup elastic lightpaths, respectively.

3. *Total Hops*–N_T:

$$H_T = \sum_{i=1}^{\overline{\mu}} H_i \tag{11}$$

where, H_i represents number of hops met along an *multi-hop elastic lightpath* ℓ_i.

4. *Total cots calculated in a simulation session–C_T*:

$$C_T = \sum_{i=1}^{\overline{\mu}} C_i \tag{12}$$

where, C_i is the cost calculated for an individual traffic demand

5. *Cost scaled to unit percentage N_T–C_u*:

$$C_u = \frac{C_T}{N_T(in\ \%\ calculation)} \tag{13}$$

6. *Hop scaled to unit percentage N_T–H_u*:

$$H_u = \frac{H_T}{N_T(in\ \%\ calculation)} \tag{14}$$

7. *Cost to hop ratio–R_H^C*:

$$R_H^C = \frac{C_u}{H_u} \tag{15}$$

Here, it is worthwhile to mention that the parameter R_H^C is inverse of the objective function defined in Sect. 3.1 and it is defined in this manner for convenient data plotting.

6.2 Simulated Numerical Results Based Algorithmic Efficiency Analysis

At the beginning under this section, 76 nodes network is considered to analyze and discuss the efficiency of the proposed algorithm $DAMEL$ based on the parameters mentioned in the Eqs. 9 to 15. Here, the numerical results obtained for these parameters under the aforesaid network at various simulation sessions are shown in Fig. 2. However, here, the network throughput N_T is represented in Gb and % calculations with its segregated parts together as $Gb(\%)(N_T^e + N_T^n)$ throughout the rest of the manuscript. Now, the network loads 7818680, 8345670 and 9155080 are considered and we focus on corresponding N_Ts and H_Ts (shown in Fig. 2a, b, respectively) obtained under MEL and $DAMEL$ algorithms. Here, the algorithm MEL produces 5888477(75.31) (3298126(42.18) + 2590351(33.13)), 6416985(76.88) (3869887(46.36) + 2547098

(30.51)) and 7105257(77.60) (4420988(48.28) + 2684269(29.32)) N_Ts at the respective network loads with hop count values 1710, 1890 and 2073, respectively, whereas, the algorithm $DAMEL$ produces 5688868(72.75) (2472261(31.61) + 3216607(41.14)), 6148255(73.66) (2949359 (35.33) + 3198896(38.33)) and 6823281(74.52) (3385548(36.97) + 3437733(37.55)) N_Ts at the same network loads, respectively. Moreover, total number of hops calculated at the aforesaid network loads are 1199, 1324 and 1452, respectively. Here, it is significantly observed that in the $DAMEL$ algorithm, an enhanced QoS is achieved by reducing H_T sharply in comparison to MEL algorithm, though a little reduction (2.5–3.5%) in network throughput is sacrificed. Here, the numerical data obtained for the parameter R_H^C under the algorithm $DAMEL$ are greater than those produced by the algorithm MEL (shown in Fig. 2e). Next, an approximately same range of network loads (as generated under 76 nodes EON) is also considered for 122 nodes EON and corresponding network throughput values (segregated into N_T^e and N_T^n), H_T are plotted and shown in Fig. 3. In this case, a more reduction in hop count is observed in comparison to corresponding values obtained under the previous considered network. Now, the largest proposed network comprising 224 nodes is considered for performance analysis and investigations. In this network, a larger incremental series of network loads (unlike the incremental N_Ls series considered under previous networks) is generated for simulating the proposed algorithm. Now, the network loads 10101545, 10843567 and 11532486 are (shown in Fig. 4) considered to focus on different aspects of the algorithm $DAMEL$. Here, the algorithm MEL produces 7911494(78.31) (2088990(20.67) + 5822504(57.64)), 8517621(78.54) (2358475(21.74) + 6159146 (56.80)) and 9078372(78.71) (2643245 (22.91) + 6435127 (55.80)) N_Ts at the considered network loads, respectively. Here, a lesser N_T^e values are observed (in comparison to the corresponding values generated under previous networks) due to large size of the network. Furthermore, the algorithm MEL produces 1369, 1489 and 1605 H_T values at these network loads, respectively. Next, the network throughput 7529943(74.54) (457597(4.52) + 7072346(70.01)), 8138097(75.03) (517238(4.76) + 7620814 (70.27)) and 8686286(75.32) (540873(4.68) + 8145413(70.64)) are produced by the algorithm $DAMEL$. Moreover, this algorithm produces 312, 331 and 357 H_Ts at the considered N_Ls, respectively (shown in the Fig. 4a, b). Here, though a 3–4% reduction in network throughput under the proposed algorithm is observed, nevertheless, the proposed algorithm is considered to be an efficient algorithm due to its capability for drastic reduction in hop counts. In other words, a huge enhancement in QoS is achieved under the proposed research study. Finally, the values obtained for the most important parameter R_H^C for both the considered algorithms are shown in Fig. 4c–e, where, a big difference in between $R_H^C(MEL)$ and $R_H^C(DAMEL)$ is shown.

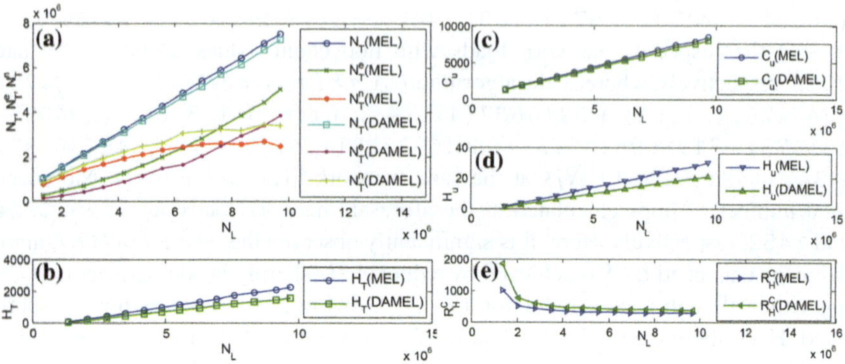

Fig. 2 N_Ts (**a**), H_Ts (**b**), C_us (**c**), H_us (**d**) and R_H^Cs (**e**) along with a series of different N_Ls obtained under 76 nodes EON

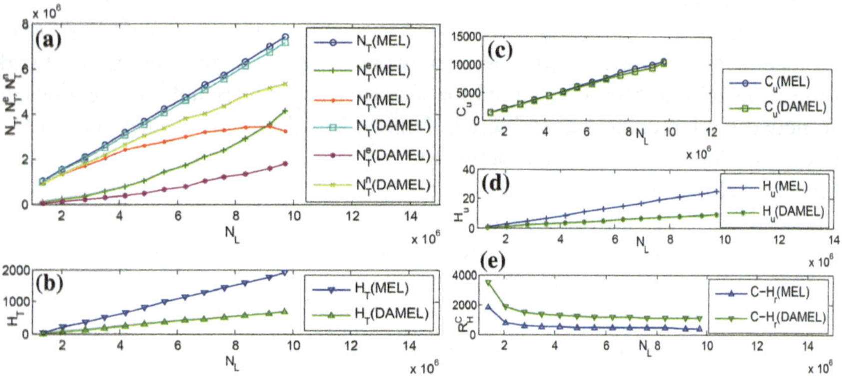

Fig. 3 N_Ts (**a**), H_Ts (**b**), C_us (**c**), H_us (**d**) and R_H^Cs (**e**) at different network loads calculated under 122 nodes EON

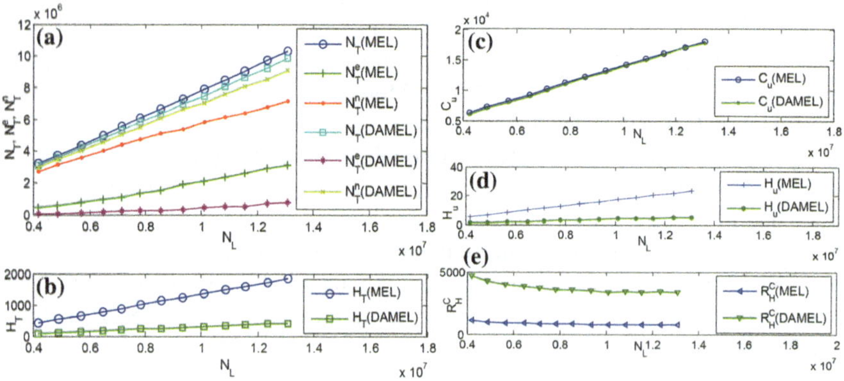

Fig. 4 N_Ts (**a**), H_Ts (**b**), C_us (**c**), H_us (**d**) and R_H^Cs (**e**) along with a series of different N_Ls obtained under 244 nodes EON

7 Conclusion

A novel concept *distance based multi-hop elastic lightpath setup* (κ) to setup a multi-hop elastic lightpath is introduced in this research works to provide an enhanced *Quality of Services* (QoS) standard. The proposed algorithm based on this concept is applied to three different EONs each with an adequate network nodes to present the proposed objective clearly. In this regard, we note note the scalability of the algorithm conforming to the recent trend in data communication under large mesh EONs. However, a linear function (i.e. an $average$ relationship) is applied in this research work for calculation of $threshold\ path\ length$. In this regard it is worth mentioning that a non-linear function may be used to calculate κ which will certainly open different possibilities to enhance network throughput. This important aspect encountered in this research work is considered to be one of the future scopes in this field of dynamic traffic grooming. A various non-linear functions may be applied to κ that possibly extends algorithmic scalability and efficiency as well.

References

1. Patel, A., Ji, P., Jue, J., Wang, T.: Survivable transparent flexible optical WDM (FWDM) networks. In: Optical Fiber Communication Conference Electronic, Los Angeles, CA, USA (2011). ISBN 978-1-55752-906-0
2. Morea, A., Rival, O.: Advantages of elasticity versus mixed data-rate schemes for restorable optical networks.In: Proceedings of European Conference and Exposition on Optical Communications (2010)
3. Gumaste, A., et al.: An autonomic virtual topology design and two-stage scheduling algorithm for light-trail WDM networks. IEEE/OSA J. Opt. Commun. Netw. **3**(4), 372–389 (2011). https://doi.org/10.1364/JOCN.3.000372
4. Gumaste, A., Das, T., Vaishampayan, R., Wang, J., Somali, A.: Extending light-trails to regional networks: multi-hop light-trails (MLT)–system design and performance. IEEE/OSA J. Opt. Commun. Netw. **4**(12), 1046–1061 (2012). https://doi.org/10.1364/JOCN.4.001046
5. Sone, Y., Watanabe, A., Imajuku, W., Tsukishima, Y., Kozicki, B., Takara, H., Jinno, M.: Highly survivable restoration scheme employing optical bandwidth squeezing in spectrum-sliced elastic optical path (SLICE) network. In: Proceedings of Optical Fiber Communication Conference, San Diego, California United States (2009). ISBN: 978-1-55752-865-0
6. Majumdar, P., Pal, A., De, T.: Extending light-trail into elastic optical networks for dynamic traffic grooming. Opt. Switch. Netw. **20**, 1–15 (2015). https://doi.org/10.1016/j.osn.2015.10.005
7. Majumdar, P., De, T.: De-multiplexing the required spectrum in a traffic demand into multiple non-adjacent granular spectrums for dynamic traffic grooming in EON. Opt. Switch. Netw. **33**, 143–160 (2018). https://doi.org/10.1016/j.osn.2018.04.001
8. Dallaglio, M., Giorgetti, A., Sambo, N., Cugini, F., Castoldi, P.: Provisioning and restoration with sliceability in GMPLS-based elastic optical networks [Invited]. IEEE/OSA J. Opt. Commun. Netw. **7**(2), A309A317 (2015). https://doi.org/10.1364/JOCN.7.00A309
9. Majumdar, P., De, T.: An alternative minimum cost route setup algorithm in a large EON for long-haul traffic under dynamic traffic grooming, Article reference: YJNCA2359. J. Netw. Comput. Appl. **140**, 65–86 (2019). https://doi.org/10.1016/j.jnca.2019.04.016

10. De, T., Jain, P., Pal, A.: Distributed dynamic grooming routing and wavelength assignment in WDM optical mesh networks. Photon Netw. Commun. **21**(2), 117–126 (2012). https://doi.org/10.1007/s11107-010-0286-7
11. Liu, M., Tornatore, M., Mukherjee, B.: Survivable traffic grooming in elastic optical networks shared protection. J. Lightwave Technol. **31**(6), 903–909 (2013). https://doi.org/10.1109/JLT.2012.2231663
12. Georges, AN., Babri, M., Kora, AD, Faye, RM., Aka, B., Lishou, C.: An efficient hybrid protection scheme with shared/dedicated backup paths on elastic optical networks. J. Digital Commun. Netw. **3**(1), 11–18 (2017). https://doi.org/10.1016/j.dcan.2016.05.001. ISSN:2352-8648
13. DAanko, M., Mikac, B., Furdek, M.: Dedicated path protection for optical networks based on function programmable nodes. J. Opt. Switch. Netw. **27**, 79–87 (2018). https://doi.org/10.1016/j.osn.2017.09.001. ISSN:1573-4277
14. Mata, J., de Miguel, I., Durán, RJ., Merayo, N., Singh, SK., Jukan, A., Chamania, M.: Artificial intelligence (AI) methods in optical networks: a comprehensive survey. J. Opt. Switch. Netw. **28**, 43–57 (2018). https://doi.org/10.1016/j.osn.2017.12.006. ISSN:1573-4277

Particle Swarm Optimization of Multi-responses in Hard Turning of D2 Steel

Rasmi Ranjan Mishra, Ramanuj Kumar, Amlana Panda, Anish Pandey and Ashok Kumar Sahoo

Abstract Good surface trait, the flexibility of the cutting process, and lesser cycle time are the significant advantages of hardened steel machining over the conventional grinding process. In this current research work, an effort has been performed to analyze the effects of rotational speed, feed rate, and depth of cutting on machining features, for example, surface trait and MRR using PSO during AISI D2 hard turned steel by coated carbide tool. The trials were planned according to L_{16} Taguchi orthogonal array. It is revealed from PSO that global best value of Ra has been found as 0.4916 μm, and correspondingly local best values are $d = 0.15$ mm, $f = 0.25$ mm/rev, and $v = 200$ m/min. Similarly, global best value of MRR has been found as 0.2849 g/s with the local best values of $d = 0.15$ mm, $f = 0.25$ mm/rev, and $v = 50$ m/min. Current optimization results confirmed that the PSO methodology is efficient and can be implemented in the future for solving multi-objective optimization problems.

Keywords PSO · Hard turning · D2 steel · Surface roughness · Material removal rate

1 Introduction

Hard turning is the most dominant cutting operation that provides more dimensional accuracy for hardened parts. To facilitate the investment associated with the installation and better machining performance of hard turning machining conditions, it is necessary to achieve the optimum condition [1, 2]. Turning of hard part steel is considered as an alternating finishing process for conventional grinding process, particularly in automotive and die & mold industry. In recent years, the machining of hardened D2 steel is enormously popular among the metal cutting industries as well

R. R. Mishra · R. Kumar (✉) · A. Panda · A. Pandey · A. K. Sahoo
School of Mechanical Engineering, KIIT Deemed to University, 751015 Bhubaneswar, Odisha, India
e-mail: ramanuj.kumar22@gmail.com

H. Das et al. (eds.), *Progress in Computing, Analytics and Networking*, Advances in Intelligent Systems and Computing 1119,
https://doi.org/10.1007/978-981-15-2414-1_24

as research communities. Further, to improve the machining performance, tooling life must be longer and surface finish should be lesser along with tool flank wear [3].

Previously, researchers have embarked on specific investigations to examine the potential of responses for specific cutting conditions. However, the current study for their assessment with factual prospective is still continued. For the hard turning of D2 steel, use of commercially available PVD-coated TiAIN cemented carbide tool (CNMG 120408) along with its significance and its multi-output parameters optimization, i.e., particle swarm optimization, is important. The available open literature shows that the optimization of responses using PSO on this insert is still lacking. In particular, for D2 steel with hardness value of 55 HRC, analysis could not be focused on critical aspects of surface roughness and material removal rate with its PSO. In light of this motivation, demand is felt that a methodical investigation is to be performed for these critical aspects. Moreover, it would be significant to notice how input process parameters sway the output parameters (surface roughness and MRR) in this hardness regime.

2 Literature Review

Particle swarm optimization (PSO) method is computationally competent; the formulation is dynamic and easy to implement. It is judged as the feasible strategy for optimizing factors to reduce the time of machining [4–6]. In general, PSO contains selection and mutation. Moreover, there is no crossover in PSO that means that PSO can have higher mobility in particles by a higher degree of exploration [7]. Xie et al. [8] formulated the prediction model with input process variables and consumption of specific cutting energy; afterwards, support vector machine (SVM) algorithm is utilized to foretell supportive surface roughness during turning practice. Manav et al. [9] optimized the input parameters for least surface roughness, lowest cutting forces, and prolonging tooling life using PSO method. During the machining of MMC, Chandrasekhar and Tamang [10] developed ANN-PSO optimization that can be easily performed in machining practice for obtaining optimum input parameters satisfying supportive surface roughness. Bouacha and Terrab [11] found that non-dominated sorting genetic algorithm shows better output performance when evaluated to the PSO-NN approach. RSM, Sugeno fuzzy model, and ANN (artifical neural network) model were developed to forecast the wear rate and hardness of the Cu-Ni-Sn spinodal alloy based on work specimen [12]. A short detail on applicability of nature-inspired optimization in cloud computing has been emphasized [13]. Various researchers have adopted various optimization methodologies to attain best possible outcomes. However, very less effort has been identified regarding the machinability aspects of hardened D2 steel in dry condition using PSO approach. The ambit of this research for hard turning operation is to find the optimal solution using PSO. The main objective is to develop a PSO-based model for prediction of important sustainable assess that were material removal rate considering an economic feature and surface roughness of D2 steel with three input parameters.

3 Experimental Scheme and Parameters

In this study, high-speed hard turning trial was designed to investigate the machining behavior of D2 hardened steel. This steel is considered as cold work tool steel affluent in carbon and chromium content together with vanadium and molybdenum appended as alloying constituents in fewer quantities. This work material is used extensively for production of punches and dies owing to its higher wear resistance, outstanding dimensional stability, and toughness.

In recent years, the multilayer coated tools have a considerable contribution to the machining industry in terms of quality of turned surface, lower tool wear, and higher MRR. Tool geometry is expected reasonably to play an effective role during hard turning. CNC turning lathe center Jyoti DX200 with Fanuc controller was used for the experiments. Taylor-Hobson (Surtonic 25) surface tester was utilized for recording the surface roughness value. Moreover, to increase the accuracy of the computed results, measurements of both the material removal rate (MRR) and the surface roughness (Ra) were performed at three different places.

The workpiece used in the trial was high carbon and chromium AISI D2 (55 HRC) cylindrical bar with a diameter of 45 mm and machining length of 175 mm. PVD-coated TiAlN cemented carbide tool (CNMG 120408) commercially available supplied by WIDIA is used for machining in this present work. WIDIA-make left-handed tool holder PCLNR2525M12 was used to accumulate the cutting tool. Before and after the machining, the weight of work surface was computed. Stopwatch was used for time measurement. The MRR was ascertained as the ratio of difference of weight and measured time. The output characteristics are surface roughness and material removal rate. The experimental figure is presented in Fig. 1. The input process parameters were all varied in four levels. Taguchi L_{16} orthogonal array was exploited for the experimental design. The input cutting parameters were selected based on the related literature survey, machine tool, cutting insert specification, and as per the tool manufacturer's recommendations. The trial was performed to identify the

Fig. 1 Experimental scheme

relation between input and output factors. Additionally, PSO approach was employed to analyze the optimal control factors.

4 Results and Discussion

The current analysis deliberated on two machining characteristics, namely, the average surface roughness and MRR. The hard turning machining using TiAIN, PVD-coated cemented carbide tool has been performed by means of L_{16} orthogonal array, and experimental results are presented in Table 1.

4.1 *Analysis of Surface Roughness*

Surface roughness is one of the significant output variables for machinability and quality point of view and also plays an important role in assessing the freshly machined component. The promising performance of the machined elements is swayed by the better superiority of their surfaces during machining procedures. In addition, it is severely affected by flank wear. The surface finish had a prevailing

Table 1 Experimental results

Run	Input variables			Output characteristics	
	d (mm)	*f* (mm/rev)	*v* (m/min)	Ra (μm)	MRR(g/s)
1	0.15	0.1	50	0.39	0.05
2	0.15	0.15	100	0.70	0.088
3	0.15	0.2	150	1.43	0.16
4	0.15	0.25	200	2.07	0.296
5	0.25	0.1	100	0.48	0.69
6	0.25	0.15	50	0.52	0.153
7	0.25	0.2	200	1.06	0.116
8	0.25	0.25	150	1.52	0.338
9	0.35	0.1	150	0.44	0.388
10	0.35	0.15	200	0.60	0.554
11	0.35	0.2	50	0.80	0.264
12	0.35	0.25	100	1.22	0.298
13	0.45	0.1	200	0.50	0.823
14	0.45	0.15	150	0.64	0.674
15	0.45	0.2	100	0.88	0.907
16	0.45	0.25	50	1.42	0.744

consequence on the turned component performance with the cost considerations. Consequently, for maximization of profit from utilizing finish turning of hardened steel by manufacturers, adequate and accurate prediction model must be generated. Ra is found to be highest (2.07 μm) at highest values of feed rate and cutting speed while it is less at lowest controllable variables of cutting speed, feed rate, and depth of cut. From results of Ra (Table 1), feed rate is the dominating agent; as with increases in feed, Ra increases [14–18]. The same trend has also been viewed by [19]. In each experimental run (Table 1), except run no. 4, the arithmetic surface roughness value is found to be less than the suggested roughness cap of 1.6 μm that signifies PVD carbide tool performed well under dry machining condition during turning of D2 hardened steel within precited range of process variables and their levels.

4.2 Analysis of MRR

MRR is generally considered as most important parameter for process efficiency. In many cases, the manufacturing productivity was termed as metal removal rate. This can be optimized by many methods. Choosing the proper cutting tool and optimal machining parameter is important for achieving maximum productivity. Combination of higher depth of cut, medium level of feed, and medium level of cutting speed condition (run no. 15) yields maximum MRR [17]. Moreover, lower combination of speed, cutting feed, and depth of cutting provide lower MRR (run no. 1).

5 Analysis Using Particle Swarm Optimization (PSO) Algorithm

Particle swarm optimization (PSO) is motivated from the behavior of fish schooling and bird flocking, and it is a population supported nondeterministic optimization technique [7]. This PSO is used to calculate the optimum and near-optimum solution of the predicament by using objective function (fitness function) $f(x) = f(x_1, x_2, x_3, \ldots x_n)$ where x_i is a population of the particles. The following equations (Eqs. 1-2) update the velocity and position of the particles:

$$v_i(t+1) = w_v \cdot v_i(t) + c_1 \cdot rand() \cdot (x_{p\text{best}} - x_i(t)) + c_2 \cdot rand() \cdot (x_{g\text{best}} - x_i(t)) \quad (1)$$

$$x_i(t+1) = x_i(t) + v_i(t+1) \quad (2)$$

where $v_i(t)$ and $x_i(t)$ are the velocity and position of the i particle at t time. The w_v is the weight function of the velocity ($0 \leq w_v \leq 1.2$). $x_{p\text{best}}$ is the best possible value obtained by the particle i at time t, and $x_{g\text{best}}$ is the global best position value obtained by the swarm. The c_1 and c_2 are the constant components, which tune the values of

x_{pbest} and $x_{gbest}\{(0 \le c_1 \le 2), (0 \le c_2 \le 2)\}$. $rand()$ is the random number range from {0, 1}.

The following objective functions provide the optimal Ra and MRR values (Eqs. 3-4):

$$\mathrm{Ra} = 0.366 - 1.708 * \mathrm{d} - 2.7341 * \mathrm{f} + 0.0067 * \mathrm{v} + 8.75 * \mathrm{d}^2 + 35 * \mathrm{f}^2 - 12.09 * \mathrm{d} * \mathrm{f} - 0.0196 * \mathrm{d} * \mathrm{v} - 0.015 * \mathrm{f} * \mathrm{v} \quad (3)$$

$$\mathrm{MRR} = 0.7073 - 0.563 * \mathrm{d} - 10.356 * \mathrm{f} - 0.0053 * \mathrm{v} + 7.275 * \mathrm{d}^2 + 17.15 * \mathrm{f}^2 + 1.696 * \mathrm{d} * \mathrm{f} - 0.0131 * \mathrm{d} * \mathrm{v} + 0.0172 * \mathrm{f} * \mathrm{v} \quad (4)$$

where

$$0.15\,\mathrm{mm} \le d \le 0.45\,\mathrm{mm}$$
$$0.1\,\mathrm{mm/rev} \le f \le 0.25\,\mathrm{mm/rev}$$
$$50\,\mathrm{m/min} \le v \le 200\,\mathrm{m/min}$$

Figure 2a, b shows the objective function value (best cost) versus the number of iteration number of Ra and MRR, respectively, using pseudocode of PSO algorithm (Fig. 3). Global best value of Ra has been found as 0.4916 μm with the local best values of d = 0.15 mm, 0.25 mm/rev, and v = 200 m/min, respectively. Global best value of MRR has been found as 0.2849 g/s with the local best values of d = 0.15 mm, f = 0.25 mm/rev, and v = 50 m/min, respectively.

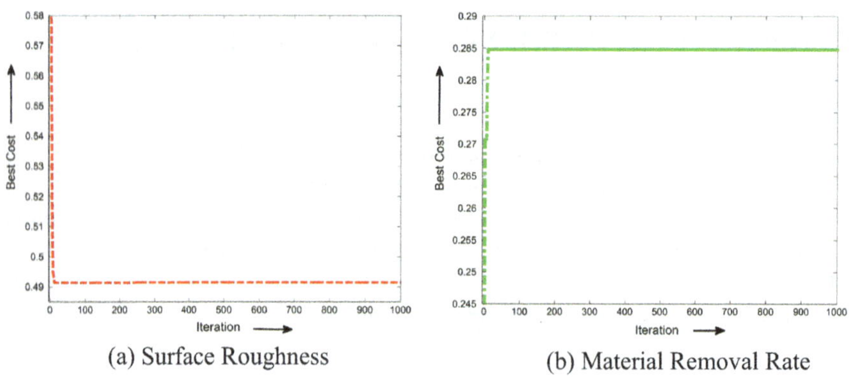

Fig. 2 a, b Objective function value (best cost) of Ra and MRR, respectively

For Every Particle
Initialization of Particle
End
Do
For Every Particle
Compute objective value
If *The objective value gives better value compare to "pbest" in the previous*
Use existingobjective value as the new "pbest".
End
Select the Particle from the best objective value of all particles as the "gbest"
For every Particle
Compute the Velocity of Particlefrom Velocity Function 1
Update Position of Particlefrom Velocity Function 2
End
Repeat the cycle until the optimum value is not achieved

Fig. 3 Pseudocode of PSO algorithm

6 Conclusion

In this research, the use of PSO approach was performed for prediction of developed model the sustainable output characteristics (MRR and surface roughness) with advantages in augmentation of product quality and quantity. The following conclusions may be illustrated:

- In most of the machining run, the arithmetic surface roughness averages (Ra) were within the suggested limit of 1.6 μm, i.e., equivalent with conventional cylindrical grinding.
- According to PSO, the global best value of Ra was observed to be 0.4916 μm with the local best values of d = 0.15 mm, 0.25 mm/rev, and v = 200 m/min, respectively.
- Likewise, global best value of MRR was found to be 0.2849 g/s with the local best values of d = 0.15 mm, f = 0.25 mm/rev, and v = 50 m/min, respectively.
- This clearly indicates that the PSO methodology for solving multi-objective optimization is to be effective for analysis of machining problems and further improve to machining efficiency.

The experimental consequences are valid inside the studied range of input process variables as well for the explicit amalgamation of the workpiece and cutting tool. Further, any additional investigation may be verified by over again performing the test. Moreover, various work materials and cutting tools require to be judged for in-depth analysis of cutting speed, feed rate, and depth of cutting on multi-performance attributes like attainable surface roughness, tooling flank wear, and cutting temperature with material characterization.

References

1. Kumar, R., Sahoo, A.K., Mishra, P.C., Das, R.K.: Comparative investigation towards machinability improvement in hard turning using coated and uncoated carbide inserts: part I experimental investigation. Adv. Manuf. **6**(1), 52–70 (2018)
2. Panda, A., Sahoo, A., Rout, R.: Multi-attribute decision making parametric optimization and modeling in hard turning using ceramic insert through grey relational analysis: a case study. Dec. Sci. Lett. **5**(4), 581–592 (2016)
3. Rajbongshi, S.K., Sarma, D.K.: Performance parameters studies in machining of AISI D2 steel with dot-textured, groove-textured & non-textured cutting tool at the flank face. Int. J. Refract Metal Hard Mater. **83**, 104970 (2019)
4. Scheffer, C., Kratz, H., Heyns, P. S., Klocke, F.: Development of a tool wear-monitoring system for hard turning. Int. J. Mach. Tools Manuf. **43**(10), 973–985 (2003)
5. Badiger, P.V., Desai, V., Ramesh, M.R., Prajwala, B.K.: Raveendra, K.: Cutting forces, surface roughness and tool wear quality assessment using ANN and PSO approach during machining of MDN431 with TiN/AlN-coated cutting tool. Arab. J. Sci. Eng. 1–13 (2019)
6. Mia, M., Królczyk, G., Maruda, R., Wojciechowski, S.: Intelligent optimization of hard-turning parameters using evolutionary algorithms for smart manufacturing. Mat. **12**(6), 879 (2019)
7. Sibalija, P.: Particle swarm optimisation in designing parameters of manufacturing processes: a review (2008–2018). Appl. Soft Comput. **84**, 1057 (2019)
8. Xie, N., Zhou, J., Zheng, B.: An energy-based modeling and prediction approach for surface roughness in turning. Int. J. Adv. Manuf. Tech. **96**(5–8) 2293–2306 (2018)
9. Manav, O., Chinchanikar, S., Gadge, M.: Multi-performance optimization in hard turning of AISI 4340 steel using particle swarm optimization technique. Mat. Today Proc. **5**(11), 24652–24663 (2018)
10. Chandrasekaran, M., Tamang, S.: ANN–PSO integrated optimization methodology for intelligent control of MMC machining. J. Instit. Eng. (India) Series C **98**(4), 395–401 (2017)
11. Bouacha, K., Terrab, A.: Hard turning behavior improvement using NSGA-II and PSO-NN hybrid model. Int. J. Advan. Manuf. Technol. **86**(9–12), 3527–3546 (2016)
12. Ilangovan, S., Vignesh, R. V., Padmanaban, R., Gokulachandran, J.. Comparison of statistical and soft computing models for predicting hardness and wear rate of Cu-Ni-Sn alloy. Prog. Comput. Anal. Netw. 559–571. Springer, Singapore (2018)
13. Nayak, J., Naik, B., Jena, A.K., Barik, R.K., Das, H.: Nature inspired optimizations in cloud computing: applications and challenges. In: Cloud Computing for Optimization: Foundations, Applications, and Challenges, pp. 1–26. Springer, Cham (2018)
14. Panda, A., Sahoo, A.K., Rout, A.K.: Investigations on surface quality characteristics with multi-response parametric optimization and correlations. Alex. Eng. J. **55**(2), 1625–1633 (2016)
15. Kumar, R., Sahoo, A.K., Das, R.K., Panda, A., Mishra, P.C.: Modelling of flank wear, surface roughness and cutting temperature in sustainable hard turning of AISI D2 steel. Proc. Manufact. **20**, 406–413 (2018)
16. Kumar, R., Sahoo, A.K., Mishra, P.C., Das, R.K.: Comparative study on machinability improvement in hard turning using coated and uncoated carbide inserts: part II modeling, multi-response optimization, tool life, and economic aspects. Adv. Manufact. **6**(2), 155–175 (2018)
17. Kumar, R., Sahoo, A.K., Mishra, P.C., Das, R.K.: Measurement and machinability study under environmentally conscious spray impingement cooling assisted machining. Measurement **135**, 913–927 (2019)
18. Kumar, R., Sahoo, A.K., Mishra, P.C., Das, R.K., Ukamanal, M.: Experimental investigation on hard turning using mixed ceramic insert under accelerated cooling environment. Int. J. Indus. Eng. Comput. **9**(4), 509–522 (2018)
19. Singh, A., Singh, H.: Effects process parameters on surface roughness and MRR in the hard turning of EN 36 steel. Int. J. Curr. Eng. Technol. **7**, 3 (2013)

Design and Performance Analysis of Wavelength Converter Using SOA for Optical WDM Network

Manoj Kr. Dutta

Abstract In all-optical Wavelength Division Multiplexing (WDM) technique, data contention is an inevitable challenge. For reliable data transmission, contention resolution is very much essential. There are different contention resolution techniques that are used in optical WDM network. In this paper, wavelength converter is used for contention resolution in optical WDM network. Here Semiconductor Optical Amplifier (SOA) is used to implement wavelength converter based on a special third-order nonlinear susceptibility phenomenon of silica fiber, called Four-Wave Mixing (FWM) technique. Realization of wavelength converter is done for two different sets of WDM networks. Performance of the wavelength converter was estimated in terms of number frequencies generated, number of new wavelength generated, its corresponding power, etc. Result shows that four-wave mixing is a very effective technique to generate new wavelengths at the output to overcome the problems due to contention in an optical WDM network. OptiSystem of Optiwave is used to design schematic layout for the wavelength converter using SOA.

Keywords Wavelength division multiplexing · Semiconductor optical amplifier · Wave length conversion · Fiber nonlinearity

1 Introduction

In the recent past, the development of different Internet-based applications like online banking, telemedicine, social networking sites, e-commerce, and many more demands huge raw bandwidth. Optical fiber is the only available transmission media which can provide the ever-increasing bandwidth requirement. A single optical fiber can provide almost 50 THz of raw bandwidth which in turn can accommodate thousands of customers effectively. Wavelength Division Multiplexing (WDM) or Dense Wavelength Division Multiplexing (DWDM) techniques are used to explore the huge available bandwidth of optical fiber [1, 2]. Although optical fiber provides very high

M. Kr. Dutta (✉)
BIT Mesra, Deoghar Campus, Jharkhand 814142, India
e-mail: mkdutta13@gmail.com

H. Das et al. (eds.), *Progress in Computing, Analytics and Networking*, Advances in Intelligent Systems and Computing 1119,
https://doi.org/10.1007/978-981-15-2414-1_25

speed optical transmission but when the signal is transmitted over the fiber, the signal suffers degradation due to various nonlinear effects name Four-Wave Mixing (FWM), Stimulated Raman Scattering (SRS), Stimulated Brillouin Scattering (SBS), Cross-Phase Modulation (XPM), Cross-Gain Modulation (XGM), etc. The signal strength within the fiber attenuates significantly due to these nonlinearities. Although the quality of the signal degrades because of different nonlinear effects, in some cases, they are very useful also. For example, four-wave mixing is useful for wavelength conversion technology, XPM is used for pulse compression, optical switching, pulse retiming, etc. However, the attenuation and loss of optical signal quality due to fiber nonlinearities can be compensated by using optical amplifiers [3, 4]. There are mainly two types of amplifiers used in WDM technology, namely, Optical Fiber Amplifier (OFA) and Semiconductor Optical Amplifier (SOA). OFA uses optical fiber that is doped with rare earth materials like (Er^{3+}) and is mainly used as in-line amplifier. Although the OFA is very useful, SOA is more commonly used for optical amplification for the following reasons. SOA is having a very wide amplification range; it is used to realize optical grating and wavelength converter and in-line amplifier, etc. Apart from the abovementioned advantages, SOA is one of the most important amplifiers which is used to implement all-optical wavelength converter using Four-Wave Mixing (FWM) technology [5, 6].

2 Wavelength Conversion Using Four-Wave Mixing

All-optical WDM technology offers very high-speed data transmission but when two or more signals wants to reserve the same wavelength at the output then data contention arises. As the information in the optical domain cannot be stored in the electrical RAM so it is very difficult to transmit the information without dropping it. To avoid this unwanted information loss, different contention resolution techniques are used in optical WDM technology. There are mainly four different technologies that are used to overcome the problems due to contention, namely, space domain, wavelength domain, time-domain contention resolution, and segmentation dropping scheme. In wavelength domain wavelength converter is used which changes the input wavelength to a new wavelength without changing the information-carrying capacity of the signal. In SOA, four-wave mixing technique is used for wavelength conversion [7–9]. Four-wave mixing is a fiber nonlinearity which arises from third-order nonlinear susceptibility of silicon optical fiber. In this process, three co-propagating waves of frequencies f_i, f_j, and f_k ($k \neq i, j$) interact with each other in silica fiber of a multichannel WDM system to generate a fourth wave of frequency given by: $f_{\rm ijk} = f_i + f_j - f$ [10].

Figure 1 shows a simple example for two waves at frequencies f_1 and f_2. When these waves co-propagate along a fiber, they mix and generate sidebands at $2f_1 - f_2$ (f_3) and $2f_2 - f_1$ (f_4). These sidebands will travel along with the original waves and will grow at the expense of signal-strength depletion. In general, for N wavelengths

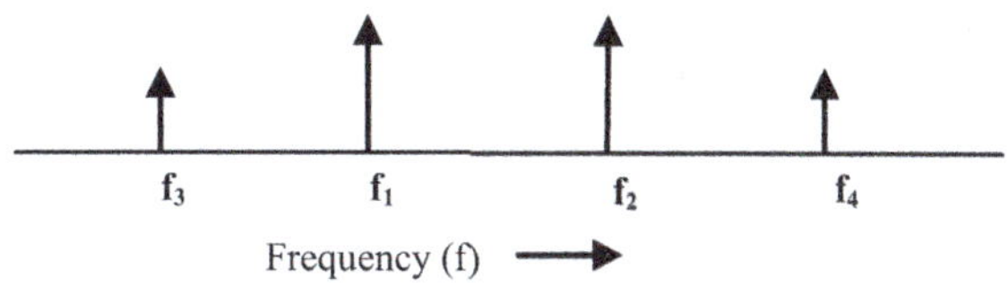

Fig. 1 FWM of two wavelengths (f_1 and f_2)

launched into a fiber, the number of generated mixing products M is, $M = \frac{N^2}{2(N-1)}$. If the channels are equally spaced, a number of new waves will have the same frequencies as the injected signals [11, 12] will be generated. Thus, the resultant crosstalk interference plus the depletion of the original signal waves can severely degrade multichannel system performance unless steps are taken to diminish it.

3 Result and Discussion

Performance of the SOA as an active medium for the use of wavelength converter was simulated using OptiSystem simulation software. Figure 2 represents the schematic layout of wavelength conversion using SOA for 2 × 1 WDM network. Figure 3 shows the frequency versus power spectrum for a 2 × 1 WDM network. Figure 4 represents the Frequencies versus Power spectrum of generated new frequencies due to wavelength conversion. Figure 5 indicates the available frequencies versus power spectrum of the output frequencies of the WDM network. The result is very interesting because it shows that at the input of the network, there was only two available wavelengths but due to the third-order nonlinearity of the silica fiber, the SOA has generated two additional frequencies which are usable at the output. If contention

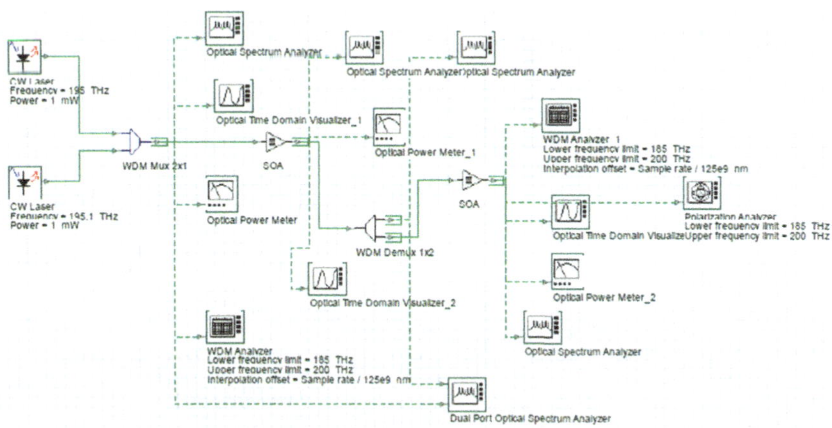

Fig. 2 Schematic layout of wavelength conversion using SOA for 2 × 1 WDM network

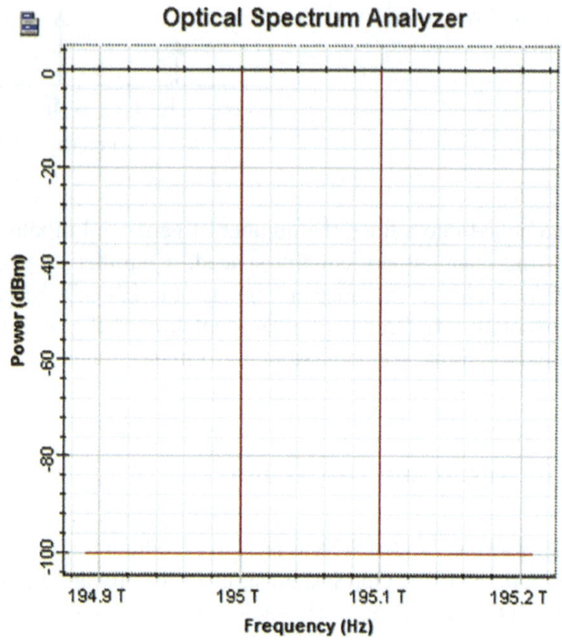

Fig. 3 Frequency versus power spectrum of input signal

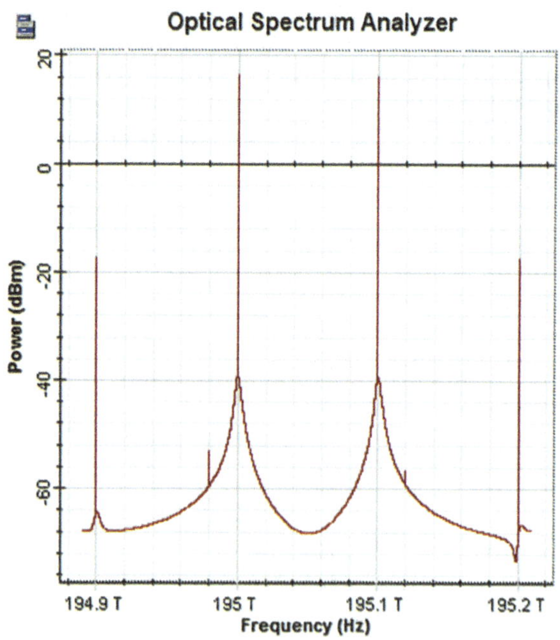

Fig. 4 Frequencies versus power spectrum of generated frequencies due to wavelength conversion

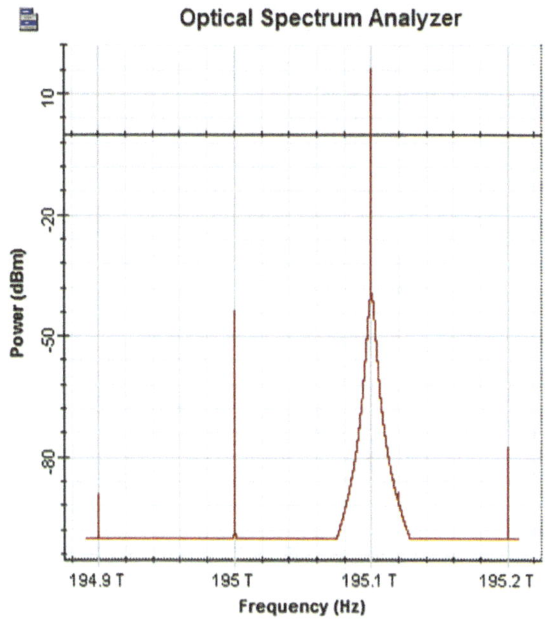

Fig. 5 Frequencies versus power spectrum of the output frequencies

occurs at the required wavelength then the information could be transmitted through the newly generated wavelengths. This technique is very useful to overcome the problems due the contention in the all-optical WDM/DWDM network. Figure 6 shows the comparative spectrum of wavelength versus power of the input wavelength and generated wavelengths due to conversion. This figure clearly verifies the generation of two new wavelengths apart from the original signal due to four-wave mixing technique. Figure 7 shows the schematic layout of wavelength conversion using SOA for 4 × 1 WDM network. Figure 8 represents the frequency versus power spectrum analysis of the input signals. Figure 9 shows the frequencies versus power spectrum presentation of the generated frequencies due to wavelength conversion. This figure ensures the generation of the new wavelengths due to the third-order nonlinearity of the silica fiber. Figure 10 shows the available frequencies at the output. Figure 11 depicts the comparative spectrum of the input frequencies and the output wavelengths generated due to the four-wave mixing. This result clearly indicates that the number of available wavelengths at the output is more than the number of wavelengths at the input.

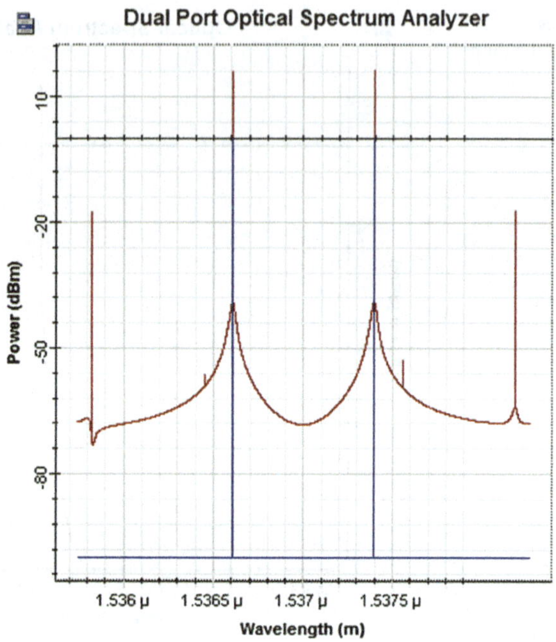

Fig. 6 Wavelength versus power spectrum of the input wavelength and generated wavelengths due to conversion

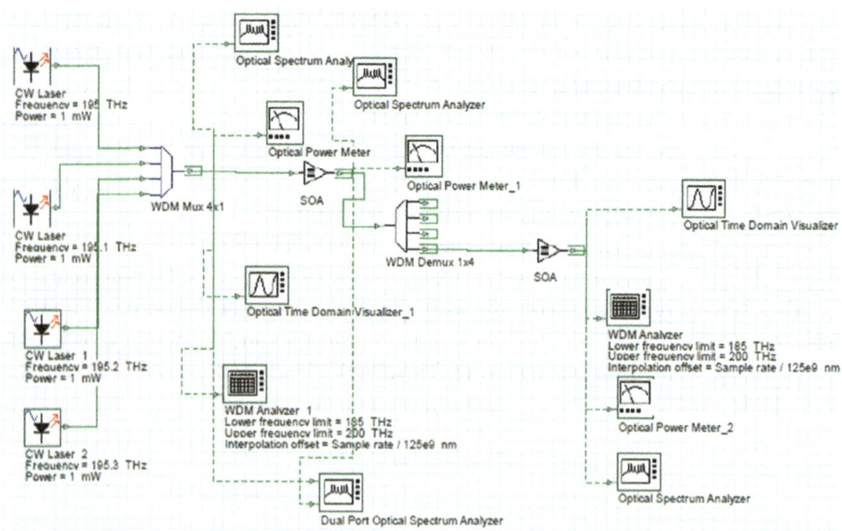

Fig. 7 Schematic layout of wavelength conversion using SOA for 4 × 1 WDM network

Fig. 8 Frequency versus power spectrum of input signal

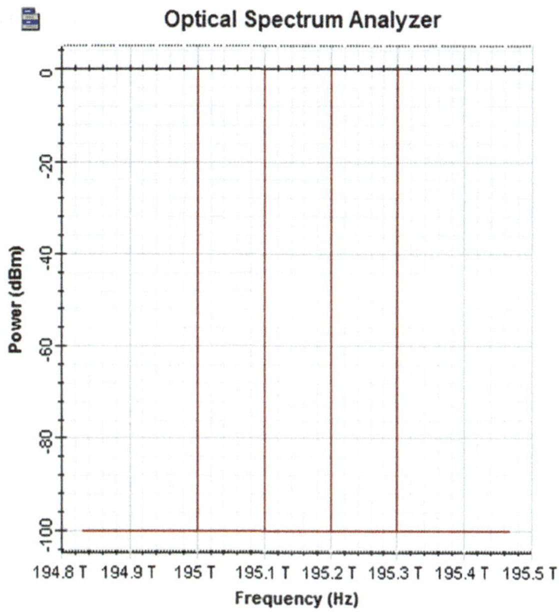

Fig. 9 Frequencies versus power spectrum of generated frequencies due to wavelength conversion

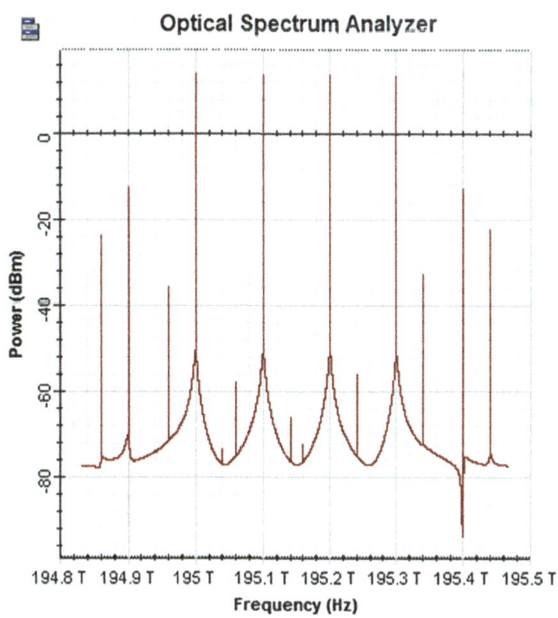

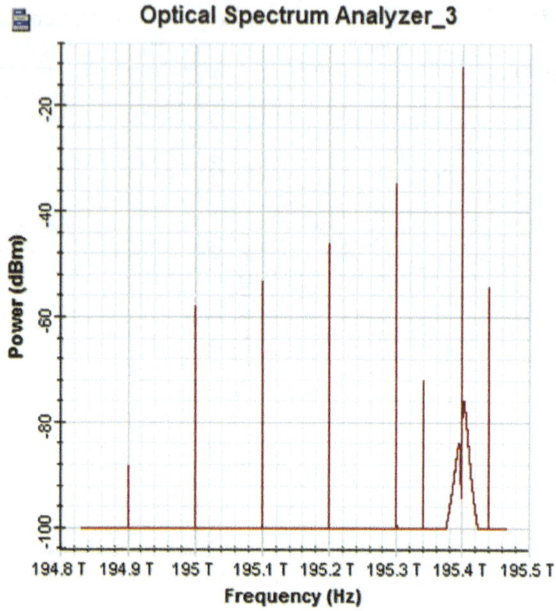

Fig. 10 Frequencies versus power spectrum of the output frequencies

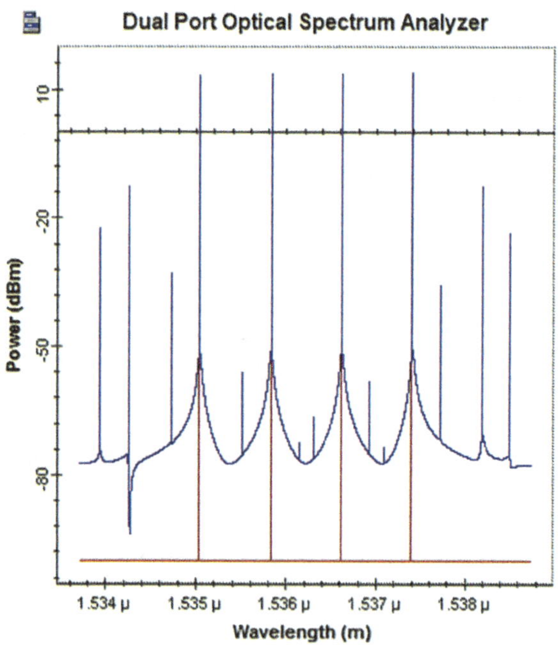

Fig. 11 Wavelength versus power spectrum of the input wavelength and generated wavelengths due to conversion

4 Conclusions

Contention due to overlapping signals at the output is a big problem in all-optical WDM network. For high-speed data transmission through optical network, contention resolution is a very important issue. There are different contention resolution techniques available nowadays. But wavelength conversion is one of the best available contention resolution techniques. When optical signal travels through optical fiber, it undergoes attenuation due to different nonlinear effects present in the optical fiber. Amplifiers are used to enhance the power of the optical signal. In this paper, Semiconductor Optical Amplifier (SOA) is used as an optical amplifier. SOA shows a very unique feature of generating new transmission wavelength by using a special type of nonlinearity called Four-Wave Mixing (FWM). FWM is caused due to the third-order nonlinearity of silicon optical fiber. In this work, schematic layout is designed for the generation of new wavelength using SOA implementing four-wave mixing technique. Results show that generation of different new wavelengths may be used for the data transmission at the time of network congestion.

References

1. Agrawal, G.P.: Fiber-Optic Communication Systems. New York. Wiley, vol. 546, pp. 226–260 (2002)
2. Osamu, A., Masateru, T., Shu, N.: FourWave mixing in optical fibers and its applications. Furukawa Rev. (2000)
3. Dutta, M.K., Chaubey, V.K.: Performance analysis of modified Optical Burst Switch ing (OBS) ring network using dummy node. Int. J. Light Elect. Optics (OPTIK) Elsevier Publication **123**, 1847–1851 (2012)
4. Byrav, R., Biswanath, M.: Wavelength conversion in WDM net working. IEEE J. Sel. Areas Commun. **16**(7) (1998)
5. Singh, S., Singh, A., Kaler, R.S.: Performance evaluation of EDFA, RAMAN and SOA optical amplifier for WDM systems. Optik Int. J. Light Electron. Opt. **124**(2), 95–101 (2013)
6. Dutta, M.K.: Design and performance analysis of EDFA and SOA for optical WDM networks: a comparative study. In: 14th IEEE India Council International Conference INDICON 2017, December 15–17, at IIT Roorkee, India (2017)
7. Schnabel, R., et al.: Polarization insensitive frequency conversion of a 10-channel OFDM signal using four-wave mixing in a semiconductor laser amplifier. IEEE Photon. Technol. Lett. **6**, 56–58 (1994)
8. Ludwig, R., Raybon, G.: BER measurements of frequency converted signals using four-wave mixing in a semiconductor laser amplifier at 1, 2.5, 5, and 10 Gbit/s. Electron. Lett. **30**, 338–339 (1994)
9. Zhou, J., et al.: Four-wave mixing wavelength conversion efficiency in semiconductor trav eling-wave amplifiers measured to 65-nm of wavelength shift. IEEE Photon. Tech. Lett. **6**, 984–987 (1994)
10. Tkach, R.W., et al.: Four-photon mixing and high-speed WDM systems. J. Lightwave Technol. **13**, 841–849 (1995)
11. Dutta, M.K., Chaubey, V.K.: Comparative analysis of wavelength conversion and segmentation based dropping method as a contention resolution scheme in Optical Burst Switching (OBS) network. Proc. Eng. Elsevier **30**, 1089–1096 (2012)
12. Singh, S.: Boost up of four wave mixing signal in semiconductor optical amplifier for 40 Gb/s optical frequency conversion. Opt. Commun. **281**(9), 2618–2626 (2008)

Comparative Study Between Star and Mesh Topology for the Application in All-Optical WDM Network

Manoj Kr. Dutta

Abstract The development of different Internet-based applications requires huge raw bandwidth and proper network management. Optical fiber along with all-optical Wavelength Division Multiplexing (WDM) network is one of the best possible solutions to overcome the bandwidth related issues. The performance of any network is heavily influenced by the choice of the proper network topology. In this paper, a comparative performance analysis between star- and mesh-based optical WDM networks is reported. The required mathematical model is developed to determine the blocking probabilities versus incoming traffic load of the said network topologies for different network parameters. Blocking probabilities of the networks are calculated for various channel capacities and different number of generated calls. The investigation further extended for the node with slow processing speed but having the same traffic load. The performance analysis has been repeated for different data processing speeds. Result shows that the mesh-based optical WDM network provides better blocking probability performance for both slow and fast networks.

Keywords Blocking probability · Star network · Mesh network · All-optical WDM network · Incoming traffic

1 Introduction

Rapid growth of different Internet-based applications, for example, Internet banking, e-commerce, telemedicine, e-learning, online gaming, etc., demands a huge raw bandwidth. Optical fiber is probably the only medium which can provide enormous bandwidth for data communication [1, 2]. Only huge available bandwidth does not fulfill the basic requirements of successful data transmission, but at the same time, communication speed and reliability is also very important aspects of information processing. All-optical Wavelength Division Multiplexing (WDM)/Dense Wavelength Division Multiplexing (DWDM) is the most useful technology to utilize

M. Kr. Dutta (✉)
BIT Mesra, Deoghar Campus, Mesra 814142, Jharkhand, India
e-mail: mkdutta13@gmail.com

H. Das et al. (eds.), *Progress in Computing, Analytics and Networking*,
Advances in Intelligent Systems and Computing 1119,
https://doi.org/10.1007/978-981-15-2414-1_26

the enormous raw bandwidth offered by a single fiber. All-optical WDM technique is implemented by using various network topologies. The most common topologies that are used in WDM technology is bus, star, ring, and mesh topologies [3, 4], etc. All the topologies have few practical advantages and disadvantages. Depending upon the associated advantages, disadvantages, and the field of applications, the network engineers select perfect network topology for optimum utilization [5, 6]. Here in this paper, two most common topologies, namely, star and mesh network, are discussed. Figure 1 shows the WDM-based star topology, and Fig. 2 shows the WDM-based mesh topology.

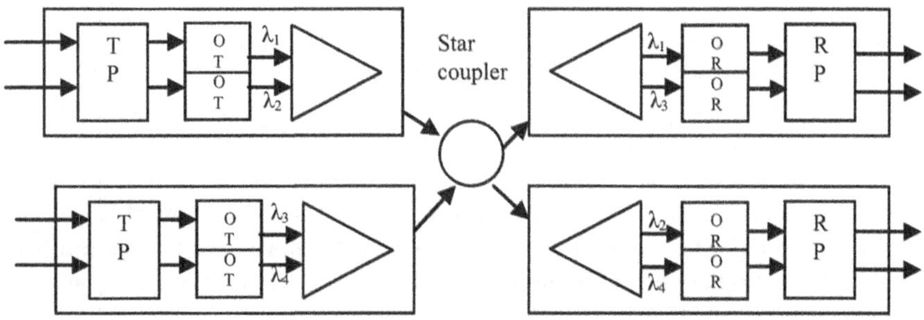

Fig. 1 Star topology

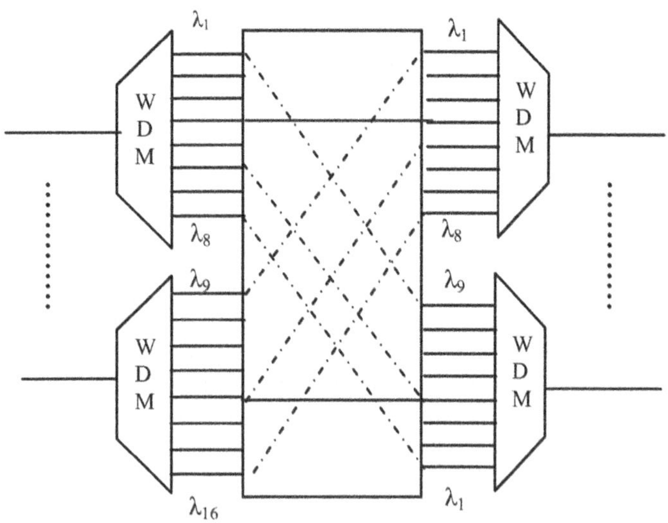

Fig. 2 Mesh topology

2 Mathematical Analysis

There are different topologies used in optical WDM networks. The most common examples are BUS, star, ring, and mesh topology, etc. [7–9]. In this paper, star and mesh topologies for the transmission of information in all-optical WDM network are discussed. In star topology, all the devices are connected to the central star coupler and the data traffic is exchanged between the nodes through the star coupler only (Fig. 1). In mesh topology, all the nodes are connected to each other individually. So if there are "n" no. of nodes, then there are "n (n–1)/2" no. of links (Fig. 2). All the topologies have their own advantages and disadvantages. Star topology reduces the unnecessary traveling of data packets through other nodes, it is very easy to add any new devices and also the faulty part can be easily be detected and repaired in case of star topology. Star topology, it is heavily dependent on central star coupler and the scalability of this topology is highly dependent on the central coupler. Mesh topology is a point to point connection so it highly secured and reliable and does not depend upon a single hub [10–13]. Though it is highly secured, there may be some redundant connections which will increase the overall cost and complexity of the network.

2.1 Star Network

In order to evaluate the performance of a star network, it is needed to derive the probabilistic evaluation of star network. The present star topology contains "n" edge nodes connected via central hub having "λ" calls generated at each node with service rate "μ" with channel capacity "C". The traffic can be expressed as

$$\rho = \frac{n\lambda}{C\mu} \tag{1}$$

The blocking probability for the star topology is derived for a case of "w" channels

$$P_{star} = \rho^{w+1}[1 - \rho]/[1 - \rho^{w+2}] \tag{2}$$

The probability that a packet coming at any edge node will be serviced by star node is $1–P_{star.}$ This makes the effective number of the data packets being routed at each node as λ $(1–P_{star})$. Net blocking probability of star topology can be written as

$$P_{B\_star} = [1 - \rho^{w+1}(1 - \rho)/(1 - \rho^{w+2})]\sum_{k=0}^{w-1}\frac{\rho^k}{k!}\mathrm{P}_0 \tag{3}$$

where,

$$p_0 = \frac{1}{1 + \frac{\rho^w\{1-(\frac{\rho}{w!})^{w+1}\}}{(1-\frac{\rho}{w})w!} + \sum_{k=0}^{w-1} \frac{\rho^k}{k!}} \tag{4}$$

2.2 Mesh Network

The present mesh topology contains "n" edge nodes connected to each other having "λ" calls generated at each node with service rate "μ" with channel capacity "C".

The traffic can be expressed as

$$\rho = \frac{[n(n-1)+2]\lambda}{2C\mu} \tag{5}$$

The blocking probability for the mesh topology is derived as for a case of w-channels:

$$P_{mesh} = \rho^{w+1}[1-\rho]/[1-\rho^{w+2}] \tag{6}$$

The probability that a packet coming at any edge node will be serviced by mesh node is $1–P_{mesh}$. This makes the effective number of the data packets being routed at each node as $\lambda\ (1–P_{mesh})$.

Net blocking probability of mesh topology can be written as

$$P_{B\_mesh} = [1-\rho^{w+1}(1-\rho)/(1-\rho^{w+2})]\left(\frac{\rho^k/k!}{\sum_{k=0}^{w-1}\rho^k/k!}\right)\left(\frac{1}{1 + \frac{\rho^w\{1-(\frac{\rho}{w!})^{w+1}\}}{(1-\frac{\rho}{w})w!} + \sum_{k=0}^{w-1} \frac{\rho^k}{k!}}\right) \tag{7}$$

The performance of a network may be evaluated on the basis of its ability to process the arriving packets efficiently.

3 Result and Discussions

The abovementioned equations are used to evaluate the blocking probabilities of mesh and star topologies for the application in all-optical WDM network. Performance of the star and mesh network was measured in terms of different network parameters, viz., channel capacity, number of calls and service rate, etc. Figure 3 represents the blocking probability versus normalized offered traffic of mesh topology for different number of generated calls at each node with Channel Capacity (CC) of 20 calls. The result shows that as the number of calls generated at each node increases, the overall blocking probability increases. The reason is that the channel capacity for a particular node is fixed but if the no of calls generated at that node increases, then the probability of blocking also increases significantly. This effect is predominant when the offered traffic is high. Figure 4 shows the same representation for Star network. The qualitative behavior of the blocking probability response is same for both the figures but the quantitative nature is different. Figure 4 shows that the overall blocking probability for star network is more than that of the mesh topology for the all values of calls generated at each node. The reason may be because the data packets for Star topology pass through a central hub. If the data rate increases then there will be a packet contention at the central hub and the blocking probability increases which in turn reduces the data processing speed.

Figures 5 and 6 depict the blocking probability of mesh network and star network, respectively, for channel capacity of 24. The graph reflects that the overall blocking probability of the network reduces for all values of call numbers. It is very interesting to note the blocking probability for the highest number of calls generated is reduced significantly comparing to the previous one for tar network. By increasing the number of output channels only by four, the blocking probability decreases by almost 15%. Figures 7 and 8 represent the blocking probability versus incoming traffic for different

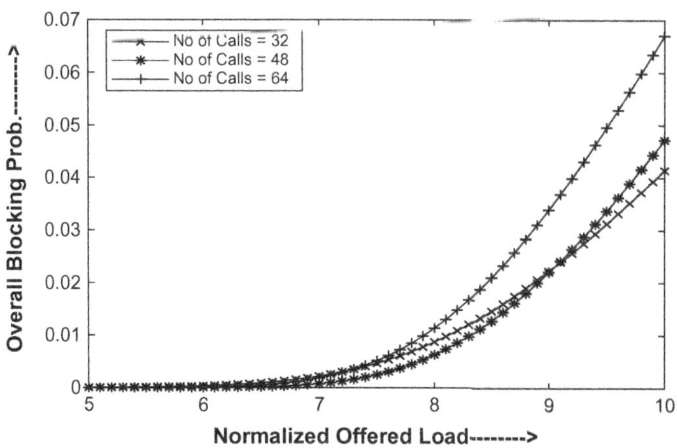

Fig. 3 Blocking prob. versus normalized offered traffic mesh topology for CC = 20

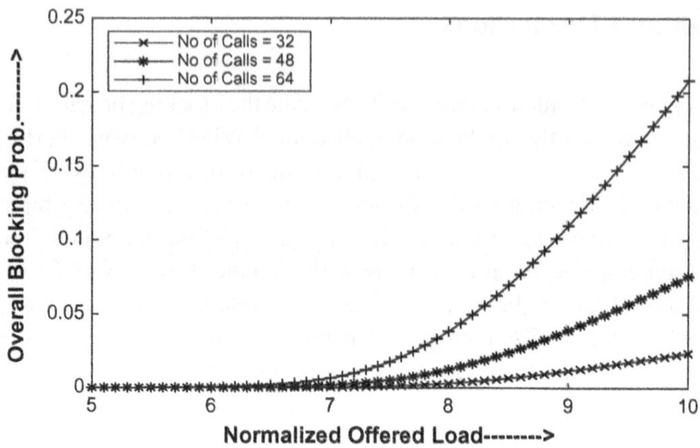

Fig. 4 Blocking prob. versus normalized offered traffic of star topology for CC = 20

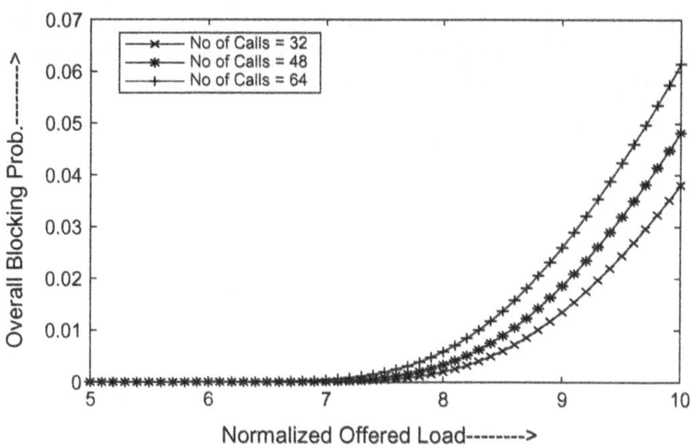

Fig. 5 Blocking prob. versus normalized offered traffic mesh topology for CC = 24

range of calls generated with channel capacity of 20 for both mesh and star topologies, respectively. However, in these cases, the processing speed of the node is reduced by keeping in the mind that the processing capacity of any node may be changed due to different practical parameters like bandwidth utilization factor, different modulation schemes and data transmission rate, etc. The graphs show that the overall behavior of the networks will remain almost same but the blocking probability changes with the incoming traffic.

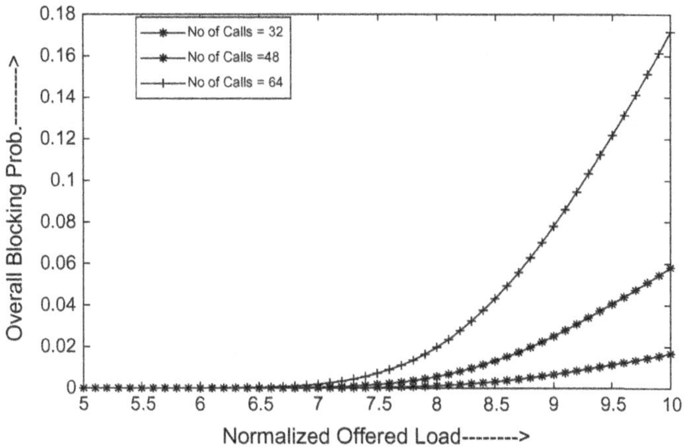

Fig. 6 Blocking prob. versus normalized offered traffic star topology for CC = 24

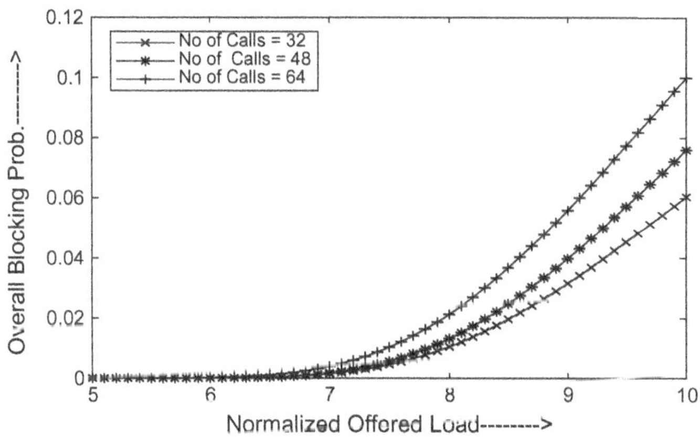

Fig. 7 Blocking prob. versus normalized offered traffic mesh topology for CC = 20 (low processing speed)

4 Conclusion

Optical fiber provides huge source of raw bandwidth but for successful data transmission, proper resource utilization technique is essential. Optical WDM technology is the most useful scheme for proper utilization of available bandwidth. WDM network is implemented using different network topologies. Here comparative performance analysis of two different most common network topologies, namely, star and mesh topologies, is considered. Performance of the two topologies is measured

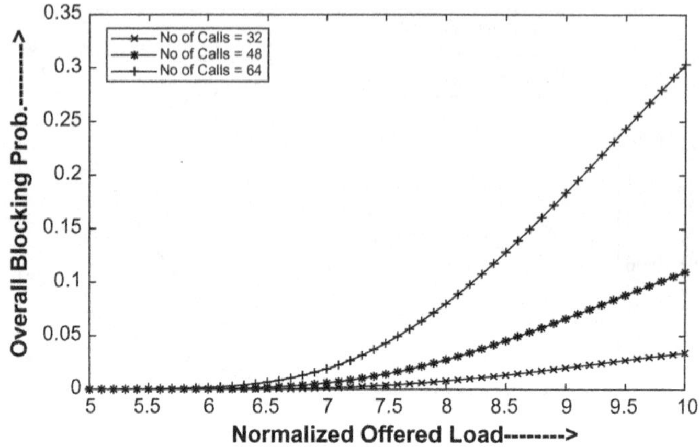

Fig. 8 Blocking prob. versus normalized offered traffic of star topology for CC = 20 (low processing speed)

using different network parameters, viz., blocking probability versus incoming traffic for different values of channel capacity and call generation rate. The investigation further extended for slow processor also. The processing speed of any node may be changed due to different factors like bandwidth utilization factor, modulation schemes and traffic flow, etc. The simulation result shows that mesh network provides better performance comparing to star topology.

References

1. Mukherjee B.: WDM optical communication networks: progress and challenges. IEEE J. Sel. Areas Commun. **18**(10) (2000)
2. Gençata, A., Mukherjee, B.: Virtual-topology adaptation for WDM mesh networks under dynamic traffic. IEEE/ACM Trans. Netw. **11**(2) (2003)
3. Krishnaswamy, R.M., Sivarajan, K.N.: Design of logical topologies: a linear formulation for wavelength-routed optical networks with no wavelength changers. IEEE/ACM Trans. Netw. **9**(2), 186–198 (2001)
4. Prasad, S., Dutta, M.K., Sarkar, R.K.: Coupled tapering/uptapering of Thirring type soliton pair in nonlinear media. Opt. Commun. **410**, 356–360 (2018)
5. Zhou, Z., Lin, T., Thulasiraman, K., Xue, G.: Novel survivable logical topology routing by logical protecting spanning trees in IP-Over-WDM networks. IEEE/ACM Trans. Netw. **25**(3) (2017)
6. Musa, M., Elgorashi, T., Elmirghani, J.: Energy efficient survivable IP-over-WDM networks with network coding. IEEE/OSA J. Opt. Commun. Netw. **9**, 207–217 (2017)
7. Dutta, M.K., Chaubey, V.K.: Modeling and performance analysis of optical WDM node architecture using SRR protocol. Int. J. Light Electron Opt. (OPTIK), **123**, 1971–1974 (2012). Elsevier Publication
8. Klaus, W., Puttnam, B.J., Luís, R.S., Sakaguchi, J., Mendinueta, J.M.D., Awaji, Y., Wada, N.: Advanced space division multiplexing technologies for optical networks. J. Opt. Commun. Netw. **9**(4) (2017)

9. Dutta, M.K.,.Chaubey, V.K.: Performance analysis of modified optical burst switching (OBS) ring network using dummy node. Int. J. Light Electron Opt. (OPTIK) **123**, 1847–1851 (2012). Elsevier Publication
10. Nayak T.K., Sivaranjan, K.N.: A new approach to dimensioning optical networks. IEEE JSAC **20**, 134–148 (2002)
11. Nayak, T.K., Sivaranjan, KN.: Dimensioning optical networks under traffic growth models. IEEE/ACM Trans. Netw. **11**(6), 935–947 (2003)
12. Feller, W.: An Introduction to Probability Theory and Its Applications. Wiley, New York (1968)
13. Leighton, T., Rao, S.: Multi commodity max-flow min-cut theorems and their use in designing approximation algorithms. J. ACM (JACM) **46**(6), 787–832 (1999)

Predicting Impact of Odia Newspaper Articles on Public Opinion

Manoj Kumar Jena and Sanghamitra Mohanty

Abstract News articles regarding a political party, a political issue, a business house, or a particular product can be judged as the feeling of writers over a time. News article can also be treated as a form of public opinion as by reading those news articles; the public opinion is influenced relating to a particular topic. So that news article can be analyzed to extract the sentiment or for opinion mining by using different supervised learning models. This paper proposes the methods to extract opinion content from Odia news article available in portal and e-paper sites using supervised learning models. A total of 500 related news articles are collected from different newspapers manually. Out of that, 350 news articles are used from training the classifier and 150 news articles are used for testing the classifier. Those collected news articles are preprocessed and then vectorized with respect to tf–idf score which is calculated using unigram and bigram representation of data at document level passed to SVM, and the results are analyzed by calculating the accuracy and F1 score.

Keywords SVM · News articles · Tf · Tf–idf · Opinion mining

1 Introduction

Newspapers are the mirror of the society, which reflect local to global issues relating to public. In the advent of web technologies, many newspapers are making their content available through e-papers and news portal. In the era of web revolution, most of the newspaper's contents are available in web and many news portals are available in various language. Common people always prefer to read news articles in their own language especially in their mother tongue. Odisha is a state located in eastern part of India and Odia is the official language of the state. The Odia language is spoken

M. K. Jena (✉) · S. Mohanty
P. G. Department Computer Science and Application, Utkal University, Bhubaneswar, Odisha, India
e-mail: manojoctmp@gmail.com

S. Mohanty
e-mail: sangham1@rediffmail.com

H. Das et al. (eds.), *Progress in Computing, Analytics and Networking*, Advances in Intelligent Systems and Computing 1119,
https://doi.org/10.1007/978-981-15-2414-1_27

by more than 40 million population of Odisha. There are more than 50 newspapers are published daily from different parts of Odisha. Most prominent dailies have their own e-paper and news portal. The news available in the portal can reach the general people quickly and influences the public opinion regarding a political party, a political issue, a business house or a particular product. The reader are ventilating their opinion in social media about a particular topic after reading those news article on that topic, which is generally influenced by opinion content of that news article. The motivation behind the work is to judge how a news article will influence public opinion on reading that particular article. The processing of available Odia news article is a challenge for the Natural Language Processing (NLP) researchers as very less resource is available for experimenting with news article in comparison to other language like English. The new machine learning techniques and text analysis tools has paved the way processing opinions. The data are collected manually and preprocessed by removing the numbers, punctuation marks, and stop words from the text. The preprocessed data is vectorized using Python and NLTK tools and then tf–idf score vector is calculated at document level using unigram and bigram data representation. The tf–idf vector passes to SVM classifier which classifies the text as positive, negative, or neutral with a confidence score.

This paper is organized as follows: Sect. 2 provides overview of related works on opinion mining. Section 3 contains elaborate explanation of the proposed method. Section 4 discusses the experimental results. Section 5 concludes and provides future scope of this proposed method.

2 Related Works

Generally, opinions are intended to be positive, negative, or neutral on a particular entity. Liu [1] defines an opinion as a quintuple, where O_i is the target object, f_{ij} is the feature of the target object O_i, h_i is the opinion holder, t_l is the time when the opinion is expressed, and so_{ijkl} is the sentiment value of the opinion expressed by the opinion holder h_i about the object O_i at time t_l. Given a set of evaluative text documents D that contain opinions (or sentiments) about an object, opinion mining aims to extract attributes and components of the object that have been commented on in each document $d \in D$ and to determine whether the comments are positive, negative, or neutral. Opinion mining has two main research directions, document-level opinion mining and feature-level opinion mining [2]. Document-level mining involves determining the document's polarity by calculating the average semantic orientation of extracted phrases. Opinion mining can be viewed as a kind of processing of natural language for tracking the attitudes, feelings or appraisal of the public about particular topic, product or services. All information available in web is of two types: facts and opinions [3]. Bruce et al. [4] have experimented the opinion mining from newspaper headline using linear SVM, tf–idf linear SVM, and SGD classifier; the comparison between the results of three classifiers has been made, and the tf–idf with linear SVM outperforms other two classifiers in terms of accuracy. However, the method proposed by van Leeuwen [5] opinion mining in newspaper articles by entropy-based word connection does not yield better result in comparison to opinion mining with SVM classifiers. Rameshbhai and Paulose [6] proposed a method for opinion mining using support vector machine using SVM but the result is not so impressive. van Leeuwen [5] has proposed algorithm classification of emotion like anger, disgust, joy, sadness, fear, and surprise from the content on SEN_VAL affective dataset. The method proposed by Scholz and Conard [7] uses integration of domain, syntactic, and lexical knowledge, and will not be applicable for resources scarce language like Odia. Jena and Mohanty [8] have proposed a descent method annotation guideline and labeling process but inter-annotator agreement was very low. Patil and Chaudhari [9] have proposed a method to find the polarity of opined sentences using N-gram-based support vector machine. The method proposed by Olsher [10] is impressive which predicts the stock market prediction by analyzing the sentimental words from newspaper article. Olsher [10] has described the role of negation and discourse relation for sentiment analysis using Hindi Sentiword Network. Balahur and Steinberger [11] have proposed the method of applying supervised opinion mining techniques on online user reviews with naïve Bayes classifier using $N = 3$ and eliminating stop words, which gives better result than $N = 1$ and $N = 2$. In this paper, the experimentation is done through Python with NLTK for calculating the tf–idf score of the documents, and SVM_{light} is used for learning from the dataset and evaluating. Jena and Mohanty [12] and Kim et al. [13] have proposed novel classification techniques for applying on different aspects like classification of diabetes mellitus disease and intrusion detection.

3 Methodology

This paper proposes a method for classification of news articles and predicting the reader's opinion on news articles, which is depicted in Fig. 1.

The whole process of opinion mining passes through six phases. Manually collected news articles are grouped as a corpus and preprocessed for removing the numbers, emails, URLs punctuations, and stop words as they have less information content by Python code. The example of news articles in Odia language is given below:

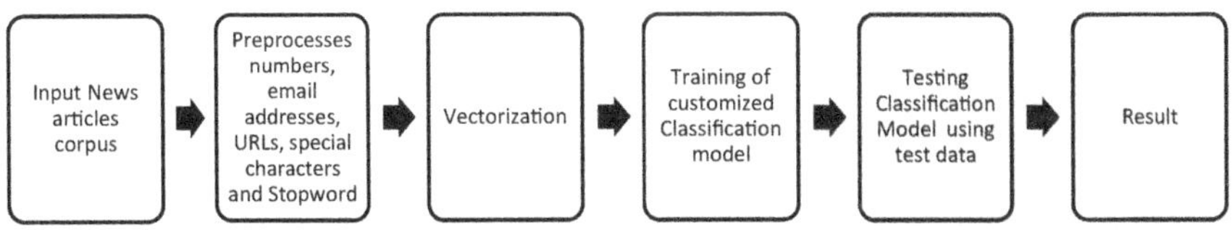

Fig. 1 Diagrammatic view of proposed approach of Opinion Classification

୨୦୧୯ ଲୋକସଭା ନିର୍ବାଚନ ଶେଷ ହୋଇଛି। କିନ୍ତୁ ଏହାକୁ ନେଇ ବିବାଦ ସରିନାହିଁ। ପଶ୍ଚିମବଙ୍ଗରେ ବିଜେପି ଏବଂ ଟିଏମସି ମଧ୍ୟରେ ହିଂସା ଜାରି ରହିଥିଲାବେଳେ ୟୁପିଏ ଅଧ୍ୟକ୍ଷା ସୋନିଆ ଗାନ୍ଧୀ ବିଜେପିକୁ ଟାର୍ଗେଟ କରିଛନ୍ତି। ନିର୍ବାଚନ ପରେ ପ୍ରଥମ ଥର ପାଇଁ ଏଠାରେ କର୍ମୀମାନଙ୍କୁ ଉଦ୍‌ବୋଧନ ଦେଇ ସୋନିଆ କହିଛନ୍ତି, ନିର୍ବାଚନରେ ଭୋଟରମାନଙ୍କୁ ବିପଥଗାମୀ କରିବାକୁ ଉଦ୍ୟମ ହୋଇଥିଲା। ଆପଣମାନେ ସମସ୍ତେ ଏହା ଦେଖିଛନ୍ତି। ସମଗ୍ରୀ ଦେଶ ବି ବୁଝିପାରିଛି ଯେ ଯାହାସବୁ ହୋଇଥିଲା ତାହା ନୈତିକ ନା ଅନୈତିକ ଥିଲା। ଏକ ନିର୍ଦ୍ଦିଷ୍ଟ ଦଳ କ୍ଷମତାକୁ ଆସିବା ଲାଗି ନୀତିନିୟମର ସମସ୍ତ ସୀମା ଟପିଗଲା। ଦେଶ ପାଇଁ ଏହାଠାରୁ ଅନ୍ୟ କିଛି ଅଧିକ ଦୁର୍ଭାଗ୍ୟଜନକ ହେବ ନାହିଁ ବୋଲି ସୋନିଆ ବିଜେପି ନାଁ ନ ନେଇ କହିଥିଲେ।
ତେବେ ଏହି କାର୍ଯ୍ୟକ୍ରମ ସ୍ଥଳକୁ ଗଣମାଧ୍ୟମ ପ୍ରତିନିଧିମାନଙ୍କୁ ଯିବାକୁ ଅନୁମତି ମିଳି ନ ଥିଲା। କିନ୍ତୁ କେତେକ ଦଳୀୟ କର୍ମୀ ସୋନିଆଙ୍କ ଉଦବୋଧନର ଭିଡିଓ କ୍ଲିପ ଶେୟାର କରିଛନ୍ତି। ନିର୍ବାଚନୀ ପ୍ରକ୍ରିୟା ଉପରେ ସୋନିଆ ପ୍ରଶ୍ନ ଉଠାଇ କହିଥିଲେ, ବିଗତ ୫ବର୍ଷ ଧରି ଭାରତର ନିର୍ବାଚନ ପ୍ରକ୍ରିୟା ଉପରେ ସନ୍ଦେହ ସୃଷ୍ଟି ହୋଇଛି। ଏକ ପୁରୁଣା କଥା ଅଛି, 'ନିଆଁ ନ ଥାଇ ଧୂଆଁ ଆସେ ନାହିଁ' ବୋଲି ସୋନିଆ କହିଥିଲେ। ଏହା ସହିତ ରାଏବରେଲି ବାସିନ୍ଦାଙ୍କ ସହିତ ତାଙ୍କର ସମ୍ପର୍କ ନୂଆ ନୁହେଁ କି ନିଜସ୍ୱ ସ୍ୱାର୍ଥ ଆଧାରିତ ନୁହେଁ ବୋଲି ସେ କହିଥିଲେ।

The cleaned text is vectorized by vectorizer code and which calculates term frequencies (tf) and inverse document frequencies (idf). As the frequently occurring word having high importance than others and also tf–idf is calculated at document level from unigram and bigram representation of cleaned text. Example of unigram and bigram representation is as follows:

News article: ତେବେ ଏହି କାର୍ଯ୍ୟକ୍ରମ ସ୍ଥଳକୁ ଗଣମାଧ୍ୟମ ପ୍ରତିନିଧିମାନଙ୍କୁ ଯିବାକୁ ଅନୁମତି ମିଳି ନ ଥିଲା। କିନ୍ତୁ କେତେକ ଦଳୀୟ କର୍ମୀ ସୋନିଆଙ୍କ ଉଦବୋଧନର ଭିଡିଓ କ୍ଲିପ ଶେୟାର କରିଛନ୍ତି। ନିର୍ବାଚନୀ ପ୍ରକ୍ରିୟା ଉପରେ ସୋନିଆ ପ୍ରଶ୍ନ ଉଠାଇ କହିଥିଲେ, ବିଗତ ୫ବର୍ଷ ଧରି ଭାରତର ନିର୍ବାଚନ ପ୍ରକ୍ରିୟା ଉପରେ ସନ୍ଦେହ ସୃଷ୍ଟି ହୋଇଛି। ଏକ ପୁରୁଣା କଥା ଅଛି, 'ନିଆଁ ନ ଥାଇ ଧୂଆଁ ଆସେ ନାହିଁ' ବୋଲି ସୋନିଆ କହିଥିଲେ। ଏହା ସହିତ ରାଏବରେଲି ବାସିନ୍ଦାଙ୍କ ସହିତ ତାଙ୍କର ସମ୍ପର୍କ ନୂଆ ନୁହେଁ କି ନିଜସ୍ୱ ସ୍ୱାର୍ଥ ଆଧାରିତ ନୁହେଁ ବୋଲି ସେ କହିଥିଲେ।

Its Unigram : 'ତେବେ' 'ଏହି' 'କାର୍ଯ୍ୟକ୍ରମ' 'ସ୍ଥଳକୁ' 'ଗଣମାଧ୍ୟମ' 'ପ୍ରତିନିଧିମାନଙ୍କୁ' 'ଯିବାକୁ' 'ଅନୁମତି' 'ମିଳିନଥିଲା'। 'କିନ୍ତୁ' 'କେତେକ' 'ଦଳୀୟ କର୍ମୀ' 'ସୋନିଆଙ୍କ' 'ଉଦବୋଧନର' 'ଭିଡିଓ' 'କ୍ଲିପ' 'ଶେୟାର' 'କରିଛନ୍ତି'। 'ନିର୍ବାଚନୀ' 'ପ୍ରକ୍ରିୟା' 'ଉପରେ' 'ସୋନିଆ' 'ପ୍ରଶ୍ନ' 'ଉଠାଇ' 'କହିଥିଲେ', 'ବିଗତ' ୫'ବର୍ଷ ଧରି' 'ଭାରତର' 'ନିର୍ବାଚନ' 'ପ୍ରକ୍ରିୟା' 'ଉପରେ' 'ସନ୍ଦେହ' 'ସୃଷ୍ଟି' 'ହୋଇଛି'। 'ଏକ' 'ପୁରୁଣା' 'କଥା' 'ଅଛି', 'ନିଆଁ 'ନଥାଇ' 'ଧୂଆଁ' 'ଆସେ' 'ନାହିଁ' 'ବୋଲି' 'ସୋନିଆ' 'କହିଥିଲେ'। 'ଏହା' 'ସହିତ' 'ରାଏବରେଲି' 'ବାସିନ୍ଦାଙ୍କ' 'ସହିତ' 'ତାଙ୍କର' 'ସମ୍ପର୍କ' 'ନୂଆ' 'ନୁହେଁ କି' 'ନିଜସ୍ୱ' 'ସ୍ୱାର୍ଥ' 'ଆଧାରିତ'' ନୁହେଁ' 'ବୋଲି' 'ସେ' 'କହିଥିଲେ'।

Its Bigram : 'ତେବେ ଏହି' 'କାର୍ଯ୍ୟକ୍ରମ ସ୍ଥଳକୁ' 'ଗଣମାଧ୍ୟମ ପ୍ରତିନିଧିମାନଙ୍କୁ' 'ଯିବାକୁ ଅନୁମତି' 'ମିଳି ନ ଥିଲା'। 'କିନ୍ତୁ କେତେକ' 'ଦଳୀୟ କର୍ମୀ' 'ସୋନିଆଙ୍କ' 'ଉଦବୋଧନର' 'ଭିଡିଓ କ୍ଲିପ' 'ଶେୟାର କରିଛନ୍ତି'। 'ନିର୍ବାଚନୀ ପ୍ରକ୍ରିୟା' 'ଉପରେ ସୋନିଆ' 'ପ୍ରଶ୍ନ ଉଠାଇ' 'କହିଥିଲେ', 'ବିଗତ ୫ବର୍ଷ' 'ଧରି ଭାରତର' 'ନିର୍ବାଚନ ପ୍ରକ୍ରିୟା' 'ଉପରେ ସନ୍ଦେହ' 'ସୃଷ୍ଟି ହୋଇଛି'। 'ଏକ ପୁରୁଣା' 'କଥା ଅଛି', 'ନିଆଁ ନ ଥାଇ' 'ଧୂଆଁ ଆସେ' 'ନାହିଁ ବୋଲି' 'ସୋନିଆ କହିଥିଲେ'। 'ଏହା ସହିତ' 'ରାଏବରେଲି ବାସିନ୍ଦାଙ୍କ' 'ସହିତ ତାଙ୍କର' 'ସମ୍ପର୍କ ନୂଆ' 'ନୁହେଁ କି ନିଜସ୍ୱ' 'ସ୍ୱାର୍ଥ ଆଧାରିତ' 'ନୁହେଁ ବୋଲି' 'ସେ କହିଥିଲେ'।

Calculation of unigram and bigram probabilities can be done accordingly.

Unigram probabilities: $P(w_i) = \text{count}(w_i))/\text{count}$ (total number of words).
Probability of word_i = Frequency of word_i in our corpus/ total number of words in our corpus.
Bigram probabilities: $P(w_i | w_{i-1}) = \text{count}(w_{i-1}, w_i)/\text{count}(w_{i-1})$.
Probability that word_{i-1} is followed by word_i = [Num times we saw word_{i-1} followed by word_i]/ [Num times we saw word_{i-1}].

So calculation tf and tf–idf vector are done for two representations. After that out of tf and tf–idf matrix of each representation, 350 rows are meant for training the SVM classifier and 150 rows are meant for testing the classifier, and the result is calculated. The SVM_{light} is used for this purpose. Basing on the result, the prediction of peoples opinion on particular news articles is predicted by calculating the confidence score.

3.1 Performance Evaluation Measure

The parameter helps in evaluating performance of supervised machine learning algorithm. Here we have used three parameters, i.e., accuracy, F1 score, and confidence of the classification. The performance measure parameters are calculated as detailed below which generally depends on the prediction as per four categories like True Positives (TP), True Negatives (TN), False Positives (FP), and False Negatives (FN).

Accuracy—Accuracy is the most intuitive performance measure, and it is simply a ratio of correctly predicted observation to the total observations. One may think that, if we have high accuracy then our model is best. Yes, accuracy is a great measure but only when you have symmetric datasets where values of false positive and false negatives are almost same.

$$\text{Accuracy} = \frac{TP + TN}{TP + FP + FN + TN} \quad (1)$$

F1 score—F1 score is the weighted average of precision and recall. Therefore, this score takes both false positives and false negatives into account. Intuitively, it is not as easy to understand as accuracy, but F1 is usually more useful than accuracy, especially if you have an uneven class distribution. Accuracy works best if false positives and false negatives have similar cost. If the cost of false positives and false negatives is very different, it is better to look at both precision and recall.

$$\text{F1 Score} = \frac{2 * \text{recall} * \text{precision}}{\text{recall} + \text{precision}} \quad (2)$$

Confidence Measure: We decided to use the word probability derived from an n-gram model as a

$$\text{Confidence measure:} \quad C(t_i) = P(t_i / t_{r-1}, \ldots, t_{rat}t) \quad (3)$$

4 Results Analysis

The snapshot of results shown in Table 1 shows that the classification of news article with a confidence that when reader will read the article his opinion about a particular

Table 1 Output snapshot of the classified sentences with calculated confidence score

News Articles	Classification	Confidence
ତେବେ ଏହି କାର୍ଯ୍ୟକ୍ରମ ସ୍ଥଳକୁ ଗଣମାଧ୍ୟମ ପ୍ରତିନିଧିମାନଙ୍କୁ ଯିବାକୁ ଅନୁମତି ମିଳି ନ ଥିଲା। କିନ୍ତୁ କେତେକ ଦଳୀୟ କର୍ମୀ ସୋନିଆଙ୍କ ଉଦବୋଧନର ଭିଡିଓ କ୍ଲିପ ଶେୟାର କରିଛନ୍ତି। ନିର୍ବାଚନୀ ପ୍ରକ୍ରିୟା ଉପରେ ସୋନିଆ ପ୍ରଶ୍ନ ଉଠାଇ କହିଥିଲେ, ବିଗତ ୫ବର୍ଷ ଧରି ଭାରତର ନିର୍ବାଚନ ପ୍ରକ୍ରିୟା ଉପରେ ସନ୍ଦେହ ସୃଷ୍ଟି ହୋଇଛି। ଏକ ପୁରୁଣା କଥା ଅଛି, 'ନିଆଁ ନ ଥାଇ ଧୂଆଁ ଆସେ ନାହିଁ' ବୋଲି ସୋନିଆ କହିଥିଲେ। ଏହା ସହିତ ରାଏବରେଲି ବାସିନ୍ଦାଙ୍କ ସହିତ ତାଙ୍କର ସମ୍ପର୍କ ନୂଆ ନୁହେଁ କି ନିଜସ୍ବ ସ୍ବାର୍ଥ ଆଧାରିତ ନୁହେଁ ବୋଲି ସେ କହିଥିଲେ।	Neutral	0.838
ଯୋଗ ଦିବସରେ ଉତ୍ତରପ୍ରଦେଶର ଜଣେ ମନ୍ତ୍ରୀ ସରକାରୀ ବାବୁଙ୍କ ହାତରେ ନିଜ ଜୋତାର ଫିତା ବନ୍ଧାଇ ବିବାଦକୁ ଆସିଛନ୍ତି। ଏହାର ଏକ ଭିଡିଓ ଭାଇରାଲ ହେବା ପରେ ମନ୍ତ୍ରୀଙ୍କୁ କଡ଼ା ସମାଲୋଚନାର ଶିକାର ହେବାକୁ ପଡ଼ିଛି।	Negative	0.816
ମନ୍ତ୍ରୀ ଜଣକ ହେଉଛନ୍ତି ଲକ୍ଷ୍ମୀନାରାୟଣ ଖାଡେହ୍। ସେ ରାଜ୍ୟରେ ସଂଖ୍ୟାଲଘୁ ବ୍ୟାପାର ମନ୍ତ୍ରୀ ଅଛନ୍ତି। ଗତ ଶୁକ୍ରବାର ଶାହାଜାହାନପୁରରେ ଏକ ଯୋଗ କାର୍ଯ୍ୟକ୍ରମରେ ସେ ମୁଖ୍ୟ ଅତିଥି ଭାବେ ଯୋଗଦେଇଥିଲେ। ଏଥିରେ ଅନ୍ୟ କେତେଜଣ ମନ୍ତ୍ରୀ, ନେତା ଏବଂ ବିଶିଷ୍ଟ ବ୍ୟକ୍ତି ସାମିଲ ହୋଇଥିଲେ। ଏହି ଅବସରରେ ଖାଡେହ୍ ଜଣେ ସରକାରୀ କର୍ମଚାରୀଙ୍କୁ ତାଙ୍କ ଜୋତାର ଫିତା ବାନ୍ଧିବାକୁ କହିଥିଲେ। ସରକାରୀ ବାବୁ ଜଣକ ତଳେ ବସି ତାଙ୍କ ଗୋଡ଼ରେ ଜୋତା ଫିତା ବାନ୍ଧୁଥିବା ଭିଡିଓରେ ଦେଖିବାକୁ ମିଳିଛି। ବିଜେପିର ଏହି ମନ୍ତ୍ରୀ ଜଣକ ନିଜ କ୍ଷମତାର ଅପବ୍ୟବହାର କରିଥିବାରୁ ତାଙ୍କୁ ବିଭିନ୍ନ ମହଲରେ ସମାଲୋଚନା କରାଯାଉଛି। କିନ୍ତୁ ସେ ଭାରତର ସଂସ୍କୃତି ଓ ପରମ୍ପରାର ଉଦାହରଣ ଦେଇ କୌଣସି ଭୁଲ କରି ନାହିଁ'	Negative	0.736

topic will be positive, negative, or neutral. The confidence is a probability of reader view about classification of the text. When confidence score is less, then the prediction about reader's motive may not be accurate. For those cases, further analysis can be made on future study (Fig. 2).

The graph is plotted to demonstrate the spread of data over the classification of positive, negative, and neutral. The 500 dataset of Odia news articles are collected from different news portal as input, but here we have demonstrated the classification in the above figure over a less text to demonstrate the spread of confidence in the hyperplane.

Table 2 depicts how accuracy and F1-score vary over selection of data representation using unigram and bigram. In the experiment, the accuracy is improved from

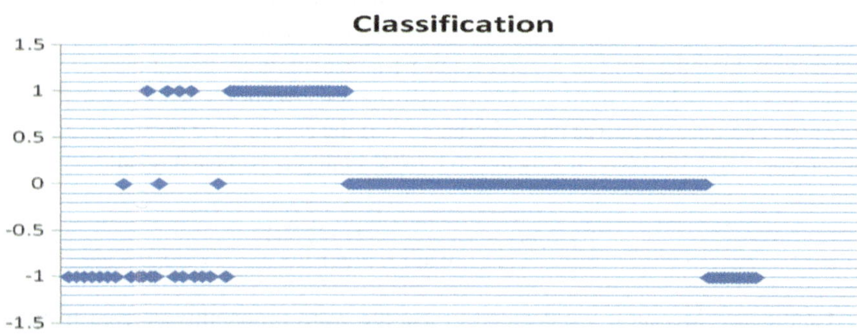

Fig. 2 The classification of datasets over the hyperplane where the data clustered at 1 are positive, –1 are negative, and 0 are neutral classification

Table 2 Results

N-gram+ tf/tf–idf score	Accuracy	F1 score
Unigram+tf	0.61	0.55
Bigram+tf	0.73	0.67
Unigram+tf–idf	0.78	0.68
Bigram+tf–idf	0.84	0.79

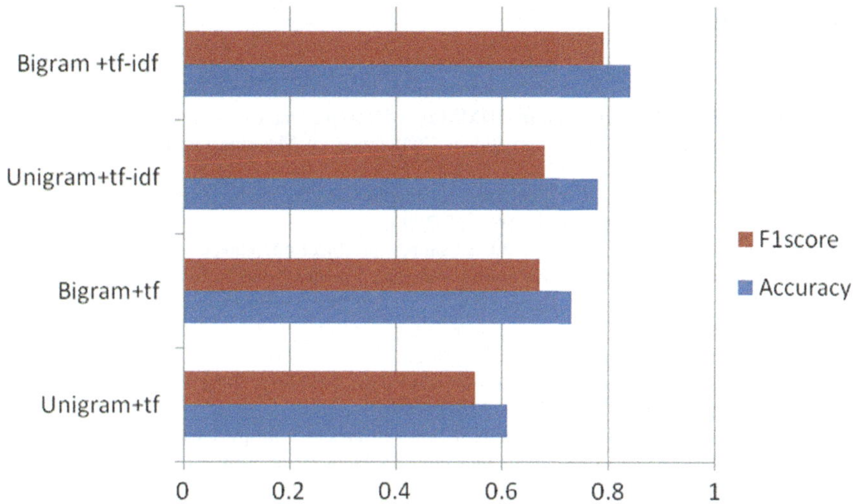

Fig. 3 The F1 score and accuracy using different approaches

0.61 to 0.73 and the F1 score is improved from 0.55 to 0.67 when term frequency is calculated from unigram representation of data and when term frequency is calculated from bigram representation of data. Similarly for when tf–idf is calculated from unigram representation of data and when tf–idf is calculated from bigram representation of data. The accuracy and F1 score for unigram+tf, bigram+tf, unigram+tf–idf, bigram+tf–idf are demonstrated in Fig. 3.

5 Conclusion

This paper makes an attempt to predict the impact of news article in Odia collected from different news portals. The impact on the reader of the article can be positive, negative, or neutral. The confidence score has been calculated to demonstrate the probability of prediction of classification impact on the reader. The accuracy and F1 score are calculated, which demonstrate the accuracy of classification of opinion. However, the result has been generated depending on tf, tf–idf, unigram, and bigram representation, which are calculated without lexical, syntactic, and semantic analysis of the news articles. So, the experimentation is to be done with large dataset with lexical, syntactic, and semantic analysis, and comparison will be done by the result with different performance measures.

References

1. Liu, B.: Sentiment Analysis and Opinion Mining. Morgan and Claypool Publishers (2012)
2. Jin, W., Ho, H.H.: A novel lexicalized HMM-based learning framework for web opinion mining. In: Proceedings of the 26th International Conference on Machine Learning, pp. 234–239. Montreal, Canada (2009)
3. Baldonado, M., Chang, C.-C.K., Gravano, L., Paepcke, A.: The stanford digital library metadata architecture. Int. J. Digit. Libr. **1**, 108–121 (1997)
4. Bruce, K.B., Cardelli, L., Pierce, B.C.: Comparing object encodings. In: Abadi, M., Ito, T. (eds.): Theoretical Aspects of Computer Software. Lecture Notes in Computer Science, vol. 1281, pp. 415–438. Springer, Berlin (1997)
5. van Leeuwen, J. (ed.): Computer Science Today. Recent Trends and Developments. Lecture Notes in Computer Science, vol. 1000. Springer, Berlin (1995)
6. Rameshbhai, J.C, Paulose, J.: Opinion on news paper headlines using SVM and NLP. Int. J. Electr. Comput. Eng. **9**(3), 2152–2163 (2019)
7. Scholz, T., Conard, S.: Opinion mining in news articles by entropy–based word connections. In: Proceedings of 2013 Conference on Empirical Methods in Natural Language Processing, pp. 1828–1839 (2013)
8. Jena, M., Mohanty, S.: Contextual opinion mining in online odia text using support vector machine. Compliance Eng. J. **10**(7), 166–169 (2019)
9. Patil, S., Chaudhari, A.: Classification of emotions from text using SVM based opinion mining. Int. J. Comput. Eng. Technol. **3**(1), 330–338 (2012)
10. Olsher, D.: Full Spectrum Opinion Mining: Integrating Domain, Syntactic and Lexical Knowledge, Cognitive Science Program, National University of Singapore
11. Balahur, A., Steinberger, R.: Rethinking sentiment analysis in the news: from theory to practice and back. In: WOMSA'09, Troyano, Cruz, Diaz(eds.), pp. 1–12 (2009)
12. Jena, M., Mohanty, S.: Document level opinion mining for odia language using n-gram based support vector machine. Compliance Eng. J. **10**(8), 430–436 (2019)
13. Kim, Y., Jeong, S., Ghani, I.: Text opinion mining to analyze news for stock market prediction. Int. J. Adv. Soft Comput. Appl. **6**(1),1–8 (2014)

An Approach to Extract Low-Grade Tumor from Brain MRI Slice Using Soft-Computing Scheme

Sangeetha Francelin Vinnarasi, J. T. Anita Rose, Jesline and V. Rajinikanth

Abstract The clinical-level assessment of brain Magnetic Resonance Imaging (MRI) varies from the visual/physical examination by a doctor to the Computer-Assisted Evaluation (CAE). Owing to its impact, a number of CAE techniques are planned to assess the brain MRI registered using different modalities. The plan of this research is to build a CAE tool by integrating the thresholding and segmentation techniques, which can effectively work with a range of MRI modalities. This work employs Brain Storm Optimization Algorithm (BSOA) and Otsu's thresholding and Watershed Segmentation (WS) to extract the tumor section from the considered two-dimensional (2D) MRI slice. The proposed CAE system is tested and validated using the benchmark MRI slices of BRATS2015 database. The examination considered the low-grade tumors of the flair, T1C and T2 modality registered pictures, and its outcome is confirmed with a qualified scrutiny with the Ground Truth (GT) given by an expert. The result of this work substantiates that this CAE tool offered superior values of Image Quality Parameters (IQPs) during the low-grade brain tumor evaluation, and hence, the developed CAE can be considered to inspect the scientific grade MRI slices.

Keywords Brain MRI · Otsu · Brain storm optimization algorithm · Watershed segmentation · Validation

---

S. F. Vinnarasi · J. T. A. Rose · Jesline
Department of Computer Science Engineering, St. Joseph's College of Engineering, Chennai 600 119, Tamilnadu, India
e-mail: fsangeetha@gmail.com

V. Rajinikanth (✉)
Department of Electronics and Instrumentation, St. Joseph's College of Engineering, Chennai 600 119, Tamilnadu, India
e-mail: v.rajinikanth@ieee.org

H. Das et al. (eds.), *Progress in Computing, Analytics and Networking*, Advances in Intelligent Systems and Computing 1119,
https://doi.org/10.1007/978-981-15-2414-1_28

1 Introduction

Recently, a considerable number of Computer-Assisted Evaluation (CAE) procedures are proposed to examine the brain abnormality based on the 2D MRI slices [1, 2]. The literature also confirms that the experimental outcomes attained with the heuristic algorithm supported techniques are good compared to the traditional and other approaches [3–6]. Further, healthier results are also attained with recent approaches like Machine Learning (ML) and Deep Learning (DL) based examination procedures [7–10]. The computation complexity and realization issues of DL are fairly high contrast to the ML method; therefore, many of brain MRI examinations are performed with a customized ML approach [9].

The ML always requires the information such as shape features, texture features, and other related attributes of the brain MRI dataset under examination. Further, the ML technique consists of different phases like preprocessing, postprocessing, feature mining, feature selection, and classification. In the proposed work, a CAE is developed to implement the essential pre- and postprocessing methodology for the considered brain MRI slices. The preprocessing is performed with Otsu's function to boost the visibility of tumor segment and the postprocessing is executed to mine the tumor. After mining the required section (tumor), a qualified evaluation among the tumor and its Ground Truth (GT) is then performed, and vital Image Quality Parameters (IQPs) are computed. In this work, the Low-Grade Tumor (LGT) of the BRATS2015 database is considered for the examination (total images = 8 volunteers × 15 slices = 120 axial view MRI slices of flair modality). The outcome of the Otsu's based CAE is then validated with the Kapur's based CAE, and the results of this work verifies that the Otsu's offered better values of average IQPs for the used image dataset. In future, this work can be tested on other view/modality images of the BRATS2015 database [11, 12].

Remaining work is presented as follows: Sect. 2 explains the tactic, Sect. 3 demonstrates the tentative results and deliberations, and the conclusion is accessible in Sect. 4.

2 Methodology

The methodology executed in this work is accessible from Fig. 1. The two-dimensional (2D) MRI slice is considered for the examination. This work

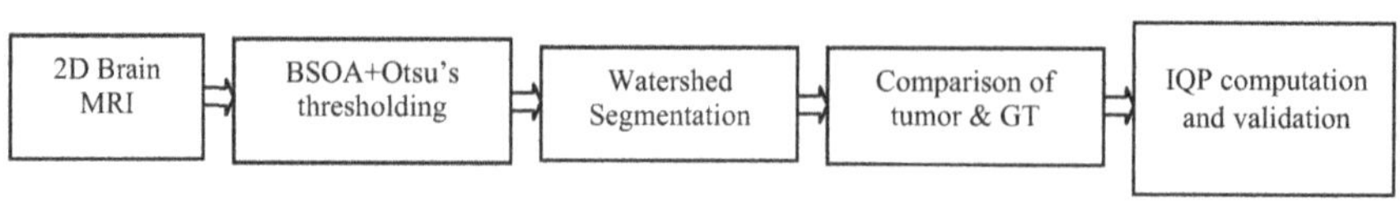

Fig. 1 Various phases in the developed CAE system

considers the flair modality slice and the initial enhancement is achieved with the multi-thresholding using the BSOA algorithm and Otsu's function. Further, the enhanced tumor segment is then mined with the WS technique. Finally, a comparison of GT and the tumor segment is done, and the IQPs like Jaccard, dice, sensitivity, specificity, accuracy, and precision are calculated. Finally, the merit of Otsu's procedure is confirmed against Kapur's technique, and the study confirmed that the IQPs of Otsu are superior to Kapur's.

The remaining part of this subdivision explains the review of the techniques applied to develop the CAE.

2.1 *Otsu's Function*

Otsu's is one of the famous preprocessing techniques usually considered to enhance the gray scale and RGB pictures [13, 14]. The multi-thresholding is widely considered schemes; which separates the test images into various sections based on chosen threshold level. Here, a three-level thresholding is implemented to separate the 2D brain MRI into background, normal brain section, and the tumor. The implemented Otsu's function helps to improve the visibility of the tumor segment by reducing the visibility of normal brain parts. Identification of an optimal threshold is a complex task and in this work, the assistance of the BSOA helped to enhance the Otsu's function.

The traditional Otsu's technique is depicted below:

The between-class variance for a bi-level problem can be presented with the following assumptions; let the task is to segregate the picture into the background (B0) and the object (B1) with the help of a chosen threshold Th.

The probability allocation for an image with L-1 thresholds, B0, and B1 is denoted as

$$B0 = \frac{P_0}{\omega_0(Th)} \cdots \frac{P_{Th-1}}{\omega_0(Th)} \text{ and } B1 = \frac{P_{Tht}}{\omega_1(Th)} \cdots \frac{P_{L-1}}{\omega 1(Th)} \tag{1}$$

where $\omega_0(Th) = \sum_{i=0}^{Th-1} P_i$, $\omega_1(Th) = \sum_{i=Th}^{L-1} P_i$ and L = 256.

The means μ_0 and μ_1 of B0 and B1 are expressed as

$$\mu_0 = \sum_{i=0}^{Th-1} \frac{iP_i}{\omega_0(Th)} \text{ and } \mu_1 = \sum_{i=Th}^{L-1} \frac{iP_i}{\omega 1(Th)} \tag{2}$$

The mean intensity (μ_T) of the picture is

$$\mu_\mathrm{T} = \omega_0\mu_0 + \omega_1\mu_1 \text{ and } \omega_0 + \omega_1 = 1$$

The maximized objective value is

$$\text{Maximize } J(Th) = \sigma_0 + \sigma_1 \tag{3}$$

where $\sigma_0 = \omega_0(\mu_0 - \mu_T)^2$ and $\sigma_1 = \omega_1(\mu_1 - \mu_T)^2$

Similar condition is obtained in trilevel thresholding, and its expressions will be

$$\text{Maximize } J(Th) = \sigma_Q + \sigma_L + \sigma_2 \tag{4}$$

2.2 *Brain Storm Optimization Algorithm*

BSOA was discussed by Shi (2011) by mimicking the decision/conversation making process in humans [15]. This technique works based on the problem-solving capabilities in humans.

This algorithm has the following steps:

Step1: Create a brainstorming people's group with a varied environment.
Step2: Initiate the algorithm to generate a variety of ideas.
Step3: Identify and choose ideas to solve the problem.
Step4: Pick the idea with the better probability.
Step5: Allow the dominant person to collect many ideas as possible to solve the given task.
Step6: Repeat the step till a better solution is attained for the chosen problem.

In this work, the BSOA is considered to identify the optimal threshold by maximizing the objective values of the Otsu's/Kapur's function. Other details on BSOA can be found in [16, 17].

2.3 *Watershed Segmentation*

Watershed Segmentation (WS) is an automated technique widely applied in image evaluation task. In this, a marker value is initiated by an operator and the algorithm extracts the entire visible pixel which resembles the marker value. The main merit of the WS is that it implements edge detection, morphological enhancement, watershed fill, and the extraction of the enhanced section. The detailed discussion on the WS can be found in [18, 19]. The earlier techniques corroborate that the outcome of WS is improved contrast to other semiautomated segmentation approaches existing in the literature.

2.4 *Image Quality Parameters and Validation*

Qualified appraisal of the extracted part and GT is carried out in pixel level to appraise the advantage of proposed system. The essential events like Jaccard, dice, accuracy, precision, sensitivity, and specificity are calculated from this assessment and based on these values; the merit of the considered approach is confirmed.

The mathematical model of these measures is presented below [20–22]:

$$\text{Jaccard} = GT \cap SEG / GT \cup SEG \tag{5}$$

$$\text{Dice} = 2(GT \cap SEG) / |GT| \cup |SEG| \tag{6}$$

$$\text{Accuracy} = \frac{T_{+ve} + T_{-ve}}{T_{+ve} + T_{-ve} + F_{+ve} + F_{-ve}} \tag{7}$$

$$\text{Precision} = \frac{T_{+ve}}{T_{+ve} + F_{+ve}} \tag{8}$$

$$\text{Sensitivity} = \frac{T_{+ve}}{T_{+ve} + F_{-ve}} \tag{9}$$

$$\text{Specificity} = \frac{T_{-ve}}{T_{-ve} + F_{+ve}} \tag{10}$$

where GT = ground truth; SEG = segmented picture; and T_{+ve}, T_{-ve}, F_{+ve}, and F_{-ve} denote true positive, true negative, false positive, and false negative, respectively.

3 Result and Discussion

This sector presents the investigational result and its discussions. Extraction of the low-grade tumor from an MRI slice is a challenging task due to its poor visibility. This work aims to propose a technique to extort the tumor segment from MRI based on the integration of thresholding and segmentation. In this work, the low-grade tumor slices of the BRATS2015 database considered for the experimental evaluation and this work is investigated with the MATLAB software.

Figure 2 presents sample trial pictures of considered database. This figure presents the test image along with the Ground Truth (GT) picture. During the examination task, this test picture is to be evaluated with the proposed technique and the tumor section is to be mined with the watershed segmentation technique. Finally, the extracted low-grade tumor is compared with GT and the essential IQPs are then calculated. Alike practice is employed for the every picture used here and the average of IQP is then used to authenticate the dominance of the proposed practice (Figs. 3, 4).

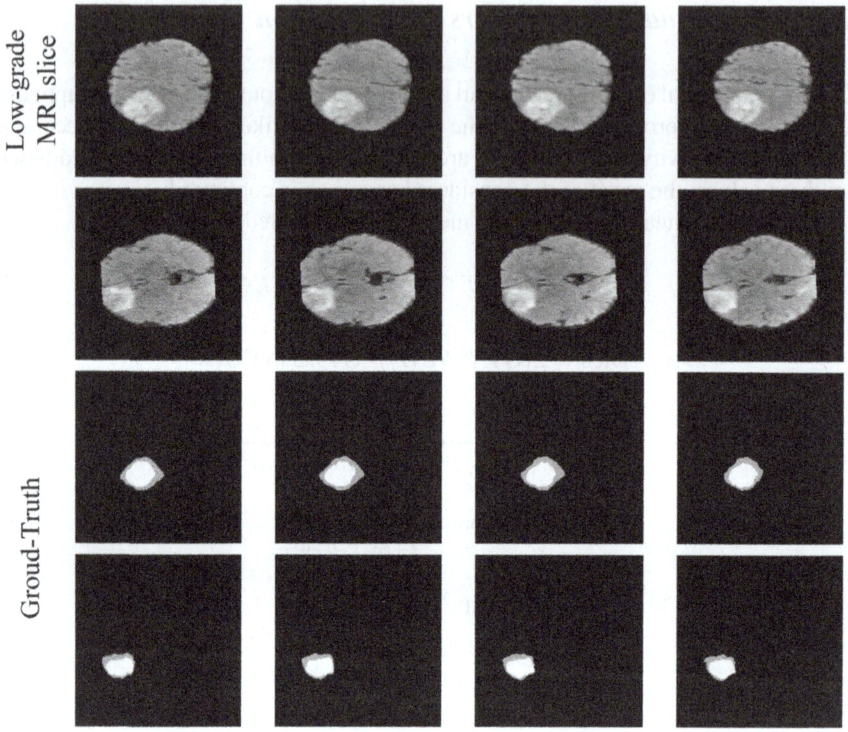

Fig. 2 Sample test MRIs and the associated GTs

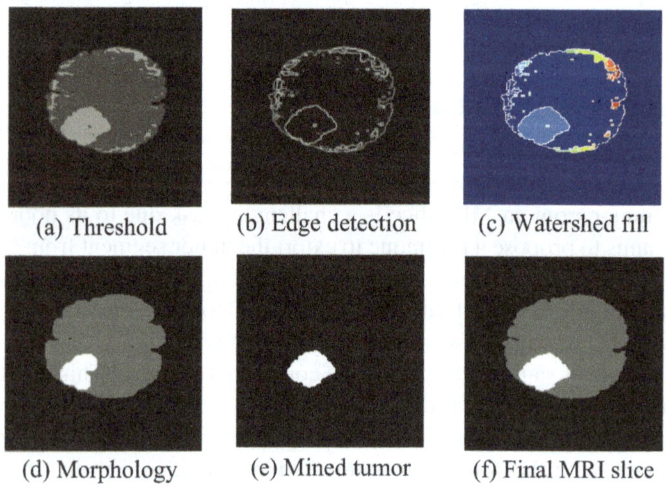

Fig. 3 Results attained with the proposed CAE system

Actual Class

Extracted class

T+ve=1722	F+ve=124	TPR=0.9914
F-ve=15	T-ve=51039	FNR=0.0086
P=T+ve+F-ve=1737	N=F+ve+T-ve=51163	TNR=0.9976
Total pixels=52900		FPR=0.0024

Fig. 4 Confusion matrix for the chosen test picture

Initially, a trilevel thresholding is implemented on the selected trial picture using the BSOA and the Otsu's function. This technique will segregate the test picture into background, brain structure, and tumor. After the segregation, the chosen WS practice is implemented to mine the tumor fragment from the MRI slice. After the extraction, the performance of the proposed technique on these MRI is evaluated with a comparative assessment. This comparison implements an appraisal linking the mined tumor regions with the GT and helps to compute the essential measures as discussed in Sect. 2.4. The performance of this approach is also validated by computing the confusion matrix, and a sample is depicted in Fig. 4, obtained for a single MRI slice. In this picture, TPR denotes true positive rate, FNR = false negative rate, TNR = true negative rate, FPR = false positive rate, P = positive pixels, and N = negative pixels. Further, the IQPs like Jaccard = 92.53%, dice = 96.12%, sensitivity = 99.14%, specificity = 99.76%, accuracy = 99.74%, and precision = 93.28% are attained for a single test image presented in Fig. 2. These values confirm that proposed practice suggests a healthier result with the Otsu's thresholding and WS mining.

The performance of Otsu's is then compared against the Kapur's based thresholding [23, 24] and WS mining and the average results of the considered images are then considered to validate the proposed technique. The results confirm that the average IQPs attained with the Otsu's and Kapur's are alike and the proposed image processing tool is the best suitable evaluation technique for the low-grade tumors.

In this work, flair modality-based two-dimensional slices (120 numbers) are considered for the evaluation and the proposed technique with Otsu's/Kapur's thresholding is implemented on every MRI slice. Further, averages of the IQPs attained with these methods are computed to identify the better procedure to evaluate the LGT. From Fig. 5, it can be observed that all the IQP values attained with this technique are greater than 90%, and also, both the threshold methods provide approximately similar results. The experimental result of this paper authenticates that proposed practice is competent in analyzing the flair modality MRI which consist low-grade tumors. In future, the recent approaches [25–31] can be employed to develop the performance of proposed system.

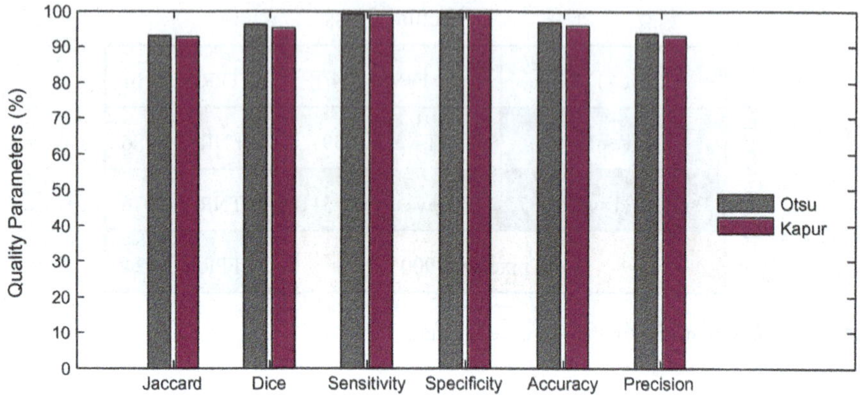

Fig. 5 An experienced analysis of the Otsu's and Kapur's based CAE system

4 Conclusion

This work aims to implement a CAE system to evaluate the low-grade tumors from the 2D brain MRI recorded using flair modality. This work employs Otsu's and BSOA-based trilevel threshold to enhance the tumor section and WS technique to mine the tumor. A comparison of the mined tumor segment and GT is then implemented to compute the essential image quality parameters to assess the efficiency of the CAE. Further, a confusion matrix-based analysis also implemented to confirm the performance of the CAE. The results attained from 120 MRI slices confirm that the proposed technique provides an average IQPs of > 90%. Further, the experimental investigation is repeated with the Kapur's threshold, and the results attained with Otsu's and Kapur's are approximately similar. These results confirm that the proposed technique is efficient in evaluating the MRI with LGT.

References

1. Rajinikanth, V., Satapathy, S.C., Fernandes, S.L., Nachiappan, S.: Entropy based Segmentation of Tumor from Brain MR Images–A study with Teaching Learning Based Optimization. Pattern Recognit. Lett. **94**, 87–94 (2016). https://doi.org/10.1016/j.patrec.2017.05.028
2. Fernandes, S.L., Rajinikanth, V., Kadry, S.: A hybrid framework to evaluate breast abnormality using infrared thermal images. IEEE Consum. Electron. Mag. **8**(5), 31–36 (2019). https://doi.org/10.1109/MCE.2019.2923926
3. Nair. M.V. et al.: Investigation of breast melanoma using hybrid image-processing-tool. In: International Conference on Recent Trends in Advance Computing (ICRTAC), IEEE, pp. 174–179 (2018). https://doi.org/10.1109/ICRTAC.2018.8679193
4. Rajinikanth, V., Dey, N., Kumar, R., Panneerselvam, J., Raja, N.S.M.: Fetal head periphery extraction from ultrasound image using jaya algorithm and Chan-Vese segmentation. Procedia Comput. Sci. **152**, 66–73 (2019). https://doi.org/10.1016/j.procs.2019.05.028

5. Satapathy, S.C., Rajinikanth, V.: Jaya algorithm guided procedure to segment tumor from brain MRI. J. Optim. **2018**, 12 (2018). https://doi.org/10.1155/2018/3738049
6. Rajinikanth, V., Satapathy, S.C.: Segmentation of ischemic stroke lesion in brain MRI based on social group optimization and Fuzzy-Tsallis entropy. Arabian J. Sci. Eng. **43**(8), 4365–4378 (2018). https://doi.org/10.1007/s13369-017-3053-6
7. Jahmunah, V., et al.: Automated detection of schizophrenia using nonlinear signal processing methods. Artif. Intell. Med. **100**, 101698 (2019). https://doi.org/10.1016/j.artmed.2019.07.006
8. Acharya, U.R., et al.: Automated detection of Alzheimer's disease using brain MRI images– a study with various feature extraction techniques. J. Med. Syst. **43**, 302 (2019). https://doi.org/10.1007/s10916-019-1428-9
9. Dey, N., et al.: Social-Group-Optimization based tumor evaluation tool for clinical brain MRI of Flair/diffusion-weighted modality. Biocybern. Biomed. Eng. **39**(3), 843–856 (2019). https://doi.org/10.1016/j.bbe.2019.07.005
10. Fernandes, S.L. et al.: A reliable framework for accurate brain image examination and treatment planning based on early diagnosis support for clinicians. Neural Comput. Appl. 1–12 (2019). https://doi.org/10.1007/s00521-019-04369-5
11. Menze, et al.: The multimodal brain tumor image segmentation benchmark (BRATS). IEEE Trans. Med. Imaging **34**(10), 1993–2024 (2015)
12. Brain Tumor Database (BraTS-MICCAI), http://hal.inria.fr/hal-00935640
13. Satapathy, S.C., Raja, N.S.M., Rajinikanth, V., Ashour, A.S. Dey, N.: Multi-level image thresholding using Otsu and chaotic bat algorithm. Neural Comput. Appl. (2016). https://doi.org/10.1007/s00521-016-2645-5
14. Raja, NSM., Rajinikanth, V., Latha, K.: Otsu based optimal multilevel image thresholding using firefly algorithm. Model Simul. Eng. **2014** (2014). Article ID 794574:17
15. Shi, Y.: Brain storm optimization algorithm. Lect. Notes Comput. Sci. **6728**, 303–309 (2011). https://doi.org/10.1007/978-3-642-21515-5_36
16. Cheng, S., Shi, Y., Qin, Q., Zhang, Q., Bai, R.: Population diversity maintenance in brain storm optimization algorithm. J Artif Intell Soft Comput Res **4**(2), 83–97 (2014)
17. Cheng, S., Qin, Q., Chen, J., Shi, Y.: Brain storm optimization algorithm: a review. Artif. Intell. Rev. **46**(4), 445–458 (2016)
18. Roerdink, J.B.T.M., Meijster, A.: The watershed transform: definitions, algorithms and parallelization strategies. Fundam. Inf. **41**, 187–228 (2001)
19. Shanthakumar, P., Kumar, P.G.: Computer aided brain tumor detection system using watershed segmentation techniques. Int. J. Imaging Syst. Technol. **25**(4), 297–301 (2015). https://doi.org/10.1002/ima.22147
20. Raja, N.S.M. et al.: Contrast enhanced medical MRI evaluation using Tsallis entropy and region growing segmentation. J. Ambient Intell. Hum. Comput. 1–12 (2018). https://doi.org/10.1007/s12652-018-0854-8
21. Dey, N., et al.: Social group optimization supported segmentation and evaluation of skin melanoma images. Symmetry **10**(2), 51 (2018). https://doi.org/10.3390/sym10020051
22. Rajinikanth, V., Dey, N., Satapathy, S.C., Kamalanand, K.: Inspection of crop-weed image database using kapur's entropy and spider monkey optimization. Adv. Intell. Syst. Comput. **1048** (2019). https://doi.org/10.1007/978-981-15-0035-0_32
23. Kapur, J.N., Sahoo, P.K., Wong, A.K.C.: A new method for gray-level picture thresholding using the entropy of the histogram. Comput. Vis. Graph. Image Process. **29**, 273–285 (1985)
24. Raja, N.S.M., et al.: Segmentation of breast thermal images using Kapur's entropy and hidden Markov random field. J. Med. Imaging Health Inf. **7**(8), 1825–1829 (2017). https://doi.org/10.1166/jmihi.2017.2267
25. Das, H., Naik, B., Behera, H.S.: Classification of diabetes mellitus disease (dmd): a data mining (DM) Approach. Adv. Intell. Syst. Comput. **710**, 539–549 (2018). https://doi.org/10.1007/978-981-10-7871-2_52
26. Sahani, R., et al.: Classification of intrusion detection using data mining techniques. Adv. Intell. Syst. Comput. **710**, 753–764 (2018). https://doi.org/10.1007/978-981-10-7871-2_72

27. Pradhan, C., Das, H., Naik, B., Dey, N.: Handbook of Research on Information Security in Biomedical Signal Processing Hershey. IGI Global, PA (2018)
28. Sahoo, A.K., Mallik, S., Pradhan, C., Mishra, B.S.P., Barik, R.K., Das, H.: Intelligence-based health recommendation system using big data analytics. In: Big Data Analytics for Intelligent Healthcare Management, pp. 227–246 (2019). https://doi.org/10.1016/B978-0-12-818146-1.00009-X
29. Dey, N., Das, H., Naik, B., Behera, H.S. (eds.): Big Data Analytics for Intelligent Healthcare Management. Academic (2019)
30. Chandrakar, P.: A secure remote user authentication protocol for healthcare monitoring using wireless medical sensor networks. Int. J. Ambient Comput. Intell. (IJACI) **10**(1), 96–116 (2019). https://doi.org/10.4018/IJACI.2019010106
31. Bhattacharya, H., Chattopadhyay, S., Chattopadhyay, M., Banerjee, A.: Storage and bandwidth optimized reliable distributed data allocation algorithm. J. Ambient Comput. Intell. (IJACI) **10**(1), 78–95 (2019). https://doi.org/10.4018/IJACI.2019010105

Development of a Semiautomated Evaluation Procedure for Dermoscopy Pictures with Hair Segment

Jesline, J. T. Anita Rose, Sangeetha Francelin Vinnarasi and V. Rajinikanth

Abstract One of the major and the life-threatening abnormality arises in skin is the skin cancer (SC). The occurrence of SC is mainly due to the exposed skin area to ultraviolet rays. The prescreening of the SC is clinically done by an experienced dermatologist with a visual check followed by the dermoscopy-related examination. This assessment will be tedious if the cancerous section is associated with the hair. The availability of the hair inside or nearer to the cancerous section may reduce the diagnosis accuracy. The aim of the proposed work is to implement a semiautomated technique to assess the dermoscopy picture with greater accuracy. Removing of the hair physically is very hard in real-time cases if it is associated with the SC segment. Hence, in this work, a Shannon's entropy and the Chan–Vese segmentation (CVS) technique is implemented to extract the suspicious skin section which is associated with the skin. The performance of CVS is then validated with level set (LS) approach. The experimental outcome authenticates that this work assists to attain healthier outcome on PH2 database. Further, this technique is also tested on DermIS database, and the attained outcome confirms that proposed technique offers better accuracy on both datasets.

Keywords Skin melanoma · Dermoscopy · Jaya algorithm · Segmentation · Validation

---

Jesline · J. T. A. Rose · S. F. Vinnarasi
Department of Computer Science Engineering, St. Joseph's College of Engineering, Chennai 600 119, Tamilnadu, India
e-mail: fsangeetha@gmail.com

V. Rajinikanth (✉)
Department of Electronics and Instrumentation, St. Joseph's College of Engineering, Chennai 600 119, Tamilnadu, India
e-mail: v.rajinikanth@ieee.org

H. Das et al. (eds.), *Progress in Computing, Analytics and Networking*, Advances in Intelligent Systems and Computing 1119,
https://doi.org/10.1007/978-981-15-2414-1_29

1 Introduction

Skin cancer (SC) is one of the life-intimidating diseases in humans, and as per the recent report by World Health Organization (WHO), around 132000 SC occurrences are recorded globally every year. Early screening and the treatment is the only measure to reduce the death rate [1]. If the SC is diagnosed in its early phase, a possible treatment can be implemented to control the disease progression.

The diagnosis of SC involves in: (i) a visual check by a dermatologist and (ii) evaluation of the suspicious section using dermoscopy. The untreated SC may raise the threat and will permit the dispersion of cancer cells to further parts through the bloodstream and hence, it should be identified in its premature stage. Recently, a numeral actions are proposed to examine the SC images recorded with a digital dermoscopy (DD) and in most of the computer-assisted methods, these DD pictures are examined using a semiautomated or automated computer algorithms, which can be used as an assisting medium for the dermatologist during the mass SC evaluation [2, 3].

In clinics, a scheduled visual inspection by an skilled dermatologist is constantly recommended to recognize the SC in its premature phase and if the possibility of SC is suspected, then a computerized screening process is suggested, in which a DD is recorded and examined with computerized algorithms. Later, the results of the computer algorithm is assessed and verified by the doctor, who implements the treatment to regulate/cure the SC.

Due to its clinical importance, a variety of regular and current SC estimation and localization measures are widely realized [2–5]. Prior works also validate that a noteworthy form of actions is considered and engaged for the independent examination of cancerous skin segment. In former research, the automated/semiautomated illness inspection methods implemented with customary and heuristic techniques are engaged to judge the infected section [5]. If the SC information is obtainable, then it may be achievable for the doctor to prepare for advanced handling procedure. The handling engages in: (i) therapeutic act to reduce the SC with enthusiastic drugs, and (ii) utilizing surgical act to eradicate the SC section.

In this paper, the grouping of multi-thresholding and a chosen segmentation methodology is executed to inspect SC based on DD images. The thresholding is implemented with Shannon's entropy (SE) and the optimization is achieved with the Jaya algorithm (JA). This work enhances the SC segment by eliminating the hair and the enhanced SC province is then mined by Chan–Vese segmentation (CVS) technique. In this work, the benchmark SC dataset "PH2" is initially utilized to judge the concert of proposed technique. This work is tested on 50 numbers of RGB scale DD images and the proposed work offered better result on all the considered images of the PH2 [6]. Later, similar technique is tested using the DermIS database [7] and better results are obtained during a comparative assessment between the extracted SC and its ground truth (GT).

The investigational work is executed with the MATLAB software, and results of this investigation confirm that the proposed procedure helps to accomplish improved result on PH2 and DermIS datasets. In future, this technique can be used to scrutinize SC using the DD images obtained from hospitals.

2 Methodology

The main aim of this research is to implement a technique to examine the SC using the DD images with considerable accuracy. Further, the proposed system should help the demonologist during the SC assessment and the treatment planning tasks. Various phases involved in the proposed examination system are depicted in Fig. 1.

The practice realized in this paper is graphically represented in Fig. 1. Initially, the test image is to be evaluated is chosen from the considered DD picture dataset. The selected picture experiences a thresholding process, which enhances the SC section and dilates the other sections and the hair segment. Later, segmentation with CV is employed to extract the SC segment. CV is a bounding-box (BB) practice initiated by the operator based on the SC segment. When the iteration value increases, the CV is allowed to converge according to the SX segment and extract a binary class picture. This picture is then compared against the GT and the vital image parameters are computed to confirm the performance of proposed system. The stages involved in the proposed system are briefly discussed in this section.

2.1 Skin Cancer Database

The proposed work considers the SC database, such as PH2 [6] and DermIS [7], which consist of RGB scaled DD pictures along with the associated GT. In order to minimize the evaluation complexity, all the test images are resized into 256 × 256 pixels. During this study, 50 images are considered from PH2 dataset and 20 images are considered from the DermIS. Some of these images are associated with artifact, like the nominal and dense air section. The infected SC segment is then extracted and evaluated using the proposed tool.

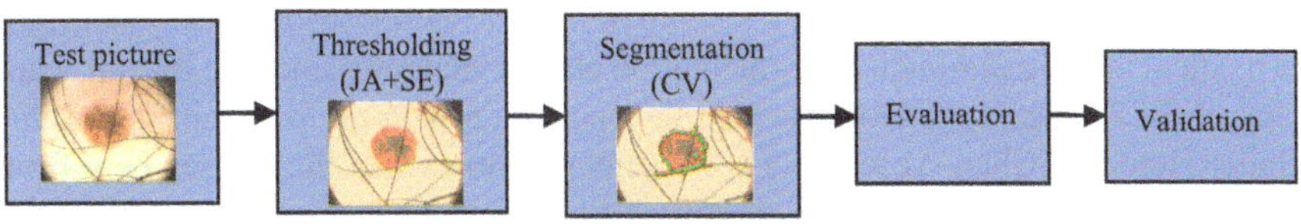

Fig. 1 Stages in the proposed DD examination system

2.2 Multi-thresholding

Thresholding is one of the accepted image preprocessing techniques extensively used to enhance the abnormal section from the test picture with assigned threshold. In this work, a three-level threshold with Shannon's entropy (SE) is considered. This thresholding scheme will provide the optimal threshold, which provides a maximized entropy value. In order to identify the maximized entropy based on the SE, this work considered the Jaya algorithm (JA), which already proved its performance for the image processing problem [8].

SE for a image of dimension M*N by pixel group (A,B) is symbolized as F (A,B), amid $A \in \{1, 2, \ldots, p\}$ and $B \in \{1, 2, \ldots, p\}$.

Let T symbolize whole pixels of image and location of every pixel {0,1,2,..., T-1} can be designated as U, as

$$F(A, B) \in U \forall (A, B) \in \text{picture} \tag{1}$$

Then, the regularized histogram is; $E = \{e_0, e_1, \ldots, e_{T\text{-}1}\}$.

In trilevel threshold, it can be framed as

$$E(th) = e_\theta(t_l) + e_d(t_2) + e_2(t_3) \tag{2}$$

where th = $\{t_1, t_2, \ldots, t_T\}$ are thresholds and $th*$ = final threshold. Other information on SE can be found in [9–11].

In this work, the optimal threshold is identified with the JA and the essential information in JA can be found in [12–15]. The aim of the JA is to arbitrarily vary the image thresholds till the SE reaches a maximal value. After getting the maximized value, the search process stops and the thresholded image goes to the next stage, where the enhanced SC segment is extracted with the CV technique. The following algorithm parameters are allocated in JA: total agents = 30, dimension = 3, iteration value = 2000, and the terminating function = maximized SE.

2.3 Segmentation

In medical image examination, segmentation plays a vital role, since the overall accuracy of the image examination system depends mainly on the performance of this phase. Here, the Chan–Vese (CV) is implemented to extract the SC segment from the thresholded picture. CV is a modified active-contour process which functions according to energy minimization impression [16–18]. In CV, the curve is permitted to recognize all the probable related pixels accessible in the picture. It is a semiautomated practice with a flexible contour, which alters its shape when the iteration rises. Lastly, the CV exploration stops after reaching the minimal energy rate and displays the section which lies within the contour.

Consider that Ψ is a surrounded position of $\Re^2$ with a border utility $\partial\psi$. Assume that $u_o : \overline{\psi} \rightarrow \Re$ is specified image and C is the curve segment. If the region inside C is Φ and outside the C is $\overline{\psi}/\phi$.

If px1 and px2 signify picture pixels within and external to C, then the energy utility will be indicated as E (e1, e2, C). For the minimized energy, $\inf_{e1,e2,C}$ E (e1, e2, C). In this paper, the CV discussed in [17] is adopted to examine PH2 and DermIS database.

2.4 *Performance Evaluation and Validation*

The advantage of medical image evaluation system is authorized by calculating the quality parameters, widely discussed in the literature [19–22]. In this paper, parameters like Jaccard (JI), dice (DC), sensitivity (SE), specificity (SP), accuracy (AC), and precision (PR) are calculated to evaluate the performance of the proposed tool on the PH2 and DermIS database.

3 Results and Discussion

The results of this procedure are accessible and conversed in this section. During this study, RGB scaled images of PH2 and DermIS datasets (256 × 256 pixels) are considered for the evaluation. Initially, the PH2 dataset is considered and the DD images are evaluated in its RGB phase. Later, the assessment is implemented for the DermIS dataset in which the RGB scaled image is evaluated in gray phase. The experimental investigation is implemented and validated using MATLAB.

Figure 2 depicts the sample test images of PH2 and its lesion segment (GT) picture. In this work, the DD images with the artifact (hair segment) are initially assessed with the proposed system. Extraction of the SC section in the presence of artifact is a challenging task and in the literature, considerable measures are already attempted by the researchers. In this work, the aim is to segment the irregular piece of DD images with better accuracy. After collecting the necessary test images from the PH2, the thresholding with JA + SE is implemented and the attained outcome is presented in Fig. 3. After enhancing the picture considerably, the CV-based mining is then implemented to mine the infected section. The segmentation results attained with this technique is shown in Fig. 4, which includes the threshold picture, initial bounding-box, converged CV, and the extracted binary picture. Later, a relative study between the binary picture and the GT is then implemented to compute the essential values discussed in Sect. 2.4 along with true positive rate (TPR), true negative rate (TNR), false positive rate (FPR), and false negative rate (FNR). Alike process is applied on all other DD images and the average of these values are considered for the validation of the implemented technique.

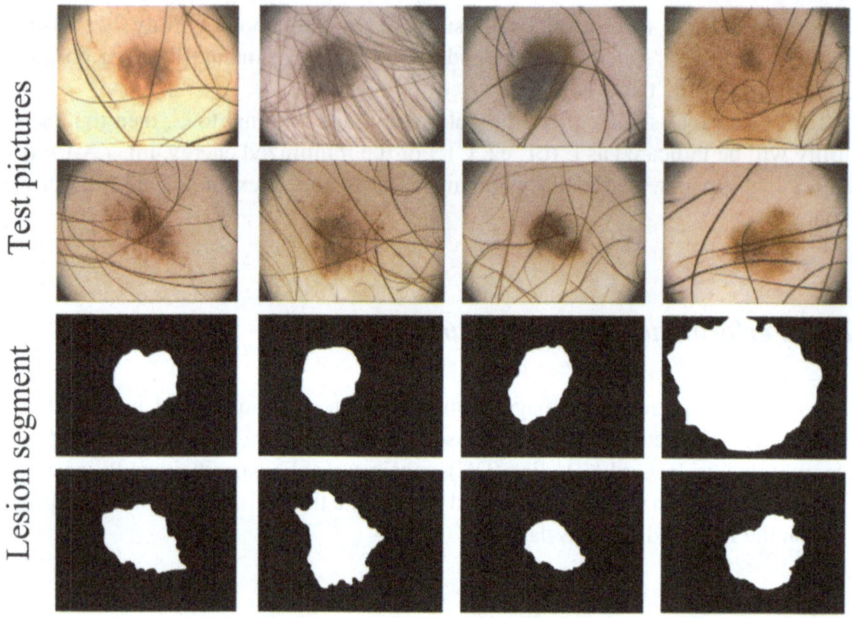

Fig. 2 Sample DD images and the GTs of PH2 database

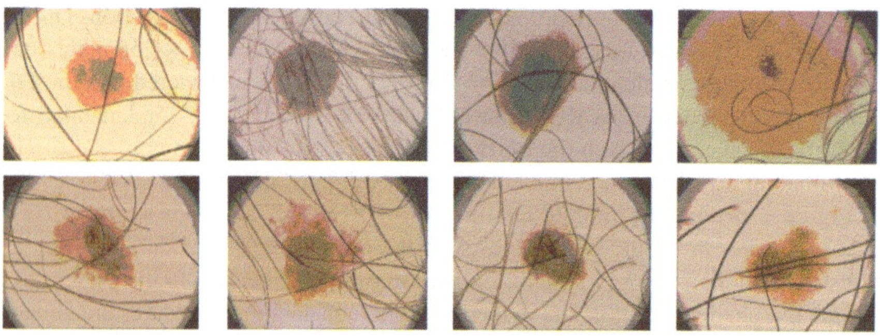

Fig. 3 Outcome attained with JA + SE thresholding process

Later, similar technique is then applied on the DermIS database and the performance values are then considered for the examination. In this, initially, the DD pictures of DermIS are converted into gray from RGB and the gray version is considered for evaluation. Figure 5 depicts the DD pictures of the DermIS database and its result. Figure 5a represents the pseudo-name, Fig. 5b, c depicts the test picture and its GT, and Fig 5d, e depicts the gray scale threshold picture and the extracted binary SC segment.

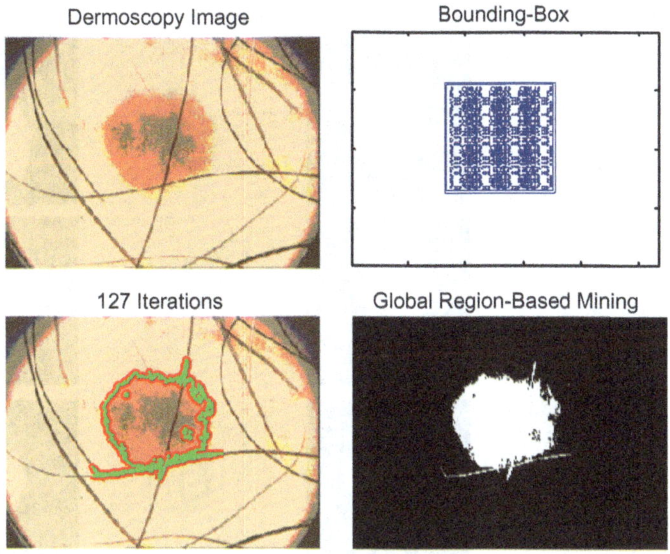

Fig. 4 Results attained with the CV segmentation

After extracting all the SC section from the database, the necessary performance events are calculated with a study among the binary and GT pictures. The performance values attained for the images of Fig. 5 is presented in Table 1. This table verifies that this system helps to attain an improved outcome to examine the DD pictures even in the gray scale form. Table 2 shows average of the performance measures attained for the considered datasets. This table also verifies that this system works fine on these two datasets and helps to attain a better performance values irrespective of the image class (RGB/gray) and the artifact. The recent procedures [23–30] can be utilized to improve the performance of skin appraisal.

4 Conclusion

The work presented in this paper aims to develop and implement a semiautomated image processing practice to assess the DD pictures of the PH2 and DermIS datasets. This work implemented JA + SE thresholding and CV segmentation to extract the SC segment with better accuracy. The threshold enhances the SC section, even though it is associated with the artifact. Later, CV extraction is executed to mine the SC region. Finally, a relative appraisal among the mined SC and GT is accomplished to mine the essential performance values. The results of this study confirm that the proposed technique works well on the RGB scaled and the gray scaled DD picture, and helps to attain better results.

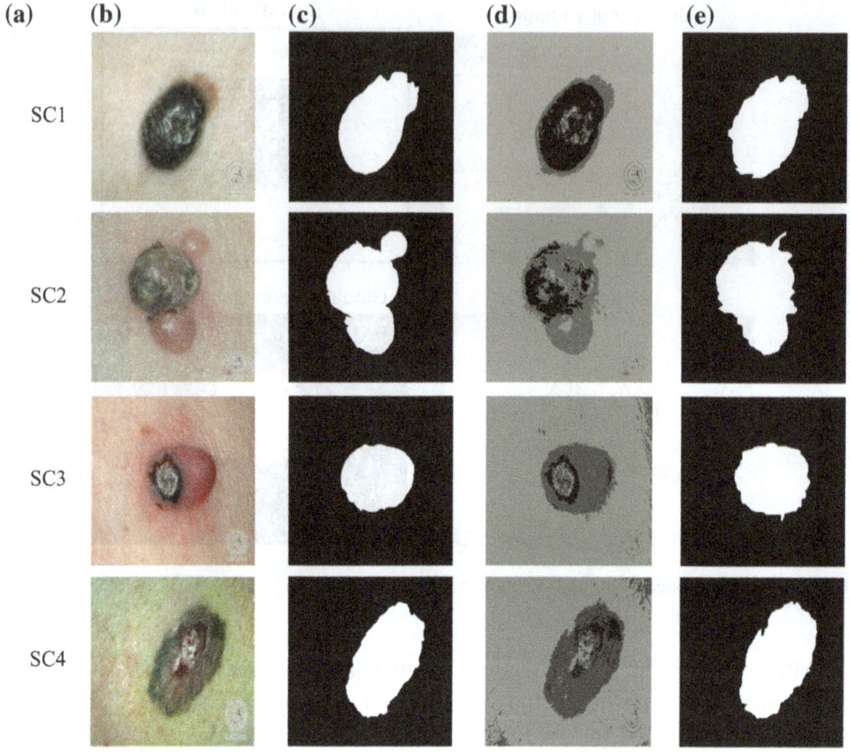

Fig. 5 Sample test images. **a** Pseudo-name, **b** sample DD image, **c** GT, **d** thresholded image, and **e** segmented binary picture

Table 1 Picture similarity parameters for the chosen DermIS images

Image	TPR	FNR	TNR	FPR	JI (%)	DC (%)
SC1	0.99	0.01	0.97	0.03	83.28%	85.01%
SC2	0.99	0.01	0.97	0.03	81.95%	84.25%
SC3	0.97	0.03	0.98	0.02	84.16%	86.03%
SC4	0.99	0.01	0.98	0.02	82.22%	83.89%

Table 2 Average of performance measures attained in this study

Dataset	JI (%)	DC (%)	SE (%)	SP (%)	AC (%)	PR (%)
PH2 (50 images)	82.19	86.28	95.36	87.38	99.04	88.38
DermIS (20 images)	83.29	85.77	96.28	88.36	98.94	87.66

References

1. https://www.who.int/uv/faq/skincancer/en/index1.html
2. Glaister, J., Wong, A., Clausi, D.A.: Segmentation of skin lesions from digital images using joint statistical texture distinctiveness. IEEE T. Biomed. Eng. **61**(4), 1220–1230 (2014)
3. Amelard, R., Glaister, J., Wong, A., Clausi, D.A.: High-level intuitive features (HLIFs) for intuitive skin lesion description. IEEE T. Biomed. Eng. **62**(3), 820–831 (2015)
4. Barata, C., Ruela, M., et al.: Two systems for the detection of melanomas in dermoscopy images using texture and color features. IEEE Syst. J. **99**, 1–15 (2013)
5. Dey, N., et al.: Social group optimization supported segmentation and evaluation of skin melanoma images. Symmetry **10**(2), 51 (2018). https://doi.org/10.3390/sym10020051
6. https://www.fc.up.pt/addi/ph2%20database.html
7. http://vip.uwaterloo.ca/demos/skin-cancer-detection
8. Satapathy, S.C., Rajinikanth, V.: Jaya algorithm guided procedure to segment tumor from brain MRI. J. Optim. **2018**, 12 (2018). https://doi.org/10.1155/2018/3738049
9. Kannappan, P.L.: On Shannon's entropy, directed divergence and inaccuracy. Probab. Theory Rel. Fields **22**, 95–100 (1972). https://doi.org/10.1016/S0019-9958(73)90246-5
10. Paul, S., Bandyopadhyay, B.: A novel approach for image compression based on multi-level image thresholding using Shannon entropy and differential evolution. In: IEEE Students' Technology Symposium (TechSym), pp. 56–61 (2014). https://doi.org/10.1109/techsym.2014.6807914
11. Rajinikanth, V., Satapathy, S.C., Fernandes, S.L., Nachiappan, S.: Entropy based segmentation of tumor from brain mr images—a study with teaching learning based optimization. Pattern Recogn. Lett. **94**, 87–94 (2017)
12. Rao, R.V.: Jaya: A simple and new optimization algorithm for solving constrained and unconstrained optimization problems. Int. J. Ind. Eng. Comput. **7**(1), 19–34 (2016)
13. Rao, R.V., More, K.C.: Design optimization and analysis of selected thermal devices using self-adaptive Jaya algorithm. Energy Convers. Manag. **140**, 24–35 (2017). https://doi.org/10.1016/j.enconman.2017.02.068
14. Rao, R.V., Saroj, A.: A self-adaptive multi-population based Jaya algorithm for engineering optimization. Swarm Evol. Comput. **37**, 1–26 (2017). https://doi.org/10.1016/j.swevo.2017.04.008
15. Rajinikanth, V., Dey, N., Kumar, R., Panneerselvam, J., Raja, N.S.M.: Fetal head periphery extraction from ultrasound image using Jaya algorithm and Chan-Vese segmentation. Procedia Comput. Sci. **152**, 66–73 (2019). https://doi.org/10.1016/j.procs.2019.05.028
16. Fernandes, S.L., Rajinikanth, V., Kadry, S.: A hybrid framework to evaluate breast abnormality using infrared thermal images. IEEE Consum. Electron. Mag. **8**(5), 31–36 (2019). https://doi.org/10.1109/MCE.2019.2923926
17. Rajinikanth, V., Satapathy, S.C.: Segmentation of ischemic stroke lesion in brain MRI based on social group optimization and Fuzzy-Tsallis entropy. Arabian J. Sci. Eng. **43**(8), 4365–4378 (2018). https://doi.org/10.1007/s13369-017-3053-6
18. Raja, N.S.M. et al.: Contrast enhanced medical MRI evaluation using Tsallis entropy and region growing segmentation. J. Ambient Intell. Hum. Comput. 1–12 (2018). https://doi.org/10.1007/s12652-018-0854-8
19. Jahmunah, V., et al.: Automated detection of schizophrenia using nonlinear signal processing methods. Artif. Intell. Med. **100**, 101698 (2019). https://doi.org/10.1016/j.artmed.2019.07.006
20. Acharya, U.R., et al.: Automated detection of Alzheimer's disease using brain MRI images–a study with various feature extraction techniques. J. Med. Syst. **43**, 302 (2019). https://doi.org/10.1007/s10916-019-1428-9
21. Dey, N., et al.: Social-Group-Optimization based tumor evaluation tool for clinical brain MRI of Flair/diffusion-weighted modality. Biocybern. Biomed. Eng. **39**(3), 843–856 (2019). https://doi.org/10.1016/j.bbe.2019.07.005

22. Fernandes, S.L. et al.: A reliable framework for accurate brain image examination and treatment planning based on early diagnosis support for clinicians. Neural Comput. Appl. 1–12 (2019). https://doi.org/10.1007/s00521-019-04369-5
23. Das, H., Naik, B., Behera, H.S.: Classification of diabetes mellitus disease (dmd): a data mining (DM) Approach. Adv. Intell. Syst. Comput. **710**, 539–549 (2018). https://doi.org/10.1007/978-981-10-7871-2_52
24. Sahani, R., et al.: Classification of intrusion detection using data mining techniques. Adv. Intell. Syst. Comput. **710**, 753–764 (2018). https://doi.org/10.1007/978-981-10-7871-2_72
25. Pradhan, C., Das, H., Naik, B., Dey, N.: Handbook of Research on Information Security in Biomedical Signal Processing Hershey. IGI Global, PA (2018)
26. Sahoo, A.K., Mallik, S., Pradhan, C., Mishra, B.S.P., Barik, R.K., Das, H.: Intelligence-based health recommendation system using big data analytics. In: Big Data Analytics for Intelligent Healthcare Management, pp. 227–246 (2019). https://doi.org/10.1016/B978-0-12-818146-1.00009-X
27. Dey, N., Das, H., Naik, B., Behera, H.S. (eds.): Big Data Analytics for Intelligent Healthcare Management. Academic (2019)
28. Chandrakar, P.: A secure remote user authentication protocol for healthcare monitoring using wireless medical sensor networks. Int. J. Ambient Comput. Intell. (IJACI) **10**(1), 96–116 (2019). https://doi.org/10.4018/IJACI.2019010106
29. Bhattacharya, H., Chattopadhyay, S., Chattopadhyay, M., Banerjee, A.: Storage and bandwidth optimized reliable distributed data allocation algorithm. J. Ambient Comput. Intell. (IJACI) **10**(1), 78–95 (2019). https://doi.org/10.4018/IJACI.2019010105
30. Rajinikanth, V., Raja, N.S.M., Arunmozhi, S.: ABCD rule implementation for the skin melanoma assesment—a study. In: IEEE International Conference on System, Computation, Automation and Networking (ICSCAN) (2019). https://doi.org/10.1109/icscan.2019.8878860

Partial Offloading for Fog Computing Using P2P Based File-Sharing Protocol

Satanu Maity and Sujoy Mistry

Abstract In the world of Internet of Things (IoT), promising paradigm such as fog computing has been emerged as an extension of cloud computing infrastructure to provide low latency and better QoS to the end-user specifically for time-sensitive applications. One of the major challenges of these applications is the task offloading, where resource-constraint end-user devices migrate the tasks to the nearby resource-rich computing environment for execution. There are several factors like latency optimization, resource management that need to be addressed for performing the task offloading. In this work, we aim to provide a new offloading technique where latency for task offloading has been optimized by taking Peer-to-Peer (P2P) technology as a basic mode of a network environment for fog computing and also taken P2P file-sharing protocol as a basic mode of offloading technique.

Keywords Fog computing · Task offloading · Delay minimization · Peer-to-Peer file-sharing protocol

1 Introduction

In the present, end-user devices like mobile devices can perform multiple real-life applications at a single instance of time. Some applications require a large amount of resources to execute those applications, but due to its resource constraint, end-user devices are unable to run such applications. Such kind of limitations have been overcome with the help of the cloud computing concept [1], where end-user devices migrates its resource-hungry task/workload to the cloud for execution. This workload migration is called task offloading process. But cloud resources are physically located in one geographical place, so to process or store the huge amount of data need to offload to a single point at every time where cloud resides is a costly affair. Most importantly, latency has been increased for such transmission/offload

S. Maity (✉) · S. Mistry
Department of Computer Science and Engineering, Maulana Abul Kalam Azad University of Technology, Kolkata, India

H. Das et al. (eds.), *Progress in Computing, Analytics and Networking*, Advances in Intelligent Systems and Computing 1119,
https://doi.org/10.1007/978-981-15-2414-1_30

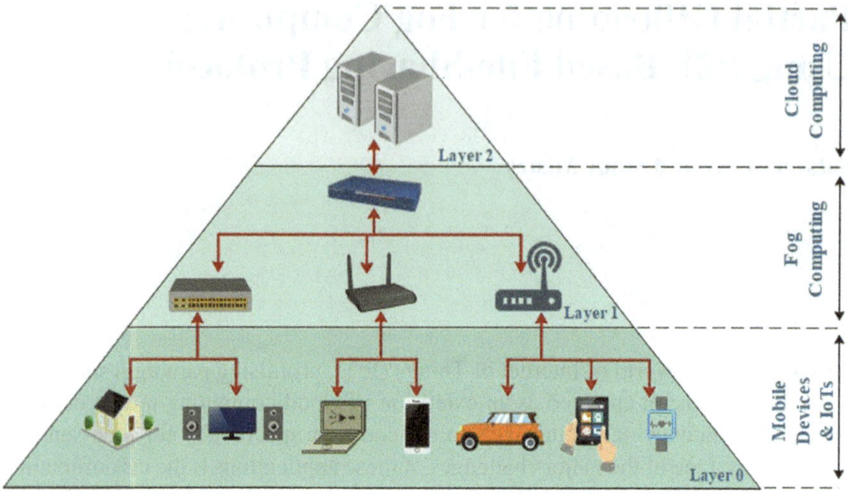

Fig. 1 Computing layered architecture

of data/workload because the bandwidth is too small and fixed from the end-user devices which are not acceptable in real-life time-sensitive applications.

To overcome these types of issues [2, 3], a new computing paradigm has been introduced named as fog computing [4], where the main concept brings the computation process more closely to the user end devices. This computing environment is prepared by a collaboration of several network edge devices (like a router, switch, etc.) named as "fog devices", which is placed in between cloud and end-user devices (Fig. 1). For this reason, the end-user devices can quickly transmit their data/task or request for a service.

Actually, the main objective of fog computing is to the growth of efficiency and trim down the latency for offloading/transmitting the workload as required in the previous computing paradigms [5]. Here in terms of "growth of efficiency" means that skilled up the computation capability to the higher level.

So far, different nodal collaboration-based fog computing architectures [6] have introduced by several researchers to optimize the delay, energy as well as increase the efficiency for task offloading purpose, whereas some works prefer to increase the provision of good quality services to the time-sensitive application rather than optimizing the different parameters related to the fog computing paradigm. After study most of these works, this has been concluded that there did not exist any concrete system which offers both at a time. Actually, it is essential to minimize the service time delay and offloading delay as well as the provision of services for the time-sensitive application.

Therefore, this works adopts three approaches to meeting the challenges as discussed. First, the SOA working approach integrates into the fog layer, by which the end-user get maximum number of services from the fog computing layer at a same time. According to SOA approach in this work, the offloaded task will remain in the fog layer for future uses purpose after initial execution until it not removes manually. So, it increases the service availability of fog computing. Second, adopt P2P nodal collaboration in the fog layer to decrease the communication delay. And finally, the third is partial offloading in the fog layer by using P2P file-sharing protocol. There are task or service containing multiple fog devices may be overloaded at the same time. At that time, a service request can arises to an overloaded fog device, then the task/service containing overloaded fog devices mutually offload the part of that service/task to a free fog device and continue the process. As an observation using a P2P-based file-sharing protocol, the partial offload approach takes less time for offloading the task rather than a simple single-source offloading method.

The remaining of this paper is organized as follows: In Sect. 2, related works to this paper are discussed. In Sect. 3, we present our proposed offloading models including system architecture of fog layer. Section 4 demonstrates the proposed system through a developed simulation. Finally, Sect. 5 concludes this paper.

2 Related Works

Fog computing paradigm is the recent technology concept, which is introduced in the broad network area by the R&D community to provide low latency and better QoS to the end-user for time-sensitive application. In fog computing paradigm, the different task offloading mechanism was proposed in several articles regarding minimizing the time-delay as well as the energy. A joint delay-efficient and energy optimization offloading system was proposed in [7], here D2D communicating fog devices were divided into two groups: high power FNs and low power FNs. Then they have presented a decision-making offloading technique for two different goals: one is node energy consumption, and other is a task processing delay. In [8], proposed a delay and energy-efficient partial offloading in distributed fog-based edge computing architecture overcentralized. Though the above papers mainly focused on applying different optimization technique jointly into a single architecture of the fog computing environment, there are some lacks in infrastructure to provide a rich amount of services to the end-users. So we have study some articles related to the provision of a huge number of services and QoS in fog computing paradigm.

The main concept behind SOA is to provide multiple and huge numbers of services to different components by application module via a different kind of communication protocols. In [9], proposed three-tier security-based SOA-Fog framework and did a case studies on geo-health data. A four-layer infrastructure combination of SOA and fog computing was proposed in [10], where the third layer provides agent-based services act as artificial intelligence to provide complex service like traffic monitoring, stream analytics, etc., whereas the second layer supports a lightweight SOA and calculates and provides services to local IoT devices in a distributed manner. These above articles of this subsection did not achieve as much as any kind of efficient model as proposed in the previous subsection.

To reduce the communication latency in the fog layer some researchers adopt Peer-to-Peer (P2P) collaboration between fog nodes. In [11], proposed a P2P-based fog network in 5G-enabled smart cities based on TSI-NFV, MANO architecture where fog nodes are communicating via Open Shortest Path First (OSPF) routing protocol and showed that a significant decrease in terms of network bandwidth usage when compared with conventional cloud solutions. The paper [12] introduce an approach named "Streamed Deployment" (SD) to make out scalability challenges in Linux container provisioning at the edge network and proposed a peer-to-peer communication model in edges of the network, at the end of this paper the proposed model tested on some real edge data.

However, some research articles had proposed several offloading approaches to increasing the quality of services for real-life applications like smart healthcare monitoring [13], traffic light management [14], smart vehicle parking system [15], etc. After studies these above papers, there is a scope to improve the service-latency and efficiency in fog computing. So in this paper, we combine both peer-to-peer collaboration among the fog devices to reduce the communication latency in the fog layer and integrate SOA in fog computing to increase the service availability as well as to provide a good quality of services to the end-users. Also, we have an introduced P2P-based file-sharing protocol in the fog layer by which we can get an advantage in partial offloading and also work in the dynamic network environment efficiently.

3 Proposed Works

3.1 System Architecture

The proposed system architecture in Fig. 2 has a fog layer with the yellow color area; consist of several heterogeneous fog devices which have some common properties like computation power, memory space, processing queue, energy, and others. All Fog Devices (FDs) are connected to each other as a P2P network fashion, as because it helps a network device can act various roles like a controller, service provider, task performer, etc., at the same time and can also transmit data to another device directly. Also, this fog computing layer is conceptually built as like Service-Oriented Architecture (SOA) works, because in the SOA a task resides in the network as a service, as long as possible until it moves out from the network manually.

The end-user devices (mobile device) symbolized by MD have considered in the above figure, which has a small wireless data transmission range denote by the dotted circle. When some fog devices come in that range then the mobile device can communicate to those fog devices and obtain the facility to get several services from the fog network as well as transmit the data/tasks. The mobile device has a primary intention to execute that uncomputable task by offloading to resource-rich devices, so if at that time some fog devices exist in its wireless network range then the mobile device request to those fog devices (called as local fog devices) for offload

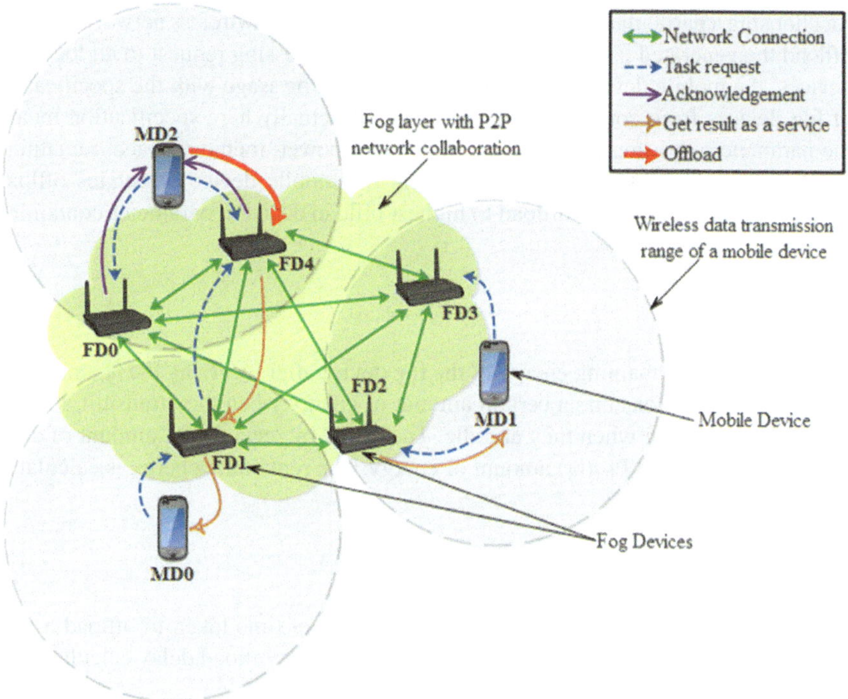

Fig. 2 P2P-based fog architecture

that task. In that case, there may arise several scenarios depend on the resource availability of the fog network, which are discussed in next subsections. (In Fig. 2, several communications types and cases are designated and shows by a particular arrow type with a unique color.)

3.2 Simple Offloading Model

Let consider a mobile device generate a task τ_i with the size of θ_i and δ_i computation requirement. But the problem is that the mobile device has not much enough resources to compute that task. According to the proposed system architecture model, a fog device has several parameters like computation power (δ_f), memory space (θ_f), processing queue (σ_f), energy (E_f), etc., and a fog device can able to perform the task τ_i as because of [$\theta_i < \theta_f$] and [$\delta_i < \delta_f$], so the mobile device wants to offload that task τ_i to the fog layer. Now consider some problems have arisen related to this offloading scenario, which are discussed below as an individual case.

Case 1. Once at a time, there may be present multiple fog devices in the wireless network range of the mobile device. So, the mobile device has to make a fair decision for choosing a particular fog device among all others in the wireless network range to offload the generated task. After sending the task processing request to all local fog devices, the mobile device gets an acknowledgment message with the specification of fog devices from corresponding fog devices. Actually here specification means the parameters of a fog device like computation power, memory space, remaining energy, etc. Depend upon those parameters the mobile device calculates offload decision parameter (α) and offload to highest offload decision parameter containing fog device.

$$\alpha = E_f^r / T_{offload} \tag{1}$$

Here E_f^r is the remaining energy of the fog device after receiving the θ_i amount of task. Fog devices consume a certain amount of energy when they transmit, receive or compute a task or when they are idle. Therefore, by receiving θ_i amount of data, fog device consumed $E_{bit}(\theta_i)$ amount of energy. The remaining energy is calculated as

$$E_f^r = E_{initial} - E_{bit}(\theta_i) \tag{2}$$

$T_{offload}$ is the offloading delay, which an amount of time taken to offload a task completely from a mobile device to a fog device. So the offload delay calculated by

$$T_{offload} = \theta_i / R \tag{3}$$

Here R is data uplink transmission rate of a mobile device and it has calculated by the help of Shannon's SNR theorem.

$$R = \omega \log_2(1 + ((P_m G_m)/(D_{mf} N_0))) \tag{4}$$

Actually, the mobile device with P_m transmission power and G_m transmission gain transmits the task to fog layer through ω bandwidth of small wireless transmission channel. This channel can be affected by noises (N_0) in a real-time environment. The noise increases in the channel with an increase in distance (D_{mf}), so as a result, the mobile device has good transmission rate at those fog devices, which are much closer to it.

Case 2. As service-oriented architecture is integrated into the proposed fog layer, there is a facility to get the result as a service from the previously deployed task by network terminal node (i.e., the mobile devices). So, if the generated offloadable task already exists in the local fog devices, i.e., that task previously offloaded to near local fog devices, then there no need to offload the task again. Simply the mobile device transmits the parameters to perform the task and get the result as a service from the local fog devices.

Case 3. Now, in this case, the task may not be present in the local fog devices but present in fog layer (i.e., currently out of the mobile wireless network range). Then local fog device can directly transfer that task perform request with parameters to that task containing fog device simply and bring back the result to the requested mobile device.

So, as a result, the network terminal devices get faster services as well as a huge number of services from the fog layer by integrating P2P network collaboration and SOA feature into the fog layer. But, in the real-time network, a task containing fog device may be overloaded or discontinue for power failure issue at dynamic runtime, so a partial task offloading model has been introduced in the next subsection.

3.3 *Partial Offloading Model*

Though the fog layer have sufficient resources to perform a mobile offloadable task but not too large as a cloud server, because there may fog devices can be overload at any time. These situation lead to degradation of the performance of the fog computing. Actually, in the proposed model, overload means either processing queue or memory space has become filled up by the number of deployed tasks. So, at this overloaded case, find a free fog device in the current fog layer for offload the task to continue the execution and control goes to that free fog device.

For SOA facility in the fog layer, a task can reside in multiple fog devices until not removes manually. But they might overload at the same time, so a free fog device can get the task from any of these overloaded fog devices. But this paper propose a mechanism to get the task from multiple fog devices as a distribution of the task as like data by the help of P2P-based file-sharing protocol; as a result, it reduces the time latency to offload the overall task from a single source. The P2P-based file-sharing protocol is a special evolution of the peer-to-peer network, here each active devices is involved in the distribution of the overall task through their upload bandwidth at the same time. That is mean; each device has transmitted some part of the overall task mutually as a requirement at the same time. So as a result, this takes less time to transmit the overall task than single-node transmission method.

As per in Fig. 3a, FD0, FD2, and FD4 are task containing overloaded fog devices and they have mutually partial offloaded the task to free fog device FD3. The overall working procedure of this proposed partial offloading model is illustrated in the sequence diagram (Fig. 3b). So, as seeing, if the availability of a task in the multiple numbers of fog devices has increased then the overall offloading time also decreases accordingly.

4 Implementation and Results

This section demonstrates that how the proposed architecture and offloading model has been implemented using Omnet++ and overall setup of the simulator. And finally, express the results with respect to the different cases and scenarios as described in the proposed work.

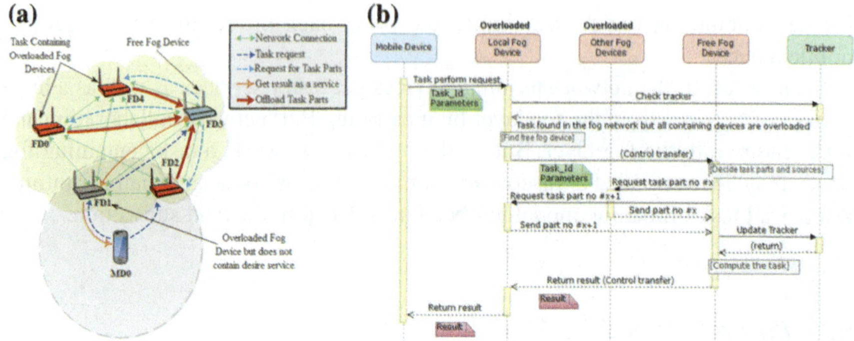

Fig. 3 Partial offloading model (**a**) and its sequence diagram (**b**)

The simulator contains mainly three type files: Network Definition File, Source Code, and Network Initialization File. The network definition file contains the overall description of a network like the specification of network devices, the topology of the network, etc., as a module. The source code contains the details of working procedure of a network device and initialization file has the regulation that how the network will initialize at the starting of simulation.

A mobile device with 40 meters of wireless network communication range generates 6 MB (maximum) of an uncomputable task which offloads or transmits as a task request to the local fog devices through a wireless uplink channel with 20 MHz bandwidth. To select a suitable fog device for offloading, the mobile device calculates offloading delay, remaining energy, and offloading decision parameter consecutively. The offloading delay relies on uplink rate, which mainly depends on the distance between the mobile device and a fog device because the noise ratio increases as the distance increases where noise ratio has considered as 0.000000000173 dBm at the unit of distance. Depend upon the several scenarios and cases in the real-life network as considering in the proposed offloading model, the time latency has formulated and computed by the simulator.

The comparative result presented in Fig. 4a shows the each and every individual cases of the simple offloading model and concludes that the offloading time is much larger than service time, so if service availability increases in the fog layer then it can reduce the offloading time of the repeated task. The results of the partial offloading model are presented in Fig. 4b which shows that the offloading time in fog network has minimized with the increases of task containing overloaded fog devices.

5 Conclusions

In this paper, partial offloading using P2P based file-sharing protocol in the SOA enabled fog computing environment is introduced. We have implemented two

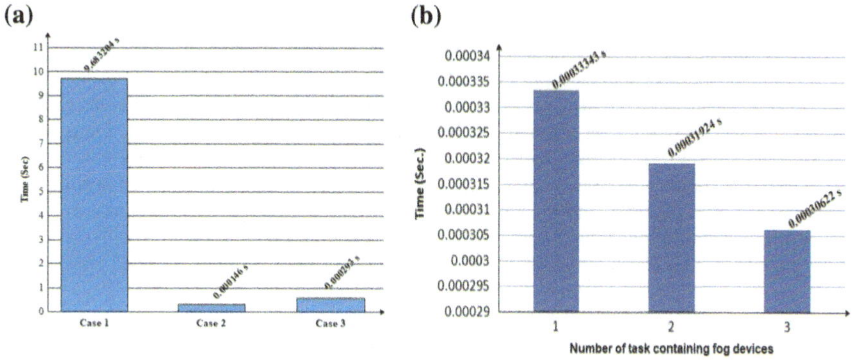

Fig. 4 Comparison of results of different offloading cases and scenarios

offloading models by considering various scenarios for minimizing the offloading delay. First, the simple offloading model is proposed for choosing a perfect fog device for initial task offloading by which it can take less time to offload from end-user devices. Second, the partial offloading model is proposed for offloading the task among fog devices using the P2P-based file-sharing protocol. This second model offered to overcome the overload problem of task/service containing fog devices and also minimize the offloading latency in the SOA-integrated fog layer. The simulation result shows that the proposed offloading models minimize the offloading delay in P2P-based fog computing as we expected. On this basis, our next footstep will be to improve the proposed system by integrating resource allocation and load balancing mechanisms. Besides this, limitation of this paper is that we do not consider the security aspect in our work, which is a vital point for real-life applications. But we would try to implement this in our architecture in the future.

References

1. Kumar, A., Ranjan, A., Gangwar, U.: An understanding approach towards cloud computing. Int. J. Emerging Technol. Adv. Eng. **2**(9) (2012)
2. Dillon, T.S., Wu, C., Chang, E.: Cloud computing: issues and challenges. In: 2010 24th IEEE International Conference on Advanced Information Networking and Applications, pp. 27–33 (2010)
3. Mishra, B.S.P., Das, H., Dehuri, S., Jagadev, A.K.: Cloud Computing for Optimization: Foundations, Applications, and Challenges, vol. 39. Springer, Berlin (2018)
4. Jalali, F., Hinton, K., Ayre, R., Alpcan, T., Tucker, R.: Fog computing may help to save energy in cloud computing. IEEE J. Sel. Areas Commun. **34**, 1728–1739 (2016)
5. Hu, P., Dhelim, S., Ning, H., Qiu, T.: Survey on fog computing: architecture, key technologies, applications and open issues. J. Netw. Comput. Appl. **98**, 27–42 (2017)
6. Mahmud, R., Buyya, R.: Fog Computing: A Taxonomy, Survey and Future Directions (2018). https://arxiv.org/abs/1611.05539

7. Bozorgchenani, A., Tarchi, D., Corazza, G.E.: An energy and delay-efficient partial offloading technique for fog computing architectures. In: GLOBECOM 2017–2017 IEEE Global Communications Conference, pp. 1–6 (2017)
8. Bozorgchenani, A., Tarchi, D., Corazza, G.E.: Centralized and distributed architectures for energy and delay efficient fog network-based edge computing services. IEEE Trans. Green Commun. Netw. **3**, 250–263 (2019)
9. Barik, R.K., Dubey, H., Mankodiya, K.: SOA-FOG: secure service-oriented edge computing architecture for smart health big data analytics. In: 2017 IEEE Global Conference on Signal and Information Processing (GlobalSIP), pp. 477–481 (2017)
10. Chakrabarty, A., Heya, T., Hossain, M., Arefin, S., Joy, K.: A novel extended-cloud based approach for Internet of Things. In: Studies in Big Data, pp. 303–331 (2018)
11. Santos, J., Wauters, T., Volckaert, B., Turck, F.D.: Fog computing: enabling the management and orchestration of smart city applications in 5G networks. Entropy **20**, 4 (2017)
12. Gazzetti, M., Reale, A., Katrinis, K., Corradi, A.: Scalable linux container provisioning in fog and edge computing platforms. In: Euro-Par 2017: Parallel Processing Workshops, pp. 304–315 (2018)
13. Al-khafajiy, M., Baker, T., Waraich, A., Al-Jumeily, D., Hussain, A.: IoT-Fog Optimal Workload via Fog Offloading. In: 2018 IEEE/ACM International Conference on Utility and Cloud Computing Companion (UCC Companion) (2018)
14. Saqib, M., Hamid, M.: FogR: A highly reliable and intelligent computation offloading on the Internet of Things. In: 2016 IEEE Region 10 Conference (TENCON) (2016)
15. Kui, X., Sun, Y., Zhang, S., Li, Y.: Characterizing the capability of vehicular fog computing in large-scale urban environment. Mobile Netw. Appl. **23**, 1050–1067 (2018)

Assessment of Fundus Images for Retinal Abnormality Screening—A Study

J. T. Anita Rose, Sangeetha Francelin Vinnarasi, Jesline and V. Rajinikanth

Abstract Screening of eye is essential to identify a range of retinal irregularities, and Fundus Imaging (FI) is one of the techniques commonly used by the ophthalmologist to record and evaluate the retinal abnormalities. Manual appraisal of FI is time-consuming, and hence, a computer-based methodology is always preferred. This work aims to develop a system to examine the retinal abnormality with the help of FI pictures. This work employs a hybrid processing scheme to enhance and extract the FI in order to extract and assess the optic-disc section. In this work, enhancement is done with firefly algorithm and Kapur's thresholding, and the extraction is implemented with level set technique. All these testing are put into practice with MATLAB software using the benchmark Rim-One FI database, and the results of this study confirmed that the anticipated technique offered enhanced outcome in extracting the disc section from the RGB scaled FI.

Keywords Fundus images · Retinal abnormality · Firefly algorithm · Optic-disc · Validation

1 Introduction

Eye is one of the vital sensory organs responsible to convert the light energy into visual picture. The disease in eye is a severe concern, and it arises due to various reasons, and the untreated eye disease will badly affect the visual system and may cause the loss in vision [1, 2].

J. T. Anita Rose · S. F. Vinnarasi (✉) · Jesline
Department of Computer Science Engineering, St. Joseph's College of Engineering, Chennai 600 119, Tamilnadu, India
e-mail: fsangeetha@gmail.com

V. Rajinikanth (✉)
Department of Electronics and Instrumentation, St. Joseph's College of Engineering, Chennai 600 119, Tamilnadu, India
e-mail: v.rajinikanth@ieee.org

H. Das et al. (eds.), *Progress in Computing, Analytics and Networking*, Advances in Intelligent Systems and Computing 1119,
https://doi.org/10.1007/978-981-15-2414-1_31

Recently, a number of disease evaluation techniques are planned and executed in the literature to detect the irregularity in eye with a number of computer-assisted techniques. Normally, the disease in eye can be assessed with a visual check by an experienced ophthalmologist and a computerized examination with the help of Fundus Imaging (FI). FI is one of the widely implemented imaging procedures to capture various parts of the eye, such as Optic-Disc (OD), Optic-Cup (OC), retina, blood vessels, etc. [3, 4]. After recording the essential section of the eye, an examination is to be implemented to detect and categorize the nature of abnormality and its severity to plan for a possible treatment. The FI can be used to assess various retinal sections, such as OD, OC, optic-nerve and blood vessels.

Recently, a range of FI examination techniques are discussed by the researchers to assess the FI to find the abnormality in retinal parts using the Computer-Based Algorithms (CBA). Because of its function and disease prediction performance, a diversity of CBA for image processing procedure is largely proposed and implemented to examine FI. Normally, evaluation of greyscale picture is fairly easy and needs lesser computational effort due to its uncomplicated pixel allocation. But the assessment of RGB image forever involves intricate computations due to its red, green and blue pixels. Hence, a selection of procedures is implemented to pre-process the RGB scale pictures to enhance the information during the evaluation [1].

The present study aims to execute a RGB picture evaluation technique to assess the considered FI pictures. The procedure in getting the FI for the evaluation of eye consists of various techniques: (i) the essential parts of eye using a specialized camera, and (ii) the recorded image is then inspected by an ophthalmologist or CBA existing in eye clinics to identify the eye abnormality to prepare for appropriate healing practice [1, 2].

In this work, examination of retinal OD using the FI is implemented using a hybrid imagine tool, which consists of a thresholding and segmentation process. The thresholding improves the visibility of the abnormal FI section and is implemented using Kapur's entropy base optimal thresholding concept. In order to discover the best threshold, the optimization process is done using the Firefly Algorithm (FA). Later, the enhanced section of FI is extracted with the Level Set (LS) practice, which works based on an adaptable contour. Finally, as evaluation of the extracted FI section, the Ground Truth (GT) is implemented, and the performance values, such as sensitivity, specificity, accuracy and precision, are then computed. Based on these values, the superiority of this tool is authenticated.

In this paper, combination of RGB scale FI multi-thresholding with the Kapur's entropy and the segmentation is realized to mine and evaluate the Optic-Disc (OD) section from the benchmark Rim-One database. This investigation is executed with the Matlab7, and the performance of proposed approach is validated by considering the quality parameters, such as Sensitivity (SE), Specificity (SP), Accuracy (AC) and Precision (PR). These values are calculated by relating the mined OD segment with the Ground Truth (GT). This database consists of five different GTs for every test image. The proposed imaging tool is very efficient and provides the better values of SE (98.21%), SP (97.55%), AC (98.04%) and PR (96.49%). In future, this process can be used to inspect other therapeutic pictures recorded with various imaging techniques.

2 Methodology

In FI assessment literature, a number of methods are existing to achieve the multi-thresholding and segmentation tasks. In this paper, the various stages presented in Fig. 1 are adopted to examine the retinal OD of the Rim-One database.

2.1 *Rim-One Database*

Rim-One is one of the famous FI datasets available in [4, 5], in which a particular retinal OD is presented in five various groups with a five number of GTs for every test picture. All these images are existing in RGB form, and the examination of this test picture is done with the proposed hybrid evaluation tool. Other details regarding the Rim-One can be found in [2, 3].

2.2 *Multi-thresholding*

Multi-thresholding is one of the commonly adopted image pre-processing schemes to enhance the visibility of the test picture. Implementation of this technique is recently considered to enhance a class of medical and classical images with various techniques. The entropy-based technique is considered widely during the medical image pre-processing, and the necessary information of this practice can be found in [6]. In this work, thresholding is applied by Firefly Algorithm (FA) and Kapur's Entropy (KE). FA is invented by Yang based on the flashing model formed by fireflies [7, 8].

The account of FA is as below:

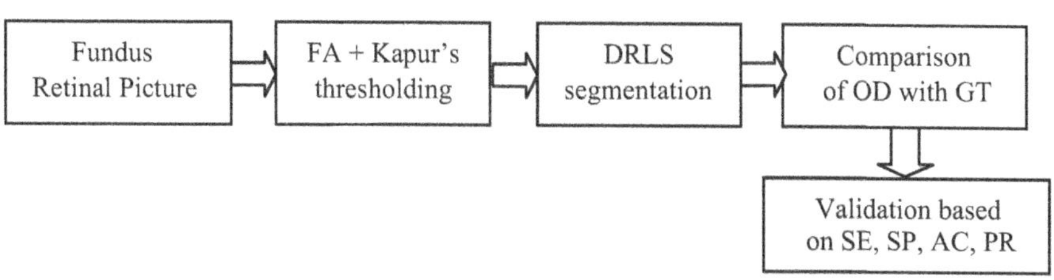

Fig. 1 Various image evaluation stages of the proposed FI examination system

$$P_{F_1}^{t+1} = P_{F_1}^{t} + \beta_0 e^{-\gamma\, d_{xy}^2}(P_{F_2}^{t} - P_{F_1}^{t}) + \alpha_1.\text{sign}\,(\text{rand} - 1/2) \oplus \text{B(s)} \quad (1)$$

where $P_{F_1}^{t+1}$ is modernized place of F_1, $P_{F_1}^{t}$ is opening background of firefly, $\beta_0 e^{-\gamma\, d_{F_1F_2}^2}(P_{F_2}^{t} - P_{F_1}^{t})$ is the striking energy among fireflies, $B(s) = A.|s|^{\alpha/2}$ is Brownian walk system, A = random variable, β = spatial supporter and α = temporal supporter. Associated particulars on FA can be established in [11–13].

In this work, the considered FA is used to discover finest threshold through the assist of KE. It can be employed to establish best threshold of test image by simply analysing its entropy. Due to its dominance, considerable studies on KE are conversed in [9, 10].

Mathematical model of the KE is defined below.

Let $Th = [t_{h1}, t_{h2}, \ldots, t_{hL-1}]$ indicate thresholds of the image, then the general entropy will be

$$\text{Cost function} = J_{\max} = J_{\text{Kapur}}(T) = \sum_{J=1}^{L} O_j^R \quad \text{for} \quad R\{1, 2, 3\} \quad (2)$$

Equation 2 shows the maximized entropy with appreciated to the selected threshold value. In multiple-thresholding assignment, the objective function is denoted as

$$\begin{aligned} O_1^R &= \sum_{j=1}^{t_1} \frac{Po_j^R}{\theta_0^R} \ln\left(\frac{Po_j^R}{\theta_0^R}\right), \\ O_2^R &= \sum_{j=t_l+1}^{t_2} \frac{Po_j^R}{\theta_1^R} \ln\left(\frac{Po_j^R}{\theta_1^R}\right), \\ &\vdots \\ O_k^R & \sum_{j=t_{k+1}}^{L} \frac{Po_j^R}{\theta_{L-1}^R} \ln\left(\frac{Po_j^R}{\theta_{K-1}^R}\right) \end{aligned} \quad (3)$$

where Po_j^R signifies likelihood distribution and $\theta_0^R, \theta_1^R, \ldots, \theta_{L-1}^R$ depicts likelihood incidence in L-levels. The main function of FA is to recognize best threshold based on maximized entropy. The FA randomly modifies thresholds of trial image till $J_{\text{Kapur}}(T)$ is achieved.

Here, FA values are chosen as follows: number of fireflies = 30, dimension of search = 3, stopping function = maximized KE and iteration value = 2500.

2.3 DRLS Segmentation

This work implemented the Level Set (LS) segmentation of Li et al. [14]. LS runs as the energy minimization process, and it is articulated as

$$\Re_e(\varphi) \triangleq \int_\Omega e(|\nabla\varphi|)dX \tag{4}$$

where e = energy concreteness charge with $e = [0, \alpha] \rightarrow \Re$.

In LS, a varied arc is endorsed to identify all the likely pixels linked to the uneven section dynamic in the picture. After recognizing all the potential pixel groups, it will mine the section which is within the converged arc for the valuation procedure. The related information on LS can be found in [15].

2.4 Assessment

The merit of the tool is confirmed by calculating the quality parameters as discussed in [16–20]. In this paper, the following values are computed to assess the performance:

$$\text{Sensitvity} = T_P \big/ (T_P + F_N) \tag{5}$$

$$\text{Specificity} = T_N \big/ (T_N + F_P) \tag{6}$$

$$\text{Accuracy} = (T_P + T_N)/(T_P + T_N + F_P + F_N) \tag{7}$$

$$\text{Precision} = T_P \big/ (T_P + F_P) \tag{8}$$

where T_N, T_P, F_N and F_P specify true negative, true positive, false negative and false positive, correspondingly.

3 Results and Discussion

This division presents the investigational outcome acquired with the proposed tool. All these results are attained using Matlab7 software. During this study, the benchmark FI pictures of the Rim-One database are considered. This dataset consists of image category, such as deep, early, moderate, normal and OHT. In every image category, 10 test pictures are used in the assessment.

Figure 2 shows the sample trial picture of deep-image category of Rim-One, which was used in preliminary assessment. Figure 2a signifies the pseudo-name, and Fig. 2b illustrates the test image. Later, a tri-level thresholding based on the KE and FA is executed, and the attained results are presented in Fig. 2c. After the thresholding, the OD section is alienated from the blood vessels and the background and this OD can be mined with the LS segmentation technique. The results attained with the LS are represented in Fig. 3. Figure 3a shows the thresholded image, and Fig. 3b, c

Fig. 2 Sample test picture of deep case: **a** pseudo-name, **b** test picture and **c** threshold picture

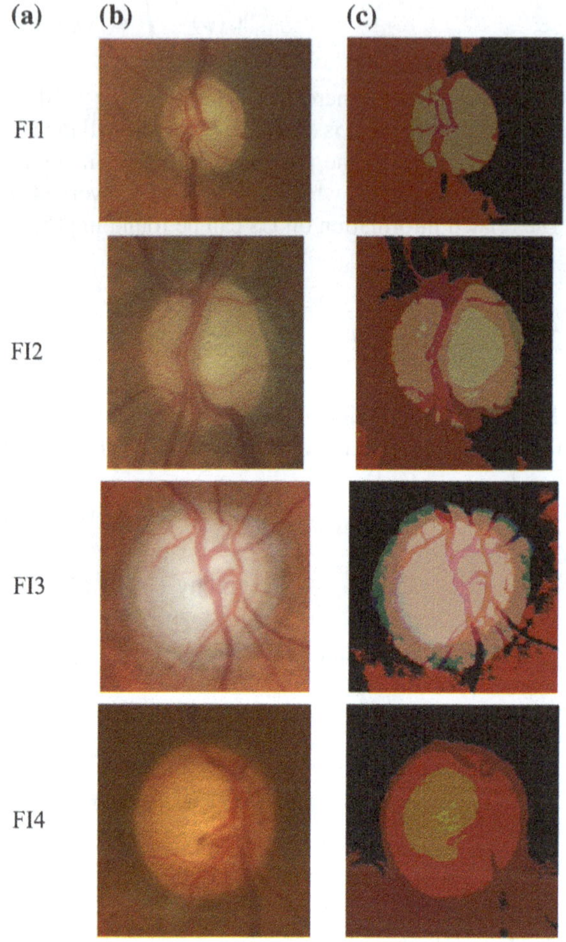

presents the initial and converged LS arc, respectively. Figure 3d presents the mined OD section, and Fig. 3e–i presents the GTs existing in the Rim-One database.

Lastly, a relative evaluation connecting the mined OD and the GTs is implemented, and the essential performance values are computed for individual GT and its values are depicted in Table 1. Finally, the average of the performance measures is computed and considered to evaluate the proposed image examination tool. Similar technique is applied for all the 10 trial pictures of the deep class, and the final mean value is considered to assess the performance.

Similar assessment is executed for the other image class of Rim-One, and Fig. 4 shows other trial test images and its image class. Figure 4a presents the pseudo-name, and Fig. 4b–d presents the sample test pictures. Similar evaluation technique is implemented for these images as discussed above, and the average of the performance values is considered to appraise the superiority of this evaluation system. The final

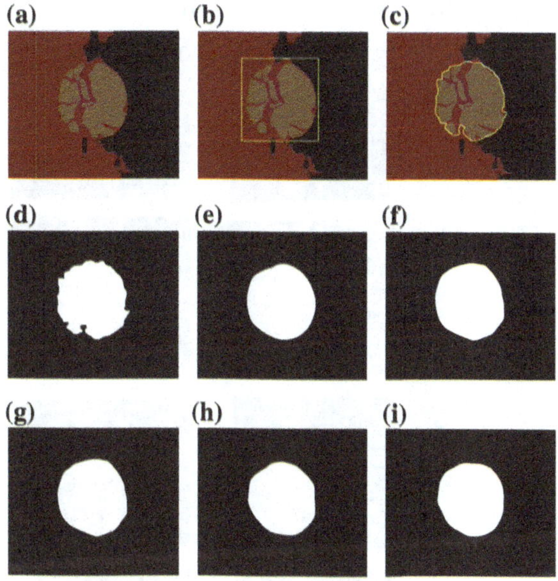

Fig. 3 Result obtained with the DRLS segmentation: **a** pre-processed FI1 picture, **b** initial bounding box, **c** converged arc, **d** extracted OD and **f–j** GT1–GT5

Table 1 Image performance values obtained for FI1 picture

Image	Sensitivity	Specificity	Accuracy	Precision
GT1	94.89	94.29	97.22	95.26
GT2	97.12	96.42	96.49	96.28
GT3	96.38	95.72	95.38	95.81
GT4	95.28	96.03	95.82	96.05
GT5	95.83	95.38	96.26	95.17
Average	95.90	95.57	96.23	95.71

outcome of the average performance values is presented in Table 2 for the every class of FI (10 pictures), and this result confirms that for each image class, the performance value attained is > 93%. This result confirms that the proposed tool with FA + KE thresholding and LS segmentation is superior in extracting the OD from the FI with better performance measures. Further, the modern methods [21–28] can also be considered to improve the outcome.

4 Conclusion

This work implemented a heuristic algorithm-based FI examination procedure based on FA + KE thresholding and LS segmentation. In this work, the aim is to mine the retinal OD with better accuracy. During the examination, Rim-One dataset with five

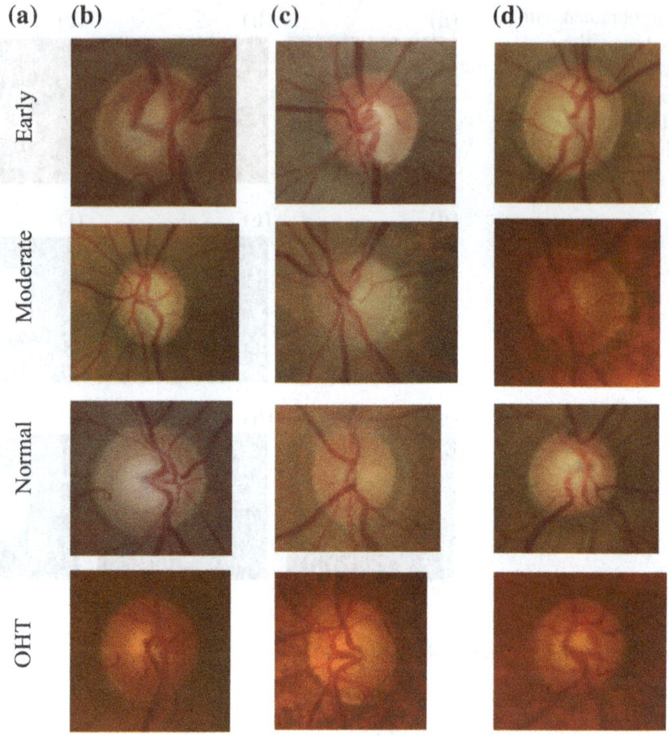

Fig. 4 Sample test pictures of various disease cases: **a** image class and **b–d** sample pictures

Table 2 Average image performance values obtained for Rim-One database

FI class	Sensitivity	Specificity	Accuracy	Precision
Deep	97.36	94.48	98.55	93.38
Early	98.21	96.29	98.70	94.17
Moderate	99.10	96.24	97.94	94.76
Normal	97.60	95.33	98.26	95.08
OHT	98.22	96.15	97.57	94.87

different image class is considered for the evaluation, and from each, class 10 pictures are considered. The main advantage of Rim-One is that it includes five numbers of GTs for every test picture, and the assessment tool should consider all the five GTs during the validation. In this work, the proposed technique is implemented on each picture of the Rim-One, and the mean of the performance is considered to evaluate the performance. The results of this study confirm that this tool is very efficient in evaluating the FI and offered an average performance value of > 93%, which confirms that the proposed work is clinically significant.

References

1. Dey, N. et al.: Optical cup to disc ratio measurement for glaucoma diagnosis using Harris corner. In: Third International Conference on Computing Communication and Networking Technologies (ICCCNT). IEEE (2012). https://doi.org/10.1109/icccnt.2012.6395971
2. Shree, T.D.V., Revanth, K., Raja, N.S.M., Rajinikanth, V.: A hybrid image processing approach to examine abnormality in retinal optic disc. Procedia Comput. Sci. **125**, 157–164 (2018)
3. Sudhan, G.H.H., Aravind, R.G., Gowri, K., Rajinikanth, V.: Optic disc segmentation based on Otsu's thresholding and level set. In: International Conference on Computer Communication and Informatics (ICCCI) IEEE, pp. 1–5 (2017). https://doi.org/10.1109/iccci.2017.8117688
4. Fumero, F., Alayon, S., Sanchez, J.L., Sigut, J., Gonzalez-Hernandez, M.: RIM-ONE: an open retinal image database for optic nerve evaluation. In: 24th International Symposium on Computer-Based Medical Systems (CBMS), pp. 1–6 (2011)
5. http://medimrg.webs.ull.es/research/retinal-imaging/rim-one/
6. Kapur, J.N., Sahoo, P.K., Wong, A.K.C: A new method for gray-level picture thresholding using the entropy of the histogram. Comput. Vis. Graph. Image Process. **29**, 273–285 (1985)
7. Raja, N.S.M., et al.: Segmentation of breast thermal images using Kapur's entropy and hidden Markov random field. J. Med. Imaging Health Inf. **7**(8), 1825–1829 (2017). https://doi.org/10.1166/jmihi.2017.2267
8. Rajinikanth, V., Satapathy, S.C., Fernandes, S.L., Nachiappan, S: Entropy based segmentation of tumor from brain MR images–a study with teaching learning based optimization. Pattern Recognit. Lett. **94**, 87–94 (2016). https://doi.org/10.1016/j.patrec.2017.05.028
9. Fernandes, S.L., Rajinikanth, V., Kadry, S.: A hybrid framework to evaluate breast abnormality using infrared thermal images. IEEE Consum. Electron. Mag. **8**(5), 31–36 (2019). https://doi.org/10.1109/MCE.2019.2923926
10. Rajinikanth, V., Dey, N., Satapathy, S.C., Kamalanand, K.: Inspection of crop-weed image database using kapur's entropy and spider monkey optimization. Adv. Intell. Syst. Comput. **1048** (2019). https://doi.org/10.1007/978-981-15-0035-0_32
11. Yang, X.S.: Nature-Inspired Metaheuristic Algorithms, 2nd edn. Luniver Press, Frome, UK (2011)
12. Rajinikanth, V., Couceiro, M.S.: Optimal multilevel image threshold selection using a novel objective function. Adv. Intell. Syst. Comput. **340**, 177–186 (2015)
13. Manic, K.S., Priya, R.K., Rajinikanth, V.: Image multithresholding based on Kapur/Tsallis entropy and firefly algorithm. Indian J. Sci. Technol. **9**(12), 89949 (2016)
14. Li, C., Xu, C., Gui, C., Fox, M.D: Distance regularized level set evolution and its application to image segmentation. IEEE Trans. Image Process. **19**(12), 3243–3254 (2010)
15. Rajinikanth, V., Lin, H., Panneerselvam, J., Raja, N.S.M.: Examination of retinal anatomical structures—a study with spider monkey optimization algorithm. Appl. Nature-Inspired Comput. Algorithms Case Stud. 177–197 (2020). https://doi.org/10.1007/978-981-13-9263-4_8
16. Jahmunah, V., et al.: Automated detection of schizophrenia using nonlinear signal processing methods. Artif. Intell. Med. **100**, 101698 (2019). https://doi.org/10.1016/j.artmed.2019.07.006
17. Acharya, U.R., et al.: Automated detection of Alzheimer's disease using brain MRI images– a study with various feature extraction techniques. J. Med. Syst. **43**, 302 (2019). https://doi.org/10.1007/s10916-019-1428-9
18. Dey, N., et al.: Social-Group-Optimization based tumor evaluation tool for clinical brain MRI of Flair/diffusion-weighted modality. Biocybern. Biomed. Eng. **39**(3), 843–856 (2019). https://doi.org/10.1016/j.bbe.2019.07.005
19. Satapathy, S.C., Rajinikanth, V.: Jaya algorithm guided procedure to segment tumor from brain MRI. J. of Optim. **2018**, 12 (2018). https://doi.org/10.1155/2018/3738049
20. Satapathy, S.C., Raja, N.S.M., Rajinikanth, V., Ashour, A.S., Dey, N.: Multi-level image thresholding using Otsu and chaotic bat algorithm. Neural Comput. Appl. (2016). https://doi.org/10.1007/s00521-016-2645-5

21. Chandrakar, P.: A secure remote user authentication protocol for healthcare monitoring using wireless medical sensor networks. Int. J. Ambient Comput. Intell. (IJACI) **10**(1), 96–116 (2019). https://doi.org/10.4018/IJACI.2019010106
22. Bhattacharya, H., Chattopadhyay, S., Chattopadhyay, M., Banerjee, A.: Storage and bandwidth optimized reliable distributed data allocation algorithm. J. Ambient Comput. Intell. (IJACI) **10**(1), 78–95 (2019). https://doi.org/10.4018/IJACI.2019010105
23. Rajinikanth, V., Raja, N.S.M., Arunmozhi, S.: ABCD rule implementation for the skin melanoma assesment–a study. In: IEEE International Conference on System, Computation, Automation and Networking (ICSCAN) (2019). https://doi.org/10.1109/icscan.2019.8878860
24. Das, H., Naik, B., Behera, H.S.: Classification of diabetes mellitus disease (dmd): a data mining (DM) approach. Adv. Intell. Syst. Comput. **710**, 539–549 (2018). https://doi.org/10.1007/978-981-10-7871-2_52
25. Sahani, R., et al.: Classification of intrusion detection using data mining techniques. Adv. Intell. Syst. Comput. **710**, 753–764 (2018). https://doi.org/10.1007/978-981-10-7871-2_72
26. Pradhan, C., Das, H., Naik, B., Dey, N.: Handbook of Research on Information Security in Biomedical Signal Processing Hershey. IGI Global, PA (2018)
27. Sahoo, A.K., Mallik, S., Pradhan, C., Mishra, B.S.P., Barik, R.K., Das, H.: Intelligence-based health recommendation system using big data analytics. In: Big Data Analytics for Intelligent Healthcare Management, pp. 227–246 (2019). https://doi.org/10.1016/B978-0-12-818146-1.00009-X
28. Dey, N., Das, H., Naik, B., Behera, H.S. (eds.): Big Data Analytics for Intelligent Healthcare Management. Academic (2019)

Generating Context-Free Group-Level Emotion Landscapes Using Image Processing and Shallow Convolutional Neural Networks

Sabyasachi Tribedi and Ranjit Kumar Barai

Abstract Emotion recognition is an integral part of any Human–Machine Interaction (HMI) system. Proper emotion recognition allows for HMI systems to choose the successive appropriate responses, given context and the emotion expressed by human(s). The advent of deep learning using Deep Neural Networks (DNN) has made incredible strides in achieving and even exceeding human accuracy in image classification and face detection. Many papers have been published mentioning the successful applications of DNNs like the Convolutional Neural Networks (CNNs), which have now become the de facto algorithm for facial image classification tasks because they combine the feature extraction and classification steps into one mathematical model. They learn the desirable features by themselves from the input images and have been demonstrated to be robust to variations in facial image data. However, there is one big bottleneck for CNNs: the models with good accuracy have many hidden layers, and hence are very deep and require heavy computing power, memory space and, of course, time to train themselves. Our paper puts forward two experimental approaches that can be extremely beneficial to the domains of CNNs and HMI systems. For the first approach, we were able to achieve a very good accuracy in emotion classification using a shallow CNN with only three and four hidden layers. This was possible only because we passed the input images through a carefully designed pipeline of image preprocessing techniques before feeding them to the CNN for training. For the second approach, we developed the interpretation of emotion landscape or the distribution of emotion classes detected in static images or videos with many people's faces visible. This is similar to the group-level emotion classification studies and publications with a distinct difference in possible applications. The rationale behind this integration is in advancing the idea of studying emotions expressed by people in a group setting and how they are mutually influential, visualize the change in the emotion distribution with time and thereby form

S. Tribedi (✉) · R. K. Barai
Department of Electrical Engineering, Jadavpur University, Kolkata, India
e-mail: stribedi15@gmail.com

R. K. Barai
e-mail: ranjit.k.barai@gmail.com

H. Das et al. (eds.), *Progress in Computing, Analytics and Networking*,
Advances in Intelligent Systems and Computing 1119,
https://doi.org/10.1007/978-981-15-2414-1_32

an emotional landscape in time, and enhance the understanding of collective sentiments non-verbally expressed through facial emotions in gatherings of known social context.

Keywords Convolutional neural networks · Shallow neural networks · Image processing · Non-verbal sentimental analysis · Social groups · Emotion recognition · Facial expressions · Facial emotion recognition · Inferring affect · Group influence

1 Introduction

Emotions are an integral and essential component of human life. Emotions can be by themselves a very strong medium of communication expressed through facial muscles, hands, body gestures and modulations of the voice. This paper focuses specifically on visual perception of emotions through a person's facial expressions. Mehrabian [1] mentions that facial expressions contribute to 55% to the effect of a speaker's message on the listener, followed by vocal modulations (38%) and message content (7%). Expressed emotions also provide information regarding the current cognitive state of the person, like boredom, interest, stress and confusion. Understanding and interpreting facial expressions contributes to the depth and dimensionality of information that can be collected from a person in various situations and contexts, apart from his or her spoken words and messages. Automatic recognition of facial expressions is becoming increasingly important in designing HMI systems that are responsive or sensitive to an interacting human's current emotional state. Mower et al. [2] has proposed an emotional classification paradigm based on Emotional Profiles (EPs). Research before the advent of deep learning extensively relied of specially engineered features [3] that effectively captured the information from facial images. Majority of the recent publications on the topic of emotion recognition from images are due to the famous Emotion Recognition in the Wild (EmotiW) Challenge where deep convolutional neural networks have won the competition consistently. The challenge consists of three separate competitions which are group-level emotion recognition, "engagement in the wild" and the audio-video sub-challenge. In context of the paper material, the objective of group-level emotion recognition is to classify an image of a group of front-facing people's perceived emotion. Gupta et al. [4] put forward a novel approach for classifying global emotions for group-level images using which the authors were able to achieve a final test accuracy of 64.83% with a ranking fourth among all challenge participants in EmotiW 2018. The approach uses two different techniques together as an ensemble predictor: CNN for obtaining a global representation of the image while an attention mechanism produced local representation by extracting individual facial features and merging them. Rassadin et al. [5] describe an approach to the EmotiW 2017 Challenge by using a pre-trained CNN for facial feature extraction and an ensemble of random forest classifiers at a validation accuracy of 75.4%. Although this paper does not refer to the first EmotiW competition for a comparative analysis, the problem statement addressed is directly derived from it. More specifically, our proposed approach has obtained inspiration from Tan, Lianzhi et al.'s publication [6] where the authors use two CNN models for learning from individual faces as well as from the whole group-level image as a whole. This approach was able to win EmotiW 2017 with 83.9 and 80.9% on the validation and testing datasets, respectively.

The motivation behind the proposed work is to find a new perspective to look at the problem domain of facial emotion classification and study the same but in scenarios of social groups. The existing literature of applications of machine learning and deep learning focuses on training and testing models on labelled images of faces. We wanted to explore the idea of identifying emotions that people express in their faces in the presence of other people, i.e. at a group-level scenario. We discuss our proposed approach of an emotional landscape for group of people in social scenario in Sect. 3.6. This is partly derived from the study of emotion inference of a group from a group-level image. Barsade et al. [7] and Kelly et al. [8] talk about the top-down and bottom-up components of emotion inference. Top-down component targets for an overall emotion of a group which is built upon the uniqueness of each of the individual members' facial expressions while the bottom-up component starts with at the group level followed by emotions of the individual members. For a static image of a group-level image, we obtain the emotion distribution of each of the detected front faces. In the EmotiW challenge, the classes defined were "positive", "neutral" and "negative". We believe that this can be too restrictive in terms of the emotion information that can be potentially gathered using the CNNs; hence, we have used seven emotion classes for our purpose: "surprise", "sad", "fear", "distress", "angry", "disgust" and "happy". The distribution of values for these seven classes for the group-level image can be found by applying an aggregation function over the same for each of the detected faces. When we consider a video of the same group, we can obtain a time-series distribution of the emotion classes values which can be visualized as a landscape with variations in the direction of time due to various individual emotions and group-inferred emotions.

The problem statements and workflow of experimentation has been defined in Sect. 2. The implementation details of the face detection, facial expression feature extraction, CNN models used for classification and the derivation of the emotional landscape from the CNN results have been briefly summarized in Sects. 3.2, 3.3, 3.4, 3.5, in Sect. 3. Section 4 contains experimentational results on the validation samples from the datasets as mentioned in Sect. 3.1. Section 5 contains our interpretation of the term "emotional landscape" from the output of the proposed pipeline. Section 6 contains concluding ideas for future prospects use cases and applications and extensions of the success of the conducted experiments.

2 Methodology Adopted

2.1 Problem Statements

The first part of this paper presents an economical and cheap way to achieve acceptable accuracy rates using CNNs with three and four hidden layers only. The CNN architecture can be abstracted as follows:

$$x^1 \rightarrow w^1 \rightarrow x^2 \rightarrow \ldots \rightarrow x^{n-1} \rightarrow w^{n-1} \rightarrow x^n \rightarrow w^n \rightarrow z \quad (1)$$

The input to the CNN is x^1, which a 3D tensor (an image) is. The parameters involved in each layer's processing are collectively denoted by a tensor w^i. Here, the CNN will be used as an image classifier, where there are seven classes or values in z. x^n is transformed to a C-dimensional vector whose i-th element encodes the posterior prediction $P(C_i|x^1)$. Note that throughout the paper, we consider the convolution operation, pooling operation and the normalization operation together as one layer in the CNN. The CNN also consist of dense layers and the final output layer. The drop out layer is not considered as a separate layer in the CNN.

The second part of this paper is dedicated to constructing a meaningful interpretation of the results obtained from the CNN. Given a group-level image of people's faces, we find z^i for each of the detected faces i in the image. z will be a column vector of length 7 in our context. These z values can be interpreted as probability percentages of each of the seven emotion classes. Aggregating the probability values across all of the detected faces for each of the seven emotion classes will give z', a column vector which provides a global context-free probability value for each class. If there are m faces detected in a single group-level image, then:

$$f\left(\begin{bmatrix} z_0^0 \\ z_1^0 \\ \vdots \\ z_5^0 \\ z_6^0 \end{bmatrix}, \begin{bmatrix} z_0^1 \\ z_1^1 \\ \vdots \\ z_5^1 \\ z_6^1 \end{bmatrix}, \ldots, \begin{bmatrix} z_0^{m-1} \\ z_1^{m-1} \\ \vdots \\ z_5^{m-1} \\ z_6^{m-1} \end{bmatrix}\right) = \begin{bmatrix} z_0' \\ z_1' \\ \vdots \\ z_5' \\ z_6' \end{bmatrix} \quad (2)$$

where f is an aggregation function such that

$$z_i' = \frac{\sum_j^{m-1} z_i^j}{\sum_i^6 \sum_j^{m-1} z_i^j} \quad (3)$$

2.2 Workflow

We obtained the JAFFE and Cohn-Kanade facial image datasets from their official websites. They can be freely downloaded after stating application use and given permission from the database authors. The CNN was trained during the individual facial images obtained from these two datasets. Every facial image used during the training and testing phase was passed through a specially designed pipeline for extracting specific or targeted regions of the face. These ROIs were then min-max normalized to centre the pixel values between 0 and 1. This helps in faster convergence during training the CNN. The output column vector from the CNN z was standardized to range of 0 to 100. So, each of the seven values in z could be interpreted as a probability value of an emotion class. During the testing phase, a grey-scaled group-level image containing multiple visible faces is passed to the preprocessing pipeline mentioned earlier. For every face i detected in the image, there were corresponding target ROIs. These ROIs was fed into the trained CNN one by one, to obtain a z vector, for m detected faces. The m output vectors was then passed to the aggregation function f to output z' which represents the global probability values for each of the seven emotion classes in the group-level image.

3 Implementation

3.1 Datasets

During the initial stages of experimentation, we used the JAFFE facial emotion image dataset from [9]. The Japanese Female Facial Expression (JAFFE) database contains 213 images of seven facial expressions (six basic facial expressions + one neutral) posed by 10 Japanese female models. Each image has been rated on six emotion classes by 60 Japanese subjects. The dataset is relatively small and was used in establishing our baseline prototype CNN model and testing the preprocessing pipeline (Fig. 1).

The Cohn-Kanade AU-Coded Facial Expression Database [10] is one of the standard datasets for research in automatic facial image analysis and emotion prediction. It has more images than the JAFFE dataset and is thereby used for the later experimental phases to tweak and tune the pipeline and the CNN models. Cohn-Kanade dataset is available in two versions. Version 1 has been used here. It includes 486 sequences from 97 posers. Each sequence begins with a neutral expression, and it proceeds to a peak expression (Fig. 2).

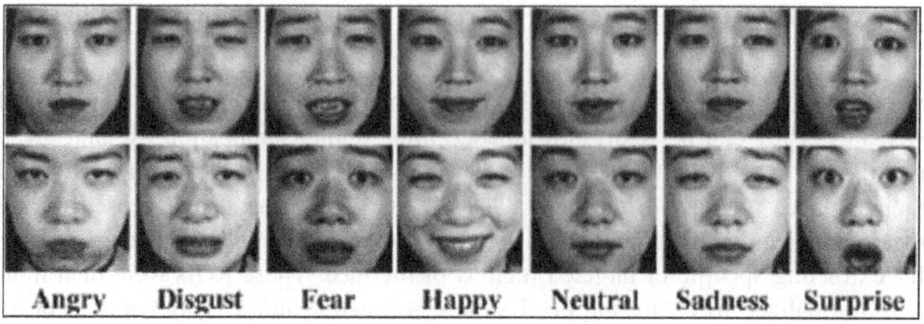

Fig. 1 Two faces from the JAFFE dataset and their six emotional expressions

Fig. 2 Sample face expressions from the Cohn-Kanade dataset

3.2 Workstation Configuration

The workstation used for preprocessing the image dataset was equipped with an i7-4790 K quad-core processor with eight threads and maximum turbo frequency reaching 4.4 GHz and 8 MB smart cache. The graphics processing unit comprises the 8 GB GDDR5 NVIDIA GeForce GTX 1080 with boost clock speed of 1733 MHz and 2560 CUDA cores. The workstation had 32 GB of DDR4 of RAM. The choice of operating system was done considering the performance and simplicity, thereby selecting Linux Mint 18.3 with Cinnamon desktop. CUDA 9.0 was used to reliably perform the training and testing on the aforementioned GPU.

We used Python 3.5.1 as our prototyping programming language and the Keras Library for Python. Keras is a high-level deep learning API which provides a simplified abstracted way to create standard neural network architectures like CNNs or RNNs along with whole set of optimizers, classifiers for neural network output layers, weight matrix initializers, etc., to choose from.

OpenCV [11] API for Python is an open-source computer vision framework which contains all the necessary tools and algorithms required to manipulate image data. OpenCV coupled with C ++ based multipurpose library called dlib and imutils (for dlib support) helped in detecting and cropping out faces and extracting the ROIs which we desired as input features to the CNN model.

3.3 Face Detection from Image

OpenCV gives access to two pre-trained face detection classifiers: the Haar cascades [12] and the Local Binary Pattern (LBP)-Cascades [13]. These classifiers have been pre-trained on many positive (images with faces) and negative (images without faces) beforehand and included as a part of OpenCV. Table 1 gives a comparative evaluation of the two classifiers.

The major differences between them appear in terms of speed and accuracy. The LBP Cascade Classifier is faster, but Haar cascades has a better accuracy rate. In our experimentation, we value accuracy over speed and hence used the Haar cascades for

Table 1 Comparative analysis of the two face classifier methods available in OpenCV [14]

Algorithm	Advantages	Disadvantages
Haar	1. High detection accuracy 2. Low false positive rate	1. Computationally complex and slow 2. Longer training time 3. Less accurate on black faces 4. Limitations in difficult lightening conditions 5. Less robust to occlusion
LBP	1. Computationally simple and fast 2. Shorter training time 3. Robust to local illumination changes 4. Robust to occlusion	1. Less accurate 2. High false positive rate

Fig. 3 A sample image from our experimental test dataset with the three faces detected and highlighted in green boxes

detecting faces from group-level images. Figure 3 gives the output of a group-level photo when passed through the Haar cascade classifier.

3.4 Targetted ROI for Feature Extraction

From the detected faces from the group-level photo, we use the dlib facial feature landmarks to extract the regions centred at the eyes, lips and the nose. We extract these specific ROIs on the basis of our assumption that these features are sufficient enough to recognize the emotion expressed by a human face. Figure 4 gives the facial landmarks as marked by dlib. Figure 5 gives the referential landmark coordinates given by the dlib facial landmark predictor, which was trained on the 68-point iBUG 300-W dataset [15]. It is important to note that there are other facial landmark detectors. A more detailed 194-point model is available which has been trained on the HELEN [16] dataset. Regardless of the dataset and landmark predictor used, dlib library helps abstract the details and allows us to focus our efforts on detecting and cropping out the ROIs we desire.

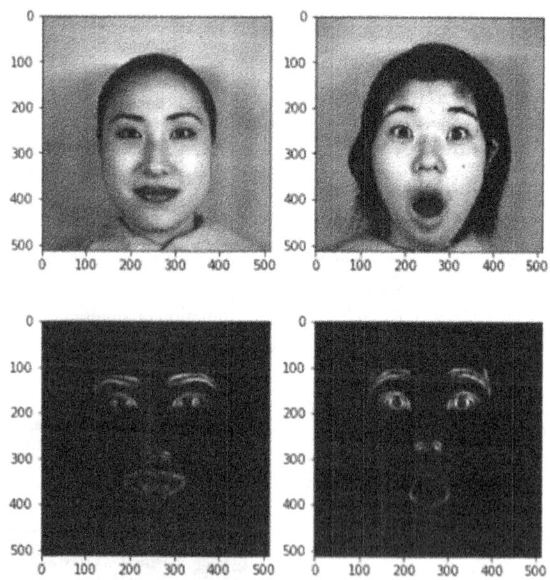

Fig. 4 ROIs as marked and extracted by dlib

Fig. 5 Visualizing the 68 facial landmark coordinates from the iBUG 300-W dataset [15]

3.5 Shallow CNNs for Identifying Emotions Expressed by Detected Faces

We used a CNN model with three hidden layers for our experimentations. The model was trained for 50 epochs. Figure 6 shows how the extracted target ROIs from each of the detected faces are passed as input to the CNN models. Our CNN models are simple in architecture and contain two repeating sequences of convolution, max-pooling and normalization operations followed by one or two densely connected neuron layers and a classifier in the last output layer. The reason why we chose to work with such a simple architecture of CNN is due to the assumption that we have already effectively reduced the image dimensions to specific regions where we want the CNN to learn from. One can also think of this as reducing the number of parameters and adjustments the network has to learn to map the input to the output classes. Increasing the number of layers implies increasing the number of parameters to be learnt. Given a reduced input feature space, this is unnecessary and wasteful [17, 18].

4 Results

An interpretation of the results of the CNN on the right of each of the figures is in due order. The CNN model had a validation accuracy of 94.28% after 50 epochs of training with a softmax cross-entropy loss of 0.48699. For the first test image Fig. 7, the pipeline as laid out in Sect. 2.2, detects six faces, and outputs the thresholded markers for each emotion class across the columns and each face across the rows. This thresholding is computed by the aggregation function f as defined in 2.1. The landscape vector below the matrix gives the distributed percentages of the emotions

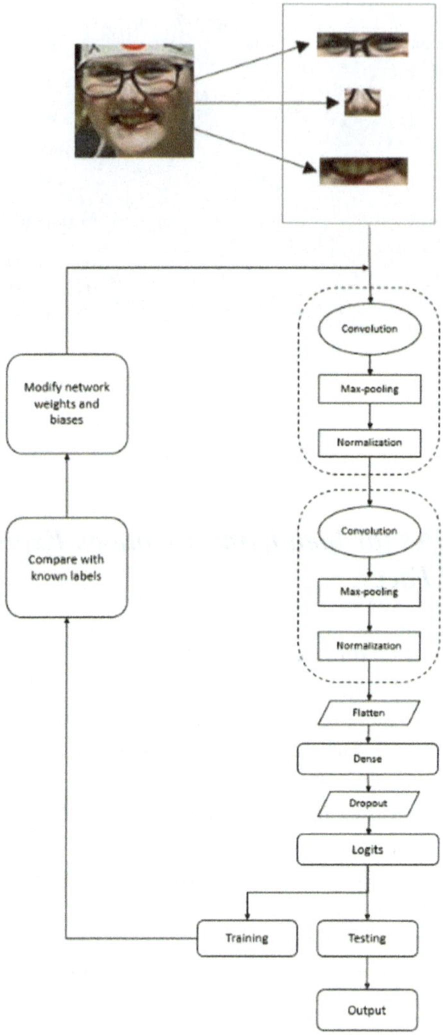

Fig. 6 The CNN architecture we used

as detected by the CNN as a whole. Similarly, for the second test image, Fig. 8, the pipeline detects three front-facing faces and outputs their thresholded values for each of the emotions. If one may note the table along with the figure, the third front-facing image from the left would actually be not “fear” but “happy” also. But given the captured facial structure of the woman, the thresholded set outputs fear as the dominant emotion, due to the structural similarity of the facial expression with a person expressing fear. The pipeline does not assume that the input image would certainly have more than one face, and this is demonstrated by the output for Fig. 9. The third test image is that of a single girl who may be expressing sadness or fear.

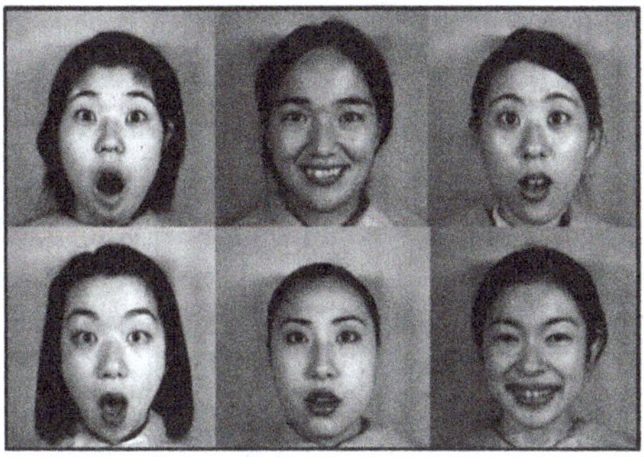

Faces detected : 6

Predicted Emotions:-

	Distressed	*Sad*	*Surprised*	*Happy*	*Fear*	*Neutral*	*Angry*
1	0.0	0.0	1.0	0.0	0.0	0.0	0.0
2	0.0	0.0	0.0	1.0	0.0	0.0	0.0
3	0.0	0.0	1.0	0.0	0.0	0.0	0.0
4	0.0	0.0	0.0	1.0	0.0	0.0	0.0
5	0.0	0.0	1.0	0.0	0.0	0.0	0.0
6	0.0	0.0	1.0	0.0	0.0	0.0	0.0

Landscape

0.0	0.0	0.67	0.33	0.0	0.0	0.0

Counter({'Surprised': 4, 'Happy': 2})

Fig. 7 Test image 01 [9]

Faces detected : 3

Predicted Emotions:-

	Distressed	*Sad*	*Surprised*	*Happy*	*Fear*	*Neutral*	*Angry*
1	0.0	0.0	0.0	1.0	0.0	0.0	0.0
2	0.0	0.0	0.0	0.0	1.0	0.0	0.0
3	0.0	0.0	0.0	1.0	0.0	0.0	0.0

Landscape

0.0	0.0	0.0	0.67	0.33	0.0	0.0

Counter({'Happy': 2, 'Fear': 1})

Fig. 8 Test image 02 [19]

Faces detected : 1

Predicted Emotions:-

	Distressed	*Sad*	*Surprised*	*Happy*	*Fear*	*Neutral*	*Angry*
1	0.0	0.0	0.0	0.0	1.0	0.0	0.0

Landscape

0.0	0.0	0.0	0.0	1.0	0.0	0.0

Counter({'Fear': 1})

Fig. 9 Test image 03 [20]

The CNN output when thresholded yields for "fear". This is due to the aggregation function *f* which takes into account the overall emotion flags for each detected face, and thresholds using a ceil operator.

5 Emotional Landscape for Group-Level Emotion Evolution

Sections 2, 3 and 4 discussed how to infer emotion class probabilities from an image of a social group of people. The approach can be directly extended for videos of social gatherings where each frame of the video can be passed to the pipeline discussed in Sect. 3 and we obtain an output vector for the seven emotion classes for every face detected across the length of the video. The output in effect is a 4D tensor where the axes are as follows: face ID, time, emotion class label, and emotion class label probability. This tensor is not a conductive way to perform further interpretations or to obtain insights regarding the prominent group-level emotions of the social gathering in the particular video. What we propose is a hint of a more visual way to present the output tensor: an emotional landscape which evolves along the video timescale. It is a 3D surface plot where each point on the surface is an accumulated function of the seven emotion classes and time. Although this idea is at a very early stage of conception, we believe that such a landscape function surface can be of great importance to the cause of developing user-friendly HMI by introducing the intelligence of understanding and predicting emotional response over a lengthy conversation with the users. The HMI would be able to predict the general emotional response to a specific question or reply and can choose to learn to decide how it should move ahead in conversation.

6 Conclusions

Our paper puts forward two main ideas of approaching emotion classification using facial expressions and allows for a more human-like inferring capability of other humans' expressions and emotional state to Human–Machine Interface (HMI)-based systems. The first idea is that if meaningful image preprocessing is performed on the input images, this allows for decreasing the depth or number of hidden layers from the CNN which learns to perform the classification task using those images. Preprocessing the input image and extracting context-specific portions to an assumption decrease the "space" where the CNN has to search for to map the input to the output classes. The second idea we have mentioned is that of the visualizing the variances of group-level emotions with passage of time as a landscape. This allows for varying interpretations of discrete classification labels which change with time and opens the scope to study and attempt to close the distance between the range of human interpretation and machine interpretation. The main limitation we have found against our proposed idea is the influence of time intervals at which the landscape is updated. Too small intervals can give misleading results because human facial expressions have a lower bound on how fast they change. In spite of this, HMI systems when equipped with the power of monitoring changes in a person's facial expressions in a social group setting can greatly enhance their decision-making potential and conducting conversations with humans.

References

1. Mehrabian, A.: Silent Messages. 1st edn. (1971)
2. Mower, E., Mataric, M.J., Narayanan, S.: A framework for automatic human emotion classification using emotion profiles. IEEE Trans. Audio Speech Lang. Process. **19**(5), 1057–1070 (2011)
3. Vaish, A., Gupta, S.: Employing Kaze features for the purpose of emotion recognition. In: Progress in Computing, Analytics and Networking, pp. 679–685. Springer, Singapore (2018)
4. Gupta, A., Agrawal, D., Chauhan, H., Dolz, J., Pedersoli, M.: An attention model for group-level emotion recognition. In: Proceedings of the 2018 on International Conference on Multimodal Interaction, pp. 611–615. ACM (2018)
5. Rassadin, A., Gruzdev, A., Savchenko, A.: Group-level emotion recognition using transfer learning from face identification. In: Proceedings of the 19th ACM International Conference on Multimodal Interaction, pp. 544–548. ACM (2017)
6. Tan, L., Zhang, K., Wang, K., Zeng, X., Peng, X., Qiao, Y.: Group emotion recognition with individual facial emotion CNNs and global image based CNNs. In: Proceedings of the 19th ACM International Conference on Multimodal Interaction, pp. 549–552. ACM (2017).
7. Barsade, S.G., Gibson, D.E.: Group Emotion: A View from Top and Bottom (1998)
8. Kelly, J.R., Barsade, S.G.: Mood and emotions in small groups and work teams. Organ. Behav. Hum. Decis. Process. **86**(1), 99–130 (2001)
9. Lyons, M.J., Akamatsu, S., Kamachi, M., Gyoba, J., Budynek, J.: The Japanese female facial expression (JAFFE) database. In: Proceedings of Third International Conference on Automatic Face and Gesture Recognition, pp. 14–16 (1998)
10. Cohn, J.F.: Cohn-Kanade AU-Coded Facial Expression Database. Pittsburgh University (1999)
11. OpenCV Project. Accessed from https://opencv.org
12. Viola, P., Jones, M.: Rapid object detection using a boosted cascade of simple features. In: Proceedings of the 2001 IEEE Computer Society Conference on Computer Vision and Pattern Recognition, 2001. CVPR 2001, vol. 1, pp. I–I. IEEE (2001)
13. Ojala, T., Pietikäinen, M., Harwood, D.: A comparative study of texture measures with classification based on featured distributions. Pattern Recognit. **29**(1), 51–59 (1996)
14. Rosebrock, A.: Facial landmarks with dlib, OpenCV, and Python (3 April 2017). Accessed from https://www.pyimagesearch.com/2017/04/03/facial-landmarks-dlib-opencv-python/
15. Sagonas, C., Antonakos, E., Tzimiropoulos, G., Zafeiriou, S., Pantic, M.: 300 faces in-the-wild challenge: database and results. Image Vis. Comput. **47**, 3–18 (2016)
16. Le, V., Brandt, J., Lin, Z., Bourdev, L., Huang, T.S.: Interactive facial feature localization. In: European Conference on Computer Vision, pp. 679–692. Springer, Berlin (2012)
17. Why are neural networks becoming deeper, but not wider? (asked July 9 2016 on stats.stackexchange.com). Accessed from https://stats.stackexchange.com/questions/222883/why-are-neural-networks-becoming-deeper-but-not-wider
18. What is the difference between a neural network and a deep neural network, and why do the deep ones work better? (asked on 20 November 2015 on stats.stackexchange.com). Accessed from https://stats.stackexchange.com/questions/182734/what-is-the-difference-between-a-neural-network-and-a-deep-neural-network-and-w
19. Image file Accessed from https://www.bfi.org.uk/sites/bfi.org.uk/files/styles/full/public/image/our-little-sister-2015-002-four-sisters-at-dining-table.jpg
20. Image file Accessed from https://i.ytimg.com/vi/B27yQMuTHWc/maxresdefault.jpg

A Deep Learning Approach to Image-Based Malware Analysis

Gurumayum Akash Sharma, Khundrakpam Johnson Singh and Maisnam Debabrata Singh

Abstract Malicious software also referred to as "Malware" is one of the serious threats on the Internet today since it has been growing exponentially over the last decade according to research, causing substantial financial trouble to various organizations. Different security companies have been proposing different techniques to defend from this threat which is a major challenge on the complexity and growing volumes. Recently, malware communities and researchers have begun to apply machine learning and deep learning model to detect potential threats. We propose a malware classification model that takes advantage of the potential of deep learning (DL) models using the convolutional neural network (CNN) and combination of machine learning classifier with CNN such as support vector machine (SVM) for classifying their families. Detection of newly released malware using such models would be possible through mathematical function. That is, $f: n \rightarrow z$, where n is the given malware and z is their corresponding malware family. Malimg dataset is used to perform the experiment which contains malware image of 25 malware families and 9339 malware samples. CNN has outperformed the CNN-SVM with a test accuracy of 97.5%.

Keywords Artificial intelligence · Deep learning · Machine learning · Malware classification · Support vector machine · Convolutional neural networks

1 Introduction

Malicious software also referred to as "Malware" is one of the serious threats on the Internet today since it has been growing exponentially over the last decade according to research, causing substantial financial trouble to various organizations. Over 350,000 new malware and potentially unwanted applications (PUA) have registered every day as per the latest report of AV-TEST Institute [7]. According to AV-TEST

G. A. Sharma (✉) · K. J. Singh · M. D. Singh
Department of Computer Science and Engineering, NIT Manipur, Imphal, India
e-mail: g_akash@nitmanipur.ac.in

H. Das et al. (eds.), *Progress in Computing, Analytics and Networking*,
Advances in Intelligent Systems and Computing 1119,
https://doi.org/10.1007/978-981-15-2414-1_33

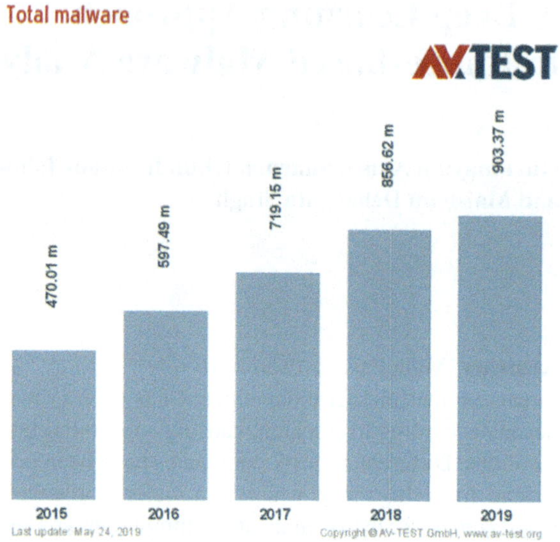

Fig. 1 Yearly survey report of malware (2015–2019) [7]

latest report, during this past 5 years, i.e., 2015–2019, it was observed that the number of malware reported has been increasing over 100 million in number each year, which is a major concern for the information security community. The detailed survey report is shown in Fig. 1.

Traditionally, pattern matching against signatures obtained from known malware was used for the detection of malware which can be easily defeated by many well-known strategies [14]. So, different security organizations have been proposing different techniques to defend from such threats which is a major challenge on the complexity and growing volumes of malware [11, 15]. In this paper, deep learning approach such as CNN and CNN-SVM will be used to perform the experiment for classification of different malware families using Malimg dataset which contains images of malware from 25 different families.

2 Related Works

Natraj et al. proposed a simple method of malware classification using a technique of visualizing and classify using image processing techniques [1, 12]. Malware binaries were converted to gray-scale images. For classification, malware code did not need to be executed or disassembled. For computing texture features, the GIST descriptor has been used which uses wavelet decomposition of an image. For classification, k-nearest neighbors with Euclidean distance have been used. For performance evaluation, bi-gram distributions were calculated from the raw data without disassembly. Classification accuracy of 0.98 was found using bi-gram distributions as a feature vector and it took 56 s to classify a sample in comparison to using the GIST feature, which took only 1.4 s for the overall classification as feature vector length used was 320 but 65 K elements were used for the distribution-based analysis using bi-grams.

Yichuan Tang proposed a deep learning model that replaced the softmax activation function and minimized cross-entropy loss with a linear support vector machine [2]. Using L2-SVMs instead of softmax has shown a significant gain on popular deep learning datasets such as MNIST and CIFAR-10. Convolutional neural network (CNN) using SVM has a lesser error rate (11.9%) as compared to conventional convolutional neural network (CNN) using softmax which has an error rate of 14.0% on MNIST and CIFAR-10 dataset.

K. Kosmidis et al. proposed an automated framework for the identification of unknown vulnerabilities using current neural network techniques such as computer vision and image classification [3]. For feature engineering, malware binaries were converted to an 8-bit vector and then gray-scale images as preparation for new training set for the machine learning algorithm. Malimg dataset was used for all experiments such as training and testing of different classification algorithms. Using the nearest centroid, the average training time of 0.218 s and testing time of 0.0211 s had an average accuracy of 0.0856 which was the least accurate result. However, using the random forest, the average training time of 1.72 s and testing time of 0.0063 s had an average accuracy of 0.0916 which is the highest accuracy result, whereas using the classification algorithm, such as decision tree, stochastic gradient, perceptron, and multilayer perceptron have accuracy results of 0.088, 0.087, 0.0905, and 0.087, respectively. However, stochastic gradient and perceptron have misclassification.

Abien Fred et al. proposed a different deep learning model using the support vector machine (SVM) classifier for the classification of malware using the Malimg dataset [4]. The classification was done on different deep learning models such as multilayer perceptron (MLP), convolutional neural network (CNN), and gated recurrent unit (GRU). Empirical evidence had shown that GRU had outperformed the other deep learning models with a predictive accuracy of ≈ 84.92%.

Jiawei et al. proposed a lightweight approach for detecting distributed denial of service (DDoS) malware in IoT environments [5]. For classifying malware families using image recognition techniques, malware binaries were converted into grayscale images and fed into fine-tuned convolutional neural networks. The converted binary images were fed into the machine learning classifier into local devices, and a suspicious file was then submitted to a remote cloud server which was then used for further classification. However, the signature matching system has a large database as it contains details of each malware sample that are not efficient for IoT devices as IoT devices have limited resources. In the case of machine learning, only a small set of training data was needed for classification once trained. A small, two-layer shallow convolutional neural network has been used to have a lightweight detection system. The proposed system can achieve an average accuracy result of 94.0% on benign and malicious classification. To obtain more representative features from malware image, a new malware image extraction method may be needed.

3 Proposed Methodology

3.1 *Machine Learning Library*

Keras which runs on top of Google TensorFlow [13] is used to implement the deep learning algorithm, with the help of other libraries such as Matplotlib [8] which is used for plotting graph, numpy [10] which is used for scientific computing, and scikit-learn [9].

3.2 *The Dataset*

In this study, the Malimg dataset would be used for the evaluation of deep learning models which consists of 9,339 malware samples from 25 different malware families [1]. The frequency distribution of malware families and their variants in the Malimg dataset are shown in Table 1.

Malimg dataset was created by Nataraj et al. [1] by converting malware binaries into an 8-bit unsigned integer composing a matrix $M \in R^{m \times n}$. The matrix can be visualized as a grayscale image having values in the range of [0, 255], with 1 representing *white* and 0 representing *black* (Fig. 2).

3.3 *Dataset Preprocessing*

For preprocessing of the dataset such as the conversion of raw pixel to numpy array and generating labels, Keras preprocessing tools have been used [1]. The Malimg dataset has 25 classes so it is converted to the binary number of 0–24 that allows the representation of categorical data to more expressive; this process is also called as one hot encoding. Categorical data cannot be applied to the machine learning algorithm directly so it is be converted to numbers as shown in Table 1.

3.4 *CNN Concepts*

3.4.1 Input/Output Volumes

CNN is usually applied to image data. Every image is a matrix of pixels values. We encode each pixel as 8 bits. Each malware image is interpreted as [0, 255]. Thus the image develops a one-dimensional structure called the input volume ($64 \times 64 \times 1$) as the image we used is a grayscale image.

Table 1 Details of Malimg dataset

No	Family	Family name	No of variants
01	Backdoor	Agent.FYI	116
02	Backdoor	Rbot!gen	158
03	Dialer	Adialer.C	122
04	Dialer	Dialplatform.B	177
05	Dialer	Instantaccess	431
06	PWS	Lolyda.AA 1	213
07	PWS	Lolyda.AA 2	184
08	PWS	Lolyda.AA 3	123
09	PWS	Lolyda.AT	159
10	Rogue	Fakerean	381
11	Trojan	Alueron.gen!J	198
12	Trojan	C2Lop.P	146
13	Trojan	C2Lop.gen!G	200
14	Trojan	Malex.gen!J	136
15	Trojan	Skintrim.N	80
16	Trojan downloader	Dontovo.A	162
17	Trojan downloader	Obfuscator.AD	142
18	Trojan downloader	Swizzor.gen!E	128
19	Trojan downloader	Swizzor.gen!I	132
20	Trojan downloader	Wintrim.BX	97
21	Worm	Allaple.A	2949
22	Worm	Allaple.L	1591
23	Worm	VB.AT	408
24	Worm	Yuner.A	800
25	Worm:autoit	Autorun.K	106

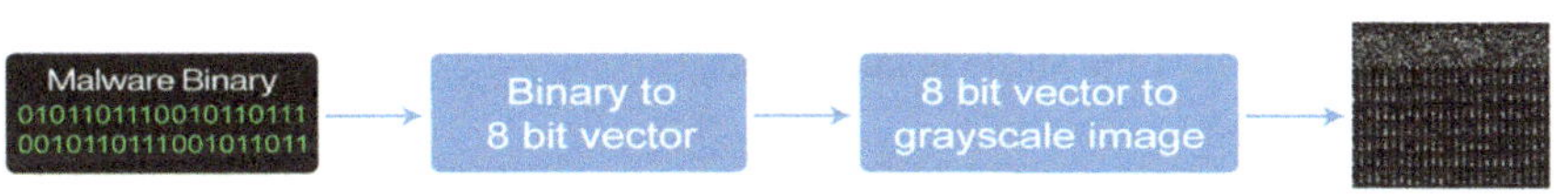

Fig. 2 Visualizing malware binary as a grayscale image

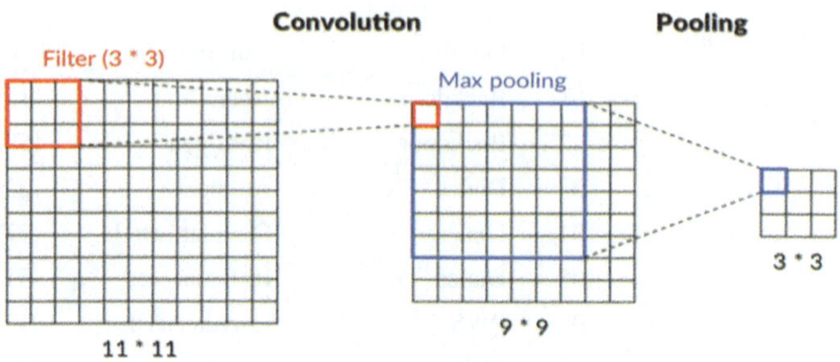

Fig. 3 Example of convolution and pooling operation diagram (source)

3.4.2 Features

A distinct and useful pattern obtained from the input data, i.e., image that helps in performing the desired image analysis is term as a feature. CNN gets the features from the input data which refers to the malware image in this study.

3.4.3 Filters (Convolution Kernels)

A filter which also refers as kernels is a small-size matrix in comparison to the input size of the image.

3.5 CNN Architecture

3.5.1 Convolution Layer

In our case, the filter size is $5 \times 5 \times 1$ and slide over input array with a stride of 1. In each step of the slide, each filter element and element of each subarea of the input array took the dot product. The result of each dot product is a scalar. Feature map is the total result of each unique position where the filter can be put on the image. For example, 64×64 unique position feature map is a $64 \times 64 \times 1$ array. We will get a smaller feature map if the stride is larger than 1.

3.5.2 The Pooling Layer

The pooling layer is used to reduce the spatial size progressively to reduce the number of features and computational complexity of the network. The main reason for the pooling layer is to avoid the model from overfitting. The most commonly used approach is max-pooling.

3.5.3 The Fully Connected Layer

Fully connected layers will perform high-level reasoning in the neural network after several convolutional and max-pooling layers. Just like regular artificial neural networks (non-convolutional), all activations in the previous layers are connected to neurons in fully connected layers.

3.6 Support Vector Machine

Vapnik originally developed support vector machines (SVMs) for binary classification [6]. To classify two classes in a given dataset with features $c \in R^m$, SVM will find the optimal hyperplane f(w, x) = w.x + b. SVM learns the parameters w and b by solving the following constrained optimization problem:

$$\text{Loss} = min\frac{1}{p}w^T w + C\sum_{i=1}^{p}\xi_i \tag{1}$$

Such that $\forall n\ w^T x_n t_n \geq 1 - \xi_n$

$$\forall n\ \xi_n \geq 0$$

where $w^T w$ is the Manhattan norm (also known as L1 norm), C is the penalty parameter (maybe an arbitrary value or a picked value using hyperparameter tuning), and ξ is the cost function.

$$loss = min\frac{1}{p}w^T w + C\sum_{i=1}^{p} max(0, 1 - y_i'(w^T x_i + b)) \tag{2}$$

This equation is known as L1-SVM, with the standard hinge loss. Its differentiable equivalent, L2-SVM (given by Eq. 3), provides more stable results [2].

$$loss = min\frac{1}{p}||w||_2^2 + C\sum_{i=1}^{p} max(0, 1 - y_i'(w^T x_i + b))^2 \tag{3}$$

where $||w||_2$ is the Euclidean norm (also known as L2 norm), with square hinge loss.

Despite intended for binary classification, SVM may be utilized for multinomial classification as well. The use of kernel tricks is one approach to accomplish this, which transforms a linear model to a nonlinear one by implementing kernel functions such as radial basis function (RBF). We then employed the one-versus-all (OvA) strategy, which treats a given class c_i as the positive class, and others as a negative class.

With OvA strategy, the L2-SVM serves as the classifier of the deep learning model in this study (CNN). That is, SVM learned the parameter weight and bias of the model.

3.7 Softmax

For classification problems using deep learning techniques, applying softmax or 1-of-K encoding at the top is the standard. For example, 25 classes in this study, the softmax layer has 25 nodes denoted by p_i, where $i = 1, 2, \ldots, 25$. p_i specifies a discrete probability distribution, therefore, $\sum_i^{25} p_i = 1$.

Let W be the weight connecting the penultimate layer to the softmax layer, h be the activation of the penultimate layer nodes, then the total input into a softmax layer, given by a, is

$$a_i = \sum_k h_k W_{ki} \tag{4}$$

Then we have

$$p_i = \frac{exp(a_i)}{\sum_j^{10} exp(a_j)} \tag{5}$$

The predicted class $\hat{i}$ would be

$$\hat{i} = arg_i\, max\, p_i = arg_i\, max\, a_i \tag{6}$$

3.8 Multiclass SVMs

Using the so-called one-versus-rest strategy, SVMs can extend for multiclass problems which seem to be the simplest way [6]. N linear SVMs will be trained independently, for N-class problems where the data from the other classes from the negative cases.

Denoting the output of nth SVM as

$$a_n(x) = w^T x \tag{7}$$

The predicted class is

$$arg_n max\, a_n(x) \tag{8}$$

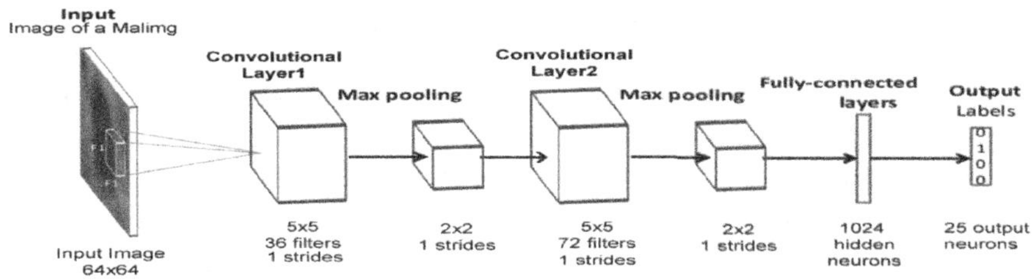

Fig. 4 CNN architecture

3.9 *Proposed CNN Model*

Convolutional neural network (CNN) has hidden layers of neurons with "learnable" parameters which make CNNs similar to feedforward neural networks. Those neurons receive input and perform a dot product then follow it with a nonlinearity such as *sigmoid* or *tanh* (Fig. 4).

The alteration proposed in the architecture design was the size of the input layer and output layer such as the input of $64 \times 64 \times 1$ and the output of 25 classes as shown in Fig. 3. And the introduction of SVM as the network classifier instead of conventional softmax function for the CNN-SVM model. Actually, Tang first presented this paradigm of combining CNN and SVM [2]. In this paper, we analyze the comparison of conventional CNN and CNN-SVM using the Malimg dataset.

4 Results and Discussion

In this study, all experiments were implemented using a personal computer with AMD Ryzen 5 2600x @ 3.5 GHz $\times$ 6, 16 GB of DDR4 RAM, and NVIDIA GeForce GTX 1050 ti 4 GB GPU.

Each deep learning (DL) model was trained on $\approx$70% of the preprocessed Malimg dataset [1]. The results for the different DL models are summarized in Table 2 where testing is done with $\approx$ 30% of the dataset. To evaluate our model like previous

Table 2 Experimental results of deep learning models

Variables	CNN	CNN-SVM
Train accuracy	97.58%	89%
Epochs	25	25
Data points	2798	2798
F1	0.96	0.86
Precision	0.96	0.83
Recall	0.97	0.89

methods, we use accuracy which indicates the percentage of malware samples labeled correctly in the test data.

We train our proposed model CNN and CNN-SVM for 25 epochs with a batch size of 256 with Malimg dataset. Table 2 shows the performance of different methods on this dataset. Our method CNN achieves the predictive accuracy of 97.58%, whereas CNN-SVM has predictive accuracy of 89%. However, the CNN model turns out to have the best performance with an accuracy of 97.58%.

Figures 5 and 6 show the experimental training and testing accuracy curves of the deep learning models (CNN and CNN-SVM) for 25 epochs (equivalent to 625 steps, since $6400 \times 25 \div 256 = 625$). The graph of learning curve for all the models shown in figures is plotted using Matplotlib [8].

Figure 7 shows the testing performance of CNN model in multinomial classification on malware families. The mentioned model has precision of 0.96, recall of 0.97, and F1 score of 0.96.

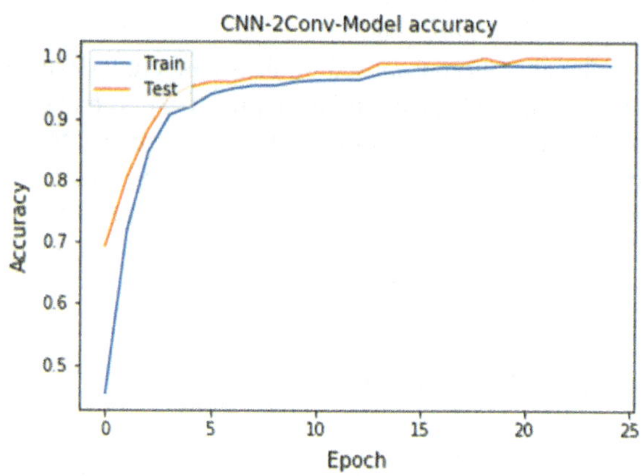

Fig. 5 Accuracy curve of CNN showing training and testing accuracy according to epoch

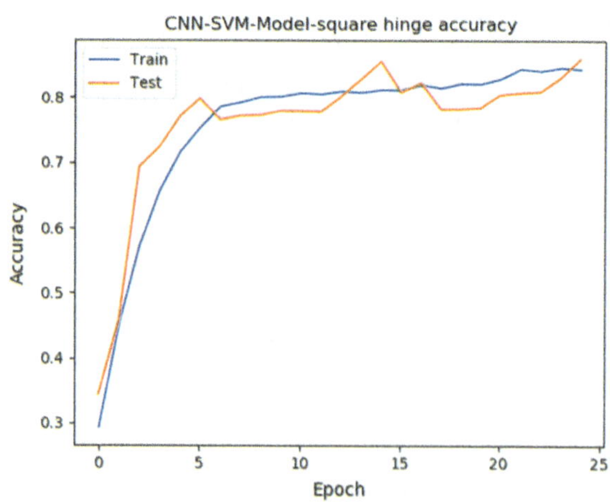

Fig. 6 Accuracy curve of CNN-SVM showing training and testing accuracy according to epoch

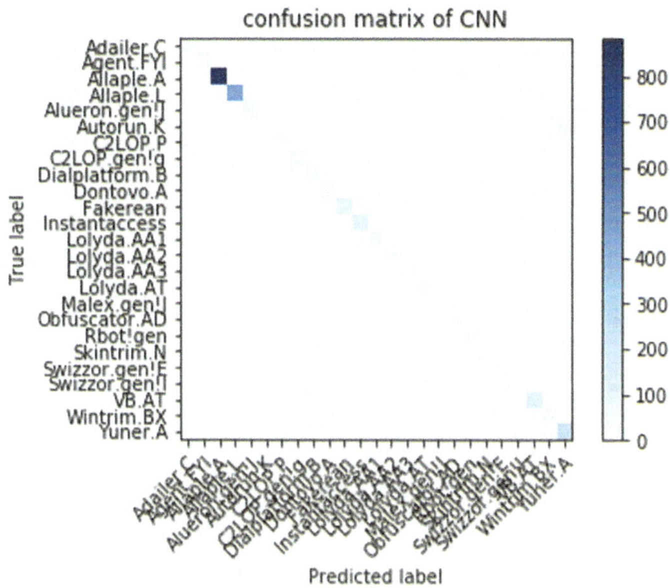

Fig. 7 The figure showing Confusion matrix of CNN

Figure 8 shows the testing performance of CNN-SVM model using SVM as final classifier. The mentioned model had a precision of 0.83, a recall of 0.89, and F1 score of 0.86.

Comparison of proposed methods with various related works done in the survey with Malimg dataset in terms of methods and accuracy is shown in Table 3, while our methods (CNN) have achieved the best performance among all the others.

5 Conclusion and Future Work

Malimg dataset is used for the experiment in this paper, which consists of malware images for the purpose of malware family classification [1]. We implemented different deep learning models with L2-SVM, and softmax as their final layer in the multinomial classification task. The experimental results show that CNN using the softmax model had the highest predictive accuracy among other deep learning models (CNN-SVM), having a test accuracy of $\approx 97.58\%$.

Improving the architectural design of the deep learning models by adding more layers and adding better non-linearities and/or using an optimized dropout may get a better result on malware classification. Other deep learning models such as MLP and RNN for image classification can be applied to experiments on malware classification.

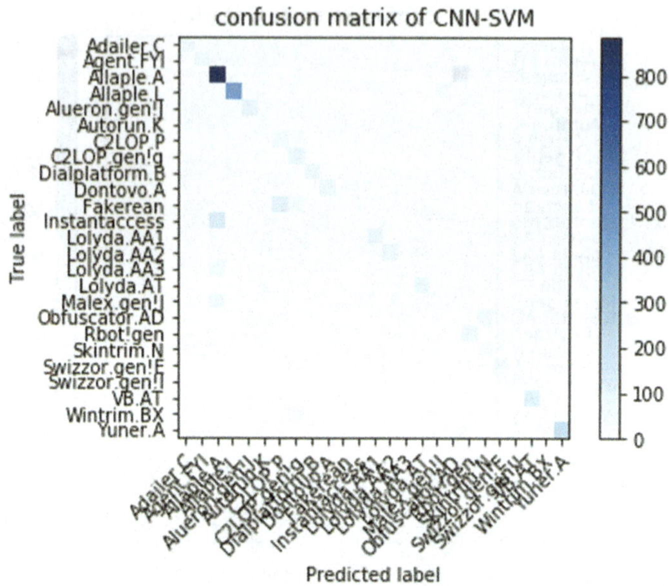

Fig. 8 The figure showing Confusion matrix of CNN-SVM

Table 3 Comparison of accuracy results for Malimg dataset

Paper	Methods	Accuracy
Kosmidis et al. [3]	MLP	≈ 87%
Agarap et al. [4]	GRU-SVM	≈ 84.92%
Agarap et al. [4]	MLP-SVM	≈ 80.46%
Agarap et al. [4]	CNN-SVM	≈ 77.23%
Proposed method	CNN	≈ 97.58%
Proposed method	CNN-SVM	≈0 89%

References

1. Nataraj, L., Karthikeyan, S., Jacob, G., Manjunath, B.: Malware Images: Visualization and Automatic Classification (2011)
2. Tang, Y.: Deep Learning using Linear Support Vector Machines (2013)
3. Kosmidis, K., Kalloniatis, C.: Machine Learning and Images for Malware Detection and Classification (2017)
4. Agarap, A.F., Pepito, F.J.H.: Towards Building an Intelligent Anti-Malware System: A Deep Learning Approach using Support Vector Machine (SVM) for Malware Classification (2017)
5. Su, J., Danilo Vasconcellos, V., Prasad, S., Daniele, S., Feng, Y., Sakurai, K.: Lightweight classification of IoT malware based on image recognition. In: 2018 IEEE 42nd Annual Computer Software and Applications Conference (COMPSAC), pp. 664–669. Tokyo (2018)
6. Cortes, C., Vapnik, V.N.: Support vector networks. Mach. Learn. (1995)
7. AVG-Test Institute: (2019). https://www.av-test.org/en/statistics/malware/. Accessed 20 May 2019

8. Hunter, J.D.: Matplotlib: A 2D graphics environment. Comput. Sci. Eng. **9**(3), 90–95 (2007). https://doi.org/10.1109/MCSE.2007.55
9. Pedregosa, F., Varoquaux, G., Gramfort, A., Michel, V., Thirion, B., Grisel, O., Blondel, M., Prettenhofer, P., Weiss, R., Dubourg, V., Vanderplas, J., Passos, A., Cournapeau, D., Brucher, M., Perrot, M., Duchesnay, E.: Scikit-learn: machine learning in Python. J. Mach. Learn. Res. **12**(2011), 2825–2830 (2011)
10. van der Walt, S., Colbert, S.C., Varoquaux, G.: The NumPy array: a structure for efficient numerical computation. Comput. Sci. Eng. **13**(2), 22–30 (2011)
11. Vinod, P., Jaipur, R., Laxmi, V., Gaur, M.: Survey on malware detection methods. In: Proceedings of the 3rd Hackers' Workshop on Computer and Internet Security, pp. 74–79 (2009)
12. Nataraj, L., Yegneswaran, V., Porras, P., Zhang, J.: A Comparative Assessment of Malware Classification Using Binary Texture Analysis and Dynamic Analysis, pp. 21–30 (2011)
13. Abadi, M., Agarwal, A., Barham, P., Brevdo, E., Chen, Z., Citro, C., Corrado, G.S., Davis, A., Dean, J., Devin, M., Ghemawat, S., Goodfellow, I., Harp, A., Irving, G., Isard, M., Jozefowicz, R., Jia, Y., Kaiser, L., Kudlur, M., Levenberg, J., ManÃl', D., Schuster, M., Monga, R., Moore, S., Murray, D., Olah, C., Shlens, J., Steiner, B., Sutskever, I., Talwar, K., Tucker, P., Vanhoucke, V., Vasudevan, V., ViÃl'gas, F., Vinyals, O., Warden, P., Wattenberg, M., Wicke, M., Yu, Y., Zheng, X.: TensorFlow: large-scale machine learning on heterogeneous systems Software (2015). Available from www.tensorflow.org
14. Gandotra, E., Bansal, D., Sofat, S.: Malware analysis and classification: a survey. J. Inf. Secur. **5**, 56–64 (2014)
15. Udayakumar, N., Saglani, V.J., Cupta, A.V., Subbulakshmi, T.: Malware classification using machine learning algorithms. In: 2018 2nd International Conference on Trends in Electronics and Informatics (ICOEI), pp. 1–9. Tirunelveli (2018)

Chlorophyll Prediction Using Ensemble Deep Learning Technique

Ashapurna Marndi and G. K. Patra

Abstract Chlorophyll is an essential component of phytoplankton and plays an important role in food chain and nutrient cycle required for survival of marine creatures. Getting suitable fishing zone is one of the commercial usages of liveliness measurement of marine ecosystem. Optimal sustainability of marine ecosystems needs an accurate prediction of chlorophyll. Dynamical models to predict the chlorophyll are challenged by the complex physical, chemical, and biological processes. Numerous researchers have attempted to address this problem using various computationally intelligent methods such as neural networks. However, normal neural networks have failed to provide a reliable forecast. This paper proposes a novel ensemble forecasting using Long Short Term Memory (LSTM) and a deep learning (DL) approach for time series data analysis. The methodology was tested to predict chlorophyll in Arabian Sea and found satisfactory result. Improved capabilities of the proposed method are also been demonstrated through various statistical analyses.

Keywords Ensemble forecasting · Long short-term memory · Artificial intelligence · Deep neural networks · Chlorophyll prediction

1 Introduction

Chlorophyll prediction plays an important role in the algal bloom predictions. The concentrations of the plant pigment, i.e., "chlorophyll-a", occur in all marine phytoplankton and provide useful proxy indicator of the amount of nutrients incorporated into phytoplankton biomass. This is because phytoplankton have predictable

A. Marndi (✉) · G. K. Patra
Academy of Scientific and Innovative Research, Ghaziabad 201002, Uttar Pradesh, India
e-mail: asha@csir4pi.in

G. K. Patra
e-mail: gkpatra@csir4pi.in

Council of Scientific and Industrial Research, Fourth Paradigm Institute, Bengaluru 560037, Karnataka, India

H. Das et al. (eds.), *Progress in Computing, Analytics and Networking*,
Advances in Intelligent Systems and Computing 1119,
https://doi.org/10.1007/978-981-15-2414-1_34

nutrient-to-chlorophyll ratios. Chlorophyll-a is the most commonly used parameter for monitoring phytoplankton biomass and nutrient status, as an index of water quality.

Increased nutrient availability, for example, from human activity such as agricultural runoff, soil erosion, discharges of sewage and aquaculture waste, etc. usually leads to a rise in chlorophyll concentrations in coastal waters because of increased phytoplankton biomass. Phytoplankton can rapidly deplete nutrients to levels which would be difficult to sample and analyze directly. Monitoring chlorophyll levels is a direct way of tracking algal growth. Amount of algal availability in a sea zone is usually proportionate to marine creature in that zone. Suitable fishing zones are detected based on extent of chlorophyll availability.

There are various conventional approaches used to solve different prediction problems. However, there are very few traditional techniques used to predict chlorophyll and in the recent years the biogeochemical models helped in assimilating the chlorophyll. Along with traditional approaches, few artificial intelligence (AI) approaches have also been attempted to predict the chlorophyll in near time. Nowadays, subfield of AI, machine learning (ML), is getting significant attention in ocean research fields. ML specifically deep learning (DL) algorithms are capable of discovering hidden important patterns from massive data leading to build insight for concise and reliable analysis. Deep learning technology has been successfully used not only in data analytics of commercial fields but also in scientific fields [1, 2].

LSTM has unique capability to utilize past learned information in best way to predict future values. It has gained significant success in predicting specifically time series data. In this experiment, LSTM technique is used to provide better solution in predicting chlorophyll for future timestamp. Further, this paper describes an enhancement upon on LSTM and found to be better than normal LSTM solving this problem. The merits of the proposed technique are demonstrated comparing with basic LSTM to predict chlorophyll in Arabian Sea. The result shows that the proposed algorithm has made significant improvement on basic LSTM that enhances the accuracy of long-term prediction of chlorophyll.

2 Related Work

Several researchers have tried to solve the problem of Chlorophyll-a (Chl-a) concentration prediction in different ways. In [3], it is attempted to use support vector machine for regression (SVR) and random forest (RF) to predict Chl-a concentration based on multiple variables and concluded that the RF model had a higher predictive ability than the SVR model. Random forest model with a sliding window strategy is used in [4] to predict Chl-a concentration in freshwater. Basic long short-term memory was used for predicting chlorophyll in 2 days and 4 days lead time based on daily measured water quality data as input. Again in [5], long short-term memory was used for forecasting air pollutant concentration. In [6], they combine ANN and generic algorithm to predict chlorophyll. In [7], it has suggested two model parameters, e.g., electrical conductivity (EC) and turbidity, and it has showed the efficiency of predictive model increased by adding them. In [8], comparison of the predictive performance was made using four types of multiple linear regression (MLR) and principal component regression (PCR) models were compared for chlorophyll-a (Chl-a) prediction. In [9], two chlorophyll predictors, ANN model and SVM models, are compared for chlorophyll prediction.

3 Methodology

3.1 Design Consideration for Model

An efficient model needs all the important parameters to be considered carefully. To solve a problem with no deterministic steps or fixed and finite input data, it is very important to consider all aspects that affect output and draw a fine balance of adjustments among them to derive conclusive result. Possible all such aspects are discussed below that can influence the outcome of this experiment:

Input Parameters

As chlorophyll concentration is influenced by various oceanographic parameters such as sea surface temperature (SST), sea surface salinity (SSS), and sea surface height (SSH) [10]. They have been considered along with chlorophyll as input to the model.

Range of Prediction

Range of prediction plays an important role for solving prediction model. Though it varies application to application, it is required to select the range of prediction carefully based on requirement of the application. However, it is obvious that prediction accuracy decreases with increase of time range of prediction. After considering usability of the application along with tolerable mismatch level, the 2-month time ahead period is been fixed for this.

Fitting Base Model

There are several variants of artificial intelligence techniques available to solve such prediction problems. The long short-term memory (LSTM) in deep learning approach is one of such techniques which can remember past important information for long time and conveniently forgets the less useful patterns. The ability to remember long back pattern leads to predict time series data efficiently. Since data used in this problem are all time series in nature, the solution in this paper is considered to be upon LSTM with suitable modification.

3.2 Long Short-Term Memory (LSTM)

The LSTM [11] model consists of forget gate (f_n), input gate (i_n), and output gate (o_n) as shown in Fig. 1. The forget gate is responsible for deciding unwanted information to be discarded. The input gate is responsible for deciding what new information is to be stored in the cell state and what new information to be added by *tanh* layer. Old cell state is updated to new one by removing the information required to be forgotten from the previous state and adding new information to the current state. Finally, the output is based on cell state but filtered through sigmoid layer and then multiply the parts of sigmoid layer output by *tanh* of cell state. Following Eqs. (1)–(5) have represented them mathematically.

$$f_n = \sigma\left(W_f I_n + U_f h_{n-1} + b_f\right) \tag{1}$$

$$i_n = \sigma(W_i I_n + U_i h_{n-1} + b_i) \tag{2}$$

$$o_n = \sigma(W_o I_n + U_o h_{n-1} + b_o) \tag{3}$$

$$C_n = f_n * C_{n-1} + i_n * \tanh(W_c I_n + U_c h_{n-1} + b_c) \tag{4}$$

$$h_n = \tanh(C_n) * o_n \tag{5}$$

Input to LSTM network is denoted by I_n. Based on values of forget gate, input gate, and output gate mentioned in Eqs. (1)–(3), cell state (C) and hidden state (h) of LSTM are being updated by Eqs. (4) and (5). $W_{f,}$ $W_{i,}$ $W_{o,}$ W_c and $U_{f,}$ $U_{i,}$ $U_{o,}$ U_c are the weight matrixes at current state and previous state and $b_{f,}$ $b_{i,}$ $b_{o,}$ b_c are the bias

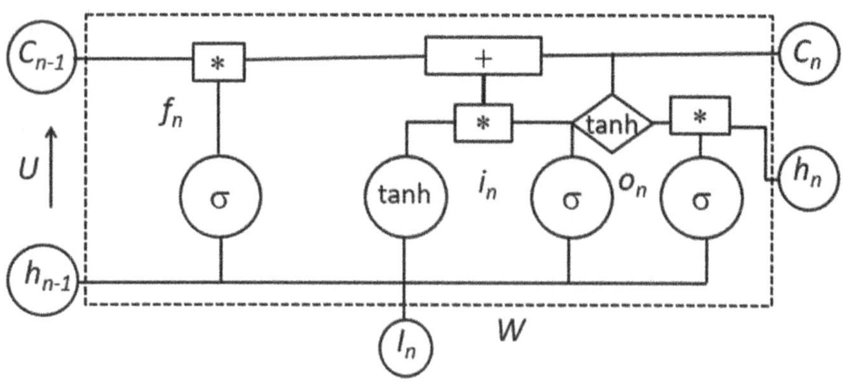

Fig. 1 Architecture of a LSTM network

vectors of forget gate, input gate, output gate, and cell state, h_{n-1} is the hidden unit of previous state, and σ is the activation function.

3.3 Proposed Model (Moving Window LSTM)

The proposed prediction algorithm is based on multilevel LSTM by treating the outputs from first level of LSTMs as the input for the second level. The first-level LSTMs take different ensemble datasets as input divided based on moving window of fixed size. The purpose of distributing input dataset into multiple time windows is that if a pattern strives out to become an impactful pattern for final output, it should be present in all windows of recurring timeframe, which is usually year in this case. Second, the time windows are made to be moving window of more than a year, by which it ensures that the impactful patterns can be in any month or covering multiple months without breaking across time boundaries of a year, though the seasonal pattern usually repeats over years. Since the algorithm is built upon moving window ensembled data with multilevel LSTM, it is named as "Moving Window LSTM" (MW-LSTM).

The whole training dataset is divided into moving window of 4 years duration and the test data is also kept for same duration. As shown in Fig. 2, the moving window period for training data starts from 2004–2007 and continues till 2009–2012 period with gap of a year in subsequent windows. The data during 2013–2016 are considered as test data. There are six LSTMs in the first level, one for each ensemble training dataset of moving window period, and are trained to output optimal result. In second level, outputs from first-level LSTMs are combined and used as inputs as shown by

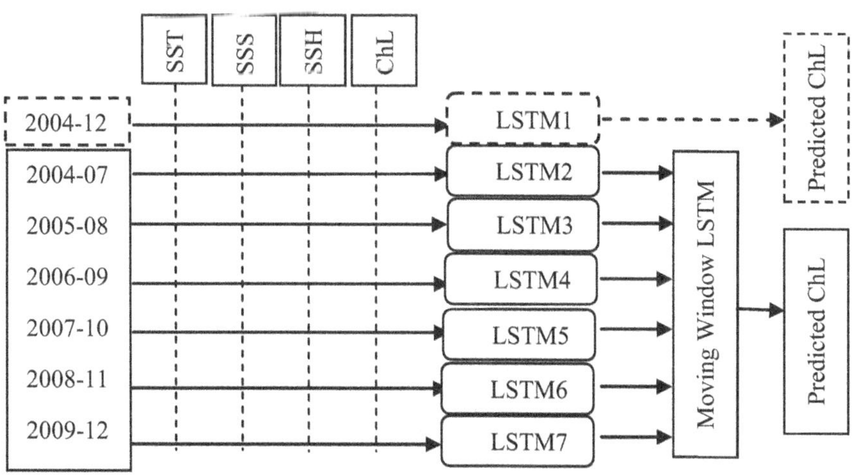

Fig. 2 Block diagram of proposed model depicting multilevel LSTMs with moving window input datasets

LSTM2 to LSTM7 in Fig. 2. In parallel, the whole input dataset 2004–2012 is also fed to a normal LSTM, i.e., LSTM1 in the figure and the output is compared with the output of second-level LSTM in the proposed model.

3.4 *Study Area and Experimental Dataset*

For this experiment, it has been chosen a portion of the Arabian Sea (Long-65E:72E and Lat-12. 5 N:15 N) as the study area for duration of 2004–2016. The whole data is divided into two sets such as data during 2004–2012 as training data and that of during 2013–2016 as testing data. The subsequent data are captured in 5 days intervals. Sea surface temperature (SST), sea surface salinity (SSS), and sea surface height (SSH) are collected from open-source live access server of Indian National Centre for Ocean Information Services Site (INCOIS) [12], and the chlorophyll data are taken from the open-source merged ocean color data [13].

3.5 *Experiment Setup*

In this experiment, LSTMs were configured with same set of hyperparameter values. In first level, all LSTMs were trained with 6 hidden layers with 50 neurons each layer. The experiment was initiated with 10 neurons at the first layer and then kept increasing by 10 more neurons in the same layer until the network gave satisfactory result. Once the number of neurons were fixed, the networks were tuned further by adding additional layer starting from second layer till optimum network was achieved. The network is treated as optimal when no further improvement was found by addition of any more neurons or layers. Thus, the LSTM configuration was fixed with 6 hidden layers and 50 neurons in each layer. Following similar approach for the second-level LSTM, it was observed that in epoch 900, it was optimized with 10 hidden layers and 50 neurons in each hidden layer.

The experiment was carried out on intel(R) Xeon(R) CPU E3-1203 v3 @ 3.30 GHz with 8 cores and 32 GB RAM.

4 Result and Discussion

Different statistical tests were carried out to find out the accuracy of predictability of the proposed method with the normal LSTM. The statistical parameters that were used to test are root mean square error (RMSE) and correlation coefficient (CC). These two parameters give a very good estimation on closeness of two time series patterns.

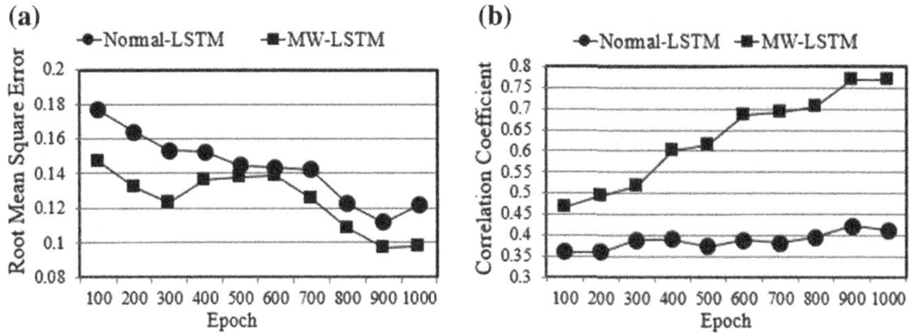

Fig. 3 For normal LSTM and MW-LSTM **a** RMSE versus epoch, **b** CC versus epoch

Figure 3a shows the RMSE of normal LSTM and proposed MW-LSTM for different number of epochs (100–1000). From this figure, it is observed that the RMSE for MW-LSTM is better compared to normal LSTM. This indicates that the average error in case of ensemble forecasting is less than the non-ensemble forecasting. However, it is not good enough to establish the superiority of the proposed method with only RMSE efficiency. Hence, the correlations between the two time series datasets were also compared. Figure 3b depicts the correlation coefficients in both the approaches, i.e., normal LSTM and proposed MW-LSTM between the two time series data that are observed and predicted for different epoch sizes. For both RMSE and CC, the values are found to be saturated after 900 epoch values and thus epoch 1000 was considered as threshold for this experiment by keeping slightly safe margin.

As shown in Fig. 4a–d, the chlorophyll values predicted using MW-LSTM are better than the normal LSTM. The outcome of this experiment is due to multiple factors such as dividing inputs into multiple ensemble datasets based on moving window period and then optimizing the output in multiple level of LSTMs. The more efficiency of this algorithm is demonstrated not only by visual graphs but also through statistical parameters such as root mean square error (RMSE) and correlation coefficient (CC) values.

5 Conclusion

Ensemble forecasting is a well-known methodology in atmospheric sciences using dynamical models. However, uses of ensemble forecasting using artificial intelligence (AI) technique are relatively new and have shown good promises. Use of ensemble forecasting using multilevel LSTMs over moving window data is novel approach giving better result. Chlorophyll prediction in ocean, especially in the Arabian Sea, is a challenging as well as important requirement for the sustainability of the marine ecosystem. This work demonstrated that using the deep learning architecture of LSTM, in an ensemble methodology, a reliable and usable prediction can be made. It has demonstrated how a set of individual LSTMs in the first layer, with

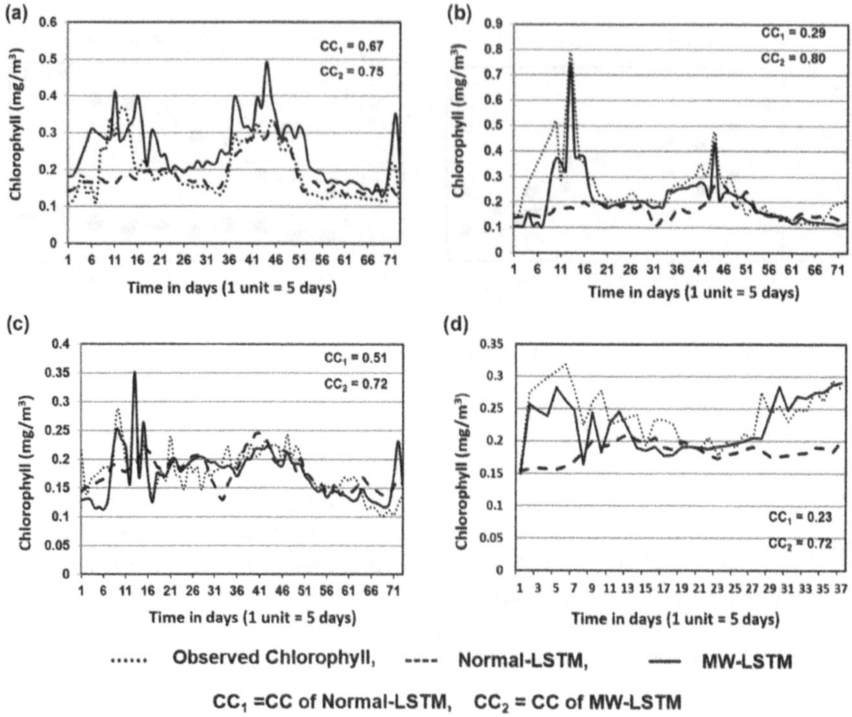

Fig. 4 Predicted results in both normal LSTM and moving Window LSTM versus observed Chl-a during **a** 2013, **b** 2014, **c** 2015, and **d** 2016

overlapping subset of data, is capable of extracting hidden important patterns that can be fed to next level of LSTM to give an optimal result, especially in this case of chlorophyll prediction. The statistical parameters such as correlation coefficient and RMSE indicate the superiority of this proposed method. In coming days, such enhancement over normal LSTM can be useful solving more complex scientific problems in variety of domains.

References

1. Das, H., Naik, B., Behera, H.S.: Classification of Diabetes Mellitus Disease (DMD): A Data Mining (DM) approach. In: Progress in Computing, Analytics and Networking, pp. 539–549. Springer, Singapore (2018)
2. Sahoo, A.K., Mallik, S., Pradhan, C., Mishra, B.S.P., Barik, R.K., Das, H.: Intelligence-based health recommendation system using big data analytics. In: Big Data Analytics for Intelligent Healthcare Management, pp. 227–246. Academic (2019)
3. Li, X., Sha, J., Wang, Z.-L.: Application of feature selection and regression models for chlorophyll-a prediction in a shallow lake. Environ. Sci. Pollut. Res. 1–11 (2018)

4. Yajima, H., Derot, J.: Application of the Random Forest model for chlorophyll-a forecasts in fresh and brackish water bodies in Japan, using multivariate long-term databases. J. Hydro Inf. **20**, 206–220 (2018)
5. Cho, H.: Deep: Learning Application to Time Series Prediction of Daily Chlorophyll-a Concentration (2018)
6. X. et al.: Long short-term memory neural network for air pollutant concentration predictions: method development and evaluation. Environ. Pollut. **231**, pp. 997–1004 (2017)
7. Lee, G., Bae, J., Lee, S., Jang, M., Park, H.: Monthly chlorophyll-a prediction using neuro-genetic algorithm for water quality management in Lakes. Desalin. Water Treat. **57**, 26783–26791 (2016)
8. Lee, G., Othman, F., Ibrahim, S., Jang, M.: Determination of the forecasting-model parameters by statistical analysis for development of algae warning system. Desalin. Water Treat. **57**, 26773–26782 (2016)
9. Cho, K.H., Kang, J.-H., Ki, S.J., Park, Y., Cha, S.M., Kim, J.H.: Determination of the optimal parameters in regression models for the prediction of chlorophyll-a: a case study of the Yeongsan reservoir. Korea. Sci. Total Environ. **407**, 2536–2545 (2009)
10. Krasnopolsky, V., Nadiga, S., Mehra, A., Bayler, E., Behringer, D.: Neural networks technique for filling gaps in satellite measurements: application to ocean color observations. Comput. Intell. Neurosci. **2016**, 9. Article ID 6156513. http://dx.doi.org/10.1155/2016/6156513
11. Hochreiter, S., Schmidhuber, J.: Long short-term memory. Neural Comput. **9**, 1735–1780 (1997)
12. ESSO—Indian National Centre for Ocean Information Services. https://las.incois.gov.in
13. NASA Ocean Color. https://oceancolor.gsfc.nasa.gov

Enabling Affordance-Guided Grasp Synthesis for Robotic Manipulation

Xavier Williams and Nihar R. Mahapatra

Abstract Empowering robots with the ability to use image data to understand complex object affordances will enable robots to intuitively interact with any given object. An affordance is an action such as scooping or swinging that an object allows. A single object may have multiple affordances that are linked to the object's functional parts. Furthermore, scooping and swinging require particular grasp configurations. This is the principle behind affordance-guided grasp synthesis. Once the affordances and the corresponding grasps are known, the robot is primed to learn the behaviors to execute affordances. This paper presents a systematic review of recent work on affordance detection and how those affordances are used to guide grasp synthesis.

Keywords Affordance detection · Functional parts · Grasping · Object manipulation · Robotic manipulation

1 Introduction

Understanding all the actions an object allows an actor is a critical step in the design of intelligent robots. With this knowledge, productivity in society will increase as robots begin to handle laborious, dangerous, or inaccessible tasks for humans [1, 2]. For example, a service robot may aid disabled people in their grasping of everyday objects, and a disaster-relief robot can help clear rubble in a hazardous environment [3].

Consider teaching a robot to throw a ball [4]. First, the robot must localize the ball and select the desired action from the set of all actions the ball affords. In this

X. Williams · N. R. Mahapatra (✉)
Department of Electrical and Computer Engineering, Michigan State University, East Lansing, MI, USA
e-mail: nrm@egr.msu.edu

X. Williams
e-mail: will3358@egr.msu.edu

H. Das et al. (eds.), *Progress in Computing, Analytics and Networking*, Advances in Intelligent Systems and Computing 1119,
https://doi.org/10.1007/978-981-15-2414-1_35

example, the desired action is throwing. The robot must first produce a stable grasp on the ball to throw it. A grasp is synthesized by using prior knowledge of successful grasps, or by using trial and error. Once the robot knows *what* it can do with the ball and how to best hold it, the robot must then learn *how* to execute the desired action. This is done either by imitating a human expert or through random attempts until the action is successful.

The above example illustrates the process by which a robot learns to execute a desired action specific to an object, or an object's affordance. An *affordance*, first defined by ecological psychologist James Gibson, refers to the ways in which an object may be used by an agent [5]. Affordances are intrinsic properties of an object's class and functional parts, and they describe what actions an object affords [6]. An object may possess multiple affordances, each requiring a unique grasp configuration. End-to-end affordance execution, as shown in Fig. 1, comprises identifying an object and its affordances, optimizing grasps for those affordances, and executing those affordances on robotic platforms. This paper seeks to present a systematic review of the first two stages of this process.

The main contributions of this paper are twofold: (a) a novel systematic overview of the state-of-the-art methods for object affordance detection and grasping and (b) a discussion of the research trends in these areas and the challenges that remain to be addressed. Thus far, there is no comprehensive survey on robotic manipulation of an object by its functional parts for affordance execution in an end-to-end manner. Conflicting terminology is prevalent in the literature describing the steps pictured in Fig. 1, so this paper offers a concise review that consolidates the disparate information.

The rest of the paper is organized as follows. Section 2 reviews methods of extracting an object's affordances. Section 3 discusses how those affordances can guide grasp synthesis. The discussions in these two sections provide a sound basis for exploring affordance execution, which is beyond the scope of this paper. Conclusions are drawn in Sect. 4.

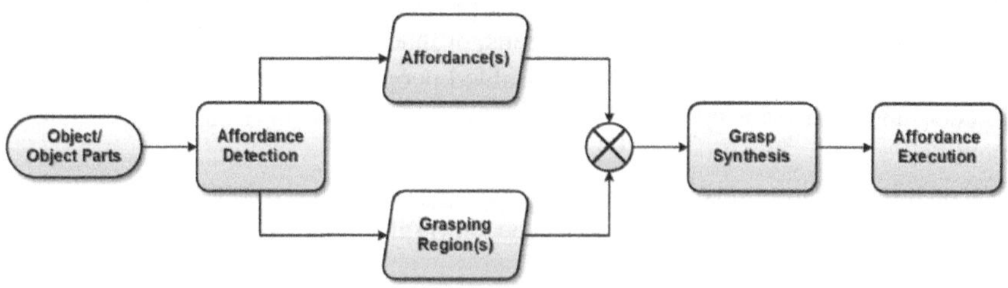

Fig. 1 End-to-end affordance execution flowchart

2 Affordance Detection

When encountering everyday objects, humans leverage prior experiences to deduce the set of actions an object affords. Affordance detection allows robots to understand objects similar to how humans do. Though we limit our focus to affordance detection with respect to manipulable objects, researchers apply this process in a variety of fields, such as autonomous driving [7, 8], sociology [9, 10], and environment understanding [11, 12].

As the concept of affordances is abstract, there are other definitions used in the literature. Su et al. regard affordances as robot-centric, e.g., the set of actions that can be completed for a given robot end-effector [13], and they use this definition to guide the types of candidate grasps for objects. This definition, however, generalizes poorly to robots with alternate end-effector designs. In contrast, we present an object-centric perspective on affordance, e.g., the set of actions available to an object, and we use this definition for the remainder of the survey.

Affordances are intrinsic properties of objects. For example, a ball affords grasping and throwing, etc. An entire object can afford a given action, but affordances are typically considered to be a function of an object's parts as shown in Fig. 2. Detection models learn to predict affordances by training on object images whose affordances are pre-labeled. Producing these labels by hand is labor-intensive, so researchers complete labeling automatically using convolutional neural networks.

Objects are segmented into parts starting from 2D images or point clouds, and those parts are assigned labels either by hand or through supervised learning. Parts are distinguished by their surface normals [15], geometric shape [16], or their spatial relations [17]. Part-based affordance detection is popular compared to object-based detection because it more easily generalizes to unknown objects. Generalization is crucial, as real-world robotic applications feature a larger variety of affordances than the limited set of learned affordances [18]. Furthermore, part-based detection has high performance in the presence of occlusions or incomplete models [14, 19].

Affordance detection models require labeled object images for training where one or more affordances are assigned to each pixel. Many authors hand-label each

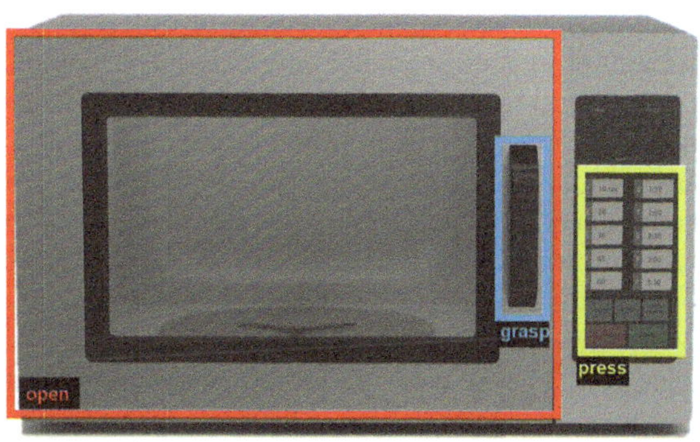

Fig. 2 Example image from RGB-D Part Affordances Dataset [14]. The color-coded regions represent the different predefined affordances

Table 1 Highlights of datasets used in affordance detection

Dataset	Affordance Classes	Object Classes	Images
Myers part affordance [14]	17		150
IIT-AAF [24]	10		8 K
COCO-tasks [25]	–	80	200 K
Cornell grasping [26]	–	280	1 K

pixel [15, 20], but, as this is labor-intensive, others look to convolutional neural networks (CNNs) for automatic labeling [21–23]. However, CNNs require an image preprocessing stage to remove variations in illuminations, and they require a larger dataset for training. To simplify this process, many publicly available datasets feature a multitude of training images with and without affordance labels, occlusions, and object classes. Some of the widely used training datasets are given in Table 1.

3 Grasp Detection

This section addresses grasping as presented throughout the literature. The prevailing definition of grasp synthesis focuses on the ability to recognize grasping points or the grasping pose for an object in any given image [27]. Grasp synthesis is classified into analytic and empirical approaches based on the use of physical feedback or prior object knowledge. We refer the reader to [28] for a more detailed review on this distinction. We instead characterize grasping according to the best grasp for an affordance, or affordance-guided grasping. We then discuss how researchers approach grasping an object with varying poses in 3D space.

3.1 Affordance-Guided Grasping

Section 2 reviewed methods focusing on object affordances themselves without considering an grasping region. Conversely, many methods extract viable grasping points on the object without explicitly considering the target's affordances. As a clarification, many studies seek to learn grasping affordances and define them as graspable areas on an object often for pick-and-place tasks [29, 30]. While this fits into our definition of affordances (an action that an object affords), we assume that an object's other affordances can be used to optimize a grasp approach. Thus, we survey methods that make a similar assumption.

Typical affordance-guided grasping algorithms begin by processing 2D or 3D data from a scene using the affordance detection methods in Sect. 2. Researchers construct a *knowledge base*, which incorporates information about an object class.

The knowledge base is represented by a graph whose nodes are object entities, such as object class, object attributes, and affordances, and whose edges are the intuitive relationships between the nodes. The higher the weight of an edge, the more likely a relation is to be true. Knowledge bases are typically implemented using neural networks. Ardón et al. use four separate convolutional neural networks (CNNs) to detect four attributes of an object (e.g., shape, texture, category, and environment) [31]. The extracted features are then used to query the knowledge base to output a corresponding affordance. The output affordance then guides the grasp synthesis process based on previous grasp knowledge. Ardón et al. extend their past framework by integrating human-defined constraints on grasp execution based on what types of affordances are socially acceptable for a given environment [32]. Knowledge bases are sensitive to noise, so researchers instead use the bag-of-words (BoW) model to classify objects [34].

While these traditional methods achieve a high grasp success rate, they are limited to operating on one object at a time. Additionally, they cannot operate in real time due to the costly training of their CNN feature detectors. The framework proposed by Madry et al. addresses these shortcomings by detecting the attributes of an object using scale-invariant feature transform (SIFT) descriptors for each attribute class (e.g., appearance, contour shape, color) [33].

3.2 Pose Estimation

An object's 6D pose (its location and orientation) is important for grasp detection, as the robot must correctly position its end-effector to form a stable grasp. 6D poses are computed using visual sensors and are therefore susceptible to occlusions. However, Zeng et al. address occlusions, cluttered environments, and sensor noise in their convolutional neural network based approach [35]. They segment and label multiple views of a scene, and then fit pre-scanned 3D object models to the resulting segmentation to acquire the 6D pose. Xiang et al. propose a similar method that uses video data to estimate 6D pose of occluded objects [36].

4 Conclusion

This paper has presented an overview of the approaches that have been developed to discover an object's affordances and affordance-guided grasping locations. This paradigm uses data to address complex real-world problems that are otherwise difficult to describe mathematically. We explained how robots detect the set of actions an object affords, and how grasps can be synthesized to take advantage of those affordances. To grasp an object, the robot either calculates the forces necessary to render the object immobile, or it recalls previous grasp examples from a remote database.

With this information, the robot is ready to execute the affordances using methods such as imitation learning, reinforcement learning, or a combination of the two.

The key challenges are as follows: (a) expanding affordance detection to handle nontraditional end-effectors and (b) design of databases that contain previous grasp attempts. As the research community tackles these and other challenges, robots will only continue to become more capable to operate in the real world. This will lead to a productivity boost in society as humans need not engage in tedious or dangerous tasks.

Acknowledgements This material is based upon work partly supported by the U.S. National Science Foundation under Grant No. 1936857.

References

1. Hassanin, M., Khan, S., Tahtali, M.: Visual affordance and function understanding: A survey (2018). arXiv:1807.06775
2. Sanchez, J., Corrales, J.-A., Bouzgarrou, B.-C., Mezouar, Y.: Robotic manipulation and sensing of deformable objects in domestic and industrial applications: A survey. Int. J. Robot. Res. **37**(7), 688–716 (2018)
3. Schwarz, M., Milan, A., Periyasamy, A.S., Behnke, S.: RGB-D object detection and semantic segmentation for autonomous manipulation in clutter. Int. J. Robot. Res. **37**(4–5), 437–451 (2018)
4. Ghadirzadeh, A., Maki, A., Kragic, D., Bjrkman, M.: Deep predictive policy training using reinforcement learning. In: 2017 IEEE/RSJ International Conference on Intelligent Robots and Systems (IROS), pp. 2351–2358 (2017)
5. Gibson, J.J.: The senses considered as perceptual systems (1966)
6. Bütepage, J., Cruciani, S., Kokic, M., Welle, M., Kragic, D.: From visual understanding to complex object manipulation. Ann. Rev. Control Robot. Auton. Syst. **2**, 161–179 (2019)
7. Wang, D., Devin, C., Cai, Q.-Z., Krähenbühl, P., Darrell, T.: Monocular plan view networks for autonomous driving (2019). arxiv:1905.06937
8. Sun, C., Vianney, J.M.U., Cao, D.: Affordance learning in direct perception for autonomous driving (2019). arXiv:1903.08746
9. Shu, T., Ryoo, M.S., Zhu, S.-C.: Learning social affordance for human-robot interaction (2016). arXiv:1604.03692
10. Shu, T., Gao, X., Ryoo, M.S., Zhu, S.-C.: Learning social affordance grammar from videos: Transferring human interactions to human-robot interactions. In: 2017 IEEE International Conference on Robotics and Automation (ICRA), pp. 1669–1676. IEEE (2017)
11. Chuang, C.-Y., Li, J., Torralba, A., Fidler, S.: Learning to act properly: Predicting and explaining affordances from images. In: Proceedings of the IEEE Conference on Computer Vision and Pattern Recognition, pp. 975–983 (2018)
12. Li, X., Liu, S., Kim, K., Wang, X., Yang, M.-H., Kautz, J.: Putting humans in a scene: Learning affordance in 3D indoor environments. In: Proceedings of the IEEE Conference on Computer Vision and Pattern Recognition, pp. 12368–12376 (2019)
13. Su, Y.-F., Liu, A., Lu, W.-H.: Improving robot grasping plans with affordance. In: 2017 International Conference on Advanced Robotics and Intelligent Systems (ARIS), pp. 7–12. IEEE (2017)
14. Myers, A., Teo, C.L., Fermüller, C., Aloimonos, Y.: Affordance detection of tool parts from geometric features. In: 2015 IEEE International Conference on Robotics and Automation (ICRA), pp. 1374–1381. IEEE (2015)

15. Lakani, S.R., Rodríguez-Sánchez, A.J., Piater, J.: Towards affordance detection for robot manipulation using affordance for parts and parts for affordance. Auton. Robots **43**(5), 1155–1172 (2019)
16. Guerin, F., Ferreira, P.: Robot manipulation in open environments: New perspectives. In: IEEE Transactions on Cognitive and Developmental Systems, pp. 1–1 (2019)
17. Indurkhya, B.: Metaphor and cognition: An interactionist approach, Vol. 13. Springer Science & Business Media (2013)
18. Chu, F.-J., Xu, R., Seguin, L., Vela, P.: Toward affordance detection and ranking on novel objects for real-world robotic manipulation. IEEE Robot. Autom. Lett. (2019)
19. Wu, Z., Song, S., Khosla, A., Yu, F., Zhang, L., Tang, X., Xiao, J.: 3D shapenets: A deep representation for volumetric shapes. In: Proceedings of the IEEE Conference on Computer Vision and Pattern Recognition, pp. 1912–1920 (2015)
20. Kokic, M., Stork, J.A., Haustein, J.A., Kragic, D.: Affordance detection for task-specific grasping using deep learning. In: 2017 IEEE-RAS 17th International Conference on Humanoid Robotics (Humanoids), pp. 91–98. IEEE (2017)
21. Nguyen, A., Kanoulas, D., Caldwell, D.G., Tsagarakis, N.G.: Detecting object affordances with convolutional neural networks. In: 2016 IEEE/RSJ International Conference on Intelligent Robots and Systems (IROS), pp. 2765–2770. IEEE (2016)
22. Srikantha, A., Gall, J.: Weakly supervised learning of affordances (2016). arXiv:1605.02964
23. Fang, K., Wu, T.-L., Yang, D., Savarese, S., Lim, J.J.: Demo2vec: Reasoning object affordances from online videos. In: Proceedings of the IEEE Conference on Computer Vision and Pattern Recognition, pp. 2139–2147 (2018)
24. Nguyen, A., Kanoulas, D., Caldwell, D.G., Tsagarakis, N.G.: Object-based affordances detection with convolutional neural networks and dense conditional random fields. In: 2017 IEEE/RSJ International Conference on Intelligent Robots and Systems (IROS), pp. 5908–5915 (2017)
25. Lin, T.-Y., Maire, M., Belongie, S., Hays, J., Perona, P., Ramanan, D., Dollár, P., Zitnick, C.L.: Microsoft COCO: Common objects in context. In: European Conference on Computer Vision, pp. 740–755. Springer, Berlin (2014)
26. Jiang, Y., Moseson, S., Saxena, A.: Efficient grasping from RGB-D images: Learning using a new rectangle representation. In: 2011 IEEE International Conference on Robotics and Automation, pp. 3304–3311. IEEE (2011)
27. Mahler, J., Liang, J., Niyaz, S., Laskey, M., Doan, R., Liu, X., Ojea, J.A., Goldberg, K.: Dex-net 2.0: Deep learning to plan robust grasps with synthetic point clouds and analytic grasp metrics (2017). arXiv:1703.09312
28. Du, G., Wang, K., Lian, S.: Vision-based robotic grasping from object localization, pose estimation, grasp detection to motion planning: a review (2019). arXiv:1905.06658
29. Kraft, D., Detry, R., Pugeault, N., Başeski, E., Piater, J., Krüger, N.: Learning objects and grasp affordances through autonomous exploration. In: International Conference on Computer Vision Systems, pp. 235–244. Springer, Berlin (2009)
30. Ten Pas, A., Platt, R.: Localizing handle-like grasp affordances in 3D point clouds. In: Experimental Robotics, pp. 623–638. Springer, Berlin (2016)
31. Ardón, P., Pairet, È., Ramamoorthy, S., Lohan, K.S.: Towards robust grasps: Using the environment semantics for robotic object affordances. In: Proceedings of the AAAI Fall Symposium on Reasoning and Learning in Real-World Systems for Long-Term Autonomy, pp. 5–12 (2018)
32. Ardón, P., Pairet, È., Petrick, R., Ramamoorthy, S., Lohan, K.S.: Learning grasp affordance reasoning through semantic relations (2019). arXiv:1906.09836
33. Madry, M., Song, D., Kragic, D.: From object categories to grasp transfer using probabilistic reasoning. In: 2012 IEEE International Conference on Robotics and Automation, pp. 1716–1723. IEEE (2012)
34. Csurka, G., Dance, C., Fan, L., Willamowski, J., Bray, C.: Visual categorization with bags of keypoints. In: Workshop on statistical learning in computer vision, ECCV, Vol. 1, No. 1–22, pp. 1–2. Prague (2004)

35. Zeng, A., Yu, K., Song, S., Suo, D., Walker, E., Rodriguez, A., Xiao, J.: Multiview self-supervised deep learning for 6D pose estimation in the Amazon picking challenge. In: 2017 IEEE International Conference on Robotics and Automation (ICRA), pp. 1386–1383 (2017)
36. Xiang, Y., Schmidt, T., Narayanan, V., Fox, D.: PoseCNN: A convolutional neural network for 6D object pose estimation in cluttered scenes (2017). arXiv:1711.00199

Towards Natural Language Understanding of Procedural Text Using Recipes

Dena F. Mujtaba and Nihar R. Mahapatra

Abstract Procedural knowledge, or how-to knowledge, is the knowledge acquired from natural language understanding of instructions in procedural text. Procedural knowledge bases containing textual descriptions of tasks in procedures have witnessed explosive growth recently. This has facilitated a significant body of work in various natural language understanding tasks. A rich source of procedural text is in the form of recipes describing food preparation procedures. The ready availability of online recipes has enabled progress in food computing, which refers to computing tasks related to recipes, such as food perception, recipe image recognition and calorie estimation, and food-oriented retrieval of recipes. However, past work on food computing has not covered the procedural knowledge inherent in recipes and the natural language understanding tasks required to uncover that knowledge. We seek to address this by presenting an overview of recent work in natural language understanding tasks in food computing and describing how this contributes to how-to knowledge and future applications.

Keywords Artificial intelligence · Natural language processing · Natural language understanding · Procedural knowledge · Food computing · Information extraction · Recipe representation

1 Introduction

Procedural knowledge, or how-to knowledge, is the knowledge acquired from natural language understanding (NLU) of instructions in procedural text. Examples of procedural text include steps in a recipe, navigation instructions, or even instructions

D. F. Mujtaba · N. R. Mahapatra (✉)
Department of Electrical and Computer Engineering, Michigan State University, East Lansing, MI, USA
e-mail: nrm@egr.msu.edu

D. F. Mujtaba
e-mail: mujtabad@egr.msu.edu

H. Das et al. (eds.), *Progress in Computing, Analytics and Networking*, Advances in Intelligent Systems and Computing 1119,
https://doi.org/10.1007/978-981-15-2414-1_36

to change a tire. Procedural knowledge bases containing textual descriptions of tasks in procedures have witnessed explosive growth recently in online how-to communities like wikiHow, eHow, and Snapguide [1]. This has enabled a large body of work in natural language understanding tasks targeting procedures.

Much work on natural language understanding tasks related to acquiring procedural knowledge has used recipes and recipe instructions, a form of procedures in the food preparation arena. Food preparation recipes have accumulated over the past decade online as a result of the widespread use of social media and online content creation. Recipe sites, such as AllRecipes.com, Epicurious.com, and Food.com, are some of the most popular sites visited (with over 50 million visits for AllRecipes.com alone for December 2015 [2]), and continue to grow in popularity as thousands of users write and share recipes.

This recipe information, alongside advances in artificial intelligence (AI), computer vision, and natural language processing (NLP) research, has enabled the creation of the field of *food computing*. First coined in [3], food computing refers to the computational approaches applied to tasks with food and recipes [4]. Food computing tasks, as described in [4], include *recipe recommendation* [5] (i.e., providing users recipes that they are likely to enjoy or are likely to meet their nutritional needs), *food perception* [6] (i.e., predicting how food will be perceived by people), and *recipe recognition* [7] (i.e., identifying food from an image).

We define two additional tasks that have not been as properly discussed in past work: (1) *recipe information extraction*, a subtask of information extraction which consists of obtaining structured knowledge from unstructured recipes, and (2) *recipe representation*, or the different forms recipes are shared and processed in. Advances in these areas have been achieved by exploiting recipe datasets and through the application of NLU to procedural text. Several food computing tasks rely on procedural text and procedural knowledge as depicted in Fig. 1.

Past Work and Our Contributions: We present an overview of NLU tasks in food computing and how this has and can contribute to procedural knowledge understanding and to the growth of robotics, AI, and other applications. To the best of our knowledge, past surveys on computational approaches and AI tasks related to food and recipes [4, 5, 8] have not discussed procedural knowledge understanding. Therefore, we seek to spur research in NLU of procedures, recipe information extraction, and other related topics.

In Sect. 2, we describe the food computing tasks that use natural language instructions and the corresponding NLU tasks. Section 3 details future challenges and applications associated with procedural knowledge understanding of recipes, along with concluding remarks and findings.

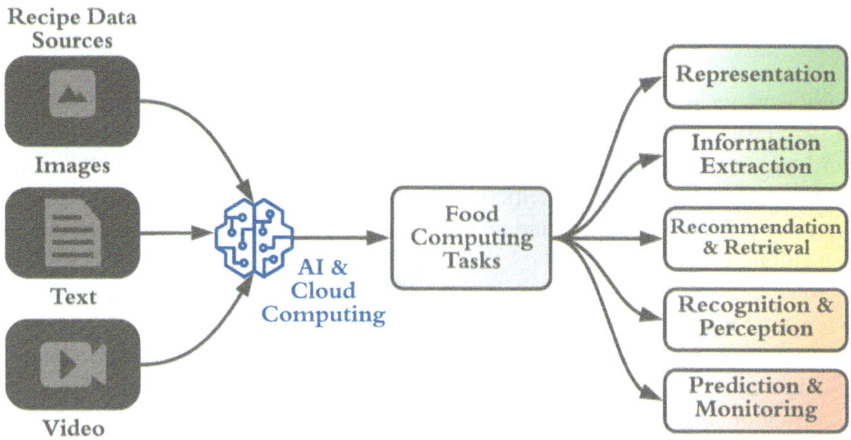

Fig. 1 Overview of food computing tasks enabled by recipe data sources coupled with AI techniques and cloud computing resources. Food computing tasks are arranged from top (green) to bottom (red) in order of most to least reliance on procedural knowledge

2 Text-Based Food Computing Tasks

Of the food computing tasks described above, there is a range of work relying on procedural text and NLU tasks. These, and other related food computing tasks, are described next.

Formal Notation for Recipes and Procedures: Consider a set of recipes in which the kth recipe is denoted by r_k and is given by [9]:

$$r_k = (\{s_k^i\}_{i=1}^{n_k}, \{g_k^j\}_{j=1}^{m_k}), \tag{1}$$

where $\{s_k^i\}_{i=1}^{n_k}$ represents the n_k natural language instructions/steps in the kth recipe and $\{g_k^j\}_{j=1}^{m_k}$ represents the m_k ingredients in the kth recipe. Furthermore, for multi-modal applications of recipes, each recipe is represented by (r_k, p_k), where p_k is the image associated with recipe r_k [9].

A recipe's sequence of instructions can be further expanded on using notation established in past work in how-to knowledge. A natural language instruction, or task, can be described as $(v, prep, o)$, where v is the verb/action needed to perform the task (e.g., stirring or pouring), o is a noun (e.g., an ingredient being modified or a tool used in the task), and $prep$ is the preposition linking the action to o (e.g., where the action occurs, or how) [1].

2.1 Information Extraction

Early work on how-to knowledge can be traced to the definition of frames developed in 1974 [10], which later led to the development of FrameNet [11] and VerbNet [12]. Furthermore, research on procedural knowledge understanding has been undertaken outside of the recipe context, such as in [1, 13, 14]. Although not specifically focusing on recipes, these approaches and representations can be used in recipe information extraction.

Recipe information extraction uses the list of instructions in a recipe to achieve a semantic understanding of procedures. Several rule-based approaches have been developed to extract part-of-speech (POS) tags, semantic roles, and dependencies amongst entities in the recipe [13, 15, 16]. Furthermore, this information has been used to ground text to pre-programmed actions in a system, such as with BakeBot, a food preparation robot that uses natural language instructions [17].

Neural network and machine learning approaches are also common methods for recipe information extraction [18–22]. Once a semantic understanding of recipe procedures has been established, a representation is used to share and visualize these findings. These are further described next in the task of recipe representation.

2.2 Representation

Advances in food computing have led to new recipe representations and visualization techniques. Several HTML and Schema-based representations have been developed and are widely used across the web. For instance, Schema.org defines a "HowTo" structure with tags for websites to use in HTML, to capture the different attributes in a procedure (e.g., preparation time, tools, and steps) [23]. Furthermore, RecipeSchema is a common HTML representation of recipe components (such as those appearing in Eq. 1) [24]. However, using recipe information extraction and procedural knowledge understanding, new representations to capture the semantics of individual instructions have been developed.

Several approaches [18, 25–28] have represented recipes as a graph. This captures ingredients and their state at each step, and the sequence of actions for execution. In addition, other representations have used images or videos alongside instructions [29, 30].

Machine interpretable representations have also been developed, an example being MILK (Minimal Instruction Language for the Kitchen), which builds on first-order logic to provide a function-based representation of recipe instructions [31, 32]. A sample instruction is shown in the code block below, where each recipe step is represented by a tuple $< M, T, S, M_d, T_d, C >$, where M is a set of ingredients, T is a set of tools, S is a set of strings, $M_d : M \times S$ is a relation between ingredients and their descriptions, $T_d : T \times S$ is a relation between tools and their descriptions, and $C : T \times I$ is a relation describing which tools contain which ingredients [31].

Similarly, TAAABLE is a case-based reasoning system for querying recipes, and has a similar representation [33]. Standardizing representation of procedures will assist in other food computing tasks, such as recipe recommendation, as described next.

```
create_ing(ing0, "strawberries")
create_ing(ing1, "blueberries")
combine((ing0, ing1)), ing2, "mixed fruit", "toss")
```

2.3 Recommendation and Retrieval

Recipe recommendation is used to provide users with meals they will enjoy or benefit from the most. There are three methods for recommending recipes: (1) direct recipe recommendation (i.e., providing the user a recipe with the best "fit"), (2) recipe modification (i.e., providing the user a recipe with slight modifications in ingredients or instructions to fit their needs), and (3) recipe generation (i.e., generation of a recipe from scratch). These tasks, along with recipe retrieval (i.e., the process of searching for recipes based on user queries [4]), are often dependent on the instructions in a recipe.

There are many factors that determine which recipes from a set to suggest to a user, such as user reviews and ratings [5, 34], shared ingredients and instructions [4, 5, 35–37], ingredient networks [38], cuisine [4], food science (e.g., looking at ingredients that fit well based on cooking methods) [4], and/or health and nutritional requirements [4, 5, 8, 39]. Furthermore, content-based algorithms, such as those presented in [34, 40], have been formulated. In these approaches, user ratings are available for a certain set of recipes, and user rating needs to be predicted for a target recipe for which the rating is unavailable. Let $rating(u_a, r_k)$ denote the rating given by user u_a to recipe r_k. For a target recipe r_t, the rating $rating(u_a, g_t^j)$ of an individual ingredient g_t^j is obtained by averaging the ratings given by user u_a across all recipes for which user u_a's ratings are available and that feature g_t^j. Then the predicted score for a target recipe r_t for user u_a is given by [34]:

$$pred(u_a, r_t) = \frac{\sum_{j=1}^{m_t} rating(u_a, g_t^j)}{m_t}. \quad (2)$$

This predicted score serves as the basis for recipe recommendations.

Although recommendations assist in meal preparation per nutrition and taste preferences and/or ratings, recipe modification and generation have the potential to provide a more tailored meal to users. Several approaches have modified recipes (e.g., their ingredients, instructions, etc.) or generated new instructions using natural language processing, deep learning, and other neural network-based approaches [4, 5, 41, 42].

Our focus in this paper is on aspects of recommendation and food computing tasks using procedure information. For surveys exploring other aspects of recipe recommendation (e.g., image or health based), see [4, 5, 8, 39, 43].

2.4 *Primarily Image-Based Tasks: Recognition, Perception, Prediction, and Monitoring*

Other food computing tasks that do not need procedural text include: recognition (i.e., recognizing the category a recipe fits in, or nutritional analysis), perception (i.e., how a user will react to a prepared recipe, and how it will affect their food choices), and prediction and monitoring (i.e., predicting popular recipes and patterns from the online food community) [4]. These tasks often rely on recipe images and computer vision research, while indirectly using recipe instruction data, such as in multimodal embeddings for recipe neural networks trained on text and image data [44].

3 Conclusion

Advancements in natural language understanding tasks using recipes have contributed to procedural knowledge understanding. Recipe instructions are an unstructured form of procedures, and developing a precise, semantic representation of these can yield improvements in multiple areas: machine interpretability of instructions, commonsense reasoning, question answering, and recipe generation and modification.

In this paper, we presented an overview of work in recipe-based NLU tasks that help with procedural text understanding. Several recipe representations and recipe information extraction methods have been developed using natural language processing and neural networks. These have the potential to improve other image-based food computing tasks, such as recipe recognition, perception, and prediction and monitoring. Applying a multi-task learning approach with recipe instructions for these food computing tasks can also increase accuracy, since a shared understanding of food preparation methods and ingredients can be leveraged (e.g., recommendation of recipes can improve while also performing recipe perception) [45]. Furthermore, this is likely to advance commonsense reasoning and other procedural knowledge understanding tasks. An example is question-answering with the RecipeQA dataset [21], which uses natural language instructions in the recipe to answer questions on the various ingredients and steps. Addressing these challenges in natural language understanding of food computing tasks can lead to better AI-based food applications, and improve the overall health and well-being of individuals by providing tailored food and nutritional options.

Acknowledgements This material is based upon work partly supported by the U.S. National Science Foundation under Grant No. 1936857.

References

1. Chu, C.X., Tandon, N., Weikum, G.: Distilling task knowledge from how-to communities. In: Proceedings of the 26th International Conference on World Wide Web, pp. 805–814. International World Wide Web Conferences Steering Committee (2017)
2. Hune-Brown, N.: Allrecipes reveals the enormous gap between foodie culture and what Americans actually cook (2016). https://slate.com/human-interest/2016/05/allrecipes-reveals-the-enormous-gap-between-foodie-culture-and-what-americans-actually-cook.html
3. Harper, C., Siller, M.: OpenAG: a globally distributed network of food computing. IEEE Pervasive Comput. **14**(4), 24–27 (2015)
4. Min, W., Jiang, S., Liu, L., Rui, Y., Jain, R.: A survey on food computing. ACM Comput. Surv. (CSUR) **52**(5), 92 (2019)
5. Min, W., Jiang, S., Jain, R.: Food recommendation: Framework, existing solutions and challenges (2019). ArXiv:1905.06269
6. Ofli, F., Aytar, Y., Weber, I., Al Hammouri, R., Torralba, A.: Is saki #delicious? the food perception gap on Instagram and its relation to health. In: Proceedings of the 26th International Conference on World Wide Web, pp. 509–518. International World Wide Web Conferences Steering Committee (2017)
7. Wang, X., Kumar, D., Thome, N., Cord, M., Precioso, F.: Recipe recognition with large multimodal food dataset. In: 2015 IEEE International Conference on Multimedia and Expo Workshops (ICMEW), pp. 1–6. IEEE (2015)
8. Theodoridis, T., Solachidis, V., Dimitropoulos, K., Gymnopoulos, L., Daras, P.: A survey on AI nutrition recommender systems. PETRA 540–546 (2019)
9. Marin, J., Biswas, A., Ofli, F., Hynes, N., Salvador, A., Aytar, Y., Weber, I., Torralba, A.: Recipe1M: A dataset for learning cross-modal embeddings for cooking recipes and food images (2018). ArXiv: 1810.06553
10. Minsky, M.: A framework for representing knowledge. MIT-AI laboratory memo 306. Massachusetts Institute of Technology (1974)
11. Baker, C.F., Fillmore, C.J., Lowe, J.B.: The Berkeley frameNet project. In: Proceedings of the 17th International Conference on Computational Linguistics, vol. 1, pp. 86–90. Association for Computational Linguistics (1998)
12. Schuler, K.K.: VerbNet: A broad-coverage, comprehensive verb lexicon (2005)
13. Yordanova, K.Y.: TextToHBM: A generalised approach to learning models of human behaviour for activity recognition from textual instructions. In: Workshops at the Thirty-First AAAI Conference on Artificial Intelligence (2017)
14. Wanzare, L.D., Zarcone, A., Thater, S., Pinkal, M.: A crowdsourced database of event sequence descriptions for the acquisition of high-quality script knowledge. In: Proceedings of the Tenth International Conference on Language Resources and Evaluation (LREC 2016), pp. 3494–3501 (2016)
15. Zhang, Z., Webster, P., Uren, V.S., Varga, A., Ciravegna, F.: Automatically extracting procedural knowledge from instructional texts using natural language processing. LREC **2012**, 520–527 (2012)
16. Schumacher, P., Minor, M., Schulte-Zurhausen, E.: Extracting and enriching workflows from text. In: 2013 IEEE 14th International Conference on Information Reuse and Integration (IRI), pp. 285–292. IEEE (2013)
17. Bollini, M., Tellex, S., Thompson, T., Roy, N., Rus, D.: Interpreting and executing recipes with a cooking robot. Exp. Robot. 481–495. Springer (2013)
18. Maeta, H., Sasada, T., Mori, S.: A framework for procedural text understanding (2015)

19. Hynes, N.: Representation learning of recipes. Ph.D. thesis, Massachusetts Institute of Technology (2017)
20. Korpusik, M., Huang, C., Price, M., Glass, J.: Distributional semantics for understanding spoken meal descriptions. In: 2016 IEEE International Conference on Acoustics, Speech and Signal Processing (ICASSP), pp. 6070–6074. IEEE (2016)
21. Yagcioglu, S., Erdem, A., Erdem, E., Ikizler-Cinbis, N.: RecipeQA: A challenge dataset for multimodal comprehension of cooking recipes (2018). ArXiv: 1809.00812
22. Yamakata, Y., Tajima, K., Mori, S.: A case study on start-up of dataset construction: In case of recipe named entity corpus. In: 2018 IEEE International Conference on Big Data (Big Data), pp. 3564–3567. IEEE (2018)
23. schema.org: https://schema.org/HowTo
24. schema.org: https://schema.org/Recipe
25. Ninomiya, A., Ozaki, T.: Learning distributed representation of recipe flow graphs via frequent subgraphs. In: Proceedings of the 11th Workshop on Multimedia for Cooking and Eating Activities, pp. 25–28. ACM (2019)
26. Mori, S., Maeta, H., Yamakata, Y., Sasada, T.: Flow graph corpus from recipe texts. LREC 2370–2377 (2014)
27. Chang, M., Guillain, L.V., Jung, H., Hare, V.M., Kim, J., Agrawala, M.: RecipeScape: An interactive tool for analyzing cooking instructions at scale. In: Proceedings of the 2018 CHI Conference on Human Factors in Computing Systems, p. 451. ACM (2018)
28. Chen, Y.: A statistical machine learning approach to generating graph structures from food recipes. Ph.D. thesis, Brandeis University (2017)
29. Xie, H., Yu, L., Li, Q.: A hybrid semantic item model for recipe search by example. In: 2010 IEEE International Symposium on Multimedia, pp. 254–259. IEEE (2010)
30. Chen, J.J., Ngo, C.W., Feng, F.L., Chua, T.S.: Deep understanding of cooking procedure for cross-modal recipe retrieval. In: 2018 ACM Multimedia Conference on Multimedia Conference, pp. 1020–1028. ACM (2018)
31. Tasse, D., Smith, N.A.: SOUR CREAM: Toward semantic processing of recipes. Technical Report CMU-LTI-08-005, Carnegie Mellon University, Pittsburgh (2008)
32. Jermsurawong, J., Habash, N.: Predicting the structure of cooking recipes. In: Proceedings of the 2015 Conference on Empirical Methods in Natural Language Processing, pp. 781–786 (2015)
33. TAAABLE: A case-based reasoning system which adapts cooking recipes. Revue des Sciences et Technologies de l'Information-Série RIA: Revue d'Intelligence Artificielle **31**(1–2), 207–235 (2017)
34. Freyne, J., Berkovsky, S.: Intelligent food planning: Personalized recipe recommendation. In: Proceedings of the 15th International Conference on Intelligent User Interfaces, pp. 321–324. ACM (2010)
35. Hong, J., Lee, H.: Culinary recipe recommendation based on text analytics. Int. J. Eng. Technol. (UAE) **7**(4), 5–6 (2018)
36. Vivek, M., Manju, N., Vijay, M.: Machine learning based food recipe recommendation system. In: Proceedings of International Conference on Cognition and Recognition, pp. 11–19. Springer (2018)
37. Wang, L., Li, Q., Li, N., Dong, G., Yang, Y.: Substructure similarity measurement in Chinese recipes. In: Proceedings of the 17th International Conference on World Wide Web, pp. 979–988. ACM (2008)
38. Teng, C.Y., Lin, Y.R., Adamic, L.A.: Recipe recommendation using ingredient networks. In: Proceedings of the 4th Annual ACM Web Science Conference, pp. 298–307. ACM (2012)
39. Trattner, C., Elsweiler, D.: Food recommender systems: Important contributions, challenges and future research directions (2017). ArXiv:1711.02760
40. Forbes, P., Zhu, M.: Content-boosted matrix factorization for recommender systems: Experiments with recipe recommendation. RecSys **11**, 23–27 (2011)
41. Salvador, A., Drozdzal, M., Giro-i Nieto, X., Romero, A.: Inverse cooking: recipe generation from food images. In: Proceedings of the IEEE Conference on Computer Vision and Pattern Recognition, pp. 10,453–10,462 (2019)

42. Lo, Y.W., Zhao, Q., Ting, Y.H., Chen, R.C.: Automatic generation and recommendation of recipes based on outlier analysis. In: 2015 IEEE 7th International Conference on Awareness Science and Technology (iCAST), pp. 216–221. IEEE (2015)
43. Vairale, V.S., Shukla, S.: Recommendation framework for diet and exercise based on clinical data: A systematic review. In: Data Science and Big Data Analytics, pp. 333–346. Springer (2019)
44. Salvador, A., Hynes, N., Aytar, Y., Marin, J., Ofli, F., Weber, I., Torralba, A.: Learning cross-modal embeddings for cooking recipes and food images. In: Proceedings of the IEEE Conference on Computer Vision and Pattern Recognition, pp. 3020–3028 (2017)
45. Liu, X., He, P., Chen, W., Gao, J.: Multi-task deep neural networks for natural language understanding. ArXiv:1901.11504 (2019)

Performance and Memory-Efficient Parallel Computing Framework for RSMT Construction

N. R. Latha and G. R. Prasad

Abstract Constructing fast and efficient Rectilinear Steiner Minimal Tree (RSMT) for modern VLSI circuit is a major issue in VLSI physical design (PD). Further, minimizing runtime and wire length is the most desired objective. The existing FLUTE-based VLSI routing model induces memory and I/O overhead. To overcome these research challenges, this paper presents a Memory-Efficient RSMT (MERSMT) construction that addresses the memory overhead problem that exists in larger nets. Further, the runtime of RSMT construction is reduced using High-Performance Computing (HPC) environment such as CPU and GPU. In order to get good trade-offs between memory efficiency and I/O efficiency, this paper presents a Performance and Memory Constraint Parallel Computation (PMCPC) algorithm. Experiments are conducted on ISPD 98 benchmarks and attained good results in terms of wire length, memory utilization, and processing time (i.e., runtime) when compared with standard VLSI routing design.

Keywords Graphical processing unit · High-performance computing · Multicore environment · Parallel computing framework · RSMT · VLSI

1 Introduction

RSMT is composed of a small set of connected pins through Steiner nodes with minimal cumulative edge size in Manhattan distance for a given set of pins. The construction of RSMT is a major issue in designing Very Large-Scale Integration (VLSI). In the field of integrated circuit physical design automation [1], many routing algorithms require the function of constructing *rectilinear Steiner minimal trees* (RSMTs). However, the RSMT construction is said to be a NP-complete deterministic problem [2]. For addressing, [3] and [4] presented lookup table-based fast and

N. R. Latha (✉) · G. R. Prasad
Department of CSE, B.M.S. College of Engineering, Bangalore, India
e-mail: latha.cse@bmsce.ac.in

G. R. Prasad
e-mail: prasad.cse@bmsce.ac.in

H. Das et al. (eds.), *Progress in Computing, Analytics and Networking*, Advances in Intelligent Systems and Computing 1119,
https://doi.org/10.1007/978-981-15-2414-1_37

accurate optimal solution for RSMT construction, namely, FLUTE. Here the nets are recursively broken into subset of nets. The FLUTE is evaluated for low degree nets, and it is suitable for VLSI design. FLUTE is also efficient for high degree nets with runtime complexity of $O(n \log n)$. However, for higher degree nets, the accuracy of RSMT construction is severely affected. This is due to the error induced during net breaking technique. Similarly, [5] presented fast lookup table-based RSMT construction which brings a good tradeoff between accuracy and the runtime complexity. However, both [6] and [5] did not address the memory constraint issue.

The modern VLSI circuit is composed of obstacle or blockages which requires efficient net breaking technique [7, 8]. FLUTE is the most widely used method among researchers [7, 9, 10]. In such design, minimizing wire length and reducing memory overhead are the most desired objectives. Further, [11] demonstrated that the methodologies designed using maze routing can similarly deal with huge scale OARSMT issues successfully. The paper presented a multi-pin maze routing that can be executed in parallel manner under High-Performance Computing (HPC) environment such as GPU environment. They came up with maze routing for obstacle avoidance RSMT construction [12] that is GPU-based maze routing method which achieved good speedup upon CPU. Further, they validated through experiments that they can address issues of obstacle avoidance RSMT construction using parallel computing environment using GPU-based design suitably.

In [13], it is shown that for creating effective multicore execution methods, one must comprehend top to bottom on different architectural and algorithmic viewpoints. The list of characteristics to be incorporated while developing a HPC parallel algorithm for HPC multicore architecture are establishing the accessible and prerequisite components of attaining good parallelization, mapping parallel threads of tasks to a conceivably huge collection of functional and processing unit, utilizing basic computational core with constrained functionalities, adapting to the restricted on-chip memory per individual processing core, and adapting to the constrained off-chip memory transmission capacity. Besides this, effective caching and synchronization method is a vital requirement. Although many algorithms have been proposed in order to solve the RSMT problem [14–16], most of these algorithms are sequential, rather than parallel [11]. As HPC platform has become widely available, developing HPC parallel algorithms allows us to exploit the computational power from these shared-memory HPC multicore systems [17]. To overcome the research problem, this work presents a HPC framework for parallel construction of memory-efficient RSMT (MERSMT) for VLSI physical design [18].

The research contribution is as described below:

- A parallel computing environment for performing memory-efficient RSMT construction for VLSI circuit is presented.
- The proposed method minimizes wire length and reduces processing time (i.e., runtime), along with improved memory utilization (i.e., consumes less memory).

The manuscript is articulated as described: Sect. 1 provides introduction to HPC environment for executing memory-efficient RSMT parallel construction for VLSI circuit and further highlights the research problem issues and challenges in presenting memory and I/O efficient routing design for VLSI circuit. In Sect. 2, literature survey for designing efficient parallel construction of RSMT is discussed. In Sect. 3, the proposed HPC platform for establishing memory-efficient routing model for VLSI circuit is presented. Experimental results and analysis is discussed in Sect. 4. Lastly, conclusion with future research direction of work is discussed.

2 Literature Survey

This section presents a survey of various existing methods presented to address RSMT construction and VLSI routing problems. In [7], a model called FOARS (FLUTE-based Obstacle Avoiding RSMT construction (OARSMT)) is presented. Their model segments a set of pins into numerous subsets using top-down technique. In [19], they showed that the global router generally decomposes the nets through RSMT. Therefore, reducing congestion and providing flexibility depends on RSMT construction. However, there is no improvement in wire length performance. In [20] and [21], a model to solve global routing problem is presented. [20] presented model, namely Global Routing Technology via Linear Programming (GRIP) and [21] presented a fast congestion-driven Steiner tree creation by adopting FLUTE. However, they did not exploit CPU and memory performance. To solve the congestion in global routing, [22] and [23] adopted game theory approach to improve runtime complexity of clustering approach for VLSI routing placement design (PD). [11] showed that maze routing can address OARSMT construction using heterogeneous High-Performance Computing (HPC) platform such as CPU and GPU. Similarly, in [24], they exploited heterogeneous computing and presented a parallel clustering approach for placement. Their model is exploited for both CPU and GPU. The model utilizes the CPU and GPU core to full extent. The outcome shows that it achieves a good speedup when compared to serial execution strategy. However, adopting GPU for processing induces high cost of deployment and their model did not consider the memory constraint [25, 26]; as a result, it increased I/O access time. Further, it is noted that in high-performance computing environment such as CPU and GPU the job execution is bounded by memory constraint; this is because most of the time the jobs wait for memory for execution. As a result, the existing VLSI routing method induces high computation cost under shared (cached) memory environment. Therefore, the future HPC framework must consider bringing good trade-offs among memory prerequisite of job and minimize runtime of VLSI routing.

3 High-Performance Computing Environment for VLSI Routing

This section presents a high-performance computing environment model for establishing a Memory-Efficient RSMT (MERSMT) routing model. First, this work describes how the memory-efficient RSMT construction is done. Then, it discusses about the parallel execution process of MERSMT under high-performance computing environment. Lastly, it presents a parallel execution model for executing sub-graphs/sub-trees with better resource utilization (i.e., minimizing execution time with better memory management).

A. ***Memory-efficient RSMT construction model***

This section describes the Memory-Efficient RSMT (MERSMT) construction. The MERSMT routing design is modeled to minimize memory usage, processing time, and more importantly the wire length. The model first divides the sub-trees based on memory availability similar to [5] and takes as input memory and spanning tree. To begin with, it computes one of the nodes as its root by computing the least overhead edge (i.e., utilizing memory-optimized spanning graph). Then, node closer to the root node is assumed to be the parent node. This is done considering child–parent relationship along the edges using depth-first search and divide-and-conquer approach is adopted to optimize memory for larger net size. Let us consider a graph $H(N, M)$, where M and N depict a set of ordered pairs of edges and nodes, respectively. Let $m = |E|$ depict the set of edges and $n = |V|$ represent the set of nodes. This work preliminarily builds and initializes spanning graph H by adding Steiner nodes α and is considered to be connected to all nodes in H as shown in Fig. 1. Then, to build a memory-optimized tree of graph H, divide–conquer approach is adopted as in Fig. 2. Lastly, the merging algorithm is presented as in Fig. 3.

B. *High-performance computing environment for parallel job execution*

The proposed high-performance computing parallel execution platform is designed using multicore environment that is composed of cache memory. In proposed high-performance computing parallel execution platform, the life span of sub-jobs may induce multiple modes. Preliminarily, entire sub-jobs in the tree/graph are in **locked mode** [27]. A sub-job will be **accessible/available** when its dependency limits are addressed. Post addressing its dependency limits, it enters into **steady/processing mode** for processing, provided memory constraint is reached/satisfied. If it failed to meet memory prerequisite, then it will go into **waiting mode**. The sub-job will be in waiting mode till the desired memory is accessible, and at same instance some will move into **accessible mode**. Considering this methodology aids in guaranteeing the memory requirement of sub-jobs for execution no matter the job queue size with respect to its corresponding processing core. The sub-job in the **execution mode** will preliminarily read input data and obtain enough memory for storing its output, and then executes the sub-job till the processing is complete. Post completion of execution of sub-job and writing its output, it goes into **completed mode**. Post completion

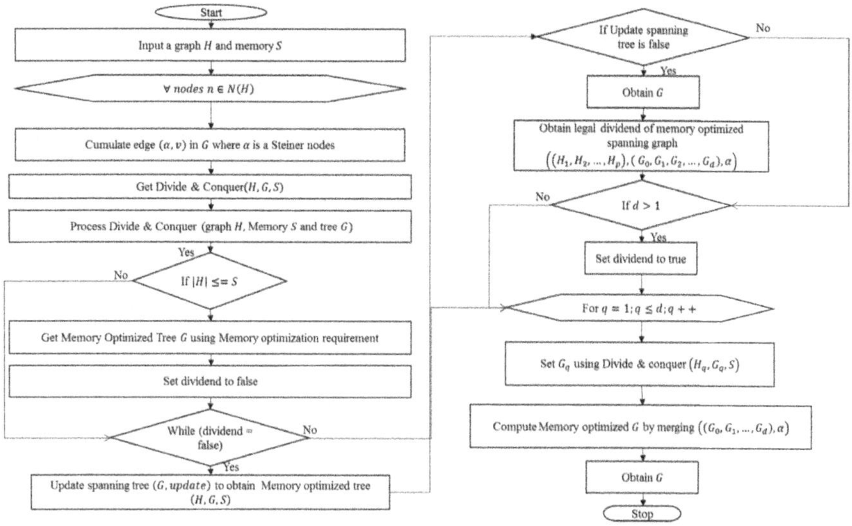

Fig. 1 Flow diagram of memory-optimized based rectilinear Steiner minimum tree construction

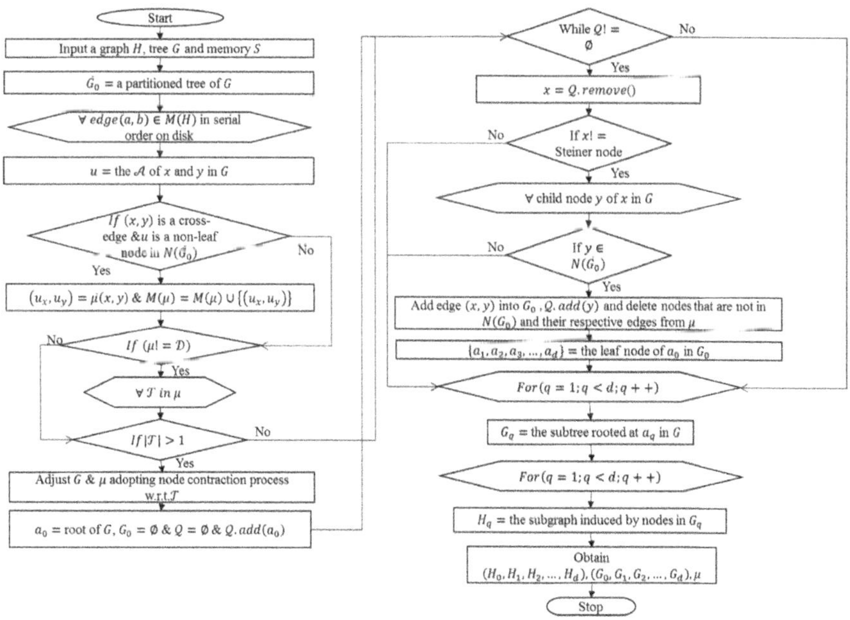

Fig. 2 Memory-optimized division algorithm

Fig. 3 Memory-optimized merging algorithm

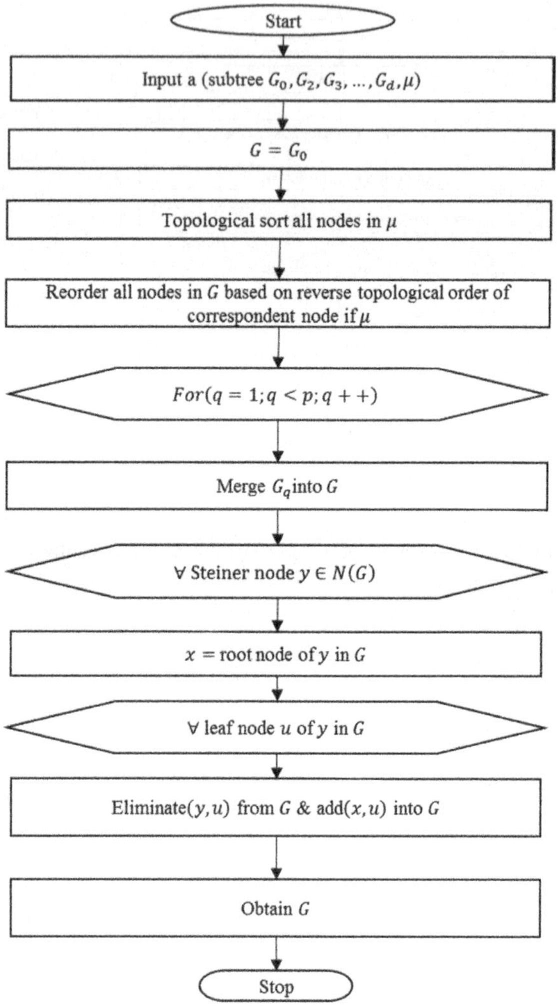

of sub-job execution, the occupied memory will be released, which can be reutilized by successive sub-jobs executions.

C. ***Performance and memory constraint parallel computing algorithm***

For any job tree to be executed under multicore environment with q cores, the makespan time of parallel execution of job tree can be bounded. Let us consider the makespan time of a job tree under single core environment be U_1 (i.e., $q = 1$) and makespan time under multicore environment be represented by U_∞ (i.e., $q = \infty$). Thus, the parallel makespan of a tree that can be bounded with respect to available processing core $q \in [1 + \infty]$ can be described as follows:

$$U_q \leq \frac{U_1}{q} + U_\infty \tag{1}$$

The above outcome is designed considering static workload size, i.e., the job size is increased as size of processing core increases. Further, the memory size is considered to be unbounded (i.e., it does not consider memory constraints). Along with using bounds of parallel jobs, the model can optimize the outcome when memory prerequisite is constraint by N. This is because the amount of processing core that can be utilized is around of $\left[i_\downarrow(N), i_\uparrow(N)\right]$. Thus, it can attain subsequent outcome corresponding to predictable parallel makespan time of the job tree. Let us consider a job tree U_n and U_∞ describing corresponding MT with sequential memory depicted as

$$N = n \tag{2}$$

and parallel MT without memory prerequisite bounding depicted as

$$N = +\infty, \tag{3}$$

then, the model can carry out bounding of predictable MT of the job tree when accessible memory is restricted by

$$N \in \left[n, N_\uparrow\right], \tag{4}$$

where $N_\uparrow$ depicts the minmax memory prerequisite of the tree computed using the following equation:

$$U_{\uparrow N} \leq \frac{lni_\uparrow(N)}{i_\uparrow(N) - i_\downarrow(N)} U_n + U_\infty \tag{5}$$

Let us assume that the amount of parallel sub-jobs is distributed in uniform nature within $\left[i_\downarrow(N), i_\uparrow(N)\right]$, for one of $q \in \left[i_\downarrow(N), i_\uparrow(N)\right]$. As a result the model posses $U_q \leq \frac{U_n}{q} + U_\infty$ as per Eq. (1). Thus,

$$\frac{\sum_{q \in \left[i_\downarrow(N), i_\uparrow(N)\right]} U_q}{i_\uparrow(N) - i_\downarrow(N)} \leq \frac{\sum_{q \in \left[i_\downarrow(N), i_\uparrow(N)\right]} \left\{\frac{1}{q} U_q\right\}}{i_\uparrow(N) - i_\downarrow(N)} + U_\infty \tag{6}$$

Since the nth frequency amount is large as the logarithm value of n, in that case, we can say Eq. (5) condition satisfies. The actual memory size utilized at an instance of execution phase depends on the order it is executed where U_n is established using maximal prerequisite of sub-job with certain contextual environment. In actual environment, it can be established that $i_\downarrow(N)$ is identical to $i_\uparrow(N)$ when N is small. Thus, $U_n \leq \frac{U_1}{i_\downarrow(N)} + U_\infty$.

For validating the proposed algorithm, our method not only addresses memory constraint but also reduces makespan time (i.e., identify shortest scheduling lengths) by utilizing both depth-first search (DFS) and breadth-first search (BFS). Using BFS aids in improving memory and parallel efficiency and DFS is contrasting to that. Further, every sub-jobs $w \in W$ has varied proportion α_w which is estimated using the following equation:

$$\alpha_w = \frac{\sum_{j \in In(w)} T_{in}(j)}{\sum_{j \in Out(w)} T_{out}(j)}. \tag{7}$$

Algorithm 1: Performance and memory constraint parallel computation (PMCPC) algorithm
Step 1. Start
Step 2. Process
Step 3. // Creation of queue
Step 4. // adding source onto queue
Step 5. Event
Step 6.while **do**
Step 7. //Remove element from queue into
Step 8. **if** **then** // can be acknowledged for processing
Step 9. //present is updated
Step 10. **else**
Step 11. //priority or selectivity is reset
Step 12. //parallelization is very restricted
Step 13. **end if**
Step 14. **if** (items in are unchanged) **then**
Step 15. **return** //not enough obtained/available
Step 16. **end if**
Step 17. **end while**
Step 18. **Event completed:**
Step 19. // freed
Step 20. For **do** //assigning out edges
Step 21. // define identifier for every edge
Step 22. For **do**
Step 23. If **then**
Step 24.
Step 25.
Step 26. **End if**
Step 27. **End for**
Step 28. **End for**
Step 29. End process
Step 30. Stop

The main aim of the proposed parallel computing framework design is to utilize the aforementioned information to give selectivity or prioritization to sub-jobs in the list that is ready to initiate, in such a manner that brings good tradeoff among memory efficiency with parallelization. The **proposed algorithm** is composed of two key events. The $PMCPC$ event is composed of a while loop which is utilized for handling $PMCPC$ event where every organized sub-job is added into queue R and sent to choose a processing core if the memory prerequisites are met. If it doesn't meet, then the organized sub-job will again enter into R by altering the selectivity parameter. The unique characteristic of proposed design is the functions $v.setq(\alpha_v)$ and $w.resetq(w, r)$, which can leverage different kinds of searching methodologies during the scheduling process. In the proposed model, the set selectivity parameter of sub-jobs nodes in R is set based on their α_w in a top-down manner (i.e.,) $\alpha_{w_1} \geq \alpha_{w_2} \geq \ldots$. Further, $degr_w$ depicts the degree of node w within $H(W, F)$, $cent_w$ depicts the centrality metric of node w in $H(W, F)$. The centrality metric describes the amount of time that node w behaves as a connector within the shortest path among nodes. Our algorithm is designed to execute job tree or graph scheduling with memory constraint that meets jobs bound prerequisite. The proposed method is hybrid in nature, i.e., it can leverage both DFS and BFS for achieving good parallelization with memory prerequisite. Post completing the sub-job execution, an event w completed happens where the memory n_w is freed and added to N and identifier is given to output edges. For every sibling of sub-job w, it is added into R with predefined selectivity provided that entire respective input edges are identified for describing that respective input data is accessible. A significant feature to be noted is that the event in **algorithm 1** runs separately and coordinates with one another by generating as well as utilizing the N memory resources through sub-jobs allotment and de-allocation.

4 Experimental Result and Analysis

This section presents the experimental evaluation of the proposed memory-efficient RSMT construction under high-performance computing environment and the outcome attained over the existing routing model. The proposed model is implemented using C ++ programing language using eclipse IDE. The GCC compiler is utilized for compiling the code. The execution platform used for analyzing the model is Centos 7.0 Operating System (OS), 2.3 GHz, Intel I-5 processor that is composed of 4 logical core with 12 GB RAM. For performance evaluation, we compare our model with existing model [5] in terms of wire length and computation time (i.e., runtime). The IBM benchmark [21] is considered for evaluation which is as shown in Table 1.

Table 1 Benchmark details of VLSI circuit

Benchmark case	Number of nets	Maximum degree	Average degree
IBM1	14111	42	3.58
IBM2	19584	134	4.15
IBM3	27401	55	3.41
IBM4	31970	46	3.31
IBM5	28446	17	4.44
IBM18	201920	66	4.06
Average	53905.33	60	3.825

Table 2 Wire Length (WL) and memory performance under HPC environment

Benchmark case	MERSMT		FLUTE [5]	
	Wire length (nm)	Memory usage (bytes)	Wire length (nm)	Memory usage (bytes)
IBM1	444307	109264	444553	398908
IBM2	527382	168457	527641	713451
IBM3	761993	180996	762276	711893
IBM4	855986	211003	856273	723607
IBM5	2809615	224661	2810816	1,188898
IBM18	2145856	1627262	2146128	6982462
Average	1257523.17	420273.83	1257947.83	1786536.5

A. ***Wire length and memory performance evaluation***

This section appraises wire length and memory performance evaluation of the proposed model and existing FLUTE model [5]. Experiments are conducted on ISPD 98 benchmark described in Table 1. The wire length and memory utilization outcome attained by proposed routing model and existing FLUTE routing model is shown Table 2. From results attained, it can be seen that an average wire length and memory usage reduction of 0.03376% and 76.4755% is achieved by proposed MERSMT over existing FLUTE routing model. Thus, the proposed model is scalable with respect to wire length in terms of wire length size minimization and memory usage reduction when compared to standard FLUTE VLSI routing model.

B. ***Runtime performance evaluation***

This section evaluates runtime performance evaluation of proposed model and the existing model [5]. Experiment is conducted on ISPD 98 benchmark described in Table 1. The runtime outcome attained by proposed routing model and existing routing model is shown in Table 3. From result attained, it can be seen that an average runtime reduction of 44.525% is achieved by proposed MERSMT over existing FLUTE routing model. Further, experiment is conducted under HPC computing

Table 3 Processing time/runtime performance under multicore environment (runtime in milliseconds)

Benchmark Case	FLUTE [5]	MERSMT	
		Single core	Multicore
IBM1	180000	90000	48000
IBM2	220000	130000	69150
IBM3	267000	163700	80100
IBM4	269000	161100	83400
IBM5	870000	410000	232000
IBM18	1880000	1090000	395000
Average	614333.3333	340800	151275

environment by varying parallel computing core size. From the results, it is seen that an average runtime reduction of 55.61% is achieved by proposed MERSMT over existing FLUTE routing model considering two processing cores, respectively. Thus, the proposed model is scalable with respect to parallel processing core in terms of runtime reduction.

5 Conclusion

This paper presented a memory-efficient RSMT construction. The MERSMT used depth-first search and divide-and-conquer approach to build a memory-efficient tree. Further, a parallel computing environment design is presented to reduce runtime of RSMT construction in parallel manner under HPC environment (i.e., both CPU and GPU). Further, this work presented Performance and Memory Constraint Parallel Computation (PMCPC) algorithms. The model allows us to exploit the computational power from these shared-memory multicore HPC systems. Experiments are conducted on ISPD 98 benchmarks for evaluating performance of proposed and existing VLSI routing method. The performance is evaluated in terms of wire length (WL), memory utilization and runtime in sequential (i.e., with single core), and parallel (i.e., using two cores). The experiment outcome shows that the proposed VLSI routing model reduces wire length size and memory utilization by 0.03376% and 76.4755%, respectively, when compared with existing VLSI routing model. Further, the proposed VLSI routing model reduces runtime by 44.525% and 55.61% when compared with existing VLSI routing model considering single and multiple cores, respectively. From overall result attained, it can be seen that proposed routing model is robust irrespective of wire length, memory, and computation performance when compared with existing VLSI routing model. However, it is important to evaluate the performance of proposed model under GPU environment which has very limited memory per core. Future work would consider performance evaluation considering higher number of cores and changing the memory size.

References

1. Chang, C.-Y., Tseng, I.-L.: A parallel algorithm for constructing obstacle-avoiding rectilinear Steiner minimal trees on multi-core systems. (2012)
2. Garey, M.R., Johnson, D.S.: The rectilinear Steiner tree problem is NP-complete. SIAM J. Appl. Math. **32**(4), 826–834 (1977)
3. Chu, C.: FLUTE: fast lookup table based wirelength estimation technique. In Proceedings IEEE/ACM International Conference on Computer-Aided Design, pp. 696–701 (2004)
4. Chu, C., Wong, Y.-C.: Fast and accurate rectilinear Steiner minimal tree algorithm for VLSI design. In: Proceedings International Symposisum on Physical Design, pp. 28–35 (2005)
5. Chu, C., Wong, Y.-C.: FLUTE: fast lookup table based rectilinear steiner minimal tree algorithm for VLSI design. IEEE Trans. Comput. Aided Des. Integr. Circ. Syst. **27**(1), 70–83 (2008)

6. Wong, Y.-C., Chris, C.: A scalable and accurate rectilinear Steiner minimal tree algorithm. In: IEEE International Symposium on VLSI Design, Automation and Test, 2008 (VLSI-DAT 2008), IEEE (2008)
7. Ajwani, G., Chu, C., Mak, W.K.: FOARS: FLUTE based obstacle-avoiding rectilinear Steiner tree construction. IEEE Trans. Comput. Aided Des. Integr. Circ. Syst. **30**(2), 194–204 (2011)
8. Huang, T., Young, E.F.Y.: An exact algorithm for the construction of rectilinear Steiner minimum trees among complex obstacles. In: Proceedings DAC (2011)
9. Hao, Z., Yea, D.-y., Guo, W.-z.: A heuristic for constructing a rectilinear Steiner tree by reusing routing resources over obstacles. **55**, 162–175 (2016)
10. Saha, P.P., Saha, S., Samanta, T.: An efficient intersection avoiding rectilinear routing technique in VLSI. In: 2013 International Conference on Advances in Computing, Communications and Informatics (ICACCI), pp. 559–562, Mysore (2013)
11. Chow, W.-K., Li, L., Young, E.F.Y., Sham, C.-W.: Obstacle-avoiding rectilinear Steiner tree construction in sequential and parallel approach. Integr. VLSI J. **47**, 105–114 (2014). https://doi.org/10.1016/j.vlsi.2013.08.001
12. Wang, R.-Y., Pai, C.-C., Wang, J.-J., Wen, H.-T., Pai, Y.-C., Chang, Y.-W., Li, J., Jiang, J.-H.: Efficient multi-layer obstacle-avoiding region-to-region rectilinear Steiner tree construction. 1–6 (2018)
13. Bader, DA., Kanade, V., Madduri, K.: SWARM: a parallel programming framework for multicore processors
14. Cai, Y., Deng, C., Zhou, Q., Yao, H., Niu, F., Sze, C.N.: Obstacle-avoiding and slew-constrained clock tree synthesis with efficient buffer insertion. In: IEEE Transactions on Very Large Scale Integration (VLSI) Systems, vol. 23, no. 1, pp. 142–155 (2015)
15. Liu, C.H., Lin, C.X., Chen, I.C., Lee, D.T., Wang, T.C.: Efficient multilayer obstacle-avoiding rectilinear steiner tree construction based on geometric reduction. IEEE Trans. Comput. Aided Des. Integr. Circ. Syst. **33**(12), 1928–1941 (2014)
16. Held, S., Spirkl, S.T.: A fast algorithm for rectilinear Steiner trees with length restrictions on obstacles. In: Proceedings of the 2014 on International symposium on physical design, pp. 37–44, ACM, (2014)
17. Panda, S.K., Panda, D.C. (2018) Developing high-performance AVM based VLSI computing systems: a study. In: Pattnaik, P., Rautaray, S., Das, H., Nayak, J. (eds.) Progress in Computing, Analytics and Networking. Advances in Intelligent Systems and Computing, vol. 710. Springer, Singapore
18. Latha, N.R., Prasad, G.R.: Wirelength and memory optimized rectilinear Steiner minimum tree routing. 1493–1497 (2017). https://doi.org/10.1109/rteict.2017.8256846
19. Ma, K., Zhou, Q., Cai, Y., Zhang, C., Qi, Z.: A Steiner tree construction method for flexibility and congestion optimization. In: 2013 International Conference on Communications, Circuits and Systems (ICCCAS), pp. 519–523, Chengdu (2013)
20. Wu, T.H., Davoodi, A., Linderoth, J.T.: GRIP: global routing via integer programming. IEEE Trans. Comput. Aided Des. Integr. Circ Syst. **30**(1), 72–84 (2011)
21. Pan, M., Xu, Y., Zhang, Y., Chu, C.: FastRoute: an efficient and high-quality global router. VLSI Des. **2012**, 608362 (2012)
22. Umair F. Siddiqi, Sadiq M. Sait, and Yoichi Shiraishi, "A Game Theory-Based Heuristic for the Two-Dimensional VLSI Global Routing Problem," Journal Of Circuits Systems And Computers, vol. 24, no. 6, 2015
23. Siddiqi, Umair F., Sait, Sadiq M.: A game theory based post-processing method to enhance VLSI global routers. IEEE Access **5**, 1328–1339 (2017)
24. Momeni, A., Mistry, P., Kaeli, D.: A parallel clustering algorithm for placement. In: Fifteenth International Symposium on Quality Electronic Design, pp. 349–356, Santa Clara, CA (2014)
25. Wang, M., Li, Z.: A spatial and temporal locality-aware adaptive cache design with network optimization for tiled many-core architectures. In: IEEE Transactions on Very Large Scale Integration (VLSI) Systems, vol. 25, no. 9, pp. 2419–2433 (2017)

26. He, S., Wang, Y., Sun, X., Xu, C.: Using minmax-memory claims to improve in-memory workflow computations in the cloud. In: IEEE Transactions on Parallel and Distributed Systems, vol. 28, no. 4, pp. 1202–1214 (2017)
27. Song, J., Li, Q., Ma, S.: Toward bounds on parallel execution times of task graphs on multicores with memory constraints. IEEE Access **7** (2019)

An Efficient Parallel Computing Framework for Over the Obstacle VLSI Routing

G. Shyamala and G. R. Prasad

Abstract Constructing a fast and efficient rectilinear Steiner minimal tree (RSMT) in the presence of obstacle/blockage is a major issue in VLSI physical design (PD). The existing method considers maze routing method for addressing obstacle-avoiding RSMT (OARSMT) problem. However, using OARSMT can induce longer wire length (WL) and routing delay. Further, in existing model, only slew constraint is considered and delay constraint are neglected. This results in wastage of routing resources (i.e. due to higher buffer size). For overcoming these problems, this work presents an efficient routing over obstacle-based RSMT construction that satisfies delay requirement. Further, we present an efficient parallel computing architecture to enhance RSMT construction (i.e. in parallel manner) under multicore environment (i.e. CPU and GPU). Experiments are conducted on small to large nets with blockages. The outcome achieved in the proposed routing method is to reduce wire length and attain better speedup when compared with sequential and existing VLSI routing model.

Keywords Graphical processing unit · Multicore environment · Parallel computing framework · RSMT · VLSI

1 Introduction

VLSI in electronic design automation is a process of placing hundreds and thousands of electronic components on one chip. The efficient algorithm for obstacle-avoiding rectilinear Steiner tree which is a part of global routing in VLSI is in demand as the number of logic circuits and the memory capacity is increasing rapidly. Placing of components in good position is essential, because this can gradually reduce the consumption of power and lower the heating of the chip, which makes the size of the

G. Shyamala (✉) · G. R. Prasad
Department of CSE, BMSCE, Bangalore, India
e-mail: shyamala.cse@bmsce.ac.in

G. R. Prasad
e-mail: prasad.cse@bmsce.ac.in

H. Das et al. (eds.), *Progress in Computing, Analytics and Networking*,
Advances in Intelligent Systems and Computing 1119,
https://doi.org/10.1007/978-981-15-2414-1_38

chips smaller and reduces the production costs. The optimization criteria that must be examined include minimizing the total wire length, total area and cost of the chip. Most algorithms focus on minimizing the wire length. The original RSMT problem assumes no obstacles in the routing region. In today's VLSI design, there can be many routing blockages like macrocells, IP blocks, and prerouted nets. Therefore, the RSMT problem with blockages, called OARSMT problem, is widely studied. In the VLSI physical design global routing phase, the suitable locations for macrocells are determined. Macrocells are the logical units that perform the desired functions of the chip. Then, the cells need to be assigned to a certain rectangular area and connected by wires which usually run rectilinearly on the chips. Hence, rectilinear Steiner trees are suited perfectly for solving this problem of routing.

In the field of integrated circuit physical design automation [1], many routing algorithms require the function of constructing *rectilinear Steiner minimal tree*s (RSMTs). However, the RSMT construction is said to have NP-complete deterministic problem [2]. Since an integrated circuit layout can contain building blocks and prerouted nets, it is desirable to consider regions that cannot be passed through by a net (or wire) as obstacles. Consequently, developing efficient algorithms for solving the *obstacle-avoiding rectilinear Steiner minimal tree* (OARSMT) problem is important and has thus received a lot of research attention in recent times [3–6]. The OARSMT [7, 8] approach can provide a possible strategy that has buffer and wires among adjacent blocks. However, it induces routing overhead and delay. [9] showed that timing constrained [10] can be addressed, provided slew constraint is met. However, existing methods are not efficient in solving global routing problem [11] as they depend on delay and wire length size. For overcoming research issues, in our previous work [12], we presented efficient routing over the obstacle-based RSMT model [13] that minimize routing overhead (i.e. reduced wire length, minimized buffer size, and minimized slew violation using Elmore model [14] rather than PERI model [15]). A comparative study of different OARSMT construction approaches is carried out in [16].

In [17] demonstrated that methodologies that are designed using maze routing can likewise deal with huge scale OARSMT issues successfully. They also presented multi-pin maze routing which can be executed in parallel manner under high-performance computing (HPC) environment such as GPU environment [18]. Despite the fact that applying GPU in different computational jobs has indicated good speedup, the utilization of GPU in EDA isn't as well known as different areas. So far it is only used in power grid analysis [19], floor planning [20] and VLSI design validation [21]. They came up with GPU-based maze routing method that can achieve good speedup upon CPU maze routing for obstacle avoidance RSMT construction. Further, they validated through experiment that they can address issues of obstacle avoidance RSMT construction using parallel computing environment using GPU-based design suitably.

For creating effective multicore execution methods, one must comprehend top to bottom of different architectural and algorithmic viewpoints. The rundown of qualities to build up a HPC methodologies for HPC multicore design incorporates the following: Establishing the accessible and prerequisite components for attaining good parallelization, mapping parallel threads of tasks to a conceivably huge collection of functional and processing unit, utilizing basic computational core with constrained functionalities, adapting to the restricted on-chip memory per individual processing core, and adapting to the constrained off-chip memory transmission capacity.

Synchronization: Implementing algorithms using multiple processing cores will require synchronization between the cores from time to time, which is an expensive operation in shared memory architectures.

Executing on multiple processing core frameworks, one must consider other features as indicated by [22]. There is a requirement to consider other multicore problems which assume significant part in high-performance computing framework. **Number of computational core**: Number of state-of-the-art environment possess two to eight processing cores incorporated on a solitary chip. Computational core generally offers features, for example, hardware multi-threading and concurrent multi-threading, which take into consideration more superior throughput and parallelization. The modern framework comes up with 100 of cores incorporated on a solitary chip such as GPU. **Caching and memory data rate**: Compared with processing power growth, the memory speed has been expanding at a much slower rate. Memory latency and data transfer capacity are essential performance parameter of most of the data-intensive jobs. Caching seriously influences the productivity of computation even on single processing core environment/frameworks. In multiple core processing environments, this will be significantly increasingly vital because of additional transmission capacity requirements. **Synchronization**: In share memory environment, executing jobs utilizing multicore would necessitate synchronization among the processing core now and again, which is a costly procedure.

Although number of methodologies are been proposed in order to address the OARSMT problem [23–26], most of these algorithms are sequential in nature, instead of parallel [17]. As inexpensive multiple processing core and virtual computational platform have become easily accessible, designing and modelling parallel methodologies aid us to exploit the computation capability from these shared memory HPC multiple processing core environment. Further, no parallel algorithm has been presented for routing over the block routing for VLSI circuit. This work presents parallel computing environment design for parallel executing construction of delay optimized over the obstacle RSMT for VLSI PD.

The research contributions are described below:

- It presented a parallel computing environment for performing delay-optimized routing over the obstacle on VLSI circuit.
- Our method minimized wire length and computation time.

The manuscript is articulated as described: Section A provides introduction of parallel computing environment for performing delay optimized over the obstacle RSMT construction and routing model for VLSI circuit, and further highlights research problem, issues and challenges in presenting efficient over the block routing for VLSI circuit under multicore environment. In section B, the proposed parallel computing environment model for establishing delay optimized over the obstacle RSMT and routing model is presented. Experiment result and analysis is discussed in Sect. 3. Lastly, the conclusion with future research direction of work is discussed.

2 Parallel Computing Environment Model for Establishing Delay Optimized Over the Obstacle Rectilinear Steiner Minimum Tree

This section presents a parallel computing environment model for establishing obstacle avoidance delay optimized rectilinear Steiner minimum tree (OADORSMT) construction. First, we describe the difference between existing and proposed routing models. Then, we discuss the parallel execution process of delay optimized over the obstacle/block RSMT under multicore environment. Lastly, we present a parallel execution model for executing sub-graphs/sub-tree with better resource utilization (i.e. minimize execution time with better memory management).

A. ***The major difference between traditional obstacle avoidance RSMT and proposed delay optimized over the obstacle RSMT model***

This section describes the major difference between existing obstacle avoidance RSMT and proposed delay optimized over the obstacle RSMT model. In Fig. 1, the existing obstacle avoidance RSMT is shown. In Fig. 2, the proposed delay optimized over the obstacle RSMT is shown. From Figs. 1 and 2, it can be seen that the existing model is composed of more delay optimizer/buffers when compared to proposed model. Thus, proposed sequential model attains better performance. The flow diagram of proposed delay-optimized routing over the obstacle under multicore environment is shown in Fig. 3. Then, a delay-optimized topology is generated to obtain influx time- and delay-resisted path at each node using flow diagram presented in Fig. 4. Post that, a delay-optimized rectilinear Steiner tree is constructed based

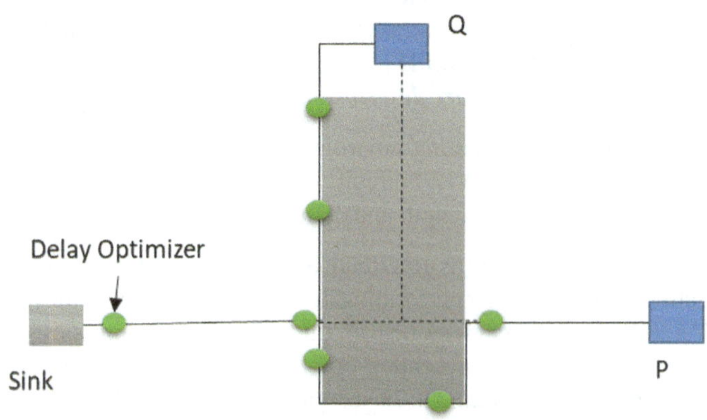

Fig. 1 Existing OARSMT construction with two sinks P and Q and one root sink

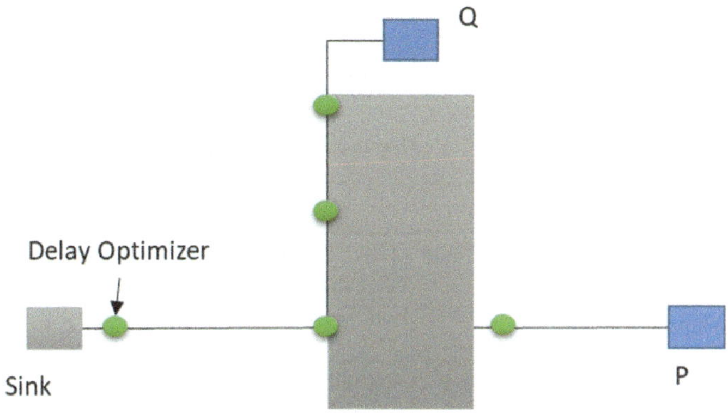

Fig. 2 Our delay optimized over the block RSMT construction with two sinks P and Q and one root sink

on the information obtained in delay optimization phase and routing over obstacle is done using flow diagram as shown in Fig. 5, respectively. More detail of notation and working process of delay-optimized routing over the block can be obtained in [12].

B. ***Multicore parallel execution model***

The proposed parallel execution platform is designed using multicore environment that composed of cache memory. The proposed model is inspired by [27], but this idea of memory bounding is applied to solve our specific problem of routing. Generally, it is difficult for designing these frameworks to potentially utilize their processing capability. This is because these frameworks possess diverse shared memory resource with multiple core; the issue of how to adjust or balance load between the processing/computing core comes when we attempt to allocate jobs to the computing core. It isn't clear how to sensibly allocate jobs to every computing core so that there won't be any load balancing issues that may affect the outcome of the framework. This work aimed at identifying technique that can adequately model the high-performance computing (HPC) multicore environment to utilize its computational capability to its full potential.

Present-day software design for parallel applications don't have knowledge or don't have capability to control maximal memory usage, making it hard for software engineers to guarantee their execution model won't come up short on memory [28, 29]. For addressing above problem, this paper presents a parallel job execution model with memory constraint problem. Let us consider a parallel job J with certain information I and a memory limit N would J be in position to finish execution within N.

- **Application partitioning**. This progression includes segmenting the task graph (TG) into progressively clear TG. Nodes with the TG are the essential processing steps in the TG. This progression is essential since we have to partition the TG

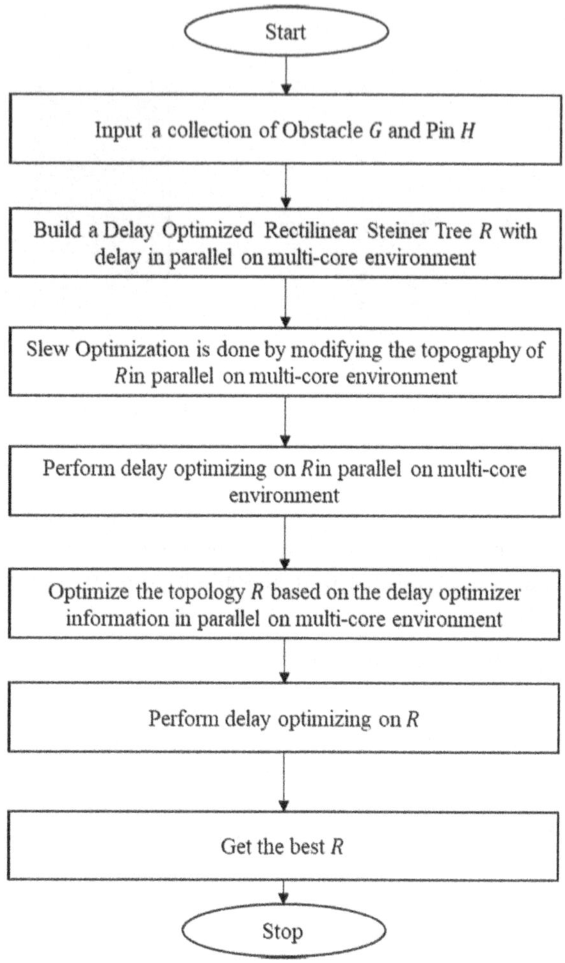

Fig. 3 Flow diagram of our delay-optimized routing over the obstacle under multicore environment

into chunks with the goal that every last one of them can be executed by one of the processing cores.

- **Task mapping**. TG is interpreted/annotated by outlining data. The interpreted TG is further mapped to the multiple processing core, job execution environment/frameworks utilizing job scheduling or mapping algorithm. This progression is very important as it decides how efficiently those computational environment can be used and how much contention happens on shared resource environment.
- **Dynamic Adaptation**. This progression includes the dynamic variation of the mapping process after certain time interim.

In proposed multicore parallel execution (MPE) platform, the lifespan of a sub-jobs may induce multiple modes. Preliminarily, entire sub-jobs in the tree/graph are in **locked mode**. A sub-job will be **accessible/available** when its dependency limits

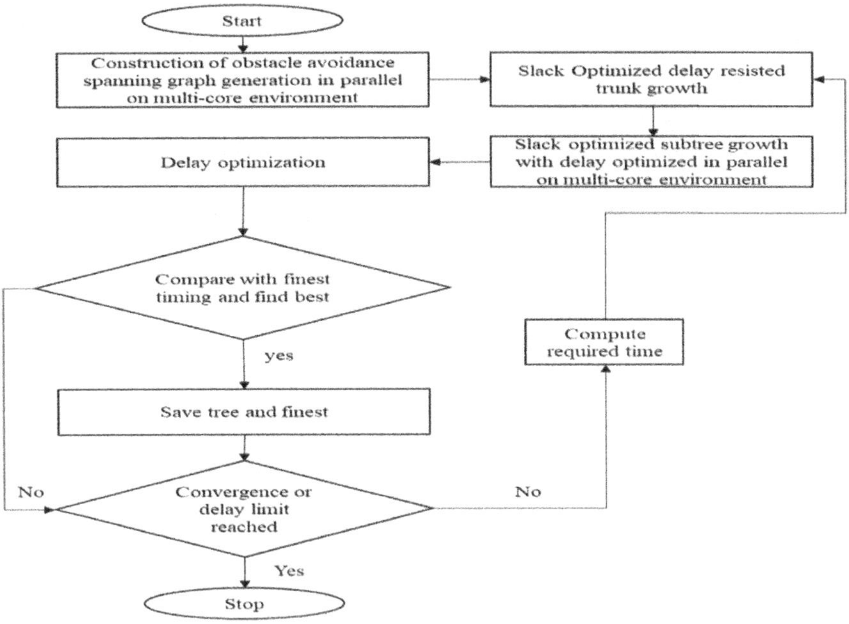

Fig. 4 Flow diagram of our delay-optimized tree creation in parallel on multicore environment

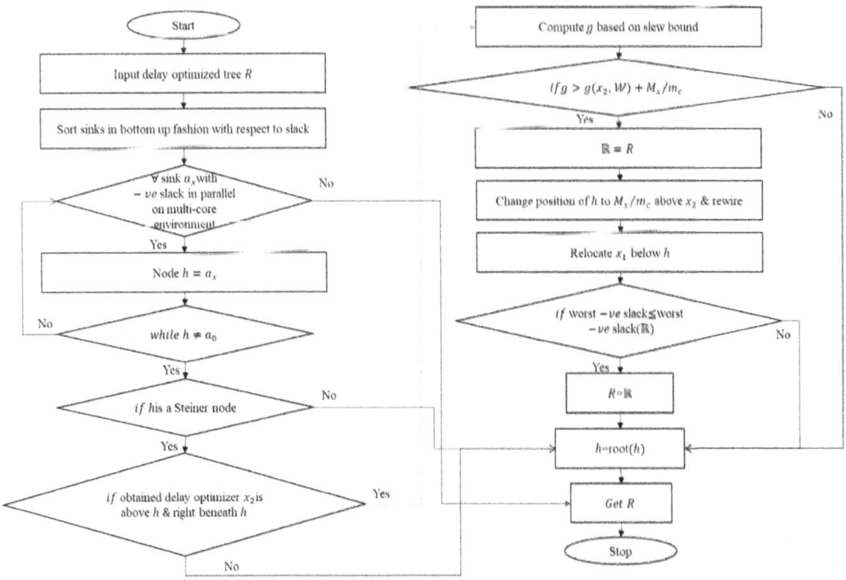

Fig. 5 Flow diagram of our path construction in parallel on multicore environment

are addressed. Post addressing its dependency limits, it enters into **steady/processing mode** for processing, provided memory constraint is reached/satisfied. If it's failed to meet memory prerequisite, then it will go into **waiting mode**. It will be in waiting mode till the desired memory is accessible, and at same instance some will move into **accessible mode**. Considering this methodology, aid in guaranteeing the memory requirement of sub-jobs for execution no matter the job queue size with respect to its corresponding processing core. The sub-job in the **execution mode** will preliminarily read input data and obtain enough memory for storing its output, and then process executing jobs till the processing is finished. Post completion of execution of sub-job and writing its output, it goes into **completed mode**. Post completion of sub-job execution, the occupied memory will be released, which can be reutilized by successive sub-job executions.

C. ***Limits modelling in Multicore parallel execution model***

This section presents a limit modelling for parallel makespan times (MT) of job trees/graphs. Here, we characterize the limit parallel job makespan times in terms of processing cores. The challenges are to identify or establish number of active cores which are bounded by memory availability during execution. Along with, to exploit maximum limit of the makespan time of job trees, however, the memory availabilities are not fixed and are dynamic in nature especially considering heterogeneous and complex characteristics of the memory prerequisite in job trees.

For overcoming the above problems, this work estimates the bounds of parallel jobs (BPJ) by estimating, for randomly designed tree and graph, the minimum and maximum limits on the peak possible amount of parallel jobs that can be processed at same instance of time within the estimated memory during the execution process. Post that, we model maximum limits on the anticipated makespan time of job trees. Lastly, a memory-bounded algorithm is modelled to enhance resource utilization/performance of job trees. Along with, cost-effective memory utilization is modelled for certain job tree by using novel metric.

D. ***Bounding of parallel jobs with enhanced resource utilization***

With the benefit of huge amount of parallel sub-jobs in a job tree, every job with varied memory prerequisite, it is practically difficult to schedule the tree to possess effective makespan time while guaranteeing efficient memory utilization outcome. Here, we analyse how bounding of maximal amount of parallel sub-jobs for scheduling methodology is optimized when memory is limited. In this work, the memory available will be assigned instantaneously, rather than reserving by the algorithm for succeeding sub-jobs. Bounding of parallel jobs aid improving resource utilization, i.e. aid in bringing good trade-offs performance among computation time and memory resources' requirement.

For meeting BPJ, we minimize and maximize limits on peak amount of parallel sub-jobs in process of processing the job tree without segmenting the memory size N. Both limits are collected by analysing every clique in the derived tree (DT) $H(W, F)$, depicted by $H_e(W, F_e)$, which is constructed by supplementing/augmenting $H(W, F)$ (i.e. an edge is supplemented among node u and w if there doesn't exist any directed path among them) and weight of node is reassigned/optimized by $Opt(w)$, $w \in W$. Let us consider that D is a clique in a well-connected component $\mathcal{D}$ of the DT, so that we can estimate the minimal for D using the following equation:

$$i_{\downarrow}(N, D) = \min_{D \in \mathcal{D}} \left\{ |T(D)| : T(D) = arg \min_{T \subseteq D\downarrow} \left\{ \left| N - \sum_{j=1}^{|T|} n_j \right| \right\} \right\} \tag{1}$$

and maximal limits for $\mathcal{D}$ using the following equation:

$$i_{\uparrow}(N, D) = \max_{D \in \mathcal{D}} \left\{ |T(D)| : T(D) = arg \min_{T \subseteq D\uparrow} \left\{ \left| N - \sum_{j=1}^{|T|} n_j \right| \right\} \right\} \tag{2}$$

where $n_j = Opt(j)$ is the memory requirement of job j in T, $T(D)$ is utilized to depict the chosen subgroup of D, whose memory prerequisite blocks/utilizes within N. However, the selection method depends on limit computation.

Using minimal and maximal limits of BPJ for every associated component, we can estimate the BPJ within memory limits N for an established tree $H_e(W, F_e)$ using the following equation:

$$BPJ = \left[i_{\downarrow}(N), i_{\uparrow}(N) \right], \tag{3}$$

where $i_{\downarrow}(N) = \min_{D \in \mathbb{E}} \left\{ i_{\downarrow}(N, D) \right\}$ and $i_{\uparrow}(N) = \min_{D \in \mathbb{E}} \left\{ i_{\uparrow}(N, D) \right\}$, in which $\mathbb{E}$ is the group of associated component of the DT.

Further, to obtain optimal computation, this work estimates the maximal limit using dynamic programming methodology, i.e. the model distributes the available memory size N between parallel sub-jobs in the graph. In precise, a allocated memory limit d, $0 < d \leq N$, we have subsequent repetition to estimate the maximal amount of parallel sub-jobs that can be processed at same instance, depicted as $\mathcal{O}[v, d]$, in process of processing the sub-graph placed at node w is estimated as follows:

$$\mathcal{O}[v, d] = \max \left\{ q_v(d), \sum_{\sum_{w \in Out(v)} n_w \leq d} \mathcal{O}[w, n_w] \right\}, \tag{4}$$

where

$$q_v(n) = \begin{cases} 0 & if\, n \leq n_v \\ 1 & Otherwise \end{cases} \tag{5}$$

is the method obtaining the amount of cores needed for node v with respect to memory size n. It shows that $\mathcal{O}[v, d]$ is accomplished by assigning the memory to sub-graphs, sub-trees or to node v, whichever offers more parallelization of sub-jobs.

From experiment analysis, it can be seen that proposed model minimizes wire length, WNS and computation time when compared with existing model, thus utilizing resources more efficiently.

3 Experimetal Result and Analysis

This section presents the experimental evaluation of proposed delay optimized over the obstacle RSMT under multicore environment outcome attained over existing routing model. The proposed model is implemented using C++ programing language using eclipse IDE. The GCC compiler is utilized for compiling the code. The execution platform used for analysing the model is Window 10 operating system, 2.3 GHz, intel I-5 processor that is composed of 4 logical core with 12 GB RAM. Further, VMware workstation 14 is installed and Centos 7.0 operating system (OS) is installed with 60 GB storage space, and memory of 4 GB is given, with 2 computing cores. For performance evaluation, we compared our model with existing model [11] in terms of wire length and computation time (i.e. runtime). Same benchmark used in [10] is used for experiment analysis. The benchmark used for experiment analysis is described in Table 1.

A. ***Wire length performance evaluation***

This section evaluates wire length performance evaluation of proposed model and existing model [10]. Experiment is conducted on benchmark described in Table 1. The wire length outcome attained by proposed routing model and existing routing model is shown in Table 2. From result attained, it can be seen that an average wire length reduction of 21.12% is achieved by proposed delay optimized over the obstacle routing model over existing routing model. Thus, proposed model is scalable with respect to wire length in terms of wire length size minimization.

B. ***Computation overhead/Runtime performance evaluation***

This section evaluates computation overhead performance evaluation of proposed model and existing model [10]. Experiment is conducted on benchmark described in Table 1. The computation overhead outcome attained by proposed routing model and existing routing model is shown in Table 2. The existing routing model [10]

Table 1 Benchmark details of VLSI circuit

Benchmark case	Number of pin	Number of obstacle
RC01	10	10
RC02	30	10
RC03	50	10
RC04	70	10
RC05	100	10
RC12	1000	10000

Table 2 Wire length (WL) and performance under multicore environment

Benchmark suite	Our approach wire length (um)	Existing approach wire length (um) [10]	Benchmark case	Proposed approach computation time with single core	Proposed approach computation time with multiple cores (2)
RC01	25632.8	29140	RC01	0.29	0.27
RC02	40040.2	42970	RC02	0.49	0.44
RC03	49162.1	62270	RC03	0.47	0.43
RC04	64424.6	73870	RC04	0.45	0.4
RC05	77424.6	87320	RC05	0.71	0.61
RC12	872277	1297170	RC12	14.33	10.07
Average	188160.22	265456.67	Average	2.79	2.036

attains an average runtime of 27.44 s for executing benchmark described in Table 1. Further, [17] presented a parallel execution of RSMT. The model attains an average runtime of 12.74 s. Further, [1] attain an average runtime of 394.36 s and 337.104 s for executing in sequential and parallel manner, respectively. The proposed model attains an average runtime of 2.79 s and 2.036 s for executing in sequential and parallel manner, respectively. The proposed model attains an average runtime performance improvement of 84.02% when compared with existing VLSI routing model [17]. The proposed model attains an average speedup of 27.025%. From overall result attained, it can be seen that the proposed model utilizes resource more efficiently when compared with existing model.

4 Conclusion

The modern VLSI design requires fast and efficient design. For overcoming research challenges, this work presented a novel routing over the blockage-based RSMT construction that reduces routing overhead (i.e. reduce buffer size) and meets delay constraint under multicore environment. Further, a parallel computing environment design is presented to improve RSMT construction in parallel manner under multicore environment (i.e. both CPU and GPU). Experiments are conducted on standard small and long nets for evaluating performance of proposed and existing VLSI routing method. The performance is evaluated in terms of wire length (WL), WNS and computation time in sequential (i.e. with single core) and parallel (multicore). The experiment outcome shows that the proposed VLSI routing model reduces wire length size by 19.63% and average runtime performance improvement of 84.02% when compared with existing VLSI routing model. Further, the proposed parallel VLSI routing model attains a computation speedup of 23.07% when compared with proposed sequential VLSI routing model. From overall result attained, it can be seen that proposed routing model is robust irrespective of wire length, a delay prerequisite constraint, with better WNS and computation performance when compared with existing VLSI routing model. Our model attained good result due to efficient memory modelling. However, some task might come with memory and delay prerequisite; how our model address brings this trade-off is a challenging considering higher number of cores, GPU with limited memory. Future work would consider performance evaluation considering higher number of cores and changing the memory size. Along with considering performance evaluation, consider varying delay prerequisite and benchmark.

References

1. Chang, C.-Y., Tseng, I.-L.: A parallel algorithm for constructing obstacle-avoiding rectilinear Steiner minimal trees on multi-core systems. (2012)
2. Garey, M.R., Johnson, D.S.: The rectilinear Steiner tree problem is NP-complete. SIAM J. Appl. Math. **32**(4), 826–834 (1977)
3. Ajwani, G., Chu, C., Mak, W.-K.: FOARS: FLUTE based obstacle-avoiding rectilinear steiner tree construction. In: Proceedings of the ISPD (2010)
4. Huang, T., Young, E.F.Y.: An exact algorithm for the construction of rectilinear Steiner minimum trees among complex obstacles. In: Proceedings of the DAC (2011)
5. Li, L., Qian, Z., Young, E.F.Y.: Generation of optimalobstacle-avoiding rectilinear Steiner minimum tree. In: Proceedings of the ICCAD (2009)
6. Li, L., Young, E.F.Y.: Obstacle-avoiding rectilinear Steiner Tree construction. In: Proceedings of the ICCAD (2008)
7. Huang, T., Li, L., Young, E.F.Y.: On the construction of optimal obstacle-avoiding rectilinear Steiner minimum trees. IEEE Trans. Comput. Aided Des. Integr. Circ. Syst. **30**(5), 718–731 (2011)
8. Liu, C.H., Kuo, S.Y., Lee, D.T., Lin, C.S., Weng, J.H., Yuan, S.Y.: Obstacle-avoiding rectilinear steiner tree construction: a steiner-point-based algorithm. IEEE Trans. Comput. Aided Des. Integr. Circ. Syst. **31**(7), 1050–1060 (2012)
9. Hu, S., Alpert, C.J., Hu, J., Karandikar, S., Li, Z., Shi, W., Sze, C.N.: Fast algorithm for slew constrained minimum cost buffering. In: Proceedings of the Design Automation Conference, pp. 308–313 (2006)
10. Lin, Y.-H., Chang, S.-H., Li, Y.-L.: Critical-trunk-based obstacle-avoiding rectilinear Steiner tree routings and buffer insertion for delay and slack optimization. IEEE Trans. Comput. Aided Des. Integr. Circ. Syst. **30**(9), 1335–1348 (2011)
11. Zhang, H., Ye, D.Y.: Key-node-based local search discrete artificial bee colony algorithm for obstacle-avoiding rectilinear Steiner tree construction. Neural Comput. Appl. **26**(4), 875–898 (2015)
12. Shyamala, G., Prasad, G.R.: Obstacle aware delay optimized rectilinear steiner minimum tree routing. In: 2017 2nd IEEE International Conference on Recent Trends in Electronics, Information and Communication Technology (RTEICT), pp. 2194–2197, Bangalore (2017)
13. Wang, R.-Y., Pai, C.-C., Wang, J.-J., Wen, H.-T., Pai, Y.-C., Chang, Y.-W., Li, J., Jiang, J.-H.: Efficient multi-layer obstacle-avoiding region-to-region rectilinear steiner tree construction. pp. 1–6 (2018)
14. Zhang, H., Ye. D.-y., Guo, W.-z.: A heuristic for constructing a rectilinear Steiner tree by reusing routing resources over obstacles. Integr. VLSI J. (2016)

15. Kashyap, C.V., Alpert, C.J., Liu, F., et al.: PERI: a technique for extending delay and slew metrics to ramp inputs. In: Proceedings of the 8th ACM/IEEE International Workshop on Timing Issues in the Specification and Synthesis of Digital Systems, ACM, pp. 57–62 (2002)
16. Shyamala, G., Prasad, GR.: A study on obstacle avoiding rectilinear steiner tree algorithms. Int. J. Innov. Eng. Technol. **4**, 120–130 (2014)
17. Chow, W.-K., Li, L., Young, E.F., Sham, C.-W.: Obstacle-avoiding rectilinear Steiner tree construction in sequential and parallel approach. Integr. VLSI J. **47**, 105–114 (2014). https://doi.org/10.1016/j.vlsi.2013.08.001
18. Panda, SK., Panda, DC.: Developing high-performance AVM based VLSI computing systems: a study. In: Pattnaik, P., Rautaray, S., Das, H., Nayak J. (eds.) Progress in computing, analytics and networking. Advances in intelligent systems and computing, vol 710. Springer, Singapore (2018)
19. Feng, Z., Li, P.: Multigrid on GPU: tackling power grid analysis on parallel SIMT platforms. In: Proceedings of ICCAD, pp. 647–654 (2008)
20. Han, Y., Roy, S., Chakraborty, K.: Optimizing simulated annealing on GPU: a case study with ic floor planning. In: Proceedings of International Symposium on Quality Electronic Design (ISQED), pp. 1–7 (2011)
21. Deng, Y.: Gpu accelerated VLSI design verification. In: Proceedings of International Conference on Computer and Information Technology (CIT), pp. 1213–1218 (2010)
22. Bader, D., Kanade, V., Madduri, K.: SWARM: a parallel programming framework for multicore processors. In: Parallel and Distributed Processing Symposium, 2007 (IPDPS 2007), pp. 1–8 (2007)
23. Cai, Y., Deng, C., Zhou, Q., Yao, H., Niu, F., Sze, C.N.: Obstacle-avoiding and slew-constrained clock tree synthesis with efficient buffer insertion. In: IEEE Transactions on Very Large Scale Integration (VLSI) Systems, vol. 23, no. 1, pp. 142–155 (2015)
24. Liu, C.H., Lin, C.X., Chen, I.C., Lee, D.T., Wang, T.C.: Efficient multilayer obstacle-avoiding rectilinear Steiner tree construction based on geometric reduction. IEEE Trans. Comput. Aided Des. Integr. Circ. Syst. **33**(12), 1928–1941 (2014)
25. Held, S., Spirkl, S.T.: A fast algorithm for rectilinear Steiner trees with length restrictions on obstacles. In: Proceedings of the 2014 on International Symposium on Physical Design. ACM, 37–44 (2014)
26. Huang, T., Young, E.F.Y.: Construction of rectilinear Steiner minimum trees with slew constraints over obstacles. In: Proceedings of the International Conference on Computer-Aided Design, ACM, pp. 144–151 (2012)
27. Song, J., Li, Q., Ma, S.: Towards bounds on parallel execution times of task graphs on multicores with memory constraints. IEEE Access **7** (2019)
28. Wang, M., Li, Z.: A spatial and temporal locality-aware adaptive cache design with network optimization for tiled many-core architectures. In: IEEE Transactions on Very Large Scale Integration (VLSI) Systems, vol. 25, no. 9, pp. 2419–2433 (2017)
29. He, S., Wang, Y., Sun, X., Xu, C.: Using minmax-memory claims to improve in-memory workflow computations in the cloud. In: IEEE Transactions on Parallel and Distributed Systems, vol. 28, no. 4, pp. 1202–1214 (2017)

Classification of Internet Traffic Data Using Ensemble Method

N. Manju and B. S. Harish

Abstract Accurate traffic classification is critical in network security and traffic engineering. Traditional methods based on port numbers and payload have proved to be ineffective in terms of dynamic port allocation and packet encapsulation. These methods also fail if the data is encrypted. In this work, we propose a ensemble method using (a) extra-tree-based feature selection, (b) support vector machines (SVMs) for classification of Internet traffic using various kernels, and finally (c) ensemble of SVM classifier using major voting technique. The method classifies the Internet traffic into broad application categories according to the network flow parameters obtained from the packet headers. We first compare three types of SVM kernels, i.e., linear, polynomial, and radial basis function (RBF) kernels. Later, we combine all the three kernels through majority voting (ensemble) method. In most of the cases, ensemble method gives better result compared with all other kernel methods.

Keywords Internet traffic · Feature selection · Ensemble · Classification

1 Introduction

In traditional method, identification of Internet applications would be through the port numbers which can be retrieved from the packet header. However, most of the advanced applications do not use the known ports. Therefore, port number-based application identification is no longer used in tradition method [1]. Further, application identification based on signature matching is commonly used approach in intrusion detection system to classify the internet traffic very accurately [2, 3]. Unfortunately, signature-based identification cannot be able to identify applications without signature and advanced version of the same applications in which the signature of the

N. Manju (✉) · B. S. Harish
Department of Information Science and Engineering, Sri Jayachamarajendra College of Engineering, Mysuru, India
e-mail: manjun007@sjce.ac.in

B. S. Harish
e-mail: bsharish@jssstuniv.in

H. Das et al. (eds.), *Progress in Computing, Analytics and Networking*, Advances in Intelligent Systems and Computing 1119,
https://doi.org/10.1007/978-981-15-2414-1_39

application has been modified. At the same time, building and maintaining database of a signature is much overpriced and takes more time. Therefore, to overcome the weakness of the port-based and signature-based approach, flow-based statistical method is used to classify the Internet traffic. To classify the Internet traffic, there are many machine learning approaches applied in the literature. Machine learning approaches present promising results that involve two phases [4].

Classification of Internet traffic has the capability to overcome network management issues for Internet Service Providers (ISPs). Traffic classification is a key part of automated system, which is capable of detecting intrusion [5, 6] patterns of denial of service (DoS), network resource allocation, and re-allocation based on user demand [7], lawful interception [8], etc.

The rest of this paper is organized as follows; Section 2 presents methodology, feature selection, and classification phases. Section 3 describes experimental results, and finally the paper is concluded in Sect. 4.

2 Proposed Methodology

2.1 Preprocessing

Standardization: Real-world data is often incomplete, inconsistent, and/or lacking in certain behaviors or trends, and is likely to contain many errors. The data present in the various columns correlate with different scales and have different units. They are scaled down to a standard range with mean value $\mu = 0$ and standard deviation $\sigma = 1$. N is the number of data points and x is the feature set. The linear transformed values are called z-scores which are computed using the equation shown below:

$$z = \frac{x - \mu}{\sigma} \tag{1}$$

where

$'\mu'$ is the mean and given by

$$\mu = \frac{1}{N}\sum_{i=1}^{N}(x_i) \tag{2}$$

$'\sigma'$ is the standard deviation and it is given by

$$\sigma = \sqrt{\frac{1}{N}\sum_{i=1}^{N}(x_i - \mu)^2} \tag{3}$$

2.2 Feature Selection—Extra Trees (Extremely Randomized Trees) Algorithm

The extra-tree algorithm constructs an ensemble of unpruned decision trees according to the fundamental top-down process. Extra trees differ from classic decision trees in the way they are built, i.e., when looking for the best split to separate the samples of a node into two groups, random splits are drawn for each of the max_features and the best split among those are chosen. When max_features is set to 1, it amounts to building a totally random decision tree. This may take many iterations to build a decision tree. Each iteration follows the following procedure:

1. Select m features randomly as a candidate set of splitting features.
2. Within each of these features F_i, with $i \in 1, \ldots, m$ draw a single random cut-point uniformly from the interval $(\min(F_i), \max(F_i))$. Evaluate the performance of the feature with the cutpoint with respect to Gini index.
3. At last, m features paired with their randomly selected cutpoints.
4. Choose splitting feature and cutpoint the pair that has the "best" performance with respect to Gini index.

Gini index is the probability of a specific variable which is wrongly classified when it is randomly chosen and it is given by

$$Gini = 1 - \sum_{i=1}^{n} (p_i)^2 \quad (4)$$

where p_i is a probability of a feature being classified to a specific class. Building decision tree would prefer choosing the feature with the least Gini index as the root node.

The major difference between random forest and extra trees is that they differ in the context of splitting of the tree. Tree splits are deterministic in case of random forests, whereas random in case of extra trees. Thus, extra tree has many advantages over the random forest like extra trees are computationally faster and give a higher performance in the presence of noisy features.

The detailed working principle of extra trees splitting algorithm [9] is presented in Algorithm 1:

Algorithm 1

Split a node(S)
Input: Local leraning subset S corresponding to the node we want to split
Output: Split[a>ac] or nothing
-if Stop split(S)=True return nothing
-otherwise, select K attributes $\{a_1,...,a_K\}$ among all constant (in S) candidate attributes
-draw K splits $\{s_1,...,s_K\}$ where s_i=Pick a random split (s,a_i), " i=1,...,K
-return a split s* such that score (s*,S)=$\max_i$ = 1,...,K score (s_i,S)

Pick random split (S,a)
Input: Subset S and attribute a
Output: Split
-Let aSmax and aSmin denote maximum and minimum value of a in S
-draw a random cutpoint ac uniformly in[aSmax, aSmin]
-return split [a<ac]

Stop split (S)
Input: Subset S
Output: Boolean
-if |S| < nmin return True
-if all attributes a are constant in S, then return True
-if the output is constant in S,then return True
-otherwise, return False

2.3 SVM Ensemble Method

To evaluate the performance of ensemble of SVM approach, we have applied a majority voting.

2.3.1 Majority Voting

Every model makes a prediction (votes) for each test instance, and the final output prediction is the one that receives more than half of the votes. If none of the predictions gets more than half of the votes, ensemble method could not make a stable prediction for the instance.

Table 1 Table shows the characteristics of datasets used for experimentation

Class no.	Class name	Standard dataset (Dataset 01) flow count	Flow distribution ratio in %	Derived subset (Dataset 02) Flow count	Flow distribution Ratio in %
1	Attack	122	00.49	1793	07.50
2	Database	238	00.95	2648	11.08
3	FTP-Control	149	00.59	3000	12.55
4	FTP-Data	1319	05.30	3000	12.55
5	FTP-PASV	43	00.17	2688	11.24
6	Mail	4146	16.67	3000	12.55
7	Multimedia	87	00.34	576	02.41
8	P2P	339	01.36	2094	08.76
9	Services	206	00.82	2099	08.78
10	WWW	18211	73.25	3000	12.55
Total number of flows		**24860**	**~100.00**	**23898**	**~100.00**

3 Experimental Setup and Results

3.1 Dataset

In the proposed experimentation, Cambridge dataset which consists of 248 features is used. The number of classes considered is 10 categories. Two variants of same dataset are used in the experimentation. One is standard dataset which is highly imbalanced and another variant is derived which is less imbalanced. These are to be called as dataset 01 and dataset 02, respectively. In dataset 01, flows are more biased toward WWW class and least biased toward multimedia class. On the other hand, dataset 02 is a normalized derived dataset (Table 1).

3.2 Results and Discussion

Extreme randomized trees use random cutpoints, and this causes the number of features selected to vary each time. The features selected are in the range of 75–82. To get a definite number of features, we used feature importance obtained by the extra trees. The feature importance are sorted in decreasing order.

It is observed from the experimentation that it reaches peak at 23 features and stays constant from 24th top features onward. Hence, we empirically decided that only 23 features are enough for effective internet traffic classification. Such as 1, 15, 22, 26, 66, 68, 72, 79, 81, 82, 83, 85, 86, 95, 97, 98, 156, 158, 165, 179, 186, 192, 245. The other features do not affect the result significantly. Since extreme random trees use random cutpoints on dataset, we obtain different feature importance each time. It is observed that only few top features with high feature importance are shuffled each time and there is no significant difference in the top 23 features. Therefore, we have selected only 23 features for the experimentation using ensemble extra trees feature selection method which gives a satisfactory result for the classification of Internet traffic.

3.2.1 Result Using Dataset 01 (Standard)

Table 2 shows the *F*-score of SVM classifier. SVM classifier is used with kernels such as linear, polynomial, and RBF on the dataset 01. Further, Table 2 shows the SVM ensemble result. Experimentation is carried out on Cambridge entry 01 standard dataset referred as dataset 01 which is highly imbalanced toward 10 categories. The result is obtained using 248 features. SVM linear kernel gives an accuracy of 99.32% compared to all other methods. Polynomial gives very less accuracy of 96.99%. The reason for degrade in the accuracy is that polynomial is not able to recognize the class 1 and class 7 effectively. Ensemble method gives approximately the same result of linear and RBF kernel. Comparison of each method in terms of accuracy is presented in Table 3.

Table 2 Result obtained using all 248 features for dataset 01 (Standard)

Class	Linear	Polynomial	RBF	Ensemble
	F-score	*F*-score	*F*-score	*F*-score
1	0.69	0.00	0.72	0.70
2	1.00	1.00	1.00	1.00
3	0.95	0.49	0.99	0.99
4	1.00	0.99	1.00	1.00
5	0.62	0.00	0.71	0.62
6	1.00	0.95	1.00	1.00
7	0.82	0.17	0.68	0.62
8	0.94	0.57	0.88	0.91
9	1.00	0.99	1.00	1.00
10	1.00	0.99	1.00	1.00

Table 3 Comparing accuracy of each method for dataset 01 using all 248 features

Classifier	Accuracy in %
SVM linear kernel	**99.32**
SVM polynomial kernel	96.99
SVM RBF kernel	99.24
SVM ensemble	99.26

Table 4 Result obtained using only 23 features for dataset 01 (Standard)

Class	Linear	Polynomial	RBF	Ensemble
	F-score	*F*-score	*F*-score	*F*-score
1	0.66	0.00	0.72	0.70
2	1.00	1.00	1.00	1.00
3	0.93	0.49	0.99	0.96
4	1.00	0.99	1.00	1.00
5	0.76	0.00	0.75	0.67
6	0.99	0.95	1.00	1.00
7	0.72	0.17	0.69	0.73
8	0.92	0.57	0.89	0.94
9	0.99	0.99	1.00	1.00
10	1.00	0.99	1.00	1.00

Table 5 Comparing accuracy of each method for dataset 01 using only 23 features

Classifier	Accuracy in %
SVM linear kernel	99.15
SVM polynomial kernel	98.09
SVM RBF kernel	99.23
SVM ensemble	**99.31**

Table 4 depicts the *F*-score of SVM classifier. SVM classifier is used with kernels such as linear, polynomial, and RBF on the dataset 01. Further, it presents the result of SVM ensemble method. Experimentation is carried out on dataset 01. The result using 23 features obtained from ensemble extra trees method in SVM ensemble classifier gives almost same accuracy of 99.31 as that of linear SVM (with 248 features) as given in Table 2. With this, it is observed that SVM ensemble gives better result with less number of features. However, ensemble gives comparable result with linear and RBF kernel. Comparison of each method in terms of accuracy is given in Table 5.

3.2.2 Result Using Dataset 02 (Derived)

Tables 6 and 8 show the *F*-Score for different kernels (Linear, polynomial, and RBF) with SVM classifier on the dataset 02 (derived dataset) and it is normalized. The flows contained by each class are almost normalized to the closest ratio of the available total flows. The result shows that RBF kernel gives better accuracy of 96.12 and 96.37% without and with using feature selection method, respectively. In ensemble method, result slightly reduced compared with RBF. The comparative analysis in terms of accuracy is presented in Tables 7 and 9. Hence, RBF kernel suits to the dataset which has less class imbalance problem.

Table 6 Result obtained using all 248 features for dataset 02 (Derived)

Class	Linear	Polynomial	RBF	Ensemble
	F-score	*F*-score	*F*-score	*F*-score
1	0.80	0.79	0.82	0.79
2	1.00	0.99	1.00	1.00
3	0.99	0.98	0.99	0.99
4	0.99	0.99	1.00	0.99
5	0.99	0.99	1.00	0.99
6	0.98	0.94	0.98	0.99
7	0.92	0.82	0.87	0.87
8	0.92	0.81	0.93	0.92
9	0.99	0.96	0.98	0.99
10	0.91	0.86	0.91	0.91

Table 7 Comparing accuracy of each method for dataset 02 using all 248 features

Classifier	Accuracy in %
SVM linear kernel	95.78
SVM polynomial kernel	92.83
SVM RBF kernel	**96.12**
SVM ensemble	95.69

Table 8 Result obtained using only 23 features for dataset 02 (Derived)

Class	Linear	Polynomial	RBF	Ensemble
	F-score	*F*-score	*F*-score	*F*-score
1	0.78	0.82	0.82	0.82
2	0.99	0.99	1.00	1.00
3	0.99	0.98	0.99	0.99
4	0.99	0.99	1.00	0.99
5	0.99	0.99	1.00	1.00
6	0.98	0.97	0.99	0.98
7	0.91	0.83	0.87	0.92
8	0.92	0.90	0.93	0.94
9	0.99	0.98	0.98	0.99
10	0.90	0.91	0.91	0.91

Table 9 Comparing accuracy of each method for dataset 02 using only 23 features

Classifier	Accuracy in %
SVM linear kernel	95.50
SVM polynomial kernel	95.52
SVM RBF kernel	**96.37**
SVM ensemble	96.29

4 Conclusion

Internet traffic classification plays a major role in network monitoring and management. To classify the Internet traffic, Cambridge dataset is used for the experimentation. First dataset is highly imbalanced (standard) and other is less imbalanced (derived). In the proposed work, following are the contributions: (a) proposed an ensemble feature selection using extra trees method, (b) explored to use an SVM classifier to classify the Internet traffic using various kernels (Linear, polynomial, and RBF), and lastly (c) an ensemble of an SVM classifier using the abovementioned kernels and applied majority voting. The result shows that polynomial kernel using SVM gives better result using 248 features on dataset 01 which is highly imbalanced. SVM ensemble gives slight improvement using 23 features for the same dataset. On the other hand, SVM using RBF kernel achieves good result on dataset 02 which is derived from the Cambridge dataset. Further, it also achieved comparable result with respect to RBF kernel by using ensemble method.

References

1. Zander, S., Nguyen, T., Armitage, G.: Automated traffic classification and application identification using machine learning. In: The IEEE Conference on Local Computer Networks 30th Anniversary (LCN'05) vol. l, pp. 250–257. IEEE (2005)
2. Vigna, G., Robertson, W., Balzarotti, D.: Testing network-based intrusion detection signatures using mutant exploits. In: Proceedings of the 11th ACM conference on Computer and communications security, pp. 21–30 (2004)
3. Shon, T., Moon, J.: A hybrid machine learning approach to network anomaly detection. Inf. Sci. **177**(18), 3799–3821 (2007)
4. Nguyen, TTT., Armitage, G., Branch, O., Zander, S.: Timely and continuous machine-learning-based classification for interactive IP traffic. IEEE/ACM Trans. Netw. (TON) **20**(6), 1880–1894 (2012)
5. Roesch, M.: Snort: lightweight intrusion detection for networks. Lisa **99**(1), 229–238 (1999)
6. Paxson, V.: Bro: a system for detecting network intruders in real-time. Comput. Netw. **31**(23–24), 2435–2463 (1999)
7. Stewart, L., Armitage, G., Branch, P., Zander, S.: An architecture for automated network control of QoS over consumer broadband links (2005)
8. Baker, F., Foster, B., Sharp, C.:Cisco architecture for lawful intercept in IP networks, (No. RFC 3924) (2004)
9. Geurts, P., Ernst, D., Wehenkel, L.: Extremely randomized trees. Mach. Learn. **63**(1), 3–42 (2006) (Springer)

Conclusion

Internet traffic classification plays a major role in network monitoring and management. To classify the Internet traffic, Cambridge dataset is used for experimentation. This dataset is highly imbalanced (standard) and other [illegible] less imbalanced. Based on the proposed work, following are the conclusions: (a) proposed an ensemble feature selection using Extra trees method, (b) explored to use an SVM classifier to classify the internet traffic using various kernels (Linear, polynomial and RBF kernels), (c) an ensemble of [illegible] kernels and applied genetic [illegible]. The results shows [illegible] RBF kernels results [illegible]

References

Toward Ameliorating K-Means Clustering Algorithm

D. Sirisha and S. Sambhu Prasad

Abstract Mining knowledge and predicting the behavior of the data have become major challenge with the advent of unprecedented escalation in the volume of the existing databases. Generally, clustering is adopted for voluminous and intricate data. In the present work, two techniques of K-means clustering, namely, K-means algorithm with random sampling (without realignment) and K-means algorithm with realignment sampling, are compared in terms of time taken and number of moves made for clustering the given data. The first one checks for any transfers between the clusters after inserting all the data. The second one is to check for any transfers between clusters for each new data inserted into cluster. The experimental results reveal that K-means clustering algorithm with realignment has performed reasonably well against K-means clustering algorithm without realignment.

Keywords Data mining · Clustering · K-means clustering algorithm

1 Introduction

The expeditious progress in the size and number of existing databases poses crucial challenges in mining knowledge from data vital in reinforcing decision-making besides predicting the behavior of the data [1]. Generally, clustering is employed keeping in view of the volume and the complexity of the data set. Clustering is often the recourse when the data set is large and complex entailing many variables and internal structures [2].

In the present work, two techniques for clustering data through K-means clustering algorithms are implemented. One is implemented by calculating the centroid of each cluster after inserting all records. Other one is implemented by realignment when a new data is inserted or transferred from one cluster to another cluster. Clustering

D. Sirisha (✉) · S. S. Prasad
Pragati Engineering College, Surampalem, India
e-mail: Sirishad998@gmail.com

S. S. Prasad
e-mail: dessp@gmail.com

H. Das et al. (eds.), *Progress in Computing, Analytics and Networking*,
Advances in Intelligent Systems and Computing 1119,
https://doi.org/10.1007/978-981-15-2414-1_40

depends on centroids selected randomly; the number of moves and time taken may get increased or decreased. If the medians are chosen as centroids, the number of moves may be reduced, but for huge data the advantage of reducing time complexity will become an overhead for sorting the data. Hence, in the current work, implementing clustering by randomly selecting the centroids is preferred. The performance of each technique is analyzed and compared with the other and conclude which one is preferable in comparison to other. The purpose of the present study is to decide which clustering algorithm is suitable for the given data in terms of time and moves.

2 The Clustering Problem

Clustering is often defined as grouping large data sets into clusters of smaller sets that exhibit similarity in some way. A good clustering technique ensures high intra-class similarity and low inter-class similarity. Clustering is considered as the most significant unsupervised learning problem, as it determines the underlying structure of unlabeled data. The quality of a cluster is usually dependent on the similarity measure employed [3]. Moreover, efficiency of a clustering algorithm can be gauged by its capability in discovering most of the hidden patterns [2].

Clustering of the data can be effectively realized by determining the distance between any pair of data items to measure their similarity. It is assumed that the input comprises a set of data items with an associated metric or distance. The distance is denoted as d, and the distance between any two data items x and y is indicated as $d(x,y)$. To identify the inherent groups in the data, a clustering algorithm attempts to determine the centroid of a group of data items. To decide the members of a cluster, mostly the clustering algorithms evaluate the distance between a data item and the cluster centroids which gives the cluster centroids and the numeral of data items in a cluster as the output [4]. The primary goal of clustering is to decide the inherent groups existing in the unlabeled data in a set [5]. However, the "best" criterion for determining the cluster is the most challenging problem [6].

Generally, K-center and K-median techniques are employed to attain desired clustering. A K-center technique gives a set of k data items $C = \{C_1, C_2, \ldots., C_k\}$ as output that indicates the center of each cluster for defining the cluster. Each data item nearest to C is placed in C and ties are resolved randomly. The cluster members can be determined by the maximum distance of a data item to its C, i.e., $max\{min\ d(x,C_i)\}$, while K-median clustering algorithm computes median for a set of k data items, describing the cluster. The quality of the cluster can be gauged by the average distance of data items to their nearest median [7].

$$\frac{1}{n}\sum_{x} \min x\, d(xk, Ck) \tag{1}$$

In K-center and K-median clustering methods, the data item assigned to each cluster is similar but only differs in the objective function. A chief constituent of any clustering algorithm is the distance measure between data items [4]. Primarily Euclidean distance metric is enough to effectively group similar data provided data is presented in the same physical units. Euclidean distance between any two data items, for instance, $x = [x_1, x_2 \ldots, x_k]$ and $y = [y_1, y_2 \ldots, y_k]$ can be determined by employing one of the following distance measures: x_i -y_i

i. Common distance:

$$\sqrt{\sum_{i=1}^{k}(xi - yi)^2} \tag{2}$$

ii. Manhattan distance:

$$\sum_{i=1}^{k}|xi - yi| \tag{3}$$

iii. Max of dimensions:

$$\max_{i=1}|xi - yi| \tag{4}$$

iv. Minkowski Metric: In situations when there is no Euclidean space, clustering becomes difficult. Minkowski metric is often preferred when a distance measure without a Euclidean space. When the dimensions of the data are high, Minkowski metric is commonly employed.

$$d_p(x_i - y_i) = \left(x\sum_{k=1}^{d}|xi, k - yj, k|^2\right)^2 \tag{5}$$

where d is the data dimensions. For Euclidean distance p is 2 and for Manhattan distance p is 1.

3 K-Means Clustering

K-means clustering technique is simple and much widely used. This technique classifies the data items based on attributes into K groups, where K is positive non-negative integer. The data is grouped by minimizing the sum of squares of distances between data item and the associated cluster centroid. Centroid of a cluster is a point that has the parametric values comprising the average parameter values of each of the data items included in the present cluster. K-means is an iterative distance-based clustering method [8]. The data can be moved among the clusters till the desired cluster is attained. K-means algorithm requires the number of clusters K to be specified in advance and, moreover, requires the definition of a cluster mean, for instance, cluster $K = \{t_1, t_2, \ldots, t_m\}$ where each tuple has only one numerical value; this cluster mean can be the centroid:

$$\frac{\sum_{i=1}^{m} ti}{m} \tag{6}$$

Usually, K-means generates good clusters, but it is computationally expensive as it requires computing the distance from a data item to its corresponding cluster centroid in each iteration. The time complexity is $O(tkn)$, where t is the number of iterations, k is the number of clusters, and n is the number of data items in each cluster. The objective of K-means technique is to minimize intra-cluster dissimilarity and maximize similarity among the data items in a cluster.

3.1 Pseudocode of the K-Means Algorithm

Input: $D = \{d_1, d_2, \ldots d_n\}$ // set of data items
K // number of chosen clusters
Output: SK // set of clusters

1. Initialize the means $m_1, m_2, \ldots m_k$ // randomly or take the first k input elements or repeat.
2. Allocate every data item to the cluster with nearest mean.
3. Compute new mean for every cluster.
4. Repeat until no or very smaller number of data items are allotted to clusters or a fixed number of iterations are carried out.
   i. Compute the cluster centroid.
   ii. Find the distance of every data item to each of the centroids.
   iii. Cluster the data items with minimal distance.

3.2 K-Means Algorithm Without Realignment

Input: $D = \{d_1, d_2, \ldots d_n\}$ // set of data items
K // number of chosen clusters
Output: SK // set of clusters

1. Choose the number of clusters K.

2. Initially, partition the data items into K clusters. The data items may be allocated either arbitrarily or as follows:
   i. Assign the first K data items as each cluster.
   ii. Allocate residual (n−K) data items to the cluster with nearest centroid.
   iii. Update the centroid of each cluster.
3. Compute the distance for every data item to the centroid of each cluster. If the data item is not present in the cluster with the nearest centroid, move the data item to that cluster and update its centroid.
4. Repeat step 3 until no data item is moved from one cluster to another.

In situations when the data items are lesser than the number of clusters, then each data item can be considered as the cluster centroid which is assigned with a cluster number. Otherwise, for each data item, compute the distance to each of the centroids and get the minimum distance. The data belongs to the nearest cluster with minimal distance. As the location of the centroid is not definite, centroid must be adjusted once the data is updated. This process repeats till none of the data items is moved to another cluster.

In this algorithm, any transfers between the clusters after inserting all the records are checked. Calculate the minimum distance for each element from the centroids. Depending upon this value the element will be placed in the respective cluster. After placing the element, the centroid is recalculated. The process continues in the similar way until no more transfers are found.

3.3 K-Means Algorithm with Realignment

1. Choose the number of clusters K.
2. Initially, partition the data items into K clusters. The data items may be assigned either randomly or systematically as follows:
   i. Assign the first K data items as each cluster.
   ii. For every (n−K) data items,
      (a) Allocate the data item to the cluster with the nearest centroid and update its centroid.
      (b) Compute the distance for every data item to the centroid of each cluster. If the data item is not present in the cluster with the nearest centroid, move the data item to nearest cluster and update its centroid.
3. Repeat step 2 till none of the data items are moved from one cluster to another.

The difference between K-means without realignment and K-means with realignment is that the first one checks for any transfers between the clusters after inserting all the records. The second one checks for any transfers between clusters for each new data item inserted into cluster.

4 Results and Discussions

To study the effect of realignment on the performance of K-means clustering algorithm, experimentations are conducted and compared with K-means clustering algorithm without realignment. In the present work, two techniques for clustering, namely, K-means algorithm with random sampling (without realignment) and K-means algorithm with realignment sampling, are implemented and are compared in terms of time taken and number of moves made for clustering the given data. The two methods are evaluated by comparing the statistics of the two algorithms and conclude which algorithm performs reasonably well. The following inputs are considered:

- A file containing records,
- Number of clusters,
- The record numbers to be taken as centroids, and
- The names of the output file series.

From the output obtained, we draw statistics for both the techniques and compare their performance. Finally, a series of output files are generated which consists of the individual clusters as each individual output file. In this way, a series of output files are obtained for both the techniques. Using these outputs, the statistics are drawn and hence performances of both the techniques are compared.

When the number of records is 100, K-means without realignment required 162 moves, while with realignment required 31 for clustering. When the number of records is increased to 500, K-means without realignment and with realignment techniques required 540 and 326 moves, respectively. For 1000, 2000, 5000, and 10000 records, the number of moves required for K-means without realignment and with realignment are {1143, 2698, 5165, 10127} and {814, 1987, 3324, 7498}, respectively. It manifests from Table 1 that K-means with realignment technique could generate clusters with lesser number of moves than K-means with realignment by 80.86%, 39.63%, 28.784%, 26.35%, 35.64%, and 25.96% for 100, 500, 1000, 2000, 5000, and 10000 number of records, respectively. Overall, K-means

Table 1 A comparative analysis of K-means without realignment and with realignment

Number of records	Without realignment		With realignment	
	Time (mm:ss:ms)	Moves	Time (mm:ss:ms)	Moves
100	00:00:125	162	00:03:844	27
500	00:02:687	637	00:55:797	148
1000	00:25:00	1418	02:19:172	430
2000	05:01:765	2698	09:28:532	1987
5000	06:01:250	5165	4:41:36	3324
10000	08:05:180	10127	18:51:22	7498

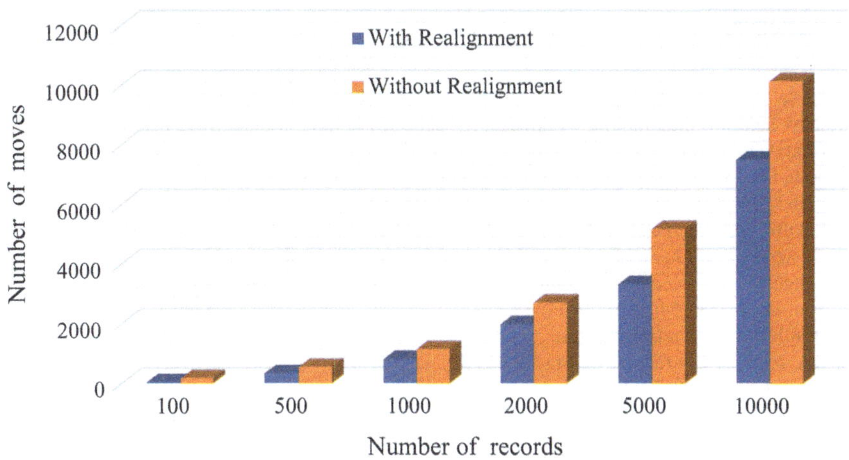

Fig. 1 A comparison of K-means clustering with alignment and without alignment

with realignment generated clusters with 29.51% lesser number of moves than K-means without realignment. Moreover, it is also evident from Table 1 that K-means without realignment has taken much lesser time than K-means with realignment (Fig. 1).

5 Conclusions

Cluster analysis is considered as the most challenging techniques as it deals with vast and multifarious data sets, characterized with varied attributes. Hence, scalability and capability to deal with diversified attributes are the major aspects for a clustering algorithm to be expedient. Moreover, the difference between two algorithms in clustering the records is manifested. Depending on the initial centroids, resulting clusters will get changed. To study the effect of realignment on the performance of the K-means clustering algorithm, experimentations are conducted on the two algorithms on a varied sized record set with eight similar attribute types. From the experimental results, it can be observed that K-means algorithm with realignment performed reasonably well compared to K-means algorithm without realignment. Overall, K-means with realignment generated clusters with 29.51% lesser number of moves than K-means without realignment.

References

1. Narayanan, R., Ozisikyilmaz, B., Zambreno, J., Memik, G., Choudhary, A.: MineBench: a benchmark suite for data mining workloads. In: 2006 IEEE International Symposium on Workload Characterization, pp. 182–188. San Jose, CA (2006)
2. Anh, DT., Thanh, LH.: An efficient implementation of k-means clustering for time series data with DTW distance. Int. J. Bus. Intell. Data Min. **10**(3), 213–232 (2015)
3. Sakthi, M., Thanamani, AS.: An effective determination of initial centroids in K-Means clustering using kernel PCA. Int. J. Comput. Sci. Inf. Technol. **2**(3), 955–959 (2011)
4. Sangalli, L.M., Secchi, P., Vantini, S., Vitelli, L.: K-mean alignment for curve clustering. Comput. Stat. Data Anal. **54**(5), 1219–1233 (2010)
5. Bradley, P., Fayyad, U.: Refining initial data items for k-means clustering. In: Proceedings 15th International Conference on Machine Learning (1998)
6. Das, H., Naik, B., Behera, H. S.: Classification of diabetes mellitus disease (DMD): a data mining (DM) approach. Prog. Comput. Anal. Netw. 539–549 (2018) (Springer, Singapore)
7. Zhang, R., Rudnicky, A.: A large scale clustering scheme for K-means. In: 16th International Conference on Pattern Recognition (ICPR'02) (2002)
8. Nepolean, G., Ganga Lakshmi, G.: An efficient K-Means clustering algorithm for reducing time complexity using uniform distribution data points. Trends Inf. Sci. Comput. (2010)

Virtual Reality-Based Driving Simulator for Testing Innovative Hybrid Automotive Powertrains

Arockia Vijay Joseph and Sridhar P. Arjunan

Abstract In the field of automotive engineering, the testing procedure of any powertrain involves a chassis dynamometer and techniques to test the performance of the powertrain. These traditional testing procedures lack in understanding the powertrain performances for different terrains. To bridge this, OEM used to test their new powertrain by on-road testing. This conventional method for testing requires more amount of time to test, validate, and re-design, and to complete the product development cycle. This paper provides a solution to the abovementioned problem by designing a virtual reality-based driving simulator (VRDS) which can wirelessly control a drive train system of a vehicle. This study has designed and enhanced a virtual and imaginary simulated reality of various tracks where the designed car can run in the required setting of the environment. The interaction between the car and the test bed is in terms of acceleration, clutch, gear position, and brake. In this study, we have tested and analyzed the performance of the hybrid engine using this real-time wireless VRDS-generated data.

Keywords Virtual reality · Hybrid · Powertrain · Test bed · Driving simulator · Unity

1 Introduction

There are various driving simulator systems [1–6] developed but very few are available in public or are open-source which make them very rigid and do not allow the user to change parameters according to the requirements. Also, there are various new ways being developed to test the automotive powertrain. The experimental research carried out by Zeng et al. [1] used three-terminal dynamometers to simulate the drive train in a test bed based on Europe NEDC condition. Through three terminals, speed and torque are varied based on the modified model, to select the operative point of

A. V. Joseph (✉) · S. P. Arjunan
Department of Electronics and Instrumentation Engineering, SRM Institute of Science and Technology, Kattankulathur 603203, Tamilnadu, India
e-mail: arockiaj1@srmist.edu.in

H. Das et al. (eds.), *Progress in Computing, Analytics and Networking*, Advances in Intelligent Systems and Computing 1119,
https://doi.org/10.1007/978-981-15-2414-1_41

the differential coupled device [1]. The study reported by Passenbrunner et al. [3] used a hydrodynamic dynamometer where a flexible shaft is used to connect the IC engine with the dynamometer. This test bench has engine speed, the dynamometer speed, and the dynamometer torque as output variables for varying inputs like the accelerator pedal position, the inlet valve position, and the outlet valve position of the hydrodynamic dynamometer [3].

Geng and Schulte [4] developed a real-time model of the powertrain which was modeled efficiently without any interface with hardware. The efficiency of the modeled system is obtained through a software simulation based on SimDriveLine [4]. Ben-gai et al. [6] used a system in such a way that the electric dynamometer to cooperate with the motor and the battery pack. The electric dynamometer uses various blocks of control modules which can simulate a real engine [6]. According to Koç et al. [5], the engine test can be performed in two different methods, namely, manual control mode and full load test mode. In the full load test mode, a PC-controlled dynamometer and a proportional controller to enable engine testing were used. Galiullin [7] developed an automated test system (ATS) for IC engines to reduce the cost and time involved in the production of the engines. The developed ATS uses fuzzy neural network-based control for the automation of the test bed [7].

Xu et al. [8] used virtual instrument technology to develop and simulate the hybrid powertrain. In this instrument, an electric dynamometer was used to simulate the work of engine and motor and control unit of the powertrain. This system works in two modes namely driving mode and load mode and operates for three different test conditions like starting, acceleration, and deceleration [8]. The selected powertrain is validated through a dynamic model using Psim and Matlab/Simulink. This study developed and implemented a fuzzy-PI double-mode controller for the powertrain on a dSPACE-based hardware-in-the-loop (HIL) test bed for real-time control [9]. Akkaya et al. [10] proposed a new powertrain system test (PST) to test and validate the interaction of all subsystems of powertrain which also involved overall quality and maturity test for the considered powertrain. This system did not test the mechanical durability, and involves four different system levels like vehicle level, powertrain level, subsystem level, and component level for testing [10].

Kouroussis et al. [11] modeled mathematically the powertrain along with gearbox, transmission, tire, and road loads and simulated the powertrain to learn its dynamics. The obtained output is validated through the measured real output of the considered real systems like Chrysler 45RFE and Aisin Warner 50–55 SN [11]. Mall et al. [12] used a 1D multibody system model of powertrain to test the dynamics for variation in acceleration and engine starts. Also, to optimize the vibration dampers of the powertrain augmented Lagrangian particle swarm optimization procedure is used. During this process of optimization, the impact of various control parameters on the mentioned process is observed [12]. Wu et al. [13] used dual-motor configured powertrain for the electric vehicles and simulated them to compare its result with the single motor. Wu et al. proved that dual-motor configuration has better efficiency when compared to single-motor configuration [13].

The main objective is to understand the method of switching between the IC engine and the electric drive varies both in mechanical and software designs. Joseph et al. [14] developed a two-wheeler hybrid powertrain where the two power sources are coupled through a chain drive mechanism [14]. However, the mentioned chain drive mechanism has a disadvantage of taking the other power source as a load, while the first power source is in operating condition and this impacts negatively on the mileage of the powertrain. Hence, this paper proposes a hybrid powertrain system with an electromagnetic switching between the power sources and an in-house developed virtual environment for the driving simulator to test the hybrid powertrain for efficiency.

2 Materials and Methods

2.1 Design of the Driving Simulator

The open-source software used to develop the simulator is Unity3D. Figure 1 shows the setup of our developed VRDS. The driving simulator consists of the following parts:

- Wheel controller,
- ABC (Accelerator, Brake, and Clutch) pedals,
- Gear shifter,
- Display monitors, and
- XBee transmitter.

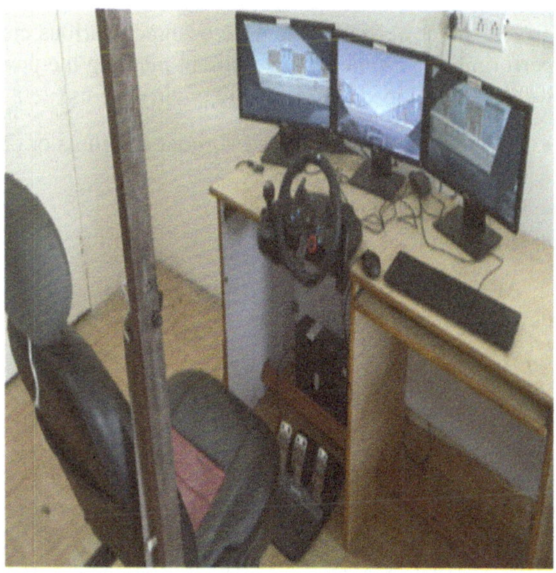

Fig. 1 Developed virtual reality-based driving simulator (VRDS)

Fig. 2 Internal view of the driver

Fig. 3 Developed different track environments for testing the powertrain

2.2 *Driving Scenarios*

To test the powertrains, we have designed various environments that would help in performing real-time testing without pitching the powertrain on-road. In this study, we have developed the following environments:

- Infinite road—Straight road without any turns or curves,
- Lake circuit,
- Desert circuit, and
- Hill circuit.

Also, the internal view of the driver is shown in Fig. 2 and the different tracks developed are shown in Fig. 3.

2.3 *Powertrain Test bed*

An elevated platform to host the engine and the motor in a fixed position was built and used for testing purpose. Hereafter, the whole setup over the elevated platform will be termed as powertrain test bed. The IC engine taken to build the hybrid powertrain is a

Table 1 Specifications of engine

Engine description	0.8 L 53.3 bhp 12 V petrol engine
Engine displacement (cc)	799
Maximum power	53.3 bhp@5678 rpm
Maximum torque	72 Nm@4386 rpm
No. of cylinders	3
No. of gears	5
Fuel system	MPFi

"RENAULT KWID 800 cc PETROL ENGINE". Table 1 mentions the specifications of the engine. The second power source is an AC motor with the specifications mentioned in Table 2.

The powertrain involves an electromagnetic-based switching between the two power sources. The prime electronic subsystems involved in the designed hybrid powertrain system are engine control system, motor control system, gear and clutch actuation system, brake system, and speed control system.

These subsystems are programmed to cooperate with each other and to respond for variations in the input signals like accelerator (A), brake (B), and clutch (C). These inputs are transmitted and received through the simulator to actuator interfacing system. Figure 4 depicts the block diagram of the simulator to actuator interfacing

Table 2 Specifications of electric motor

Power	3.7 kW/5 HP
Voltage	415 V
RPM	2865
Efficiency	85%
Connection type	DELTA

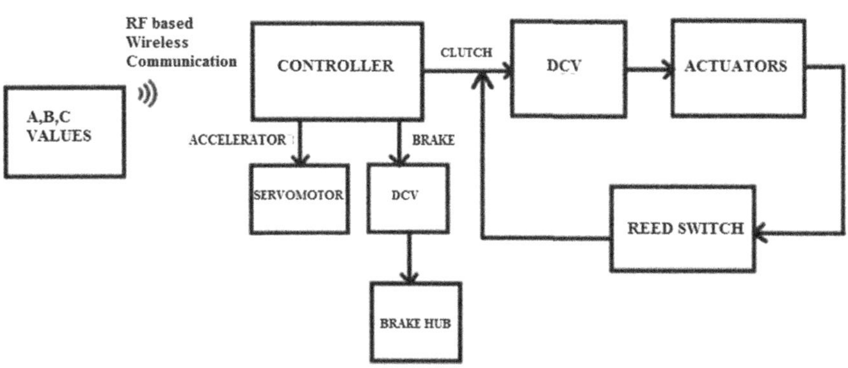

Fig. 4 Block diagram of the simulator to an actuator interfacing system

system. The variations in the ABC (Accelerator, Brake, and Clutch) pedals are measured as variable resistance with a range of 0–100 Ω and the gear position input acts as a step input with variable step size for each of the different gear positions. The designed wireless communication system, a part in the interfacing system as shown in Fig. 4, transmits the variable data from the VRDS to the hybrid powertrain test bed.

These simulator inputs are encoded into a byte format before they are transmitted using the coordinator XBee module connected to the simulator. The encoded data received by the router XBee module was connected to the microcontroller. The encoded data was decoded by the microcontroller, and then three different actuators were controlled according to the received data. The transmitted data and the received data of throttle position and gear positions are plotted and represented as in Figs. 5 and 6.

For switching the gears, two actuators are employed and one more actuator is used for the engagement and disengagement of the mechanical clutch. The two actuators used for gear shifting are gear positioner and gear selector.

The relay switches AC supply to the direction control valve (DCV) which supplies the pressure to the actuators according to the signal initiated by the controller. The positions of the pistons are noted by the reed switches and fed back to the controller.

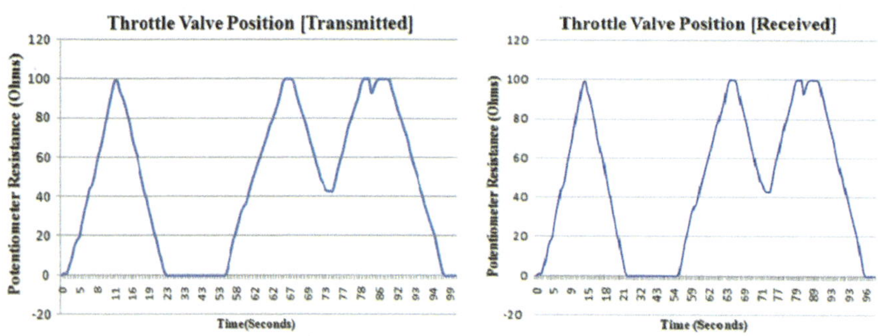

Fig. 5 Acceleration signals transferred wirelessly from simulator to the test bed

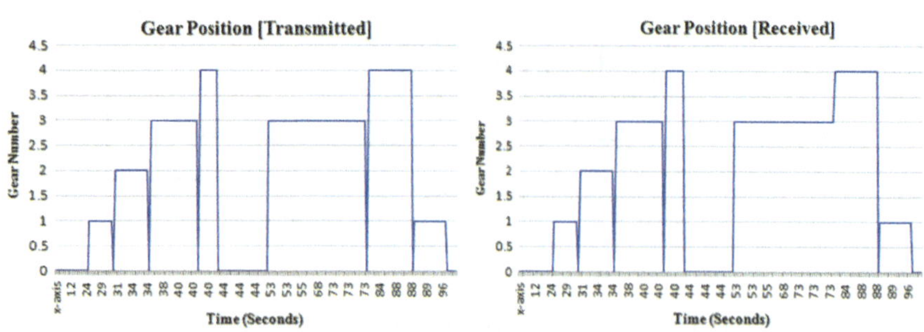

Fig. 6 Gear shifter signals transferred wirelessly from simulator to the test bed

For braking, the signal from the brake pedal is given to the relay, and this relay will initiate the DCV and the pressure is directly supplied to the brake.

3 Control Methodology

3.1 Switching Mechanism

The hybrid powertrain system on the elevated test bed has an electromagnetic-based switching system between the IC engine and the electric motor. The developed system has one electromagnetic clutch coupled to the electric motor and a mechanical clutch coupled to the IC engine.

For testing purpose, the fifth and reverse gears are disabled. The electromagnetic clutch engages the first and second gear during the low speeds, thus driving the gearbox using a chain drive, whereas the mechanical clutch engages the third and fourth gears during higher speeds and drives the gearbox directly. This engagement order of clutches and gear is mentioned in Table 3. This seamless switching is carried out using a designated customized controller. The switching mechanism is based on the gear position.

4 Results

This system is proven to show up the following results upon testing under the engine mode/normal mode and hybrid mode.

While the system is operated on both modes, the power transmitted from the motor or engine passes to the wheel through the transmission system. The transmission system was isolated from the power source when any of the two available clutches was engaged. The fuel efficiency is computed upon comparing the fuel consumption under both modes of operation.

The system is tested, by driving in the VRDS infinite road track for a period of 5 min and repeated for four times to count 20 min during which the simulator sends the corresponding signals to the hybrid powertrain system, under both engine and hybrid modes. The transferred data is received by the test bed controller and executed at the engine with a delay range of 10–30 s. It is observed that the usage

Table 3 Engagement order of clutches and gears

Gear	Mechanical clutch	Electromagnetic clutch	Driving source
1,2	Disengaged	Engaged	Motor
3,4	Engaged	Disengaged	Engine

Table 4 Mileage results

For infinite road track	Driving time (min)	Distance traveled (m)	Fuel consumed (ml)	Mileage (Kmpl)
Engine mode	20	8030.49	480	16.73
Hybrid mode	20	8030.49	360	22.3

Table 5 Parameters considered for computing the distance

Parameters	Values
Time for each run (min)	5
Engine RPM	1600
Radius of the tire (m)	0.255
Total no. of gears considered	4

of pneumatic actuators causes the delay. Upon measuring the fuel consumption, the below calculations are performed, and the results are tabulated in Table 4.

4.1 Mileage

The distance is calculated using the below formula:

$$\sum_{i=1}^{4}(\text{Engine RPM} * \text{Time of each run} * 2\pi r * \text{gear ratio})$$

The parameters considered for calculating the distance are given in Table 5.

4.2 System Efficiency

The efficiency of the developed hybrid powertrain with patented switching system is calculated as follows:

$$\text{System efficiency } (\%) = \frac{\text{Hybrid mileage(Kmpl)} - \text{Engine mileage(Kmpl)}}{\text{Engine mileage}} \times 100$$

The system efficiency was increased to 33.29% using the above formula. Thus, the developed hybrid powertrain system is proved to provide 33.29% more efficiency than the normal IC powertrain.

4.3 *Video*

The test run of the developed hybrid powertrain and the virtual driving simulator can be seen in this link, https://www.youtube.com/watch?v=TttnBicAF8w.

5 Conclusion

This study has developed and tested a hybrid powertrain system with an electromagnetic switching. The efficiency of the hybrid powertrain was tested using our developed virtual reality-based driving simulator (VRDS). We have increased the power efficiency using a hybrid powertrain when compared with the considered IC powertrain.

Acknowledgements We acknowledge SRM Institute of Science & Technology for supporting this research under Selective Excellence Program (SEP).

References

1. Zeng, X., Nie, L., Wang, Q.: Experimental study on the differential hybrid system hybrid electric vehicle (International workshop on automobile, power and energy engineering). Procedia Eng. **16**, 708–715 (2011)
2. Mikulski, M., Wierzbicki, S.: The concept and construction of the engine test bed for experiments with a multi-fuel CI engine fed with CNG and liquid fuel as an ignition dose. J. KONES Powertrain Transp. **19**(3), 289–296 (2012)
3. Passenbrunner, T.E., Sassano, M., del Re, L.: Optimal control of internal combustion engine test benches equipped with hydrodynamic dynamometers. In: 7th IFAC Symposium on Advances in Automotive Control, The International Federation of Automatic Control (2013). 978-3-902823-48-9/2013
4. Geng, S., Schulte, T.: Real-time models of hybrid electric vehicle powertrains. In: 7th IFAC Symposium on Advances in Automotive Control, The International Federation of Automatic Control (2013). 978-3-902823-48-9
5. Koç, T., Karayel, D., Boru, B., Ayhan, V., Cesur, I., Parlak, A.: Design and implementation of the control system of an internal combustion engine test unit. Adv. Mech. Eng. **2014** (2014)
6. Ben-gai, S., Li-ming, C., Yan-min, X.: Design of light hybrid electric vehicle powertrain test system. In: Fifth International Conference on Intelligent Systems Design and Engineering Applications (2014). 978-1-4799-4261-9/14
7. Galiullin, L.A.: Automated test system of internal combustion engines. In: IOP Conference Series: Materials Science and Engineering, vol. 86, p. 012018 (2015)
8. Xu, Y., Guo, K., Chen, L.: In: IOP Conference Series: Earth Environment Science, vol. 64, p. 012086
9. Li, Y., Xu, X., Sun, X., Xue, H., Jiang, H., Qu, Y.: Theoretical and experimental analytical study of powertrain system by hardware-in-the-loop test bench for electric vehicles. Int. J. Vehicle Syst. Modell. Test. **12**(1/2), 44–71 (2017)
10. Akkaya, F., Klos, W., Schwämmle, T., Haffke, G., Reuss, H.: Holistic testing strategies for electrified vehicle powertrains in product development process. World Electr Veh. J. **9**, 5 (2018). https://doi.org/10.3390/wevj9010005

11. Kouroussis, G., Dehombreux, P., Verlinden, O.: Vehicle and powertrain dynamics analysis with an automatic gearbox. Mech. Mach. Theory **83**, 109–124 (2014)
12. Mall, P., Fidlin, A., Krüger, A., Groß, H.: Simulation based optimization of torsional vibration dampers in automotive powertrains. Mech. Mach. Theory **115**, 244–266 (2017)
13. Wu, J., Liang, J., Ruan, J., Zhang, N., Walker, P.D.: Efficiency comparison of electric vehicles powertrains with dual motor and single motor input. Mech. Mach. Theory **128**, 569–585 (2018)
14. Joseph, A.V., Kumar, G.J.R.: Switching control in hybrid vehicle system based two wheeler. Int. J. Autom. Smart Technol. **7**(1), 15–19 (2017)

Distance-Based Clustering Protocol (DBCP) in Wireless Sensor Network

Takhellambam Sonamani Singh and Ajoy Kumar Khan

Abstract The primary missions of the Wireless Sensor Network (WSN) are to grow network lifetime and to utilize each node energy effectively. To grow the network lifetime, communication routing protocols play a very essential role. The goal of this protocol is to shun the data transmissions and data reception unnecessarily. In the paper, Distance-Based Clustering Protocol (DBCP) for WSN that is an improvement of the Stable Election Protocol (SEP) is being introduced. The new protocol objective is to raise the stable lifetime of the WSN and ultimately to elongate its lifetime based on the method of taking membership during cluster formation to minimize energy dissipation among the sensor networks. Our method introduces an independent node, which is nearer to the sink and such independent nodes can transmit their own sensor data after compressing directly to the sink. Finally, the model result illustrates that DBCP outperformed better than the existing system.

Keywords Wireless sensor network · Network lifetime · Clustering

1 Introduction

WSNs are generally designed or configured to ensure a large areas of applications such as environmental monitoring, medical diagnosing, military awareness, infrastructure protection, and temperature sensing [1]. A WSN is collection of an array of very small and low complex diverse devices known as sensor nodes which are interconnected through a communication link. The WSN is made up of four units [2]. They are sensing, communication, power supply, and processing unit. The sensor node senses the quantities that are available in the monitoring field and then translates them into data to reach finally to the Base Station (BS). BS is a sink node

T. S. Singh (✉)
Manipur Institute of Technology Takyelpat, Imphal 795004, India
e-mail: sona.cse2013@gmail.com

A. K. Khan
Assam University Silchar, Silchar 788011, India
e-mail: ajoy.kumar.khan@aus.ac.in

H. Das et al. (eds.), *Progress in Computing, Analytics and Networking*,
Advances in Intelligent Systems and Computing 1119,
https://doi.org/10.1007/978-981-15-2414-1_42

which interacts with the remote controller node by satellite links or Internet [1, 2]. The nodes in WSN are equipped with an inadequate power source. The lifespan or the longevity of the network is truly dependent on how fast the sensor nodes energy is depleted. When the battery dies, no node in the network is any longer useful. So, researchers have now been motivated to begin developing a lot of efficient energy routing mechanisms to lengthen the sensor nodes lifetime and also to protect energy of node in the sensing field. SEP [3] for WSN is a widespread clustering protocol. It is built on heterogeneity of network. Here, nodes of energy higher are considered as advanced nodes and advanced nodes' probability to be CH is more as related to that of normal nodes. SEP has few limitations as the selection of CH among sensor nodes is not vibrant which affects that the far-away nodes from the CH node or sink node will die first. As a result, Distance-based Clustering Protocol (DBCP) for WSN is being introduced. The main objective of this protocol is to consider a method of taking membership during cluster formation so that it will minimize energy dissipation among the sensor nodes. Our goals are to achieve reliability and scalability by interconnecting multiple sensor nodes in the WSN in addition to dynamic utilization of energy for achieving longer network time. Our proposed model introduces an independent node which is closer to the sink and that independent node can transmit their own sensing data after compressing directly to the sink. So, we study the effect of independent node in WSN for the above protocol in heterogeneous environment. The other parts of the finding are arranged as under. In Sect. 2, we highlighted the concerned work with respect to our system. In Sect. 3, we highlighted the SEP system description. Section 4 describes the DBCP system description. Section 5 shows the performance evaluation of DBCP protocol by simulation through a certain parameters and compared it with SEP protocols. Finally, Sect. 6 presents concluding and future scopes.

2 Related Work

Heinzelman et al. [1] proposed the microsensors which are designed with small power hardware and small power radiofrequency. This paper proposes LEACH, a clustering-centered routing protocol that operates on selection of CHs randomly to expand equally the energy load among the nodes. Shepard [2] discussed the two parameters of the transmitted signal necessary for accepting the system performance, i.e., transmitted signal's power level and bandwidth. Duarte-Melo et al. [3] stated that there are three common types of communications in WSN, i.e., Clock-driven, query-driven, and event-driven. The paper examines sensors which are equipped with different battery power levels in a clock-driven network. Kalpakis et al. [4] discussed the wireless networks that contains hundreds of low-cost nodes which may be positioned in physical environment for gathering data. Data is collected from each node, and then data aggregation is done to reach it to the sink. Tan et al. [5] introduced two algorithms, i.e., power-efficient data gathering and aggregation protocol and power-efficient data gathering and aggregation protocol-power aware.

It is built upon minimum spanning tree in which one of the two is the power-aware version of the other. Conner et al. [6] offered two energy-conserving protocols which work together to enable sensor nodes to preserve energy by sleeping. The paper discusses its various areas of applications. Akyildiz et al. [7] illustrated the sensor network ideas which are feasible by convergence of micro-electro-mechanical system technology, wireless communications, and digital systems. Mhatre et al. [8] discussed homogeneous versus heterogeneous LEACH system. A single-hop homogeneous network system ensures that relatively complete utilization of overall network energy distributed among all nodes but each sensor node must be designed with necessary hardware capabilities to act as a CH in this network. In heterogeneous sensor network, all sensor nodes except CH may be deployed with simple hardware devices which facilitate communication range short but the nodes near to the margin of the cluster disburse more energy than those near the CH. Lindsey et al. [9] proposed a system called power-efficient data gathering and aggregate in sensor information system. In this paper, every node interacts with neighbor nodes only and then transmits to BS, thus dropping spent of energy per round. For MSLEACH [10], we break the rule that only the CH has to report to the sink.

3 The SEP Protocol

The key features of SEP are as follows:

1. It is a protocol that works in heterogeneous environment to lengthen the time interval until first node dies. It is decisive for numerous applications where the feedback is consistent.
2. It is created on weighted election probabilities for each node to be CH on residual energy basis.
3. It generates larger stable region for large extra energy values.

The main objective of SEP is to preserve the system stability, especially when the nodes are heterogeneous by using a dynamic election process of CH. SEP attempts to elect advanced nodes as CH in most of the time and finally improved the stable region.

SEP generates clusters by employing distributed algorithms where nodes make independent decisions. This is for weight of optimal probability p_{opt}. The weight is equal to initial energy of each node divided by the initial energy of the node (normal).

Suppose m= part of total n numbers of nodes, which are deployed with α time more energy than other nodes. These powerful nodes are called advanced nodes and remaining $(1-m) \times n$ part as normal nodes. Probability of normal nodes and advanced nodes to become CHs are as shown in (1) and (2):

$$P_{\text{nrm}} = \frac{p_{\text{opt}}}{1 + \alpha \cdot m} \tag{1}$$

and

$$P_{\mathrm{adv}} = \frac{P_{\mathrm{opt}} \times (1 + \alpha)}{1 + \alpha \cdot m} \tag{2}$$

where P_{opt} = optimal probability of each node to become CH,

$T(S_{\mathrm{nrn}})$ is the threshold for normal nodes, and $T(S_{\mathrm{adv}})$ is the threshold for advanced nodes.

Thus for normal nodes, we have

$$T(S_{\mathrm{nrn}}) = \begin{Bmatrix} \dfrac{P_{\mathrm{nrn}}}{1 - P_{\mathrm{nrn}} r \bmod \left(\dfrac{1}{P_{\mathrm{nrn}}}\right)} & \text{if} \quad S_{\mathrm{nrn}} \in G \\ 0, \quad \text{otherwise} & \end{Bmatrix} \tag{3}$$

where r is the round (current), G is the set of normal nodes which have not become CHs inside the last $1/P_{\mathrm{nrm}}$ rounds of the epoch, and $T(S_{\mathrm{nrm}})$ is the threshold function to a population of n (1−m) normal nodes. This ensures that each normal node will become a CH exactly once every $1/P_{\mathrm{opt}} \times (1{+}\alpha{\cdot}m)$ rounds per epoch and the average number of CH that are normal nodes per round per epoch is equal to $n \times (1-m) \times P_{\mathrm{nrm}}$.

Similarly, for advanced nodes, we have

$$T(S_{\mathrm{adv}}) = \begin{Bmatrix} \dfrac{P_{\mathrm{adv}}}{1 - P_{\mathrm{adv}} r \bmod \left(\dfrac{1}{P_{\mathrm{adv}}}\right)} & \text{if} \quad S_{\mathrm{adv}} \in G \\ 0, \quad \text{otherwise} & \end{Bmatrix} \tag{4}$$

where r is the round (current), G is the set of advanced nodes which have not become CHs inside the last $1/P_{\mathrm{adv}}$ rounds of the epoch, and $T(S_{\mathrm{adv}})$ is the threshold function to a population of $n{\cdot}m$ advanced nodes.

Steady Phase

Here, data transmission commences. Non-CH forwards data depending on signals strength of advertisement.

All nodes except CH can be switched off unless the CH node allocates TDMA slot to each node. The CH then combined all the data from the member nodes along with itself and finally forwards to the sink.

4 The Proposed Methodology

The main goal of existing SEP is to utilize more energy in advanced nodes by electing them as CH more than the normal nodes. Our proposed system attempts to minimize energy dissipation during cluster formation and data transmission.

In our proposed system, we break the rule that only the CH has to forward data to the sink. In setup phase of cluster formation, each node that picks itself as a CH for the round (current) shows advertisement message to the remaining sensors. All CHs transmit advertisement message using same energy level. Non-CH nodes should retain receivers on at this phase to get the message. After this, they decide which cluster to join for the round by choosing CH that needs communication energy at the minimum. The proposed system adds an intermediate step of comparing two distances before deciding which CH has to be chosen by a non-CH node. After identifying the nearest CH of a non-CH node for dissipating minimum intra-cluster communication energy, we compare the distance of the sink and the non-CH, and distance of that non-CH and the nearest CH before taking membership to that particular CH. If the distance of non-CH node and sink is shorter, then non-CH node is independent of that CH and will not be the member of any CH for the current round. Such nodes are independent nodes. Otherwise, that non-CH node acts as a member of the CH. Independent nodes for a round are eligible to send data for itself to the sink with compressing its own sense data.

4.1 Architecture of Proposed DBCP System

We can increase the network diameter beyond certain level but cannot determine the location of the nodes which are distributed in network field randomly. As the role of being CH nodes is rotated for each round among the sensor nodes, some non-CH nodes which are far from the selected CH but nearer to sink may take membership to a CH which leads to more energy consumption to the member nodes for transmission and the CH loads more energy for reception and aggregation of the received data. Hence, we determine the distance of each non-CH node and its nearest CH for each round before taking membership to the CH. The non-CH nodes will not take membership to any CH if it is more nearer to the sink, and it will act as an independent node which can communicate to the sink directly (Fig. 1).

The above architecture shows 50 sensor nodes being dispersed randomly in the sensor field. A 0.14% of sensor nodes are advanced nodes and others as normal nodes. For this current round of the above network, 6 clusters are formed having 7 nodes including CH. After setup phase of cluster formation, we find eight independent nodes for this current round of the above network field.

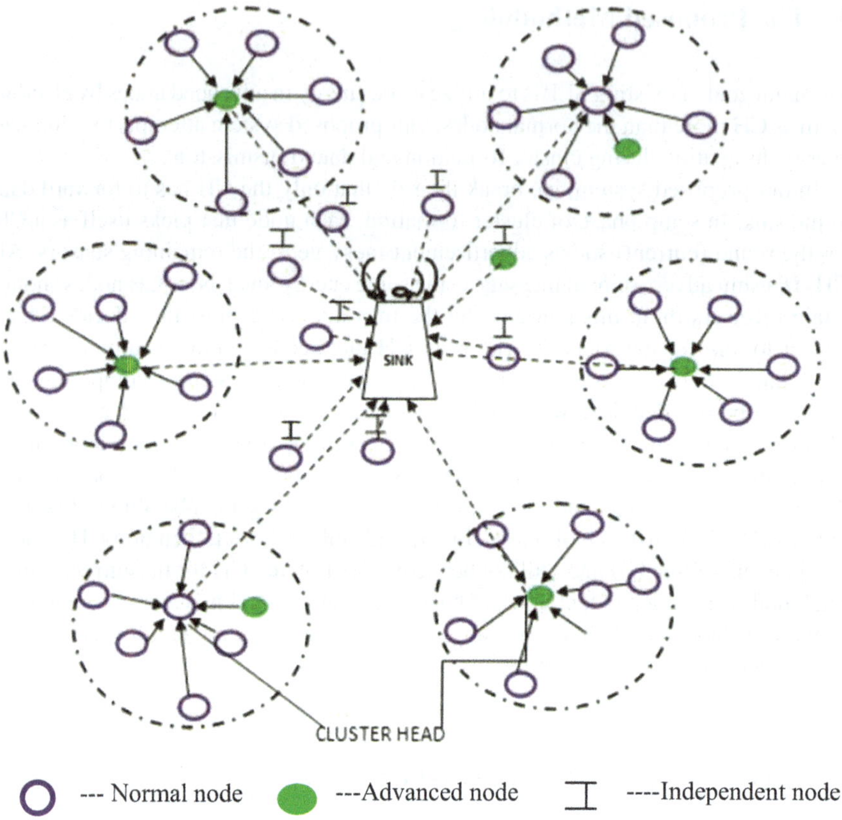

Fig. 1 Architecture of proposed (DBCP) network system

5 Implementation

Figures 2 and 3 show the initial distribution of the network for the two routing protocols SEP and DBCP. Here, 100 nodes are dispersed at random in 100 * 100^2 m. The sink or BS denoted by X and Y is deployed at the middle of field (0.5, 0.5), and its number of round is 5000. In these two protocol systems, 20% of nodes are found to have more initial energy than other nodes called advanced nodes. In this case, 20 advanced nodes have 2 J of initial energy in 100 nodes, whereas remaining 80 normal nodes have 0.5 J of initial energy.

Advanced nodes are represented by plus symbol (+), and the normal nodes are represented by circle symbol (o). In Figs. 2 and 3, all the nodes are found to be alive in the network.

Figures 4 and 5 show the total number of nodes that stayed alive above 4000 rounds for SEP and DBCP. All the normal nodes drain out their energy after 4000 rounds in SEP. Out of 20 advanced nodes, 6 advanced nodes remain alive in SEP system and

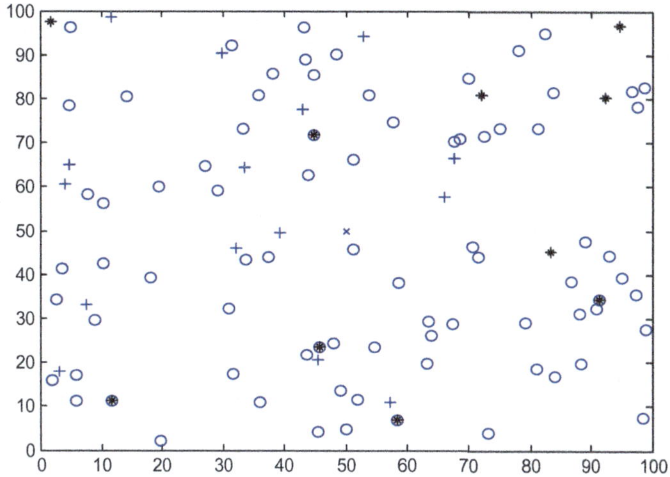

Fig. 2 Initial setup phase of SEP system

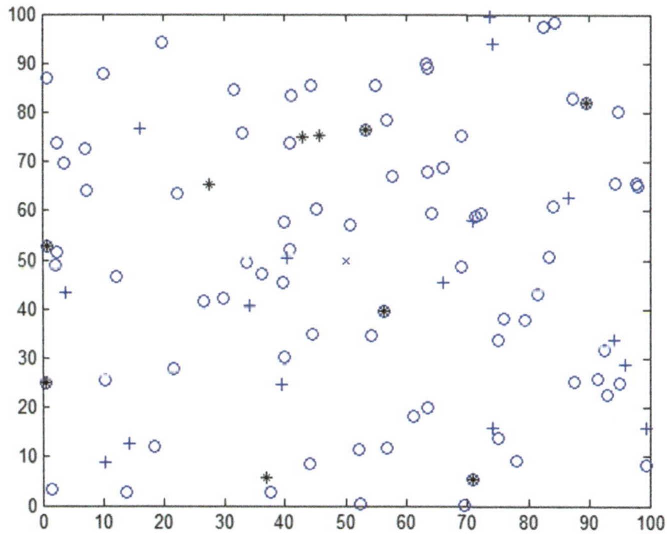

Fig. 3 Initial setup phase of DBCP system

16 advanced nodes with 2 normal nodes still remain alive in DBCP system. Dead nodes in the network are indicated by the red diamond shape. We further compare the SEP and DBCP systems.

Figure 6 shows a comparison between SEP and DBCP systems in terms of alive nodes per round. The graph illustrates that alive node numbers per round in most of the rounds in DBCP are greater than those of the SEP system.

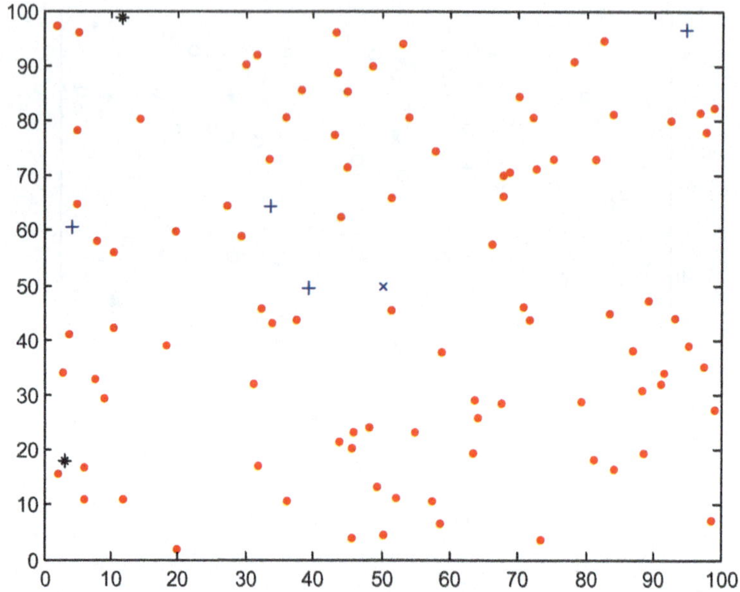

Fig. 4 Simulation result after 4000 rounds in SEP system

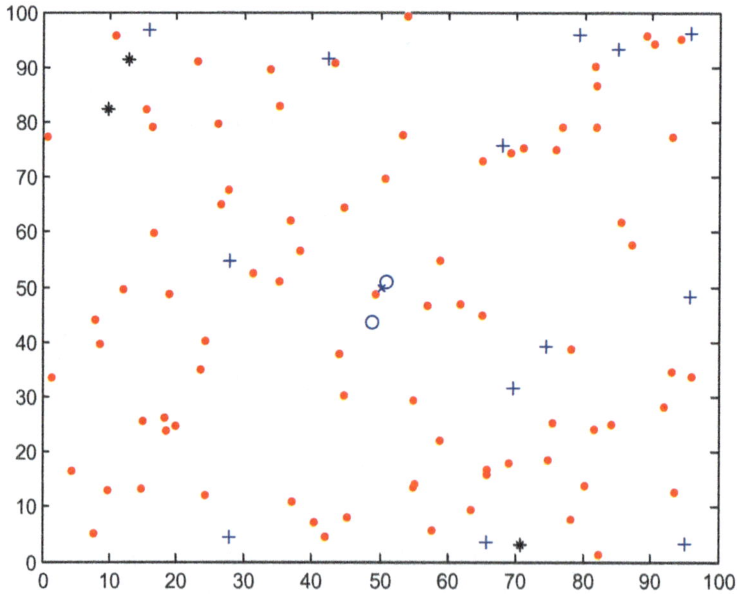

Fig. 5 Simulation result after 4000 rounds in DBCP system

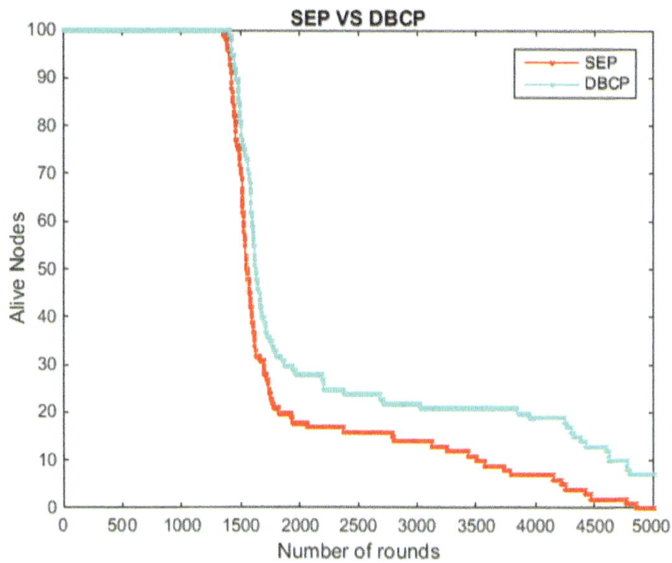

Fig. 6 No. of alive nodes versus no. of rounds in SEP and DBCP systems

Figure 7 shows a comparison between SEP and DBCP systems in terms of residual energy per round. The graph shows that DBCP uses less energy in most of the rounds than those of the SEP systems. Hence, our DBCP system performed better.

6 Conclusion and Future Scope

In our findings, we discussed DB–SEP which is an energy-aware clustering protocol used in non-homogeneous WSN depending upon distances and compared it to the existing SEP protocol. In DBCP, each node in the network field choose itself autonomously as a CH built on its initial energy and introduce an independent node by comparing minimum distance of the non-CH nodes and BS with the distance that of non-CH and its nearest CH. Hence, we can minimize energy dissipation for transmission of data in the WSN. Results from this model indicate that DBCP provides better results for efficiency in energy and network lifetime. In future, we would consider by changing residual energy of advanced nodes and different data aggregation techniques like chain-based that can be used among all the independent nodes in WSN.

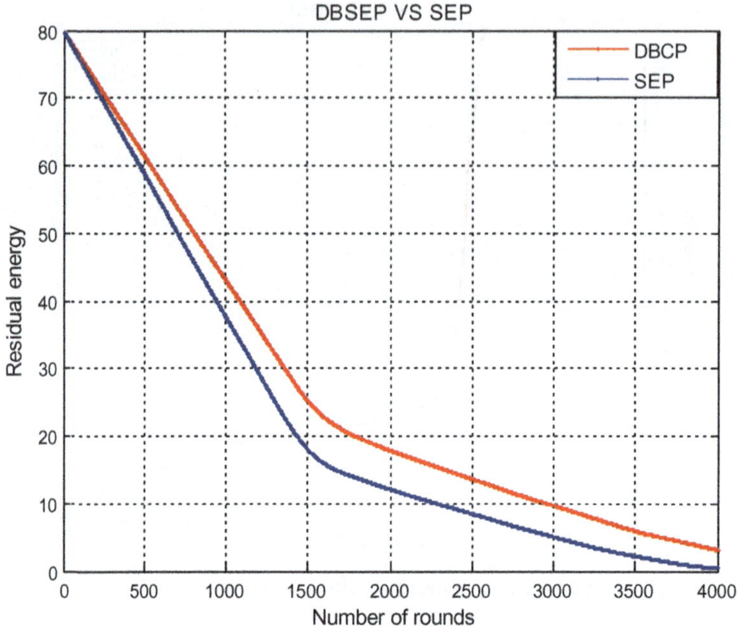

Fig. 7 Residual energy versus number of rounds in SEP and DBCP systems

References

1. Heinzelman, W.R., Chandrakasan, A., Balakrishnan, H.: Energy-efficient communication protocol for wireless microsensor networks. In: Proceedings of the 33rd Annual Hawaii International Conference on System Sciences, 10 pp. IEEE (2000)
2. Shepard, T.J.: A channel access scheme for large dense packet radio networks. ACM SIGCOMM Comput. Commun. Rev. **26**(4), 219–230 (1996)
3. Duarte-Melo, E.J., Liu, M.: Analysis of energy consumption and lifetime of heterogeneous wireless sensor networks. In: IEEE GLOBECOM'02, Global Telecommunications Conference 2002, vol. 1, pp. 21–25. IEEE (2002)
4. Kalpakis, K., Dasgupta, K., Namjoshi, P.: Efficient algorithms for maximum lifetime data gathering and aggregation in wireless sensor networks. Comput. Netw. **42**(6), 697–716 (2003)
5. Tan, H.Ö., Körpeoğlu, I.: Power efficient data gathering and aggregation in wireless sensor networks. ACM Sigmod Rec. **32**(4), 66–71 (2003)
6. Conner, W.S., Chhabra, J., Yarvis, M., Krishnamurthy, L.: Experimental evaluation of synchronization and topology control for in-building sensor network applications. In: Proceedings of the 2nd ACM International Conference on Wireless Sensor Networks and Applications, pp. 38–49. ACM (2003)
7. Akyildiz, I.F., Su, W., Sankarasubramaniam, Y., Cayirci, E.: A survey on sensor networks. IEEE Commun. Mag. **40**(8), 102–114 (2002)
8. Mhatre, V., Rosenberg, C.: Homogeneous versus heterogeneous clustered sensor networks: a comparative study. In: 2004 IEEE International Conference on Communications (IEEE Cat. No. 04CH37577), vol. 6, pp. 3646–3651. IEEE (2004)

9. Lindsey, S., Raghavendra, C.S.: PEGASIS: Power-efficient gathering in sensor information systems. In: Proceedings, IEEE Aerospace Conference, vol. 3, pp. 3–3. IEEE (2002)
10. Singh, T.S., Soram, R., Khan, A.K.: Distance based multi single hop low energy adaptive clustering hierarchy (MS LEACH) routing protocol in wireless sensor network. In: 2016 IEEE 6th International Conference on Advanced Computing (IACC), pp. 613–617. IEEE (2016)

Feeder Transit Service Efficiencies with Vehicle-to-Passenger (V2P) Communication

Shailesh Chandra, Mamta Kumari and R. Thirumaleswara Naik

Abstract This paper provides a novel technique of evaluating routing efficiencies that can result from vehicle-to-passenger (V2P) communication. Examples of public transport vehicles that can leverage the advantages of V2P communication include demand-responsive transit (DRT)—which is a form of flexible transit system often used as a feeder transport for main transit line. Special case has been presented for a grid street network system found predominantly in residential or downtown areas where feeder transit is frequently used. With an assumption of uniform random spatial distribution of passengers over a grid street network, efficiency improvements of as much as 73–98% can be expected with V2P deployment for variable grid network side length of 300–2000 ft, respectively, for higher number of passengers served. These efficiency improvements have been validated through several simulation experiments with a varied number of grid street network systems.

Keywords Transit systems · Grid street · Block length · Simulation · Efficiency

1 Introduction

Street systems are an integral part of our transportation infrastructure and hence, form important ingredient for most of the Intelligent Transportation System (ITS)-related studies. A well-connected street system (reflective of the infrastructure) allows easy

S. Chandra (✉)
Department of Civil Engineering and Construction Engineering Management, California State University, 90840 Long Beach, CA, USA
e-mail: shailesh.chandra@csulb.edu

M. Kumari
College of Business, California State University, 90840 Long Beach, CA, USA
e-mail: mamta.kumari@student.csulb.edu

R. T. Naik
Department of Mechnical Engineering, Indian Institute of Science, 560 012 Bengaluru, Karnataka, India
e-mail: rtnaik@iisc.ac.in

H. Das et al. (eds.), *Progress in Computing, Analytics and Networking*,
Advances in Intelligent Systems and Computing 1119,
https://doi.org/10.1007/978-981-15-2414-1_43

vehicle navigation [1] and thus assists in improved vehicle-to-passenger (V2P) interactions—particularly for flexible forms of public transit systems. Global Positioning System (GPS) is widely used in establishing the V2P communication. Communication using GPS has often been described under vehicular ad hoc network (VANet) studies, wherein vehicles can interact with the surroundings and passenger location of service requests [2]. Other smart communication and system technologies have also been developed [3]. However, functionality of GPS is not always reliable [4]. There are failures associated with the functionality of GPS, and these failures are often attributed to weak GPS signals resulting in location of passengers being not determined properly. These potential failures can be overcome if information regarding passenger coordinate locations can be gathered through alternate methods—with the goal to improve routing. Currently, there are several such alternatives in the form of location-based services [5]. However, just like GPS, there are possible failures associated with these location-based tools or devices. When it comes to V2P communication, the vehicle's interaction with the built environment, especially with streets, plays a critical role in location-based services. Current literatures that describe this critical role that V2P communication can play in improving public transport service operations are sparse.

V2P communication is also found to be widely utilized in emerging systems, such as in case of autonomous ground vehicles [6]—wherein the exact location of a potential passenger/vehicle on a street network needs to be determined in advance to allow for efficient service. This is particularly relevant when a passenger pick-up/drop-off is involved through a vehicle having dynamic routes over the streets. Thus, V2P communication could become very useful for flexible type of transit systems such as demand-responsive transit (DRT) which serves passengers placed randomly across the street network in a shared-ride mode [7]. In fact, a DRT system, though not driverless, is specifically designed to serve passengers door-to-door on request, thereby providing a quick and convenient connection to the main transit line, such as metro bus, light rail, etc. [8]. The passengers are often required to make an advance service request by phone call or Internet. The DRT shuttle operates from a common depot or a major transit transfer point to serve randomly distributed passengers across the streets using a suitable scheduling algorithm [9]. However, as a DRT vehicle equipped with V2P communication system begins to serve passengers once all the service requests are received, a major challenge exists in understanding the role street network can play in improving the efficiency of the vehicle.

Grid forms of street systems are much easier to model and relatively easier for vehicular navigations as compared to curvilinear or cul-de-sacs (or dead ends). These forms of streets can be found in plenty all over the United States and elsewhere in the world [10]. In this paper, we model the V2P communications using random passenger location requests over a grid street network system having square (or regular) blocks.

2 Methodology

Efficiency of DRT is evaluated based on minimum travel distance taken by the transit bus starting from the terminal, picking up all passengers at their designated service request locations, and back to the terminal. As the V2P communication exists, the efficiency in picking up passengers can be assessed by noting the bounds on the average closest distance between two passengers on the grid street network. Thus, on an average, each passenger location for service requests is located at mid-block of a street link in the network (see Fig. 1 for example). We call this average street-based distance as impedance (denoted by $I_{\text{imp}}^{L,P}$). Subsequently, we investigate maximum efficiency in routing that could be possible by calculating the lower bound on $I_{\text{imp}}^{L,P}$—with considerations to the effects of block lengths or block spacing of the grid street network. Knowledge of desired block spacing will help transportation planners and engineers in designing efficient grid street systems for vehicles utilizing V2P communication system particularly for transit vehicles such as the DRT.

2.1 Lower Bound Solution

We derive the expression for the lower bound on the average closest street-based distance (i.e., average smallest possible value of $I_{\text{imp}}^{L,P}$) over a grid network system. The distance is derived between two random spatially distributed passengers under the existing V2P communication between the DRT vehicle and the passengers who

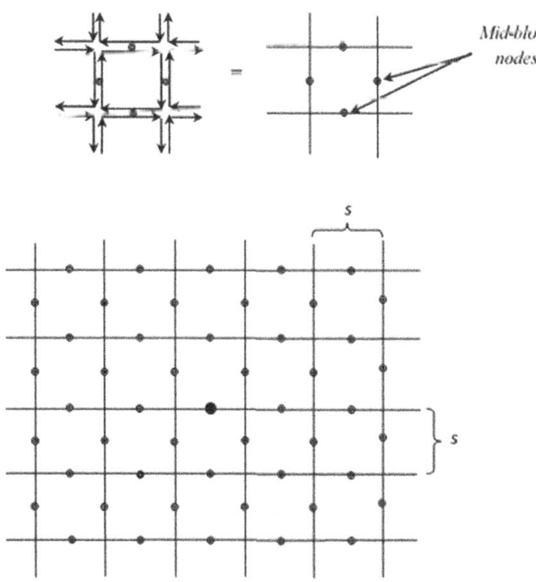

Fig. 1 Representation of the grid street network system

have already made the service request for pick-up. Consider the grid street network system with block spacing, s, as shown in Fig. 1. The mid-block nodes are the average potential location points of vehicles on street links and are designated as on-demand nodes in rest of the paper. The expected closest distance between the vehicle and a passenger (the lower bound for impedance, $I_{\text{imp}}^{L,P}$) can be written as

$$I_{\text{imp}}^{L,P} = \sum_{n=1}^{n=n^*} (ns + tV)P\{\Omega_A = a\} \tag{1}$$

where $P\{\Omega_A = a\}$ is the probability of spotting a number of passengers at the on-demand node n for the random number Ω_A and n^* is the total number of mid-block nodes in the given grid network. V is the mean speed of the vehicle movement on the network. Simplification of (1) is possible when the spatial probability $P\{\Omega_A = a\}$ of passengers is known and is mathematically tractable to obtain closed-form expressions of (1). We assume passengers can be distributed over the whole area instead of just on the streets—since there is flexibility of the passengers to walk to the closest street whenever there is a need to use a vehicle. This is similar to passenger request procedure adopted for any demand-responsive transit system [11]. We first show some simplified forms for passenger distributions spread over the whole service area containing the grid street system. Uniformly and randomly scattered passengers over an area follow a spatial Poisson distribution [12]. Assuming that the number of passengers is Ω_A within the area A is a Poisson random variable, its distribution is given by

$$P\{\Omega_A = a\} = \frac{(\sigma A)^a}{a!} e^{-\sigma A} \quad \text{where,} \quad a = 0, 1, 2, 3, \ldots \tag{2}$$

For a zero count of passengers within the area A, it would imply $a = 0$, and the following expression is obtained:

$$P\{\Omega_A = 0\} = e^{-\rho A} \tag{3}$$

Each passenger is assigned to the closest mid-block node for further being served by a vehicle. Thus, using simple geometry, if a vehicle is at node Z (see Fig. 2), the location of a passenger within an assigned area A_n decides how far the nth closest mid-block stop is located on the grid street system. For example, the first closest mid-block node from Z is at a street-based distance of "s" in the shaded area A_1 as shown in Fig. 2. The second closest distance to a mid-block node from the node Z is $2s$ in the area designated as A_2 and so on. An example has been shown using shaded area A_2 of Fig. 2 for illustration. The arrows show the shortest path direction of travel to all the four closest nodes enclosed within the shaded area A_2 which are at a street-based distance of ($3s/2 + s/2 = 2s$) from node Z (Note: the area is analyzed for quarter section and can be expanded fourfold to capture the whole grid area surrounding node Z). Observing the geometry of each of the quarter sections in

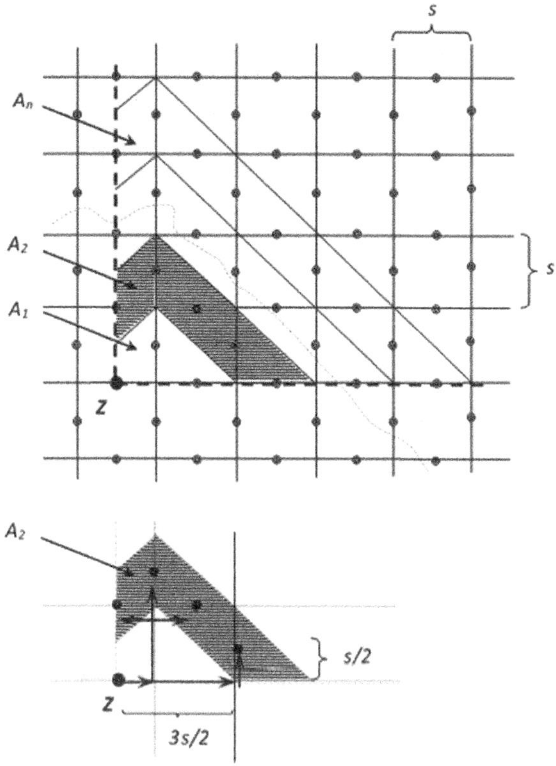

Fig. 2 Location of the closest vehicle/passenger at a mid-block node from a known node Z

Fig. 2, the following expressions hold: $A_1 = \frac{s^2}{2} + \frac{3s^2}{8}$; $A_2 = \frac{(2s)^2}{2} + \frac{3s^2}{8} + \frac{s^2}{2}$; or, in general for the nth demand area,

$$A_n = \frac{1}{2}(ns)^2 + \frac{3s^2}{8} + (n-1)\frac{s^2}{2} \tag{4}$$

Therefore, using Eq. (1), the expression of $I_{\text{imp}}^{L,P}$ can be written as

$$\begin{aligned} I_{\text{imp}}^{L,P} &= \sum_{n=1}^{n\approx\infty} (ns + tV)P\{\Omega_A = 0\} = \sum_{n=1}^{\infty} (ns + tV)e^{-4\rho A_n} \\ &= \sum_{n=1}^{\infty} (ns + tV)e^{-2\rho s^2\left(n^2+(n-1)+\frac{3}{4}\right)} = \sum_{n=1}^{\infty} (ns + tV)e^{-2\rho s^2\left(n^2+n-\frac{1}{4}\right)} \\ &= \left(se^{\rho s^2}\right)\left(\sum_{n=1}^{\infty} ne^{-2\rho s^2\left(n+\frac{1}{2}\right)^2} + \sum_{n=1}^{\infty} \frac{tV}{s} e^{-2\rho s^2\left(n+\frac{1}{2}\right)^2}\right) \end{aligned} \tag{5}$$

The variable n in Eq. (5) being small compared to the exponential terms and $\left(\frac{1}{e^{4\rho s^2}}\right) \in (0, 1)$, we have by Szablowski [13],

$$I_{\text{imp}}^{L,P} \approx \left(se^{\rho s^2}\right)\frac{1}{2}\left[\sqrt{\frac{\pi}{2\rho s^2}}\left(1+2\sum_{n\geq 1} e^{-\frac{\pi^2 n^2}{2\rho s^2}}\right) - 1\right] + \left(se^{\rho s^2}\right)\frac{tV}{2s}\left[\sqrt{\frac{\pi}{2\rho s^2}}\left(1+2\sum_{n\geq 1} e^{-\frac{\pi^2 n^2}{2\rho s^2}}\right) - 1\right] \tag{6}$$

And further ignoring the summation terms, since $\left(2\sum_{n\geq 1} e^{-\frac{\pi^2 n^2}{2\rho s^2}}\right) \ll 1$ $I_{\text{imp}}^{L,P}$ can be further simplified as

$$I_{\text{imp}}^{L,P} \approx \frac{se^{\rho s^2}}{2}\left(1+\frac{tV}{s}\right)\left(\sqrt{\frac{\pi}{2\sigma s^2}} - 1\right) \tag{7}$$

The expression for $I_{imp}^{L,P}$ in (7) should hold only for $\rho \leq \frac{1.57}{s^2}$ (using $\pi = 3.14$) as right-hand side should be a positive quantity. Some existing limits from literature can be found for the right-hand side of (7) by setting block length s and stop time t at the intersection to be too small, which gives $\lim_{t,s\to 0}\left[I_{\text{imp}}^{L,P}\right] \approx \frac{0.63}{\sqrt{\rho}}$. The latter value can be verified with exactly similar one obtained for a rectilinear-based metric assumption for distance between two uniformly distributed points inside a large area as derived by Quadrifoglio et al. [12]. The lower bound solution in (7) is more accurate for a very high vehicular density across the area. The availability of this kind of street-based distance information is much more superior to the usual Euclidean metric adapted to measure distances between two passenger locations on a network.

3 Simulation Experiments and Results

Simulation experiments are performed to validate the lower bound obtained using analytical derivations. The experimental setup consists of coding four different grid street networks with varying block sizes of s (in feet) = 300, 667, 1000, 1333, and 2000 within a rectangular grid street system of length, $L = 8000$ ft and width, $W = 4000$ ft. The block sizes selected such that all possible range of grid network system is covered for simulation—from being too dense network with block size of 300 ft to too sparse network with block size of 2000 ft.

Passengers are uniformly random distributed across the entire service area containing the street network and assigned to the closest on-demand node in a rectilinear manner. The outputs consist of average closest street-based distance among the vehicles/passengers. To accomplish this task over a wide number of simulations, help of a scheduling algorithm is sought. In this paper, we use a very widely used "insertion heuristic" in the field of vehicle scheduling to get the outputs [14].

Using the insertion heuristic, the objective function (namely, minimization of average shuttle travel distance between passenger locations) was obtained within reasonable computational times. At lower block length values, the computational time and solution convergence are noted to be impractical with the algorithm since several stops are created with an increased number of intersections. Hence, block length values of less than 300 ft are not considered for the simulation. The simulations are carried out with 25 replications for every demand setting and using each of the five different grid street systems with various block length values. The maximum seating capacity of DRT shuttle is usually less than 10 passengers. Therefore, the number of passengers used for simulation is varied from 2 to 12 passengers (with an increment of 2 passengers). Corresponding travel times are converted to time units by multiplying it by the speed $V = 20$ mph of the vehicle to reach another closest vehicle/passenger located on an on-demand node. An intersection stop time $t = 2$s is also assumed. The entire simulation is carried out by coding the algorithm and the networks using a simulation software.

4 Results

The results of the simulation experiments are reported in Tables 1 and 2. The notations used are LB = lower bound distance impedance in (7), and Sim = distance impedance using simulation. The outputs are the average distance traveled by the DRT shuttle in "feet" between passenger demand locations.

Observing the outputs from Tables 1 and 2, the results of simulation experiments lie above the lower bound values. The main aim of the simulation was to demonstrate the validation of the analytically derived closed-form expressions for the bounds on average impedance between uniformly distributed passengers on a given grid network

Table 1 Efficiency improvement comparisons for variable block lengths and number of passengers = 2 and 6

Block length (in feet)	Number of passengers				Efficiency improvement (%)	
	2		6		Number of passengers = 2	Number of passengers = 6
	LB	*Sim*	*LB*	*Sim*	*LB*	*Sim*
2000	1991	4955	974	4609	60	79
1333	2147	4701	1137	4326	54	74
1000	2261	4672	1210	4325	52	72
667	2431	4588	1317	4189	47	69
300	2904	4842	1576	4569	40	66

Table 2 Efficiency improvement comparisons for variable block lengths and number of passengers = 8 and 12

Block length (in feet)	Number of passengers				Efficiency improvement (%)	
	8		12		Number of passengers = 8	Number of passengers = 12
	LB	*Sim*	*LB*	*Sim*	*LB*	*Sim*
2000	709	4649	108	4326	85	98
1333	955	4298	725	4005	78	82
1000	1024	4220	806	3933	76	80
667	1119	4070	887	3847	73	77
300	1348	4498	1079	4040	70	73

system. We achieve the aim from the outputs reported in Tables 1 and 2. Note that the lower bound analytical values are only applicable when the grid street network has smaller block lengths which creates large number of route options on the network. In other words, as the block length increases the grid mesh size becomes larger at which resulting lower bound distances obtained using (7) may not be applicable. This is very well reflected from the values of Tables 1 and 2.

Tables 1 and 2 also present efficiency improvement for the passenger counts of 2, 6, 8, and 12. The percentage increase in efficiency is observed to be lower with reduction in block length—as noted in the last two columns of Tables 1 and 2. As noted in Table 1, the percentage improvement in efficiency varies from 60 to 79% for the block size of 2000 ft, for passenger counts equal to 2 and 6, respectively. From Table 2, the percentage improvement in efficiency for the DRT shuttle is 85–98% for the block size of 2000 ft, for number of passengers equal to 8 and 12. These percentages are lower for lower values of block lengths. It is also observed that the percentage improvement in efficiency is the lowest for smallest size of block length equal to 300 ft selected for the analysis. For the same block size of the grid street network system, the percentage improvement in efficiency increases with an increase in passenger counts served by the DRT shuttle. These findings are presented in Tables 1 and 2.

5 Conclusions

Closed-form expressions are derived for the lower bound average street-based distances between passengers on a grid street network system important for ITS-related V2P communication systems. These bounds on distance-based impedance are an improvement over the Euclidean-metric-based distances that are usually employed by location-based services. For the example case of grid street network with length = 8000 ft and width = 4000 ft, it is also observed that the percentage improvement in efficiency is the lower for smaller size of block length used in the grid street system.

For the same block size of the grid street network system, the percentage improvement in efficiency increases with an increase in passenger counts served by the DRT shuttle. These findings are very useful for transportation planners and practitioners who can use their own parameter settings to derive and estimate lower bounds to control efficiency improvements for DRT service. This certainly makes this work versatile not only for ITS but across many other system planning areas involving grid network systems. Our future work consists of extending the findings from this research to non-uniform passenger distribution over grid street network systems and estimating efficiency improvements in routing.

References

1. Chandra, S.: Safety-based path finding in urban areas for older drivers and bicyclists. Transport. Res. part C Emerg. Technol. **48**, 143–157 (2014)
2. Rout, C., Panigrahi, A., Badjena, J. C., Pradhan, C., and Das, H.: An approximation solution to the NP-complete joint problem in multi-radio WMN. In: Smart Intelligent Computing and Applications, pp. 385–396. Springer, Singapore (2020)
3. Blum, J., Eskandarian, A., Hoffman, L.: Challenges of inter-vehicle ad hoc networks. IEEE Trans. Intell. Transp. Syst. **5**(4), 347–351 (2004)
4. Boukerche, A., Oliveira, H.A.B.F., Nakamura, E.F., Loureiro, A.A.: Localization systems for wireless sensor networks. IEEE Wirel. Commun. (Special Issue on Wireless Sensor Networks) **14**, 6–12 (2007)
5. Boukerche, A., Oliviera, H., Nakamura, E., Loureiro, A.: Vehicular ad hoc networks: a new challenge for localisation-based systems. Comput. Commun. **31**(12), 2838–2849 (2008)
6. Luettel, T., Himmelsbach, M.: Autonomous ground vehicles concepts and a path to the future. IEEE Trans. Robot. (2012)
7. Dial, R.B.: Autonomous dial-a-ride transit introductory overview. Transport. Res. Part C Emerg. Technol. **3**(5), 261–275 (1995)
8. Koffman, D.: Operational experiences with flexible transit services: a synthesis of transit practice. TCRP synthesis, vol. 53. Transportation Research Board, Washington, DC
9. Chandra, S., Quadrifoglio, L.: Critical street links for demand responsive feeder transit services. Comput. Ind. Eng. **66**(3), 584–592 (2013)
10. Kostof, S.: The City Shaped. Little, Brown, Boston (1991)
11. Long, J., Gao, Z., Orenstein, P., Ren, H.: Control strategies for dispersing incident-based traffic jams in two-way grid networks. IEEE Trans. Intell. Transp. Syst. **99**, 1–13 (2011)
12. Quadrifoglio, L., Hall, R.W., Dessouky, M.: Performance and design of mast services: bounds on the maximum longitudinal velocity. Transport. Sci. **40**, 351–363 (2006)
13. Szablowski, P.J.: Discrete normal distribution and its relationship with Jacobi-theta functions. Stat. Probab. Lett. **52**, 289–299 (2001)
14. Campbell, A.M., Savelsbergh, M.: Efficient insertion heuristics for vehicle routing and scheduling problems. Transport. Sci. **38**(3), 369–378 (2004)

For the same block-sized grid street network system, the percentage [illegible] efficiency increases [illegible] an increase in passenger counts served by the DRT shuttle. These findings [illegible] useful for transportation planners and practitioners who can use their own parameter settings to derive and examine [illegible]. Although it may [illegible] for DRT services. This certainly makes this work [illegible] not only for TPS but across other [illegible] areas involving grid network systems. Our future work consists of extending the findings from this research to non-uniform passenger distribution over grid street networks and such [illegible].

[illegible]

An NLP-Based Cryptosystem to Control Spread of Fake News Through Social-Media

Arghya Ray and Pradip Kumar Bala

Abstract The era of digitalization has seen a transition from traditional methods of providing services to digital means. The advancement of technological innovations has not only impacted the human lifestyle but also has aided the growth of various social channels for people to communicate with one another. But the rise of social media platforms has also led to the spread of fake news. Fake news spread fast and can affect the brand image of various product/service-based organizations. To reduce the effect of fake news, we have proposed a method to temporarily encrypt the message sent from a consumer using the company's tag for a fixed time, say, 1 day, in Twitter, hence giving the company the time to respond to the concerns raised by the customer. Once the response is received, or the stipulated time ends, the message will be made visible to the public automatically.

Keywords Data analytics · Fake news · Knapsack cryptosystem · Natural language processing · Twitter

1 Introduction

The era of digitalization and penetration of Internet has helped companies to advertise their products or services to consumers within minutes. While companies need to focus on consumer engagement for building a long-lasting and successful relationship between brand and consumer, fake news about the brand can result in the loss of brand image [1]. Fake news is referred as the viral posts that look like actual reports and are intentionally made by certain people to misguide the readers [2]. Such fake news results in loss of loyal customers and defamation of the company within minutes. The presence of social platforms enables people to stay "just an app" away from any news going on around the world and people often share articles shared by people in

A. Ray (✉) · P. K. Bala
Information Systems Area, Indian Institute of Management Ranchi, Ranchi, India
e-mail: arghya.ray16fpm@iimranchi.ac.in

P. K. Bala
e-mail: pkbaba@iimranchi.ac.in

H. Das et al. (eds.), *Progress in Computing, Analytics and Networking*,
Advances in Intelligent Systems and Computing 1119,
https://doi.org/10.1007/978-981-15-2414-1_44

their social network without much efforts to check the authenticity of the message they share. These activities can lead to spread of fake news rapidly. Through this paper, we propose a method to prevent the spread of fake news using encryption.

The cipher block chaining (in short, CBC) encryption is a technique for encrypting a block of textual data. The specialty of this algorithm is that the input is actually the output of the operations performed on the preceding encoded text block. The second advantage is that the input block for encryption at each step bears no fixed relationship with the original text. The knapsack problem is an NP-complete problem in combinatorial optimization. Knapsack problem is used here since they help to solve various encryption problems in industries where public keys are involved.

In this article, we try to provide a solution for product/service-based organizations present on social media platforms, like, Twitter, Facebook, LinkedIn, etc., to avoid such situations through a cryptosystem which will enable such organizations to respond to the issues raised by the audience before the message is posted on the social platform. This will give them the buffer time to reduce fake news and defamation.

The article is divided into the following sections. Section 2 following this discusses the literature review followed by the proposed method. It is followed by the results, discussion, implications, and conclusions.

2 Review of Existing Literature

In this era of data analytics, there has been advancement in almost all aspects including data analytics and networking [3]. There has been improved optimizations [4], intelligent health recommendations [5], etc. But, with the advancement of technological innovations and the social media platforms, there has also been an increase in the spread of fake news. Network security now plays an important role not only for privacy concerns but also for protecting the image of individuals and organizations from fake social issues.

The 0–1 knapsack problem is a popular technique to identify the weights associated with each item for filling the knapsack of a predefined capacity. The main objective is to maximize the weight of the items without increasing the knapsack's capacity. This whole idea is summarized below:

$$\text{Max} \cdot \sum_{i=0}^{n} a_i x_i \tag{1}$$

$$\text{Subject to: } \sum_{i=0}^{n} \text{we}_i x_i \leq \text{WE} \tag{2}$$

$x_i = 1$, when the item is taken; $x_i = 0$, when the item is discarded.
where

'a' value of the item 'i',

‘we’ weight of item ‘I’,
‘WE’ knapsack’s maximum capacity, and
‘*n*’ number of items.

The subset sum problem is an advanced knapsack problem. Here, a group of integers is taken from a vector V. Another binary vector X is created, so that V * X = S (S = sum of the subset of elements in V) [6]. A super-increasing knapsack vector (s) and another vector M are created by modular multiplication and permutation. M is the public key of the encryption process which is also needed for decryption [7].

Twitter is a micro-blogging site where users can post their views. These are called tweets. Users often mark their posts with a particular labels or “hashtags”. This can help other users to identify a certain message [8, 9]. Customers posting comments using ‘@’ against a company name get displayed instantly and hence if a consumer is posting rumors, it gets posted and can misguide others.

Over the years, researchers have been working on various techniques to improve the security aspect during usage of internet. Some of the techniques used by earlier researchers involve decision tree approach by use of C4.5 [10], cipher block chaining (CBC) method [11], Merkle–Hellman knapsack cryptosystem (MHKC) [12], etc. CBC encryption accepts the plain text in a number of blocks and encrypts the blocks simultaneously, Merkle–Hellman knapsack cryptosystem (hereby referred to as MHKC) is another encryption technique. The tweets consist of blocks of text. Hence, we felt that CBC and MHKC are suitable for solving our purpose. A combinatorial model using CBC and MHKC was proposed by Ray et al. [13] which have been used in this study.

3 Proposed Method

In order to reduce the impact of fake messages, we propose certain steps using the combinatorial cryptosystems to encrypt messages posted by customers so that organizations can read the message and only after proper reply, the message will be visible to all. If the company fails to address the query/issue within a stipulated time, say 1 day, the message will be shown to all. This will prevent spread of rumors.

The proposed model is prepared such that the message posted by the customers is encrypted using the combinatorial cryptosystem. The unique key is dependent on the previous output and hence helps to generate a more powerful ciphertext. Performing the operations in reverse order helps to get the original message back. Figure 1 demonstrates the encryption process. The steps involved are described below:

Step 1: A message is extracted from Twitter from an Indian e-commerce website “Snapdeal”.
Step 2: An initial vector (referred to as IV) and initial key (referred to as IK) (here the user_id) (each 7-bit long) is chosen.
Step 3: Twitter message is divided into blocks (p_1, p_2, …, p_n) of equal length. Each block contains 7 bits each.

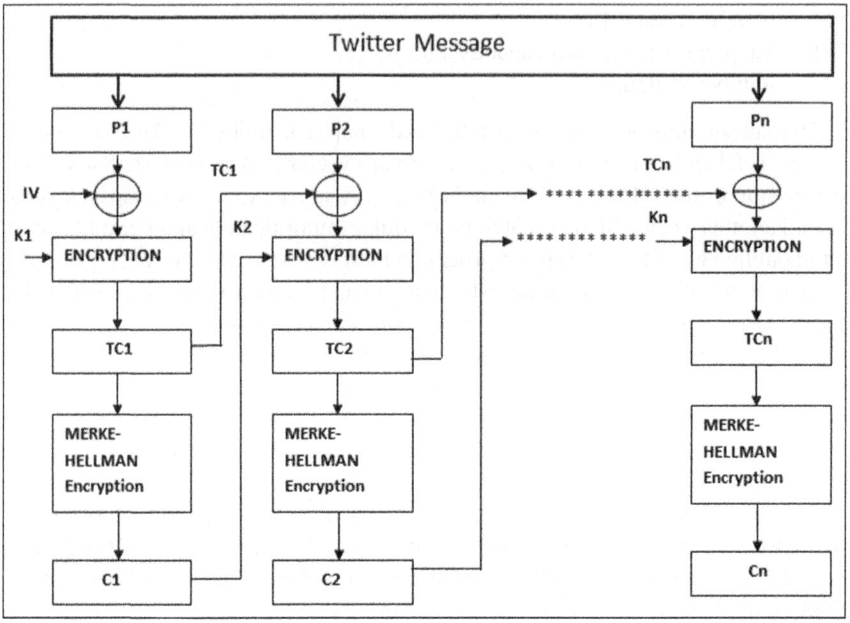

Fig. 1 The encryption process (Adapted from, *Source* Ray et al. [13])

Step 4: Each block is encrypted by using the IV and IK. This generates a temporary message (TM).
Step 5: A super-increasing sequence of positive integers is chosen $s = (s_1, s_2, s_3, \ldots, s_n)$.
Step 6: Two secret numbers (an integer 'a' and its co-prime 'r') are chosen. The integer 'a' should be greater than the sum of all numbers 's'. 's' and 'a' and 'r' together form the private key.
Step 7: $b_i = r * s_i \bmod (a)$.
Step 8: TM is now encrypted through MHKC. This generates the ciphertext blocks $(C_1, C_2, \ldots, C_n)$.

Figure 2 shows the decryption process. The steps involved in the decryption process are as follows:

Step 1: The encrypted message is broken down to get the ciphered message and the IV and IK.
Step 2: Similar to encryption, now the blocks of encrypted message are decrypted simultaneously.
Step 3: Each block is initially decrypted using MHKC.
Step 4: The blocks are again decrypted using the IV and IK.
Step 5: Steps 3 and 4 are repeated for all the ciphered blocks.
Step 6: The output generates the plaintext blocks $(p_1, p_2, \ldots, p_n)$.

The encrypted text can only be decrypted by the person who has the private key.

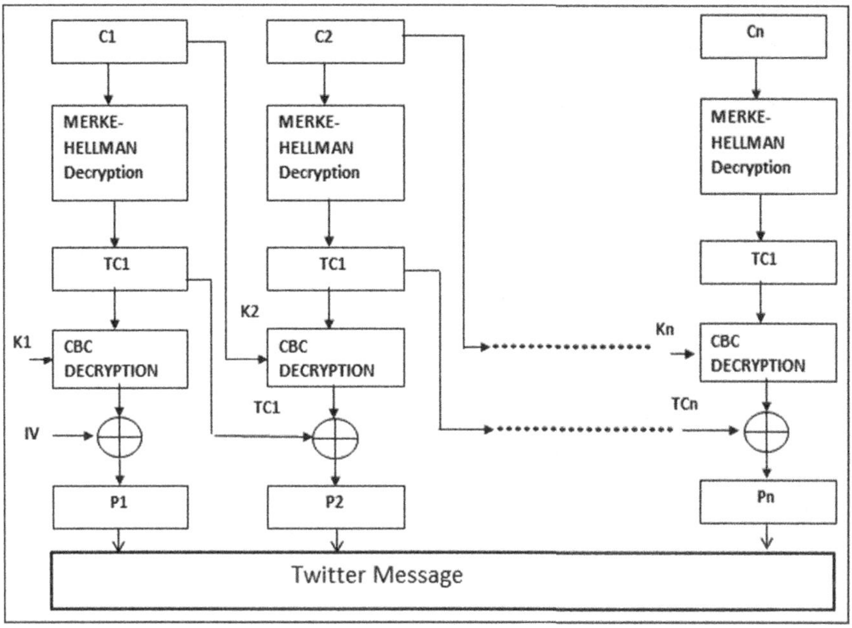

Fig. 2 Steps involved in the decryption process (Adapted from, *Source* Ray et al. [13])

$$M = \sum_{i=0}^{n} p_i^* b_i \tag{3}$$

4 Results

For developing the cryptosystem and explaining the results, Java platform BlueJ version 1.3.5 and RStudio Version 1.0.136 were used. Using RStudio, the authors have extracted tweets where "Snapdeal" was used after the period "2014-08-20" and we got 4160 records (shown in Fig. 3). We take any tweet as sample input, and two 7-bit binary values or in short numeric codes—one for the IV (this one the company can provide while registering) and one for the IK as inputs (which can be based on the customer user id) and we get the ciphertext as output. For our demonstration purpose in this paper, we have used a small text "Snap" which the customers use often. We have not shown automated code extraction process here, since we are more interested to understand the process. Hence, we have demonstrated the output with just a word used frequently by customers.

This encryption output shown in Fig. 4 shows the input message "Snap" (we can automate the process in real time and pass the entire line as input). User or the

	text	favorited	favoriteCc	replyToSN	created
1	@Snapdeal_Help I request you to kindly tak	FALSE	0	Snapdeal_	9/27/2018 7:09
2	@snapdeal order id 23576643628 & sub	FALSE	0	snapdeal	9/27/2018 7:08
3	@Snapdeal_Help Pathetic customer care wh	FALSE	0	Snapdeal_	9/27/2018 7:07
4	What should you climb next?	FALSE	0	NA	9/27/2018 7:06
5	@AjayKum67550010 Thanks for sharing the	FALSE	0	AjayKum6	9/27/2018 7:04
6	@Snapdeal_Help Why are you not	FALSE	0	Snapdeal_	9/27/2018 7:02
7	eCommence Marketplace is so fast, Setup y	FALSE	1	NA	9/27/2018 7:01
8	@yatirajpanchal As informed, replacement	FALSE	0	yatirajpan	9/27/2018 7:00
9	@Snapdeal_Help I had provided my	FALSE	0	Snapdeal_	9/27/2018 6:58
10	@Snapdeal_Help Hey, Please explain here,	FALSE	0	Snapdeal_	9/27/2018 6:58
11	@MohdHusayn Please do not respond to an	FALSE	0	MohdHus	9/27/2018 6:56

Fig. 3 Data extracted from Twitter

Fig. 4 Output of the encryption process

```
BlueJ: Terminal Window - Cryptosystem
Options
Enter the plain text
Snap
Enter the Initialization Vector
0100100
Enter the key
1000011

 *****************

The cipher text is..
301608553678
```

company executive enters an initial code and a secret key (this also can be automated during the registration process). We get the ciphertext. This is done to maintain the privacy of the user. This message is then transmitted to the concerned company issue solving editor.

At the company's site, the message will get decrypted only when the company's employee uses the correct initialization vector. The key and the other processes will be automated. The decryption process is given in Fig. 5.

We find that we have received the original message back. So when a customer enters a code, and a sentiment analysis through machine learning tells the system that the customer has posted some negative comments about the company, the message will automatically get encrypted and will be just be visible to the customer and the company (provided the company's employee enters the right key). The encryption is done to not only protect the customer privacy, but also to avoid false allegations/concerns. In case the company fails to reply within a stipulated time, the

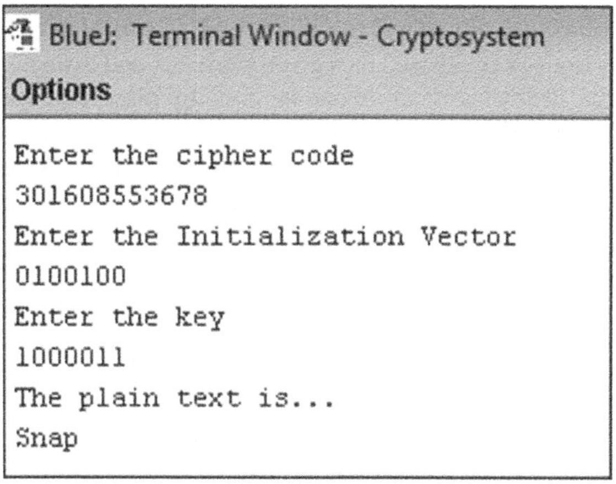

Fig. 5 Output of the decryption process

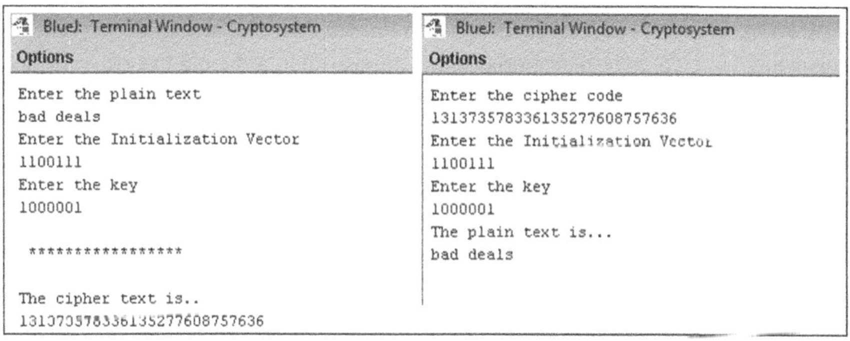

Fig. 6 The encryption and decryption processes of another block of text

message will be automatically made visible to the public. Another simulation is shown in Fig. 6. It is to be noted that the ciphertext here is denoted by numbers. It can be converted into alphabets as well which we are actually used to. Also, the IV and IK can be embedded in the encrypted message. We have just shown these figures for demonstration purpose only.

5 Discussion

This study provides a new avenue to tackle the growing issues related to fake messages. Information security plays an important role in every sector and even biomedics [14]. The usage of encryption techniques will hamper the spread of messages for some time but will help brands or companies from falling into the trap of falsity.

To achieve this objective, this study has utilized an NLP-based (web mining) method to extract tweets from Twitter based on a given keyword, and extract sentences which they feel are fake. This classification can be done by the use of various classifiers like Naïve Bayes, etc. A combinatorial cryptosystem is used to encrypt the assumed fake message. Checks are done to verify if it is actually a fake message or not. If not, this message can be decrypted after a certain period of time. If the company fails to verify within the given period of time, the message will automatically be available online again.

It is to be noted that the combination of the two cryptosystems helps to improve the strength of the key because of its combination with a super-increasing sequence. The CBC helps in chaining the blocks together, thus making the encrypted message difficult to break. In future, this method can be tried with much stronger methods like DES, AES, etc. to see if it can improve the performance.

5.1 Implications

This study has two main managerial implications. First, reducing the effect of fake news can help the companies take preventive measures when they see fake news related to them. Second, this study will help managers get some time to strategize accordingly and reduce the impact of the fake message. This will also help them find out if there is any service gap that they can rectify.

The theoretical implications of this study are as follows: First, this study will help researchers in management domains to look into proper use of encryption techniques for handing various safety issues related to big data. Second, this study contributes to the existing literature on fake news affecting brand images.

6 Conclusion and Future Directions

This paper explains how to reduce issues faced by organizations from fake news in this digital era. The double encryption-based cryptosystem helps in establishing the secrecy of the message posted by a customer and also helps companies to address the concerns within a fixed time period. The whole proposed system has been demonstrated using data extracted from Twitter of the Indian e-commerce Snapdeal. In future, researchers can enhance this model for preventing spread of fake news during natural disasters which leads to death of many innocent people from all around the globe. This model can also help in preventing lynching cases due to spread of fake news about certain person. Additionally, researchers can also work on developing a machine learning algorithm which can detect fake news. A simple way is to use the Naïve Bayes technique. Naïve Bayes has been used in various contexts like detecting diabetes mellitus [15]. Some other classification techniques that can be followed is PPO-based backpropagation learning MLP [16].

Acknowledgements We would like to thank the authors whose papers we have cited for helping us with the idea of the proposed model. I would also like to thank Twitter and Snapdeal whose data we have used for our study.

References

1. Ravaglia, V., Brivio, E., Graffigna, G.: Engaging consumers via Twitter: three successful communicative strategies. Digital marketing and consumer engagement: concepts, methodologies, tools, and applications (2018)
2. Tandoc Jr., E.C., Lim, Z.W., Ling, R.: Defining 'fake news'. Digit. J. **6**(2), 137–153 (2017)
3. Pattnaik, P.K., Rautaray, S.S., Das, H., Nayak, J.: Progress in computing, analytics and networking. In: Proceedings of ICCAN (2017)
4. Nayak, J., Naik, B., Jena, A.K., Barik, R.K., Das, H.: Nature inspired optimizations in cloud computing: applications and challenges. In: Cloud Computing for Optimization: Foundations, Applications, and Challenges, pp. 1–26. Springer, Cham (2018)
5. Sahoo, A.K., Mallik, S., Pradhan, C., Mishra, B.S.P., Barik, R.K., Das, H.: Intelligence-based health recommendation system using big data analytics. In: Big Data Analytics for Intelligent Healthcare Management, pp. 227–246. Academic Press (2019)
6. WolframMathWorld: http://mathworld.wolfram.com/SubsetSumProblem.html
7. Merkle, R., Hellman, M.: Hiding information and signatures in trapdoor knapsacks. IEEE Trans. Inf. Theory **24**, 525–530 (1978)
8. Weng, L., Flammini, A., Vespignani, A., Menczer, F.: Competition among memes in a world with limited attention (2012)
9. Simon, T., Goldbergb, A., Adinia, B.: Socializing in emergencies—a review of the use of social media in emergency situations. Int. J. Inf. Manage. **35**, 609–619 (2015)
10. Sahani, R., Rout, C., Badajena, J.C., Jena, A.K., Das, H.: Classification of intrusion detection using data mining techniques. In: Progress in Computing, Analytics and Networking, pp. 753–764. Springer, Singapore (2018)
11. Bellare, M., Kilian, J., Rogaway, P.: The security of cipher block chaining. Adv. Cryptol. CRYPTO **94**, 341–358 (1994)
12. Lagarias, J.C.: Performance analysis of Shamir's attack on the basic Merkle-Hellman knapsack cryptosystem. International Colloquium on Automata, Languages, and Programming, pp. 312–323. (1984)
13. Ray, A., Padhmavathi, B., Anjum, A., Bhat, S.: Improvement of CBC encryption technique by using the Merkle-Hellman Knapsack Cryptosystem. Intelligent Systems and Control (ISCO), pp. 340–344 (2013)
14. Pradhan, C., Das, H., Naik, B., Dey, N.: Handbook of Research on Information Security in Biomedical Signal Processing, pp. 1–414. IGI Global, Hershey, PA (2018)
15. Das, H., Naik, B., Behera, H. S.: Classification of Diabetes Mellitus Disease (DMD): A Data Mining (DM) Approach. In: Progress in Computing, Analytics and Networking, pp. 539–549. Springer, Singapore. (2018)
16. Das, H., Jena, A.K., Nayak, J., Naik, B., Behera, H.S. A novel PSO based back propagation learning-MLP (PSO-BP-MLP) for classification. In: Computational Intelligence in Data Mining, vol. 2, pp. 461–471. Springer, New Delhi (2015)

Analysis of Proactive Simulated Topology Reconfiguration for WDM Networks

P. Liyaz, B. Surendra Reddy, K. Manoj Kumar, A. Basi Reddy and P. M. D. Ali Khan

Abstract The existing approaches for reconfiguration followed reactive type, reconfiguring only after the change in traffic or after the occurrence of failure. The reactive type of reconfiguration algorithms suffers from poor Quality of Service (QoS) performance since there is an exchange off between QoS parameters. In this exploration work, another methodology called, Proactive simulated Topology Reconfiguration for WDM networks, is proposed so as to solve the shortcomings in the existing approaches. Further, most of the existing approaches use Poisson traffic model which fails to represent the real network traffic from WAN or LAN. To solve this problem, a self-similar traffic model with Gaussian distribution is proposed which is used throughout the validation of the proactive VTR. An efficient traffic prediction model using the tool of wavelet transform is derived to compute the future traffic on links of the network with Gaussian traffic distribution. This new prediction model shows that the prediction error ratio less than 4%. The computational cost in terms of number of changes and number of iterations for the proactive reconfiguration is also less compared with the reactive approaches.

Keywords WDM optical networks · Virtualization · QoS

P. Liyaz (✉) · B. S. Reddy · K. M. Kumar · A. B. Reddy · P. M. D. A. Khan
Department of Computer Science and Engineering, Sri Venkateswara College of Engineering (SVCE), Karakambadi Road, Tirupati 517507, Andhra Pradesh, India
e-mail: liyaz.liyaz2009@gmail.com

B. S. Reddy
e-mail: suri1253@gmail.com

K. M. Kumar
e-mail: kandalamanojkumar@gmail.com

A. B. Reddy
e-mail: basireddy.a@gmail.com

P. M. D. A. Khan
e-mail: pathan.md@gmail.com

H. Das et al. (eds.), *Progress in Computing, Analytics and Networking*,
Advances in Intelligent Systems and Computing 1119,
https://doi.org/10.1007/978-981-15-2414-1_45

1 Introduction

One of the serious issues in the system designing today is gigantic interest for increasingly more transfer speed. The unexpected development of Internet drastically expands the interest for information transmission limit. Expansion of Internet-based services in day-to-day life generates the demand for high bandwidth in the network backbone, and this demand will grow further along with new real-time multimedia applications [1]. The corresponding enormous increase in network bandwidth due to Wavelength Division Multiplexing (WDM) has finely tuned the need for faster switching at the core of the network. Currently, there is no other technology, which can meet this demand rather than WDM technology. With the advancement of optical systems and the utilization of WDM innovation, another and most likely, a pivotal achievement is being come to in system development.

Lately, the WDM methods and Optical fiber innovations together acquired an unrest fast information correspondence system [2], which are presently ready to fulfill the high transfer speed needs of information traffic. WDM innovation has risen as capable innovation for spine systems and Wide Area Networks (WAN) with Terabits every subsequent data transfer capacity viewed as a potential contender for the cutting edge wide territory spine systems [2]. In these kinds of systems, steering and exchanging are done in the optical area as opposed to electronic space. WDM innovation has risen as a promising innovation for the spine systems and WAN with every subsequent data transfer capacity. As the reconfiguration is a multi-objective optimization problem, which is computationally hard to solve, several heuristic algorithms are available in the literature.

The Internet Protocol (IP) over WDM systems will be the correct possibility for the future age web arranges as they give colossal transfer speed fulfilling exponentially developing information traffic [3]. In these types of networks, IP payload is directly carried over WDM physical network, forming a virtual topology in the IP-layer They are system configuration issues, mechanical issues and system operational issues. Casually, this examination business locales the issues on Topology Engineering, which puts the system data transfer capacity where the traffic necessities emerge. Interestingly, traffic designing is to put the system traffic where transfer speed is accessible. Be that as it may, topology reconfiguration and traffic designing both improve arrange use and traffic execution.

The work is systematized as follows. Section 1 describes the need for proactive VTR and its challenges. Section 2 reviews the literature in the field of WDM optical networks. Section 3 states the proposed work followed by its description of methodology. The simulated outcomes and entire discussion are made in the Sects. 4 and 5 conclude the approach.

2 Literature Survey

The greater part of the current reconfiguration methodologies pursued responsive sort, reconfiguring simply after the adjustment in rush hour gridlock. The traffic-based reconfiguration system for the consistent topologies in huge scale IP-over WDM systems is depicted by Yongbing et al. (2005). The creators Brzezinski and Modiano (2005) have read calculations for the joint WDM reconfiguration system & IP-layer steering in IP-over-WDM systems [4]. The structuring calculations can work dependent on the greatest weight booking, and arrive at asymptotically ideal throughput. Akgun et al. (2006) researched the VTRP of optical WDM organizes by taking the traffic preparing factor into thought. The VTRP considering traffic preparing was figured with new factors and imperatives by a blended whole number straight program (MILP).

An online versatile reconfiguration approach was proposed to pursue the dynamic changes in rush hour gridlock designs without the earlier learning. They consider an adjustment component for reconfiguration [5]. The proposed calculation upgrades the virtual topology as per a normal traffic design. Toward the part of the bargain time frame normal light path burdens are determined. Here the creators consider a solitary advance methodology where reconfiguration is performed through changing the topology adaptively with single light path expansion or erasure. A higher and lower watermark level is utilized to discover when to reconfigure the system by including or erasing the light paths [5]. In this basic methodology, reconfiguration calculation goes about as a system controller that directs the system to manage traffic changes for an ideal topology. A focal director is in charge of gathering traffic data and light path change. The issue of VTR for the traffic versatile WDM systems fathomed with activating component (Baldline et al. 2001). The versatile reconfiguration plans depicted above are utilizing lower and higher limit esteems with the supposition of the traffic vacillations are smooth and pursue an example. However, they neglect to adjust with sudden traffic changes, following no example.

The traffic models depicted (Gencata et al. 2002; Baldline et al. 2001; Yongbing et al. 2005; Brzezinski et al. 2005) in the writing pursued Poisson forms on the grounds that such procedures are scientifically tractable [6]. In any case, the ongoing investigations (Paxson and Sally Floyd 1995) have demonstrated that Poisson procedure neglects to speak to both neighborhood wide region system traffic, and the appropriation of bundle bury entries plainly contrasts from exponential. Such sort of systems can be better demonstrated utilizing measurably self-comparative procedures, which have vastly different hypothetical properties than Poisson forms because of the way that there is no characteristic length for a "burst" yet traffic blasts show up on a wide scope of time scales.

The greater part of the methodologies accessible in writing expected that the traffic request is known ahead of time, and dependent on the realized traffic design another topology is planned [7]. Thus, reconfiguration is activated because of some occasions instead of thinking it as an online procedure. In addition, the accentuation has stayed on topology structure and reconfiguration calculations instead of reconfiguration activating and topology relocation systems. In the main arrange, the reconfiguration is constrained to a couple of changes so as to accelerate the reconfiguration procedure and decrease the reconfiguration cost. In separated administrations, parcels are stamped distinctively dependent on the administration prerequisites [7]. A heuristic algorithm for virtual topology reconfiguration with successive approximations is given by Easwaran Baskaran et al. (2007). The authors Seungyeon You et al. (2007) proposed Virtual Topology Reconfiguration controller for WDM networks. The reconfiguration is problematic to the system under activity. In this way, it desires to think about the interchange between the exhibition of new consistent topology & the expense of topology production [8].

The past research works demonstrate that self-comparative system traffic can detrimentally affect arrange execution, particularly on QoS parameters. Consequently, there is a need to address the issue of traffic displaying to improve QoS execution for the future age virtual worlds utilizing IP-over-WDM innovation.

3 Proposed Work

The virtual topology reconfiguration issue considered in this examination work is multi-target issue, which considers not just the QoS goals of IP-over-WDM systems, yet in addition execution destinations, for example, AWHT and number of changes in virtual topology. The new approach devised in this research work is a proactive one, which estimates the future traffic demands and measures the corresponding QoS metrics, based on which the reconfiguration is activated. The mere changes in the traffic demands or in the physical topology or any failure of any of the network component will not trigger the reconfiguration [9]. This problem assumes wavelength continuity constraint since flexible wavelength conversion in all nodes of the WDM network increases the network cost. Given a physical topology T1 which is enhanced for system state S1 and traffic request frameworks, the proactive virtual topology reconfiguration issue is to discover when and how to change the system topology into T2 which is ideal for the future system state S2 with insignificant disturbance and maximal QoS. The learning about the future state S2 is inaccessible at the hour of reconfiguration. The proactive VTR issue includes the accompanying five sub-issues.

i. Traffic modeling sub-issue
ii. QoS modeling sub-issue
iii. Lightpath addition/deletion sub-issue
iv. Lightpath routing and wavelength assignment sub-issue and
v. Traffic routing sub-issue.

Traffic modeling subproblem models the future traffic using normal traffic distribution and wavelet transform [10]. The QoS modeling subproblem, models QoS parameters for VTR problem The DCA limitation expresses that light ways on a similar fiber connection must be relegated with the particular hues. The traffic steering sub-issue is a directing of traffic on a virtual topology. Our concern detailing considers the SONET traffic streams unto the limit of light ways. The QoS model found in literature for the IP-over-WDM networks assumes Poisson distribution. As it fails to represent the accurate model, a new QoS model using Gaussian distribution is devised. The proactive VTR approach proposed needs future traffic demands as input. Hence, the real-time traffic patterns are modeled and predicted in advance using Gaussian distribution and Wavelet Transform, respectively [11]. Further, efficient heuristic algorithms for proactive VTR are also devised to study the performance of the proposed approach. The new proposed approach monitors the traffic and predicts the future traffic between all source destination pairs using normal traffic distribution. Considering the existing virtual topology and its performance metrics, QoS and also the predicated traffic, the reconfiguration is triggered. The most congested lightpath is chosen to be deleted and a new lightpath is established with maximum traffic flow and reroute the traffic through it. This step also repeated until reaching the maximum number of changes or till deleting all underutilized lightpaths. Since this is proactive reconfiguration, if there is any degradation of network performance or QoS, then the reconfiguration heuristics will be triggered. The following two different proactive reconfiguration algorithms are presented.

i. Blocking Parameter Driven VTR Algorithm.
ii. Differential Traffic Driven VTR Algorithm (Fig. 1).

The blocking parameter-driven VTR expects to decrease parcel misfortune or blocking likelihood [12]. This heuristic calculation predicts the future traffic and gauges parcel misfortune and henceforth the blocking likelihood. This calculation acknowledges weighted jump check (WHT) for all s-d matches as info. In the event that the blocking parameter is more noteworthy than the limit, VTR calculation is called and emphasizes till arriving at the negligible blocking [12].

Fig. 1 Heuristic approach of proactive VTR

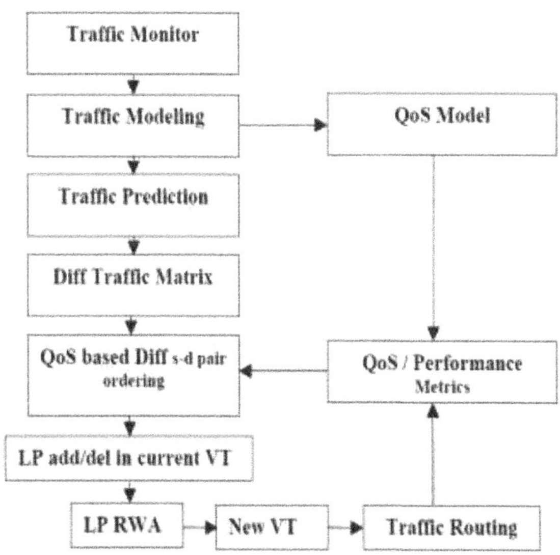

4 Experimental Analysis

The performance study involves the reconfiguration of logical or virtual topologies viewed by Internet Protocol for carrying data. While considering the problem of virtualization in the network layer, it is essential to consider the QoS parameters related to the network layer, especially on IP. The QoS parameters associated with the virtualization are

i. Reconfiguration latency
ii. Blocking probability,
iii. Throughput,
iv. Resource utilization.

The reconfiguration process is repeated by varying the percentage of traffic change and the corresponding variation in the Average Weighted Hop Count for the Topology (AWHT) is plotted in Fig. 2. From this chart, it is seen that the AWHT is getting decreased after half change in rush hour gridlock. The decrease in AWHT is because of the presence of more light paths and the ideal way taken by the RWA calculation [13]. The blocking probability measured for different virtual loads of the dynamic network with Gaussian traffic model is plotted in Fig. 3. As the proactive approach is driven by critical blocking probability and predicted future traffic demands, lightpaths are satisfying most of the dynamic traffic changes and hence the blocking probability is minimal compared to the reactive approaches [14].

The system dormancy estimated by differing levels of progress in rush hour gridlock for dynamic system with Gaussian traffic model. From this chart, it is seen that the system inactivity is insignificant contrasted with the receptive methodology. The improvement in network throughput is due to the minimal blocking of connection requests [14]. In other words, most of the traffic loads are carried by the available reconfigured lightpaths, which increases throughput of the network. Comparing with reactive approach, the proactive approach optimizes the number of lightpaths for increasing the traffic load and hence the network Utilization is considerably improved. It is also to be noted that after reaching 50% traffic load, the

Fig. 2 AWHT versus percentage of traffic change

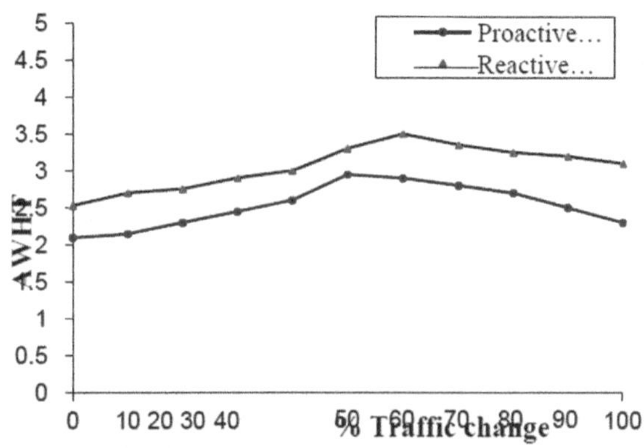

Fig. 3 Blocking probability versus virtual link load

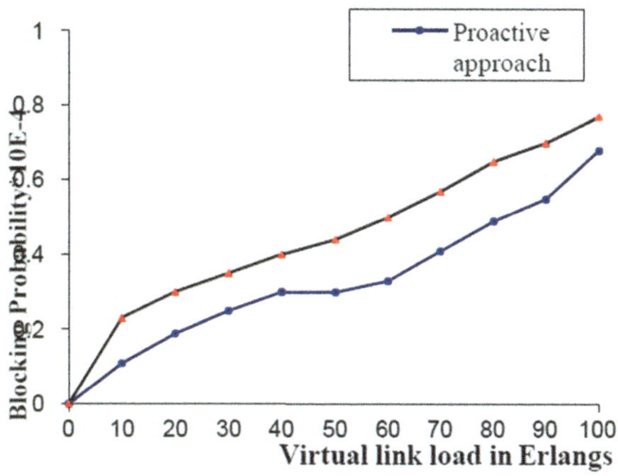

utilization traffic, is the predicted traffic value. It is computed decreases. This is due to the creation of more lightpaths for increased load but not all are fully utilized. This section analyzes the computational complexity of the proactive VTR. A divide and conquer type of algorithm is used to perform the traffic series estimation using wavelet transform model, which involves a runtime of O(log N) for N(N−1) elements [15]. Therefore, for small N a runtime of O(N) is expected. It should be noted that the initialization for the algorithm has a runtime that grows as a polynomial in the number of nodes (N) in network. But once the divide and conquer binary tree is completed the reconfiguration stage of the algorithm is running faster than any other algorithm.

5 Conclusion

This research work proposed a new approach called Proactive Simulated Topology Reconfiguration for WDM networks with Gaussian Traffic modeling to satisfy QoS requirements. The proposed proactive approach of reconfiguration performs well for single hop as well as multi-hop IP-over-WDM optical networks. The proposed heuristic methodology was approved for NSFNET with 14 hubs utilizing recreation. The QoS metrics observed are blocking probability: 0.68×10^{-4}, latency: 0.39×10^{-4}s and throughput: 0.95 which are much better than the reactive approaches found in literature. The simulation results are also correlating with the analytical results. The performance of the proactive reconfiguration heuristics is studied through several metrics namely Average Weighted Hop Distance (AWHD), Average Weighted Packet Delay (AWPD), Reconfiguration Cost (RC) and Reconfiguration Gain (RG), which are comparatively better than reactive approaches. The Virtual Topology Utilization (VTU) approaches near unity (0.9799) which shows the efficient utilization of the virtual topology resulted from the proactive approach. Further, the new proactive approach yields the reconfigured topology near optimal and hence this new approach avoids frequent and unnecessary reconfigurations.

References

1. Basi Reddy, A., Liyaz, P., Surendra Reddy, B., Manoj Kumar, K., Sathish, K.: Advanced spatial reutilization for finding ideal path in wireless sensor networks. In: Singh, M., Gupta, P., Tyagi, V., Flusser, J., Ören, T., Kashyap, R. (eds.) Advances in Computing and Data Sciences. ICACDS 2019. Communications in Computer and Information Science, vol. 1046. Springer, Singapore (2019)
2. Sinha, S., Rammohan, N., Murthy, C.S.R.: Dynamic virtual topology reconfiguration algorithms for groomed WDM networks. Photonic Netw. Commun. **9**(2), 181–195 (2005)
3. Chaturvedi, Sushil, Shrivastava, Manish, Goel, Aditya: Examining of reconfigurations and rerouting operations: WDM networks. Int. J. Inform. Technol. Knowl. Manag. **5**(1), 69–72 (2012)
4. Takagi, H., Zhang, Y., Jia, X., Takagi, H.: Reconfiguration heuristics for logical topologies in wide-area WDM networks. In: Proceedings of IEEE Global Telecommunications (GLOBECOM 02), pp. 2701–2705. Taipei, Taiwan (2002)
5. Jany Shabu, S.L., Manoj Kumar, K.: Preserving user's privacy in personalized search. Int. J. Appl. Eng. Res. (IJAER) **9**(22), 16269–16276 (2014)
6. Manoj Kumar, K., Vikram, M.: Disclosure of user's profile in personalized search for enhanced privacy. Int. J. Appl. Eng. Res. (IJAER) **10**(16), 36358–36363 (2015)
7. Kandala, H., Tripathy, B.K., Manoj Kumar, K.: A framework to collect and visualize user's browser history for better user experience and personalized recommendations. In: Satapathy S., Joshi, A. (eds.) Information and Communication Technology for Intelligent Systems (ICTIS 2017)—Volume 1. Smart Innovation, Systems and Technologies, vol. 83. Springer (2018)
8. Yuvaraju, B.V., Venkiteela, L., Kumar, K.M.: A presumptuous encephalon electronic brain computing in gaming gadgets. In: 2018 International Conference on Soft-computing and Network Security (ICSNS). Coimbatore, India (2018)
9. Kumar, K.M., Kandala, H., Reddy, N.S.: Synthesizing and imitating handwriting using deep recurrent neural networks and mixture density networks. In: 2018 9th International Conference on Computing, Communication and Networking Technologies (ICCCNT), pp. 1–6. Bangalore (2018)
10. Devi, K.J., Moulika, G.B, Sravanthi, K., Kumar, K.M.: Prediction of medicines using LVQ methodology. In: 2017 International Conference on Energy, Communication, Data Analytics and Soft Computing (ICECDS), pp. 388–391. Chennai (2017)
11. Bhavana, G., Kumar, K.M.: Cardiac disease monitoring based on ZigBee technology. In: 2017 International Conference on Energy, Communication, Data Analytics and Soft Computing (ICECDS), pp. 392–395. Chennai (2017)
12. Praveen Kumar, R., Manoj Kumar, K., Tejasree, S., Aswini, R.: Review on cost effective and dynamic security provision strategy of staging data items in cloud. Res. J. Pharm. Biol. Chem. Sci. (RJPBCS) **7**(6), 1592–1597 (2016)
13. Razvi, S.A., Neelima, S., Prathyusha, C., Yuvasree, G., Ganga, C., Kumar, K.M.: Implementation of graphical passwords in internet banking for enhanced security. In: 2017 International Conference on Intelligent Computing and Control Systems (ICICCS), pp. 35–41. Madurai (2017)
14. Manoj, K., Sandeep, T.S., Sudhakar Reddy, D.N., Alikhan, P.M.D.: Genuine ratings for mobile apps with the support of authenticated users' reviews. In: 2018 Second International Conference on Green Computing and Internet of Things (ICGCIoT), pp. 217–221. Bangalore, India (2018)
15. Rajendran, P.K., Asbern, A., Manoj Kumar, K., Rajesh, M., Abhilash, R.: Implementation and analysis of MapReduce on biomedical big data. Indian J. Sci. Technol. (IJST), ISSN/E-ISSN: 0974-6846 / 0974-5645 **9**(31), 1–6 (2016)

Arabic Text Extraction by Convolutional Neural Network Through Robust Dataset Design

Mosin Hasan, Bhavesh Tanawala, N. Dhruti, Ketan Bhalerao and Dhruvkumar Patel

Abstract Text extraction from images plays a vital role in modern-day applications like verifications of checks, vehicle's number plate scanning, finding evidences of crime cases, etc. we proposed a methodology of extracting Arabic characters from a given image. The process has been divided into three phases. In the first phase, the given image containing Arabic text to be extracted has been subjected to various image molding subprocesses like cleaning, blurring, binarization, and segmentation to make the image compatible for recognition phase. Available Dataset for Arabic has limitation of one form and hence we created dataset which fits real-world scenario (second phase). The final phase describes the Convolutional Neural Network (CNN) model used for recognizing the Arabic characters. The achieved accuracy of the trained model was about 95% for extraction of Arabic characters.

Keywords Arabic language · Text extraction · Image segmentation · Arabic dataset · Convolutional neural networks · Character recognition

M. Hasan (✉) · B. Tanawala · N. Dhruti · K. Bhalerao · D. Patel
Computer Engineering Department, BVM Engineering College, Anand, Gujarat, India
e-mail: mihasan@bvmengineering.ac.in

B. Tanawala
e-mail: bhavesh.tanawala@bvmengineering.ac.in

N. Dhruti
e-mail: dhrutipatel5501@gmail.com

K. Bhalerao
e-mail: iamketan14@gmail.com

D. Patel
e-mail: dhruv814@gmail.com

H. Das et al. (eds.), *Progress in Computing, Analytics and Networking*,
Advances in Intelligent Systems and Computing 1119,
https://doi.org/10.1007/978-981-15-2414-1_46

1 Introduction

1.1 Importance of Text Extraction

Extraction of words of languages like English is comparatively easy as the letters are generally written with certain space in the middle which makes the entire process straightforward for accurate results. However, for languages like Arabic, the above processes and algorithms cannot be applied since the Arabic letters are jointly written which makes the extraction process challenging. Also, not much research has been done in recognition of Arabic characters in an image due to which an initiative was undertaken to develop a new methodology for efficient extraction of Arabic letters from an image.

Text recognition from an image plays a really important role in variety of applications such as postal address reading, check verifications in banks, office automation, and many more. In addition to such common and traditional applications, there is a large interest in searching scanned documents that are available on the internet and for searching handwritten manuscripts. Other possible applications are preparing efficient libraries, text recognition over maps, recognizing vehicle license plates, using it as first phase in text readers for visually impaired people and understanding filled forms. Arabic text extraction has been one of the quiet unnoticed areas in terms of research and algorithmic development because of unique features of Arabic language which make the process of text extraction comparatively complex as compared to other languages.

1.2 Characteristics of Arabic Language

Arabic is an ancient language written in reverse order of English and is cursive in nature. It has an ancient flavor and is spoken by approximately 400 million people over the globe [1]. Moreover, it is also used by more than triple the previous number of Muslims all over the world as a second language, for it is the language in which the Holy Quran was revealed [1]. That is, Arabic is being used by more than 1.5 billion people [1]. It has 28 basic letters. A single letter in Arabic language can be written in four different forms depending upon the position. Vertical overlapping is also possible between letters of a single word. Arabic letters can be of different widths and heights. Diacritics can also be constructed over a single word. Entire meaning of a full word can get changed due to improper diacritic. Experts of Arabic can easily read the text correctly even without the presence of diacritics (Fig. 1).

Since every Arabic character can be written in four different forms, the extraction method must be able to differentiate the 28 letters as well as each of the four forms of a particular letter. Thus, in order to extract Arabic texts, the recognition algorithm has to differentiate between approximately 110 classes. All these unique characteristics make the overall process of extraction difficult in implementation [1] and

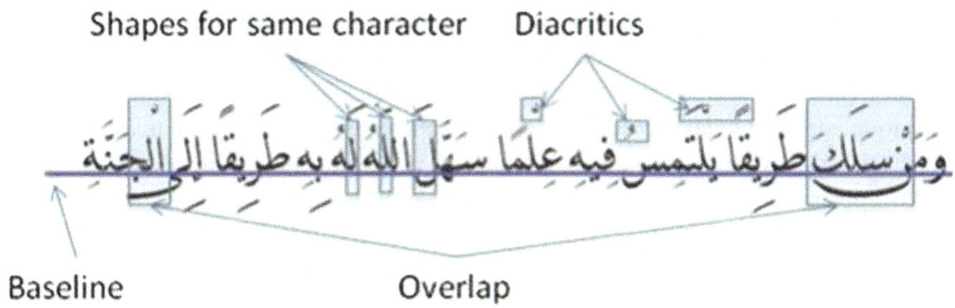

Fig. 1 Arabic sentence indicating some characteristics of Arabic language [15]

ا ب ت ث ج ح خ د ذ ر ز س ش ص ض ط ظ ع غ ف ق ك ل م ن ه و ي

Fig. 2 Basic Arabic characters [15]

thus, a better methodology is required for the extraction of such a complex textual language. Although certain datasets of standalone form of Arabic characters were readily available, the lack of rest of the three forms acted as a major barrier in getting the required accuracy (Fig. 2).

2 Existing Research in English and Arabic Text Extraction

The approach of finding candidate lines of text through multi-scale wavelet features and the structural information was proposed by Zhan and Wang [2]. Morphological Component Analysis (MCA) algorithm was utilized by Hoang and Tabbone [3] for effective text extraction from graphical document images. Maximum Color Difference (MCD) and Boundary Growing Method (BGM) for the detection of multi-oriented handwritten scene text from video was practiced by Shivakumara et al. [4]. Moreover, a segmentation technique was proposed by Gopalan and Manjula [5].

Khelifi and Zaghden in [6] proposed an approach for text extraction to categorize and index documents from documents having different properties. Text regions were identified based on their font size. SVM classifier, fractal descriptors were used to perform this task. To make image same in properties, different color layers were extracted. The results were binarized, followed by a required grouping of texts. Paraag Agrawal [7] proposed a scheme for the extraction of textual areas from an image using Globally Matched Wavelet filters with Fisher classifiers. They have used these filters to segment the document images and classify them into text, background, and picture components. The obtained results were improved by further subjecting them to Markov random fields. Samarabandu and Liu [8]'s dealt with printed document images as well as with scene text. The proposed method was based on characteristics like edge strength, density and the orientation variance. The output of the process was later converted to a feature map. Text detection was achieved through multi-scale edge detector and text localization stages were tackle through morphological dilation operator. Strouthopoulos [9] used a combination of an adaptive color reduction (ACR) technique and a page layout analysis (PLA) approach for making the text extraction process insensitive to variations in font, color, or size of the text inside the image.

An approach involving a set of sequential algorithms for text extraction and enhancement of image using cellular automata was proposed by Sahoo and Kumar [10] in which the Luminance-based algorithm was used to convert the image into gray scale image. A combination of artificial neural network along with Nonnegative Matrix Factorization for detecting complex texts in images was proposed by Jung and Kim [11].

A block methodology has been proposed by [7] for extracting simple Arabic text in pre-binarized images. As per the survey conducted by [12], even though a large number of algorithms have been proposed in the literature, no single method can provide satisfactory performance in all the applications due to the large variations in character font, size, texture, color, etc.

However, Al-Muhtaseb [1] proposed a method stating the use of Markov models for extraction of Arabic texts of different fonts like Akhbar, Naskh, Andalus, etc. Since Arabic is cursive, the segmentation process has been a spot of concern in this area. Moreover, pytesseract emerged as an important python utility for extracting Arabic text from image. However, in case of complex images, the results were not very efficient [13]. Suggested a method of recognition of printed Arabic text using decision tree learning techniques and features of global word [14].

The recognition algorithm fails to identify certain Arabic characters due to their cursive nature and thus the efficiency witnessed at the end of recognition phase was hardly 40–50% which made the model inefficient for using it in practical applications like signature and check verifications [12]. Moreover, the raw image did not provide sufficient and appropriate features for the recognition model due to which the overall extraction turned out to be inefficient. The simple artificial neural networks set with random weights took a long time to train due to lack of proper datasets [12].

3 Proposed Methodology

3.1 Architecture of Proposed Model

Followings are the steps of our proposed model.

1. Dataset will be passed to CNN models with various tweaks will be identify for efficient accuracy.
2. New input image will be passed for preprocessing.
3. CNN model will extract features from it and using final dense layer and softmax layer characters will be classified.

3.2 Image Preprocessing

A raw image cannot be used as an input to any recognition algorithm. The image needs to be appropriately processed and segmented before it is fed to recognition phase. The binarization would reduce the possible pixel values which can prove beneficial during features extraction phase. Image blurring via Gaussian process will be useful for determining the ROI for appropriate segmentation.

Many readymade APIs like EAST were trained models in order to determine the required ROI but failed at certain instances over a wide series of images containing some Arabic text. The primary reason for inappropriate segmentation in case of Arabic text is its cursive nature and availability of diacritics above and below the connected letters. Following have been the expected outputs of image preprocessing phase:

1. Cleaned image via blurring will remove the unwanted noise in the input image which will result in better accuracy of text extraction.
2. Binarized image will limit the pixels to only 2 values (0 or 1) based on which the required features can be extracted.
3. Dilation of image can make the characters merged so that image segmentation can be efficiently carried out.
4. Contouring process can make segments over desired ROI which are ready to be inputted to the CNN algorithm (Figs. 3, 4, 5, 6).

The formed segments are then stored as individual images and can be inputted to recognition algorithm for required text's extraction.

Fig. 3 Sample image (raw) [16]

Fig. 4 Negative binarization over sample image

Fig. 5 Sample image after dilation

Fig. 6 Segments of sample image

3.3 Dataset Preparation

Arabic datasets required for training the CNN model were indeed scarce and hence it was decided to make one. The prepared dataset consists of images of handwritten Arabic characters with all of the four forms of every Arabic character which were then subjected to negative binarization. Thus, the foreground color of the dataset images is white and the background color is black.

All the four forms of every Arabic letter were written on paper and then separate image of every form of a character was obtained through the cropping algorithm. Thus, separate images of every handwritten Arabic character in all of its four forms was obtained which was later negatively binarized so that appropriate training of CNN model can take place by counting the number of white pixels. Following Image shows a screenshot of one of the Arabic characters—"*Baa*":-

In the below figure, each of the four forms has been written for 14 times. The standalone form is at the extreme left, followed by middle form, a row later. The fifth column shows the extreme left form of Arabic letter, followed by extreme right form, a row later. Similarly, the other Arabic letters were recorded on a paper and cropping algorithm was applied on the images to split out all the four forms in separate folders.

The cropping algorithm scanned the image from left to right and from top to bottom. Due to equal spacing of every writing space inside the image, the required Arabic forms easily separated out from each other. Following figure shows the separated and negatively binarized form of letter—"Baa":- Total such 28 character and for every 20 sheets having 52 characters. Comprising 29,120 images were collected (Figs. 7 and 8).

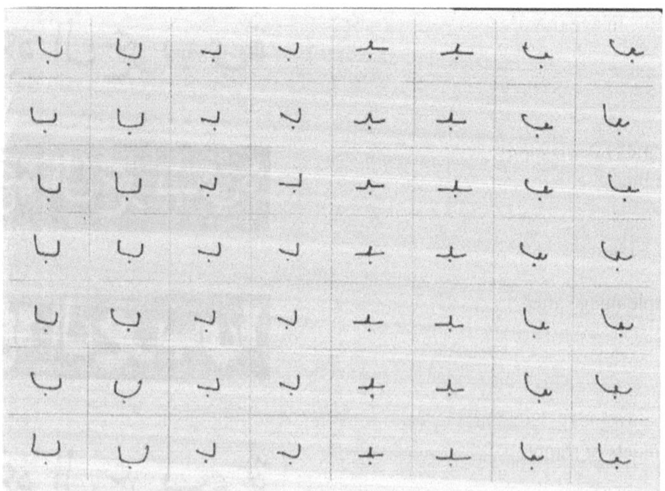

Fig. 7 All Four Forms of Arabic letter–Baa

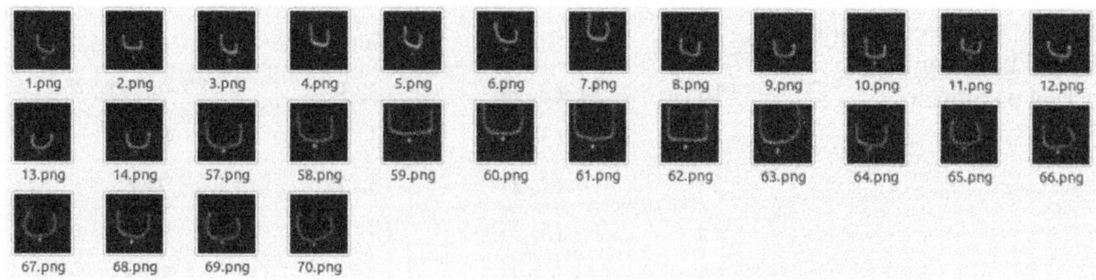

Fig. 8 Separated Standalone form of letter "Baa" through cropping algorithm

4 Classification Through Convolutional Neural Networks

Extracting the text from images captured from varied backgrounds, the idea to use Hidden Markov model failed as it requires probabilities which highly depends on background of images. Hence, a generalized model which can extract text being indifferent to the background images was decided to be trained. Thus, most pertinent and efficient way to recognize the object in given condition is performed by Convolutional Neural Network. It has achieved great triumph for its classification and recognizing techniques. Convolutional Neural Networks (CNNs) comprised of neurons that self-optimize through learning. CNNs are comprised of three types of layers. These are Convolutional layers, pooling layers and fully connected layers. CNN trained with neural network backpropagation algorithm. CNN has the ability to learn from high-dimensional complex inputs, nonlinear mappings from very large number of data. It automatically learns salient features and trains the network by manipulating the weights.

CNN architecture in our model: The flow of training model

Input layer → Convolutional layer → Batch Normalization layer → Max pooling layer → Dropout layer → Fully Connected Layer → Output Layer (Softmax)

These layers automatically find out features from images and train the model based on features that image consists of. It uses various filters to find out features in images. To increase the efficiency of model, we used batch normalization layer, dropout layer, and optimizer.

Efficiency:

Batch Normalization is used to reduce the changes in the distributions of internal nodes of a deep network, in the course of training, as Internal Covariate Shift. Eliminating it offers a promise of faster training. Batch Normalization, that takes a step towards reducing internal covariate shift, and in doing so dramatically accelerates the training of deep neural nets. [BN]Mini-batches are formed and each batch is normalized and backpropagated. It adds only two activation function to retain the property of CNN. Dropout Layer removes over fitting in model. This layer used to remove some nodes that may cause over fitting in model to increase the training speed.

Optimizer is the method which is straightforward to implement, is computationally efficient, has little memory requirements, is invariant to diagonal rescaling of

Table 1 Observed accuracies of CNN by changing its internal parameters

Filters	Strides	Kernel size	Accuracy
16	(2, 2)	(3, 3)	94.52
16	(3, 3)	(3, 3)	89.25
32	(2, 2)	(3, 3)	94.87
32	(3, 3)	(3, 3)	93.28
16	(2,2)	(2, 2)	89.33
32	(2, 2)	(2, 2)	92.84
16	(2, 2)	(4, 4)	91.87
32	(2, 2)	(4, 4)	94.25

the gradients, and is well suited for problems that are large in terms of data and/or parameters. This helped the CNN model to gain 93% accuracy to classify and recognize the Arabic characters. The following table depicts the behavior of CNN model by variations in its internal parameters (Table 1).

It was observed that tweaking the parameters led to variations in accuracy of classification. Excess variations in one parameter only decreased the model accuracy.

5 Conclusion

The overall accuracy of the CNN model for a variety of images was close to 95%. This work concludes that CNN is one of the best algorithms for extracting the required text from image and can be prominently used in recognition of cursive languages like Arabic whose extraction turns out to be complex by the use of Markov models and dense neural networks. By tweaking a single CNN parameter, the accuracy of model can get decreased and thus, all the other parameters are required to be modified proportionally. However, the proposed approach has certain restrictions, one classical Arabic language while neglecting the other calligraphies which require another specialized dataset having the same writing font and patterns. Moreover, the diacritics not considered. Also, for certain characters resembling almost similar in shapes (like Baa and Taa), the extraction algorithm for certain images gave flipped results.

References

1. Al-Muhtaseb, H.A.: Arabic text recognition of printed manuscripts efficient recognition of offline printed Arabic text using hidden markov models, bigram statistical language model, and post-processing (n.d.)
2. Zhan, Y., Wang, W.: A robust split-and-merge text segmentation approach for images. In: International Conference on Pattern Recognition, pp. 1002–1005 (2006)

3. Hoang, T.V., Tabbone, S.: Text extraction from graphical document images using sparse representation (2010)
4. Shivakumara, P., Dutta, A.: A new method for handwritten scene text detection in video. In: International Conference On Frontiers In Handwriting Recognition, pp. 16–18 (2010)
5. Gopalan, C., Manjula, D.: Text region segmentation from heterogeneous images. Int. J. Comput. Sci. Netw. Secur. 108–113 (2008)
6. Khelifi, B., Zaghden, N.: Unsupervised categorization of heterogeneous text images based on fractals. In: Proceedings of ICPR, pp. 1–4 (2008)
7. Paraag Agrawal, R.V.: Text extraction from images. In: IJCSET (2012)
8. Samarabandu, J., Liu, X.: Multiscale edge-based text extraction from complex images. In: International Conference on Multimedia and Expo, pp. 1721–1724 (2006)
9. Strouthopoulos, N.P.: Text extraction in complex color documents. Pattern Recogn. **35**, 1743–1758 (2002)
10. Sahoo, G., Kumar, T.: Text extraction and enhancement of binary images using cellular automata. Int. J. Autom. Comput. 254–260 (2009)
11. Jung, K., Kim, K.I.: Text information extraction in images and video: a survey (2012)
12. Gundu, N.K., Jadhav, S.M.: Text extraction from image and displaying its related information. Int. J. Sci. Res. Publ. (2014)
13. Amin, A.: Recognition of printed Arabic text based on global features and decision tree learning techniques (2000)
14. Moradi, M., Mozaffari, S.: Farsi/Arabic text extraction from video images by corner detection. In: 6th Iranian Conference on Machine Vision and Image Processing (2010)
15. Sumathi, C.P., Santhanam, T.: A survey on various approaches of text extraction in images. Int. J. Comput. Sci. Eng. Surv (IJCSES) (2012)
16. Diab, A.: Reconstruct Arabic sentences to be used in applications that don't support Arabic. https://github.com/mpcabd/python-arabic-reshaper
17. Saudagar, A.K., Mohammed, H.V.: Arabic character extraction and recognition using traversing approach. Int. Arab J. Inf. Technol. (2018)
18. Sunil Kumar, R.G.: Text extraction and document image segmentation using matched wavelets and MRF model. IEEE Trans. Image Process. (2007)
19. Alves, W.A.L., Hashimoto, R.F.: Text regions extracted from scene images by ultimate attribute opening and decision tree classification. In: 23rd Sibgrapi Conference on Graphics, Patterns and Images (2014)
20. Mukhopadhyay, S.: Improving the efficiency of tesseract OCR through superresolution. https://edu.authorcafe.com/academies/7609/improving-the-efficiency-of-tesseract-ocr-throughsuperresolution

Cryptanalysis of A New Cryptosystem of Color Image Using a Dynamic-Chaos Hill Cipher Algorithm: A Chosen Ciphertext Attack

Vadlamudi Naveen Kumar and N. Ravi Shankar

Abstract This paper carries out a chosen ciphertext attack on the image encryption proposed by Said Hraoui et al. [1]. The scheme proposed by Said Hraoui et al. uses PWLCM to generate key matrix and other confidential information. But, the whole encryption algorithm boils down to a simple Affine-Hill Cipher where only one component is generated dynamically for each pixel encryption. Since the relationship between the plaintext and the ciphertext is linear, the cipher suffers from the same weaknesses as that of the Hill Cipher. Particularly, this cipher exhibits its vulnerabilities in the face of an all zero plaintext. This paper exploits that weakness and breaks the cipher.

Keywords PWLCM · Hill Cipher · Chaotic maps · Image cryptography · Cryptanalysis · Chosen-plaintext attack

1 Introduction

Growing dependence on the Internet in every sphere of activity necessitated the incorporation of strong measures for the security and privacy of the information. Cryptography is widely used for ensuring the confidentiality of information that is exchanged over the Internet, be it in the form of Textual data or multimedia. Cryptographic algorithms are classified as private key algorithms and public key algorithms depending on the number of keys involved, in encryption and decryption. They are also classified as block ciphers and stream ciphers depending on the unit of encryption (one byte/a group of bytes).

Hill Cipher [2, 3] proposed by Lester Hill is the first encryption algorithm, reported in the literature, that encrypts multiple Characters at a time. This cipher, though exhibits very good diffusion property, is vulnerable to known-plaintext attack as the relationship between plaintext and the ciphertext is linear in nature. There are

V. N. Kumar · N. Ravi Shankar (✉)
Geethanjali College of Engineering and Technology (Autonomous), Keesara, Cheeryal, Medchal, Hyderabad, India

H. Das et al. (eds.), *Progress in Computing, Analytics and Networking*, Advances in Intelligent Systems and Computing 1119,
https://doi.org/10.1007/978-981-15-2414-1_47

numerous modifications proposed by various researchers to the Hill Cipher algorithm [4–11].

One such algorithm [1] proposed by Said Hraoui et al. is of particular interest in this paper. Hill Cipher is also used in encrypting multimedia data like images [12–16]. Chaotic cryptography, as a method for ensuring confidentiality, is gaining popularity. Chaos theory is used more in image encryption than encrypting plaintext data. Chaos theory as an offshoot of nonlinear dynamics is found suitable, by many researchers, for multimedia encryption owing to its high sensitivity to the initial conditions. A minute change in the initial conditions results in wide dissimilarity in the output due to the inherent nature of chaotic maps. This property makes chaos theory suitable for cryptography. There are a number of chaotic systems reported in the literature [17–24]. Piece Wise Linear Chaotic Maps (PWLCM) [25] is one such chaotic system that generates binary sequences. This paper carries out a chosen-plaintext attack on [1] which uses a combination of Hill Cipher and PWLCM. The scheme proposed by Said Hraoui et al., (henceforth referred to as *the target algorithm*), is vulnerable to the chosen-plaintext attack, as demonstrated in the subsequent sections.

Section 2 of this paper describes the target algorithm and its vulnerabilities, which forms motivation to this research. Section 3 demonstrates the cryptanalytic attack with the chosen plaintext. Section 4 proposes modifications to the target algorithm to remove its vulnerabilities. Finally, conclusions are dealt with in Sect. 5.

2 The Target Algorithm

The algorithm proposed by Said Hraoui et al. broadly consists of two components, PWLCM, and the Hill Cipher. Image is encrypted pixelwise, i.e., the R, G, B components of each pixel are transformed by applying the affine Hill Cipher algorithm. The key in the target algorithm comprises four components, namely, A, a 3×3 Key matrix, as used in the Hill Cipher, generated through PWLCM, the Key Matrix A is generated as outlined in [1].

$$\mathrm{A} = \begin{pmatrix} 1 & p & 0 \\ q & 1+pq & 0 \\ \alpha & \beta & \lambda \end{pmatrix} \tag{1}$$

The Components p, q, α, β, λ are generated as described in [1]. The pixel encryption is governed by the relation.

$$C = (AP + X_0) \bmod 256 \tag{2}$$

where P and C contain original and encrypted R, G, B values, of the pixel, respectively, governed by the encryption algorithm mentioned in (2).

The vector X is dynamically generated for each pixel using the recurrence relation

$$X_i = (TX_{i-1} + Y) \bmod 256 \tag{3}$$

$$X_0 = \begin{pmatrix} X_0^0 \\ X_0^1 \\ X_0^2 \end{pmatrix} \tag{4}$$

$$\mathrm{Y} = \begin{pmatrix} Y_0 \\ Y_1 \\ Y_2 \end{pmatrix} \tag{5}$$

$$X_1^0 = (X_0^0 T_{11} + X_0^1 T_{12} + X_0^2 T_{13} + Y_0) \bmod (256) \tag{6}$$

$$X_2^0 = (X_1^0 T_{21} + X_1^1 T_{22} + X_1^2 T_{23} + Y_1) \bmod (256) \tag{7}$$

$$X_3^0 = \left(X_2^0 T_{31} + X_2^1 T_{32} + X_2^2 T_{33} + Y_2\right) \bmod (256) \tag{8}$$

A close study of the target algorithm reveals that all the components of the key are either generated randomly or through PWLCM. However, once these components are generated, they form the permanent elements of the key used for encryption of data. Thus, irrespective of the complex mechanisms adopted, the target algorithm essentially boils down to an affine transformation, wherein X is the only component that is dynamically generated through another affine transformation for each pixel.

An all zero first pixel $[000]^T$, when given as input to the algorithm given in (2), yields a ciphertext equivalent to X_0 only. An all zero second pixel yields X_1 and so on. Thus, we can find as many $X_i s'$, as we want. From these $X_i s'$ it is possible to ascertain the values of T and Y. Similarly a pixel with RGB properties as $[100]^T$ gives away the first column, $[010]^T$ reveals the second column, $[001]^T$ exposes the third column of the key matrix A, when we know X, Y and T (as seen from (2)).

3 The Cryptanalytic Attack

This section demonstrates how the vulnerabilities of the target algorithm are exploited to break the cipher. For this purpose, the following values are assumed for the key elements.

$$\mathrm{A} = \begin{bmatrix} 232 & 244 & 0 \\ 238 & 95 & 0 \\ 88 & 136 & 144 \end{bmatrix} \quad \mathrm{T} = \begin{bmatrix} 1 & 13 & 17 \\ 7 & 11 & 19 \\ 3 & 5 & 31 \end{bmatrix} \quad X_0 = \begin{bmatrix} 5 \\ 7 \\ 9 \end{bmatrix} \quad \mathrm{Y} = \begin{bmatrix} 231 \\ 255 \\ 241 \end{bmatrix}$$

Firstly an all zero pixel, when encrypted using the target algorithm, results in

$$C = \left[\begin{bmatrix} 232 & 244 & 0 \\ 238 & 95 & 0 \\ 88 & 136 & 144 \end{bmatrix} * \begin{bmatrix} 0 \\ 0 \\ 0 \end{bmatrix} + \begin{bmatrix} 5 \\ 7 \\ 9 \end{bmatrix}\right] = \begin{bmatrix} 5 \\ 7 \\ 9 \end{bmatrix} \tag{9}$$

Now X_0, through recursive relationship given by (3), is transformed into X_1 as given below

$$X_1 = [TX_0 + Y] \bmod 256 \tag{10}$$

$$X_1 = \left[\begin{bmatrix} 1 & 13 & 17 \\ 7 & 11 & 19 \\ 3 & 5 & 31 \end{bmatrix} * \begin{bmatrix} 5 \\ 7 \\ 9 \end{bmatrix} + \begin{bmatrix} 231 \\ 255 \\ 241 \end{bmatrix}\right] \bmod (256) = \begin{bmatrix} 11 \\ 13 \\ 15 \end{bmatrix}$$

$$X_2 = \begin{bmatrix} 154 \\ 248 \\ 36 \end{bmatrix} X_3 = \begin{bmatrix} 125 \\ 137 \\ 243 \end{bmatrix} X_4 = \begin{bmatrix} 124 \\ 86 \\ 130 \end{bmatrix} X_5 = \begin{bmatrix} 99 \\ 187 \\ 209 \end{bmatrix} \tag{11}$$

It can be readily noticed that, as given in (6), that

$$11T_{11} + 13T_{12} + 15T_{13} + Y_0 \equiv 154 (\bmod 256) \tag{12}$$

$$154T_{11} + 248T_{12} + 36T_{13} + Y_0 \equiv 125 \ (\bmod 256) \tag{13}$$

$$125T_{11} + 137T_{12} + 243T_{13} + Y_0 \equiv 124 \ \bmod (256) \tag{14}$$

$$124T_{11} + 86T_{12} + 130T_{13} + Y_0 \equiv 99 \ \bmod (256) \tag{15}$$

Here the superscripts (0, 1, 2) of X denote the (R, G, B) values of each pixel. The subscripts of X signifies the position of that pixel in the chosen plaintext. The four Eqs. (12)–(15) contain four unknowns. These are essentially a system of four linear congruences. Solving these equations gives the key elements T_{11}, T_{12}, T_{13} and Y_0 as

$$T_{11} = 1 \quad T_{12} = 13 \quad T_{13} = 17 \quad Y_0 = 231$$

Similarly, the elements of the second row of T and Y, as given in equation in (7), satisfy the following condition.

$$11T_{21} + 13T_{22} + 15T_{23} + Y_1 \equiv 248 \ \bmod (256) \tag{16}$$

$$154T_{21} + 248T_{22} + 36T_{23} + Y_1 \equiv 137 \ \bmod (256) \tag{17}$$

$$125T_{21} + 137T_{22} + 243T_{23} + Y_1 \equiv 86 \ \bmod (256) \tag{18}$$

$$124T_{21} + 86T_{22} + 130T_{23} + Y_1 \equiv 187 \mod (256) \tag{19}$$

From these equations, we get

$$T_{21} = 7, T_{22} = 11, T_{23} = 19, Y_1 = 255$$

The third row of T and Y are obtained from (8)

$$11T_{31} + 13T_{32} + 15T_{33} + Y_2 \equiv 36 \mod (256) \tag{20}$$

$$154T_{31} + 248T_{32} + 36T_{33} + Y_2 \equiv 243 \mod (256) \tag{21}$$

$$125T_{31} + 137T_{32} + 243T_{33} + Y_2 \equiv 130 \mod (256) \tag{22}$$

$$124T_{31} + 86T_{32} + 130T_{33} + Y_2 \equiv 209 \mod (256) \tag{23}$$

Solving the equations gives

$$T_{31} = 3, T_{32} = 5, T_{33} = 31, Y_2 = 241$$

Thus, we obtain the crucial key elements T, Y and hence the X of the key. The next step is to find the elements of A, which completes the cryptanalysis. A second chosen plaintext in which the first pixel, is of the form $[100]^T$ when applied in (2) gives

$$C = \left[\begin{pmatrix} 232 & 244 & 0 \\ 238 & 95 & 0 \\ 88 & 136 & 144 \end{pmatrix} * \begin{pmatrix} 1 \\ 0 \\ 0 \end{pmatrix} + \begin{pmatrix} 5 \\ 7 \\ 9 \end{pmatrix} \right] \mod (256 = \begin{bmatrix} 237 \\ 245 \\ 97 \end{bmatrix} \tag{24}$$

This is same as

$$\begin{pmatrix} A_{11} + X_0^0 \\ A_{21} + X_0^1 \\ A_{31} + X_0^2 \end{pmatrix} \mod (256)$$

Since we already know X_0, obtaining the values of the first column of the key matrix is a trivial task. Thus, we get the first column A as

$$\begin{bmatrix} 237 \\ 245 \\ 97 \end{bmatrix} - \begin{bmatrix} 5 \\ 7 \\ 9 \end{bmatrix} = \begin{bmatrix} 232 \\ 238 \\ 88 \end{bmatrix} \tag{25}$$

Similarly the second and third column elements of the key matrix, can be obtained by taking $[010]^T$ and $[001]^T$ as the second and the third plaintext pixels. With the help of two chosen plaintexts, we can find all the elements of the key of the cipher. Thus, the cipher is broken.

4 Strengthening the Target Algorithm

The major vulnerability exhibited by the target algorithm is the linear relationship between the plaintext and the ciphertext. An all zero plaintext neutralizes the interaction of the key matrix A with the plain text and gives away the elements of X_i. If the algorithm can be strengthened against the all zero plaintext, its performance can certainly be improved. To offset this problem, two vectors Z and X are employed instead of X alone. The encryption function is modified as

$$
\begin{aligned}
& \mathrm{K} = 0 \\
& \text{for i in range(image\_height):} \\
& \quad \text{for j in range(image\_width):} \\
& \qquad C = (A * (P + Z_k) + X_K) \bmod 256 \\
& \qquad X_{K+1} = (T X_K + Y) \bmod 256 \\
& \qquad Z_{k+1} = (\mathrm{S} Z_k + \mathrm{W}) \bmod 256 \\
& \qquad \mathrm{K} = \mathrm{K} + 1
\end{aligned}
$$

where Z is a (3×1), matrix generated dynamically through a recursive relationship.

$$Z_i = (\mathrm{S}Z_{i-1} + \mathrm{W}) \bmod 256 \tag{26}$$

Here Z_0 is taken randomly. S and W, 3×3 and 3×1 vectors, respectively, are generated through PWLCM as was the case with T and Y.

A random value, when added before multiplication with the key matrix, will strengthen the algorithm against all zero chosen-plaintext attack. This modification, outlined above, enhances the security of the cipher without any degradation of its performance, as the additional operations incorporated into the cipher are column vector addition and affine transformation.

5 Conclusion

This study analyzes the strengths and weaknesses of the target algorithm proposed by Said Hraoui et al. [1]. This scheme exhibits a fatal deficiency in the face of an all zero plaintext, and thus cannot withstand rigorous cryptanalytic attacks. The algorithm is broken with the help of two chosen plaintexts (chosen images). The first one comprises five pixels whose R, G, B component values are zeroes. Second chosen image contains three pixels in which the first pixel has R, G, B values of $[100]^T$, the second pixel has $[010]^T$ and the third pixel $[001]^T$. The composition of the first chosen plaintext helps in retrieving the components T, Y, and X. The second chosen plaintext reveals the elements of the key matrix A.

Addition of Z to P, before interaction with the key matrix A overcomes the vulnerability exhibited by the target algorithm in the wake of an all zero plaintext. This strengthens the target algorithm considerably.

References

1. Hraoui, S., Gmira, F., Fouad Abbou, M., Jarrar Oulidi, A., Jarjar, A.: A new cryptosystem of color image using a dynamic-chaos Hill Cipher algorithm. In: Second International Conference on Intelligent Computing in Data Sciences (ICDS 2018), Procedia Computer Science, vol. 148, pp. 399–408 (2019)
2. Hill, L.S.: Cryptography in an algebraic alphabet. Am. Math. Mon. **36**(6), 306–312 (1929)
3. Hill, L.S.: Concerning certain linear transformation apparatus of cryptography. Am. Math. Mon. **38**(3), 135–154 (1931)
4. Sastry, V.U.K., Ravi Shankar, N., Durga Bhavani, S.: A modified Hill Cipher involving interweaving and iteration. Int. J. Netw. Secur. **11**(2), 51–56 (2010)
5. Sastry, V.U.K., Ravi Shankar, N., Durga Bhavani, S.: A large block cipher involving key dependent permutation, interlacing and iteration. Cybern. Inf. Technol. **13**(3), 50–63 (2013)
6. Sastry, V.U.K., Ravi Shankar, N.: Modified Hill Cipher with interlacing and iteration. J. Comput. Sci. (Science Publications) **3**(11), 854–859 (2007)
7. Sastry, V.U.K., Ravi Shankar, N.: Modified Hill Cipher for a large block of plaintext with interlacing and iteration. J. Comput. Sci. (Science Publications) **4**(1), 15–20 (2008)
8. Sastry, V.U.K., Ravi Shankar, N., Durga Bhavani, S.: A modified playfair cipher involving interweaving and iteration. Int. J. Comput. Theory Eng. **1**(5), 594–598 (2009)
9. Sastry, V.U.K., Ravi Shankar, N., Durga Bhavani, S.: A modified playfair cipher for a large block of plaintext. Int. J. Comput. Theory Eng. **1**(5), 590–594 (2009)
10. Sastry, V.U.K., Ravi Shankar, N., Durga Bhavani, S.: A generalized playfair cipher involving intertwining, interweaving and iteration. Int. J. Netw. Mobile Technol. **1**(2), 45–53 (2010)
11. Sastry, V.U.K., Ravi Shankar, N., Durga Bhavani, S.: A blending of a generalized playfair cipher and a modified Hill Cipher. Int. J. Netw. Mobile Technol. **2**(1), 35–43 (2011)
12. Megantara, R., Rafrastara, F., Mahendra, S.: A combination of Hill CIPHER-LSB in RGB image encryption. Kinetik: Game Technol. Inf. Syst. Comput. Netw. Comput. Electron. Control **4**(3), 241–248 (2019)
13. Handoko, L.B., Umam, C., Anindita, A.S.: Upload file security on the server using LSB and Hill Cipher. J. Appl. Intell. Syst. **4**(1), 28–38 (2019) (e-ISSN: 2502–9401, p-ISSN: 2503-0493)
14. Essaida, M., Akharraza, I., Saaidiabet, A., Mouhiba, A.: Image encryption scheme based on a new secure variant of Hill Cipher and 1D chaotic maps. J. Inf. Secur. Appl. **47**, 173–187 (2019)
15. Abd El Hafez Bakr, M., Amr Mokhtar, M., El Sherbini Takieldeen, A.: Elliptic curve cryptography modified Hill Cipher dependent on circulant matrix. Int. J. Ind. Electron. Electr. Eng. (IJIEEE) **6**(1), 24–29 (2018)
16. Rachmawati, D., Budiman, M.A., Wardhono, M.I.: Hybrid cryptosystem for image security by using Hill Cipher 4 × 4 and ElGamal elliptic curve algorithm. In: 2018 IEEE International Conference on Communication, Networks and Satellite (Comnetsat), Medan, Indonesia, pp. 49–54 (2018)

17. Li, C., Luo, G., Li, C.: An image encryption scheme based on the three-dimensional chaotic logistic map. Int. J. Netw. Secur. **21**, 22–29 (2019)
18. Luo, Y., et al.: A novel chaotic image encryption algorithm based on improved baker map and logistic map. In: Multimedia Tools and Applications 1–21 (2019)
19. Li, X., Mou, H., Yan, B., Gao, L.: A reversible data hiding with quality controllable scheme for encrypted-compressed image. Wuhan Univ. J. Nat. Sci. **24**, 383 (2019). https://doi.org/10.1007/s11859-019-1410-x
20. Gupta, R., Pachauri, R., Singh, A.K.: Image encryption method using dependable multiple chaotic logistic functions. IJISP **13**(4), 53–67 (2019). 19 Sept 2019. https://doi.org/10.4018/ijisp.2019100104
21. Yi, X.: Hash function based on chaotic tent maps. IEEE Trans. Circuits Syst. II Express Briefs **52**(6), 354–357 (2005). https://doi.org/10.1109/TCSII.2005.848992
22. Liu, J., Yang, K., Wang, S.: Coupled chaotic tent map lattices system with uniform distribution. In: 2010 2nd International Conference on e-Business and Information System Security (EBISS), pp. 1–5 (2010)
23. Pareek, N.K., Patidar, V., Sud, K.K.: Image encryption using chaotic logistic map. Image Vision Comput. **24**(9), 926–934 (2006)
24. Kuperin, Yu.A., Pyatkin, D.A.: Two-dimensional chaos: the baker map under control. J. Math. Sci. **128**(2), 2798–2802 (2005)
25. Li, S., Chen, G., Mou, X.: ON the dynamical degradation of digital piecewise linear chaotic maps. Int. J. Bifurc. Chaos **15**(10), 3119–3151 (2005)

Counterfeit Product Detection Analysis and Prevention as Well as Prepackage Coverage Assessment Using Machine Learning

Aradhana Behura, Ashutosh Behura and Himansu Das

Abstract There is no deny in the fact that duplicate product jeopardizes the luck of businesses worldwide over violating patent rights and causing immense commercial wound. Hence, it has to change into basic for firms to ensure their disgraces opposed to duplication. Existing technologies for electrical forged find encompass the applying of extra security looks prefer Watermark Technology to the emblem stock itself that raises the charges of one's merchandise. In this paper, a reliable method for counterfeit prediction is proposed. This method can be used by customer to predict counterfeit for their daily need items available in the market. The item may be medicine, detergent or food packet. Now the development of new package of the product always comes with the risk of counterfeiting, sometimes that could affect our health, company reputation and goodwill. We have used the concept of invisible and visible watermarking present in item itself to provide authenticity of the product. This is a low cost solution that help enterprises and consumers identify the authenticity of products.

Keywords Boosting · Product detection · Feature extraction · Discrete Wavelet Transform (DWT)

1 Introduction

Due to the rapid information firing in the modern era, a great shape of multimedia contents has been digitalized and their duplication made easy without any discount

A. Behura (✉)
Department of Computer Science and Engineering, VSSUT, Burla 768018, Odisha, India
e-mail: aradhanabehura@gmail.com

A. Behura (✉)
Department of Computer Science and Engineering, KIIT University, Bhubaneswar 768018, Odisha, India
e-mail: ashutosh.behura@gmail.com

H. Das (✉)
School of Computer Engineering, KIIT Deemed to be University, Bhubaneswar, Odisha, India
e-mail: das.himansu2007@gmail.com

H. Das et al. (eds.), *Progress in Computing, Analytics and Networking*,
Advances in Intelligent Systems and Computing 1119,
https://doi.org/10.1007/978-981-15-2414-1_49

in excellence through each felony and unauthorized distribution channels. In addition, the information generated due to code affirmation can be used by the product owner to scrutinize regions wherein duplicate products are supplied, which includes the possibility of identifying noncompliant supply chain operators. There are many popular technologies, which can be used for the consumer to apprehend whether or not the product is duplicate. Watermark is considered as identifiers, inserted in the original package image [1, 2]. This is a trend in today's years and it is because of the boom diploma of sophisticated packaging of the products is a completely effective technique in opposition to counterfeiting which permits the tracing of the duplicates within the supply and distribution chain. The drawback of packaging is that items inside the packet may be extracted from the bundle and substituted with some other goods [2, 3]. In these scenarios, sealing can mitigate the risk of duplicacy. Encryption [4] a very popular method to convert the data into an unrecognized format so that confidentiality can be preserved during transmission over the Internet and in the market. To ensure copyright safety, digital watermarking [2, 5] embeds hidden statistics into the package that can not be removed and its detection affirms the ownership of products [6]. In this paper first we make a database then train [1, 5, 7] it for feature extraction. After that, we proposed an algorithm to confirm whether a product is counterfeit or not.

2 Literature Review

As products move through the supply chain, numerous channels are involved and each channel is a door for counterfeit products to enter into market find their way to the customers [1]. Although some paths are secure but the probability of entering counterfeit products into market still exists. The range of products being counterfeited is continuously expanding not only for traditional retailers but also for e-tailers [4, 8]. The need of the hour is to develop the best methods for counterfeit detection, prevention, building a response mechanism in case counterfeit is detected and finally punishing or warning the defaulters. Blind image [9] watermarking is one of the popular methods which delineates the difference among the patent issues and redistribution of the information and it is a nice process for data authentication [2, 5]. Wavelets [10] are used to cut off fine details in images [6]. It helps us to identify among the significant and non- significant bits of the images. An image encryption scheme based on a combination of Kolmogorov flows with a fast shift-register-based pseudorandom number generator was proposed in reference [7]. Although these existing techniques operating on block cipher offered greater security and speed, but cannot withstand loss compression. To tackle this menace, an image encryption technique that performs compression [3, 6] and encryption both was proposed in reference [3]. This scheme encrypts image [9] by manipulating the Huffman coding tables. To achieve this, it selects many different Huffman tables employing them alternatively and the selection of particular Huffman tables and its order are used as a secret key. This method is computationally efficient but cannot resist chosen-plaintext attack. An encryption scheme was proposed in reference [6] which employs wavelet transform to decompose the image into several sub-bands and encryption at each level is achieved by random permutation. To provide high security without visual degradation, many security mechanisms are provided. In the next phase, the scrambled image is changed into an unrecognized form using secret keys which are generated with the help of logistic map [2]. Intrusion detection mechanisms [11] are important in current year. Duplicate product not only hazardous for our health, it also affects the brand name of our country.

3 Methodology

Training mechanism used to train the dataset, so it is easy for a customer to predict whether the package the product is original or duplicate by some intruder.

3.1 The Proposed Algorithm Consists of Three Stages

1. **Feature Selection**:

Measurement of the pixel is used to know the intensities of pixels. If the weight of the lesser region is increased, then the illumination areas are varied a lot. We also check the edge features, line features, point and diagonal features of the image. The important task is to find out the structures, surrounded by the ten thousand accessible, would better differentiate among the positive and negative samples, and how to conglomerate these into an efficient learning machine (Fig. 1).

2. **Apply AdaBoost Training method**:

Boosting helps to combine weak learner to form a strong one. At the end, we have got combined D1, D2, and D3 stumps to form a strong prediction having complex rule in assessment to individual vulnerable learner which is shown in Fig. 2. It helps to reduce miscalculation error. We take total amount of loads of all positive and negative images. If the dataset was misclassified, then error (x_j) is 1; otherwise, it is 0. We have to minimize overall miscalculation of classifier. If required, weak classifier weights are updated. It provides a better classification of large weights and poor classification of small weights. In this case, the selection of one feature is affected by the selections made for previous features which are shown in Figs. 3 and 4.

3. **Cascading Classifiers**:

Various types of positive and negative samples are trained for a particular product images of the same size. After the classifier is trained, it can be applied to an area of an image and detect the original image to search for a product package image in the

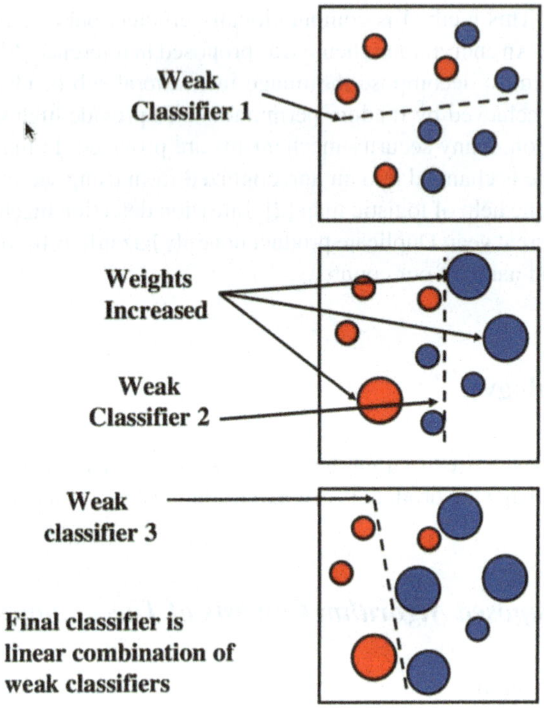

Fig. 1 Graphical representation of AdaBoost training

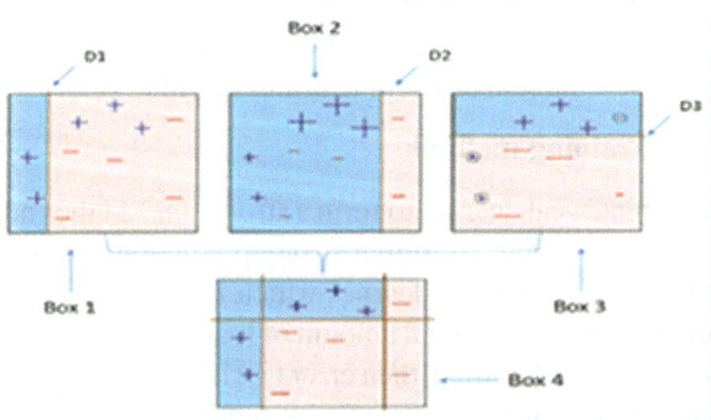

Fig. 2 AdaBoost training mechanism

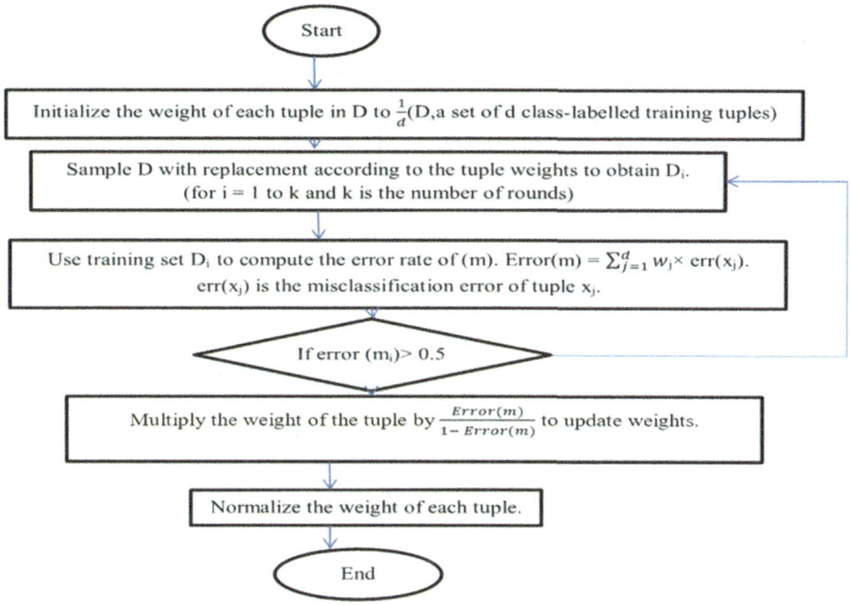

Fig. 3 AdaBoost training process (Phase-1)

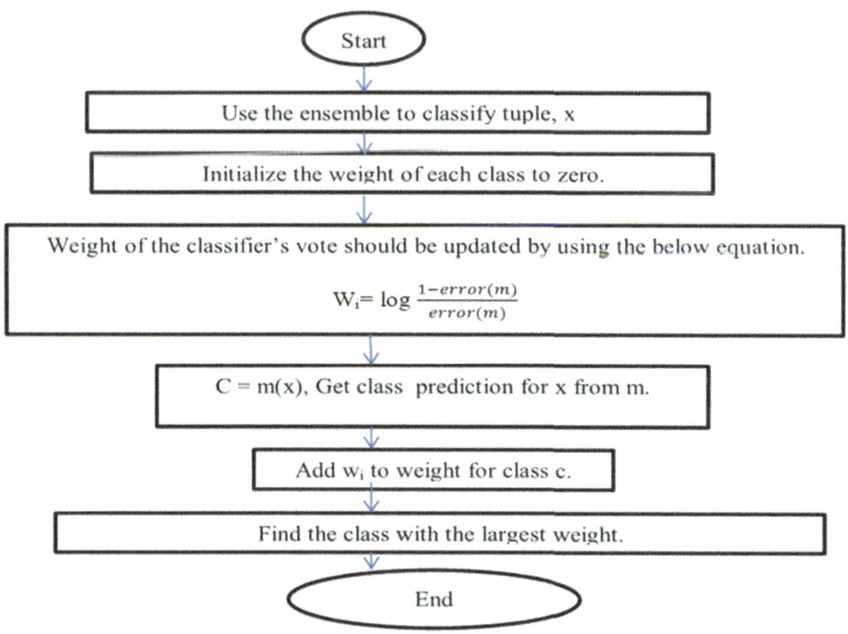

Fig. 4 AdaBoost training process (Phase-2)

entire frame, the space can be moved athwart the package image and check its every location for the classifier. Every phase must be legalized for original package images and can harvest more false positives package images. Initially, if the trained stages are very fast, we have eliminated many rectangles not containing package images quickly. Every phase produces weak learners in its training stage and train them as a group.

3.2 *Testing of Trained Model*

(i) If the scaling factor of the package image is 1.02, it means we are using a proper phase for resizing, i.e., diminish size by 2%, we develop the chance of a matching size with the prototype for prediction is found, while it is very affluent.
(ii) minNeighbours lay down how many neighbors each contestant rectangle should have to preserve it. This constraint will affect the eminence of the detected original cover image. In the end, we conclude that higher value consequences in less detections but with better excellence of the images.
(iii) minSize: If the package size is smaller then that are also ignored.
(iv) maxSize: It is defined as the maximum possible package image size. If the cover image is extremely larger then that are ignored.

We need to test out different scales, starting from 1, while keeping the minNeighbours as 0. This will display multiple detected objects as shown in Fig. 9. To enhance the pixel quality of the image, we use gamma correction mechanism which is shown in Fig. 5.

In case the object is not detected, we try a lower scale. In case of false positives we use a higher scale. This way we fine-tune the parameter until all the boxes are more or less around the desired object. Now, we set the minNeighbours as somewhere around. This process should be repeated for multiple examples of the same object to improve detection (Fig. 6).

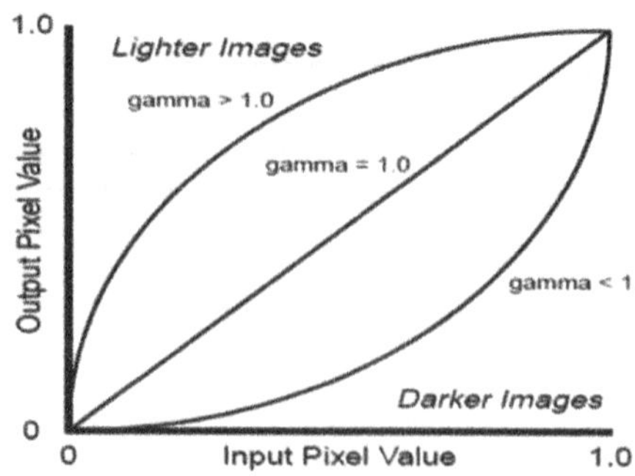

Fig. 5 Gamma correction mechanism

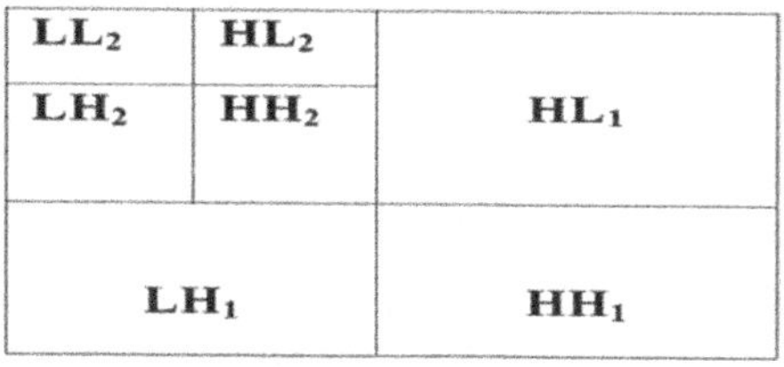

Fig. 6 Discrete Wavelet Transform (DWT)

$$F = \prod_{i=1}^{K} f_i.$$

$$D = \prod_{i=1}^{K} d_i.$$

3.3 Prevention of Counterfeit Product

We take a watermark and detect its edge to eliminate unnecessary part of the image. For edge detection of the logo, we propose a hybrid edge detection algorithm. We hybridize sobel edge detection algorithm by using ant colony optimization algorithm. Ants can odor pheromones; while selecting their track, they have a tendency to indicate the paths noticeable by strong pheromone deliberation. By using its fitness value, we can detect edge efficiently. After embedding, we used gamma correction to enhance the quality of the image and the result is shown in Figs. 10 and 14.

1. **Invisible watermark**

    1.1. **Embedding algorithm in the product cover image**
    Step 1. We take the main cover image input pixel value is 250 randomly.
    Step 2. Convert the input cover image pixel value into binary form, i.e.: 11111010.
    Step 3. Then take the watermark image pixel value suppose 155 and its binary value is 10011011.
    Step 4. Then make the two least significant bit of the cover image value 250 to zeros i.e.: 11111000.
    Step 5. Take the two most significant bits of the watermark value of the image is 155 i.e.: 10.
    Step 6. Finally, place the two MSB of the watermark image at the two LSB of the main cover image to produce a watermarked image.

$$11111000 + 00000010 = 11111010 \text{ i.e. } 250.$$

Step 7. Actually, the watermarked image is nearly equal to main cover image.

1.2. **Watermark extraction process**

Step 1. Take a watermarked image.

Step 2. Extract the two least significant bits of the watermarked image.

2. **Visible Watermark**

Visible digital watermarking is a simple process in which logo is embedded into main package image to identify the ownership. Hence a robust visible watermarking techniques should maintain details of the stuffing, while certifying that the embedded logo is difficult to be detached, so that no one could alter or manipulate watermarked cover image illegitimately. We have used discrete wavelet transform technique to compress the redundant bit from the cover image of the package which is shown in Fig. 9. In the low resolution part we can find significant part of the image. The visible watermarking algorithm is presented in Figs. 7 and 8 and the result is shown in Fig. 12.

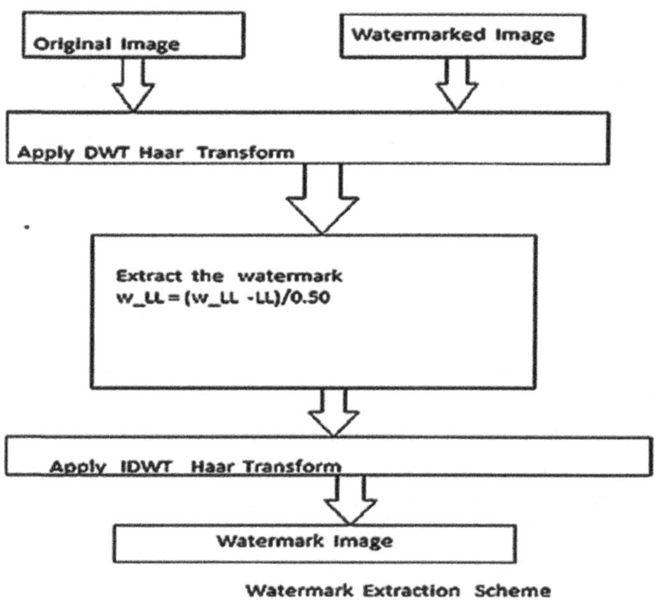

Fig. 7 Extraction algorithm

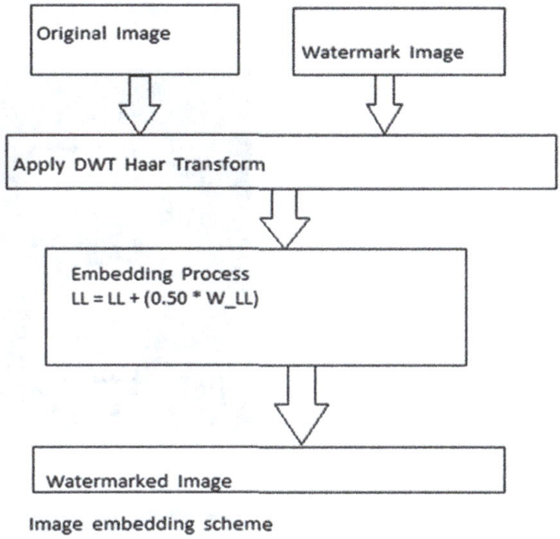

Fig. 8 Embedding algorithm

4 Result and Discussion

Recommended scrutiny is based on equating the eminence of the processed image to the quality of its image. We have taken many parameters to measure the performance of the algorithm. Intruders are using various types of attacks to manipulate or to change the originality of the product (Figs. 9, 10, 11, 12, 13 and 14).

Fig. 9 False object detection

Fig. 10 Package detection

Fig. 11 Watermark

Fig. 12 Visible watermark detection

Fig. 13 Original image

Fig. 14 Package detection

5 Presentation Scrutiny

The chief aim is to insert a watermark in the package to maintain own integrity and to provide high resistance and it is used to check the robustness of the technique.

1. **Structural Similarity Index Measure (SS)**:
   Its value exists between the decimal values 1 and −1. If it is equal to 1, it means the images are identical sets.

$$\mathrm{SS(x, y)} = \frac{(2uxuy + c3)(2kxy + c2)}{((ux \times ux) + (uy \times uy) + c3)((kx \times \mathrm{kx}) + (\mathrm{ky} \times \mathrm{ky}) + \mathrm{c2})} \times 100 \quad (1)$$

ux and uy = average of x and y, kxy = covariance of x and y. c3 and c2 = Two variables to stabilize the division with weak denominator.

2. **Error rate (E)**:

$$E = \frac{\text{Incorrect classifier image samples}}{\text{Total number of image samples}} \times 100 \quad (2)$$

3. **Accuracy (A)**:
We can measure accuracy by substacting error rate from the total dataset.
4. **Prevalence (P)**:

$$P = \frac{\text{True positive image samples}}{\text{Total number of image samples}} \times 100 \quad (3)$$

5. **Peak Signal to Noise Ratio (PSNR)**:
This constraint measures the quality of encrypted [10] image. It is defined as ratio of amount of significant signal information to noise. High PSNR value of the image provides high security.
6. **Sensitivity (S)**:
This [6] is the ratio of correct classified positive image sample with respect to the true positive image samples.

$$S = \frac{\text{Correct classified positive image samples}}{\text{True positive image samples}} \times 100 \quad (4)$$

Table 1 indicates the values of different parameters of the algorithm for training, testing, and total data. It can be observed that this algorithm is giving quite impressive results. Table 2 indicates the PSNR value of the encrypted product cover image.

Table 1 Performance measurement

All (%)	Training data	Testing data	Total data
Accuracy	97.56	95.33	98
Specificity	99.46	99.26	99.56
Prevalence	59.11	52.67	62.5
Sensitivity	96.24	94.08	97.07
Average mean square error	0.2244	0.0067	0.3033
Error rate	2.44	0.67	2.45

Table 2 Attack analysis of product image

Cover image	No attack	Salt and Pepper noise attack	Rotation attack	Translation attack
Surf excel	32.13	28.11	27.23	27.35
Tide surf	39.17	35.06	35.98	36.72

Table 3 Correlation measurement

Image	Horizontal analysis	Vertical analysis	Diagonal analysis
Surf	−0.0002	−1.0005	0.0017

7. **Correlation enactment of image**:
   Entropy amount of the image is 7.9693. This indicates that the entropy values of the encoded images gained from the recommended system are nearer to 8 which designate the data in the suggested algorithm is enhanced randomized and less conceivable to divulge the data. To verify the right to use of information to interlopers, which is important to confirm that snarled and original package image do not have any similarities. So we carefully chosen combines of two neighboring pixels horizontally, vertically and diagonally consequently correlation measurement, which are measured in appropriate fashion and discussed in Table 3.

6 Conclusion

In this paper, the suggested procedure, which is efficient that a customer and a product producing company can able to know whether a product (let us take an example surf cover image) is counterfeit or not. For security purpose, we insert two watermarks, one of them is visible another is invisible by using discrete wavelet transform. Hence the intruder cannot able to alter the package image. Invisible and visible watermarked image resembles the cover image without any loss of detail between cover image and watermarked image. The proposed method withstands all the image processing attacks and in various noises. Simultaneously, this algorithm also helps manufacturers to minimize consumer risk by preventing the sale of counterfeit/tampered goods.

References

1. Najafi, E.: A robust embedding and blind extraction of image watermarking based on discrete wavelet transform. Department of Mathematics, Faculty of Science, Urmia University, Iran. Math. Sci.**11**, 307–318 (2017). 10.1007/s40096-017-0233-1. Springer
2. Averbuch, A., Lazar, D., Israeli, M.: Image compression using wavelet transform and decomposition. IEEE Trans. Image Process. **5**(1) 2017
3. Nandi, S., Roy, S., Nath, S., Chakraborty, S., Karaa, W.B.A., Dey, N.: 1-D group cellular automata based image encryption technique. In: 2014 International Conference on Control, Instrumentation, Communication and Computational Technologies (ICCICCT) (2014)
4. Horn, C., Blankenburg, M.: Automated Detection of Counterfeit Consumer Goods using Product. IWF, Technische University at Berlin, Pascalstrasse 8, 10587 Berlin

5. Drews Jr., P.L.J., Nascimento, E.R., Botelho, S.S.C., Campos, M.F.M.: Underwater depth estimation and image restoration based on single images. IEEE Comput. Graph. Appl. **36**(2), 24–35 (2016)
6. Yeung, M.M., Mintzer, F.: An invisible watermarking technique for image verification. IEEE Trans. Consum. Electron. **39**, 93–103 (2007)
7. Gonzalez, R.C., Woods, R.E.: Digital Image Processing, 2nd edn. Pearson International Edition, London, 2008.
8. Lin, P.-Y., Chen, Y.-H., Chang, C.-C., Lee, J.-S.: Contrast-Adaptive Removable Visible Watermarking (CARVW) mechanism. Image Vis. Comput. (2013)
9. OECD: The Economic Impact of Counterfeiting and Piracy, OECD, Paris (2008). www.oecd
10. Ancuti, C.O., Ancuti, C., De Vleeschouwer, C., Bekaert, P.: Color balance and fusion for underwater image enhancement. IEEE Trans. Image Process. (2018)
11. Sahani, R., Rout, C., Badajena, J.C., Jena, A.K., Das, H.: Classification of intrusion detection using data mining techniques. In: Progress in Computing, Analytics and Networking, pp. 753–764. Springer, Singapore (2018)

Recommender System Using K-Nearest Neighbors and Singular Value Decomposition Algorithms: A Hybrid Approach

Rounick Palit and Rajdeep Chatterjee

Abstract Recommender systems have found their importance in many parts of our life, from helping us decide what product to buy next to helping us discover songs of different genres. This article provides some basics of recommender systems and how to implement them using content and collaborative-based filtering methods. The model takes a matrix of user ratings for books as input and top recommendations are generated. It also discusses hybrid or mixed recommender systems and how they can provide better recommendations to users. It identifies areas that need enhancement for future implementation.

Keywords Recommender system · K-Nearest Neighbors · Singular Value Decomposition · Correlation

1 Introduction

There is growing number of content on Internet each and every day. And with the increasing number of people adopting smartphones and Internet, content consumption is on the rise. People want to look at interesting videos or songs or want to read interesting books. And content creators want to serve these people with quality content. But catering to each and every individual's taste is really difficult. That can be achieved using recommender systems. Recommender systems are being used in more areas than we can imagine. Netflix uses recommender systems to learn their viewers' tastes and keep suggesting them with interesting movies and series. Amazon uses

R. Palit (✉) · R. Chatterjee (✉)
KIIT University, Bhubaneswar 751024, Odisha, India
e-mail: rounickpalit@outlook.com

R. Chatterjee
e-mail: cse.rajdeep@gmail.com

H. Das et al. (eds.), *Progress in Computing, Analytics and Networking*,
Advances in Intelligent Systems and Computing 1119,
https://doi.org/10.1007/978-981-15-2414-1_50

recommender systems to suggest a buyer similar products that other users have also bought and liked. Spotify has one of the most advanced recommendation engines that suggest their users with songs that they want to listen to next, which makes it a very popular music and podcast streaming service. In [1], a hybrid memory and model-based approach was discussed to recommending movies. In [2], B. M. Khan, A. Mansha, F. H. Khan, S. Bashir gave us a survey of collaborative filtering-based online recommendation systems.

Despite such advances in this field, book recommendation continues to be an area that has not been explored that much. It is an area where there are still a lot of unexplored parameters, different factors that affect a reader's mind when he or she picks up the book to read. This was our main motivation behind this study. We wanted to explore the different algorithms used for building recommendation systems and fine-tune the model specifically for book recommendations. We experimented with three different algorithm and the top N recommendations were generated.

Section 2 discusses the related work done on recommender systems so far. In Sect. 3, a brief background has been discussed so as to get the reader acquainted with the terminologies. Section 4 contains the experimental setup used for performing this study and a detailed analysis of our result. Finally, Sect. 5 concludes the research and discusses the future work.

2 Related Works

There is a lot of research work that has been done on collaborative and content-based filtering algorithms, most of them focusing either on movie recommendations, product recommendations, or article recommendations. In [3], C. Pereira, S. Iyer, and C. Raut discussed a product recommendation system based on cosine similarity algorithm. In [4], S. Kathait, S. Tiwari, and P. Singh proposed a hybrid recommender system that suggests articles based on short-term and long-term interests of users. In [5], D. Jannach, M. Zanker, A. Felfernig, G. Friedrich discussed a new approach for evaluating user similarity metrics without using any of the standard similarity functions. In [6], J. Leskovec, A. Rajaraman, J. D. Ullman mentioned a concept called "The Utility Matrix". According to them, the users have preferences for a certain number of items. Those preferences need to be extracted from the data. Finally, the data is represented as a utility matrix, in which for each user–item pair, the value represents the magnitude of preference of the user for that item. The matrix is assumed to be sparse, meaning that most of the entries in it are "unknown". Figure 1 shows the top 10 rows of a sample utility matrix.

```
book_id  1    2    3    4    5    6    ...  9994  9995  9996  9997  9998  9999
user_id                                 ...
1        0.0  0.0  0.0  5.0  0.0  0.0  ...  0.0   0.0   0.0   0.0   0.0   0.0
2        0.0  0.0  0.0  0.0  0.0  0.0  ...  0.0   0.0   0.0   0.0   0.0   0.0
3        0.0  0.0  0.0  3.0  0.0  0.0  ...  0.0   0.0   0.0   0.0   0.0   0.0
4        0.0  5.0  0.0  4.0  4.0  0.0  ...  0.0   0.0   0.0   0.0   0.0   0.0
6        0.0  0.0  0.0  0.0  0.0  0.0  ...  0.0   0.0   0.0   0.0   0.0   0.0
7        0.0  0.0  0.0  0.0  3.0  0.0  ...  0.0   0.0   0.0   0.0   0.0   0.0
8        0.0  0.0  0.0  3.0  3.0  0.0  ...  0.0   0.0   0.0   0.0   0.0   0.0
9        4.0  4.0  4.0  0.0  5.0  0.0  ...  0.0   0.0   0.0   0.0   0.0   0.0
10       0.0  0.0  0.0  5.0  5.0  0.0  ...  0.0   0.0   0.0   0.0   0.0   0.0
11       0.0  0.0  0.0  3.0  0.0  0.0  ...  0.0   0.0   0.0   0.0   0.0   0.0

[10 rows x 6414 columns]
```

Fig. 1 Top 10 rows of the Utility Matrix

3 Background

3.1 Machine Learning

Machine Learning (ML) is a small subset of Artificial Intelligence (AI) that uses statistical algorithms and provides systems the ability to make decisions on their own without being programmed explicitly to do so.

3.2 Recommendation System

A recommendation engine sorts the data with the help of algorithms and generates item recommendations relevant to the users. It captures the past behavior of the user and predicts items that the user might like in the future. There are two main types of recommendation systems: content based and collaborative based.

3.3 Content-Based Filtering

Content-based recommendation system concentrates on the attributes of the items and produces recommendations based on the similarity between those items. Pearson's Correlation is a very popular method that is used to calculate the similarity.

- Pearson's Correlation: It tells us the value of the correlation between two items. If the correlation is higher, the similarity will be more. Pearson's correlation is calculated using the formula given in (1). This method is mostly used when our data is subject to user bias or the rating scales of users are different.

$$sim(p, q) = \frac{\sum(r_{pi} - \bar{r}_p)(r_{qi} - \bar{r}_q)}{\sqrt{\sum(r_{pi} - \bar{r}_p)^2}\sqrt{\sum(r_{qi} - \bar{r}_q)^2}} \tag{1}$$

3.4 Collaborative-Based Filtering

Collaborative Filtering (CF) generates recommendations based on the attitude of users toward the items. This algorithm has the ability to do feature learning automatically. There are two types of CF approaches: Memory-based CF and Model-based CF.

Memory-Based CF approach can be divided into two main sections: user–item filtering and item–item filtering.

- User–item filtering: In this method, a particular user is chosen, and similar users are found based on the similarity of the ratings, and then those items are recommended that similar users also liked.
- Item–item filtering: In this method, an item is chosen, and then those users are found who liked that item, and then other items are found that those users or similar type of users also liked.

In Model-based CF approach, CF models are developed using various machine learning algorithms to predict user ratings of unrated items. This can again be subcategorized into Clustering Based, Matrix Factorization Based, and Deep learning based. Out of these three approaches, deep learning-based algorithm has the best accuracy overall.

4 Experimental Setup and Result Analysis

For this experiment, we used the Goodreads dataset from Kaggle[1]. The dataset consisted of a total 59,76,479 ratings from 53,424 users for 7877 books. After trimming and processing the data, the final version is shown in Table 1. Figure 2 gives us a histogram of ratings versus count of ratings.

Due to the sheer large number of ratings available in the dataset and considering the amount of time it would have taken to train the models, we took the top 10,00,000 user ratings as a standard for all algorithms.

All the code executions were done on the same machine and it had the following specifications:

- Intel Core i7—7700HQ @ 2.80 GHz
- NVIDIA GeForce GTX 1050 Ti 4GB DDR5 VRAM
- 16 GB DDR4 RAM
- Windows 10 Operating System

The code was written on Spyder editor. Programming language used was Python 3.7. We tried three algorithms and after analyzing we got the most effective one out of those. First, we tried Pearson Correlation algorithm. After that, we tried K-Nearest

[1]https://www.kaggle.com/zygmunt/goodbooks-10k.

Table 1 Top 5 items of the table

user_id	book_id	user_rating
1	258	5
11	258	3
143	258	4
242	258	5
325	258	4

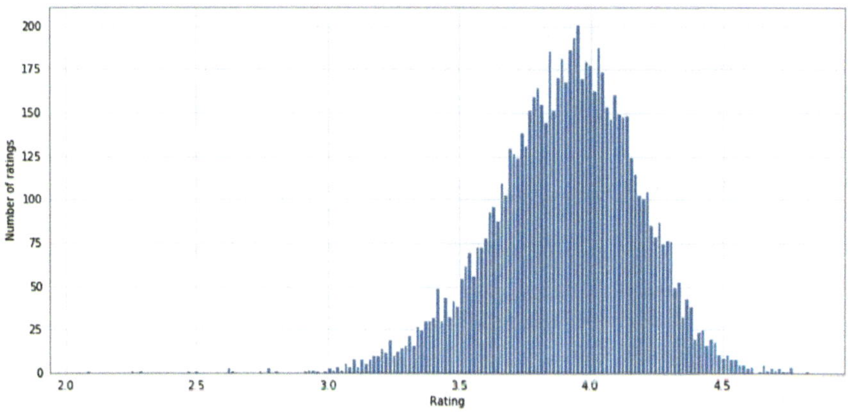

Fig. 2 Rating versus ratings count

Neighbors (KNN) algorithm, and then lastly Singular Value Decomposition (SVD) algorithm.

Let us analyze the results for Pearson Correlation algorithm given in Table 2. The list of recommendations were generated for the movie "Harry Potter and the Goblet of Fire".

For the sake of this experiment, we chose to only display those recommendations that have at least 100 user ratings. The first movie in each of these lists was the one based on which the rest of the recommendations have been fetched. So it had a perfect correlation. But the rest of the movies were sorted according to their correlation score in the descending order. But, a huge drawback of this algorithm was that it would only recommend items that are of similar type. It would not recommend books that the user has not read previously. So, if a user read only romantic books, for example, in the past, then the engine would only recommend books of the romantic genre. For example, in Table 2, we can clearly see that most of the recommendations shown to the user are from "Harry Potter" series. So, it proved to be a very limited way to build an engine.

To overcome this drawback, we needed an algorithm that would also consider the behavior of users as well. So we created a model using KNN algorithm and trained it with our dataset.

Table 2 Top recommendations based on Pearson Correlation

Title	Correlation	Number of ratings
Harry Potter and the Goblet of Fire	1.000000	4163
Harry Potter Boxset	0.788166	422
Harry Potter and the Half-Blood Prince	0.716292	3981
Harry Potter and the Prisoner of Azkaban	0.703594	4180
Harry Potter and the Order of the Phoenix	0.683845	4051
Harry Potter and the Chamber of Secrets	0.670286	4172
This is Where I Leave You	0.641210	146
Harry Potter and the Sorcerer's Stone	0.637437	4707
Harry Potter and the Deathly Hallows	0.619126	4016
Secrets of the Dragon Sanctuary	0.611922	138

Table 3 Recommendations based on KNN algorithm

Book ID	Title
815	No Country for Old Men
8	The Catcher in the Rye
65	Slaughterhouse-Five or The Children's Crusade
13	Nineteen Eighty-Four
4	To Kill a Mockingbird
5	The Great Gatsby
28	Lord of the Flies
14	Animal Farm: A Fairy Story
32	Of Mice and Men
1319	All the Pretty Horses

Here based on the same book, we fetched top 10 recommendations as shown in Table 3. We were no longer getting the books from the "Harry Potter" series, rather the recommendations were from a broad array of genre. We chose K value as 20, and used cosine similarity method for nearest neighbor search.

Finally, we trained the same dataset over SVD algorithm and got our top 10 recommendations which are shown in Table 4. These recommendations were based on a user that had read "Harry Potter and the Goblet of Fire" before and rated it. We trained our model using the same dataset, and it provided us with a list of book names that the user might like along with their predicted rating scores. Based on the recommendations, the user would have quite definitely liked "Harry Potter and the Philosopher's Stone", but he might have also liked "Twilight", "To Kill a Mockingbird", "The Great Gatsby", and so on.

We can see in Table 5 that the Root Mean Square Error (RMSE) score was 0.9032 with a Standard Deviation (SD) of 0.0069 which meant our models had a good

Table 4 Recommendations based on SVD algorithm

Title	Predicted rating
Harry Potter and the Philosopher's Stone	5.0
Twilight	4.89435797825888
To Kill a Mockingbird	4.81196333808558
The Great Gatsby	4.80950000353788
The Hobbit or There and Back Again	4.80706662161423
The Catcher in the Rye	4.78082147391311
Angels & Demons	4.77913179939264
Pride and Prejudice	4.73783697128662
The Kite Runner	4.71992985554299
Nineteen Eighty-Four	4.70085919856097

Table 5 SVD Model Evaluation

	Fold 1	Fold 2	Fold 3	Fold 4	Fold 5	Mean	SD
RMSE	0.8953	0.8973	0.9108	0.9120	0.9005	0.9032	0.0069
MAE	0.7042	0.7037	0.7165	0.7142	0.7069	0.7091	0.0053
Fit time	57.70	59.80	61.35	61.88	60.88	60.32	1.48
Test time	1.29	1.54	1.37	1.65	1.57	1.48	0.13

accuracy. But it had a major problem. Training the model was very slow and computationally expensive. Also, in case of both KNN and SVD algorithms, they suffered from popularity bias. So, they only recommended books that were popular. Another major drawback here was Cold Start. Over time, new books will get added to the database. And these books will be unrated. So, the recommender would fail to recommend these books because they had either none or very little interactions with the user.

In [7], R. Burke mentioned about a hybrid or mixed type of recommendation system that would be a mixture of two or more algorithms. As apparent from the name, two or more algorithms could be combined to generate top N recommendations. Here, if we combined the top N recommendations of KNN and SVD algorithm to create a new list of recommendations for the user, it would be the intersection of the two lists. In Tables 3 and 4, we see that books like "To Kill a Mockingbird", "The Great Gatsby", and "The Catcher in the Rye" appear in the top recommendations of both the algorithms, which means they can surely be recommended to the user.

Our model still suffered from Cold Start problem. For a new book, it would be really difficult to predict its rating without any existing ratings. In [8], Z. Wen discussed a trick to overcome this problem by filling all the unrated books' ratings with the mean rating of that book. So if we fill all the unrated books' ratings with the mean rating for that book, our Utility Matrix will become less sparse, leading to better predictions.

Huge companies like Netflix and Amazon use their own version of hybrid type of recommendation system. This helps to overcome any drawback that could be caused due to one type of algorithm and ultimately increasing the accuracy.

5 Conclusion

There could be various different ways to implement a recommender system. First, we tried content-based filtering method. But due to the drawbacks like recommending only those items that are of the same type, this method alone could not produce good results. After that we tried KNN and SVD algorithms that come under CF method. These algorithms improved upon the problems that we faced in content-based filtering. But SVD, being computationally slow and expensive, could not be used with large amount of data. This also meant that it was not scalable.

There are factors that can be taken into account while predicting recommendations, for example, how many pages did the user read of that book before he or she got disinterested in it. Also, RMSE scores are not a proper way to evaluate when it comes to recommender systems because we are trying to predict what a user might or might not like to read in the future. The recommendations that we provide may also convince the user to try a completely different genre that he has never read before. There can be a lot of hit and miss in this field. So in the future, we plan to address these issues and try to take into account all these factors when implementing a recommender system.

References

1. Pennock, D.M., Horvitz, E., Lawrence, S., Giles, C.L.: Collaborative filtering by personality diagnosis: a hybrid memory- and model-based approach. In: Proceedings of the Sixteenth Conference on Uncertainty in Artificial Intelligence (UAI-2000), pp. 473–480. Morgan Kaufman, San francisco (2000)
2. Khan, B.M., Mansha, A., Khan, F.H., Bashir, S.: Collaborative filtering based online recommender system: a survey. In: International Conference on Information and Communication Technologies (2017)
3. Pereira, C., Iyer, S., Raut, C.: Recommendation system based on cosine similarity algorithm. Int. J. Recent. Trends Eng. Res. **3**(9) (2017). ISSN: 2455-1457
4. Kathait, S.S., Tiwari, S., Singh, P.K.: Intelligent recommendation system. Int. J. Adv. Res., February (2017)
5. Jannach, D., Zanker, M., Felfernig, A., Friedrich, G.: Recommender Systems—An Introduction. Cambridge University Press (2011)
6. Leskovec, J., Rajaraman, A., Ullman, J.D.: Mining of Massive Datasets. Cambridge University Press (2011)
7. Burke, R.: Hybrid Recommender Systems: Survey and Experiments. User Modeling and User-Adapted Interaction, November (2017)
8. Wen, Z.: Recommendation System Based on Collaborative Filtering. Technical Report, CS229, Stanford University, USA, December (2008)

Dynamics in Fractional Calculus: A Computational Approach

Pinakadhar Baliarsingh, Birupakhya Prasad Padhy, Laxmipriya Nayak and Snigdha Samantaray

Abstract Recently, Fractional Derivatives (FDs) and Fractional Differential Equations (FDEs) are extensively used in modelings of most dynamic processes in the physical world involving biological and nonbiological materials. Normally, the nonuniform and violating natures of the FDs are more helpful for describing the dynamic behaviors of nature. In the present note, we provide some inconsistent and violating behaviors of well known Riemann–Liouville fractional derivatives.

Keywords Riemann–Liouville fractional derivatives · Caputo fractional derivatives · Fractional derivatives

1 Introduction

The theory of Fractional Calculus (FC) usually deals with the notion of derivatives and integrations of a function in the case of arbitrary orders. Due to its numerous applications in various fields of science and engineering, the theory of FC has been considered as one of the emerging sources of research. The theory initially developed mainly as a pure theoretical field of mathematics, recently, besides mathematics, the idea has been developed through other fields such as mechanics, electrical and electronics, chemistry, biology, and many other dynamic models for real-life problems.

P. Baliarsingh · B. P. Padhy (✉) · L. Nayak · S. Samantaray
Department of Mathematics, School of Applied Sciences,
Kalinga Institute of Industrial Technology, Deemed to be University, Patia, Bhubaneswar 751024, India
e-mail: birupakhya.padhyfma@kiit.ac.in

P. Baliarsingh
e-mail: pb.math10@gmail.com

L. Nayak
e-mail: laxmipriyamath@gmail.com

S. Samantaray
e-mail: samantaraysnigdha1995@gmail.com

H. Das et al. (eds.), *Progress in Computing, Analytics and Networking*,
Advances in Intelligent Systems and Computing 1119,
https://doi.org/10.1007/978-981-15-2414-1_51

The theory of FC is dynamic in nature, therefore the differential equations involved, are quite sufficient enough to capture and describe the dynamic and uncertain characteristics and behaviors of the physical world and the nature itself. Recently, Fractional-order Differential Equations(FDEs) are increasingly used to model the problems in control of dynamic systems, viscoelasticity, electrochemistry, diffusion processes, physical and biological processes, and many others. Besides, the theory of FDEs has been frequently used in fractional-order models involving non-differentiable problems in fractal engineering such as non-differentiable solution of LC circuit, the heat conduction equation on Cantor sets, the damped wave equation in fractal strings, the perturbation solution of the oscillator of free damped vibrations etc. and many more. For more detail of fractional derivatives and difference operators one may refer to [1–11].

As certain formulas involved in fractional derivatives and integrations have various forms and not compatible with each other, therefore the mathematical results involved have no clear visualization as that of the classical theory of calculus. This leads to come across several violations and dynamics in the theory of FC. For instance, the formulas based on Leibniz rule, Chain rule, Exponent rules, etc for fractional derivatives are found to be inconsistent and violated. The primary objective of this note is to discuss some of these inconsistent and violated results.

The most violations and dynamic natures in fractional calculus are due to the fact that the fractional derivative of a constant function need not be zero. For instances, in Riemann–Liouville sense, the fractional derivative of a constant function is never zero whereas it is exactly zero in other such as Caputo and Grunwald–Letnikov sense. Some of the deviations or violating behaviors based on Leibniz and Chain rules are also found in [12–16]. This violating and nonuniform behavior of fractional derivatives provides enough ingredients for the entire dynamic structure of FC.

2 Definitions and Preliminaries

Let $\mathbb{N}$ be the set of all natural numbers and α be a positive real. Then the initial definitions of fractional derivatives of order α due to Riemann–Liouville and Caputo are given bellow:

(i) The *Riemann–Liouville* fractional derivatives of order α and $n-1<\alpha\leq n$ is given by

$$({}_a^{RL}D_t^{\alpha}f)(t)=\frac{1}{\Gamma(n-\alpha)}\frac{\mathrm{d}^n}{\mathrm{d}t^n}\int_a^t (t-x)^{n-\alpha-1}f(x)\mathrm{d}x,\quad (t>a),$$

(ii) The *Caputo* fractional derivative of order α, is given by

$$({}_a^{C}D_t^{\alpha})(t)=\frac{1}{\Gamma(n-\alpha)}\int_a^t (t-x)^{n-\alpha-1}f^{(n)}(x)\mathrm{d}x,\quad (t>a),$$

where $\Gamma(\alpha)$, $(\alpha \notin \mathbb{Z}^- = \{-1, -2, -3, ...\})$ denotes the Euler gamma function with $\Gamma(\alpha+1) = \alpha\Gamma(\alpha)$ and $f^{(n)}(x) = \frac{\mathrm{d}}{\mathrm{d}x} f(x)$.

In fact, the relation between Riemann–Liouville and Caputo fractional derivatives is as follows:

$$({}_a^{RL}D_t^{\alpha} f)(t) = ({}_a^{C}D_t^{\alpha} f)(t) + \sum_{k=0}^{n-1} \frac{t^{k-\alpha}}{\Gamma(k-\alpha+1)} f^{(k)}(0).$$

Now we provide some properties of above fractional derivatives:

Let $f(t)$ and $g(t)$ be a continuous and differentiable function in whole real line. Then

P1: **Linearity**
The fractional differential operators given by Riemann–Liouville and Caputo are linear, i.e.,

$${}_a^{RL}D_t^{\alpha} [\lambda_1 f(t) + \lambda_2 g(t)] = \lambda_1({}_a^{RL}D_t^{\alpha} f)(t) + \lambda_2({}_a^{RL}D_t^{\alpha} g)(t),$$

and

$${}_a^{C}D_t^{\alpha} [\lambda_1 f(t) + \lambda_2 g(t)] = \lambda_1({}_a^{C}D_t^{\alpha} f)(t) + \lambda_2({}_a^{C}D_t^{\alpha} g)(t),$$

where $\lambda_1, \lambda_2 \in \mathbb{R}$, the set of all real scalars.

P2: **Backward compatibility**
If α is an integer, then proposed fractional derivatives provide same results as ordinary derivatives, i.e.,

$$({}_0^{RL}D_t^{\alpha} f)(t) = ({}_0^{C}D_t^{\alpha} f)(t) = \frac{\mathrm{d}^{\alpha}}{\mathrm{d}t^{\alpha}} f(t).$$

P3: **Identity**
The zero order derivatives of a function returns the original function, i.e.,

$$({}_a^{RL}D_t^{0} f)(t) = ({}_a^{C}D_t^{0} f)(t) = f(t).$$

P4: **Exponent rule**
For $\alpha, \beta > 0$, the following results do not hold in general

$$({}_a^{RL}D_t^{\alpha})({}_a^{RL}D_t^{\beta} f)(t) = ({}_a^{RL}D_t^{\beta})({}_a^{RL}D_t^{\alpha} f)(t) = ({}_a^{RL}D_t^{\alpha+\beta} f)(t)$$

and

$$({}_a^{C}D_t^{\alpha})({}_a^{C}D_t^{\beta} f)(t) = ({}_a^{C}D_t^{\beta})({}_a^{C}D_t^{\alpha} f)(t) = ({}_a^{C}D_t^{\alpha+\beta} f)(t).$$

P5: **Consistency**
Indeed, fractional derivatives of a constant function is exactly zero in Caputo sense, whereas those are nonzero in Riemann–Liouville sense.

P6: **Leibniz rule**
The well known formula for Leibniz rule due to [1] is consistent only for the integer $\alpha = 1$ and may not hold for fractional values of α in Riemann–Liouville sense, i.e.,

$$({}_a^{RL}D_t^\alpha fg)(t) = ({}_a^{RL}D_t^\alpha f)(t) \cdot g(t) + f(t) \cdot ({}_a^{RL}D_t^\alpha g)(t) \text{ does not hold in general.}$$

P7: **Chain rule**
For any integer α, the Chain rule is consistent in both Caputo and Riemann–Liouville sense but do not hold for a proper fraction α, i.e.,

$$({}_a^{RL}D_t^\alpha f(g))(t) = ({}_a^{RL}D_{g(t)}^\alpha f(g))(t)({}_a^{RL}D_t^\alpha g)(t)$$

and

$$({}_a^{C}D_t^\alpha f(g))(t) = ({}_a^{C}D_{g(t)}^\alpha f(g))(t)({}_a^{C}D_t^\alpha g)(t)$$

do not hold for a proper fraction α.

P8: **Geometrical interpretation**
The geometrical interpretations of derivatives of integer orders are well captured and visualized, whereas those of fractional derivatives are dynamic in nature.

3 Main Results

In this section, we provide the proof of the properties of fractional derivatives mentioned above. For this we take the help of following theorems:

Theorem 1 *Let α be positive real, $f(t) = (t-a)^\beta$ for $\beta \in \mathbb{R} \setminus Z^-$ and $t > a$. Then*

$$({}_a^{RL}D_t^\alpha f)(t) = ({}_a^{C}D_t^\alpha f)(t) = (t-a)^{\beta-\alpha}\frac{\Gamma(\beta+1)}{\Gamma(\beta+1-\alpha)}.$$

Proof of P1, P2, P3: Linearity, backward compatibility, identity properties are clear from the definitions, hence omitted.

Proof of P4: We provide the following counter example for the inconsistence of P4. For this we first consider the fractional derivatives in Riemann–Liouville sense.

Example: Consider the function $f(t) = t^{\alpha-1}$, for all $\alpha > 0$. Then clearly, taking $a = 0$ in Theorem 1, we have

$$({}_0^{RL}D_t^\alpha f)(t) = t^{-1}\frac{\Gamma(\alpha)}{\Gamma(\alpha-\alpha)} = 0.$$

For each $\beta \notin \mathbb{N} \cup \{0\}$ such that $(\alpha - \beta) \notin \mathbb{Z}^- \cup \{0\}$, and from the linearity property, one can derive that

$$({}_0^{RL}D_t^{\beta})({}_0^{RL}D_t^{\alpha} f)(t) = 0. \tag{1}$$

Since $(\alpha - \beta) \notin \mathbb{Z}^- \cup \{0\}$, we now get

$$({}_0^{RL}D_t^{\beta} f)(t) = t^{\alpha-1-\beta} \frac{\Gamma(\alpha)}{\Gamma(\alpha - \beta)}.$$

Finally, for $\beta \notin \mathbb{N} \cup \{0\}$, we derive that

$$({}_0^{RL}D_t^{\alpha})({}_0^{RL}D_t^{\beta} f)(t) = t^{\alpha-1-\beta-\alpha} \frac{\Gamma(\alpha)}{\Gamma(\alpha - \beta)} \frac{\Gamma(\alpha - \beta)}{\Gamma(\alpha - \beta - \alpha)} = t^{-1-\beta} \frac{\Gamma(\alpha)}{\Gamma(-\beta)} \neq 0. \tag{2}$$

Combining (1) and (2), we complete the proof of the first two equality and for the last part we may use the similar techniques in $f(t) = t^{\beta-1}$.

Proof of P5: From the definition, the fractional derivatives of a constant function $f(t) = \lambda$ is zero, but this does not happen in the case of Riemann–Liouville derivatives. A consequence of Theorem 1, we have the following remark:

Remark 1 *Let* $0 < \alpha < 1$ *and* $f(t) = \lambda$. *Then*

$$({}_0^{RL}D_t^{\alpha} f)(t) = \frac{\lambda t^{-\alpha}}{\Gamma(1 - \alpha)}.$$

Proof of P6: The well known Leibniz rule for fractional derivatives due to [1] is inconsistent. To show it we use contradiction method. Suppose the following identity holds for $a = 0$.

$$({}_0^{RL}D_t^{\alpha} fg)(t) = ({}_0^{RL}D_t^{\alpha} f)(t) \cdot g(t) + f(t) \cdot ({}_0^{RL}D_t^{\alpha} g)(t).$$

Now, replacing $g(t) = 1$ in above expressions, we have

$$\begin{aligned}
({}_0^{RL}D_t^{\alpha} f)(t) &= ({}_0^{RL}D_t^{\alpha} f \cdot 1)(t) \\
&= ({}_0^{RL}D_t^{\alpha} f)(t) + f(t) \cdot ({}_0^{RL}D_t^{\alpha} 1)(t) \\
&= ({}_0^{RL}D_t^{\alpha} f)(t) + f(t) \cdot \frac{t^{-\alpha}}{\Gamma(1 - \alpha)}
\end{aligned}$$

Therefore, for any $\alpha > 0$ and continuous function $f(t)$, we conclude that

$$({}_0^{RL}D_t^{\alpha} f)(t) = \frac{t^{-\alpha}}{2\Gamma(1 - \alpha)} f(t),$$

which is not always true, hence the above identity for Leibniz rule is inconsistent. In fact, the corrected and generalized formula of Leibniz rule for fractional derivatives is as follows:

$$({}_a^{RL}D_t^\alpha fg)(t) = \sum_{k=0}^{\infty} \binom{\alpha}{k} ({}_a^{RL}D_t^{\alpha-k} f)(t)({}_a^{RL}D_t^k g)(t).$$

An application of this formula, we can approximate the fractional derivatives of a function $f(t)$ by Taylor type series expansion as

$$\begin{aligned}({}_0^{RL}D_t^\alpha f)(t) = ({}_0^{RL}D_t^\alpha 1 \cdot f)(t) &= \sum_{k=0}^{\infty} \binom{\alpha}{k} ({}_0^{RL}D_t^{\alpha-k} 1)(t)({}_0^{RL}D_t^k f)(t) \\ &= \sum_{k=0}^{\infty} \binom{\alpha}{k} \frac{t^{-\alpha+k}}{\Gamma(1-\alpha+k)} ({}_0^{RL}D_t^k f)(t) \\ &= \sum_{k=0}^{\infty} \binom{\alpha}{k} \frac{t^{-\alpha+k}}{\Gamma(1-\alpha+k)} \frac{\mathrm{d}^k}{\mathrm{d}t^k} f(t).\end{aligned}$$

Proof of P7: We demonstrate the result by taking the following counter example:

Example: Consider the functions $f(t) = t$ and $g(t) = t^2$. Then clearly, $f(g(t)) = f(t^2) = t^2$ and for $\alpha > 0$, we have the left hand side as

$$({}_0^{RL}D_t^\alpha f(g))(t) = ({}_0^{RL}D_t^\alpha t^2)(t) = \frac{2t^{2-\alpha}}{\Gamma(3-\alpha)}. \tag{3}$$

But the right hand side can be written as

$$\begin{aligned}({}_0^{RL}D_{g(t)}^\alpha f(g))(t)({}_0^{RL}D_t^1 g)(t) &= ({}_0^{RL}D_{t^2}^\alpha t^2)(t)({}_0^{RL}D_t^1 t^2)(t) \\ &= ({}_0^{RL}D_p^\alpha p)(p)2t \\ &= \frac{2tp^{1-\alpha}}{\Gamma(2-\alpha)} = \frac{2t^{3-2\alpha}}{\Gamma(2-\alpha)},\end{aligned} \tag{4}$$

where we have considered a new variable $p = t^2$. Using (3) and (4), we finish the proof.

Proof of P8: The geometrical interpretation of fractional derivatives are still an open problem to the field of fractional calculus. In this context, we consider a function $f(t) = 2$ and with the help of MATLAB we plot the graphs of the fractional derivatives of $f(t)$ for different orders. In Fig. 1, we present the green color plots for the fractional derivatives of $f(t)$ with integer order. Red and blue color plots signify the fractional derivatives with positive and negative fractional orders, respectively. It is seen that the plots do not exhibit the uniform behaviors.

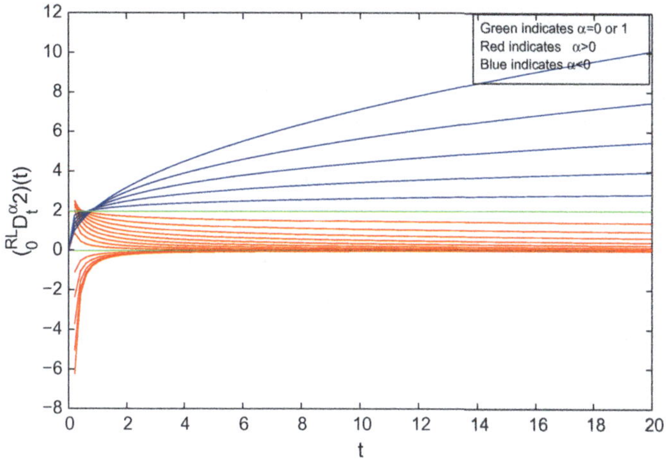

Fig. 1 Plot of the fractional derivatives $({}_0^{RL}D_t^{\alpha}2)(t)$ verses t

Conclusion: The present note states that the several properties for fractional derivatives is only valid for integer order and may deviate for fractional cases. So many attempts have been taken in order to get a unified or generalized formulas which can easily accumulate in both classical and fractional cases. But due to the nonlinearity of the fractional differentiation, the entire attempts could not get the success and still make an open problem with the existing field.

References

1. Jumarie, G.: On the representation of fractional Brownian motion as an integral with respect to $(dt)^{\alpha}$. Appl. Math. Lett. **18**(7), 739–748 (2005)
2. Jumarie, G.: On the solution of the stochastic differential equation of exponential growth driven by fractional Brownian motion. Appl. Math. Lett. **18**(7), 817–826 (2005)
3. Liu, C.: On the local fractional derivative of everywhere nondifferentiable continuous functions on intervals. Commun. Nonlinear Sci. Numer. Simul. **42**, 229–235 (2017)
4. Tarasov, V.E.: Quantum Mechanics of Non-Hamiltonian and Dissipative Systems. Elsevier Science, New York (2008)
5. Atangana, A., Baleanu, D.: New fractional derivatives with nonlocal and non-singular kernel theory and application to heat transfer model. Therm. Sci. **20**(2), 763–769 (2016)
6. Baleanu, D., Diethelm, K., Scalas, E., Trujillo, J.J.: Fractional Calculus: Models and Numerical Methods. World Scientific Publishing, Singapore (2012)
7. Baliarsingh, P., Dutta, S.: On a spectral classification of the operator Δ_{ν}^{r} over the Sequence Space c_0. Proc. Natl. Acad. Sci. India, Sect. A, Phys. Sci. **84**(4), 555–561 (2014)
8. Baliarsingh, P., Dutta, S.: On certain Toeplitz matrices via difference operator and their applications. Afr. Mat. **27**(5–6), 781–793 (2016)
9. Baliarsingh, P.: Some new difference sequence spaces of fractional order and their dual spaces. Appl. Math. Comput. **219**(18), 9737–9742 (2013)

10. Baliarsingh, P., Dutta, S.: A unifying approach to the difference operators and their applications. Bol. Soc. Paran. Mat. **33**(1), 49–57 (2015)
11. Baliarsingh, P., Dutta, S.: On the classes of fractional order difference sequence spaces and their matrix transformations. Appl. Math. Comput. **250**, 665–674 (2015)
12. Osler, T.J.: A further extension of the Leibniz rule to fractional derivatives and its relation to Parseval's formula. SIAM J. Math. Anal. **3**(1), 1–16 (1972)
13. Baliarsingh, P.: On a fractional difference operator, Alex. Eng. J. **55**(2), 1811–1816 (2016)
14. Baliarsingh, P., Nayak, L.: A note on fractional difference operators. Alex. Eng. J. **57**(2), 1051–1054 (2018)
15. Tarasov, V.E.: On chain rule for fractional derivatives. Commun. Nonlinear Sci. Numer. Simul. **30**(1), 1–4 (2016)
16. Tarasov, V.E.: No violation of the Leibniz rule. No fractional derivative. Commun. Nonlinear Sci. Numer. Simul. **18**(11), 2945–2948 (2013)

Privacy Preserving in Collaborative Filtering Based Recommender System: A Systematic Literature Review

Srishti Raj, Abhaya Kumar Sahoo and Chittaranjan Pradhan

Abstract Recommender systems solve the information overload problem by filtering data on the basis of the user's preferences, interest, or previous behavior regarding an item. Data filtering techniques employed are content-based (based on the user's past behavior), collaborative (based on the behavior of users that are alike to the active one), or hybrid (a combination of filtering techniques). Due to its versatility, the most popular technique used in the recommender systems is collaborative filtering. However, the privacy of the user is at risk because malicious users can attack the targeted user or the recommendation server may reveal the personal data of users' to other parties or misuse the data for targeted advertising. The existing works mostly employ encryption or randomizations based methodologies, but often sacrifice privacy for accuracy and accuracy for privacy.

Keywords Collaborative filtering · Perturbation · Privacy · Randomization · Recommender system

1 Introduction

The massive expansion of the world wide web in recent years has presented users with a vast variety of choices in several fields like e-commerce, movies, music, documents, etc. This in turn has rendered the selection of a few relevant items from thousands of available items as a time and effort consuming process. However, when these choices available on the web are provided with ratings, ranks, reviews, opinions, complaints, remarks, feedback, and comments (regarding a product, event, individual, and services), filtering of choices and finding the right ones for the current user become

S. Raj · A. K. Sahoo (✉) · C. Pradhan
School of Computer Engineering, KIIT Deemed to Be University, Bhubaneswar, India
e-mail: abhayakumarsahoo2012@gmail.com

S. Raj
e-mail: srishti.raj.doll19@gmail.com

C. Pradhan
e-mail: chitaprakash@gmail.com

H. Das et al. (eds.), *Progress in Computing, Analytics and Networking*,
Advances in Intelligent Systems and Computing 1119,
https://doi.org/10.1007/978-981-15-2414-1_52

easier and less time taking. The applications that perform these tasks are known as recommender systems. They provide recommendations based on the ratings and summary of the relevant text about items that can be used for decision-making [1].

Recommender systems are profitable to the service providers as well to the users. They significantly reduce the time required for finding a suitable item by the active user through their curated list of recommendations lest the user is lost in a variety of categories and options. When users get benefits of a recommender system, helping them find relevant items in a short span of time, they return and often repeat their transactions. This in turn enhances the revenue of the business (utilizing the recommender system) and hence proves beneficial to the service providers. It brings loyalty to customers as well as helps improve the decision-making process and quality of services. Therefore, the need to use efficient and accurate recommendation techniques within a system that will provide relevant and dependable recommendations for users cannot be over-emphasized. Also, the massive growth of e-commerce sites and online businesses is enhancing the requirements of a robust recommendation system [2]. We provide privacy to the recommender systems to hide the user's personal information like ratings or specific items that have been rated. Apart from accuracy, privacy is a major metric for recommender systems and the challenge is to maintain the high privacy of the users and simultaneously provide them with accurate recommendations. Different privacy-preserving techniques are used in different applications of a recommender system. In this paper, we focus on both accuracy and privacy factors. We examine some of the proposed works utilizing randomization techniques in a chronological order for privacy preservation in collaborative filtering and identify the area that still holds an immense scope of improvement. In this paper, the Sect. 2 describes about recommender system and its basic concepts. Section 3 shows the different privacy-preserving collaborative filtering methods. Section 4 explains the comparison study among different privacy-preserving collaborative filtering methods. Section 5 concludes the paper and presents the future scope in the domain of recommender system.

2 Background

Recommender system takes user–item matrix as input and uses different types of filtering techniques such as collaborative filtering, content-based filtering, and hybrid filtering. Recommender systems address the issue of information overload by generating suitable information from a massive quantity of dynamically produced information in accordance with the taste, interest, or past behavior of the user regarding an item. On the basis of the user's profile, recommender system acquires the capability to predict if or not a specific user would favor an item, which is shown in Fig. 1.

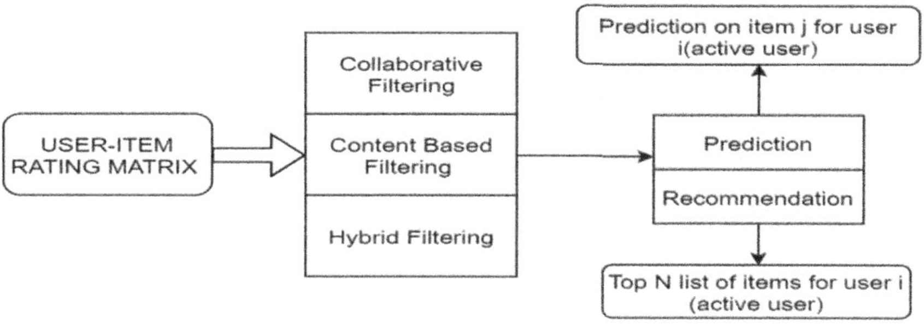

Fig. 1 Block diagram of a recommender system

2.1 Different Phases of Recommender System

To build an efficient Recommender system, different phases should be followed. These phases consist of input and output phases along with learning phase. Figure 2 shows the phases of recommendation process. This process broadly consists of three phases; information collection phase, learning phase, and prediction/recommendation phase.

Information collection phase collects appropriate user information and creates a profile or model for each user including the user's meta data, behavior, or attributes of the assets accessed by him for the task of prediction. The system requires to learn as much as possible from the user to make sensible recommendations straight from its inception [1]. Recommender systems depend upon various kinds of input that can be classified into explicit feedback where users express their interest for an item or implicit feedback where the system infers a user's choices indirectly by monitoring his behavior or hybrid feedback attained by employing a mix of the other two feedback methods [1]. Learning Phase implements a learning algorithm to filter and utilize the characteristics of the user from the collected feedback of the previous phase [1]. Prediction/Recommendation phase recommends or predicts what type of items the user might like. It can be done either via memory-based or model-based techniques directly applied on the collected data set in the first phase of the recommendation process or can be achieved based on the previous activities of the user recorded by the system [1, 10].

Fig. 2 Phases of recommendation process

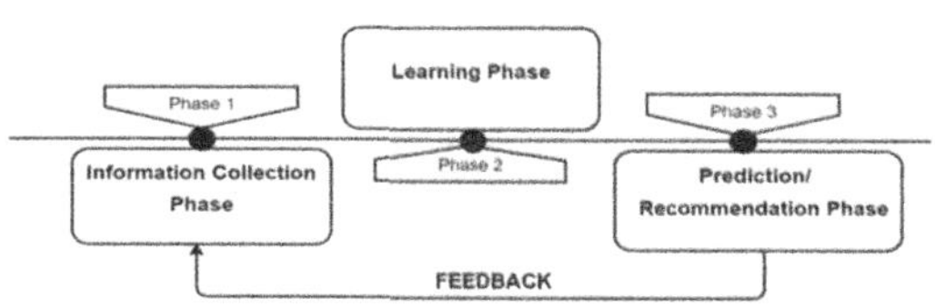

2.2 Recommendation Filtering Techniques

Data collected in the previous phases needs to be filtered for the process of recommendation. Filtering can either be *content-based filtering* which utilizes the information about past choices [2] made by the user and the meta data of items or *Collaborative Filtering (CF)* which functions by creating a user–item matrix that specifies the choice of items made by the users or *hybrid filtering* which is a combination of one or more filtering techniques. CF can either be *model based or memory based.* Recommendations generated by CF can either be a prediction or recommendation.

2.3 Evaluation Metrics for Recommendation Algorithms

Recommendation algorithms are classified as statistical or decision support accuracy metrics. Statistical accuracy metrics assess the precision of the data filtering process by comparing the predicted ratings straight with the actual user rating. Mean Absolute Error (MAE), Root Mean Square Error (RMSE), and Correlation are the widely used statistical accuracy metrics. Decision support accuracy metrics being popularly used are reversal rate, weighted errors, Receiver Operating Characteristics (ROC), and Precision–Recall Curve (PRC), Precision, Recall, and F-measure. These measures assist users in choosing items of excellent grade from all the accessible items at hand [10, 11].

2.4 Privacy for Recommender System

Commonly users do not mind to reveal some of their private data for the sake of receiving useful recommendations; however, they are concerned that such systems might gather too much of their personal information and misinterpret their preferences, or employ the data to fulfill objectionable motives. The recommender system can become a security threat involving an unacceptable collection of data, targeted marketing or promotions, and bias toward certain users. Other system users might become an adversary trying to gain access to the personal data of a targeted user by exploiting the outputs of the recommender. Also, some external entities might attempt to access the gathered information by the recommender or interrupt the communication of the system and its consumers [3, 10].

3 Different Privacy-Preserving Collaborative Filtering Methods

Privacy-preserving collaborative filtering methods are used to enhance the privacy level with better accuracy of recommender system. The following methods are explained below.

3.1 Nonnegative Matrix Factorization (NMF) Based Privacy-Preserving Recommendation Method

Li et al. [4], proposed an algorithm based on the concept of Nonnegative Matrix Factorization (NMF) for privacy preservation in collaborative filtering. The algorithm uses random perturbation techniques to disguise ratings of items and items that have already been rated. Arbitrary numbers generated either through Gaussian or normal distributions are added to the original numbers. These disguised data are then sent to the servers for computations instead of the original ones. The steps to obscure data of a recommender system having n users and m items are explained. (a) The server selects the distribution (uniform or Gaussian) and parameters (α, μ, σ), and informs each user. Every user generates uniform random values within a range of $[-\alpha, \alpha]$ where α is a constant number when the distribution is decided to be a normal distribution, or creates random values having a normal distribution with mean $\mu = 0$ and variance $\sigma 2$ when the distribution is Gaussian. (b) The empty cells of the user's ratings vector are filled by his ratings' mean and z-scores are calculated. (c) n random numbers duj, are generated by each user u either from a uniform or Gaussian distribution. The random numbers are added to the z-score values and disguised z-scores are generated. $z'uj = zuj + duj, j = 1, 2, \ldots, m$, where m is the total number of items. (d) At last, every user communicates these values to the recommendation server which creates the new disguised user–item matrix R'. (e) Utilizing the concept of NMF, the matrix R' is now factorized into W and H and the rating of item j by the user i is predicted as

$$P_{ij} = \bar{r}_{\mathrm{i}} + \mathrm{sgn}\big(R'(i, j)\big) \cdot \sigma_i \cdot [W(i) \cdot H(j)]. \tag{1}$$

where $\bar{r}_i$ is the mean value, σ_i is the standard deviation of user i, and $\mathrm{sgn}(R'(i, j))$ is the signature of the element in ith row and jth column of R'.

3.2 *Naive Bayesian Classifier (NBC) Based Privacy-Preserving Recommendation Method*

NBC might be employed in CF by utilizing the classification of ratings into two classes; like and dislike. In this scheme, all users are treated as a feature. Let "a" be an active user with ratings as the class labels and D that maintains the ratings of the other users against the feature/attribute values, where the attributes are independent of the class labels. Then, the conditional probability of a specific item belonging to class *j*, where *j* ε {like, dislike} provided its n attributes, can be given by

$$P(\text{class}\, j | f1, f2, \ldots, fn) \alpha\; p(\text{class}\, j) \Pi \text{in}\; P(fi | \text{class}\, j) \tag{2}$$

where *fi* is the feature of target item (*q*) for user "a", and *p*(class*j*) and *P*(*fi*|class*j*) are obtained from the training set. The calculation of conditional probabilities is completed and q is allocated to the class having the largest probability. In this paper, Bilge et al. [5] aim to enhance NBC-based recommendations by creating a subset of data being used for online recommendations through two preprocessing techniques applied to the already disguised data set D.

Neighborhood formation: The total number of items in a neighborhood has a direct impact on the efficiency of recommendations. An optimum number of neighbors N is selected to be the neighbors of an item, where value of N is obtained experimentally, and N is a constant satisfying the condition $N < m$. Here, Tanimoto coefficient is used to find the neighbors of an item [5].

Increasing density preprocessing method: A predetermined optimum percentage of the empty cells is filled to stabilize the performance of the system. The method proceeds as follows: (1) The server decides on an experimentally calculated performance and accuracy parameter p, supposed to provide the best results. This parameter is closely related with the density of D. (2) An R percent (uniformly randomly selected from the range $(0, p)$ of the empty ratings' cells (V) is selected. (3) V number of personalized predictions for the selected empty ratings' cells using PPNBC is estimated. (4) Lastly, all chosen empty cells are filled with the calculated personalized ratings [5].

3.3 *Distributed Anonymization Based Privacy-Preserving Recommendation Method*

Zhifeng et al. [6] proposed a locally distributed data scheme enabling the users to independently anonymize their ratings' data without viewing others' data, which is shown in Fig. 3. This scheme assumes three major roles: the authorized user u_a, a group of collaborative users $\{u_c\}$, $u_c \in U_{co}$, and the unauthorized user u_h.

Let $U := u_a \cup u_c \cup U_{co}$ denote the set of all users. Every user of the system is allocated a unique integer c, $c \in [0, |U| - 1]$, for the user id and every item a unique

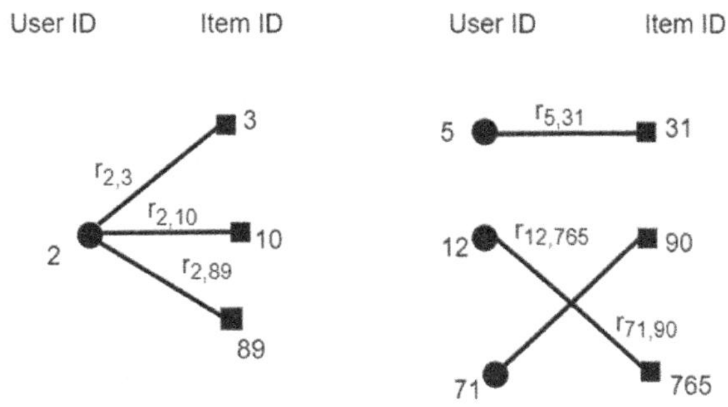

Fig. 3 An anonymized bipartite review graph of user 2 [6]

integer o, o ∈ [0, |*O*| − 1], where *O* is the set of all items. In this scheme, a user needs to anonymize his local data individually using anonymized bipartite review graphs and anonymization map.

3.4 Random Perturbation Based Privacy-Preserving Recommendation Method

Polatidis et al. [7] utilize a multilevel method to safeguard the ratings' privacy. Multiple levels are used and at each level, the rating is perturbed from a different value range, thus protecting the privacy of the user fairly and also retaining an acceptable accuracy. The overall steps followed in this algorithm are (a) Number of levels of perturbation and the range of random values, say [low, high] to be generated at each level is decided experimentally based on the data set and other conditions at the client side. (b) At each level, random numbers are generated from the range specified previously. For every rating, a perturbed rating is created by adding the rating and a random number together. If the perturbed rating falls below low, the lower limit of the range; value of low is assigned to it. Similarly, if the perturbed rating crosses the higher limit of the range, high; it is assigned the value of high. (c) When all the levels are exhausted, the new set of disguised ratings is released to the server by all users [7].

3.5 Fuzzy Based Random Perturbation Method for Privacy Preservation in Recommender System

Manju et al. [8] propose a methodology for privacy preservation using fuzzy theory, randomization, and ant-colony based clustering. The proposed methodology is pictorially represented in Fig. 4. In this method, the user–item rating matrix is subjective leading to uncertainty about the user's behavior and preferences. Hence, the

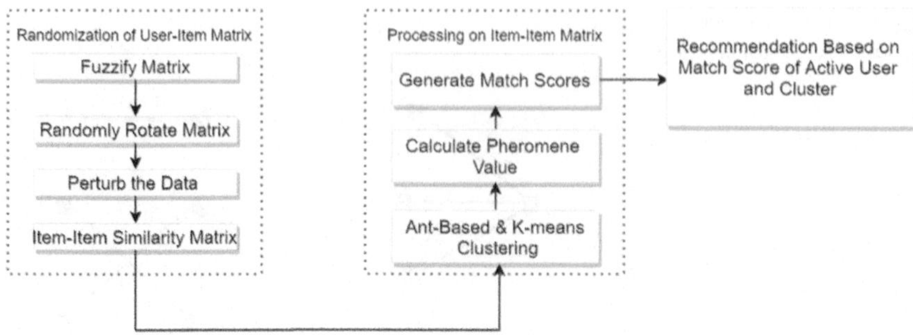

Fig. 4 Recommendation process followed for the fuzzy data set [8]

fuzzy theory can be used to depict the user–item rating matrix. After the generation of the fuzzified user–item rating matrix, a random rotation perturbation of the matrix follows. This helps the system to conceal the real interests of every user from the adversary. Cosine similarity measure is applied to the perturbed matrix and an item–item similarity matrix is derived. Items in the item–item similarity matrix are clustered via the ant-colony clustering algorithm for the first iteration and the cluster centroid is obtained. This initial cluster's centroid is improved with the help of *k*-means clustering algorithm. All these steps are carried out offline. The pheromone value $P(t)$, on every cluster is calculated. When an active user logs in, Euclidean Distance is used and similarity between the active user and the cluster center is obtained. A match score is generated and the best matched item cluster for the active user is identified and the items of the cluster not yet seen by the active user are recommended.

Where $P(t)$ is the deposited pheromone of the cluster, SM is the similarity, I_n is an active user item, and I_c is the item cluster center. Finally, pheromone values are updated after each recommendation and the updated value can be used in the future [8].

4 Comparison Study Among Different Privacy-Preserving Collaborative Filtering Methods

We compare different privacy-preserving based collaborative filtering methods along with advantages and disadvantages which are shown in Table 1.

Our survey tries to include the most novel issues that have not been explained carefully in the existing papers. We find the most outstanding features of this survey.

It summarizes different privacy-preserving methods to recommender system along with their advantages and disadvantages. This paper shows different techniques based on collaborative filtering to find out accuracy of recommender model and measure the privacy level toward user data. It finds the recent challenges toward recommender

Table 1 Comparison of the existing methodologies

Literature, Year	Methodology	Advantages	Disadvantages
Nonnegative Matrix Factorization (NMF) Based Privacy-Preserving Recommendation Method [4], 2009	1. Random perturbation technique to disguise ratings 2. NMF for prediction	1. Better in large data 2. Performance better than SVD-based ones	1. Since better with large data, might end up using the whole original data set 2. All empty cells filled which affect originality of data
Naive Bayesian Classifier(NBC) Based Privacy-Preserving Recommendation Method [5], 2010	1. Offline neighborhood formation and increasing density 2. NBC for prediction	1. Able to predict from small amount of training data 2. More truthful predictions	1. Affects efficiency negatively. Improvements required strengthening the overall system 2. Performance through preprocessing
Distributed Anonymization Based Privacy-Preserving Recommendation Method [6], 2013	1. Users' anonymize their own data individually 2. Data stored as bipartite graphs and not matrix 3. Weighted sum of votes applied for prediction [9]	1. Higher level of privacy than perturbation based schemes	1. Page Rank, HITS are still focused on unipartite graphs 2. Much harder to implement than classical matrix-based solution
Random Perturbation Based Privacy-Preserving Recommendation Method [7], 2017	1. Multiple levels of perturbation 2. Pearson correlation for prediction	1. Perturbation range created randomly leaving no information about actual perturbation range 2. Privacy of users is well protected	1. Accuracy is decreased 2. Several experiments need to be conducted to decide the number of levels and perturbation range

system, i.e., enhancement of privacy level, diversity, stability, and accuracy of model to avoid cold start problem and shillong attack.

5 Future Work and Conclusion

The privacy preservation of the user's data is a major concern in the recommendation process as their critical information might be revealed to the public or malign users via the recommendation server. Any measure taken to protect the privacy of user's rating results in a degradation of accuracy and vice versa. One of the most effective methods to protect the privacy in collaborative filtering systems is to perturb the rating before being provided to the recommendation server. Some other existing works utilize encryption techniques. Systems that provide users with reliable recommendations without revealing their sensitive data is the demand of the near future. In the future, we would aim to implement the proposed methodologies and investigate the employability of other similarities, clustering, and deep learning methods and compare the result with the existing ones. Optimization of different algorithms and methods of recommender systems would be another goal. By using bio-inspired approaches, we can build an efficient recommender system.

References

1. Isinkaye, F.O., Folajimi, Y.O., Ojokoh, B.A.: Recommendation systems: principles, methods and evaluation. Egypt. Inform. J. **16**(3), 261–273 (2015)
2. Bobadilla, J.E.S.U.S., Serradilla, F., Hernando, A.: Collaborative filtering adapted to recommender systems of e-learning. Knowl. Based Syst. **22**(4), 261–265 (2009)
3. Knijnenburg, B.P., Berkovsky, S.: Privacy for recommender systems: tutorial abstract. In: 11th ACM Conference on Recommender Systems, RecSys' 17, pp. 394–395. Association for Computing Machinery (ACM) (2017)
4. Li, T., Gao, C., Du, J.: A NMF-based privacy-preserving recommendation algorithm. In: 2009 First International Conference on Information Science and Engineering, pp. 754–757. IEEE (2009)
5. Bilge, A., Polat, H.: Improving privacy-preserving NBC-based recommendations by preprocessing. In: 2010 IEEE/WIC/ACM International Conference on Web Intelligence and Intelligent Agent Technology, vol. 1, pp. 143–147. IEEE (2010)
6. Luo, Z., Chen, S., Li, Y.: A distributed anonymization scheme for privacy-preserving recommendation systems. In: 2013 IEEE 4th International Conference on Software Engineering and Service Science, pp. 491–494. IEEE (2013)
7. Polatidis, N., Georgiadis, C.K., Pimenidis, E., Mouratidis, H.: Privacy-preserving collaborative recommendations based on random perturbations. Expert Syst. Appl. **71**, 18–25 (2017)
8. Manju, S., Thenmozhi, M.: Privacy preserving collaborative filtering approach for recommendation system. In: 2018 International Conference on Wireless Communications, Signal Processing and Networking (WiSPNET), pp. 1–5. IEEE (2018)
9. Breese, J.S., Heckerman, D., Kadie, C.: Empirical analysis of predictive algorithms for collaborative filtering. In: Proceedings of the Fourteenth Conference on Uncertainty in Artificial Intelligence, pp. 43–52. Morgan Kaufmann Publishers (1998)
10. Sahoo, A.K., Mallik, S., Pradhan, C., Mishra, B.S.P., Barik, R.K., Das, H.: Intelligence-based health recommendation system using big data analytics. In: Big Data Analytics for Intelligent Healthcare Management, pp. 227–246. Academic Press (2019)
11. Sahoo, A.K., Pradhan, C., Barik, R.K., Dubey, H.: DeepReco: deep learning based health recommender system using collaborative filtering. Computation **7**(2), 25 (2019)

A Topology Based Address Selection Technique for Network Virtualization Environment

Kumari Soni, Navin Mani Upadhyay, Juli Singh and Ankur Srivastava

Abstract Network Virtualization is a concept about the distribution of resources for easy access, accounting, rapid connection sharing, and release. Therefore, to make this possible, it is necessary to manage modern data centers effectively. However the resources from computer and storage have virtualized by applying service providers and infrastructure providers techniques which is quite effective, but a comprehensive solution for the network virtualization has not been developed yet. Since the selection of source address and the destination address is very difficult. So this paper will analyze the essential components of network virtualization which have point out into six processing steps of simple and unique observation. One is specifying the network-related requirements, such as source address, destination address and another one is, processing over the unconventional path in a methodical and adaptable manner. This paper introduces a topology-based address selection technique. In complex network architecture like Distributed Overlay Virtual Ethernet network, selection of the correct node is quite difficult from which any data will pass through. At last, this paper demonstrates the scalability and the efficiency of the solution.

Keywords Network virtualization · Address selection algorithm · VPNS · DOVE · Virtual nodes · Virtual link

---

K. Soni · J. Singh · A. Srivastava
Ashoka Institute of Technology and Management, Varanasi, Uttar Pradesh 221007, India
e-mail: soni.suman2232@gmail.com

J. Singh
e-mail: mail2juli@gmail.com

A. Srivastava
e-mail: ankur.saurabh@gmail.com

N. M. Upadhyay (✉)
Indian Institute of Technology (BHU), Varanasi, Uttar Pradesh 221005, India
e-mail: navinmupadhyay.rs.cse17@iitbhu.ac.in

H. Das et al. (eds.), *Progress in Computing, Analytics and Networking*,
Advances in Intelligent Systems and Computing 1119,
https://doi.org/10.1007/978-981-15-2414-1_53

1 Introduction

Since the Internet is almost ossified and there are lots of Band-Aids and makeshift solutions. So there is a need for new architecture, and it is hard to come up with a one size fit all architecture because almost it is impossible to predict what future might unleash. So all the architectures created till today are open and expandable architectures and they are the testbed for future networking architectures and protocols [1]. According to the services and applications like large-scale data sharing, multimedia services, parallel accessing of USB devices, internet etc., the main requirement is having a source address (senders IP) and a destination address (receivers IP). In different operation systems, our current internet needs modifications, and this new architecture is referred to as Network Virtualization or future internet [2, 3]. Apart from being a part of a more complex network, the management, as well as configuration of these networks, have not been changed yet either conceptually or physically. From the last few decades, the involvement of researchers has been increased who deals with different protocols as, address and port. The lower level of a physical node has derived from the upper level of nodes and transport technologies, which is not directly reflected by the network administrator [4, 5]. Since the Network virtualization environment is a collection of multiple heterogeneous Virtual Networks (VNs) from different service providers, which shares the Substrate network resources. The traditional Internet Service Providers (ISPs) are decomposed into two independent dependencies. The first one is Infrastructure Providers (InPs), who manage the physical network and provides the resources whereas the second is Service Provider (SPs), who creates a virtual network by aggregating resources form multiple Infrastructure Providers. [6, 7]. For example, in order to select a source address from the lower level of virtual network, the user has to configure routing rules and strong control over different levels shown in Fig. 1.

In these prospects, the network virtualization plays a very important role, according to robust routing, efficient search, scalability, decentralization, fault tolerances, trust and authentications, performance etc. Basically, network virtualization is a pro-

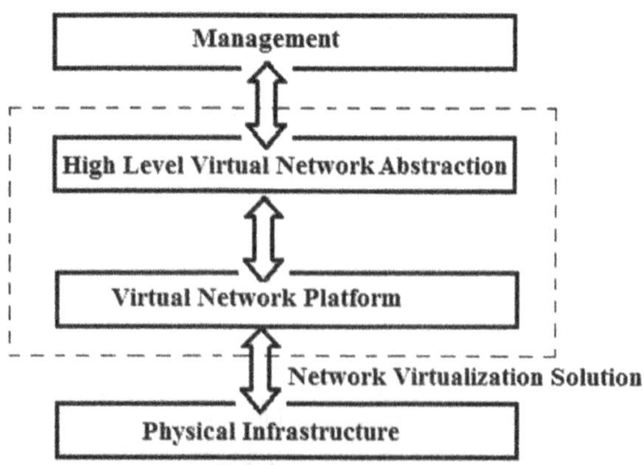

Fig. 1 Reference architecture of network virtualization

cess of combining the two different types of resources such as hardware and software. The term Network virtualization has two distinct components which are used in two distinct aspects also, the first is virtualization of hardware on the network entities and the second one is the virtual network. The virtualization of hardware on the network is used to provide virtual server capacity on existing hardware using some software like XEN, VMware or virtual Box. A comprehensive network virtualization solution must consist of two necessary components [1, 8] as Fig. 1 shows. In any Network virtualization platform there must be a multiple but independent nodes and isolated paths which belongs to multiple layers of virtual network, so that every node can have a define conventional networks, topologies, technologies, and virtualized data centers where lower level to upper level all are dynamically connected [9].

2 Challenges in Address Selection

Consider a network architecture, connected through a Physical Network Controller (PNC) to operate servers and web applications. In that architecture, at the physical end, all node has some topology, which is associated with some PNC, designed for such nodes. To ensure such policies, the configuration of physical network infrastructure will be needed at every level of architecture. The maximum steps required to configure are based on technology in low-level network control management. There is no concern with the functionality of any node at any service level of the three-tier network architecture. The final configuration depends on the lower level of physical network topology and moreover, the connectivity or virtual link is directly managed by the configuration of lower-level nodes. A similar kind of network has shown in Fig. 2 based on the above scenario. Based on the proposed design, the application needs three Virtual Networks, where PNC must be connected. In most of the cases, all physical ports, which are defined in configuration rules are connected to a pre-defined network. For example in the above figure, All virtual networks carried their endpoint from physical network node1, node2, and node3 to the virtual network2 as New InP, and databases. The process is strongly independent and avoids most of the topologies which affect the configuration.

In the functionality of deployed network, virtual nodes and physical port policies, all new virtual nodes have their own properties according to the physical endpoint. All endpoints must be correctly configured by network administrator [10, 11]. Any physical node which is misconfigured can be reconfigured for the respective virtual node, as in Fig. 2 where the new virtual network can be configured to route between virtual Node1, Node2, and virtual Node3. Every routing path and their endpoints are part of the network control (High-Level Network Abstraction) and do not relate directly to the functionality of the physical node. In this approach network service providers and infrastructure providers have the same service provided by PNC. This paper introduced the terms of different policy domains for address selection and such criteria depending on physical nodes property [12–14]. The address selection

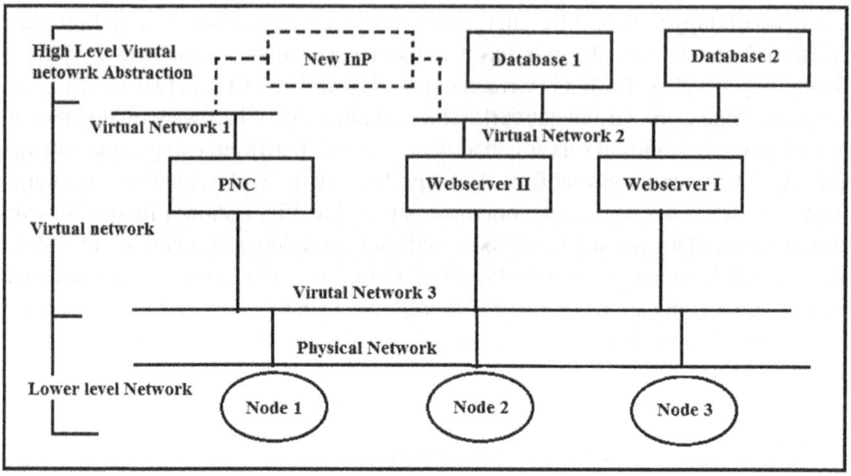

Fig. 2 Physical network controller in network virtualization

approach decouples the network management from physical node creation to virtual node validation scheme to ensure the correctness of prior network deployment or migration to a new physical infrastructure.

3 Proposed Algorithm

The proposed algorithm is divided into two distinct algorithms such as Source Address Selection Algorithm and Destination Address Selection Algorithm. The selection will depend on various parameters such as: address type, scope, non-deprecated addresses, outgoing interfaces, matching label, and public addresses over temporary addresses. While determining destination address selection algorithm the algorithm preferences are: matching scope, non-deprecated addresses, home addresses over address, matching label, higher precedence, prefer native transport, smaller scope, longest matching prefix, leave order unchanged and finally, it avoids unusable destinations. So the proposed algorithm is as follows:

- *Step 1*: Select Source Address and Corresponding Destination address sequentially.
- *Step 2*: Convert them to binary format. That will be a 4×16 matrix. This matrix is known as Combination matrix.
- *Step 3*: Obtain Transpose of that matrix. That will be an 8×8 matrix.
- *Step 4*: Select any middle four Rows. That will be a 4×8 matrix.
- *Step 5*: Transpose the obtained matrix and pair them with two separate rows. That is 8×4 matrixes.
- *Step 6*: Add Row1 with Row2 and Row3 with Row4 and store both of them into Result-1 and Result-2 respectively with a simple adding operation.

- *Step 7*: Subtract Result-1 from Result-2.
- *Step 8*: if (Result-element < 0) Then, if (Result-element is more than 1), Then apply FCFS and select first address position as path.
  Else: Select the position as path. otherwise, sort the subtracted Result-elements into ascending order and select the high magnitude as path.

3.1 Iteration of Proposed Algorithm

Based on the above algorithm, the two different selected addresses are as follows: source address: 192.168.69.16 and destination address: 192.168.1.86. After applying the proposed algorithm to obtain data path to maintain the traffic in the network virtualization environment. The steps are:

Step 1: Select Source Address and Corresponding Destination address sequentially. Source address: 192.168.69.16 and Destination address: 192.168.1.86. *Step 2*: Convert them to binary format. That will be 8×8 matrix followed by source and destination address having 8-bit each. Write source in top-down approach but the destination address in bottom-up approach, i.e., source address:192.168.69.16 and destination address: 86.1.168.192 in binary representation shown below

Source address Destination address
11000000
10101000
01000101
00010000
01010110
00000001
10101000
11000000

Step 4: Select any middle four Rows. That will be 4×8 matrix. Suppose we select

01000101
00010000
01010110
00000001

Step 5: Transpose the obtained matrix and pair them with two separate rows. That is 8×4 matrixes

Row1	Row2	Row3	Row4
0	0	0	0
1	0	1	0
0	0	0	0
0	1	1	0
0	0	0	0
1	0	1	0
0	0	1	0
1	0	0	1

Step 6: Add Row1 with Row2 and Row3 with Row4 and store both of them into Result-1 and Result-2, respectively, with a simple adding operation.

(Row1 + Row2)	(Row3 + Row4)
0	0
1	1
0	0
1	1
0	0
1	1
0	1
1	1

Step 7: Subtract Result-1 from Result-2 and store into result-element (i).

Result-element (i)
0
0
0
0
0
0
−1
0

Now apply Step 8 and select the 6th position as path for sending the data to destination.

Since the architecture becomes more complex depending on the number of nodes at every level. The packet processing is difficult at every level by the switch. So this paper introduced an algorithm to packet processing in network virtualization path architecture shown in Fig. 3.

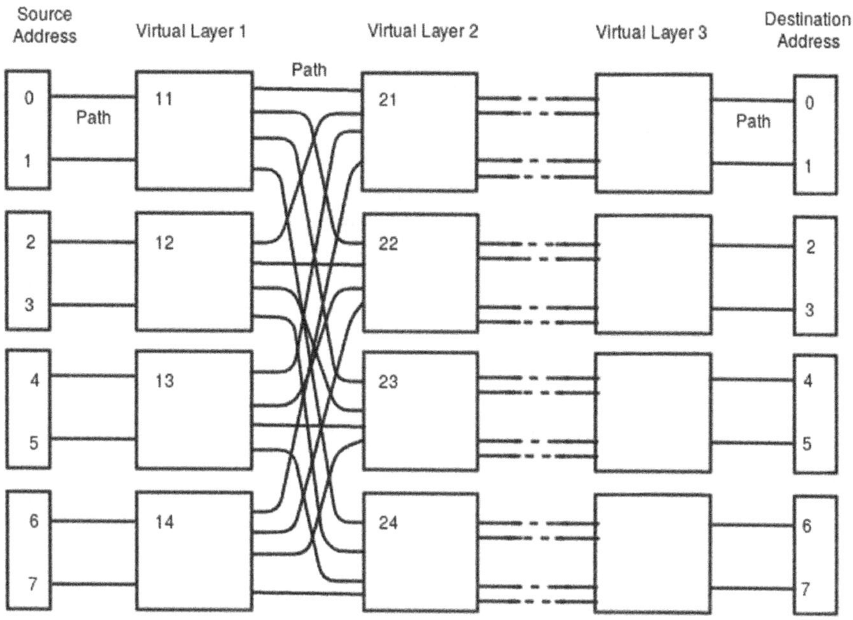

Fig. 3 Network virtualization path architecture

The above packet_processing algorithm intercept an endpoint node selected by the virtual network for the data packet, the hosting Node must acquire policy information about associated sending data packets, referred to as packet processing policy, either from the physical network or from the Physical Network Controller [15, 16]. However, the data required encapsulation because sending and receiving of source and destination addresses may require Quality of Services based policies.

Figure 4 shows that the packet will pass through the endpoint nodes, PNCs are responsible for security policies. The intermediary switch has configured to use all protocols of PNC, applied for configuration support. For example, Virtual extensible Local Area Network (VXLAN), Stateless Transport Tunneling (STT), Generic Routing Encapsulation (GRE) [6, 18], shown in Fig. 4. PNC at Destination node, encapsulates the header provided by SPs at PNC sender endpoint, and then delivers it to the destination endpoint. In the case of Virtual Machine migration, the destination PNC through lower level of the controller, provided by Infrastructure Provider.

Apart from the Network Virtualization path architecture, the controlled architectures are successfully supported by a large number of nodes and their defined policies. For example, to support Quality of Services in the infrastructure it needs a quality PNC, to reduce heavy traffics. Similarly, The supported policies necessarily depend on the specific physical infrastructure, and configuration [16, 17, 19].

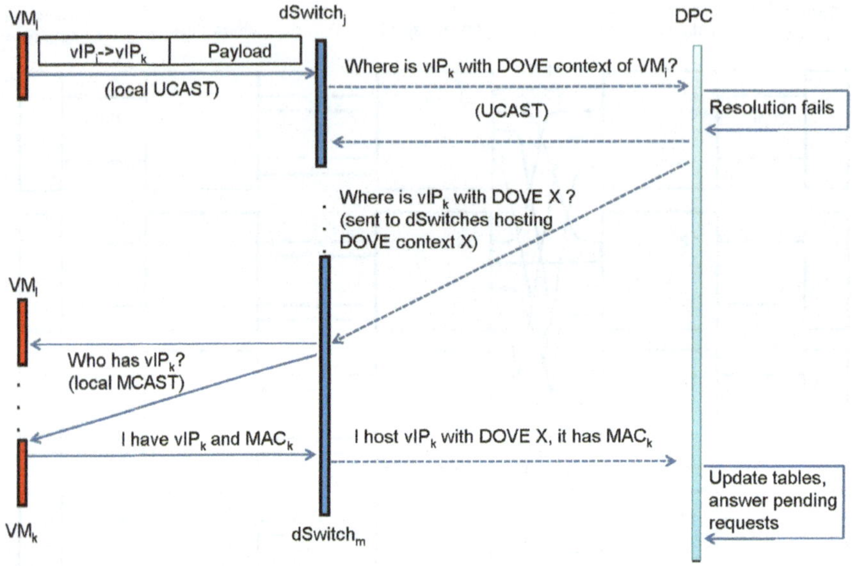

Fig. 4 Policy request-response cycle by PNC [17]

4 Correctness of Proposed Algorithm

The experimental results shown in this paper are based upon planet Lab simulator [11–13] and on Omnet++ [2, 3, 10] based network simulator. These Simulators have operated upon VM-level, modeling the switch, as well as network appliances, such as adapter, different operation over packets such as queuing, buffering, scheduling, and mapping. Network Virtualization path architecture is mapped on a lower level of common physical network node provided by the infrastructure provider. For the evaluation of Path selection according to the virtual node existing on a higher level of the virtual network, we have investigated the packet loss shown in Fig. 5a, Load generator shown in Fig. 5b, the avg. Latency measurement graph shown in Fig. 5c, and the avg. throughput graph calculation shown in Fig. 5d. Based on these measurements we have observed that it works well for all the parameters. We also measure the performance of a variable number of virtual nodes inherited by a lower level of the physical network node. The measured results are in good agreement with existing network provided by ISPs.

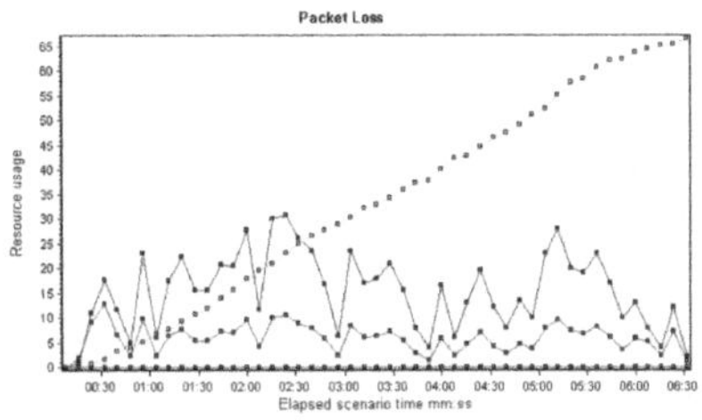

(a) Network Virtualization based on SPs and InPs Per Group

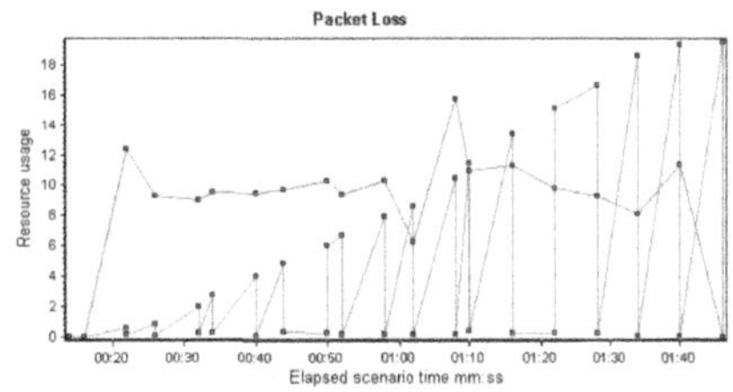

(b) Network Virtualization based Load Generator

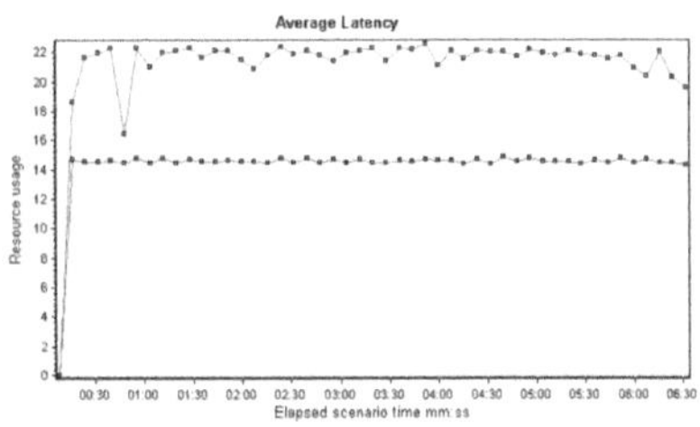

(c) Network Virtualization based Average Latency Graph

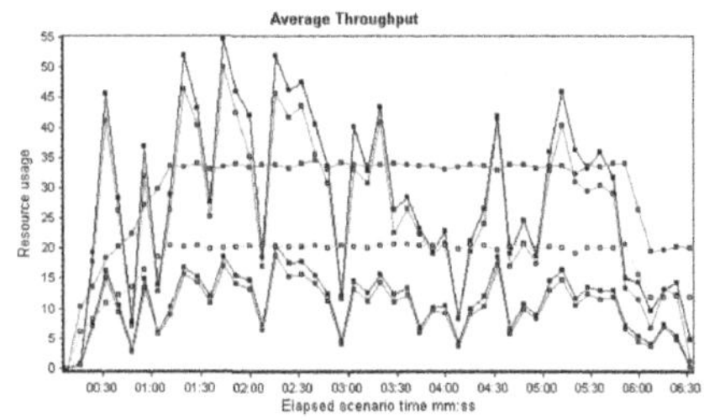

(d) Network Virtualization based Average Throughput Graph

Fig. 5 Validating the proposed algorithm on different parameters in network virtualization environment

5 Conclusion

In this paper, we have presented a complete network virtualization solution for the selection of source address and destination address. We have proposed an optimization-based algorithm to create a better network blueprint to determine the specification and policies of network functionality. The life cycle of any virtual node independently depends on the physical node which is managed by the network administrator who controls the PNC. This paper has a major breakthrough in achieving the separation between service provider and infrastructure provider. We believe our proposed algorithm has a major concern with the network virtualization and virtualization administration (PNC). In addition, we have presented the network virtualization path architecture to serve as a platform carrying out a task of selection of source and destination address node according to specified network policies traced by a higher level of virtual nodes. Since traffic management is a very big problem in network virtualization environment because it degrades the performance of the network. By using of Address Selection Algorithm we make on traffic management architecture shown in this paper and it works successfully in case of other traffic management algorithms. In every layer, it calculates the length of the packet, finds the path number, and then it stores the remaining path numbers for further use at the next level of layers.

References

1. Mattos, D.M.F., Velloso, P.B., Duarte, O.C.M.B.: An agile and effective network function virtualization infrastructure for the internet of things. J. Internet Serv. Appl. **10**(1) (2019)
2. Al-Fares, M., Loukissas, A., Vahdat, A.: A scalable, commodity data center network architecture. In: Proceedigs of ACM SIGCOMM, Global Grid Forum, pp. 63–74 (2008)
3. Peterson, L., Shenker, S., Roscoe, T., Anderson, T.: Overcoming the Internet impasse through virtualization. IEEE Comput. Mag. **38**, 34–41 (2005)
4. Bays, L.R., Oliveira, R.R., Barcellos, M.P., Gaspary, L.P., Madeira, E.R.M.: Virtual network security: threats, countermeasures, and challenges. J. Internet Serv. Appl. **6**(1) (2015)
5. Kim, W.: Toward network function virtualization for cognitive wireless mesh networks: a TCP case study. EURASIP J. Wirel. Commun. Netw. **1** (2015)
6. Gao, L., Rexford, J., Feamster, N.: How to lease the internet in your spare time. ACM Comput. Commun. Rev. **114–121** (2006)
7. Boutaba, R., Chowdhury, N. M. K.: A survey of network virtualization. David R. Cheriton School of Computer Science, University of Waterloo, Technical Report CS-08-25 (2008)
8. Huh, J-H., Seo, K.: Design and test bed experiments of server operation system using virtualization technology. Human-Centric Comput. Inf. Sci. **6**(1) (2016)
9. Prez, C., Ordua, J.M., Soriano, F.R.: A nested virtualization tool for information technology practical education. SpringerPlus **5**(1) (2016)
10. Taylor, D., Turner, J.: Diversifying the internet. In: IEEE Global Telecommunications Conference (GLOBECOM'05), vol. 2 (2005)
11. Gao, L., Rexford, J., Feamster, N.: How to lease the internet in your spare time. SIGCOMM Comput. Commun. Rev. **37**(1), 61–64 (2007)
12. Boutaba, R., Chowdhury, N.M.K.: Network virtualization: state of the art and research challenges. IEEE Commun. Mag. **47**(7), 20–26 (2009)
13. Zhang-Shen, R., Rangarajan, S., Rexford, J., Zhu, Y.: Cabernet: connectivity architecture for better network services. Workshop on Re-architecting the Internet, ACM SigComm (2008)
14. Al-Fares, M., Loukissas, A., Vahdat, A.: A scalable, commodity data center network architecture. In: Proceedings of ACMSIGCOMM Conference on Data Communication, Seattle, WA (2008)
15. Barabash, K., Cohen, R., Hadas, D., Jain, V., Recio, R., Rochwerger , B.: A case for overlays in virtualization. In: 3rd Workshop on Data Center, Converged and Virtual Ethernet Switching, pp. 30–37 (2011)
16. Upadhyay, N.M., Soni, K., Arvind, K.: Security management in network visualization environment by using modified root management algorithm for creation of virtual node and virtual link. IJCA **162**(6), 37–43 (2017)
17. Birke, R., Crisan, D., Barabash, K., Levin, A., DeCusatis, C., Minkenberg, C., Gusat, M.: Partition/aggregate in commodity 10g ethernet software-defined networking. In: IEEE 13th international conference on high performance switching and routing (HPSR), pp. 7–14 (2012)
18. Wang, L., Zhang, F., Hou, C., Aroca, J.A., Liu, Z.: Incorporating rate adaptation into green networking for future data centers. In: 12th international symposium on network computing and applications, pp. 106–109 (2013)
19. Bethanabhotla, D., et al.: User association and load balancing for cellular massive MIMO. Information Theory and Applications Workshop (ITA) **1–10** (2014)

A Model for Prediction of Paddy Crop Disease Using CNN

Ritesh Sharma, Sujay Das, Mahendra Kumar Gourisaria, Siddharth Swarup Rautaray and Manjusha Pandey

Abstract The agriculture industry is the most important industry for society as it serves the most important need of life. But the plant diseases in agriculture lead to a decrease in productivity and hence it is very important to prevent, detect, and get rid of the diseases. Image processing and deep learning are nowadays the buzzwords in the IT industry and their applications in the agriculture industry can enhance decision making in various aspects of the agriculture industry. Paddy crop is one of the most demanding crops especially in South Asia. This paper proposes a predictive model using CNN for classification and prediction of disease in paddy crop. Paddy crop diseases are very fatal and can affect the crops severely if it is not taken care in the initial stages. The proposed model will improve the decision making using CNN in case of various diseases in paddy crop for prediction of diseases in initial stages and prevention of mass loss in productivity of the whole yield.

Keywords Convolutional neural networks · Image processing · Paddy crop diseases

R. Sharma (✉) · M. K. Gourisaria · S. S. Rautaray · M. Pandey
School of Computer Engineering, KIIT Deemed University, Bhubaneshwar, India
e-mail: sharmaritesh3312@gmail.com

M. K. Gourisaria
e-mail: mkgourisaria2010@gmail.com

S. S. Rautaray
e-mail: siddharthfcs@kiit.ac.in

M. Pandey
e-mail: manjushafcs@kiit.ac.in

S. Das
School of Electronics Engineering, KIIT Deemed University, Bhubaneshwar, India
e-mail: dasujoy1999@gmail.com

H. Das et al. (eds.), *Progress in Computing, Analytics and Networking*,
Advances in Intelligent Systems and Computing 1119,
https://doi.org/10.1007/978-981-15-2414-1_54

1 Introduction

Agriculture was started during the Neolithic age and is going strong until today. But due to harsh changes in climatic conditions and malpractices by humans the condition of agriculture is worsening day today. As in human beings, the diseases in plants and crops are also fatal and can destroy agriculture if not diagnosed in the early stages. There are number of diseases that prove to be very fatal and cannot only destroy the crop but can also destroy the crops surrounding it. But the diagnosis is not always possible due to lack of education and expertise. So, the experts are trying their best to provide the services to the farmers to get rid of these kinds of diseases.

Since the last few years, the IT industry is also playing a vital role in agriculture as it provides efficient solutions to the various aspects of the agriculture. Machine learning, deep learning, image processing, artificial intelligence, and many more tools are providing very efficient solutions to the problems in agriculture like fighting food scarcity and empowering small farmers by receiving useful data, making yield prediction, coping with climate change, and providing images of crops and land.

Image processing and deep learning are the new trends in agriculture as far as the IT industry is concerned. Image processing and finding out the information about the disease that the plant might be suffering from is one of the major techniques in deep learning domain and then taking precautions is one way to prevent the disease to grow or transmit to another plant.

We have presented our approach to predict the disease that the plant might be suffering from by introducing a model based on Convolutional Neural Network (CNN). We have chosen CNN because it works effectively with images. We have taken data set which contains the images of various infected leaves of a rice plant. Our model processes the image followed by image augmentation and then trains on it. Then with the help of the trained images, it predicts what kind of disease the given plant leaf might be suffering from. We have done all our works in Spyder IDE. We have used the Python programming language and Keras library tool to create this model.

This paper is organized as follows: Sect. 1 discusses the introduction, Sect. 2 gives a detail about the related work in the field of crop disease prediction, Sect. 3 explains about the Proposed Model, Sect. 4 gives the Implementation detail, Sect. 5. Contains the Pseudo Code, Sect. 6 Depicts the Result achieved, the conclusion has been detailed in Sect. 7 along with the details of future work.

2 Related Works

Suraksha et al. [1] propose a technique for prediction paddy crop disease using data mining and image processing. In this paper, they have proposed a model for image processing followed by feature extraction and data mining techniques.

Rajmohan et al. [2] propose a technique for smart paddy crop disease prediction using deep CNN and SVM classifier. In this paper, they have proposed a model for image processing followed by feature extraction and SVM classifier for classification. In which they have selected 250 images from which 50 are used for training and rest for testing. In this paper, they have developed a mobile app which clicks the image, zooms, and crops it then uploads the image and the person receives a notification.

Barik [3] proposes a technique for region identification of Rice Diseases using image processing. In this paper, they have presented a model for the identification of disease and region of infection using image processing and classification techniques like SVM classifier and Naive Bayes classifier. Prediction is done and the severity of various diseases is found and then it is classified into different categories.

Jagan Mohan et al. [4] present a technique for disease detection of plants using canny edge detection. In this paper, they have proposed a model that uses canny edge detection algorithm to track the edge and then get the histogram value to detect the disease. It is concluded that this model periodically monitors the cultivated field. The diseases are detected using edge detection in early stage and machine learning is used for training which takes proper decision regarding the diseases.

Badage [5] presents a technique for disease detection in paddy crops which uses Scale Invariant Feature Transform for extracting the features then the features are taken and SVM and K Nearest Neighbors are used to recognize the image.

Dhaygude et al. [6] presented a technique for prediction of diseases in rice plant. This consists of four steps that are color transformation for the RGB image is created then the RGB is converted to HSI then the green pixels are masked and removed using specific threshold value and after that the image is segmented and the useful features are extracted and the texture statistics are computed from SGDM matrices. After that, the presence of disease in the leaf is evaluated.

3 Proposed Model

The objective of this research is to recognize the type of disease that a rice plant is diagnosed with. Three families of diseases are bacterial, viral, and fungal. We have taken three diseases into consideration that are brown spot, leaf smut, and bacterial leaf blight.

3.1 Paddy Crop Disease Modeling

Paddy crops mainly suffer from three types of diseases which we are using in our dataset for training and testing the data. The detailed discussions of those diseases are

3.1.1 Brown Spot (Cochliobolus miyabeanus)

These are circular brown spots that appear on the paddy leaf. There are also other symptoms by which it can be identified like the death of seedlings, death of large areas of leaf, brown or black spots on grains. It is a type of fungal infection. All over South and South East Asia is does 5% yield loss [7]. Treating seeds with chemicals can also be proved helpful as it decreases the chance of the infection [8].

3.1.2 Bacterial Leaf Blight (Xanthomonas oryzae pv. Oryzae)

This disease usually occurs in the leaf of a rice plant and can easily be found out by looking at the yellow and white strips on the leaves. This disease can be recognized by looking at the youngest leaf which will be pale yellow in color if the plant is suffering from bacterial leaf blight. This disease can be prevented by avoiding the use of excess nitrogen fertilizer and by plowing stubble and straw into the soil after harvesting the crop [9].

3.1.3 Leaf Smut (Ustilaginoidea virens)

Leaf smut is a widely spread but a minor disease that occurs in paddy. It can be identified by looking at the leaf as the leaf will be covered by fungus all over. This disease is caused by Entyloma oryzae [10]. This disease can be controlled by doing clean cultivation and growing resistant varieties [11].

3.2 Process Flow Model

See Fig. 1.

4 Implementation

Neural Networks are a set of complex algorithms that are designed just like the human brain, which helps us recognize different patterns or images. The core idea behind the framework is Convolutional Neural Network (CNN). We are using the CNN for classification as well as for prediction of the diseases in the rice plant. There are different steps in the model which are defined below.

Fig. 1 Process flow model

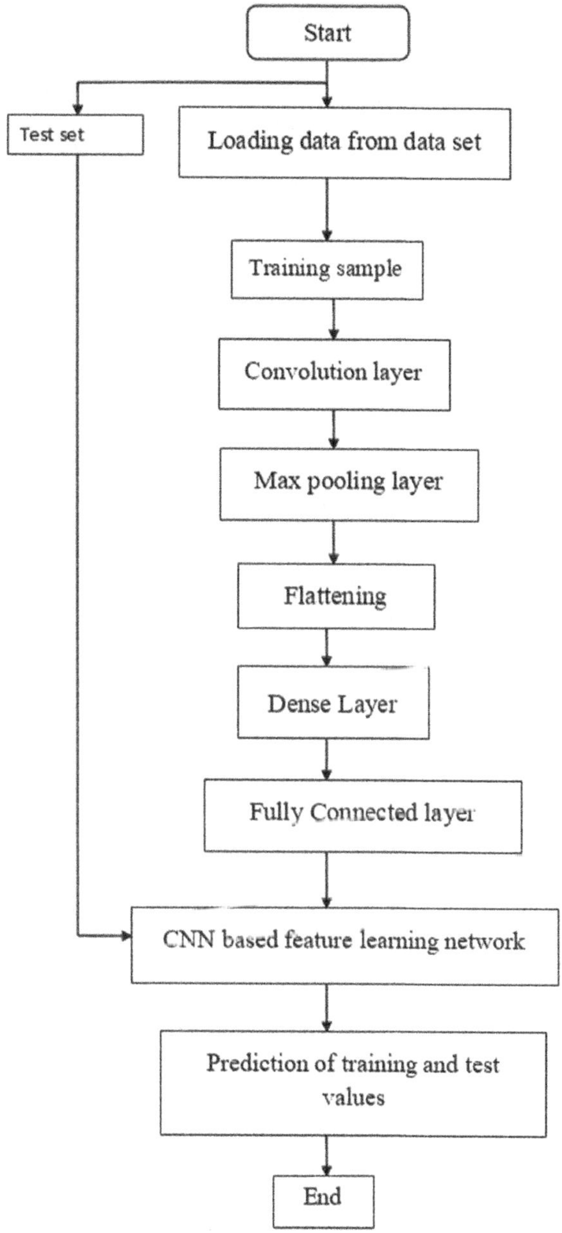

4.1 Image Generation and Augmentation

Neural networks processes images using CNN and the CNN model interprets the image in the form of matrices and all the operations are performed on the matrix formed by the image. The images are generated and augmentation is performed. Images are augmented to increase the amount of training data. We have performed the horizontal flip, vertical flip, shearing, and brightening of images in the augmentation process.

4.2 Convolution Step

The model gets an image as its input in the form of matrices. In a CNN model, matrices play a key role. In the convolution step, a filter or kernel matrix is taken and convolution is performed with the input image matrix by sliding the filter over the input image [12]. Let us consider

x Input
a^k After convoluted image
k Index of kernel (weight filter)
W Kernel (weight filter)
b Bias
E Cost function.

For Forward Propagation:

$$a_{ij}^{(k)} = \sum_{i=0}^{m-1} \sum_{j=0}^{n-1} W_{st}^{(k)} x_{(i+s)(j+t)} + b^{(k)} \tag{1}$$

For Backward Propagation:

$$\frac{\partial E}{\partial W_{st}^{(k)}} = \sum_{i=0}^{M-m} \sum_{j=0}^{N-n} \frac{\partial E}{\partial a_{ij}^k} \frac{\partial a_{ij}^k}{\partial W_{st}^{(k)}} = \sum_{i=0}^{M-m} \sum_{j=0}^{N-n} \frac{\partial E}{\partial a_{ij}^k} x_{(i+s)(j+t)} \tag{2a}$$

$$\frac{\partial E}{\partial b^{(k)}} = \sum_{i=0}^{M-m} \sum_{j=0}^{N-n} \frac{\partial E}{\partial a_{ij}^k} \frac{\partial a_{ij}^k}{\partial b^{(k)}} = \sum_{i=0}^{M-m} \sum_{j=0}^{N-n} \frac{\partial E}{\partial a_{ij}^k} \tag{2b}$$

4.3 Pooling Step

Pooling is one of the most important building blocks of a CNN model. The main objective of applying pooling is to reduce the spatial size of an image. Pooling is performed on each depth dimension; therefore the depth of the image remains intact. There are three kinds of pooling: Max pooling, Min pooling, and Average pooling. In our model, we have implemented the max-pooling technique [13].

Forward Propagation:

$$a_{ij} = \max(0, x_{(i+s)(j+t)}) \tag{3}$$

Backward Propagation:

$$\frac{\partial E}{\partial x_{(i+s)(j+t)}} = \frac{\partial E}{\partial a_{ij}^{(k)}} \frac{\partial a_{ij}^{(k)}}{\partial x_{(i+s)(j+t)}} = \begin{Bmatrix} \frac{\partial E}{\partial a_{ij}^{(k)}} & (a_{ij}^{k} = x_{(i+s)(j+t)}) \\ 0 & (Otherwise) \end{Bmatrix} \tag{4}$$

4.4 Fully Connected Layers

The output of the pooling step is a matrix that should be mapped into a data structure which is feasible for classification and prediction and the vector is used for that. A fully connected layer is used to map a matrix to a one-dimensional vector. A matrix gets converted to a vector using a flatten() function and some linear operations are performed in the hidden layers using the Dense() function. In one of the Dense layer, we have used 'ReLU' as an activation function and 'Softmax' for the other Dense layer.

ReLU activation function:

$$\mathrm{Re}LU(x) = \max(0, x) \tag{5}$$

Softmax activation function:

$$\sigma(z)_i = \frac{e^{z_i}}{\sum_{j=1}^{K} e^{z_i}} \quad for\ i = 1, \ldots, K\ and \quad z = (z_1, \ldots, z_K) \in R^K \tag{6}$$

Forward Propagation:

$$a_{ij} = \mathrm{Re}LU(x) = \max(0, x_{ij}) \tag{7}$$

Backward Propagation:

$$\frac{\partial E}{\partial x_{ij}} = \frac{\partial E}{\partial a_{ij}} \frac{\partial a_{ij}}{\partial x_{ij}} = \begin{cases} \frac{\partial E}{\partial a_{ij}} & (a_{ij} \geq 0) \\ 0 & (Otherwise) \end{cases} \tag{8}$$

4.5 Disease Prediction

After training the data set on the training set and classifying the disease on test set we have predicted the disease present in a random paddy crop image using the predict() function and the results were obtained in the form of a binary vector with three classes. If any disease is present in the crop then the corresponding element in the output vector will be 1 and all the other are 0 [14].

5 Algorithm of the Model

Input: A dataset with images of infected paddy crop leaves.
Output: A predicted value of disease.

Start
Step1. Import necessary libraries
Step2. Add a 2-dimensional convolution layer using conv2D() in the neural network
Step 3. Add a 2-dimensional max pooling layer using maxpooling2D() in the neural network
Step 4. Add a flattening layer using Flatten()
Step 5. Add a fully connected layer with an activation function of ReLu
Step 6. Add a dense layer using dense() with an activation function of Softmax
Step 7. Perform image augmentation using ImageDataGenerator() by Horizontal flip, vertical flip, brightening, shearing
Step 8. Load the images as training and test sets and load them in different Variables
Step 9. Fit the data and train the model on training dataset using fit_generator()
Step10. Test the accuracy of the model on the test set.
Step 11. Predict the disease on a random image and print the output.
Stop

Fig. 2 Bacterial leaf blight

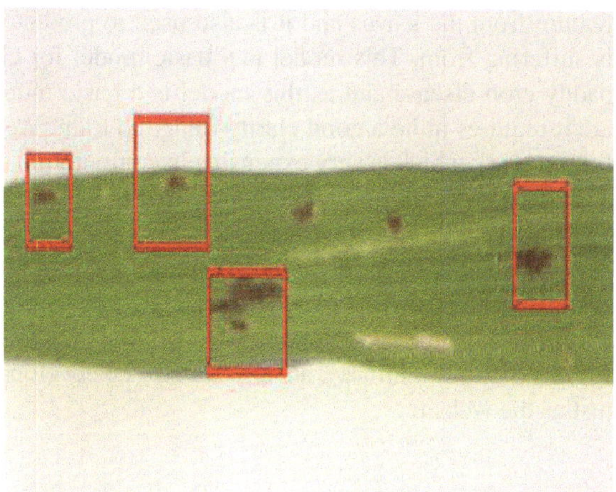

Fig. 3 Brown spot

6 Result

An accuracy of 90.32% is seen in test set and an accuracy of 93.58% is seen in the training set. The diseases in the paddy plant can be recognized as (Figs. 2, 3, 4).

7 Conclusion and Future Work

Agriculture plays a very important role in the lives of living beings as it provides food to all the living organisms. As it is important not only for humans to survive and also for animals so it must be protected and means should be created to protect agriculture from different types of diseases. Convolutional Neural Network is used to extract the

Fig. 4 Leaf smut

feature from the leaves and it is also used to predict what type of disease the plant is suffering from. This model is a basic model for classification and prediction of paddy crop disease and as this model is a basic model it contains some flaws that are it requires quite a good clarity image to train. We are using the CNN technique for our work which is very expensive in computation so a machine with higher GPU and processor with a large dataset of images will work more effectively.

There is a lot of scope for this model to deploy on apps and websites. A website can be made just to predict which disease the plant is suffering from. The person will have to upload a well-clicked image of the plant leaf in the website and immediately the predicted result will come along with the precaution measures as well as a cure for this disease. Similarly, an app can also be developed to function in the same way just as the website.

References

1. Suraksha, I.S., Sushma, B., Sushma, R.G., Susmitha, K., Uday Shankar, S.V.: Disease prediction of paddy crops using data mining and image processing techniques. Int. J. Adv. Res. Electr. Electron. Instrum. Eng. 5(6) (2016)
2. Rajmodhan, R., Pajany, M., Rajesh, R., Raghuraman, D., Prabu, U.: Smart paddy crop disease identification using deep convolutional neural network and SVM classifier. Int. J. Pure Appl. Math. 118(15) (2018)
3. Barik, L.: A survey on region identification of rice disease using image processing. Int. J. Res. Sci. Innov. 5(1) (2018)
4. Jagan Mohan, K., Balasubramanian, M., Palanivel, S.: Detection and Recognition of Diseases from Paddy Plat Leaf, International Journal of Computer Applications Vol. 144 No. 12(2016)
5. Badage, A.: Crop disease detection using machine learning: indian agriculture. Int. Res. J. Eng. Technol. 5(9) (2018)
6. Dhaygude, S.B., Kumbhar, N.P.: Agricultural plant leaf disease detection using image processing. Int. J. Adv. Res. Electr. Electron. Instrum. Eng. 2(1) (2013)

7. Rice knowledge bank. http://www.knowledgebank.irri.org/training/fact-sheets/pest-management/diseases/item/brown-spot
8. Plant village. https://plantvillage.psu.edu/topics/rice/infos
9. Rice knowledge bank. http://www.knowledgebank.irri.org/training/fact-sheets/pest-management/diseases/item/bacterial-blight?category_id=326
10. Louisiana plant pathology. https://www.lsuagcenter.com/NR/rdonlyres/41C7A6F2-E14A-4713-BFE1-724F8BE27509/75781/pub3115LeafSmutLOWRES.pdf
11. Rice knowledge management portal. http://www.rkmp.co.in/content/leaf-smut-2
12. Jaswal, D., Sowmya, V., Soman, K.P.: Image classification using convolutional neural networks. Int. J. Adv. Res. Technol. 3(6) (2014)
13. Scherer, D., Muller, A., Behnke, S.: Evaluation of pooling operations in convolutional architectures for object recognition. In: International Conference on Artificial Neural Networks (ICANN) (2010)
14. Ullah, I.: Data mining algorithms and medical sciences. Int. J. Comput. Sci. Inf. Technol. (IJCSIT) 2(6) (2010)

Testbeds, Attacks, and Dataset Generation for Big Data Cluster: A System Application for Big Data Platform Security Analysis

Swagata Paul, Sajal Saha and R. T. Goswami

Abstract A big data cluster consists of number of network-connected computers. A big data cluster offers a huge data store and processing power. End users submit both data and application to the cluster. All the computers called nodes in the cluster work together to give the result from the data. During data processing, lots of process run on different nodes and exchange data. The data exchange is done via regular network protocols. During processing, one or multiple computers may not participate well due to its bad hardware or operating system health. Some computers may receive known network attack like DOS and thus slow down the performance of the cluster. Some other computers may receive unknown attacks generated by the big data job itself. Therefore, the system requires a mechanism to detect such nodes under attack or the nodes generating attacks and isolate thereafter. To detect this attack, we need to analyze the cumulative network traffic of all the nodes in the cluster. Therefore, we must collect such network traffic of all the nodes participating in data processing job simultaneously. This work is to present an efficient testbed for external or internal attack generation and dataset creation for different attacks. The proposed architecture captures network traffic from all nodes of the cluster and stores them for attack detection in near future.

Keywords Big data cluster · Attack detection framework · Big data security analysis

1 Introduction

Hadoop [1] is a distributed data store and processing platform. We have used Hadoop 3.1.1 deployed through Hortonworks Data Platform [2] and Apache Ambari [3] for

S. Paul (✉) · R. T. Goswami
Techno International New Town, Kolkata 700156, West Bengal, India

S. Saha
The Assam Kaziranga University Koraikhowa, Jorhat 785006, Assam, India
e-mail: sajalsaha@kazirangauniversity.in

H. Das et al. (eds.), *Progress in Computing, Analytics and Networking*, Advances in Intelligent Systems and Computing 1119,
https://doi.org/10.1007/978-981-15-2414-1_55

our experiments. To detect any external or internal attack to a big data cluster, we must look into both incoming and outgoing network traffics in every node in the cluster. So we need to capture network packets in all the nodes in the cluster. In this work, the focus is to present an efficient testbed for attack generation and network packet capture from all nodes of the cluster, and store them for further analysis. To accomplish the capture, we have used dumpcap utility of Wireshark Software [4] on all nodes. All such captured packets are stored temporarily on the cluster nodes as .pcapng file format [5]. This research is described in Sects. 2–8. Section 2 is related to past works. Section 3 shows generalized Hadoop architecture. Section 4 shows the details of testbed architecture. Experimental setup is given in Sect. 5. The framework used to generate attack, start and stop capture, and collect packets is shown in Sect. 6. Sections 7 and 8 represent Results and Discussions and Conclusion, respectively. The system we developed can be used on any big data platform including Hadoop using Linux as host operating system.

2 Related Work

Preparing a testbed and generation of dataset for computer network security analysis has been explored in many research studies. Considering a big data cluster consisting of 100 s or 1000 s of computers engaged in processing huge amount of data, it is almost impossible to trace which computer is in attack, especially when the attack is generated from an internal cluster node. The work [6] presents a wonderful method by generating process signature to detect attack on a big data cluster using older version of Hadoop which uses JobTracker and TaskTracker. Unfortunately, these two components have been replaced by ResourceManager and NodeManager in newer version of Hadoop (version 3.1.1). Moreover, the work does not create any dataset for attack detection. The works presented in [7, 8] showed that completion time increases in a MapReduce Job while attack was generated. However, a spark streaming application, based on MapReduce framework, running on a Hadoop cluster does not end until it is explicitly killed. Therefore, the methods will detect this as attack. Every day, new applications are developed by big data application developers, so new JAR files are created. Sometimes, they are not always Java applications. The code may be written in Scala, Spark, Python, etc. The work presented in [9] uses run-time assembly code from JARs to detect attack. It will fail for a new Java application as well as for Scala, Spark, and Python codes. In [10, 11], the authors show many tools to simulate different network-based attacks like DoS, DDoS, etc., and generate attack dataset. Each such tool is used for different attacks and is not applicable for a cluster of computer. The project Apache Metron [12, 13], a cybersecurity framework for big data, is still in incubation stage and is not so promising as attack detection framework. Hadoop can be configured to works in secure mode with Kerberos [14] which provides the features—authentication, authorization of services, web consoles, etc. [15]. But Kerberos has its known drawbacks [16]. The article [17] shows that more than 5000 clusters are open to Internet with no security. This describes about

Hadoop Distributed File System (HDFS) open port number 50070 and not other components. The article [18] by Cloudera shows recent open areas where people can work to protect a big data cluster. Different recent applications on big data like healthcare recommendation [19, 20] does not tell about data security or attack on the system. The research work presented in [21] shows how to build a big data firewall using cloudlets. The firewall works near source of data and not inside the cluster during operation. The performance improvement scheme for big data application [22] does not consider the attacks on big data system. Applications on geospatial big data, biomedical signal processing, etc. are growing without considering the attacks on the system [23–26]. The goal is to provide a platform which can generate datasets for big data cluster security analysis and create example datasets for future research on big data cluster security analysis and attack detection.

3 Big Data Architecture

Along with Hadoop ecosystem as a big data processing framework, the other emerging platforms are Ceph, DataTorrent RTS, Disco, Google BigQuery, HPCC, Hydra [27], etc. All such platforms can process big data using the cluster computers even if deployed on cloud. The generalized and simplified framework is shown in Fig. 1a. The Hadoop processes on a cluster are mainly Java applications. They can be classified into two sets. The first set of processes represents HFDS, MapReduce, YARN, ZooKeeper, HBase, Hive, Spark, etc. These are big data service (we call these as BDS). The second set of processes is the big data applications submitted by end users. We call them big data process (we call these as BDP). There exists many similar BDSs on a set of nodes and some unique BDS on other set of nodes. All BDSs may or may not be required for a BDP. A set of BDSs works in parallel to complete a BDP. In Fig. 1a, a BDS named ResourceManager and a set of Slave

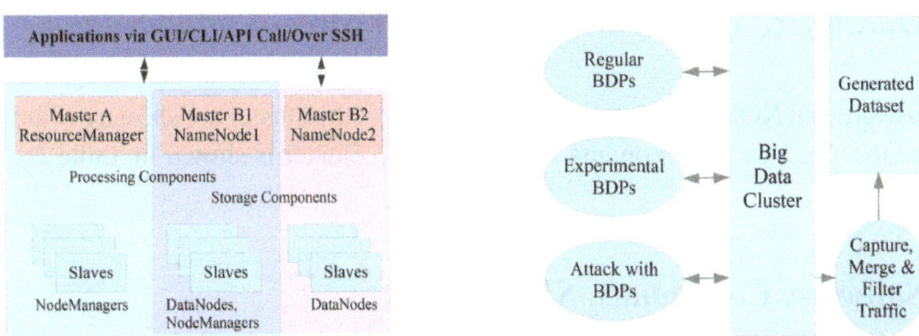

(a) Generalized Big Data Framework

(b) Testbed, Attack Generation and Dataset Creation.

Fig. 1 Big data framework and proposed architecture

BDSs named NodeManagers offers a big data processing component called YARN. BDPs communicate with one or multiple BDSs to complete a data processing job.

4 Testbed Architecture: With Attack Generation and Dataset Creation

The generalized architecture for the testbed, attack generation, and dataset creation is shown in Fig. 1b. The cluster specification is given in Table 2b. For our experiment, along with regular BDPs, we submit other two types of processes to the cluster—(a) experimental BDPs and (b) attack processes. These are launched from attack launcher (AL) node. While the cluster runs BDPs, we capture the network traffic from all the nodes in the cluster. The capture operation is initiated from capture center (CC) node. The captured traffic is stored temporarily on cluster nodes as .pcapng file. CC node then collects all such network traffic. The traffic data is then merged and filtered for further investigation on attack detection. The results are the dataset generated from experimental processes.

5 Experimental Setup

The following setup was considered for our experiment to generate network traffic in all nodes of the cluster. Configuration of each part is given in the following subsections. The designated node called capture center communicates with all other hosts via SSH to start the traffic capture, stop the capture operation, and collect the traffic data.

5.1 Hardware, System, and Cluster Software—Configuration

The hardware and NOS configurations of the nodes of the cluster are shown in Table 1. Configuration of basic components for the Hadoop cluster is shown in Table 2.

5.2 Software, Commands, Shell Scripts

We have used (a) hping3 tools [28] to generate some known attacks like DOS, ICMP flooding, etc., (b) Java code and TestDFSIO benchmark tools to generate huge file I/O operation, and (c) YARN REST API, etc. to generate network activity inside

Table 1 Big data cluster—hardware and NOS

Sl	Item	Detailed description
1	Operating system	Ubuntu 18.04 64 bits
2	CPU	Intel-Core i5
3	Primary memory	Type—DDR3, Size—8 GB (cluster allocation: 6 GB)
4	Secondary memory	Type—SATA, Size—1 TB. A 700 GB partition is mounted on the directory/data
5	LAN details	Network switch—1 Gbps, Network adapter—1 Gbps
6	IP Address range	192.168.10.171–192.168.10.198
7	OS Hostnames	cseoslab001, cseoslab002, … cseoslab028
8	OS Hostname: NodeName	cseoslab001:N1, cseoslab002:N2, … cseoslab028:N28

Table 2 Big data cluster—Hadoop basic configuration details

Sl	Item	Detailed description
1	Ambari manager	Ver 2.7.x, on cseoslab001
2	Hadoop file system	Ver 3.1.x, block size: 128 MB, total storage size: 13.7 TB, replication factor: 3
3	NameNode—HDFS	Count: 1, Hostname: cseoslab002
4	DataNodes—HDFS	Count: 25, Hostnames: cseoslab004 to cseoslab028
5	YARN	Ver 3.1.1
6	ResourceManager—YARN	Count: 1, Hostname: cseoslab002
7	NodeManagers—YARN	Count: 24, Hostnames: cseoslab005 to cseoslab028
8	Cluster memory—YARN	Size: 142 GB
9	Apache spark	Ver 2.3.2
10	Apache Hive	Ver 3.1.0 Hostname: cseoslab004
11	Client library	On all nodes: cseoslab001 to cseoslab028

the cluster and capture all such activity for further research toward the detection of attacks on the nodes of big data cluster. Table 3 shows the details.

6 Attack Generation and Data Capture

In the big data cluster specified in Table 2, along with regular BDPs, we submit experimental BDPs and attack process to the cluster. We have developed a socket program pair named capture agent and capture manager. Capture agents are server daemon that runs on all the nodes in the cluster. Capture manager is a client process that runs on CC node. The attack launcher process runs on AL node. It sends attack as well as experimental BDPs to the cluster.

Table 3 Experimental BDPs and attack generation

Sl	BDPs and/or attack	Description
1	DOS attack	Packet are generated and send to ResourceManager BDS of YARN using hping3 command. Sample packet generation command is *sudo hping3 -d 122 -c 5000 -w 64 -S -p* $p *–rand-source –flood* $h [where $p = 50470, $h = cseoslab002]
2	Busy HDFS	Attack generated by a Java code populating big files. Thus it creates huge data transfer over network and make it busy
3	Make YARN busy	YARN REST APIs has been used via shell script. It runs spark streaming job and never terminates until explicitly killed. This consumes all possible YARN containers
4	HBase loading	We loaded 1 GB data file 1gb.csv to table table1gb using HBase ImportTsv

In Fig. 2, the capture agents are marked as 'A', the capture manager is marked as 'M', and attack launcher is marked as 'L'. The operation sequence is as follows—(a) Process 'M' Send start-capture() message to all 'A' processes running on cluster nodes; (b) Process 'A' starts the network traffic capture operation using dumpcap; (c) The process 'L' sends experimental BDPs and/or attack process to the cluster and notify 'M'; (d) After a certain period process 'M' sends stop-capture() message to all 'A' processes to stop capture; (e) Process 'M' then can send get-data() message to all 'A' processes to collect the traffic data in .pcapng format; (f) After receiving all .pcapng files, Wireshark's mergecap utility is used to create a merged single .pcapng file. (g) If required we can filter unwanted traffic; (h) Next, we use Wireshark TShark utility to create a .csv file from .pcapng.

This .csv file represents the dataset for the corresponding experiment or attack. The dataset contains all the communications among nodes in the cluster during big data job executions. If attack is generated, the dataset contains attack packets also. Thus for each experiment, we have one .csv file, which can be used as dataset for

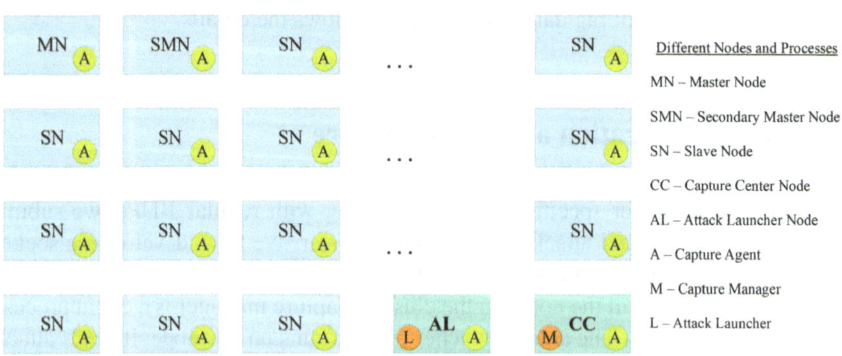

Fig. 2 Attack generation and data capture processes on different nodes in the cluster

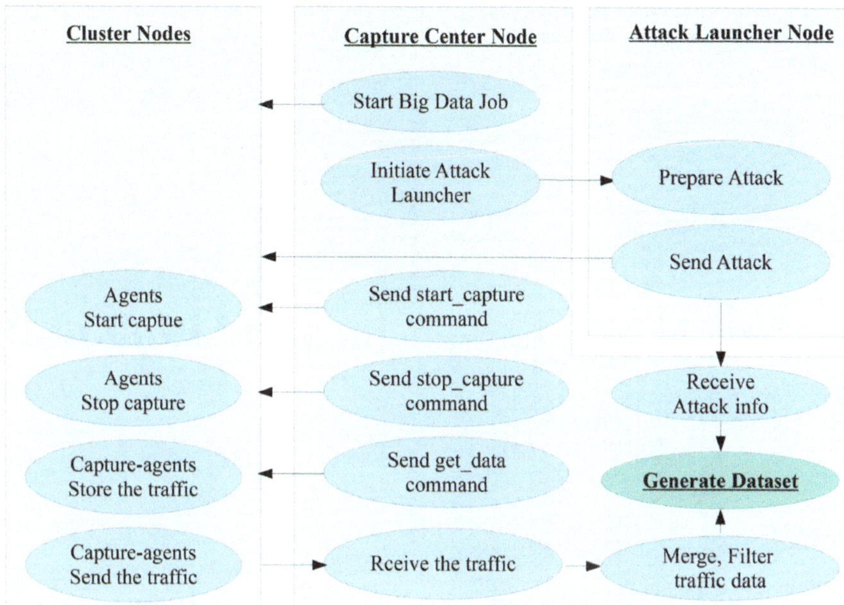

Fig. 3 Attack generation and data capture flow

security analysis. The flow of operations on cluster nodes, capture center node, and attack launcher node is shown in Fig. 3.

7 Results and Discussion

All the outcomes of the experiments are recorded in a .csv file. We store historical experimental details. Each .csv file along with the specification given below forms a complete dataset for future analysis. The structure of a dataset, which is generated in CC node and in AL node in the cluster, consists of the following fields:
exp_id, //Unique id of the experiment; exp_name, //Name of the experiment; exp_desc, //Experiment description; attack_type //Type of attack; attack_desc, //Attack description; cluster_id, //Name of the cluster; input_data //File etc used; input_data_spec, //Specification of input data; script_command, //Commands or scripts used; attack_script_command, //Commands or scripts used; experiment_start_time; experiment_end_time; capture_start_time; capture_end_time; csv_file, //Path of .csv; exp_op //experiment output if any.

We have visualized the dataset to analyze the pattern of the attack toward determination of source of attack. In experiment No. 2, in Table 3, we have used Java code to create large files in HDFS. Input to the program is *Path, Nsize, NFiles*. This generates NFiles number of files with Nsize GB file size in the directory Path. The

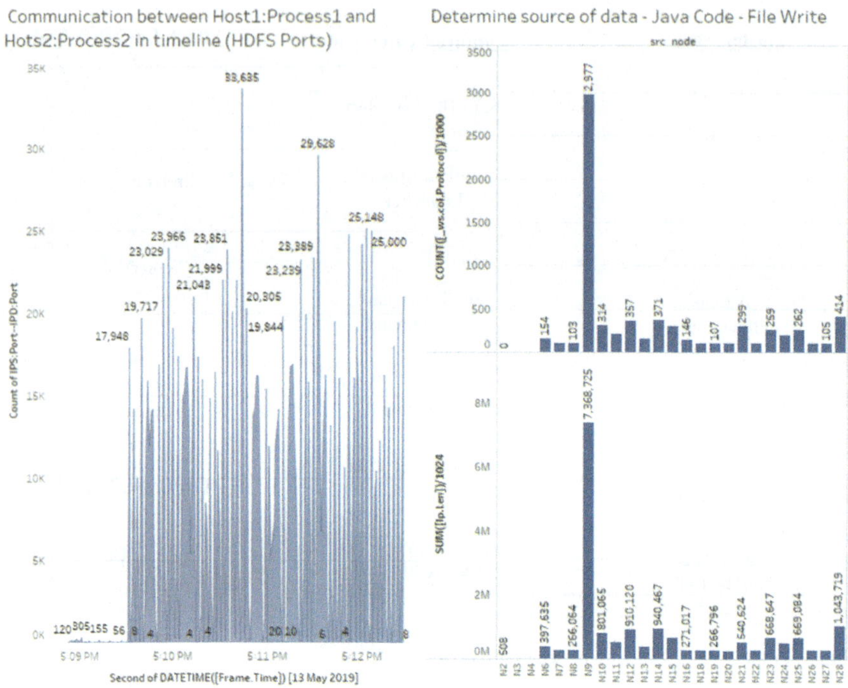

Fig. 4 Search the node sending huge data to other processes in the cluster during huge file write

program was executed on a node in the cluster. This is an insider attack to make the system busy. During the file generation operation, we have captured the traffic and the dataset is generated. The visualization in the left side of Fig. 4 shows that there are huge number of communication between two processes on different hosts. We considered only HDFS processes by applying the filter on HDFS port numbers 50070, 50470, 8020, 9000, 50075, 50475, 50010, 1019, 50020, 8485, and 50090. In the right side of Fig. 4, we tried to determine which node in the cluster is generating more traffic. We found that node N9 is generating huge traffic. Therefore, we may say node N9 is the source of attack. However, more investigation is required to confirm the result. The impact of data capture by Wireshark is very low. It does not consume additional resource from the OS of the node of the Hadoop cluster. Depending on requirement, we capture continuous network traffic or sometimes we capture in a periodical way. We have used Tableau version 2019.3 to generate the visualizations of the pattern of application behavior or attack.

8 Conclusion

In this paper, we studied the frameworks or tools available for testbeds architecture for attack generation, network capture, and dataset generation. We introduced a system application to implement the same. We have used Java Codes, Shell Scripts, and open-source or free third-party tools like Wireshark and hping3 to accomplish the job. The tools put a little impact on the Hadoop cluster performance as separate nodes are being used to send attack and process the network traffic.

We have used different experimental processes for dataset generation. We have done a little analysis toward attack detection. For future research, we will generate more datasets for known and unknown external and/or internal attacks. We will move forward to developing a feature-rich system for dataset generation for big data platform attack detection.

Acknowledgements We have conducted this research at Big Data Lab, Techno International New Town (Formerly known as Techno India College of Technology), Kolkata. We received professional version of Tableau through their Tableau Academic Programs.

References

1. Cutting, D.: The apache™ hadoop®. https://hadoop.apache.org, May 2019
2. Hortonworks: Apache hadoop ecosystem and open source big data projects. https://hortonworks.com/ecosystems, May 2019. Accessed 8 July 2019
3. Ambari: The apache ambari project. https://ambari.apache.org, May 2019
4. Wireshark: Wireshark—the world's foremost and widely-used network protocol analyzer. https://www.wireshark.org, July 2019. Accessed 10 Sept 2019
5. Wireshark: Pcapng file format. https://www.wireshark.org/docs/dfref/f/file-pcapng.html, March 2019. Accessed 10 Sept 2019
6. Aditham, S.K., Ranganathan, N.: Systems and methods for detecting attacks in big data systems. http://www.freepatentsonline.com/y2019/0089720.html, March 2019
7. Glenn, W., Yu, W.: Cyber attacks on mapreduce computation time in a hadoop cluster. Big Data Concepts, Theories, and Apps, pp. 257–279. Springer, Berlin (2016)
8. Huang, J., Nicol, D.M., Campbell, R.H.: Denial-of-service threat to hadoop/yarn clusters with multi-tenancy. In: 2014 IEEE International Congress on Big Data, pp. 48–55. IEEE (2014)
9. Aditham, S., Ranganathan, N.: A system architecture for the detection of insider attacks in big data systems. IEEE Trans. Dependable Secur. Comput. **15**(6), 974–987 (2017)
10. Alzahrani, S., Hong, L.: Generation of ddos attack dataset for effective ids development and evaluation. J. Inf. Secur. **9**(04), 225 (2018)
11. Haider, W., Hu, J., Slay, J., Turnbull, B.P., Xie, Y.: Generating realistic intrusion detection system dataset based on fuzzy qualitative modeling. J. Netw. Comput. Appl. **87**, 185–192 (2017)
12. Metron: Apache metron big data security. https://metron.apache.org
13. Metron: Apache metron. https://hortonworks.com/apache/metron, May 2019
14. MIT: Kerberos - network auth. protocol. https://web.mit.edu/kerberos/
15. Hadoop: Hadoop in secure mode. https://hadoop.apache.org/docs/current/hadoop-project-dist/hadoop-common/SecureMode.html, July 2019
16. MIT: Kerberos drawbacks. https://web.mit.edu/rhel-doc/5/RHEL-5-manual/Deployment_Guide-en-US/ch-kerberos.html

17. Millman, R.: Thousands of hadoop clusters still not being secured against attacks. https://www.scmagazineuk.com/thousands-hadoop-clusters-not-secured-against-attacks/article/1475302
18. Yoder, M., Acharya, S.: Protecting hadoop clusters from malware attacks. https://blog.cloudera.com/blog/2018/11/protecting-hadoop-clusters-from-malware-attacks, November 2018
19. Dey, N., Das, H., Naik, B., Behera, H.: Big Data Analytics for Intelligent Healthcare Management. Academic, London (2019)
20. Sahoo, A.K., Mallik, S., Pradhan, C., Mishra, B.S.P., Barik, R.K., Das, H.: Intelligence-based health recommendation system using big data analytics. In: Big Data Analytics for Intelligent Healthcare Management, pp. 227–246. Elsevier, Amsterdam (2019)
21. Panigrahi, C.R., Tiwary, M., Pati, B., Das, H.: Big data and cyber foraging: future scope and challenges. In: Techniques and Environments for Big Data Analysis, pp. 75–100. Springer, Berlin (2016)
22. Reddy, K.H.K., Das, H., Roy, D.S.: A data aware scheme for scheduling big data applications with savanna hadoop. In: Networks of the Future, pp. 377–392. Chapman and Hall/CRC, Boca Raton (2017)
23. Barik, R.K., Dubey, H., Misra, C., Borthakur, D., Constant, N., Sasane, S.A., Lenka, R.K., Mishra, B.S.P., Das, H., Mankodiya, K.: Fog assisted cloud computing in era of big data and internet-of-things: systems, architectures, and applications. In: Cloud Computing for Optimization: Foundations, Applications, and Challenges, pp. 367–394. Springer, Berlin (2018)
24. Barik, R.K., Kandpal, M., Dubey, H., Kumar, V., Das, H.: Geocloud4gi: Cloud sdi model for geographical indications information infrastructure network. In: Cloud Computing for Geospatial Big Data Analytics, pp. 215–224. Springer, Berlin (2019)
25. Das, H., Barik, R.K., Dubey, H., Roy, D.S.: Cloud Computing for Geospatial Big Data Analytics: Intelligent Edge, Fog and Mist Computing, vol. 49. Springer, Berlin (2018)
26. Pradhan, C., Das, H., Naik, B., Dey, N.: Handbook of Research on Information Security in Biomedical Signal Processing. IGI Global (2018)
27. Gupta, B.: 10 hadoop alternatives that you should consider for big data. https://www.analyticsindiamag.com/10-hadoop-alternatives-consider-big-data
28. Sanfilippo, S.: hping3 package description. http://www.hping.org, March 2019

Energy-Efficient Optimization-Based Routing Technique for Wireless Sensor Network Using Machine Learning

Aradhana Behura and Manas Ranjan Kabat

Abstract Wireless Sensor Systems (WSNs) are utilized to gather the information and to propel them to the base station and this domain is also used to design and analyse routing problems in the network. Besides, the corresponding vitality in the sensor systems is a crucial challenge to keep away from loss of packets or packet drop, quick power consumption and compelling to degradation in node performance while increasing the delay in packet delivery across the network. In order to have an effective routing decision, there is an outrageous need to check the usage of power of the nodes by maximizing the overall performance of the network using the improved machine learning techniques. Along these issues, balancing the loads while enhancing the utilization of cluster heads and nodes is one of the essential issues for bunching sensor nodes. For this problem, we have proposed a hybrid C-means donkey–smuggler optimization technique to achieve the overall routing performance in WSNs. We have validated our proposed technique against performance metrics such as packet delivery ratio, network lifetime, energy utilization and delay. Based on the experiments conducted, our proposed method outperforms better and achieves an efficient network performance.

Keywords C-means clustering · Donkey–smuggler algorithm · Routing

1 Introduction

The structure of WSN must deal with provoking issues as for interconnection and correspondence. The current conventions intended for wired, remote and sensor systems can't be utilized straightforwardly for upgrading the presentation of clustering-based systems because of the heterogeneous gadgets including modest gadgets and bigger household gadgets. In addition, WSN can be utilized alongside our vehicles utilizing vehicular ad hoc systems and they can be utilized for information gathering directing and a measure of insight can be presented in WSN for powerful co-appointment

A. Behura (✉) · M. R. Kabat
Department of Computer Science and Engineering, Veer Surendra Sai University of Technology, Burla 768018, India

H. Das et al. (eds.), *Progress in Computing, Analytics and Networking*, Advances in Intelligent Systems and Computing 1119,
https://doi.org/10.1007/978-981-15-2414-1_56

and correspondence. The insight can be presented in WSN gadgets through the use of AI systems and delicate processing draws near. In such a situation, the sensors present in the WSN can utilize the guidelines framed by deep learning calculations for settling on successful choices on activities to be performed. In addition, portability of gadgets is likewise permitted in WSN gadgets and henceforth the arrangement of principles for versatility the executives, vitality improvement and insightful steering are the significant difficulties to be tended to in the plan of WSN-based systems. In a WSN framework, the human-usable items to be specific deodorizers and brilliant vehicles can be made keen by composing standard-based programming strategies so such gadgets can react adaptably dependent on the choice settled on by the choice director. A surmising motor can likewise be intended for successful basic leadership as for development in nature of administration in WSN condition. Be that as it may, the information gathering, information portrayal, information stockpiling and correspondence should all together collaborate for powerful working of the sensor networking application. This can be accomplished by creating AI and principle-based methodology for making information gathering and correspondence shrewdly. Choosing the Cluster Head (CH) properly among the hubs can reduce the vitality use and draw out the life expectancy of the WSN. In addition, the majority of the scientists concentrated on CH choice in grouping and bunch-based steering. Be that as it may [1, 2] and [3–5], energy utilization is the most significant factor on cluster development and directing just a couple of researchers has focused on group course of action for viable steering this past [6]. In Low-Energy Adaptive Clustering Hierarchy (LEACH) [7, 8], groups are enclosed for every round dependent on the space between the hub and the CH, and the hubs ignore the other parts that affect the performance use and the framework life expectancy. In this convention, energy is displayed and the sensor node having the most noteworthy vitality and the least good ways from the part hubs are considered for the choice of CHs. At whatever point the energy accessible in the CH turns out to be not exactly any of the accessible bunch individuals in the system at that point, another CH is chosen utilizing similar criteria. Every one of these works endeavoured to decrease the vitality utilization and gave systems to ideal vitality utilization. Notwithstanding, because of the vulnerability occurring because of the development of hubs, the greater part of this works had the confinement as far as performance advancement. In [9], training is performed using convolution neural network which takes much time and it increases the time complexity. Along these lines, the system is prepared occasionally and the course revelation procedure utilizes the principles acquired from preparing for performing compelling steering. Additionally, vulnerability is dealt with in this work by the powerful utilization of fuzzy guidelines for ideal cluster development and furthermore to achieve group-based directing. By observing the real-life situation in donkey–smuggler algorithm, to avoid congestion and to find out shortest path in the networks become easy for faster routing. Thus, the proposed method must consume less energy and provide earlier routing and less time complexity by using optimized hybrid c-means donkey–smuggler algorithm.

2 Literature Review

Energy-efficient productive plan of sensor systems based on WSN framework is perplexing because of the vitality imperatives present in the sensor hubs. In such a situation, power preservation during the information accumulation and coxswaining procedure is a significant plan that is likewise utilized for assessing the exhibition of sensor network. To make routing effectiveness, various procedures dependent on bunching have been proposed in the writing [1, 2, 8, 10, 11]. Drain [1–6, 9–11] is a significant leading convention among the group-based directing strategies produced for viable directing in WSN because of its vitality productivity through bunching. This accomplishes directing over and done with the cluster heads which are chosen occasionally dependent on the system settings. In [3, 12, 24], separated from utilizing remaining power, it gives node to productive revolution of cluster heads so as to improve the system lifespan. In grouping, these heads endure some extra burden for data social affair, conglomeration and correspondence using the base station. In the past, numerous analysts utilized fuzzy relation for dealing with the vulnerability issue in numerous applications. In [13], another hand-off determination calculation was proposed by applying fuzzy relation-based basic leadership model. In [1], the creators utilized fuzzy standards for choosing appropriate group takes and for steal away the routing access over and done with the cluster heads. This prototype gave enhancement in execution up to a specific degree. Nonetheless, the exactness is to be improved promote for settling on progressively precise choices. The fuzzy guidelines can be extended by thinking about extra number of traits. Consequently, in [4], the creators built up a fuzzy-based steering model by considering the factors to be specific separation and vitality. Be that as it may, the energy proficiency issue couldn't be unravelled completely because of the idea of the hubs. A few structure rules to keep up QoS decency are additionally referenced. Blockage control calculations are ordered relying upon three criteria to be specific the identification of clog, giving notice on blockage to the sink hub just as the neighbour hubs and the alteration of information stream rate. Here, the vehicle conventions are part into blockage governor and unwavering quality assurance conventions in a less difficult and solidified gathering of present conventions. Researchers [12–19] proposed another and efficient energy directing convention which is likewise a disseminated and dynamic bunching-based convention that utilizes three stages to be specific probabilistic CH race process, use of fuzzy relation for basic leadership and the arrangement of on interest grouping. Their prototypical spotlights on all sequences and thus it achieves grouping consistently alike the channel convention. The neural system is prepared with system follow information got from the over a significant time span interchanges. The past information is utilized for the underlying preparing. Loads are balanced by utilizing the present information and by applying fuzzy standards that are terminated by the fuzzy derivation framework. The cluster individuals are powerfully refreshed dependent on the good ways from the group heads and the vitality accessibility. For this reason, the routing devoured in various hubs and the steering examples are broke down utilizing the convolution neural system. The calculation which was presented in [2] is a vitality proficient group-based steering calculation that chooses the cluster heads haphazardly by utilizing the likelihood esteems so as to perform bunch-based directing viably. Be that as it may, notice changes the CHs all the more consistently over the sensor arrange in different emphases among littler bunch ranges. Besides, every hub in Notice calculation can turn into a CH through revolution strategy utilizing its very own likelihood esteems when the calculation is hearing no bunch head announcement from the neighbour hubs. The fundamental preferred position of Notice is that the CHs are chosen dependent on the turn arrangement. Additionally, the choice expert cess has been improved by applying rules all the more viably. However, the vast majority of the above talked about calculations took a shot at CH choice utilizing fuzzy relation and just a couple dealt with group development. In [4], on the off chance that the additional amount of individuals link the CH, at that point the CH acquires overweight and this drops its energy in all respects rapidly. In [20, 21] built up a matrix station-based briefest way directing calculation for the anonymous aeronautical frameworks. The creators could accomplish improved system execution dependent on the nature of administration measurements. Disregarding the accessibility of every one of these mechanisms in the writing on clever cluster created directing for sensor network, the vast majority of the current frameworks utilized just the grouping procedures dependent on separation. Anyway in a sensor arrange, energy proficiency is a significant constraint that must be well thought out for upgrading the system life span. In this manner, a deep learning-based strategy has been proposed [9] for successfully grouping the hubs which aides in framing compelling bunch. Additional test in the plan of directing calculations for systems is the basic leadership access in the navigation calculations. Here, fuzzy principles are utilized for settling on increasingly productive choices to give an ideal incentive in the course disclosure route [8, 22]. This situation and subsequently the current directing conventions which were anticipated for routing in remote sensor systems are not ready to give ideal power utilization arrangement. Be that as it may, the proposed energy productive methodology for cluster arrangement and group-based directing gives ideal outcomes by decreasing the vitality utilization and postponement. Additionally, the directing calculation created in this examination work is expanding the bundle conveyance proportion by maintaining a strategic distance from hub disappointments. This is accomplished by the powerful observing of energy levels and in this way maintaining a strategic distance from the cluster drops happening because of hub disappointments by vitality. At long last, it is seen that the proposed work furnishes ideal outcomes concerning the improvement as for quality of service in WSNs for energy-efficient routing. Convolution neural network used to train the sensor dataset [9] and neuro-fuzzy inference system is used for testing purpose which is shown in Fig. 1. Cluster head energy, distance between cluster head and the base station, distance between cluster head and node and cluster head degree have induced in neuro-fuzzy inference system. Using rule-based protocol, neuro-fuzzy rule manager and decision manager have decided the data routing module [9]. Shadowed Type-2 fuzzy logic is used to provide efficient routing protocol [23]. It helps to avoid collision among the sensor nodes [5, 7]. ARIMA protocol is used for efficient data collection in wireless sensor network [2].

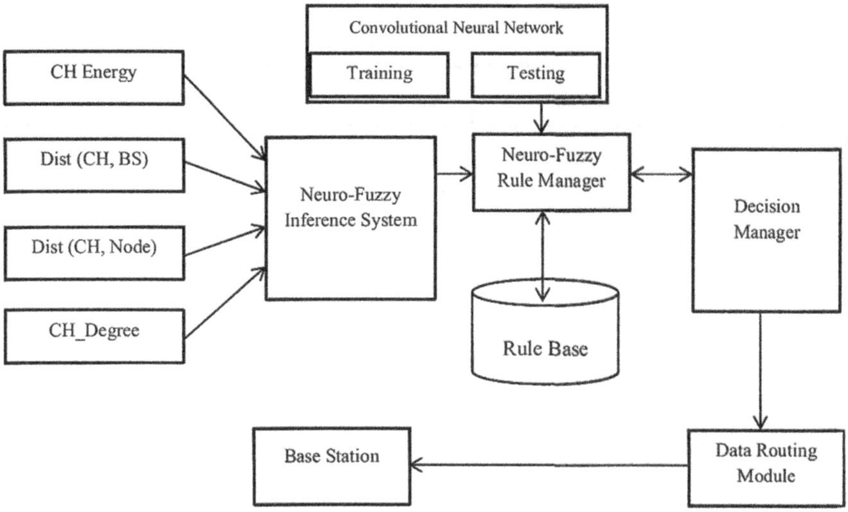

Fig. 1 Fuzzy rule-based routing system architecture

3 Donkey–Smuggler Algorithm

The smuggler decided the destination of the donkey (best solution) and he has the ability to change the destination or to set the shortest path according to the best solution. When the smuggler found that the sign of the congestion or overloading, then he sets two channels to reduce the traffic in the network. The schematic representation is shown in Figs. 2 and 3 [12].

4 Proposed Work

4.1 Cluster Head Degree

Group individuals are significant for adjusting the heap in a cluster. For that reason, we presented another calculation strategy for choosing the cluster head. At this juncture, the individuals earlier unite with the cluster head; this deliberates the tally of the individuals which are as of now existent in the separate group. In addition, it ascertains the group highest cluster degree utilizing Eq. (1). In the event that the CH degree is high, at that point there is exceptionally less shot for a hub to join as a bunch part with that CH.

$$CH_{\text{degree}} = \frac{\text{Total number of nodes in the cluster area}}{\text{Total number of nodes in the coverage area}} \quad (1)$$

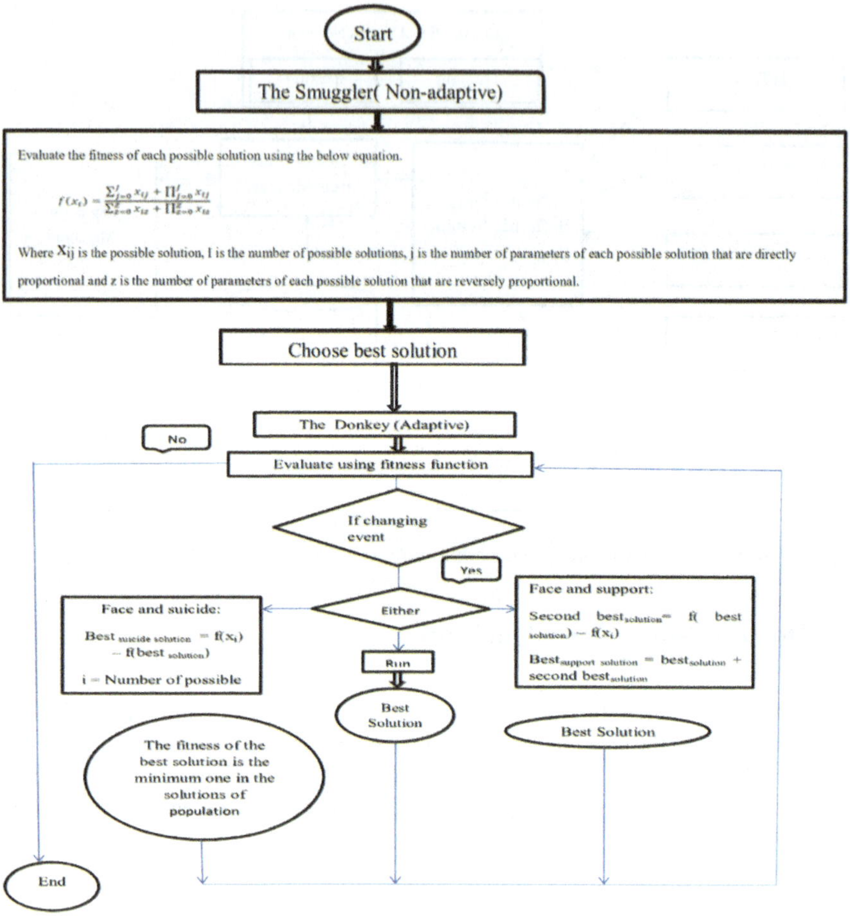

Fig. 2 Flow chart of donkey–smuggler algorithm

4.2 *An Optimized Hybrid DAS C-means Clustering Algorithm*

It is obvious that similarity of two data points residing in the same cluster will have a high value, and two points belonging to two different clusters will be dissimilar in nature. To reduce the distance, avoid packet loss, and consume less energy, we propose this hybrid algorithm.

Step 1: Decide the number of clusters and define the appropriate level of fuzziness cluster.
Step 2: By using DAS, find initial cluster centre.
Step 3: Find nth extreme samples from each cluster centre.
Step 4: Evaluate mean for nth extreme samples for each cluster.

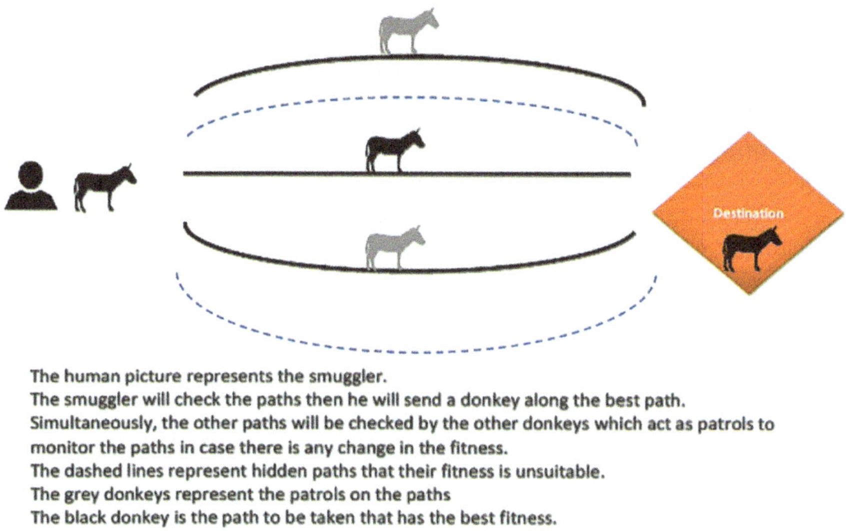

Fig. 3 Graphical representation of algorithm [12]

Step 5: Inverting the coordinate of each calculated means then find closest data to each cluster centre by using various pairwise distances.
Step 6: When the iteration ends, find cluster centre and corresponding data to each cluster or change the type of the distance.

4.3 *Proposed Routing Algorithm*

In this work, a bunch-based directing calculation that utilizes optimized hybrid c-means donkey–smuggler algorithm has been proposed for performing vitality effective steering. The means of the proposed steering calculation are as per the following:

Stage 1: Use the energy levels and area (xxi, yyi) of sensor hubs Si, $I = \{1, 2, 3, \ldots, 7, n\}$
Stage 2: Drive "DATA" wallets to entire neighbour hubs from the Base Station (BS) and discover the separations of hubs from the BS and between the hubs.
Stage 3: Demand the hybrid *C*-means donkey–smuggler optimization bunching calculation to frame *c* groups (centre of mean value) by gathering the hubs dependent on separation.
Stage 4: Utilizing BS as the controller, play out the cluster head determination for all groups by allowing for the separations in addition to energy stages of hubs.
Stage 5: Carry out course disclosure by discovering the briefest way as of every hub to the BS over and done with the cluster head among the nodes.

Stage 6: Direct information gathered through hubs over and done with the cluster heads utilizing the most limited way establish in stage 5, and then furthermore by spread on clustering standards.
Stage 7: Gather information at Base Station (BS).
Stage 8: In the event that vitality levels of at least half hubs are depleted, at that point STOP.
Stage 9: Verify if cluster head pivot if important.
Stage 10: If it is yes go to stage 4. Otherwise go to stage 6.

Utilizing the calculation, information gathered by the sensor hubs are directed to the BS occasionally and the calculation is ended at whatever point the vitality level of half of the hubs are depleted and has just under 11% of the first energy stage.

5 Results

The projected group development convention was tried through reproductions utilizing the MATLAB programming. The parameters that have been utilized in this work for the re-enactment are introduced in Table 1 (Figs. 4, 5, 6 and 7).

6 Conclusion

In this paper, another routing calculation for efficient cluster-based sensor arranges that optimized hybrid c-means donkey–smuggler algorithm-based grouping approach for performing cluster-based directing so as to improve the network execution. In this methodology, the cluster arrangement in WSNs used the vitality demonstrating for proficiently directing the parcels through the utilization of AI utilizing hybrid clustering system with donkey–smugger algorithm for weight alteration, and henceforth the system lifetime is drawn out. In addition, we considered four parts in particular lingering vitality of the CH, space between the CH and the sink hub, space between the sensor hub and the CH and the level of the CH which

Table 1 Parameters

Parameters	Value
Simulation area	100×100 m^2
Number of sensor nodes	100
Initial energy of nodes	0.5 J
E_{elec}	50 nj/bit
ε_{fs}	10 bit/pj/m^2
E_{mp}	0.0013 bit/pj/m^4
Packet size	4096 bits

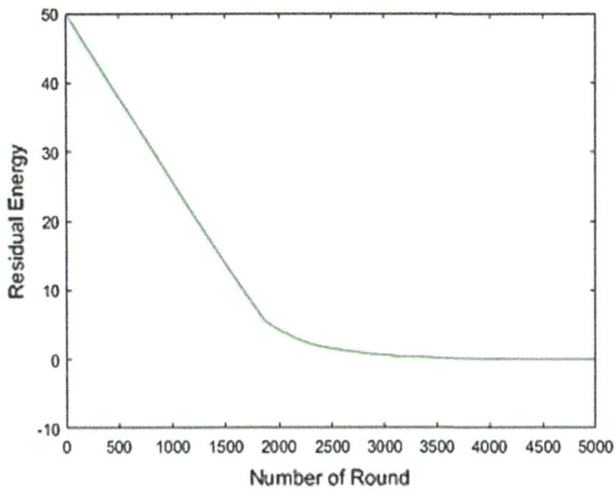

Fig. 4 Residual energy

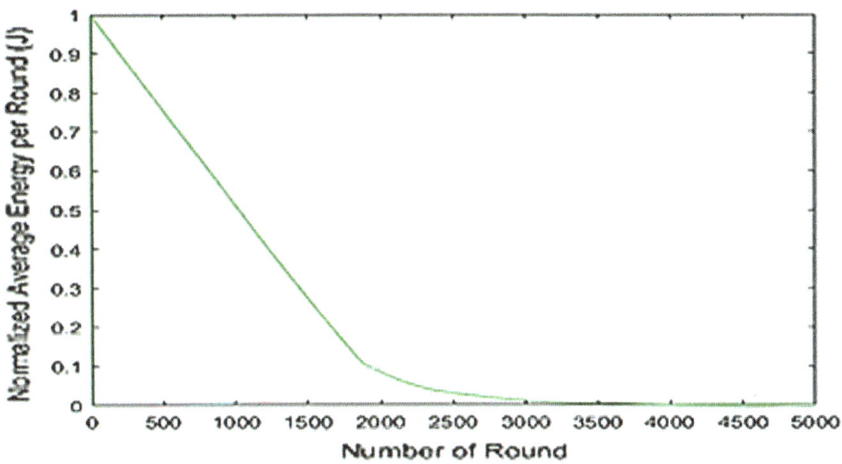

Fig. 5 Normalised average energy per round

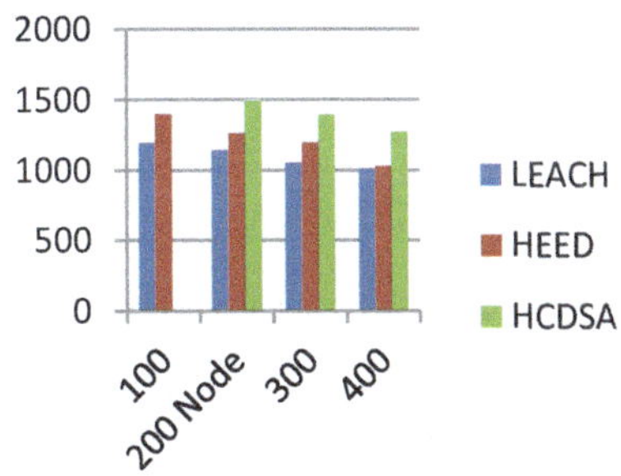

Fig. 6 Network lifetime versus number of sensor nodes

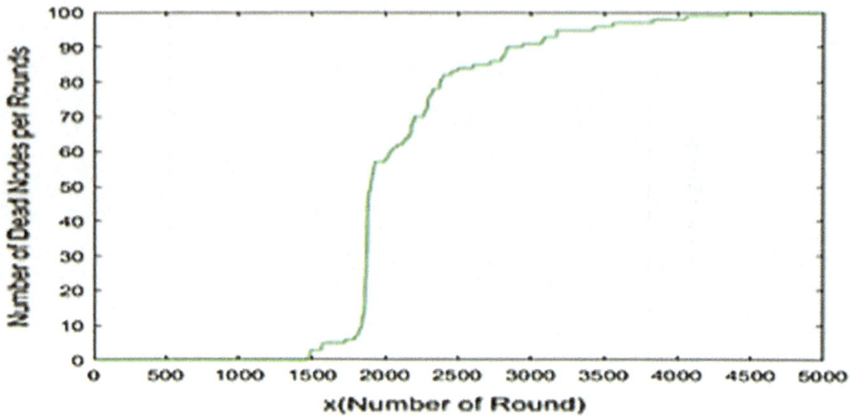

Fig. 7 Number of dead nodes per rounds

are significant elements for the strength use and system life expectancy. We have assessed the proposed calculation utilizing reproductions with MATLAB in which the previously mentioned segments were utilized as optimized hybrid C-means donkey–smuggler algorithm (HCDSA). The yield estimation of HCDSA was utilized to decide the CH for the hub to join as a part. From the outcomes acquired in this work, it has been seen that the proposed convention performed better when this is contrasted and channel, clustering and notice as far as energy use and framework life expectancy because of the utilization of clustering guidelines got through learning and by applying cluster-based directing. One confinement of this work is the suspicion that all hubs are trustful hubs which isn't constantly conceivable.

References

1. Mishra, B.B., Dehuri, S., Panigrahi, B.K., Nayak, A.K., Mishra, B.S.P., Das, H. (eds.): Computational Intelligence in Sensor Networks. Springer (2019)
2. Diwakaran, S., Perumal, B., Devi, K.V.: A cluster prediction model-based data collection for energy efficient wireless sensor network. J. Supercomput. **75**, 3302–3316 (2019). https://doi.org/10.1007/s11227-018-2437-z
3. Lee, J.S., Cheng, W.L.: Fuzzy-logic-based clustering approach for wireless sensor networks using energy predication. IEEE Sens. J. **12**(9), 2891–2897 (2012)
4. Izadi, D., Abawajy, J., Ghanavati, S.: An alternative clustering scheme in WSN. IEEE Sens. J. **15**(7), 4148–4155 (2015)
5. Bagci, H., Yazici, A.: An energy aware fuzzy approach to unequal clustering in wireless sensor networks. Appl. Soft Comput. **13**(4), 1741–1749 (2013)
6. Mhemed, R., Aslam, N., Phillips, W., Comeau, F.: An energy efficient fuzzy logic cluster formation protocol in wireless sensor networks. Proc. Comput. Sci. **10**, 255–262 (2012)
7. Lin, Q., Song, H., Gui, X., Wang, X., Su, S.: A shortest path routing algorithm for unmanned aerial systems based on grid position. J. Netw. Comput. Appl. 215–224 (2018)

8. Palattella, M.R., Dohler, M., Grieco, A., Rizzo, G., Torsner, J., Engel, T., Ladid, L.: Internet of things in the 5G era: enablers, architecture, and business models. IEEE J. Sel. Areas Commun. **34**(3), 510–527 (2016)
9. Thangaramya, K., Kulothungan, K., Logambigai, R., Selvi M., Ganapathy, S., Kannan, A.: Energy aware cluster and neuro-fuzzy based routing algorithm for wireless sensor networks in IOT. Comput. Netw. (2019)
10. Das, H., Naik, B., Pati, B., Panigrahi, C.R.: A survey on virtual sensor networks framework. Int. J. Grid Distrib. Comput. **7**(5), 121–130 (2014)
11. Logambigai, R., Kannan, A.: Fuzzy logic based unequal clustering for wireless sensor networks. Wirel. Netw. **22**(3), 945–957 (2016)
12. Shamsaldin, A.S., Rashid, T.A., Al-Rashid Agha, R.A., Al-Salihi, N.K., Mohammadi, M.: Donkey and smuggler optimization algorithm: a collaborative working approach to path finding. J. Comput. Des. Eng. (2019). https://doi.org/10.1016/j.jcde.2019.04.004
13. Selvi, M., Logambigai, R., Ganapathy, S., Sai Ramesh, L., Khanna Nehemiah, H., Arputharaj, K.: Fuzzy temporal approach for energy efficient routing in WSN. In: Proceedings of the International Conference on Informatics and Analytics, ACM, pp. 1–5 (2016)
14. Ari, A.A.A., Yenke, B.O., Labraoui, N., Damakoa, I., Gueroui, A.: A power efficient cluster-based routing algorithm for wireless sensor networks: honey bees swarm intelligence based approach. J. Netw. Comput. Appl. **69**, 77–97 (2016)
15. Bajaber, F., Awan, I.: An efficient cluster-based communication protocol for wireless sensor networks. Telecommun. Syst. **55**, 387–401 (2014)
16. Xie, G., Pan, F.: Cluster-based routing for the mobile sink in wireless sensor net- works with obstacles. IEEE Access **4**, 2019–2028 (2016)
17. El-said, S.A., Osamaa, A., Hassanien, A.E.: Optimized hierarchical routing tech- nique for wireless sensors networks. Soft. Comput. **20**, 4549–4564 (2016)
18. Anand, K., Ganapathy, S., Kulothungan, K., Yogesh, P., Kannan, A.: A rule based approach for attribute selection and intrusion detection in wireless sensor networks. Proc. Eng. **38**, 1658–1664 (2012)
19. Heinzelman, W.R., Chandrakasan, A., Balakrishnan, H.: Energy-efficient communication protocol for wireless microsensor networks. In: Proceedings of thirty-third IEEE Annual Hawaii International Conference on System Sciences, pp. 1–10 (2000)
20. Lin, H., Wang, L., Kong, R.: Energy efficient clustering protocol for large-scale sensor networks. IEEE Sens. J. **15**(12), 7150–7160 (2015)
21. Logambigai, R., Ganapathy, S., Kannan, A.: Energy-efficient grid-based routing algorithm using intelligent fuzzy rules for wireless sensor networks. Comput. Electr. Eng. **68**, 62 75 (2018)
22. Betzler, A., Gomez, C., Demirkol, I., Paradells, J.: CoAP congestion control for the internet of things. IEEE Commun. Mag. **54**(7), 154–160 (2016)
23. Chatterjee, K., De, A., Chan, F.T.: Real time traffic delay optimization using shadowed type-2 fuzzy rule base. Appl. Soft Comput. J. **74** (2019)
24. Parasakthi, S., Mohan Kumar, P.: EART: enhancing an energy aware routing protocol in cluster based wireless sensor networks. Sens. Lett. **13**, 611–617 (2015)

8. [illegible]: Internet of things in the … data centers, architecture, and business models. IEEE J. Sel. Areas Commun. 34(4), 510–527 (2016)
9. Thangaramya, K., Kulothungan, K., Logambigai, R., Selvi, M., Ganapathy, S., Kannan, A.: Energy aware cluster and neuro fuzzy based routing algorithm for wireless sensor networks in IoT. Comput. Netw. (2019)
10. Das, H., Naik, B., Pati, B., Panigrahi, C.R.: A survey on virtual sensor networks framework. Int. J. Grid Distrib. Comput. 7(5), 121–130 (2014)
11. Logambigai, R., Kannan, A.: Fuzzy logic based unequal clustering for wireless sensor networks. Wirel. Netw. 22(3), 945–957 (2016)
12. [illegible]
[illegible]

Energy-Efficient Resource Scheduling in Fog Computing Using SDN Framework

Mahmoud Al Ahmad, Sudhansu Shekhar Patra and Rabindra K. Barik

Abstract Fog computing is emerging as a powerful technology and computing paradigm which enables the providers to execute the services in the edge network. Both sensor computing and mobile computing are selected as the fog nodes, as well as the cloud nodes, as and when needed for their resource provisioning. The networking policy is more dynamically controlled by the Software Defined Network (SDN), which is an intelligent technology in the network that controls the forwarding devices. It is a challenge to realize the energy efficiency for the computing nodes in the fog-assisted cloud environment. In this paper, we focus on the energy-saving issues in Fog Server (FS) selections for the Fog Node (FN) in the fog-assisted cloud computing environment. In view of this challenge, Fog Node (FN) scheduling method for energy efficiency in fog-assisted cloud computing is projected. We analyze the energy influencing factors and then design energy-efficient FN selection algorithms based on dynamic programming scheme. Consequentially, the experimental evaluation studies and comparative analysis are performed to validate the effectiveness and efficiency of our proposed scheme. In this paper, the dynamic programming scheme is used for optimizing the selection of FNs for better energy efficiency. The experimental results illustrate that the proposed dynamic programming technique achieves good energy-saving effect. We perform the experiments with ifogsim and the results depict that the proposed algorithm can be used for minimizing the energy consumed by the fog system while maintaining the SLA parameters.

Keywords Sensor computing · SDN · Fog computing · Fog node scheduling · Energy efficiency

---

M. A. Ahmad
School of Computer Engineering, KIIT Deemed to be University, Bhubaneswar, India
e-mail: Alahmad.ayla@gmail.com

S. S. Patra (✉) · R. K. Barik
School of Computer Applications, KIIT Deemed to be University, Bhubaneswar, India
e-mail: sudhanshupatra@gmail.com

R. K. Barik
e-mail: rabindra.mnnit@gmail.com

H. Das et al. (eds.), *Progress in Computing, Analytics and Networking*,
Advances in Intelligent Systems and Computing 1119,
https://doi.org/10.1007/978-981-15-2414-1_57

1 Introduction

In the current scenario and the years to come, the increasing energy demand by the data centers has become a challenging problem for the developing, as well as, to the developed countries. There is an article in New York Times regarding energy efficacy noted that the energy going to be consumed only in the data centers is close to 30 billion watts worldwide, approximated by 30 nuclear power plants [1]. Research by an Italian researcher depicted that the overall energy consumption by the Internet will rise approximately by 50% in the next 5 years as per the current demand. Therefore, the network energy consumption by the data centers cannot be ignored, and we should concentrate on the energy-saving techniques on the networks. This becomes a sensitive issue and has to put focus on it. Data centers plays a vital role in the network as they are the computing, storage, infrastructure, software, and many more service providers. To make the data centers reliable, many redundant switches and servers are used. The Internet of Things (IoT) and the sensor networks are going to be used in all areas of the society, generate a promising volume of data flows on the Internet. Cloud computing, a promising distributed computing model is used to store and process a huge volume of data originated by the IoT sensors and helps end-users with many reliable services [2, 3]. On the other hand, there is a large amount of network bandwidth consumed by the smart devices and so decreases the performance by the burden to the cloud data centers [4–7]. Because of that certain delay-sensitive services in the IoT cannot be processed timely. In addition to this many real time requirements of edge devices with mobility, location awareness, etc., are not met. To overcome all these technological difficulties a new computing paradigm known as fog computing has been proposed by Cisco in 2012, in which they extended the traditional cloud computing technology paradigm to the edge of the network [8, 9]. Fog computing provides a virtualized platform, which is placed between the sensor devices and cloud servers, which has the capacity of computing, as well as storage services, at the edge of the Internet closer to the end-users [6, 10, 11]. Fog computing is more appropriate for terminal users. Due to the dynamic nature of the resources and uncertainty of them scheduling is a prime technology that has to be in the attention of the researchers and has to be resolved in fog environment [9, 12]. Because of the massive data in fog computing, scheduling faces many challenges and special attention has to be given by the researcher's community.

Because of the high energy consumption, the operating cost of the fog-assisted cloud systems has increased drastically, and also has a bad effect on the climate. Cloud data centers increase carbon emissions and electricity costs [13, 14]. The concept of fog/edge computing offers a technique where storage and computing are closer to the devices which generate data at the edge of the networks [12, 15]. Software Defined Network (SDN) is a promising next generation architecture, which provides packet forwarding option to a logically centralized controller which is called the control plane and the routers have only the data plane with them.

In Network Function Virtualization (NFV), the role of a hypervisor is to create many a number of virtual fog nodes that uses resources of a fog server which enables

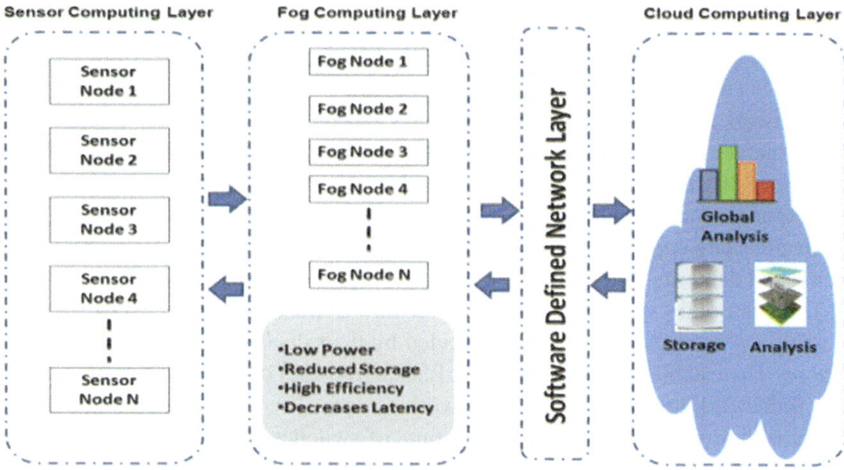

Fig. 1 Fog computing architecture with sensor computing, SDN and cloud computing layer

to handle the heavy traffic in the network using virtualization at the edge nodes. SDN is an important network technology that makes the Fog nodes more intelligent, analytic, and efficient. SDN decouples the forwarding plane from the control plane. The control plane acts as an intermediary between the two, i.e., the data plane and forwarding plane. In the data plane the designer able to write decision-making algorithms, policy management techniques, and route discovery algorithms, etc. Figure 1 depicts the architecture of fog computing with sensor computing, cloud computing, and SDN.

The energy consumption can be minimized at the data centers by energy-saving models at SDN or by energy-saving models at fog layer or by cloud layer. In this paper, we focus on fog server selection on an overloaded host for fog Node placement and optimize the energy consumption.

The remainder of this paper is organized as follows: related work is explained in Sect. 2, the energy-saving fog node scheduling problem is elaborated in the Sect. 3, experimental results are analyzed in Sects. 4, and 5 presents the conclusions of the present work.

2 Related Work

Energy efficiency of data center networks is often used as a by-product of other optimization strategies in cloud and fog servers. The major focal point in data centers is on energy-saving FN placement. The networks are treated as the side trouble in terms of providing power savings. In this case, power savings schemes in the networks are normally achieved by traffic locality or traffic consolidation [16, 17]. For example, the frameworks are presented in [13, 14, 18, 19], minimize the number of active racks in the data centers networks by consolidating FNs in a fewer number of racks.

FN placements are identified as one of the routing problems [20] and it combines with network optimization. A distributed flow scheduling scheme [21] is determined on the networking component within the data center network. To get the trade-off among the optimization in the network, still, there is some lack of research and experimental studies made on. It basically concentrates on the aim of giving application developers that tries to program the network a lucid overview of the merits and demerits of their respective algorithmic choices.

There are other studies that deploy Software Defined Networks (SDN) in order to apply the energy-aware changes decided by their decision frameworks or applications. The establish work in this era, ElasticTree [22] designs a centralized decision framework to turn off the idle network devices. The main disadvantage of this model is that the multi-minute booting time of switches makes handling appropriately bursty traffic more difficult. Differently, it uses in the simulation the sleeping mode of switches. This method takes a much shorter time, i.e., around 1 s to turn them back on, while still providing a significant amount of power-saving. There are similar approaches that put idle network devices into the sleeping mode or turn them off by implementing heuristic scheduling algorithms [15, 23, 24]. It mostly provides the traffic matrix as an input to the scheduling algorithm which is not always a realistic scenario. Contrarily, it focuses on the real time flow requests while keeping the scalability quality requirement in mind [25].

3 Energy-Saving Fog Node Scheduling Algorithm

The goal of the fog layer is the minimization of the number of active fog servers by migrating the fog nodes by making a fewer number of fog servers. Let A be a fog node placement matrix defined by

$$A_{\mathrm{ij}} = \begin{cases} 1, \text{ if fog node } j \text{ is placed on fog server } i \\ 0, \text{ elsewise} \end{cases}$$

F_i^r is the available resource type r on fog server i, whereas N_j^r is the demand by the fog node j for the available resource type r. Let a binary variable FS_i can be defined with value 1 if fog server i is on, elsewhere 0.

$$\text{Minimize} \sum_i FS_i \tag{1}$$

$$\text{subject to} \sum_{j=1} N_j^r * A_{\mathrm{ij}} \leqslant F_i^r \; \forall i, r \tag{2}$$

$$\sum_{i} A_{ij} = 1 \ \forall j \tag{3}$$

$$\mathrm{FS}_i >= A_{\mathrm{ij}} \ \forall i, j \tag{4}$$

where $1 <= i| <= |F|, 1 <= j <= |N|$ and $r \in R$.

The objective function (1) minimizes the number of fog servers turned on. The constraint (2) defines the summation of the resource demands installed on a given fog server by the fog nodes cannot be greater than the available capacity of the fog server. Constraint (3) ensures each fog node should be placed on exactly one fog server. Equation (4) has the variables FS_i and A_{ij} maintains a fog server turned on or off as it is in use or not.

The tasks are first assigned to the fog nodes and then the fog nodes are assigned to the fog server. So tasks assignment and fog nodes assignment are the two well-known steps in the resource allocation problem. The allocation of tasks is to decide which fog node should be assigned to run the task. In this paper, we mainly focus on the allocation of fog nodes. Fog node scheduling has four major steps

I. To detect the fog server which is currently overloaded;
II. To detect the underloaded fog server;
III. To select the suitable fog node which has to be migrated from the overloaded fog server;
IV. Place the fog nodes are selected in step-III and all the fog nodes from the underloaded fog server to the appropriate fog server.

In this paper, we work on step-III which uses the policy of static CPU threshold which helps to detect the overloaded as well as underloaded fog servers, and MBFD [19] as the placement algorithm for fog node.

A. Fog Node selection algorithm based on Greedy

The basic idea of the Greedy algorithm is to select a fog node with the minimum memory in the remain fog node list. So, we first sort the fog Node list on overloaded fog servers in ascending order based on fog node's memory, then add the fog node with minimum memory to migration list, after that check whether the physical node is in underloaded status, or not, if not repeat this process until physical node is in underloaded status, finally return the selected migration fog node's. Figure 2 shows an example of the migration fog nodes selected by the greedy algorithm.

We can see that fog node selected by greedy algorithm are the set {1, 2, 3, 4, 5} is, but fog node 7 is a more optimal solution. So the greedy algorithm is giving the local optimized solution is not the best method to solve this problem. But the time complexity is $O(n)$ where n is the number of fog nodes running on the fog server.

B. Fog node selection algorithm based on dynamic programming

Assume that overloaded fog server k has n fog nodes running in it, $N(i, j)$ represents the total minimum memory of the first i $(1 <= i <= n)$ fog nodes running in fog server with CPU threshold j $(1 <= j <=$ upper Threshold X $C_{\mathrm{k\_total}})$, we can obtain the following state transition equation:

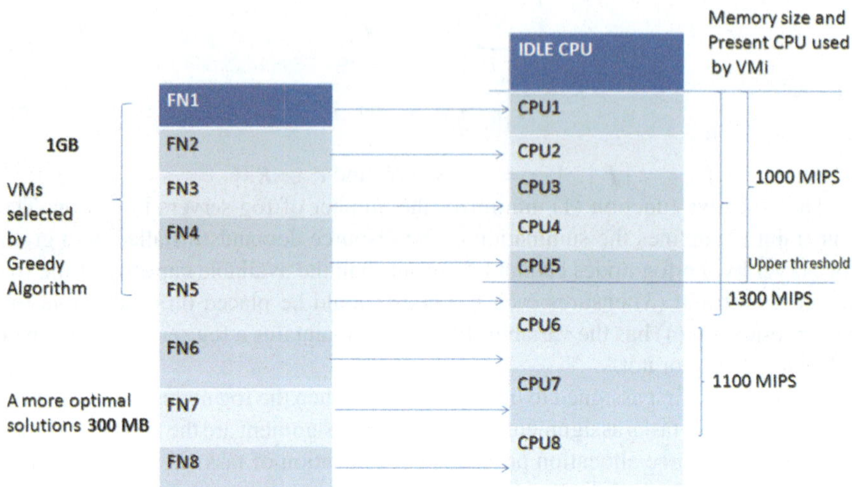

Fig. 2 Migration Fog nodes selected by Greedy Algorithm

$$N(i, j) = \begin{cases} 0,\ i = 0 \\ 0,\ j = 0 \\ N(i-1, j),\ j < c_i \\ \max\{N(i-1, j),\ N(i-1, j-n_i) + n_i\},\ j > c_i \end{cases}$$

When there are no fog nodes running on fog server k ($i = 0$), or the fog server k can't supply any CPU capacity for fog node ($j = 0$), N (i, j) equals 0, there are no fog nodes need to migrate. When the CPU resource required by fog node i is greater than the current CPU resource required by fog node i is greater than the current CPU threshold (j IMPS) on fog server k, then N (i, j) equals to N ($i-1, j$), which means the minimum total memory of the first j fog nodes reserved in fog server k with CPU threshold j is equivalent to minimum total memory of the first $i-1$ fog nodes. On the other hand, if the CPU resource requirement of fog node j can be satisfied fog server k ($j > c_i$), there are the following two cases

I. Fog node i is selected as a reserved fog node
II. Fog node i has not been selected as reserved fog node, we should choose the greater one as N (i, j)

Dynamic programming algorithm can be implemented in a bottom-up or top-down manner. Since the top-down implementation may require a relatively large system stack space and uses stack overflows, we may use bottom-up approach to implement the algorithm.

Algorithm : Dynamic programming to select the fog node to migrate from overloaded fog server
Input : overloaded fog server
Output : migration fog node list

```
fogNodeReservelist = NULL
get fogNodeList from host
fogNodeNum = fogNodeList.size()
thresholdCapacity = upperThreshold X hostCpuTotal
define array M[fogNodeNum + 1, thresholdCapacity +1]
N[i,0]=0
N[0,j]=0
for i= 1 to fogNodeNum
        fogNodeCPU= fogNodeList[i].getCurrentUsedCpu()
        fogNodeMem = fogNodeList[i].getCurrentUsedMemory()
        for j= 1 to thresholdCapacity
            if j < fogNodeCpu then
                    N[i,j]=N[i-1,j]
            else
                    N[i,j]=max{N[i-1,j],N[i-1,j-fogNodeCpu]+fogNodeMem}
            endif
        endfor
endfor
i= fogNodeNum, j=thresholdCapacity
while ((i>0) && (j>0))
        fogNodeCpu=fogNodeList[i].getCurrentusedCpu()
        fogNodeMem=fogNodeList[i].getCurrentusedMemory()
        if (j < fogNodeCpu) then
                i=i-1
                continue
        endif
        if N[i,j] equals N[i-1, j-fogNodeCpu] + fogNodeMem then
                add fogNodeList[i] to fogNodeReserveList
                i=i-1
                j=j-fogNodeCpu
        else
                i=i-1
        endif
endWhile
fogNodeMigrationList = fogNodeList.RemoveAll(fogNodeReserveList)
return fogNodeMigrationList
```

4 Experiment and Result Analysis

We create a fog node center with the following energy-saving model, we created four categories of fog nodes, the fog nodes are equal to the number of the task of every workload data, the organization of each fog server and fog node is depicted in Table 1.

Table 1 Configuration of each fog server and fog node

Parameter	Fog server	Fog node
No. of PE	2	1
Performance of each PE	2000 or 3000 MIPS	1000 or 500 MIPS
RAM size	4 GB	870/1740/613 MB
Bandwidth	1 Gbit/s	100Mbit/s
Storage size	1 TB	2.5 GB

We maintain the CPU utilization upper Threshold mesaued to 50% to 100%, and the power consumption, SLA violation rate, numbers of fog node migration, fog node selection mean time between greedy and dynamic programming in Table 2 noted. Figure 3 shows the energy consumption for different workloads by setting the CPU

Table 2 Various performance parameter using Greedy and Dynamic Programming Algorithm for different CPU utilization upper threshold

Parameters	Greedy technique	Dynamic programming
CPU threshold value: **0.5**		
Energy (KWh)	248.07	223.32
Fog node selection mean time	0.00056	0.00232
Number of fog node migration	38360	29944
Overall SLA violation	0.0012	0.0015
CPU threshold value: **0.55**		
Energy (KWh)	234.4	211.62
Fog node selection mean time	0.00047	0.00217
Number of fog node migration	33704	26806
Overall SLA violation	0.0001	0.0013
CPU threshold value: **0.6**		
Energy (KWh)	223.4	201.94
Fog node selection mean time	0.00043	0.00284
Number of fog node migration	31814	25286
Overall SLA violation	0.0009	0.0012
CPU threshold value: **0.65**		
Energy (KWh)	214.92	194.35
Fog node selection mean time	0.00043	0.00283
Number of fog node migration	29982	24169
Overall SLA violation	0.0008	0.0011
CPU threshold value: **0.7**		
Energy (KWh)	207.32	186.93
Fog node selection mean time	0.00042	0.00296
Number of fog node migration	29398	23338
Overall SLA violation	0.0008	0.0011
CPU threshold value: **0.75**		
Energy (KWh)	200.4	179.52
Fog node selection mean time	0.00039	0.00329

(continued)

Table 2 (continued)

Parameters	Greedy technique	Dynamic programming
Number of fog node migration	28182	22869
Overall SLA violation	0.0007	0.0011
CPU threshold value: **0.8**		
Energy (KWh)	191.73	173.27
Fog node selection mean time	0.00047	0.00354
Number of fog node migration	26634	22359
Overall SLA violation	0.0007	0.0012
CPU threshold value: **0.85**		
Energy (KWh)	184.68	166.45
Fog node selection mean time	0.00042	0.00408
Number of fog node migration	26029	21565
Overall SLA violation	0.0008	0.0014
CPU threshold value: **0.9**		
Energy (KWh)	177.32	159.77
Fog node selection mean time	0.00048	0.00391
Number of fog node migration	24492	20610
Overall SLA violation	0.0009	0.0017
CPU threshold value: **0.95**		
Energy (KWh)	171.43	140.05
Fog node selection mean time	0.00075	0.0053
Number of fog node migration	47289	32037
Overall SLA violation	0.0025	0.0032
CPU threshold value: **1**		
Energy (KWh)	163.23	131.22
Fog node selection mean time	0.00091	0.0061
Number of fog node migration	45517	29132
Overall SLA violation	0.0054	0.0073

utilization threshold which equals 0.8. We perform experiments with ifogsim and results drawn from the experiment proofs that the proposed algorithm can effectively reduce the power consumption by maintaining the SLA needs.

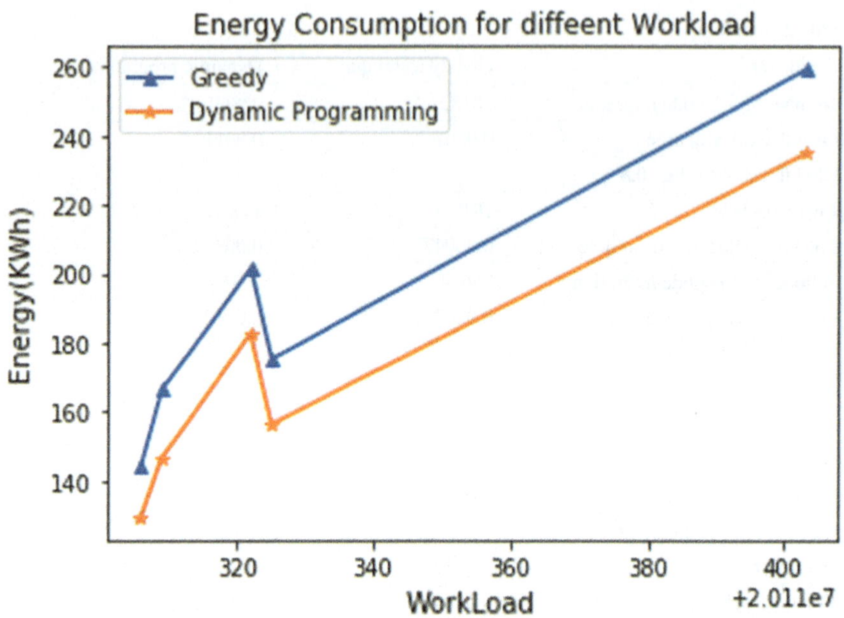

Fig. 3 Energy consumption for different workload

5 Conclusion

In this paper, we propose the architecture of sensor node with fog computing and SDN. The proposed architecture helps to support real time data delivery, high level of scalability, as well as mobility, by bringing the computing closer to the edge devices. For sensor node, fog computing is the appropriate platform due to the capability of faster analysis and decision-making. Along with SDN uses logically centralized control plane implementing sophisticated mechanisms for traffic control, as well as resource management. In the future growth of IoT environment, such a network gives a vital significance to address the increasing capacity demands where a large number of Internet-connected terminal devices are expected. The proposed architecture addresses the benefits of energy-saving by combining SDN and fog environment in one system and adapting to each other. In this paper, a dynamic programming scheme is used to optimize the selection of fog nodes for better energy efficiency. The experimental results show that the proposed dynamic programming technique achieves the good energy-saving effect.

References

1. Data-Center-Issue-Paper-final826, http://anthesisgroup.com/wpc-Ontent/uploads/2014/08/Data-Center-Issue-Paper-final826.pdf
2. Yu, L., Cai, Z.: Dynamic scaling of virtual clusters with bandwidth guarantee in cloud datacenters. In: IEEE INFOCOM 2016-The 35th Annual IEEE International Conference on Computer Communications, pp. 1–9, IEEE (2016, April)
3. Armbrust, M., Fox, A., Griffith, R., Joseph, A.D., Katz, R., Konwinski, A., Lee, G., Patterson, D., Rabkin, A., Stoica, I.: A view of cloud computing. Int. J. Comput. Technol. **4**, 50–58 (2013)
4. Lei, Y., Shen, H., Cai, Z., Ling, L., Pu, C., Lei, Y., Shen, H., Cai, Z., Ling, L., Pu, C.: Towards bandwidth guarantee for virtual clusters under demand uncertainty in multi-tenant clouds. IEEE Trans. Parallel Distrib. Syst. **29**, 450–465 (2018)
5. Lei, Y., Shen, H., Sapra, K., Lin, Y., Cai, Z.: CoRE: cooperative end-to-end traffic redundancy elimination for reducing cloud bandwidth cost. IEEE Trans. Parallel Distrib. Syst. **28**, 446–461 (2017)
6. Das, H., Jena, A.K., Badajena, J.C., Pradhan, C., Barik, R.K.: Resource allocation in co-operative cloud environments. In: International Conference on Computing, Analytics and Networking, Springer, India (2018)
7. Kar, I., Parida, R.N.R., Das, H.: Energy aware scheduling using genetic algorithm in cloud data centers. In: International Conference on Electrical, Electronics, and Optimization Techniques, IEEE (2016)
8. Bonomi, F., Milito, R., Zhu, J., Addepalli, S.: Fog computing and its role in the internet of things. In: Proceedings of the First Edition of the MCC Workshop on Mobile Cloud Computing, Helsinki, Finland, pp. 13–16 (17 August 2012)
9. Das, H., Barik, R.K., Dubey, H., Roy, D.S.: Cloud Computing for Geospatial Big Data Analytics: Intelligent Edge, Fog and Mist Computing. Springer
10. Barik, R.K., Dubey, H., Mankodiya, K., Sasane, S.A., Misra, C.: GeoFog4Health: a fog-based SDI framework for geospatial health big data analysis. J. Ambient. Intell. Hum.Ized Comput. **10**(2), 551–567 (2019)
11. Barik, R.K., Priyadarshini, R., Dubey, H., Kumar, V., Mankodiya, K.: FogLearn: leveraging fog-based machine learning for smart system big data analytics. Int. J. Fog Comput. (IJFC) **1**(1), 15–34 (2018)
12. Barik, R.K., Tripathi, A., Dubey, H., Lenka, R.K., Pratik, T., Sharma, S., Mankodiya, K., Kumar, V., Das, H.: MistGIS: optimizing geospatial data analysis using mist computing. In: International Conference on Computing, Analytics and Networking, Springer, India (2018)
13. Mishra, B.S.P., Das, H., Dehuri, S., Jagadev, A.K.: Cloud Computing for Optimization: Foundations, Applications, and Challenges, vol. 39. Springer (2018)
14. Pattnaik, P.K., Rautaray, S.S., Das, H., Nayak, J. (eds.): Progress in computing, analytics and networking. In: Proceedings of ICCAN, vol. 710. Springer (2017)
15. Barik, R.K., Kandpal, M., Dubey, H., Kumar, V., Das, H.: Geocloud4GI: cloud SDI model for geographical indications information infrastructure network. In: Cloud Computing for Geospatial Big Data Analytics: Intelligent Edge, Fog and Mist Computing, Springer (2019)
16. Chiang, M., et al.: Clarifying fog computing and networking: 10 questions and answers. IEEE Commun. Mag. **55**(4), 18–20 (2017)
17. Ye, K.J.: Power management of virtualized cloud computing platform. Chin. J. Comput. **35**(6) (2012)
18. Ruan, S.L., Caiwu, L.U.: Global scheduling method for energy consumption optimization for virtualized data center. Comput. Eng. (2013)
19. Belogazov, A., Abawajy, J., Buyya, R.: Energy-aware resource allocation heuristics for efficient management of data centers for Cloud computing. Futur. Gener. Comput. Syst. **28**(5), 755–768 (2012)
20. Cheng, C.L., Yu, P., Zhang, D.Y.: Energy saving resource scheduling algorithm in cloud environment. Syst. Eng. Electron. **35**(11), 2416–2423 (2013)

21. Calheiros, R.N., Ranjan, R., Beloglazov, A., et al.: CloudSim: a toolkit for modeling and simulation of cloud computing environments and evaluation of resource provisioning algorithms. Softw. Pract. Exp. **41**(1), 23–50 (2011)
22. Sun, X., Ansari, N.: EdgeIOT: mobile edge computing for the internet of things. IEEE Commun. Mag. **54**(12), 22–29 (2016)
23. Gai, K., Qiu, M., Zhao, H., et al.: Dynamic energy-aware cloudlet-based mobile cloud computing model for green computing. J. Netw. Comput. Appl. **59**(C), 46–54 (2016)
24. Beloglazov, A., Buyya, R.: Optimal online deterministic algorithms and adaptive heuristics for energy and performance efficient dynamic consolidation of virtual machines in Cloud data centers. Concurr. Comput. Pract. Exp. **24**(13), 1397–1420 (2012)
25. Goswami, V., Patra, S.S., Mund, G.B. (2012, March). Performance analysis of cloud with queue-dependent virtual machines. In: 2012 1st International Conference on Recent Advances in Information Technology (RAIT), pp. 357–362, IEEE

Industrial Automation: Case Study—Vision Based Live Object Monitoring System

S. Shishira, R. Roopalakshmi, Sithu D Sudarsan and Nilabja Ash

Abstract In the current era of *Industry 4.0* automation plays a vital role towards enhanced productivity. In this paper case study of a metal rod manufacturing plant is considered which currently employs proximity sensors for the automation process at the inspection unit. However, field sensors are frequently damaged because of their nature of proximity towards object recognition resulting in production downtime. As a solution vision based automation is presented in this paper by incorporating image analytics in three stages namely Image acquisition and pre-processing, Segmentation, Feature extraction and analysis at real-time. Case study results carried out at live object monitoring demonstrates the capability of the proposed framework in an open environment with changing lighting conditions.

Keywords Industry 4.0 · Computer vision · Image analytics · Object detection

1 Introduction

In the current era of *Industry 4.0* automation plays a vital role. Industrial automation is the process of making production operations with increased efficiency. Specifically, production efficiency is improved by using speed-up techniques which results in intelligent manufacturing solutions in terms of improved product quality, higher

S. Shishira (✉) · S. D. Sudarsan
ABB Corporate Research, Bengaluru, India
e-mail: shishira.s@in.abb.com

S. D. Sudarsan
e-mail: sudarsan.sd@in.abb.com

R. Roopalakshmi
Alvas Institute of Engineering and Technology, Mijar, India
e-mail: drroopalakshmir@aiet.org.in

N. Ash
Business Line, Process Industry, Bengaluru, India
e-mail: nilabja.ash@in.abb.com

H. Das et al. (eds.), *Progress in Computing, Analytics and Networking*,
Advances in Intelligent Systems and Computing 1119,
https://doi.org/10.1007/978-981-15-2414-1_58

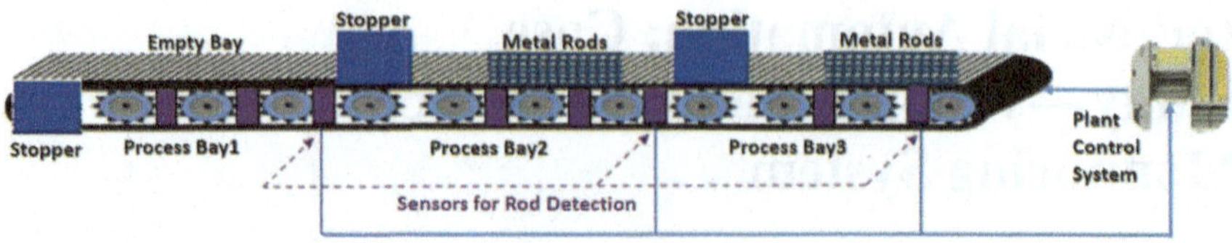

Fig. 1 Inspection unit of a manufacturing plant

productivity and wastage reduction. Further in the current manufacturing field incorporating *Industry 4.0* standard [1, 2] is considered as one of the greatest challenge focused towards automating the whole of industrial manufacturing process with minimal or no human intervention. A recent US study report on industrial automation indicates that productivity in the nation's factories is doubled with automation [3]. Also, the annual productivity growth with automation is estimated to raise globally by 0.8 to 1.4% [4] and market growth is estimated to reach $321.93 billion by 2024 [5].

In recent years innovative solutions in the direction of industrial automation is significantly enhanced by incorporating different technologies including computer vision [6], robotics [7], additive manufacturing [8] and machine learning techniques [9] in the automation process. Further it is found that computer vision is evolving as one of the key component in industrial automation; the applications of computer vision in industrial automation are enormous [10]. For example, global research and markets 2019 report indicates that the market growth for computer vision technology is valued at $9.12 billion in 2018 and is expected to reach $14 billion by 2024 [11].

In this paper, a vision-based real-time monitoring system is proposed to automate the production process at inspection unit of a manufacturing plant. The inspection unit of the plant is as shown in Fig. 1. The inspection unit is divided into set of blocks, having a conveyor belt that continuously rolls from block 1 to block n referred as a process bay; where the manufactured objects (metal rods) enter the inspection unit in batches. Each process bay is associated with a stopper to avoid collision with next incoming batch of rods. The stopper is raised/active when the block is occupied, blocking the next incoming batch. As the objects roll out from the process bay, the bay is empty, and the stopper is deactivated. The stopper operation is automated using proximity sensors. Specifically, proximity sensors are installed on the process bay, which will sense the occupancy of rods on the bay and accordingly operate the stopper signaling the control system of inspection unit. However, the major bottleneck is the sensors are intensely placed close to the conveyor for effective sensing; they tend to damage when improper shaped or defective rods hit the proximity sensors. As a consequence the manufacturing process is delayed until the array of damaged sensors are replaced. To overcome these issues a *vision-based live object monitoring system* with *Existence Detection algorithm* is presented as a replacement to the proximity sensors in this paper. The main contributions of this paper are:

1. Vision based framework as a solution for the automation of batch processing at metal industrial environment.

2. Visual perceptual feature extraction from low resolution images towards decision making for conveyor speed control mechanisms at real time.
3. Real time object monitoring in open environment with varying environmental factors with the proposed Existence Detection (ED) Algorithm.
4. Vision based automation as a replacement to proximity based automation.

The paper is organized as follows: Section 2 details on the review of related work, Sect. 3 describes the vision based framework with ED algorithm. Section 4 details on the experimental setup and results and Section 5 concludes the paper.

2 Related Work

Keisuke et al. [12] proposed a framework which is capable of capturing the tilt, distance and contact of a fragile object using high-speed high-precision proximity sensor equipped in a high-speed robot hand. Satoshi et al. [13] proposed object detection mechanism with tactile and proximity sensors by employing self-capacitance measurement on the curved surface for human collaboration robots; the sensor usage is limited with robot application. Hiraoki et al. [14] proposed a net-structure proximity based sensor system, which covers large sensing area with less responsive time suitable for real time applications; but is limited only to robot hand systems and human machine interface systems. Ren et al. [15] proposed a robotic conveyor tracking system using visual feedback, wherein a robotic arm is used to grasp the objects from a conveyor of a production line using specified tracking strategy. Though it works better for smaller dimensional objects but not suitable for lengthy and multi dimensional objects like metal rods. Recently in 2018 Delail et al. [16] discussed particle filtering-based object tracking techniques which are applicable under changing illuminations. However targets of smaller dimension and individually separated are tracked; but effectively fails in tracking fine grained overlapped objects and objects in batches.

To summarize, in the existing literature object detections are focused with sensors (proximity and vision) being placed very close in the object's vicinity ranging from millimeters to centimeters. Further smaller dimension objects such as bottles, mugs are mostly considered in the automation process, where less focus is given for large dimensional objects like lengthy rods which are approximately 12 to 15 meters in length. Detection of lengthy rods on a conveyor requires the placement of vision sensors at farther distance running in meters. In other perspective object detection and tracking using robot operated environment are popular in the existing literature in which measurements and estimations are fed to the operations of robots for picking and grasping of objects. However, automation of industrial systems in non-robotic environment is also essential. Also in the current literature, object detections are in a closed environment within the operating premises; in a constant and fixed light source with artificial illumination, whereas detections at open environments like open sheds/mills with varying illumination of natural light/skylight/twilight and artificial light; along with changing environmental features like humidity, temperature needs to be explored and is challenging for image processing.

In the present literature, conveyor systems continuously roll at constant speed, with fixed object dimensions, whereas a conveyor with varying speed, and varying object dimensions in batches at real time is less focused and object occupancy influenced conveyor speed controller is a critical task. This paper proposes a framework using vision which can overcome the problems of existing system by means of tracking larger dimension objects at varying environmental conditions and also in batches at real time with varying conveyor speed. More specifically to detect and track the objects in batches on the conveyor at real time an **Existence Detection (ED) Algorithm** is proposed in this paper.

3 Proposed Framework

3.1 Image Acquisition and Pre-processing

The block diagram of the proposed system is as shown in Fig. 2. Real time chromatic video, captured from the monitoring camera in the inspection unit is input to the proposed framework. Image acquisition is a process of extracting individual frames from the video for further processing. The extracted image includes the complete view of the process bay, the objects of interest in the input image appears only on the conveyor belt, hence conveyor belt is the region of interest and other regions from the camera view obtained are to be excluded. The system is being pre-calibrated with respect to the position of the camera and the distance from the conveyor belt. Hence, Region of Interest (ROI) module excludes the unwanted regions from the image; followed with scaling where in the cropped ROI with RGB features with 3 channels is converted to a gray scale image, black and white features with a single channel, effectively the size of the image is drastically reduced in this module.

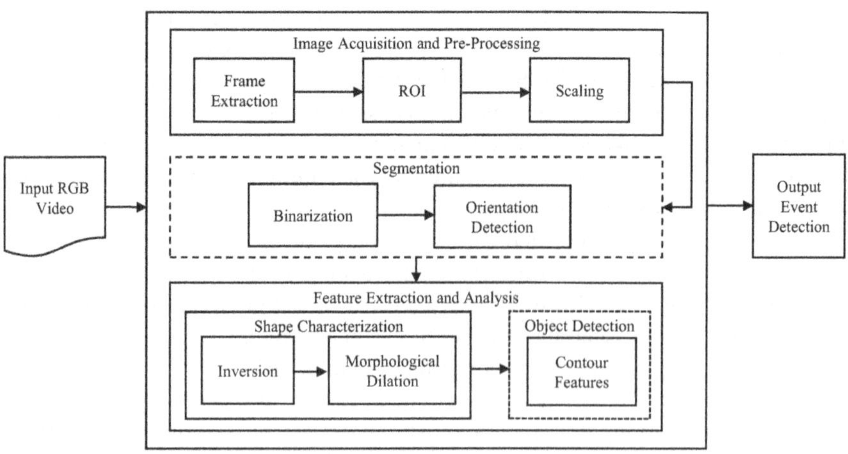

Fig. 2 Block diagram of the proposed vision-based object detection framework detailing step by step image processing operations

3.2 Segmentation

A clear separation of the foreground objects from its background is one of the crucial phase in image processing. Thresholding based segmentation with binarisation [17] is used to extract foreground objects—metal rods from its background—conveyor belt in the image; which is based on observed distribution of pixel values in finding optimal threshold.

The threshold value T estimated by iterating through all possible threshold values which minimizes the weighted within class variance. The algorithm [17] finds an optimal threshold value where in the sum of foreground and background spreads are minimum. More specifically taken the pixel intensity range of t from 0 to 255, calculating the probability for each t followed with the calculations of weighted average $\omega_i(t)$, mean $\mu(t)$, variance $\sigma_i^2(t)$ and weighted within class variance $\sigma_\omega^2(t)$ [17]. t with the minimum class variance among all the class variance is selected as the final minimum threshold T.

Existence Detection (ED) Algorithm The chromatic video sequence W with n frames, where $W = \{I_i \mid i \in [1, n]\}$, I_i is the ith extracted frame of the video sequence W; represented as a two dimensional image function $I(m, n)$ with m rows and n columns, where (x,y) are the discrete co-ordinates of the image I_i with $x = 0, 1, \ldots, m-1$ and $y = 0, 1, \ldots, n-1$, with the image size of $(m * n * 3)$ pixels considering the RGB channels. Let $V_{k,l}$ denote the intensity value of the pixel of image I_i at kth row and lth column where $k <= m-1$ and $l <= n-1$. The image I_i is pre-processed by selecting the region of interest followed with scaling. P_i represents the resultant pre-processed image which is a two dimensional image function with $P(p, q)$ where $p \in [0, m-1]$ and $q \in [0, n-1]$, with image size of $(p * q)$ pixels with a single channel. Further P_i undergoes segmentation with threshold T with minimum class variance resulting in a binary image B_i with each of its pixel having an intensity value of either 0 or 255.

The stopper operation at the process bay relates with bay occupancy. The stopper of a process bay is activated when the complete batch is arrived on the bay and deactivated only when the complete batch has left the bay, accounting to this two control points are identified at the entry and exit of every process bay, one at the entry of the bay CP_1 and other at exit CP_2. The control points are quadrilateral in nature, and could be a trapezoid, parallelogram or a rectangle dependent on the camera view. $CP_i \subset B_i$ and considering a rectangular polygon, with r $rows$ and s $columns$, CP_i is a set of all pixel intensity values in the rectangular region. CP_i is given by the Eq. 1 where (a, b) and (c, d) are the top left and bottom right discrete co-ordinates of the rectangular polygon. $CPThresh_i$ is pre-determined from an occupied bay image I_i, along with R_i determines bay status as explained in the algorithm detailed in Table 1.

$$CP_i = \{x \mid x \in V_{i,j}\ ,\ \text{where}\ a \leq i \leq c,\ b \leq j \leq d\} \quad (1)$$

3.3 *Feature Extraction and Analysis*

Shape Characterization In this module shape of the segmented objects is identified and analyzed. Since the conveyor belt and metal rods are of same color intensity, they are grouped into single intensity cluster in the segmented binary image, and edges of rods into a different cluster; hence the image follows an inversion process for the individual metal rod detection. The inverted image further undergoes a morphological operation of dilation to connect the broken edges of rod in the binary image.

Table 1: Existence Detection Algorithm

Input: B - Binarized segmented image
Output:Bay Status, Stopper status

1: B_i is the ith binarized image of the input video stream W,with $(p \times q)$ rows and columns.
2: Extract the Control Points CP_1, CP_2 from B_i
3: Consider a Mask M_i such that sizeof(M_i) = sizeof(B_i) with all pixel intensity values equal to zero.
4: $MaskedImage_i = fillPoly(M_i, CP_{icoordinates}, 255)$
5: $CP_i = bitwiseAnd(B_i, MaskedImage_i)$
6: Calculate SI_i, Sum of the intensity values of CP_i,given by

$$SI_i = \sum_{i=a}^{c} \sum_{j=b}^{d} V_{i,j}$$

$$ASI_i = r * s * 255$$

$$R_i = SI_i / ASI_i$$

7: If $R_1 > CPThresh1$ and $R_2 > CPThresh2$ then *BayStatus= Empty* and *Stopper= OFF* otherwise *BayStatus= Occupied* and *Stopper= ON*
8: If $R_1 < CPThresh1$ and $R_2 > CPThresh2$ then *BayStatus= OccupiedEntry* and *Stopper= OFF*
9: If $R_1 < CPThresh1$ and $R_2 < CPThresh2$ then *BayStatus= OccupiedCenter* and *Stopper= ON*
10: If $R_1 > CPThresh1$ and $R_2 < CPThresh2$ then *BayStatus= OccupiedExit* and *Stopper= ON*

Object Detection Contours are used to analyze the shape of segmented foreground objects. After smoothing of edges in the dilation module, a topological analysis is performed on the dilated binary image to determine the shape of the objects. Further, contour information of the foreground objects is used to derive insights about the objects like height and width, angle of deviation, perimeter and area. Bounding rectangles are used to highlight the detected metal rods. Contour features are grabbed and analyzed with the pre-defined measures of metal rods under experimentation. With the contour features recognized and analyzed the objects are identified as metal rods.

4 Experiments and Results

The experimental setup of the proposed system is as shown in Fig. 3. The video camera is mounted above the conveyor platform. CCTV based video camera with 30 fps, 1280 × 720 resolution is interfaced to a Windows 10 machine with 16GB RAM, through a network switch. The system continuously reads the camera stream using RTSP protocol. The video captured from the camera is input to the proposed system, which performs various image processing activities along with applied analytics. The output of the proposed system is continuously recorded in a log file with time stamp. The proposed framework is implemented in Python3 using OpenCV. Videos of metal rods with different modularities including dimensions, background color, camera projections and frame rate are captured and verified with the proposed framework. The snapshots of the images S1 to S6 shown in Fig. 4 represent the different modularity of metal rods with varying background color, rod spacing and orientation.

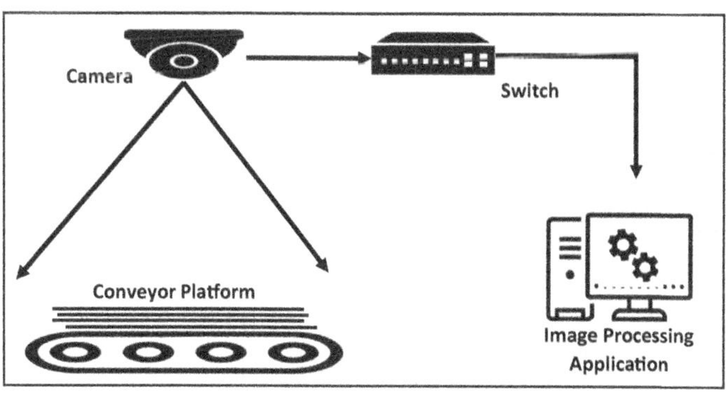

Fig. 3 Experimental setup of the proposed framework

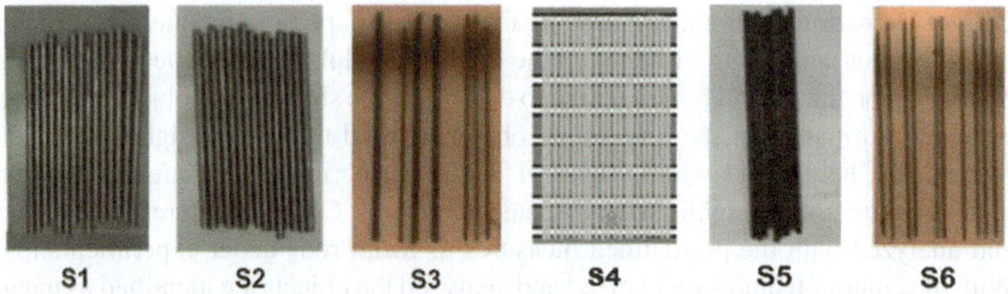

Fig. 4 Image snapshots with different modularities

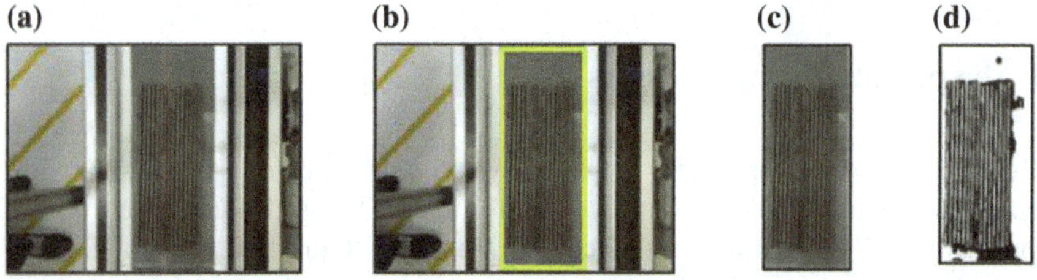

Fig. 5 Image acquisition with pre-processing and segmentation

4.1 Image Acquisition and Pre-processing

The sample video of the metal rods on a conveyor imitating inspection unit is captured with a frame rate of 30 fps with continuous batches of rod's entering and exiting the bay. The images/frames are grabbed from the video stream which are of approximately 350 to 420 kb. The snapshot of the image extracted from the video stream is shown in Fig. 5a. To remove the unwanted region of interest in the image captured, region-based cropping is employed, the selected ROI is highlighted as shown in Fig. 5b. The resultant cropped image with ROI is as shown in the Fig. 5c. Further the image is converted to a gray-scale image with a single channel, wherein the size is being drastically reduced to approximately 150 kb.

4.2 Segmentation

To extract the foreground objects from the background of the image different segmentation algorithms are experimented including OTSU, Adaptive Mean, Adaptive Gaussian Mean, Background Subtraction techniques with Mixture of Gaussian and K-Nearest Neighbors, and Canny Edge Detection technique. Since the process is real time based, the segmentation process should be quick enough in extracting the foreground, and not involve too many filtering or noise removal techniques to arrive at a clear segmented image. Specifically, the extracted ROI of the image is tested

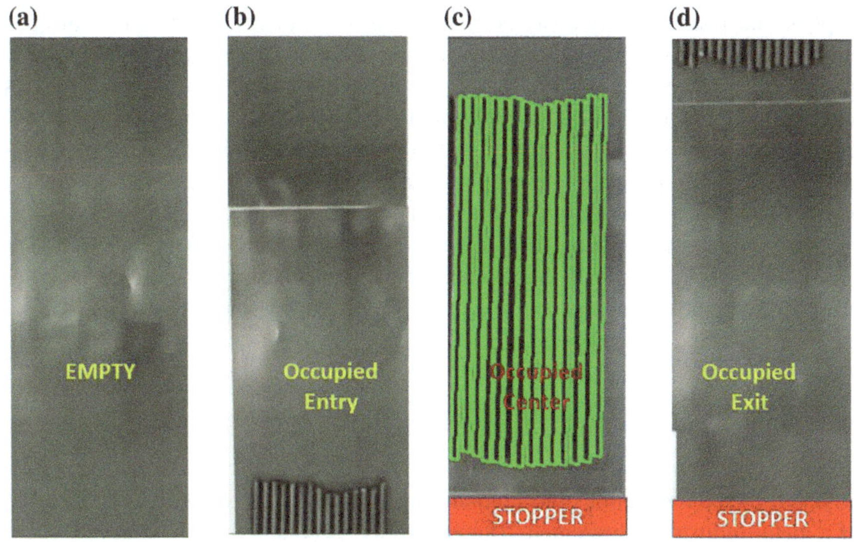

Fig. 6 ED Algorithm: Bay occupancy, orientation of bars with stopper status (highlighted with a red rectangle)

with segmentation techniques mentioned above, and it is found that OTSU clearly segments the foreground objects with the average execution time of 5 μs. Adaptive mean techniques require an additional noise removal filter before to segmentation, and execution time of background subtraction techniques and canny edge detector is extensively high and not acceptable for real time. It is found that OTSU binarization is best in obtaining a clear segmented image effectively with least processing time of 5 μs and effectively extracting the foreground with no additional noise filtering.The resultant segmented binarized image as depicted in Fig. 5d is further processed for the orientation detection with the ED algorithm as described in Sect. 3. Bay and stopper status with rod occupancy is depicted in Fig. 6.

4.3 Feature Extraction and Analysis

With the bay status occupied-center the segmented image is further processed for feature extraction and analysis. The bar dimensions are pre-defined, and with the identified contour, its contour features are analyzed with bar features like height, width and angle of deviation and if the contour features are satisfying the acceptance level of bar features the objects identified are recognized as metal rods. The experimental results shown indicates the efficiency of the proposed framework and ED algorithm in terms of providing better results including segmentation, feature extraction and analysis and object detection. The proposed framework is deployed at plant and tested continuously at real time for consecutive 1 month in an open environment with varying lighting conditions with natural, twilight, artificial light, wherein CCTV cameras where used for the application with a frame rate of 25 fps and an image resolution of 1M pixel. Parallel processing is employed in monitoring four process bay simultaneously with the overall execution time of 10 m secs per frame. The proposed framework successfully detects the metal rod occupancy at each Bay and accordingly logs the status at real time which also helped in the stopper automation resulting in continuous operation. The stopper automation is successfully carried out with the bay occupancy identified with the image analytics proposed which resolved the problems with the proximity sensors.

5 Conclusion

A Vision-Based Real-Time object monitoring framework for the process automation of a metal industry is presented. The experimental results indicate that bay occupancy, orientation of metal rods on the conveyor platform are successfully detected which helped in stopper automation. The proposed system positively replaces proximity sensors with vision sensors and operates favourably at varying environmental conditions, leading to continuous operation at inspection unit resulting in reduced downtime which effectively increases the production rates. In future, the proposed framework can be enhanced for the identification of defective rods which can speed up the inspection process and eliminate manual inspection. Also the framework can be further inspected with other large dimensional objects like wooden logs at conveyor platforms.

References

1. Bassi, L.: Industry 4.0: hope, hype or revolution?. In: 3rd IEEE International Forum on Research and Technologies for Society and Industry, pp. 1–6. IEEE, Modena (2017)
2. Cisneros-Cabrera, S., Sampaio, P., Mehandjiev, N.: A b2b team formation microservice for collaborative manufacturing in industry 4.0. In: IEEE World Congress on Services, pp. 37-38. IEEE, San Francisco, CA (2018)
3. A future that works: Industrial Automation and Productivity, www.eletimes.com
4. A future that works: Automation, Employment and Productivity, www.mckinsey.com/mgi
5. Global Industrial Automation Market Will Reach USD 321.93 Billion By 2024, https://www.globenewswire.com
6. Fernandes, A.O., Moreira, L.F.E., Mata, J.M.: Machine vision applications and development aspects. In: 9th IEEE International Conference on Control and Automation, pp. 1274-1278. IEEE, Santiago (2011)
7. Guo, S., Diao, Q., Xi, F.: Vision based navigation for omni-directional mobile industrial robot. J. Procedia Comput. Sci. **105**, 20–26 (2017); **22**, 562–569 (2016)
8. Bhate, D.: Design for Additive Manufacturing Concepts and Considerations for the Aerospace Industry. SAE (2019)

9. Semeniutaa, O., et al.: Towards increased intelligence and automatic improvement in industrial vision systems. J. Procedia CIRP **67**, 256–261 (2018)
10. Malamas, E.N., Petrakis, E.G.M., Zervakis, M.: A survey on industrial vision systems, applications and tools. J. Image vis. Comput. **21**, 171–188 (2003)
11. The Machine Vision Market is Expected to Reach USD 14.0 Billion by 2024 From USD 9.9 Billion in 2019, https://www.globenewswire.com
12. Koyama, K., et al.: High-Speed High-Precision Proximity Sensor for Detection of Tilt, Distance, and Contact. J. IEEE Robot. Autom. Lett. **3**, 3224–3231 (2018)
13. Tsuji, S., Kohama, T.: Tactile and proximity sensor using self-capacitance measurement on curved surface.In: International Conference on Industrial Technology (ICIT), pp. 934-937. IEEE, Toronto, ON (2017)
14. Hasegawa, H., et al.: Net-structure proximity sensor: high-speed and free-form sensor With analog computing circuit. J. IEEE/ASME Trans. Mechatron. **20**(6), 3232–3241 (2015)
15. Luo, R.C., Liao, C.: Robotic conveyor tracking with dynamic object fetching for industrial automation. In: 15th International Conference on Industrial Informatics, pp. 369–37. IEEE, Emden (2017)
16. Al Delail, B., et al.: Robust likelihood model for illumination invariance in particle filtering. J. IEEE Trans. Circuits Syst. Video Technol. **28**(10), 2836–2848 (2018)
17. Otsu, N.: A threshold selection method from gray-level histograms. J. IEEE Trans. Syst. Man Cybern. **9**(1), 62–66 (1979)

Performance Enhancement of Distribution Network by Optimal Placement of Multiple Capacitors Using FKBC

Rudresh B. Magadum and D. B. Kulkarni

Abstract The outdated existing electrical infrastructure and shortage in power generation causes complex operation and poor performance of electrical distribution networks. This results in low voltage profile, huge T&D losses, load shedding, and poor power quality of supply. To address these problems several researchers are working across the globe. The many techniques, algorithms, devices, online analysis of power system softwares are introduced to overcome these issues by taking quick corrective measures. In this paper, enhancement of voltage values and minimization of total power loss are taken as key objectives. The placement of multiple capacitors is used for compensation of required reactive power using fuzzy knowledge-based controllers. The proposed methodology is tested on IEEE-15 bus and IEEE-33 bus radial networks. The obtained results are satisfactory in improving the overall enhancement of the voltage profile.

Keywords Capacitor placement · Distribution network · Energy loss · Voltage profile · Fuzzy knowledge-based controller · Radial distribution networks

Nomenclature

CPI	Capacitor Placement Index
FKBC	Fuzzy Knowledge based Controller
B	Best
Be	Better
P	Poor
G	Good
VB	Very Best

R. B. Magadum (✉) · D. B. Kulkarni
Electrical and Electronics Engineering Department, KLS Gogte Institute of Technology, Belagavi, Karnataka, India

D. B. Kulkarni
e-mail: dbkulkarni@git.edu

H. Das et al. (eds.), *Progress in Computing, Analytics and Networking*,
Advances in Intelligent Systems and Computing 1119,
https://doi.org/10.1007/978-981-15-2414-1_59

RDS	Radial distribution Network
Ik	Current flowing in *k*th line
Rk	Resistance of the *k*th line
EMTP	Electro Magnetic Transient Programming
BAV	Below average
AV	Average
AAV	Above average
P_{Loss}	Power loss
P_{new}	Power loss after capacitor connection
P_{old}	Base case power loss
p.u.	Per unit system
FIS	Fuzzy Interface System
N	Total number of transmission lines
RES	Renewable Energy Sources
LFA	Load Flow Analysis

1 Introduction

The distribution network plays a vital role in supplying power to the consumers and it operates at lower voltage causing more power losses in the network [1]. The distribution network alone contributes 15% of the total generation as loss, i.e., active power and reactive power losses [2]. These losses can be reduced by compensating an appropriate amount of reactive power at optimal node points with the help of optimal siting of shunt capacitor with less investment [3, 4].

In the past few years, lots of research papers are published on the optimal siting of the capacitor in the RDS. The improvement of voltage profile is achieved by ant colony search for the best placement location of a capacitor with network reconfiguration [5]. The shark smell optimization technique is implemented for enhancement of voltage profile with the improvement of the network efficiency by optimal siting of capacitors in radial distribution systems [6]. The PSO technique has been used for optimal sizing and placement of capacitor in RDS to minimize the total power losses with improvement of the voltage profile of the system. This paper is focused on the optimal placement of multiple capacitors in RDS to improve the voltage profile and minimize the total power losses. The obtained results have been compared to the exhaustive LFA for IEEE fifteen bus radial network. The proposed FKBC algorithm gives acceptable improvements in voltage profile with the least complexity.

The main motivation for the placement of shunt capacitors is to correct the power factor by supplying the required reactive power, which leads to improving the voltage profile with minimizing the energy loss [7, 8]. The important objectives of shunt capacitor placements are to reduce the total power loss by correcting the power factor [9, 10], enhancement of the voltage profile with the improvement of quality of the power supply and also reduction of loading burden on the transmission lines by injecting required reactive power [12–17].

Fuzzy knowledge-based controller is used for optimal siting of shunt capacitors for minimization of energy loss with maximization of the voltage profile. The proposed technique is tested on IEEE-15 bus and IEEE-33 bus systems using MATLAB and MiPower software tools. The obtained result shows significant improvements in the system's performance. The rest of this paper is organized as follows: Sect. 2 gives methodology Sect. 3 provides results and analysis of optimal siting of shunt capacitor and Sect. 4 concludes the paper.

2 Methodology

This work discusses optimal siting shunt capacitor issues in distribution networks. The objectives are,

a. Minimize the total power loss in the network.
b. Improve the voltage profile.

- Enhancement of Voltage profile.

$$\text{Vmin} \leq \text{Voltage} \leq \text{Vmax}$$
$$V_{\text{min}} = \text{Minimum voltage in p.u.}$$
$$V_{\text{max}} = \text{Maximum voltage in p.u.}$$

- Minimization of total power loss

$$P_{\text{Loss}} = \sum_{k=1}^{N} I_K^2 R_k \tag{1}$$

Figure 1 shows the flow chart for the optimal siting of the capacitor. The proposed FKBC algorithm is tested on IEEE-15 bus and IEEE-33 bus test systems and both are tested on MATLAB and MiPower software packages. MiPower software is a strong power system simulation tool, which can be used to carry out various analysis like load flow analysis, transient analysis, short circuit, EMTP, harmonic analysis, etc. The power loss and voltage at each node are computed by LFA using the NR method. Depending on the voltage and power loss values the ranges of fuzzy membership function will be selected for inputs and outputs. Depending on the membership functions fuzzy rules are developed for the analysis. The CPI is computed for all the busses in the test network with the help of appropriate voltage and power loss at each node. The optimal siting of the capacitor is decided by finding the weakest point in the network, i.e., the largest CPI value among all the busses in the network. With the

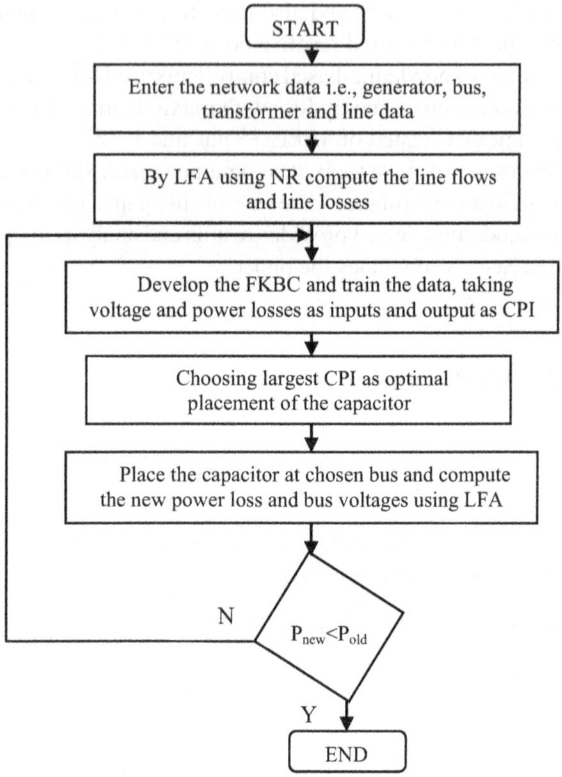

Fig. 1 Flow chart of FKBC algorithm for optimal siting of the capacitor

capacitor, LFA is carried out to know the effectiveness of the capacitor in improving the voltage profile with significant suppression of the power loss.

3 Results and Discussions

Figure 2 shows the single line diagram of IEEE -15 bus distribution network consisting of 14 loads contributing a total load of 1.2264 MW of real power and 1.2510 MVar of reactive power. The entire load is supplied by one generator with a specified voltage of 1.0 p.u., to supply the power from the generator to load lines are connected in the network.

Table 1 shows the percentage loading of each transmission line, power loss, and power flow in each of the branches. In the base case, LFA is carried out with given network data as shown in Table 2. From the obtained LFA results for IEEE 15 bus RDS it can be observed that total loss in the network is 61.5 kW with minimum voltage magnitude in the network 0.9463 p.u. at node 13 as shown in Table 2. The exhaustive LFA is carried out considering each node as a candidate bus for the

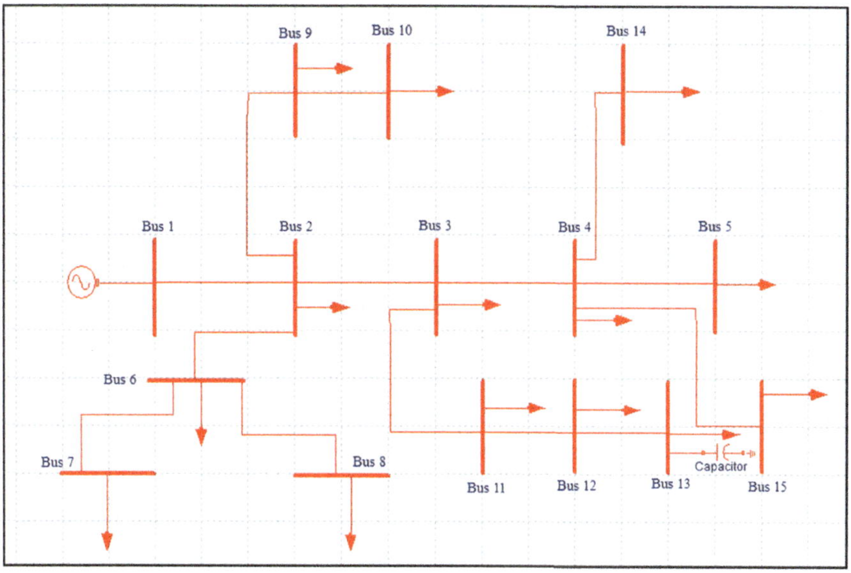

Fig. 2 Single line diagram of IEEE 15 bus system

Table 1 Power loss and power flow in the IEEE-15 bus network

From node	To node	Forward		Power loss		Percentage loading
		kW	kVAr	kW	kVAr	
1	2	1288	1304	37.6	32.5	52.6
2	3	735	748	11.2	11	30.9
3	4	397	405	2.4	2.4	17
4	5	44	45	0.1	0	1.9
2	9	115	117	0.5	0.3	4.8
9	10	44	45	0.1	0	1.9
2	6	356	361	5.7	3.9	15
6	7	140	143	0.4	0.3	6
6	8	70	71	0.1	0.1	3
3	11	257	261	1	1.5	11
11	12	115	117	0.6	0.4	4.9
12	13	44	45	1.3	0	1.9
4	14	70	72	0.2	0.1	3
4	15	140	143	0.4	0.3	6

Table 2 Voltages in p.u. after placement of capacitor at various nodes

Nodes	Base case	Node-2	Node-4	Node-8	Node-12	Node-13	Node-15
1	1.0000	1.0000	1.0000	1.0000	1.0000	1.0000	1.0000
2	0.9730	0.9739	0.9739	0.9739	0.9739	0.9739	0.9739
3	0.9584	0.9593	0.9601	0.9593	0.9601	0.9601	0.9601
4	0.9527	0.9536	0.955	0.9536	0.9544	0.9544	0.955
5	0.9517	0.9526	0.954	0.9526	0.9534	0.9534	0.954
6	0.9600	0.9609	0.9609	0.9609	0.9609	0.9609	0.9609
7	0.9577	0.9586	0.9586	0.9586	0.9586	0.9586	0.9586
8	0.9587	0.9596	0.9596	0.9596	0.9596	0.9596	0.9596
9	0.9697	0.9706	0.9706	0.9715	0.9706	0.9706	0.9706
10	0.9686	0.9695	0.9695	0.9705	0.9695	0.9695	0.9695
11	0.9517	0.9526	0.9534	0.9526	0.9543	0.9543	0.9534
12	0.9476	0.9485	0.9493	0.9485	0.9502	0.9514	0.9493
13	0.9463	0.9472	0.948	0.9472	0.9489	0.9510	0.9480
14	0.9504	0.9513	0.9527	0.9513	0.9521	0.9521	0.9527
15	0.9502	0.9511	0.9525	0.9511	0.9519	0.9519	0.9531
P_{loss} in kW	61.5	59	57.3	58.7	57.1	56.8	57.1

capacitor placement. Out of all results, few samples of exhaustive LFA results are shown in Table 2. After placing the capacitor at node-2 the power loss reduced from 61.5 kW to 59 kW with a minimum voltage magnitude of 0.9472 p.u. at bus 13. Similarly, after integrating the capacitor at bus-4, power loss was reduced to 57.3 kW with 0.948 p.u. voltage values. From Table 2 it can be observed that acceptable step up in the voltage profile with minimization of the total energy loss is achieved. Out of all nodes, bus number 13 can be considered as an optimal location with significant development of efficiency and voltage profile of the network.

The placement of the capacitor at bus 13 gives a total loss of 56.8 kW with minimum voltage 0.9510 p.u. The exhaustive LFA is complex and it is suitable for small networks. For a large network, soft computing techniques are more reliable and less time consuming for termination of the programs. Hence in this program FKBC is used for optimal placement of capacitor (Fig. 3).

The power loss and voltage are taken as inputs and output is taken as CPI. For input and output, five membership functions are chosen for more accuracy as shown in Fig. 4. The range for the voltage and CPI is selected as 0 to 1 p.u. and for the power loss it is chosen as 0–37 kW.

Figure 5 shows FIS for optimal placement of the capacitor where the centroid method is chosen for defuzzification. Table 3 shows the rules developed for capacitor placement. Five membership functions are chosen for both the inputs, due to which $5 * 5 = 25$ rules are deployed in the fuzzy tool. Out of formulated twenty-five rules few of them are as shown below,

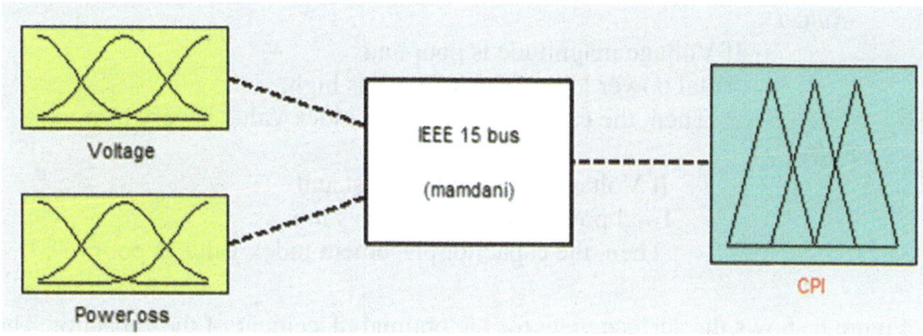

Fig. 3 FIS for capacitor placement [11]

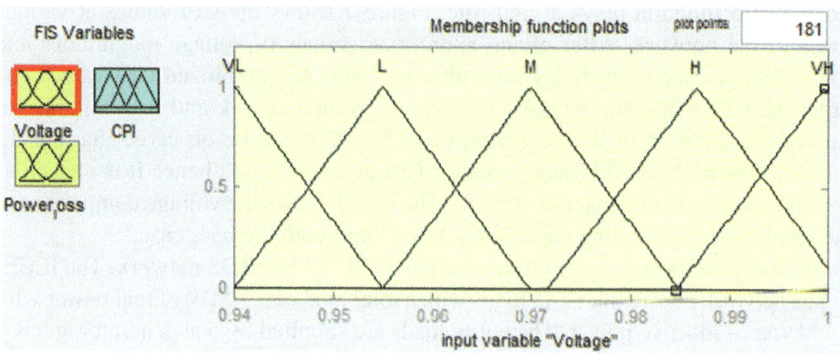

Fig. 4 Membership functions [11]

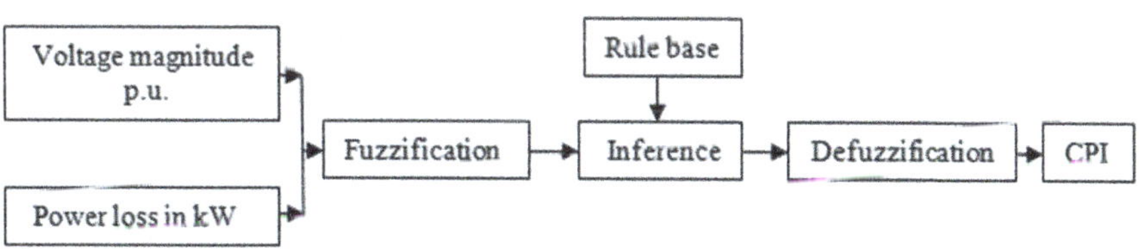

Fig. 5 Fuzzy interface system for optimal placement of capacitor [11]

Table 3 Fuzzy rules table

AND	Power loss in kW					
Voltage magnitude in p.u.		**L**	**BAV**	**AVG**	**AAV**	**H**
	P	G	Be	**B**	**B**	**B**
	G	G	G	Be	**B**	**B**
	Be	P	G	G	Be	Be
	B	P	P	P	G	G
	VB	P	P	P	P	G

Rule-1

If Voltage magnitude is poor and
Total power loss of the network is high,
Then, the capacitor placement index value is very best.

Rule-2

If Voltage magnitude is best and
Total power of the network is low,
Then, the capacitor placement index value is poor.

Figure 6 shows the surface view for the optimal placement of the capacitor. The fuzzy surface viewer purely depends on the ranges chosen for each membership function and rules developed. For more accuracy, the selection of ranges for each membership function plays a vital role. Figure 7 shows the CPI values at various nodes of the network. After giving appropriate values of voltage magnitudes and power loss at each node fuzzy algorithm will give the optimized CPI value. The larger the CPI value, the weakest is the node in the network and it is suitable for the optimal location of the capacitor. From Fig. 7 it can be observed that bus 13 is having the highest CPI value amongst CPI at other busses hence it is chosen as optimal placement of capacitor (Fig. 8). The Fig. 9 shows the voltage comparison at each node after connecting capacitor at bus 13 and with the base case.

Similarly, proposed algorithm is tested on IEEE-33 bus RDS network. The IEEE-33 bus network consisting of 32 loads with a total load of 3.7 MW of real power with 2.3 MVAr of reactive power. The entire loads are supplied by one generator at bus 1 with its specified voltage of 1.0 p.u. With LFA the total losses in the network were 202 kW with the lowest voltage magnitude of 0.9131 p.u. at bus 18.

Figure 9 shows the CPI values at each bus. For the optimal siting of first capacitor in IEEE-33 bus system the appropriate inputs are given. Depending on those values fuzzy algorithm will optimize the output value. The large value of CPI indicates a more suitable location for the performance enhancement of the network. Node 30 is having the highest CPI, i.e., 0.919 hence it is selected for capacitor placement. For placement of the second capacitor, the new values of power loss and voltages are

Fig. 6 Fuzzy surface viewer

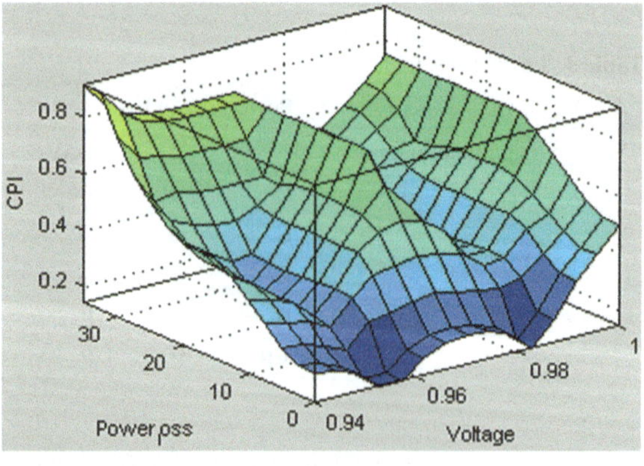

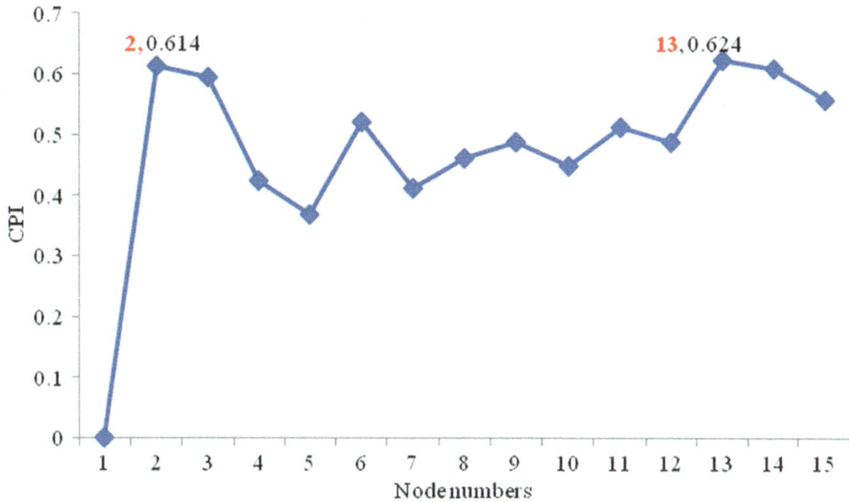

Fig. 7 CPI value at various nodes

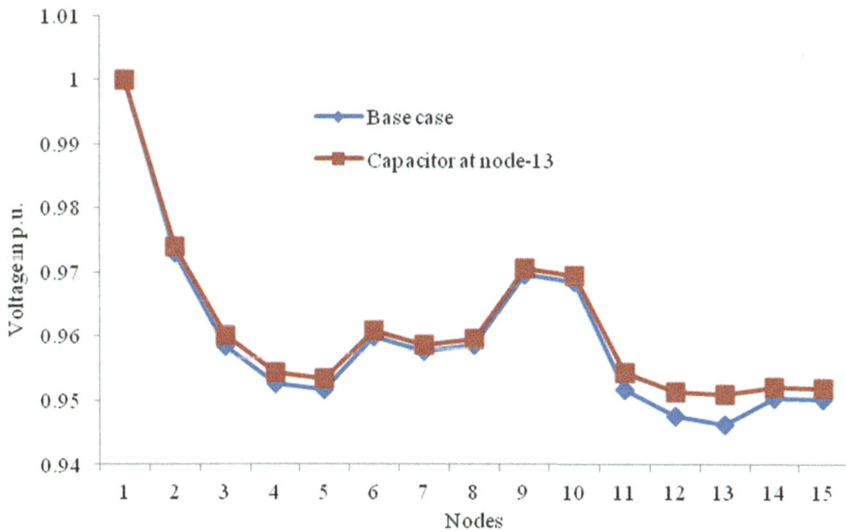

Fig. 8 Voltage comparison with and without capacitor

computed by LFA using the NR method. Figure 10 shows the new CPI values for the second capacitor placement after the integration of the first capacitor. Node 12 is having the highest CPI value, i.e., 0.927 hence it is the most appropriate location for second capacitor placement. Figure 11 shows the voltage comparison under various conditions.

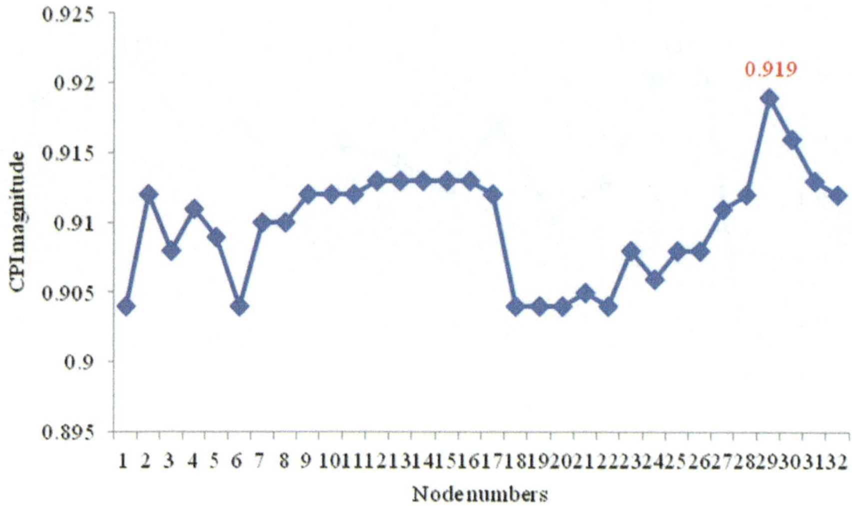

Fig. 9 CPI value for first capacitor placement

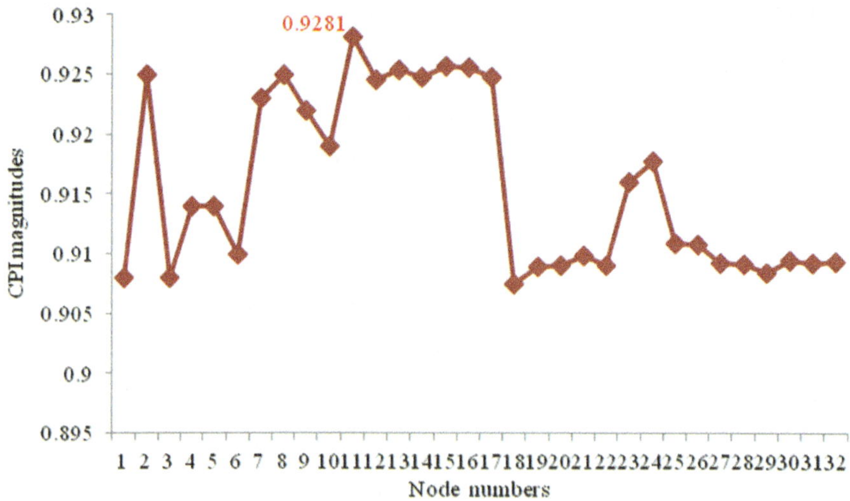

Fig. 10 CPI value for second capacitor placement

The optimal placement capacitor by FKBC gives an impressive improvement in the voltage profiles with significant power loss reduction. The proposed methodology is tested on both IEEE 15 bus and IEEE 33 bus system. The obtained results of both test cases, it is observed that acceptable improvement of the voltage profiles.

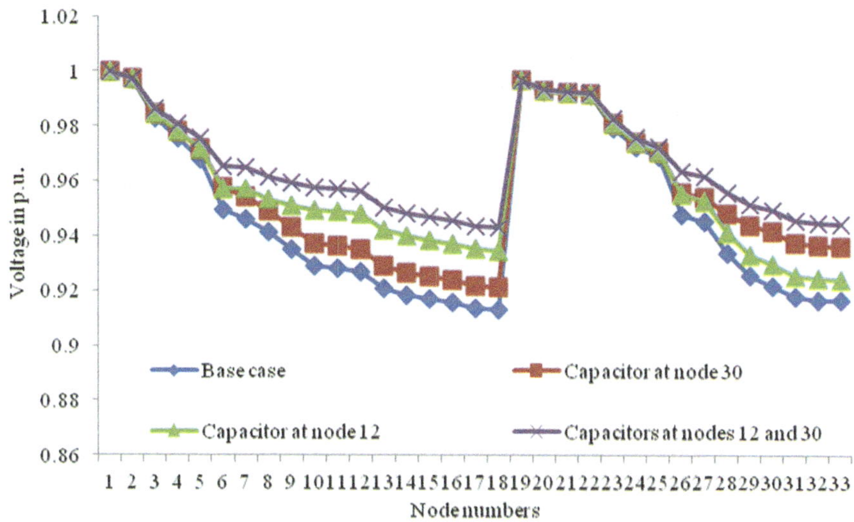

Fig. 11 Voltage profile of IEEE-33 bus system at various test cases

4 Conclusion

The total power loss minimization with a satisfactory step up in the voltage profile is achieved by the optimal placement of multiple capacitors using FKBC technique. The proposed methodology is tested on IEEE-15 bus and IEEE-33 bus systems. The results obtained from FKBC are validated with exhaustive LFA. The FKBC can be used for any number bus system for multiple placements of a capacitor with the least complexity. The obtained results are satisfactory in terms of improvement in the voltage profile and minimization of power loss.

References

1. Magadum, R.B., Kulkarni, D.B.: Optimal placement of capacitor to enhance the efficiency of the distribution network. IJITEE **8**(9), 2877–2881 (2019)
2. Gnanasekaran, N., Chandramohan, S., Sathish Kumar, P., Mohamed Imran, A.: Optimal placement of capacitors in radial distribution system using shark smell optimization algorithm. Ain Shams Eng. J. **7**, 907–916 (2016)
3. Dixit, M., Kundu, P., Jariwala, H.R.: Optimal allocation and sizing of shunt capacitor in distribution system for power loss minimization. In: IEEE Students' Conference on Electrical, Electronics and Computer Science (2016)
4. Xu, Y., Dong, Z.Y., Wongm, K.P., Liu, E., Yue, B.: Optimal capacitor placement to distribution transformers for power loss reduction in radial distribution systems. IEEE Trans. Power Syst. **28**(4), 4072–4079 (2013)
5. MiPower user manual by Power research Development and Consultants, Bangalore
6. Ramalinga, R.M., Murthy, K.V.S., Ramachandra, R.K.: Direct search algorithm for capacitive compensation in radial distribution systems. Int. J. Electr. Power Energy Syst. **42**, 24–30 (2012)

7. Chang, C.-F.: Reconfiguration and capacitor placement for loss reduction of distribution systems by ant colony search algorithm. IEEE Trans. Power Syst. **23**(4) (2008)
8. Farahani, V, Vahidi, B., Askarian Abyaneh, H.: Reconfiguration and capacitor placement simultaneously for energy loss reduction based on an improved reconfiguration method. IEEE Trans. Power Syst. **27**(2) (2012)
9. Srinivasa Rao, R., Ravindra, K., Satish, K., Narasimham, S.V.L.: Power loss minimization in distribution system using network reconfiguration in the presence of distributed generation. IEEE Trans. Power Syst. **28**(1) (2013)
10. Magadum, R.B., Kulkarni, D.B.: Optimal placement and sizing of multiple distributed generators using fuzzy logic. In: 5th IEEE International Conference ICEES-2019, pp. 88–85 (2019)
11. Magadum, R.B., Kulkarni, D.B.: Power loss reduction by optimal location of DG by using fuzzy logic. In: IEEE Conference ICSTM-2015, pp. 338–343 (2015)
12. Sukraj, K., Yuvaraj, T., Hariharan, R.: Optimal allocation of shunt capacitor in the radial distribution network using bio inspired bat algorithm. Int. J. Pure Appl. Math. **119**(12), 15901–15917 (2018)
13. Bagheri Tolabi, H., Ali, M.H., Rizwan, M.: Simultaneous reconfiguration, OPTIMAL Placement of DSTATCOM, and photovoltaic array in a distribution system based on fuzzy-ACO approach. IEEE Trans. Sustain. Energy **6**(1), pp. 210–218 (2015)
14. Magadum, R.B., Kulkarni, D.B.: Minimization of power loss with enhancement of the voltage profile using optimal placement of distribution transformer and distributed generator. In: 8th IEEE Conference ICCSP-2019, pp. 392–395 (2019)
15. Magadum, R.B., Kulkarni, D.B.: Improvement of voltage profile by using line reconfiguration and distribution transformer placement. In: IEEE International Conference ICEET-2016, pp. 330–335 (2016)
16. Das, H., Mishra, S.K., Roy, D.S.: The topological structure of the Odisha power grid: a complex network analysis. IJMCA **1**(1), 012–016 (2013)
17. Das, H., Jena, A.K., Rath, P.K., Muduli, B., Das, S.R.: Grid computing-based performance analysis of power system: a graph theoretic approach. In: Intelligent Computing, Communication and Devices, pp. 259–266. Springer, New Delhi (2015)

Control of Home Appliances and Projector by Smart Application Using SEAP Protocol

E. Udayakumar, T. Kanagaraj, G. Venkata Koti Reddy, K. Srihari and S. Chandragandhi

Abstract In the current scenario, to utilize Android application to control home apparatuses. Android application, which you can use to control electrical machines with directions. Directions are sent by means of Bluetooth to Arduino Leonardo. Along these lines, there is no compelling reason to get up to turn ON or OFF the gadget while viewing a film or doing some work. Despite what might be expected, controlling home machines and projector requires a major exertion in programming. The proposed strategy is making a brilliant application for controlling the home apparatuses and projector utilizing an Arduino Leonardo. Crafted by a keen application is to produce order without anyone else's input each n-times to the gadget. Here, Secure Efficient Authentication Protocol (SEAP) convention is utilized for a productive verifying reason. Arduino bolsters increasingly number of sensor for cutting edge future reason. In this work, an answer is found for filling this opening in the proposed system, definitely just as they existed embedded in Android condition. The effort handles the Bluetooth HC-05 advancement and comprehends a library for sharp contraptions directing. The proposed design is tentatively assessed in a conceivable situation of home computerization and projector control.

Keywords Secure efficient authentication protocol · Bluetooth · PBNFCP · LCD

1 Introduction

The extraordinary failure cost remote frameworks are boundless: from the mechanical condition, where they were, for the most part, abused previously, to our homes, for encompassing helped living and home robotization. The home computerization

E. Udayakumar (✉) · T. Kanagaraj
Department of ECE, KIT-Kalaignarkarunanidhi Institute of Technology, Coimbatore, Tamilnadu, India

G. Venkata Koti Reddy · K. Srihari
Department of CSE, SNS College of Engineering, Coimbatore, Tamilnadu, India

S. Chandragandhi
Department of CSE, JCT College of Engineering and Technology, Coimbatore, Tamilnadu, India

H. Das et al. (eds.), *Progress in Computing, Analytics and Networking*,
Advances in Intelligent Systems and Computing 1119,
https://doi.org/10.1007/978-981-15-2414-1_60

innovation makes human life agreeable and rich. That is the reason today every individual needs a brilliant home. This innovation gives the control of the home gadgets just as a protected and insightful collaboration between workforce inside and outside of the room. The control framework gives the status worried ON or OFF the chose instrument to the house proprietor [1]. This framework is likewise intended to help, offer help, and to satisfy the requirements of an older individual in the home. The framework is planned to control electrical machines in the house and successful association to the guest with moderately ease structure, easy to use interface just as simplicity of establishment.

The motivation behind this task is to increase a superior comprehension on how Bluetooth correspondence works, more inside and out, as far as sending and accepting sequential information between different gadgets. This task will actualize the plan, hardware, programming, graphical UI (G.U.I), and undertaking the executive's aptitudes so as to develop a framework that enables the client to control numerous yields remotely utilizing Bluetooth innovation.

2 Related Work

An insignificant exertion and clear remote-controlled home robotization structure are demonstrated using the Arduino board, Bluetooth module, mobile phone, ultrasonic sensor, and dampness sensor. A PDA application is used in the supported structure which empowers the customers to control up to 18 contraptions including home mechanical social occasions and sensors using Bluetooth improvement. Nowadays, most of the standard home robotization systems are typical for remarkable purposes while the proposed structure is an extensively pleasing home computerization structure [2]. Which can without a lot of a stretch be recognized in the existing home utilizing huge information. The proposed structure has a more prominent number of features than standard home robotization systems, for instance, an ultrasonic sensor is used for water a level area and soil soaked state sensor is used for changed plant water framework structure. This paper in like manner depicts the hardware and programming working of structure, future work, and degree. The proposed model of home robotization structure is done and had a go at gear and it gave the exact and foreseen results utilizing enormous information.

The novel validation convention for MANETs requiring more grounded security. The convention chips away at a two-level system engineering with customer hubs and verification server hubs and supports dynamic participation. We utilize an outer enrollment conceding server (MGS) to furnish more grounded security with dynamic participation. In any case, the outside MGS in our convention [3] is semi-online as opposed to being on the web, i.e., the MGS can't start an association with a system hub, however, any system hub can speak with the MGS at whatever point required. To guarantee efficiency, the convention utilizes symmetric key cryptography to actualize the verification administration [4]. Be that as it may, to accomplish capacity versatility, the convention utilizes a pseudo arbitrary capacity (PRF) to tie the mystery key of a customer to its character utilizing the mystery key of its server. Likewise, the convention has an efficient server repudiation instrument alongside an efficient server re-task system, which makes the convention powerful against the server hub bargain.

Approval show expects a noteworthy activity in the short-expand remote exchanges for the Near Field Communication (NFC) development. Due to the basic thought of remote correspondence frameworks, there are a couple of sorts of security vulnerabilities. Starting late, a pen name NFC show (PBNFCP) has been proposed to withstand the security traps found in the current prohibitive insurance sparing security show (CPPNFC). In any case, this paper further assesses PBNFCP and exhibits that regardless of all that it fails to stay away from the ensured security properties, for instance, emulate ambushes against an adversary [5], who is a harmful selected customer having a generous pseudonym contrasting private key.

In order to crush these security detriments, this paper proposes a SEAP for NFC applications using lifetime-based pseudonyms. The proposed SEAP is duplicated for the formal security affirmation using the comprehensively recognized AVISPA (Automated Validation of Internet Security Protocols and Applications) mechanical assembly by big data. The entertainment results show that SEAP is secure. The exhaustive security and execution examination shows that the proposed SEAP is secure and successful when stood out from the related existing check shows for NFC applications [6]. The design is tentatively done to coordinate cell phone with top-notch sensors and correspondence frameworks, effectively overseen by straightforward application under Android.

3 System Design

Home apparatuses and projectors are controlled in a verified way by utilizing SEAP (Secure proficient Authentication Protocol) convention. MIT (Massachusetts Institute of Technology) application innovator is utilized to make an application for controlling the home apparatuses by utilizing Arduino Leonardo. The work misuses the Bluetooth innovation and understands a library for remote overseeing. Crafted by a savvy application is to produce direction without anyone else each n-times to the gadgets. Here, SEAP convention is utilized for a proficient verifying reason. Arduino underpins progressively a number of sensor for cutting edge future reason. Arduino board is associated with the PC that perceives the Arduino board as a mouse or a console. Along these lines, the slide can be changed and the power supply is constrained by Arduino. All procedures are done through an application in a verified way. LCD show is utilized here to demonstrate the condition of the power supply to the home machines [7]. The power supply unit is a wellspring of steady DC supply voltage. The required DC supply is acquired from the accessible AC supply after amendment, filtration, and guideline.

The Block diagram of the proposed system is shown in Fig. 1. The 230 V as supply is changed over into 12 V air conditioning supply through the transformer. The yield

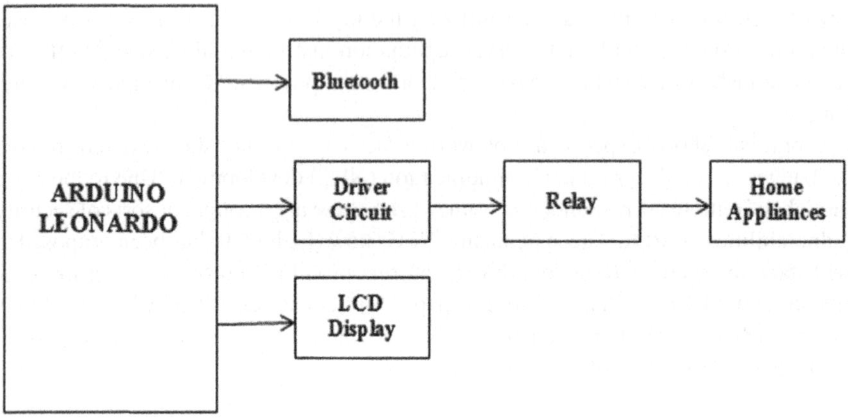

Fig. 1 Block diagram of the proposed system

of the transformer has a similar recurrence as in the info air conditioning power this air conditioner power is changed over into DC power supply. This changed over DC power supply has the swell of substance and for the typical activity of the circuit; the swell substance of the DC supply ought to be as low as would be prudent. Since the swell [8] substance of the power supply will lessen the life of the circuit. The channel is utilized to decrease the swell substance of the DC power supply. The channel is only the enormous worth capacitance.

3.1 Liquid Crystal Display

LCDs have materials that solidify the properties of the two liquids and valuable stones. Rather than having a melting point, they have a temperature kept running inside which the particles are almost as versatile as they would be in a liquid yet are amassed in an orchestrated structure [9] like a diamond. An LCD contains two glass sheets, with the liquid jewel material sand witched in them. The internal surface of the glass plates is secured with clear anodes which portray the character, pictures or guides to be indicated polymeric layers are accessible in the terminals and the liquid jewel, which makes the liquid molecules to keep up a described heading point. One each polarizer is stuck outside the two glass plates. This polarizer would turn the light pillars experiencing them to a positive edge, in a particular bearing. Exactly when the LCD is in the off state, light pillars are rotated by the two polarizers and the liquid valuable stone, to such a degree, that the light bars leave the LCD with no bearing, and thus the LCD appears to be direct.

3.2 SEAP Protocol

A basic action in the short-go remote exchanges for the NFC progression. It is capable when stood out from the related existing confirmation shows for NFC applications. The convention chips away at a two-level system design with customer hubs and confirmation server hubs and supports dynamic enrollment. We utilize outer participation conceding server to give more grounded security dynamic enrollment. In any case, the outside MGS in our convention is semi-online as opposed to being on the web [10], i.e., the MGS can't start an association with a system hub, however, any system hub can speak with the MGS at whatever point required. To guarantee efficiency, the convention utilizes symmetric key cryptography to actualize the confirmation administration. Be that as it may, to accomplish capacity adaptability, the convention utilizes a pseudo arbitrary capacity (PRF) to tie the mystery key of a customer to its character utilizing the mystery key of its server. Moreover, the convention has an efficient server repudiation instrument alongside an efficient server re-task system, which makes the convention powerful against the server hub bargain.

3.3 MIT App Inventor2

A versatile application is a bit of programming produced for portable figuring gadgets, for example, advanced cells, and tablet gadgets. There are numerous devices accessible for creating versatile applications, yet you will utilize Google (presently MIT) Application [10] Inventor. Application Inventor was structured so that those with next to zero programming background could create Mobile Applications for Android-based telephones.

4 Results and Discussion

This section manages the all out equipment and programming depiction based home robotization. It portrays the general working procedure of the framework. The equipment parts are picked with care. So that the total framework can fit for the brilliant application. The segments are picked based on particulars, highlights, and cost.

The underlying phase of the framework is demonstrated where the parts stay inert. In this express, the machines are in the OFF state. OFF state alludes to the apparatuses which are in turned off. Figure 2 delineates the structured application. The high and low flag shows whether the apparatuses are ON/OFF. As indicated by the versatile inclusion the client can switch on in any direction. For example, if the client needs to ON the fan or light he can tap on to the catch present on the application.

This segment manages the different yields which are gotten at different condition. From Fig. 3, it is seen that the client sends the order to control the fundamental

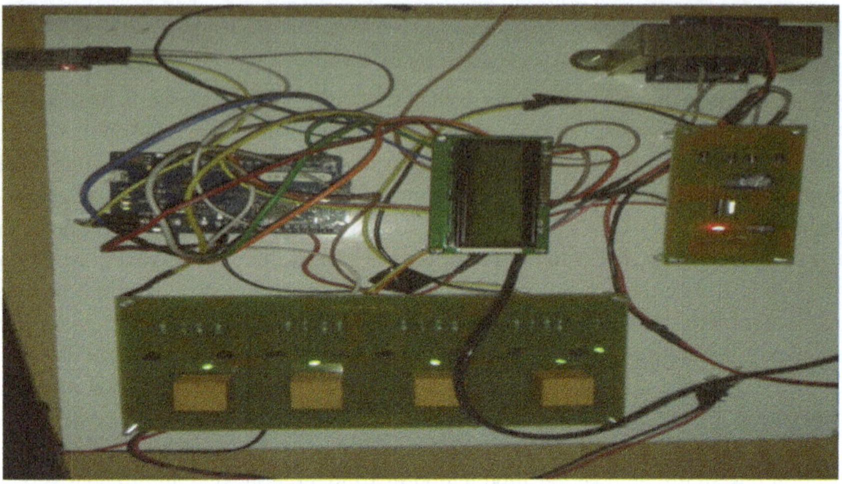

Fig. 2 Overall circuitry of the system

apparatuses at home. Here, the client attempts to ON the light, fan, and projector, he taps the [11] particular ON catch on the screen to execute his attractive yield as appeared. The ON catch sends a sign to Bluetooth to the equipment containing the separate module. Utilizing the Bluetooth HC-05 module and Arduino Leonardo, the sign are passed from the transmitter of the Bluetooth to the collector of the Arduino and the other way around, the projector slides are changed all together consequently.

5 Conclusion

Home apparatuses and projectors are controlled in a verified way by utilizing SEAP convention. MIT application creator is utilized to make an application for controlling the home apparatuses by utilizing Arduino Leonardo. The work misuses the Bluetooth innovation and understands a library for remote overseeing. Crafted by a brilliant application is to create direction independent from anyone else each n-times to the gadgets. Here, SEAP convention is utilized for a proficient verifying reason. Arduino underpins an increasing number of sensors for cutting edge future reasons. Arduino board is associated with the PC that perceives the Arduino board as a mouse or a console. Thus, the slide can be changed and the power supply is constrained by Arduino. All these procedures are done through an application in a verified way. LCD show is utilized here to demonstrate the condition of the power supply to the home apparatuses. The data stored in the cloud can be used to automate the ON/OFF timing of the appliances or machineries in the home or industry by analyzing the data regarding the usage of appliances. The mobile application will ask for permission to switch ON/OFF the device according to the timing. Thus, the devices in the home

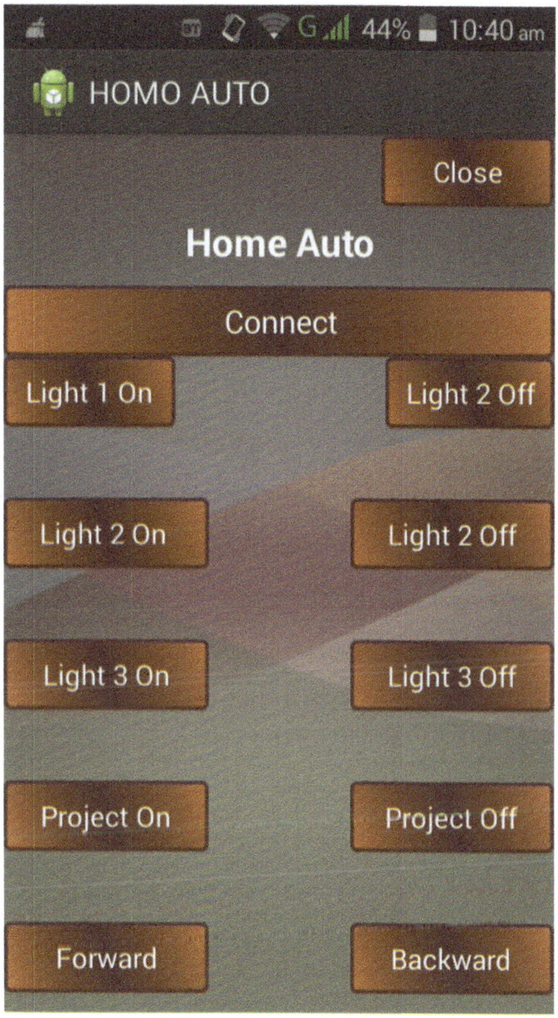

Fig. 3 Designed application

switches ON/OFF easily with the signal touch on your mobile, even the ON/OFF time can be snoozed as the application asked your approval.

References

1. Varaiya, P.P., et al.: Smart operation of the smart grid: risk-limiting dispatch. In: Proceedings of IEEE, pp. 40–57 (2011)
2. Ferrari, P., Flammini, A., et al.: A bluetooth sensor net with web interface. IEEE Trans. Instrum. Meas. 2359–2363 (2005)

3. Sarfaraj, H., et al.: IoT access & analysis of WSN protocols with low power host connectivity. In: 2015 International Conference on Smart Tech and Management for Computing, Communication, Controls, Energy and Materials, pp. 372–375 (2015)
4. Dey, N., Das, H., et al.: Big Data Analytics for Intelligent Healthcare Management. Academic Press (2019)
5. Kanagaraj, T., et al.: Foot pressure measurement by using ATMEGA164 microcontroller. Adv. Nat. Appl. Sci. (AENSI Journals) **10**(13), 224–228 (2016)
6. Vetrivelan, P., et al.: A Neural Network Based Automatic Crop Monitoring Robot for Agriculture. The IoT and the Next Revolutions Automating the World, pp. 203–212. IGI Global (2019)
7. Piyare, R., Tazil, M,: Bluetooth home automation system using cell phone. In: 2011 IEEE 15th International Symposium on Consumer Electronics, pp. 192–195 (2011)
8. Srihari, K., et al.: Automat battery replace of robot. Adv. Nat. Appl. Sci. 33–38 (2015)
9. Pavithra, D., Balakrishnan, R.: IoT monitoring and control system for home automation. In: Global Conference on Communication Technology, pp. 169–173 (2015)
10. Zhang, W.: Study of IoT's app in digital for Agriculture construction. In: 2011 International Conference on Electrical & Control Engineering, pp. 2578–2581 (2011)
11. Mandula, K., et al.: Mobile home automation IoT. In: International Conference on Control Instrumentation Communication and Computational Technologies (2015)
12. Rinaldi, S., Ferrari, P., et al.: Intro to the measurement virtual-eGateway of domestic energy management. In: 2015 IEEE International Instrumentation and Measurement Technology Conference, pp. 1819–1824, Pisa, Italy (2015)
13. Colombo, A., et al.: Flexible indoor localization & track on a wearable platform & sensor data fusion. IEEE Trans. Instrum. Meas. **63**, 864–876 (2014)
14. Oh, S., et al.: A reliable communication strategy for real-time data dissemination in WSN. In: 2012 IEEE 26th International Conference on Advanced Information Networking & Applications, pp. 26–29, 817–823 (2012)
15. Ferrari, P., et al.: On the implement & performance assessment of a wireless HART distributing packet analyzer. IEEE Trans. Instrum. Meas. **59**(5), 1342–1352 (2010)
16. Vetrivelan, P., et al.: Design of smart surveillance security system based on wireless sensor network. Int. J. Res. Stud. Sci. Eng. Technol. **4**(5), 23–26 (2017)
17. Aripin, N., et al.: Voice of home appliance for Android. In: 2014 Electrical Power, Electronics, Communications, Controls and Informatics Seminar, pp. 142–146, Malang (2014)
18. Asadullah, M., Ullah, K.: Smart home automation system using bluetooth technology. In: 2017 International Conference on Innovations in Electrical Engineering and Computational Technologies, pp. 1–6, Karachi (2017)
19. Mandula, K., Parupalli, R., et al.: Mobile automation using IoT. In: 2015 International Conference on Control Instrumentation Communication and Computational Technologies, pp. 340–343 (2015)
20. Waradkar, G., et al.: Automated room light controller with visitor counter. Imp. J. Inter Res. **2**(4) (2016)
21. Pranay, P., et al.: A survey on smart home system IoT. In: International Conference on Computation of Power Energy Inform & Communication, pp. 330–335 (2015)
22. Pattnaik, P.K., Rautaray, S.S., et al.: Progress in Computing, Analytics and Networking: Proceedings of ICCAN 2017, vol. 710. Springer (2018)
23. Pradhan, C., Das, H., Naik, B., Dey, N.: Handbook of Research on Information Security in Biomedical Signal Processing, pp. 1–414. IGI Global, Hershey, PA (2018)

A Study on Securing Non-financial Data Using Blockchain Mechanism

Abhyarthna Sontakke, Shivam Shadangi, Abhavya Verma and Roshni Pradhan

Abstract Blockchain is a sequence of blocks that contains information. It is originally designed intended to timestamp digitally notary documentation. It was developed by a group of researchers and later it was discovered by Nakamoto Satoshi. Block chain is a distributed public ledger which permits to approach and records the data which are difficult to eradicate. It is a technology intended for industries and companies to cooperate with faith as well as lucidity in work in a safe way. The well-known use of blockchains is a cryptographic currency like bitcoin along with some other uses. In Blockchain, each block contains data, hash of the data, and hash of the previous block. Hash is a fingerprint that identifies a block and block of data containing within it. After the block is created the hash is calculated and it is connected to the previous hash like a chain or necklace. Blockchain in a very simple language is like a garland that secures data with high cryptocracy.

Keywords Blockchain · Hash · Bitcoin · Cryptography · Containers · Ledger · Timestamp · Cryptocrurrency

1 Introduction

The blockchain technology introduced by Nakamoto Satoshi in the year 2009 has the underlying security mechanism like bitcoin. He is the founder of bitcoin cryptocurrency. Many industries are planning to shift while some have already implemented

A. Sontakke · S. Shadangi · A. Verma · R. Pradhan (✉)
Kalinga Institute of Industrial Technology, Deemed University, Bhubaneswar, Odisha, India
e-mail: roshni.pradhanfcs@kiit.ac.in

A. Sontakke
e-mail: 1606243@kiit.ac.in

S. Shadangi
e-mail: 1606301@kiit.ac.in

A. Verma
e-mail: 1606242@kiit.ac.in

H. Das et al. (eds.), *Progress in Computing, Analytics and Networking*,
Advances in Intelligent Systems and Computing 1119,
https://doi.org/10.1007/978-981-15-2414-1_61

blockchain technology to enhance the security of the data or the organization. Many researchers have termed this adaptation to secure nonfinancial information as "bitcoin 2.0".

Before the use of blockchain, upholding the level of security while expanding the network was a huge challenge. As the number of devices in the traffic increased it was difficult to maintain the level of security in the existing security mechanism. This increased the expenditure of the industries because they had to maintain an infrastructure for security purpose. As blockchain is growing, data becomes larger in volume, the loading and computing will also getting difficult. It takes a lot of time to synchronize as the data continuously increases in volume. It creates a problem to the client during computation in the system [1]. However, the problem with the existing system remains the same as a large number of systems relied on a handful number of systems. If the attacker is able to identify and attack the system or bypass the security measures implemented, then the entire organization is at risk. This limitation is eliminated with the help of blockchain technology.

Blockchain is a distributed governance that includes distributed consensus request validation and a peer-to-peer model for data storage. In order to get access to any information, the request needs to be validated by a large number of devices in the network. This becomes more and more challenging for the attacker as the network strength increases because the attacker needs to compromise a great number of devices in a small interval of time without being detected. Even if a device gets compromised, only a small part of the data gets compromised which will make no sense to the attacker. The lost data can be easily retrieved by analyzing the remaining portions of the entire data. In the case of a geographically distributed system, the applicable cyber laws can vary. Hence, it could be incorporated as a consensus parameter.

2 Blockchain Technology

2.1 Evolution of Blockchain

Blockchain technology is designed for the purpose of back-end security mechanisms like digital cryptocurrency bitcoin. Blockchain technologies are not only a single-handed technique, but also contains cryptography, mathematics, algorithm and economic model, peer-to-peer network model, and distributed consensus algorithm to solve traditional distributed database synchronization problem. It is an integrated architecture which has multi-field infrastructure construction [2–4]. There have been various versions of cryptocurrencies and blockchain mechanisms. The evolution can be divided into three categories based on their popularity and acceptances are as follows.

Blockchain using Distributed Ledger Technology
It is also known as blockchain version 1.0. Bitcoin foundation is based on the concept of Distributed Ledger Technology (DLT). DLT works by converting the centralized database into a peer-2-peer database. Validation is required to access information stored at the specified node. Validation is done by consensus mechanism which is elaborately discussed in the following subsections.

Integration of Smart Contracts in Blockchain
This version of the blockchain is called Ethereum and is also known as Blockchain Version 2.0. Bitcoin suffered from the problem of scalability of the network. The process of generating a bitcoin involved mining technology. In mining, the users have to solve complex cryptographic puzzles which require an enormous amount of graphic processors. It consumes electricity in a huge amount. Ethereum extended the concept of blockchain by enabling developers to set a path for distributed applications based on blockchain. The operation was based on "Smart Contracts" programs that determine the set of rules to be followed in a blockchain network [5–7].

Permissioned Blockchain Networks
Bitcoin and Ethereum are non-permission networks which creates problem if the system enters in a deadlock condition. In this regard, the peers participate in every consensus mechanism without knowing each other. This leads to high latency as well as stability issues. Further public consensus mechanism provides incorrect result in some few cases. Privacy issues are also associated as all the parameters of the transaction are visible to all nodes participating in the transaction. This lead to the evolution of blockchain version 3.0. Efforts are being made to adapt Blockchain to protect nonfinancial industries and consumers. Hyperledger Fabric is one such example. It is a permissioned blockchain mechanism hosted by Linux Foundation and nurtured by IBM Corporation.

2.2 *Blockchain Consensus Mechanism*

Blockchain has got a wide area of popularity in different industrial applications. However, there also exist some difficulties and challenges that need to be faced. Blockchain enables the distribution of data throughout the nodes of the network. The consensus mechanism is responsible for validation of the access of distributed data which plays an important role in deciding the security of data [8, 9].

Proof-of-Work (PoW)
Proof-of-Work is the oldest consensus technology or mechanism used in bitcoin and Ethereum. In Ethereum, it is used to certain extent. The leader of the data block is decided by the amount of work done by the device. This mechanism helps to counter flooding attacks like Denial of Service (DoS) and Distributed Denial of Service

(D-DoS) attack. In order to obtain complete control over the network, the attacker needs to compromise at least 51 percent of the participants in the network which is difficult to achieve [1, 10]. The popularity of bitcoin resulted in a huge consumption of electricity and so the developers started working on designing of other consensus mechanisms that are more eco-friendly. Due to the low probability of a successful generation, this makes it unpredictable which worker computer in the network will be able to generate the next block [11].

Proof of Stake (PoS)
Proof of Stake is an evolved version of the previous consensus aimed at the financial and nonfinancial data. Proof of Stake doesn't need expensive computing power. With Proof of Stake, the resource that's compared is the amount of bitcoin a miner holds—someone holding 1% of the bitcoin can mine 1% of the "Proof of Stake blocks" [12]. Every participant is assigned a certain weight denoting the importance of the vote in the consensus. The participant with the highest weight has the responsibility of ensuring fair execution of the consensus as well as protecting the network from various attacks.

Proof-of-Elapsed Time (PoET)
Evolution of Proof-of-Work architecture aims to counter the ceaseless use of resources. This is used in Hyperledger Sawtooth. After every transaction, the participant has to wait for a random amount of time to initiate the next transaction. The one who completes the block earlier becomes the leader. The attacker is unable to continuously send messages due to variable waiting period and so this mechanism prevents flooding and efficiently tackles various network attacks.

Practical Byzantine Fault Tolerant (PBFT)
Blockchain implementation is generally on a large-scale network. Blockchain aims to protect the whole system and not individual systems. There is a high possibility of the system in the network being compromised and showing faulty behavior. Such behavior is called Byzantine fault. The consensus mechanism that identifies anomaly behavior and function normally is termed as Byzantine fault tolerance mechanism.

3 Work Objective

The objective is to design a network which can be used to secure the nonfinancial data. The designed network aims to solve the following concerns:

1. The network should adaptable.
2. Despite heavy computation, the network should be able to respond in real time.
3. The network should work in a heterogeneous environment.
4. The network must be scalable.
5. It should handle the tasks uniformly as the devices with different capacities are required to work together.

6. The network should have a dynamic storage mechanism. There should be sufficient space for new data. The devices with different storage capacities should operate in a way to avoid deadlock state.

4 Tools and Technologies Used

4.1 Platforms

Multiple programming languages and tools in conjunction are used to show the progress in technology. The system consists of the inter-communicating component, i.e., hardware and software combination [13]. The platforms used in the proposed model for implementation purpose is as follows.

Ubuntu 16.04 LTS
Ubuntu 16.04 which is also known as Xenial Xerus. It is a long-term release version of the Ubuntu operating system. It is based on Linux and developed by Canonical Ltd. It supports a wide range of applications, compilers, tools, binaries, and systems because of being open-sourced. It serves as a basic platform for the proposed model.

Hyperledger Fabric
Hyperledger serves as a platform for implementing blockchain mechanism in the proposed model. Some of the variants of Hyperledger are Hyperledger Burrow, Hyperledger Iroha, Hyperledger Sawtooth and Hyperledger Indy. Hyper ledger is an authenticated blockchain infrastructure that provides a commutable building block along with a description of part plays in between nodes of building block, the implementation of Smart Contracts, also form consensus and the subscription facility.

4.2 Software Components

This is a cross-platform implementation. Hence our model uses a number of software components are as follows.

GIT
Linus Torvalds constructed GIT in the year 2005 for the development of the Linux Kernel along with other kernel innovators. The goals consist of data probity, speed, and assist for distributed network and nonlinear workflow. It is a directory on each computer, which is a complete repository along with the full version. It also provides tracking abilities and complete history. It is an open-source software under GNU.

Go Programming Language
Google Engineers Griesemer Robert, Pike Rob, and Thompson Ken designed Go Programming Language. Go is non-dynamically typed, compiled, and syntactically common to C. Non-dynamic typing, run time efficient, readability, usability, high-performance networking, and multiprocessing are some of the features in Go programming language.

Docker
Docker is developed by Docker, Inc. in the year 2013. It is a computer's set of instructions used for performing operating system-level virtualization. The components of the Docker are as follows:

1. Docker Daemon
2. Docker Registry
3. Docker Objects
    (i) Docker Container
    (ii) Docker Image
    (iii) Docker Service

Most important contrivances of docker are the following:

I. Docker compose
II. Docker Swarm.

5 Establishing of Connection

Algorithm 1: Network Creation and Configuration Algorithm

1. Create a workspace.
2. Copy bin folder into fixtures.
3. Make a file named crytoconfig.yaml.
4. Create Organization Domain.
5. Create another file configtx.yaml.
6. Create artifacts folder.
7. Create Environment variable.
8. Create Genesis block.
9. Create Channel.
10. Create Anchor peer.
11. Create a file docker-compose.yaml.
12. Run the network.
13. Close the network.

Algorithm 2: Initialization Algorithm of Fabric SDK

1. Access the workspace folder.
2. Create a file config.yaml.
3. Create a blockchain folder.
4. Create a new file setup.go.
5. Create Main file.
6. Satisfy dependencies.
7. Create make file.
8. Execute the network.

6 Securing of Nonfinancial Data

Blockchain enables to have full control over the identity of data by storing and processing the digital identity of data [14, 15]. Hence it provides a higher level of certitude to everyone about the identity. It consumes less time and makes security check operations much simpler [16]. In this paper, work is done to supervise the identity and personal data of each and every individual. It improves privacy and safety across the digital age at the same time. It will also help the government to digitize the land registry along with maintaining authenticity. In the energy sector, it will solve problems with the use of blockchain-based smart contracts coupled with smart meters. This will enhance the prediction result in energy consumption patterns. It can also add one more layer of certitude. It eliminates the administration charge for energy transitions and utility charge for purchasers. There are many use case of blockchain technologies like protection of intellectual property, traceability in supply chain, identity certification, insurance, international payments, Internet of Things, patient's privacy in medical treatment or prediction market [17, 18], etc.. Blockchain will prevent the attackers from intruding into the system and changing the nonfinancial information since it requires large amount of work to send multiple requests and compromise at least 51% of the system in a very small span of time.

7 Conclusion

Blockchain will provide transparency to the entire flow process thus creating a trustworthy connection allowing everyone to participate by creating a favorable environment. Blockchain will also reduce the expenditure of industries in preventing denial of service and distributed denial of service as the companies need not establish security workplaces in different geographical areas. With the use of blockchain, the companies will be able to secure their financial data as well as the nonfinancial data of the employees thus help in maintaining proper management of the whole organization. It helps in real-time stock monitoring and identification of exact location and quantity of stocks thus providing quality monitoring.

References

1. Karame, G.: On the security and scalability of bitcoins blockchain. In: Proceedings of the 2016 ACM SIGSAC Conference on Computer and Communications Security—CCS16 (2016). https://doi.org/10.1145/2976749.2976756
2. Garay, J., Kiayias, A., Leonardos, N.: The bitcoin backbone protocol: analysis and applications. In: Advances in Cryptology—EUROCRYPT 2015 Lecture Notes in Computer Science, pp. 281–310 (2015). https://doi.org/10.1007/978-3-662-46803-6_10
3. Gervais, A., Karame, G.O., Capkun, V., Capkun, S.: Is bitcoin a decentralized currency? IEEE Secur. Priv. **12**(3), 54–60 (2014). https://doi.org/10.1109/msp.2014.49
4. Nakamoto, S.: Bitcoin: a peer-to-peer electronic cash system (n.d.)
5. Chang Lin, I., Chun Liao, T.: A survey of blockchain security issues and challenges (n.d.)
6. Kosba, A., Miller, A., Shi, E., Wen, Z., Papamanthou, C.: Hawk: the blockchain model of cryptography and privacy-preserving smart contracts. In: 2016 IEEE Symposium on Security and Privacy (SP) (2016). https://doi.org/10.1109/sp.2016.55
7. Watanabe, H., Fujimura, S., Nakadaira, A., Miyazaki, Y., Akutsu, A., Kishigami, J.: Blockchain contract: securing a blockchain applied to smart contracts. In: 2016 IEEE International Conference on Consumer Electronics (ICCE) (2016). https://doi.org/10.1109/icce.2016.7430693
8. Bonneau, J., Miller, A., Clark, J., Narayanan, A., Kroll, J.A., Felten, E.W.: SoK: research perspectives and challenges for bitcoin and cryptocurrencies. In: 2015 IEEE Symposium on Security and Privacy (2015). https://doi.org/10.1109/sp.2015.14
9. Heilman, E., Kendler, A., Zohar, A., Goldberg, S.: Eclipse attacks on bitcoin's peer-to-peer network (n.d.)
10. Eyal, I., Sirer, E.G.: Majority is not enough: bitcoin mining is vulnerable. In: Financial Cryptography and Data Security Lecture Notes in Computer Science, pp. 436–454 (2014). https://doi.org/10.1007/978-3-662-45472-5_28
11. Gervais, A., Karame, G.O., Wüst, K., Glykantzis, V., Ritzdorf, H., Capkun, S.: On the security and performance of proof of work blockchains. In: Proceedings of the 2016 ACM SIGSAC Conference on Computer and Communications Security—CCS16 (2016). https://doi.org/10.1145/2976749.2978341
12. King, S., Nadal, S.: Peer-to-peer crypto-currency with proof-of-stake (n.d.)
13. Solapure, S.S., Kenchannavar, H.: Internet of things: a survey related to various recent architectures and platforms available. In: 2016 International Conference on Advances in Computing, Communications and Informatics (ICACCI) (2016). https://doi.org/10.1109/icacci.2016.7732395
14. He, Q., Guan, N., Lv, M., Yi, W.: On the consensus mechanisms of blockchain/DLT for internet of things. In: 2018 IEEE 13th International Symposium on Industrial Embedded Systems (SIES) (2018). https://doi.org/10.1109/sies.2018.8442076
15. Zyskind, G., Nathan, O., Pentland, Sandy, A.: Decentralizing privacy: using blockchain to protect personal data. In: 2015 IEEE Security and Privacy Workshops (2015). https://doi.org/10.1109/spw.2015.27
16. Courtois, N.T., Bahack, L.: On subversive miner strategies and block withholding attack in bitcoin digital currency (n.d.)
17. Luu, L., Narayanan, V., Zheng, C., Baweja, K., Gilbert, S., Saxena, P.: A secure sharding protocol for open blockchains. In: Proceedings of the 2016 ACM SIGSAC Conference on Computer and Communications Security—CCS16 (2016). https://doi.org/10.1145/2976749.2978389

18. Tsai, W.-T., Blower, R., Zhu, Y., Yu, L.: A system view of financial blockchains. In: 2016 IEEE Symposium on Service-Oriented System Engineering (SOSE) (2016). https://doi.org/10.1109/sose.2016.66
19. Lin, J., Yu, W., Zhang, N., Yang, X., Zhang, H., Zhao, W.: A survey on internet of things: architecture, enabling technologies, security and privacy, and applications (n.d.)

Appraisal of Breast Ultrasound Image Using Shannon's Thresholding and Level-Set Segmentation

R. Ifan Roy Thanaraj, B. Anand, J. Allen Rahul and V. Rajinikanth

Abstract As per the statement of the World Health Organization (WHO), breast malignancy is one of the major impacting cancers among women. The availability of the modern disease diagnostic systems and treatment procedure will assist to improve the survival rate. Even though considerable modalities are available to record the breast abnormality, ultrasound imaging technique is frequently used in clinics to record the abnormalities. This study aims to propose a computer-assisted procedure to examine the Breast Ultrasound Image (BUI). The proposed work implements an integration of Shannon's Entropy Thresholding (SET) to improve the visibility of the image and Level-Set Segmentation (LSS) to extort the abnormal division. The proposed scheme is a semiautomated approach, which aims to mine the suspicious section from the BUI. The extracted suspicious segment is then compared against a ground truth and the essential performance measures are computed to justify the performance of LSS. The overall performance of LSS is then compared and validated with other methods, such as active-contour (AC) and Chan–Vese (CV) and the results of this study confirmed that performance measures attained with LSS, AC, and CV are roughly similar.

Keywords Breast malignancy · Ultrasound image · Thresholding · Segmentation · Assessment

1 Introduction

Cancer is one of the deadliest diseases with a high mortality rate as per the recent report of the World Health Organization (WHO) [1]. As per the WHO, in 2018 alone, 9.6 million estimated deaths occurred globally due to cancer. Breast and lung cancer are the leading causes of these deaths (2.09 million). Recently, substantial awareness

---

R. Ifan Roy Thanaraj · B. Anand · J. Allen Rahul · V. Rajinikanth (✉)
Department of Electronics and Instrumentation, St. Joseph's College of Engineering, Chennai 600119, Tamilnadu, India
e-mail: v.rajinikanth@ieee.org

H. Das et al. (eds.), *Progress in Computing, Analytics and Networking*,
Advances in Intelligent Systems and Computing 1119,
https://doi.org/10.1007/978-981-15-2414-1_62

programs are implemented to support; (i) Cancer prevention, (ii) Early diagnosis, and (iii) Detection and Treatment process.

Breast Cancer (BC) is one of the universal cancers widely affect women and premature finding can help to control the mortality rate. The research work on BC confirms that early diagnosis may help to provide a possible treatment process to cure the disease with various procedures ranging from medication to surgery [2]. The detection of the BC mainly depends on the imaging techniques and after the detection; it can be confirmed by an experienced doctor with or without the help of biopsy. Normally, the examination of BC is carried with biomedical-imaging procedures implemented using the modalities, such as mammogram [3], thermal imaging [4, 5], and ultrasound [6–9]. Mammogram is one of the oldest modalities broadly used to screen the breast, to detect the abnormality. Thermal imaging and ultrasound are recent imaging methodologies, usually used to screen the breast part due to its noninvasive nature and cost-effectiveness. The recent works confirm the use of the Breast Ultrasound Imaging (BUI) technique to examine the abnormalities in breast [10]. In this approach, various regions of the breast are recorded with the help of Ultrasound Scanning Device (USD) and then these imageries are studied manually or with the assist of computer software. The manual evaluation procedure is time consuming and also to be carried with the help of an experienced doctor. The alternate approach involves a computerized examination technique, in which the assessment of BUI is done with a computer-based approach, which extracts the irregular segment from the picture for further assessment. The success of the computer-based assessment depends mainly on the technique implemented to mine the irregular slice from the considered BUI.

In recent years, hybrid image examination methods are extensively adopted to asses a range of image modalities, like ultrasound [6–9], thermal images [11], and magnetic resonance images [12–15]. In this method, the integration of thresholding and segmentation techniques is applied to extort the interested image segments from the trial pictures. Further, the hybrid imaging technique helped to attain better results on a class of gray and RGB scaled pictures as confirmed in the literature [16]. Hence, to evaluate the BUI, the combination of thresholding and segmentation is implemented to extract the suspicious section. Thresholding is implemented by maximizing Shannon's Entropy [17, 18] value and the extraction is implemented with Level-Set Segmentation (LSS) [19, 20]. The earlier studies confirmed that these techniques offered a superior effect in contrast to further techniques available in the literature. Here, a three-level threshold is applied based on the Spider Monkey Algorithm (SMA), which helps to group the BUI pixels into three groups, like background, breast tissue, and suspicious breast region. The enhanced suspicious section is then mined with the LSS and its pixel group is then compared against ground truth (GT) existing in the considered image catalog. The comparison helps to compute the essential quality parameters; Jaccard, Dice, accuracy, precision, sensitivity, and specity. Further, for every BUI, a confusion matrix is also developed with the pixel-level information like the True Positive (TP), True Negative (TN), False Positive (FP), and False Negative (FN). This matrix also authenticates that, implemented LSS works fine on the considered image catalog. Finally, the performance of LSS is validated with the

Active-Contour and Chan–Vese mining procedures and the outcomes confirm that the performance measures attained in the considered mining methods are approximately similar. This confirms that the implemented hybrid technique works well on the considered BUI.

Other sections of this work are discussed as follows. Sect. 2 gives the related existing approaches, database information, and methodology is discussed in Sect. 3, Sect. 4 presents the results and discussion and the conclusion is presented in Sect. 5.

2 Former Study

Breast abnormality was widely studied by the researchers, due to its consequence. The work of Raja et al. [11] proposes a hybrid image evaluation tool for breast thermal images (BTI) based on Kapur's entropy and Markov-Random-Field-based segmentation and achieved matter result [11]. Rajinikanth et al. [5] employed Fuzzy-Entropy thresholding and Active-Contour based segmentation of BTI and attained better outcome to examine the ductal carcinoma in situ [5]. The recent research by Fernandes et al. [4] proposed a detailed evaluation procedure for the BTI [4]. All these methods implement a hybrid technique to examine the BTI. The recent work of Nair et al. [3] employs an image examination method to assess the breast abnormality using the mammogram [2]. Further, a considerable number of evaluations of the BUI are also discussed by Chucherd and Makhanov [10] and Rodtook and Makhanov (2013) [9, 10]. Further, segmentation of the infected section from the BUI is also discussed by Kirimasthong et al. (2017), Rodtook et al. (2018) and Keatmanee et al. (2019) [6–8]. All these techniques implement a possible segmentation technique to extract and evaluate the suspicious breast section.

3 Methodology

The method executed is shown in Fig. 1. In this work, the BUI dataset obtained from [2] is considered for the assessment. This dataset consists of clinical-grade ultrasound image along with its Ground Truth (GT). Later, three-level thresholding is implemented with SMA and Shannon's entropy, which will improve the visibility of the suspicious section. Finally, the enhanced section is mined with an LSS approach.

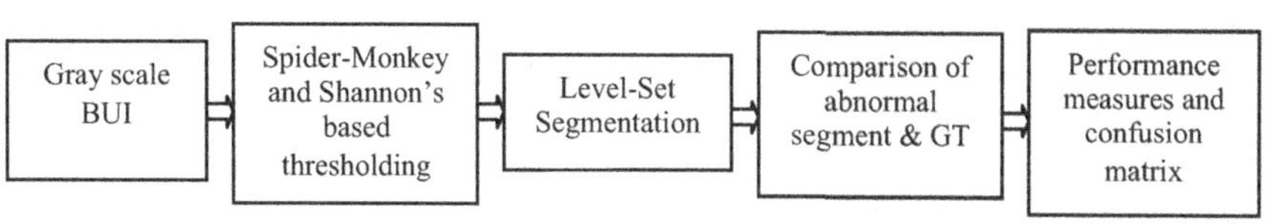

Fig. 1 Various assessment phases in this system

Finally, the extracted section is compared with GT and the essential performance values are calculated to authenticate the presentation of this system.

The remaining part of this subdivision discusses the summary of the scheme proposed in this work.

3.1 Database

The essential test images are obtained from [2]. This dataset is a clinical-grade BUI and already considered by the searchers to test the performance of their developed computer-based evaluation tool. This dataset consists of the benign and the malignant class BUI with its related GT offered by an expert member. The binary GT from the existing GT is extracted using ITK-Snap tool [21, 22], since the binary GT only helps to compute the essential performance measures based on its pixel value.

3.2 Thresholding

Thresholding is the proven picture preprocessing practice, extensively considered to improve trial pictures based on the preferred threshold level. Here, a three-level threshold is chosen to separate the BUI into background, normal section and the tumor. In this work, the maximization of Shannon's entropy based on SMA is implemented and the enhanced image is then considered to extract the tumor section.

The essential details of SMA can be found in recent articles [23–25]. In this work, the SMA subjectively varies with the thresholds, until the entropy of the picture is maximized. The details of Shannon's Entropy can also be found in [26].

3.3 Segmentation

Segmentation is a widely considered image handling system used to extort the required division from the trial picture. This approach is considered to mine the anomalous sector from the BUI (benign/malignant). The proposed work implements Level-Set-Segmentation (LSS) [27], a semiautomated approach works based on the bounding box practice. In this method, initially, the operated initiates a bounding box around the section to be extracted and the box will converge toward the border of the abnormal segment as per the increase in the iteration. After identifying the required section, this bow will extract this section and provides a binary picture (background with a value '0' and tumor section with a value '1'). The segmentation performance of LSS is then compared with the related methods, such as Active-Contour (AC) [19] and Chan–Vese (CV) [12] existing in the literature.

3.4 Computation of Performance Measures

Comparative evaluation of mined piece and GT is carried out in pixel level to appraise the advantage of the proposed system. The essential actions, like Jaccard, Dice, accuracy, precision, sensitivity, and specificity are calculated from this assessment and based on these standards; the eminence of the system is established.

The mathematical model of these measures is presented below [23, 24];

$$\text{Jaccard} = GT \cap SEG/GT \cup SEG \tag{1}$$

$$\text{Dice} = 2(GT \cap SEG)/|GT| \cup |SEG| \tag{2}$$

$$\text{Accuracy} = \frac{T_{+ve} + T_{-ve}}{T_{+ve} + T_{-ve} + F_{+ve} + F_{-ve}} \tag{3}$$

$$Precision = \frac{T_{+ve}}{T_{+ve} + F_{+ve}} \tag{4}$$

$$\text{Sensitivity} = \frac{T_{+ve}}{T_{+ve} + F_{-ve}} \tag{5}$$

$$\text{Specificity} = \frac{T_{-ve}}{T_{-ve} + F_{+ve}} \tag{6}$$

where GT = ground truth, SEG = segmented image, T_{+ve}, T_{-ve}, $F_{+ve,}$ and F_{-ve} denotes true positive, true negative, false positive, and false negative correspondingly.

4 Result and Discussion

The proposed work is experimentally realized using the MATLAB software. During this study, 20 numbers of benign class and 20 numbers of malignant class BUI are considered for the examination and the average performance measures are considered during the validation of the proposed LSS with other related segmentation techniques.

Figure 2 presents the sample test images (malignant/benign) attained from [2]. Figure 2a depicts the grayscale version of BUI, Fig. 2b, c shows the actual and the mined binary GTs correspondingly. Initially, the proposed threshold is implemented on the test BUI and then the tumor section is extracted using the LSS. The LSS is capable of extracting the tumor accurately and presents a binary form of the image as the outcome.

Figure 3 presents the results attained for a malignant class BUI, in which Fig. 3a shows the thresholded picture, Fig. 3b, c present the trace made by the LSS and the segmentation outcome, respectively. Later, the comparison of the binary GT with the extracted tumor is performed and the essential performance measures are then

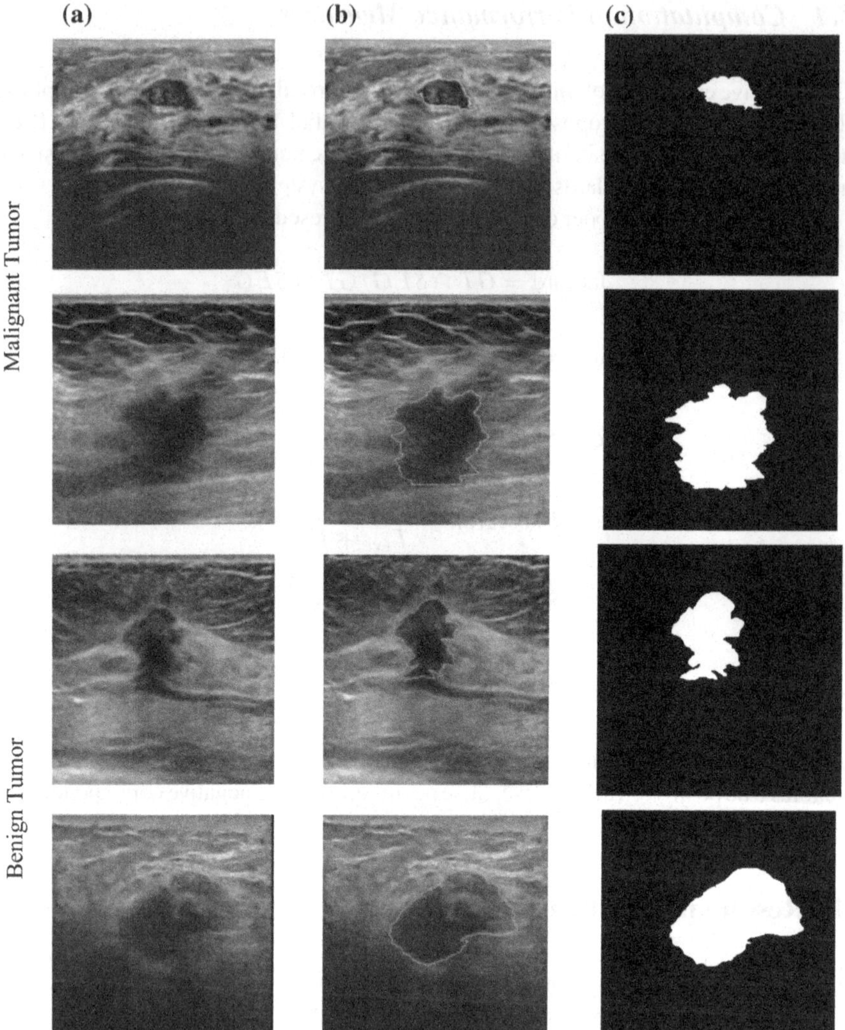

Fig. 2 Sample BUI considered for the examination. **a** Test image. **b** Existing GT. **c** Binary GT extracted using ITK-Snap tool

computed during this assessment. The attained value of performance measures of the sample test BUI discussed in Fig. 3 is considered for the demonstration and these values are considered for the confusion matrix shown in Fig. 4. From this outline, it can be confirmed that the proposed BUI evaluation system works efficiently and offers superior values of the performance measure. The alike practice is then applied for all additional images (benign and malignant) existing in the considered database and the results are summarized to compute the average performance values. A similar

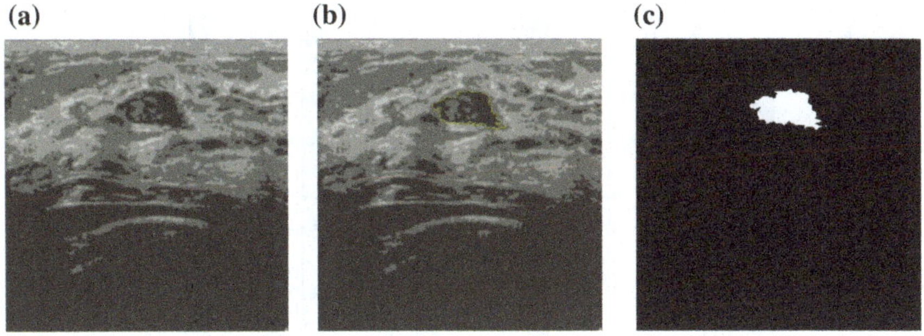

Fig. 3 Results attained with the proposed practice. **a** Threshold picture. **b** Converged boundary of LSS. **c** Extracted binary GT using LSS

Actual Pixels

Recognized pixels

<table>
<tr><td>T+ve=5203</td><td>F+ve= 1006</td><td>Accuracy=
99.52%</td><td rowspan="2">Jaccard=
80.39%</td></tr>
<tr><td>F-ve= 263</td><td>T-ve= 255667</td><td>Precision=
83.80%</td></tr>
<tr><td>Positive=5466</td><td>Negative=256673</td><td>Sensitivity=
95.19%</td><td rowspan="2">Dice=
89.13%</td></tr>
<tr><td colspan="2">Total pixels=262139</td><td>Specificity=
99.61%</td></tr>
</table>

Fig. 4 Sample confusion matrix attained for a chosen BUI

technique is then repeated by replacing the LSS with the AC and CV and these results are also tabulated separately to validate the performance of LSS.

Figure 5 presents the comparison of the average values of the performance measures attained with the SMA and Shannon's thresholding and the segmentation with LS, AC, and CV. From these results, it can be noted that the average measures attained with these techniques are superior. Further, all the considered segmentation methods offer approximately similar average values and the differences found in LS, AC, and CV are negligible. This confirms that the proposed technique is very efficient in appraising the considered BUIs. In the future, a feature extraction and classification procedure can be employed to organize the BUIs into benign/malignant class. The recent procedures may help to get better results [28–32].

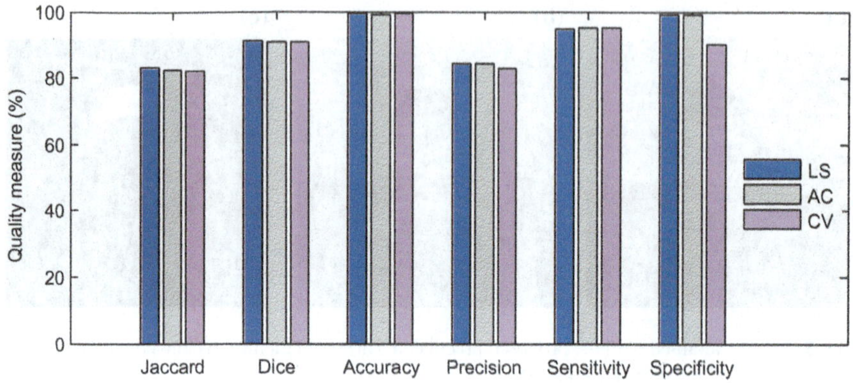

Fig. 5 Average performance measure attained with the considered segmentation methods

5 Conclusion

This work implements a hybrid image examination system to evaluate the tumor section of the BUI. This work implements Spider Monkey Algorithm assisted Shannon's three-level thresholding to separate the test picture into background, normal section and tumor. Later, an LSS is implemented to mine the tumor region with better correctness and the performance of the LSS is then validated with AC and CV. The results of this work confirm that the proposed technique helps to attain better performance measures, T_{+ve}, T_{-ve}, $F_{+ve,}$ and F_{-ve} values. The average value attained for all the 40 test images (20 malignant and 20 benign) also confirms that the results attained are superior. Further, this study also confirms that the segmented results of LSS, AC, and CV are approximately similar. In the future, this system can be used to estimate the clinically obtained BUIs.

References

1. https://www.who.int/health-topics/cancer
2. http://www.onlinemedicalimages.com/index.php/en/
3. Nair, M.V., et al.: Investigation of breast melanoma using hybrid image-processing-tool. In: International Conference on Recent Trends in Advance Computing (ICRTAC), pp. 174–179. IEEE (2018). https://doi.org/10.1109/ICRTAC.2018.8679193
4. Fernandes, S.L., Rajinikanth, V., Kadry, S.: A hybrid framework to evaluate breast abnormality using infrared thermal images. IEEE Consum. Electron. Mag. **8**(5), 31–36 (2019). https://doi.org/10.1109/MCE.2019.2923926
5. Rajinikanth, V., et al.: Thermogram assisted detection and analysis of ductal carcinoma in situ (DCIS). In: International Conference on Intelligent Computing, Instrumentation and Control Technologies (ICICICT), pp. 1641–1646. IEEE (2018). https://doi.org/10.1109/ICICICT1.2017.8342817

6. Keatmanee, C., Chaumrattanakul, U., Kotani, K., Makhanov, S.S.: Initialization of active contours for segmentation of breast cancer via fusion of ultrasound, Doppler, and elasticity images. Ultrasonics **94**, 438–453 (2019). https://doi.org/10.1016/j.ultras.2017.12.008
7. Rodtook, A., Kirimasthong, K., Lohitvisate, W., Makhanov, S.S.: Automatic initialization of active contours and level set method in ultrasound images of breast abnormalities. Pattern Recogn. **79**, 172–182 (2018)
8. Kirimasthong, K., Rodtook, A., Chaumrattanakul, U., Makhanov, S.S.: Phase portrait analysis for automatic initialization of multiple snakes for segmentation of the ultrasound images of breast cancer. Pattern Anal. Appl. **20**, 239–251 (2017)
9. Rodtook, A., Makhanov, S.S.: Multi-feature gradient vector flow snakes for adaptive segmentation of the ultrasound images of breast cancer. J. Visual Commun. Image Represent. **24**(8), 1414–1430 (2013)
10. Chucherd, S., Makhanov, S.S.: Sparse phase portrait analysis for preprocessing and segmentation of ultrasound images of breast cancer. IAENG Int. J. Comput. Sci. **38**(2), 146–159 (2011)
11. Raja, N.S.M., et al.: Segmentation of breast thermal images using Kapur's entropy and Hidden Markov random field. J. Med. Imaging Health Inform. **7**(8), 1825–1829 (2017). https://doi.org/10.1166/jmihi.2017.2267
12. Rajinikanth, V., Dey, N., Kumar, R., Panneerselvam, J., Raja, N.S.M.: Fetal head periphery extraction from ultrasound image using Jaya algorithm and Chan-Vese segmentation. Procedia Comput. Sci. **152**, 66–73 (2019). https://doi.org/10.1016/j.procs.2019.05.028
13. Satapathy, S.C., Rajinikanth, V.: Jaya algorithm guided procedure to segment tumor from brain MRI. J. Optim. **2018**, 12 (2018). https://doi.org/10.1155/2018/3738049
14. Rajinikanth, V., Satapathy, S.C.: Segmentation of ischemic stroke lesion in brain MRI based on social group optimization and Fuzzy-Tsallis entropy. Arab. J. Sci. Eng. **43**(8), 4365–4378 (2018). https://doi.org/10.1007/s13369-017-3053-6
15. Raja, N.S.M., et al.: Contrast enhanced medical MRI evaluation using Tsallis entropy and region growing segmentation. J. Ambient Intell. Humaniz. Comput. 1–12 (2018). https://doi.org/10.1007/s12652-018-0854-8
16. Dey, N., et al.: Social group optimization supported segmentation and evaluation of skin melanoma images. Symmetry **10**(2), 51 (2018). https://doi.org/10.3390/sym10020051
17. Jahmunah, V., et al.: Automated detection of schizophrenia using nonlinear signal processing methods. Artif. Intell. Med. **100**, 101698 (2019). https://doi.org/10.1016/j.artmed.2019.07.006
18. Acharya, U.R., et al.: Automated detection of Alzheimer's disease using brain MRI images—A study with various feature extraction techniques. J. Med. Syst. **43**, 302 (2019). https://doi.org/10.1007/s10916-019-1428-9
19. Dey, N., et al.: Social-group-optimization based tumor evaluation tool for clinical brain MRI of Flair/diffusion-weighted modality. Biocybern. Biomed. Eng. **39**(3), 843–856 (2019). https://doi.org/10.1016/j.bbe.2019.07.005
20. Fernandes, S.L., et al.: A reliable framework for accurate brain image examination and treatment planning based on early diagnosis support for clinicians. Neural Comput. Appl. 1–12 (2019). https://doi.org/10.1007/s00521-019-04369-5
21. http://www.itksnap.org/pmwiki/pmwiki.php
22. Yushkevich, P.A., Piven, J., Hazlett, H.C., Smith, R.G., Ho, S., Gee, J.C., Gerig, G.: User-guided 3D active contour segmentation of anatomical structures: significantly improved efficiency and reliability. Neuroimage **31**(3), 1116–1128 (2006)
23. Rajinikanth, V., Lin, H., Panneerselvam, J., Raja, N.S.M.: Examination of retinal anatomical structures—A study with spider monkey optimization algorithm. In: Applied Nature-Inspired Computing: algorithms and Case Studies, pp. 177–197 (2020). https://doi.org/10.1007/978-981-13-9263-4_8
24. Rajinikanth, V., Dey, N., Satapathy, S.C., Kamalanand, K.: Inspection of crop-weed image database using Kapur's entropy and Spider monkey optimization. In: Advances in Intelligent Systems and Computing, p. 1048 (2019). https://doi.org/10.1007/978-981-15-0035-0_32

25. Bansal, J.C., Sharma, H., Jadon, S.S., Clerc, M.: Spider monkey optimization algorithm for208 numerical optimization. Memetic Comput. **6**(1), 31–47 (2014)
26. Kannappan, P.L.: On Shannon's entropy directed divergence and inaccuracy. Probab. Theory Rel. Fields **22**, 95–100 (1972)
27. Chen, Y., Chen, G., Wang, Y., Dey, N., Sherratt, R.S., Shi, F.: A distance regularized level-set evolution model based MRI dataset segmentation of brain's caudate nucleus. IEEE Access **7**, 124128–124140 (2019)
28. Chandrakar, P.: A secure remote user authentication protocol for healthcare monitoring using wireless medical sensor networks. Int. J. Ambient Comput. Intell. (IJACI) **10**(1), 96–116 (2019). https://doi.org/10.4018/IJACI.2019010106
29. Bhattacharya, H., Chattopadhyay, S., Chattopadhyay, M., Banerjee, A.: Storage and bandwidth optimized reliable distributed data allocation algorithm. J. Ambient Comput. Intell. (IJACI) **10**(1), 78–95 (2019). https://doi.org/10.4018/IJACI.2019010105
30. Das, H., Naik, B., Behera, H.S.: Classification of diabetes mellitus disease (dmd): a data mining (DM) approach. Adv. Intell. Syst. Comput. **710**, 539–549 (2018). https://doi.org/10.1007/978-981-10-7871-2_52
31. Sahani, R., et al.: Classification of intrusion detection using data mining techniques. Adv. Intell. Syst. Comput. **710**, 753–764 (2018). https://doi.org/10.1007/978-981-10-7871-2_72
32. Pradhan, C., Das, H., Naik, B., Dey, N.: Handbook of Research on Information Security in Biomedical Signal Processing Hershey. IGI Global, PA (2018)

Optimal Scheduling of Household Appliances Using GOA for Electricity Cost Saving with Solar PV

Kumari Kasturi, Sushil Kumar Bhoi and Manas Ranjan Nayak

Abstract Exponential increase in energy demand and environmental crisis causes the need for an efficient improvement in energy consumption. The following apex report describes the grasshopper optimization algorithm (GOA) to minimize the cost function in a smart home environment due to which peak demand of microgrid can be reduced. The proposed optimization technique is used to find the most suitable time to use home appliances as per the time of use (TOU) electricity pricing. All home appliances have different energy demand peaks. The proposed method is further used to compare the results with and without the use of photovoltaic (PV) modules. Finally, the results obtained show the efficiency of the grasshopper optimization algorithm (GOA) to achieve optimal scheduling for home appliances and to reduce the daily electricity cost.

Keywords Smart home · Energy consumption · Scheduling · Photovoltaic (PV) · Grasshopper optimization algorithm (GOA)

1 Introduction

A smart home is one which uses communication technology and energy management methods to operate its appliances wisely to reduce consumption cost. So the

K. Kasturi (✉)
Department of Electrical Engineering, Siksha O Anusandhan University, Bhubaneswar 751030, India
e-mail: kumari.kasturi1986@gmail.com

S. K. Bhoi
Department of Electrical Engineering, Goverment College of Engineering, Kalahandi, Bhawanipatna, India
e-mail: sushilkumarbhoi@gmail.com

M. R. Nayak
Department of Electrical Engineering, CAPGS, Biju Patnaik University of Technology, Rourkela, India
e-mail: manasnk72@gmail.com

H. Das et al. (eds.), *Progress in Computing, Analytics and Networking*,
Advances in Intelligent Systems and Computing 1119,
https://doi.org/10.1007/978-981-15-2414-1_63

management of smart home appliances has dragged more attention of researchers nowadays. To achieve this goal, all appliances in a house should act "smart" or more automated [1, 2]. Smart optimization models are required to solve issues like an exponential increase in energy demand and serious concern on environmental impacts. The optimization of energy resources can be handled by managing energy efficiency at the microgrid level by providing appropriate scheduling for home appliances. The integration of a solar photovoltaic (PV) can achieve further improvement in schedule plans of appliances.

For reduction in energy bills, many methods have been proposed as per the time of use pricing [3, 4]. As per the supply and demand, TOU pricing is varied throughout the day [5]. In [6], scheduling for four home appliances is formulated and the result is compared between two optimization techniques. The supercity of the proposed single-objective optimization method can be verified from the result. The mixed-integer linear problem is formulated in convex programming framework to model appliances for a single home in [7]. The authors analyze cost minimization considering user dissatisfaction. In [8], an hourly energy schedule for electric vehicles, heaters, water heater, and pool pump is presented. Particle swarm optimization is used as an optimization tool for finding optimal scheduling for considered appliances.

The main contribution of the proposed method is as follows:

(a) Application of novel grasshopper optimization algorithm (GOA) optimization method for determination of optimal scheduling of home appliances to minimizing the residential electricity bill.
(b) Present the scheduling of home appliances for two days of the year which are chosen as 15th May for the summer season and 15th December for the winter season.
(c) Shows the profit gain by the use of the surplus of energy from the PV to get preferred use time for home appliances.

In Sect. 2 the system modeling is described. In Sect. 3 problem formulation is defined, whereas Sect. 4 describes GOA. Section 5 shows the results and analysis of the proposed problem. Finally, Sect. 6 gives the conclusion.

2 System Model

2.1 Energy Consumption Modes of Home Appliances

In a smart home environment, the use of energy should be done wisely to reduce energy consumption costs. So, we have to use smart appliances i.e. power consumption time of that appliance can be scheduled to a preferred working period to get low consumption cost. To get more benefits, optimal scheduling is required, so that usages of appliances can be done at low electricity price hours.

Table 1 Appliances and power consumption patterns

Appliances	Daily power (W)	Energy consumption patterns
Heater 1	1100	Preferred hours: 7 a.m.–9 a.m.: 300 Wh, 10 a.m.: 200 Wh
Washing machine	500	Preferred hours: 12 p.m.: 500 Wh
Iron	400	Preferred hours: 9 a.m.: 500 Wh, 1 p.m.: 300 Wh
Dishwasher	400	Preferred hour: 12 p.m.–2 p.m.: 400 W
Heater 2	800	Preferred hour: 9 a.m.: 500 Wh, 2 p.m.: 300 Wh

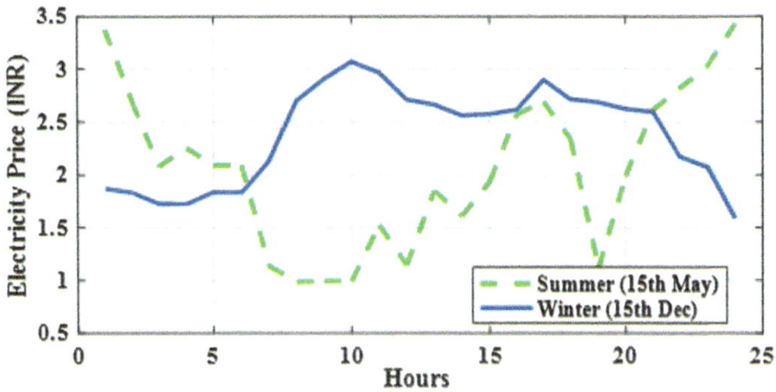

Fig. 1 Hourly electricity price

The power consumption of the ith appliance in a time slot $t \in T = \{1, \ldots, 24\}$ can be defined as:

$$P_i(t) = (P_i(1), P_i(2), \ldots, P_i(24)) \in R^{24}, \, for \; i \in H \,\&\, t \in T \qquad (1)$$

where H is the home appliances set defined as $H = \{1,2,3, \ldots, m\}$, m is the number of home appliances.

Here, five home appliances are considered. Appliances and their power Consumption patterns are shown in Table 1. As per the consumption of energy, the electricity price is also varying during the day. Hourly electricity price for 2 days is considered as shown in Fig. 1. One is chosen as 15th May for the summer season and another one is 15th December for the winter season.

2.2 *Modeling of PV*

MPPT and an inverter are connected to the output terminals of the PV arrays at the time of PV output power calculation. Global horizontal irradiance (GHI) ($E_{ir}(t)$) and the real time temperature (*Tem*) [9] are used for calculation of PV output.

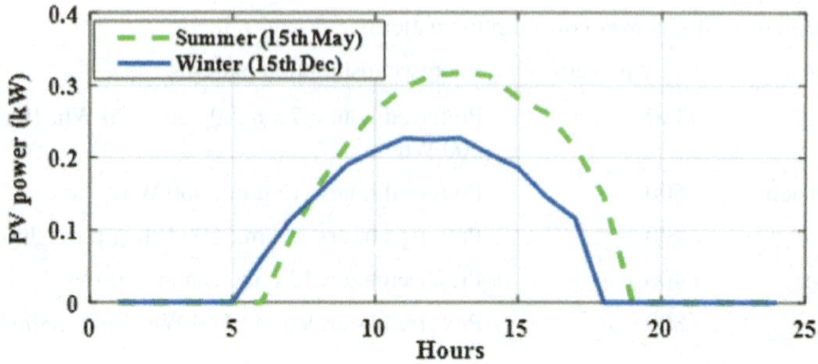

Fig. 2 PV module power output

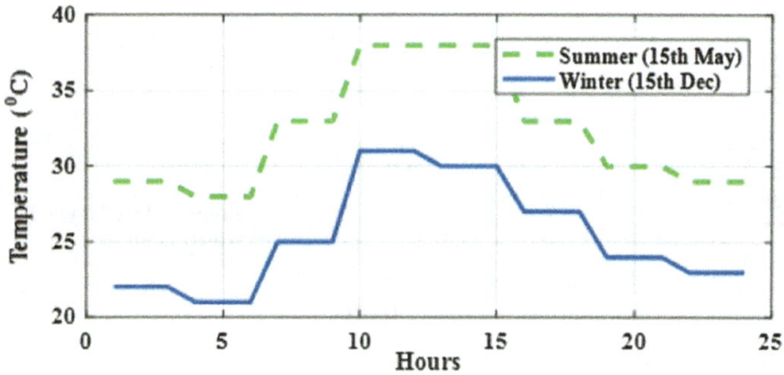

Fig. 3 Real-time temperature

$$P_{PV}(t) = d \times \eta_{PV} \times \eta_{inv} \times P_{PV}^{rated} \times \frac{E_{ir}(t)}{E_{ir}^{std}} \times [1 + \beta(Tem(t) - Tem_{std})] \quad (2)$$

where, d (=89%) is the power de-rating coefficient by shading, cable and switching losses, and dust accumulation on the array, η_{PV} (=15%) is the PV efficiency, η_{inv} (=97%) is the inverter efficiency, P_{PV}^{rated} (=5 kW) is the rated PV power output, E_{ir}^{std} and T_{std} are the GHI and the temperature at standard test conditions respectively and β is the temperature coefficient of the PV cell. The PV module power output and real time temperature for two days are shown in Figs. 2 and 3 respectively.

3 Problem Formulation

The problem of finding the optimal scheduling for home appliances has the main objective of minimizing the total electricity consumption cost in order to benefit the consumer within the framework of system operational constraints.

3.1 Objective Function

3.1.1 Minimization of Electricity Consumption Cost

The objective function is defined as

$$f_{obj} = \min\left(\sum_{t=1}^{24}\sum_{i=1}^{5} ec(t)(P_i(t) - P_{PV}(t))\right) \tag{3}$$

where $ec(t)$ is the 24-h electricity price.

3.2 System Operational Constraints

The constraints imposed to solve the proposed problem are the follows:

$$\sum_{i\in H} ec(t) \times P_i(t) \leq L_i(t)\ \forall t \in T \tag{4}$$

$$P_i^{\min} \leq P_i(t) \leq P_i^{\max} \tag{5}$$

$$\pm P_{grid}(t) = P_i(t) - P_{PV}(t) \tag{6}$$

$$P_{PV}^{\min} \leq P_{PV}(t) \leq P_{PV}^{\max} \tag{7}$$

where $L_i(t)$ highest electric consumption load at time t, $P_i^{\min}$ and $P_i^{\max}$ are the lower and upper bounds of appliance's power, respectively, $P_{PV}^{\min}$ and $P_{PV}^{\max}$ are the lower and upper bounds of PV power respectively. +ve and −ve sign are used for delivering and consuming power with the grid, respectively, $\pm P_{grid}(t)$ is delivering and consuming power with the grid.

4 Grasshopper Optimization Algorithm (GOA)

The swarming behavior of grasshopper is simulated by the help of the grasshopper optimization algorithm. The position of grasshopper provides the solution to the optimization problem. X_n indicates the of grass hopper at nth location.

$$X_n = S_n + G_n + A_n \tag{8}$$

where S_n: social interaction, G_n: *gravity* forces on *n*th grasshopper, and A_n: wind advection. The social interaction component of grasshopper can be visualized as

$$S_n = \sum_{j=1, j \neq n}^{N} S(d_{np})\hat{d}_{np} \tag{9}$$

where d_{np}: distance of *n*th and *p*th grasshopper. It can be calculated as $d_{np} = |x_p - x_n|$, *s*: strength of social forces, and $\hat{d}_{np} = \frac{x_p - x_n}{d_{np}}$ is the unit normal vector from *n*th grasshopper to the *p*th grasshopper. Here x_n and x_p are constants.

The function S is the social force which can be formulated as

$$S(r) = fe^{\frac{-r}{l}} - e^{-r} \tag{10}$$

where *f* shows the intensity of attraction and *l* is the attractive length scale. For large distances the function s, is unable to produce strong forces between the grasshoppers. So to solve this problem the distance needs to be mapped or normalized.

Now consider the "G" in Eq. (8)

$$G_n = -g\hat{e}_g \tag{11}$$

where *g* is a constant for gravity and $\hat{e}_g$ is the unit vector in the direction to the center of the earth. Further, the A component in (8) can be realized as

$$A_n = -u\hat{e}_w \tag{12}$$

where *u* indicates the drift and $\hat{e}_w$ is a unit vector in the direction of the wind.

Finally, we can write Eq. (24) with all components as

$$x_i = \sum_{j=1, j \neq 1}^{N} s(|x_p - x_n|)\frac{x_p - x_n}{d_{np}} - g\hat{e}_g + u\hat{e}_w \tag{13}$$

where *N* denotes the number of grasshoppers.

A stochastic algorithm needs to be performed for exploration and exploitation effectively to find an accurate approximation of the global optimum. Special parameters are used further to show the exploration and exploitation of optimization.

$$x_i^d = c\left(\sum_{j=1, j \neq 1}^{N} c\frac{ub_d - lb_d}{2} s(|x_p - x_n|)\frac{x_p - x_n}{d_{np}}\right) + \widehat{T}_d \tag{14}$$

where ub_d is the maximum limit in the *d*th dimension, lb_d indicates the minimum limit in the *d*th direction. $S(r) = fe^{\frac{-r}{l}} - e^{-r}$, $\widehat{T}_d$ shows *d*th dimension in the target, and "*c*" is a decaying coefficient which minimizes the comfort area, repulsion area, and attraction area. Here the gravity is not considered and for wind direction, it is assumed to be towards the target, i.e., $\widehat{T}_d$. The decrease in the attraction or repulsion among the grasshoppers is formulated by the inner "c". It is proportional to the number of cycles. The decrease in the search area is realized by the outer "c". We can update the "c" by the help of the equation:

$$c = c_{mx} - l\frac{c_{mx} - c_{mn}}{L} \tag{15}$$

where c_{mx} denotes the maximum value, c_{mn} denotes the minimum value, l denotes the current iteration, and L indicates the number of iterations.

5 Result and Analysis

Electricity consumption cost for each home appliance is given in Table 2. The analysis is done for two days of the year which are 15th May and 15th December, respectively. Without optimization technique total cost is 2.98 INR whereas when GOA optimization is applied it reduces to 2.47 INR without PV integration. With integration of PV, electricity consumption is reduced to 2.20 INR and 2.33 INR for 15th May and 15th December, respectively, which is more benefit to consumers.

Figures 4 and 5 show the optimal scheduling of five home appliances. There is an effective reduction in electricity consumption cost if all home appliances will operate as per this schedule. So proposed GOA optimization scheduling can tackle energy demand and can also reduce total electricity bills in a smart home environment effectively.

Table 2 Cost comparison

Home appliances	Electricity consumption cost (INR) (without optimization)	Electricity consumption cost (INR) (with optimization)	Electricity consumption cost (INR) (with optimization and integration of PV)	
			15th May	15th December
Heater 1	0.99	0.96	0.89	0.91
Washing machine	0.55	0.46	0.44	0.45
Iron	0.32	0.27	0.19	0.20
Dishwasher	0.48	0.15	0.15	0.19
Heater 2	0.64	0.63	0.53	0.58
Total	2.98	2.47	2.20	2.33

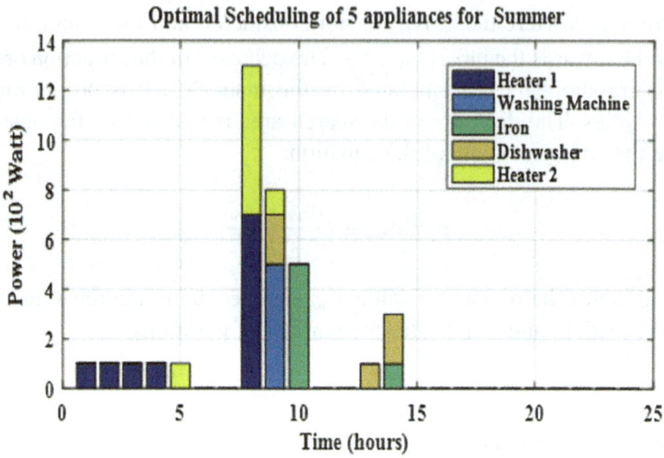

Fig. 4 Scheduling of 5 appliances for 15th May without PV integration

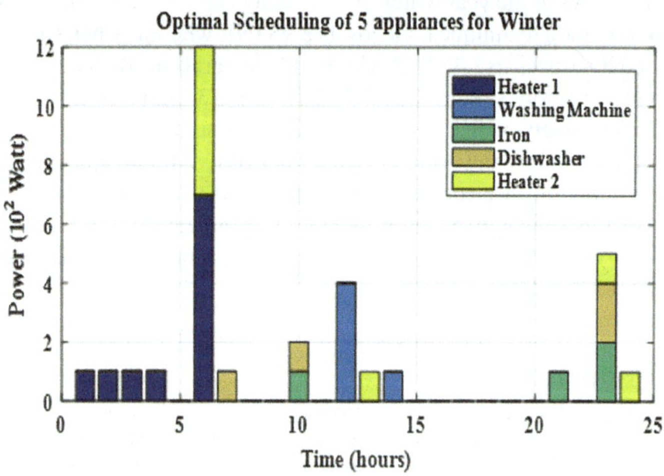

Fig. 5 Scheduling of 5 appliances for 15th December without PV integration

Figures 6 and 7 show the optimal scheduling of five home appliances with the integration of PV. The generated PV power output is used to supply the five home appliances first and the rest portion of PV power output is to be fed back to the utility grid as surplus power which is shown as negative power in the figures.

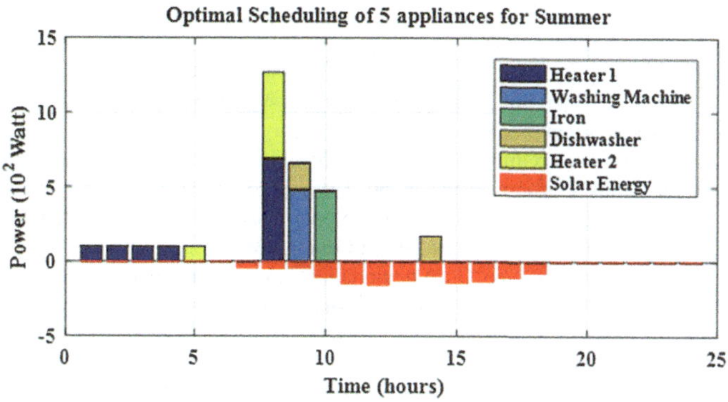

Fig. 6 Scheduling of 5 appliances for 15th May with PV integration

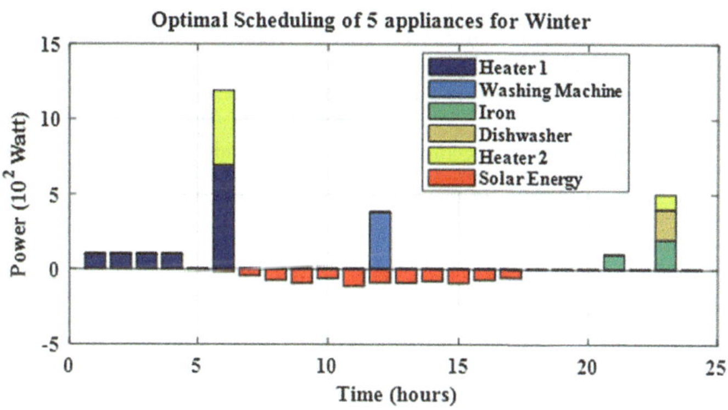

Fig. 7 Scheduling of 5 appliances for 15th December with PV integration

6 Conclusion

In this paper, a novel and efficient GOA optimization method is implemented. The proposed method is more effective to get optimal scheduling for smart home appliances. Proposed scheduling is implemented with the integration of PV so that consumers can get more profit. and The analysis is done on two days of the year which are 15th May and 15th December with to respect different electricity prices and PV power output.

References

1. Han, D.M., Lim, J.H.: Smart home energy management system using IEEE 802.15. 4 and zigbee. IEEE Trans. Consum. Electron. **56**, 1403–1410 (2010)
2. Han, D., Lim, J.: Design and implementation of smart home energy management systems based on zigbee. IEEE Trans. Consum. Electron. **56**, 1417–1425 (2010)
3. Agnetis, A., De Pascale, G., Detti, P., Vicino, A.: Load scheduling for household energy consumption optimization. IEEE Trans. Smart Grid **4**, 2364–2373 (2013)
4. Ozturk, Y., Jha, P., Kumar, S., Lee, G.: A personalized home energy management system for residential demand response. In: IEEE 4th International Conference on Power and Energy systems Engineering, pp. 1241–1246 (2013)
5. Erol-Kantarci, M., Mouftah, H. T.: TOU-aware energy management and wireless sensor networks for reducing peak load in smart grids. In: IEEE 72nd Vehicular Technology Conference, pp. 1–5 (2010)
6. Erol-Kantarci, M., Mouftah, H.T.: Wireless sensor networks for cost efficient residential energy management in the smart grid. IEEE Trans. Smart Grid **2**, 314–325 (2011)
7. Tsui, K.M., Chan, S.C.: Demand response optimization for smart home scheduling under real-time pricing. IEEE Trans. Smart Grid **3**, 1812–1821 (2012)
8. Pedrasa, M.A.A., Spooner, T.D., MacGill, I.F.: Coordinated scheduling of residential distributed energy resources to optimize smart home energy services. IEEE Trans. Smart Grid **1**, 134–143 (2010)
9. Nayak, C.K., Nayak, M.R.: Technoeconomic analysis of a grid-connected PV and battery energy storage system considering time of use pricing. Turk. J. Electron. Eng. Comput. **26**, 318–329 (2018)

A Systematic Approach to Enhance the Forecasting of Bankruptcy Data

Udjapana Mahapatra, Sasmita Manjari Nayak and Minakhi Rout

Abstract Several models have been developed for the forecasting of bankruptcy dataset but still, this is an active research area without which it may lead to a severe financial crisis. This paper focuses on the preprocessing phase which is very much essential in this domain to enhance the performance of the prediction model. We have filled out the missing values of the feature by the means of corresponding feature vector and then used oversampling technique SMOTE, normalization in the preprocessing phase is then applied to the transformed data set to five different popular classifiers such as random forest, decision tree, K-nearest neighbor (K-NN), logistic regression, and artificial neural network (ANN) to see the effect of these preprocessing steps in the prediction performance of all these five classifiers.

Keywords Bankruptcy prediction · SMOTE · Normalization · Random forest · Logistic regression · K-NN · Decision tree · ANN

1 Introduction

The problem of bankruptcy is a financial situation in which cash flows are not sufficient to pay the debt. It is the most challenging and active research area related to financial distress and crisis. Several methods based on statistical methods, machine learning techniques, and hybrid techniques have already been developed and implemented to predict the companies bankruptcy and financial crisis. Most of the studies reveal that the machine learning approach is more suitable than statistical techniques. Still, this research area needs more attention to develop intelligent models along with appropriate data preprocessing techniques which can predict the bankruptcy prior

U. Mahapatra · S. M. Nayak · M. Rout (✉)
School of Computer Engineering, KIIT Deemed to be University, Bhubaneswar, Odisha, India
e-mail: minakhi.rout@gmail.com

U. Mahapatra
e-mail: Udjam2000@gmail.com

S. M. Nayak
e-mail: sasmita.lina@yahoo.com

H. Das et al. (eds.), *Progress in Computing, Analytics and Networking*,
Advances in Intelligent Systems and Computing 1119,
https://doi.org/10.1007/978-981-15-2414-1_64

to the crisis so that it allows the investors, financial institutions, shareholders, and enterprises to take some preventive steps before the crisis.

Machine learning approaches such as random forest, bagging, boosting, and SVM are compared with statistical approaches such as discriminant analysis and logistic regression and the study [1] reveals that 10% more accuracy has been obtained in comparison to other statistical models. A new boosting approach along with feature selection has been proposed in [2] and compared with some well-accepted machine learning models and the study reveals that the FS-boosting model can be used as an alternative to the prediction of bankruptcy. Instead of classifying whether the sample is bankrupt or non-bankrupt, in [3] the author has used intellectual capital to predict the going concern doubt (GCD) of corporate using hybrid random forest and rough set theory. A collection of intelligent techniques such as MLP, random forest, SVM, logistic regression, and regression trees were considered in [4] with the t-statistics feature selection technique. The reduced data set is applied to the ensemble model comprises of the mentioned techniques and the validation is carried out with the ROC curve which reveals that the proposed model is superior to other models reported prior to the studies. Extreme gradient boosting [5] for learning an ensemble of the decision tree has been used with synthetic feature value on polish data. If the financial ratio is measured based on the Markov for discrimination (MFD) [6] model then it can able to extract the sequential information from the time series of financial ratios help to improve the prediction accuracy. Oversampling technique, synthetic minority oversampling technique (SMOTE), borderline-SMOTE, adaptive synthetic sampling, SMOTE with edited nearest neighbor (SMOTE + ENN) [7] are applied and compared. Out of these five sampling techniques SMOTE + ENN with random forest is performing better on Korean mixed data set and achieving 84.4%. class imbalance problem complicates the procedure of learning of the model. Hence we couldn't be able to correctly address the situation. Thus a hybrid approach using oversampling technique and cost-sensitive learning (HAOC) [8] has been designed for the prediction of the Korean bankruptcy dataset and performing well than the existing approaches. Spanish bankruptcy dataset [9] is taken into consideration for investigation and various oversampling and undersampling techniques are applied on it to see the impact of it on the individual classifiers and ensemble learners. The study reveals that the performance of the Logistic Regression and DTBagging with oversampling technique and C4.5 and random forest with undersampling techniques are performing superior to other models in terms of G-Mean and area under the curve (AUC) measures. Imbalance issues are also addressed in a dataset collected from the Altares database which contains income statements and balance sheets of French firms [10]. Both oversampling and undersampling techniques are considered with well-accepted statistical learners and machine learning based models.

From the literature survey, we have found that the prediction performance is getting deteriorate if missing values are not handled at the beginning stage then due to the imbalanced nature of the data. Secondly, it consists of 64 components or features with different ranges which may lead to a problem when applying it a model for prediction similarly we can also go for feature reduction techniques to reduce the computational complexity without compromising the accuracy. Fourth, as the data is imbalance we should choose the performance measure in such a way that it should also focus on minority class prediction as accurate as possible so that it can helpful for giving warning before the crisis occurs. Details of all the aspects are described in detail in the rest of the section of this paper and as follows: Sect. 2 describes the proposed workflow of this study where the dataset description followed in the next Sect. 3 and the experimental studies and results are carried out in the Sect. 4. Finally, the conclusion of this work and further work carried out in this domain is explained in Sect. 5.

2 Proposed Model and Workflow

SMOTE Algorithm

Imbalanced data distribution happens when observations or data samples in one of the class are much higher or lower than the other classes. As machine learning algorithms tend to increase accuracy by reducing the error, they do not consider the class distribution. SMOTE (synthetic minority oversampling technique) [11] is one of the most commonly used oversampling methods to solve the imbalance problem by randomly increasing minority class. SMOTE synthesizes new minority instances between existing minority instances (i.e., bankrupt class). SMOTE algorithm is used to generate new samples for minority class data. SMOTE algorithm combines a certain observation with k similar minority class samples to generate a new sample according to the following calculation:

$$X_{\text{new}} = X + r \times (X_k - X)$$

X_{new}, X, and X_k, respectively, are newly generated sample, the original sample, and one of the k-nearest neighbors to the original sample. Through the repetitive sample generation process, the SMOTE algorithm effectively creates new samples of minority class by combining a certain sample with k similar minority class observations multiplied by Gaussian random distances until both the numbers of the minority class and majority class become the same. The proposed model is represented in Fig. 1.

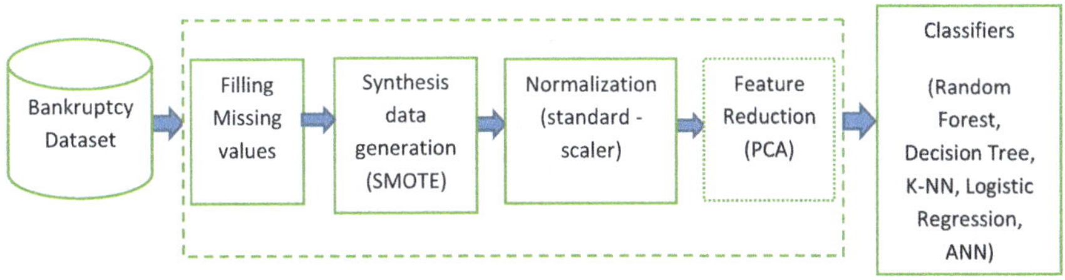

Fig. 1 Block diagram of the proposed workflow

Table 1 Details of the dataset

Year	Total instances	Non-bankrupt instances	Bankrupt instances
First	7027	6756	271
Second	10173	9773	400
Third	10503	10008	495
Fourth	9792	9277	515
Fifth	5910	5500	410
Total	43405	41314	2091

3 Dataset Selection

The dataset is about the bankruptcy prediction of Polish companies. The data was collected from the Emerging Markets Information Service (EMIS). EMIS is a database containing information on emerging markets around the world including the Polish one [1]. In Poland, since 2012, many companies in the manufacturing sector went bankrupt. The research sample consists of both bankrupt and non-bankrupt/operating companies data. The bankrupt companies were analyzed in the period 2000–2012, while the still operating companies were evaluated from 2007 to 2013. The dataset consists of 64 features. Detailed information about the total instances of bankrupt and nonbankrupt in each year is represented in Table 1.

4 Simulation Result and Analysis

The experiment was done on the datasets by considering three major aspects. The first one is with or without the synthesis sample generation technique (SMOTE) to address the effect of the imbalance factor of the dataset on the prediction. The second one is through the dataset that consists of 64 different features and the ranges of each of the feature are different, so to address this fact, we have used standard scaler normalization technique and the third aspect is the feature reduction using PCA to see the effect of lossless dimensionality reduction with respect to prediction. Finally, we have combined some popular classifiers used in this domain to model fusion classifiers for the prediction of class labels. To validate the model, we have considered F1-score as the major performance measure due to the imbalanced nature of the dataset because F1-score plays an immense role in determining the performance of minority class in an imbalanced dataset. If the F1-score of minority class increases for a particular classifier that means that classifier is performing well. Other performance measures are also included in this study to validate the performance of the proposed model also. We have investigated the work in the following phases:

(I) Without SMOTE and Normalization but with and without PCA.
(II) With SMOTE and without Normalization but with and without PCA.

(III) Without SMOTE with Normalization and with and without PCA.
(IV) With SMOTE, normalization and with and without PCA.

I Without SMOTE and normalization but with and without PCA

The given imbalanced dataset is neither normalized nor synthesized in this method. Different classifiers like the random forest, decision tree, *k* nearest neighbors, logistic regression and neural networks were fitted with the training data with and without PCA. The ROC curve, F1 score, and accuracy graph are represented in Figs. 2 and 3, respectively. In an ROC curve, the true positive rate (Sensitivity) is plotted in function of the false positive rate (100-Specificity) for different cutoff points of a parameter. Each point on the ROC curve represents a sensitivity/specificity pair corresponding to a particular decision threshold. The area under the ROC curve (AUC) is a measure

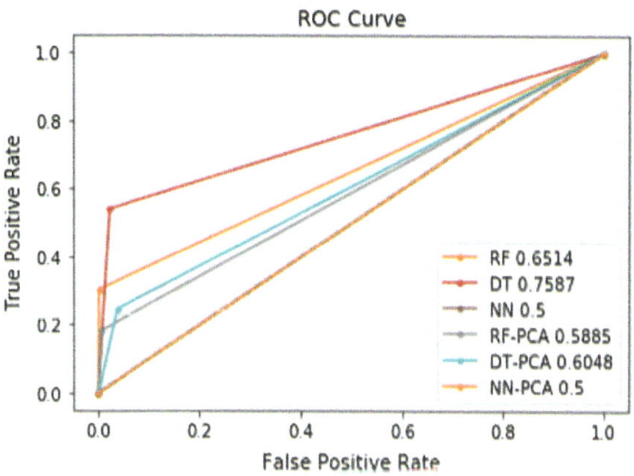

Fig. 2 ROC curve obtained using random forest, decision tree, and neural network without SMOTE and normalization but with and without PCA

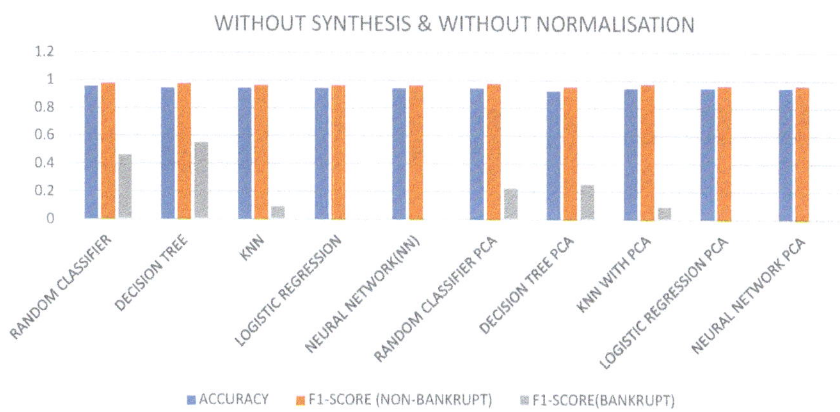

Fig. 3 F1 score (bankrupt and non-bankrupt) and accuracy chart obtained through all 5 classifiers without SMOTE and normalization but with and without PCA

of how well a parameter can distinguish between two groups (bankrupt and non-bankrupt). As it is a highly imbalanced dataset the classifiers do not perform well on the raw data, as a result of which it is not having a good ROC score which implies the model cannot distinguish the classes well. From the above graph, it can be seen that the F1 score of the bankrupt class decreases when we apply PCA on it. Hence, PCA is not working well with unprocessed data.

II **With SMOTE and without normalization but with and without PCA**

Though in the above method (I), the accuracy is quite good it is because of the fact that the models predict the non-bankrupt class or majority class well as non-bankrupt class is having a good F1 score. Hence it has good accuracy but we need a model that can predict the minority class as well as the majority class accurately. In this method, we are synthesizing the imbalanced data by using SMOTE oversampling technique. It generates virtual training records by linear interpolation for the minority class. After the oversampling process is done the data is reconstructed and several classification models were applied to the processed data.

The ROC curve and F1 score graph are obtained and presented in Figs. 4 and 5 respectively. From the ROC curve, it can be observed that the area under the curve for different classifiers has improved. From Fig. 5, it can be found that PCA does not work well even after synthesizing the data.

III **Without SMOTE with normalization and with and without PCA**

In method (II), PCA does not work well with the synthesized data. So in this method, we are applying the classifier models on the normalized data to check whether PCA performs well on that or not. The ROC curve and F1 score graph are represented in Figs. 6 and 7 respectively. From the ROC curve, it is clear that after applying normalization the ROC score is improved and better than the score of method (I). This is because of the fact that in method 1 we were not normalizing the data, as a result of which different features where having values in different scales and ranges. This leads to the overpowering of one feature over others. This leads to the ignorance

Fig. 4 ROC curve obtained using random forest, decision tree, and neural network with SMOTE without normalization but with and without PCA

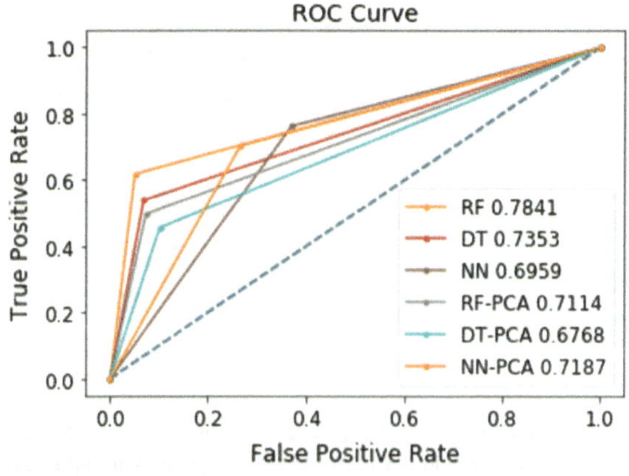

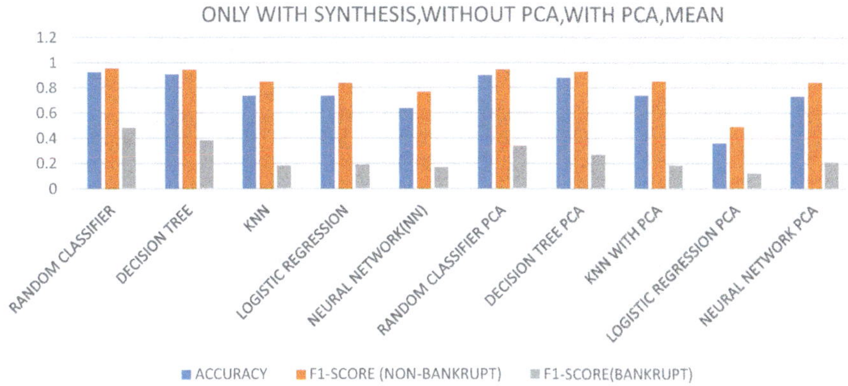

Fig. 5 F1 score (bankrupt and non-bankrupt) and accuracy chart obtained through all 5 classifiers with SMOTE without normalization but with and without PCA

Fig. 6 ROC curve obtained using random forest, decision tree, and neural network without SMOTE with normalization and with and without PCA

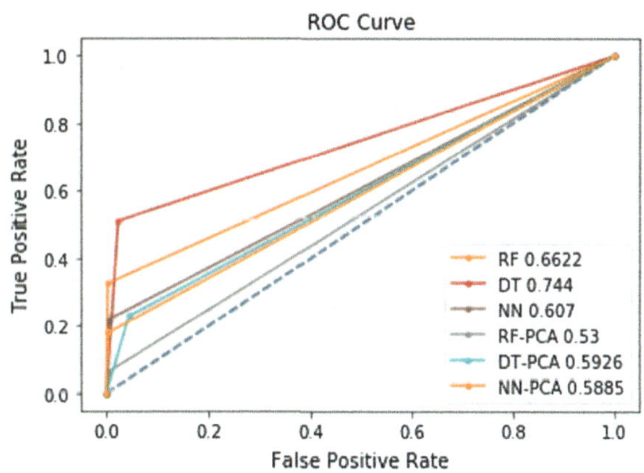

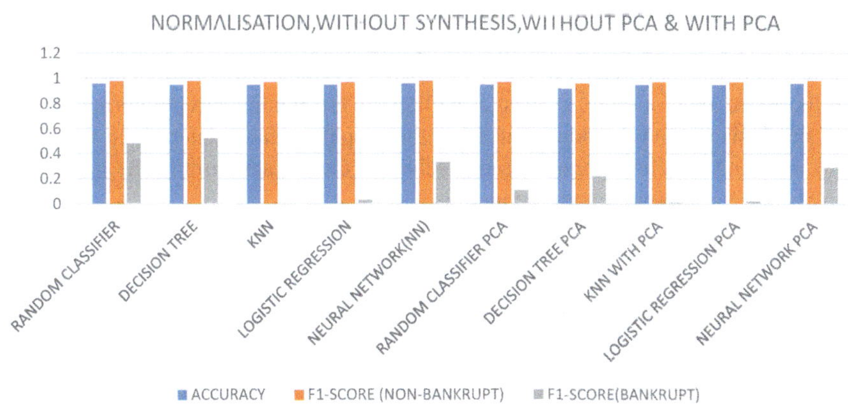

Fig. 7 F1 score (bankrupt and non-bankrupt) and accuracy chart obtained through all 5 classifiers without SMOTE with normalization and with and without PCA

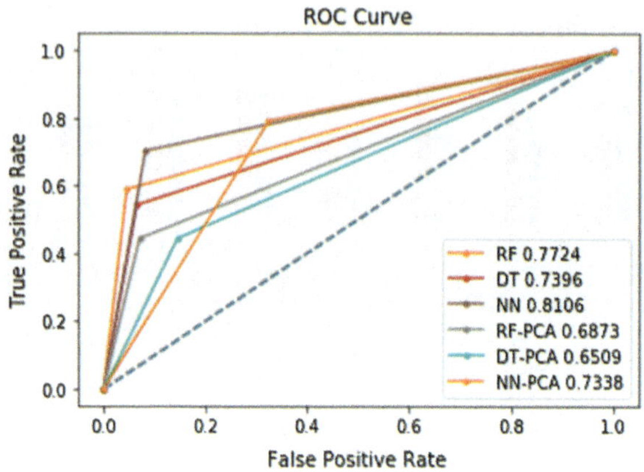

Fig. 8 ROC curve obtained using random forest, decision tree, and neural network with SMOTE, normalization and with and without PCA

of some feature which may have more importance than the other features that are taken into consideration. As here we have normalized the data using standard scalar. So, the ROC curve is improving. But in Fig. 7, again PCA doesn't work well even after normalizing the data.

In this method, it is also noticed that K-NN and logistic regression gives a very low F1 score for minority class with and without PCA.

IV **With SMOTE, normalization and with and without PCA**

From method II and III, it is clear that when we are either synthesize the data or normalize it using feature scaling, the ROC scores for all classifier models are improving the prediction of the bankrupt and non-bankrupt classes. So in this method, we combined both SMOTE and normalization techniques for preprocessing the data and then applied all the 5 classifiers on the transformed dataset. The ROC curve and the graph are represented in Figs. 8 and 9, respectively. From Fig. 8, we can see the ROC score for all models is improved as well as from Fig. 9, it is clearly visible that the F1 score has increased a lot for minority class as well. As in this method, proper normalization and synthesis of data are done. But from the graph, we have determined that even after applying different preprocessing techniques and transforming the data, still, PCA is not working well with the models.

5 Conclusion and Future Directions

From the above experiment, it is clear that the preprocessing steps have a great impact on the prediction of the bankruptcy data set. In this preprocessing phase, we have first filled the missing values by the mean of that respective feature vector and then applied the synthesis data generation technique to address the imbalanced

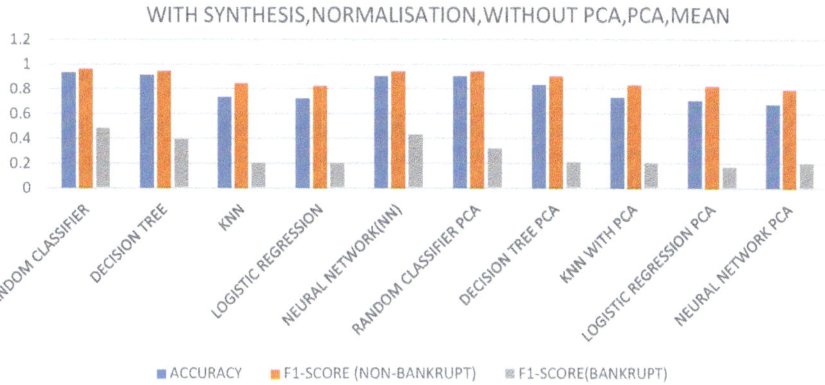

Fig. 9 F1 score (bankrupt and non-bankrupt) and accuracy chart obtained through all 5 classifiers with SMOTE, normalization and with and without PCA

nature of the data, then applied normalization to handle a variety of ranges of data values of the features of the dataset. Then we have applied dimensionality reduction technique in a thought that it improves the prediction performance but we observed from the result analysis it is not performing well in any manner in this data set. K-NN and Logistic Regression are not performing well in this domain. This work can be further investigated with other sampling techniques similarly it can be further applied to other machine learning and deep learning based classifiers. One can also focus on other suitable dimensionality reduction techniques to reduce the computational cost. Though the prediction of bankruptcy data is a crucial one, it can be useful to consider ensemble or fusion classifier instead of a single classifier.

References

1. Barboza, F., Kimura, H., Altman, E.: Machine learning models and bankruptcy prediction. Expert Syst. Appl. **83**, 405–417 (2017)
2. Wang, G., Ma, J., Yang, S.: An improved boosting based on feature selection for corporate bankruptcy prediction. Expert Syst. Appl. **41**(5), 2353–2361 (2014)
3. Yeh, C.C., Chi, D.J., Lin, Y.R.: Going-concern prediction using hybrid random forests and rough set approach. Inf. Sci. **254**, 98–110 (2014)
4. Chandra, D.K., Ravi, V., Bose, I.: Failure prediction of dotcom companies using hybrid intelligent techniques. Expert Syst. Appl. **36**(3), 4830–4837 (2009)
5. Zięba, M., Tomczak, S.K., Tomczak, J.M.: Ensemble boosted trees with synthetic features generation in application to bankruptcy prediction. Expert Syst. Appl. **58**, 93–101 (2016)
6. Volkov, A., Benoit, D.F., Van den Poel, D.: Incorporating sequential information in bankruptcy prediction with predictors based on Markov for discrimination. Decis. Support Syst. **98**, 59–68 (2017)
7. Le, T., Lee, M., Park, J., Baik, S.: Oversampling techniques for bankruptcy prediction: novel features from a transaction dataset. Symmetry **10**(4), 79 (2018)
8. Le, T., Vo, M.T., Vo, B., Lee, M.Y., Baik, S.W.: A hybrid approach using oversampling technique and cost-sensitive learning for bankruptcy prediction. Complexity (2019)

9. Sisodia, D.S., Verma, U.: The impact of data re-sampling on learning performance of class imbalanced bankruptcy prediction models. Int. J. Electr. Eng. Inform. **10**(3), 433–446 (2018)
10. Veganzones, D., Séverin, E.: An investigation of bankruptcy prediction in imbalanced datasets. Decis. Support Syst. **112**, 111–124 (2018)
11. Sun, J., Lang, J., Fujita, H., Li, H.: Imbalanced enterprise credit evaluation with DTE-SBD: decision tree ensemble based on SMOTE and bagging with differentiated sampling rates. Inf. Sci. **425**, 76–91 (2018)

Leveraging Blockchain as a Solution for Security Issues and Challenges of Paperless E-Governance Application

Ambica Sethy and Abhishek Ray

Abstract DigiLocker, a government of India initiative, provides a free platform to the citizens of India to store and access important personal documents. This platform is implemented by using various open source technologies to deliver a mass solution and contributes back to the ever-growing Indian community. DigiLocker is a weapon for paperless E-governance. It is a platform to issue and verify all types of certificates and documents digitally. Thus gradually eliminate the use of physical documents. In this paper, we have proposed a framework to secure individual information using secured verification method of blockchain-based validation technique. The proposed framework ensures proper validation and verification with a unique hash key. Additionally, Argon2 encryption mechanism is used for securing the hash key in the proposed framework. This will enable a timeline for visibility of the uploaded documents by any authorized person/organization and reduce fraudulent use of documents.

Keywords Aadhaar authentication · Argon2id · Blockchain security · DigiLocker · Data analytics · Ethereum

1 Introduction

A citizen desires to cautiously keep files like Birth Certificate, Medical Documents, Passport, PAN (Permanent Account Number) Card, Voter ID Card, Ration Card, BPL (Below Poverty Line) Card, Degree Certificate, License, and all [1]. It is quite difficult for a person to keep all the relevant original hard copies with him/her, which he/she could provide at the time of requirement. In order to make

A. Sethy (✉) · A. Ray
School of Computer Engineering, Kalinga Institute of Industrial Technology, Deemed to be University, Bhubaneswar, India
e-mail: ambica.sethy@gmail.com

A. Ray
e-mail: arayfcs@kiit.ac.in

H. Das et al. (eds.), *Progress in Computing, Analytics and Networking*, Advances in Intelligent Systems and Computing 1119,
https://doi.org/10.1007/978-981-15-2414-1_65

it more handy with online a citizens should have some limited rights, such as uploading documents and use it for all type of verification purpose. For Indian citizens' empowerment, Govt. of India has taken multiple initiatives to bring rapid growth among the nation. To facilitate this, we need digital technologies to be used by citizens in day-to-day life. DigiLocker is one of the digital technologies which can empower Indian citizens for storage of personal and official documents like Birth Certificate, Medical Documents, Passport, PAN (Permanent Account Number) Card, Voter ID Card, Ration Card, BPL (Below Poverty Line) Card, Degree Certificate, Driving License, University certificates, and the e-documents issued by various issuer departments of government of India [2]. This DigiLocker application is so essential because it provides dedicated personal storage space, can reduce use of physical documents, provides authenticity of e-documents. Furthermore, citizens get access to the documents issued by government. DigiLocker can reduce administrative expenses of government departments and agencies and its easy for citizens to receive services through this digital technology. As currently UIDAI is connected with DigiLocker, the citizen can share personal details and data with agencies. But this process generates huge amount of Data taking into consideration the number of citizens of India and possible threat related to improper use of personal data, and it becomes essential to keep citizen data secure with authenticated secured access [3].

Moreover, the Department of Electronics and Information Technology (DeitY) is focused more on mobile governance and the implementation of DigiLocker to store Id and Certificates on public Cloud Storage can be device-friendly and capable of being accessed over smart Android-based mobile phones. Current DigiLocker Framework lags with Security Concerns as only Aadhaar Authentication is not sufficient for the mass Indian Community. Important thing concerning Aadhaar is that a centralized identification database containing demographic and biometric information of over one thousand million human beings is an apparent honey-pot for hackers. This is not always the only concern, although. All the information leaks that have been suggested within the beyond needed to do with how Aadhaar-based authentication basically works as Aadhaar information is attached to a 12-digit Aadhaar number. There is no doubt that scalability and storage capacity of the cloud will be a boost to the ever-growing databases of e-governance projects. We have outlined few developments in implementing security aspects of blockchain in the E-governance initiatives while in the proposed areas it still needs to be applied like DigiLocker is one of the applications [4].

1.1 Problem with Traditional System

Document storage with centralized services is at risk of information breaches and may even cause different attacks by intruder, in case the provider is compromised. Government has been more trust-able medium through which a user can reveal their personal information's or files without any threat fear. The files are also at risk of

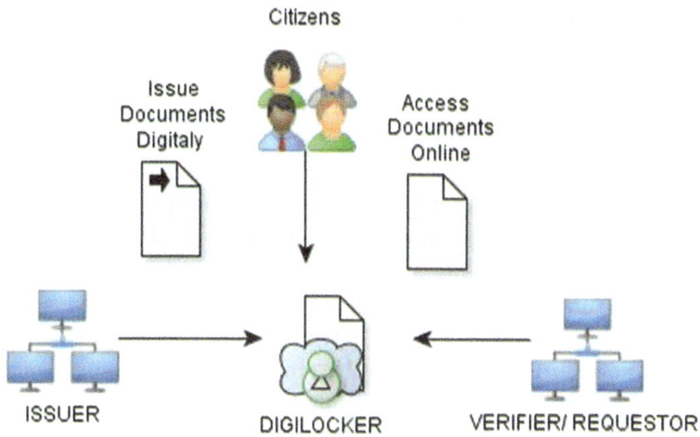

Fig. 1 Working of DigiLocker system

hacking from insiders who are able to acquire the protected key. Digital locker tool device in India is a great initiative taken by using India towards e-file and e-financial system. For the research study it will cause some auxiliary data, which is firmly used for specific purpose. Using these reasons the books, magazines, news papers and websites are get updated [5].

Aadhaar is the world's biggest sole identification system that holds all information of Indian residents. Aadhaar collects name, date of birth, gender, address of residents in India and stores them against the corresponding biometric data. With such massive amount of personal data stored in a centralized database, the anxiety was more as shown in Fig. 1.

This Aadhaar number does not hold any meaningful information itself but a service provider can use this Aadhaar number and authenticate against information provided by you to verify your identity.

Aadhaar User Agency (AUA) is the way to identify a person with their exact identity for that it requires the user to provide their private data to the AUA.

An AUA with a harmful target may keep a secret record of all individual's demographic and biometric details with Aadhaar number and commit fraudulent activities [6].

A person should not disclose their Aadhaar number to an AUA, but you will still be required to provide your personal information. So, keep in mind, while communicating about the safety worries of Aadhaar, we are speaking about security for the duration of the distinctive levels of implementation which includes the Aadhaar authentication way in addition to the centralized database. Because of the spacious procedure of the similar data of a person by different organizations to provide a variety of services to the perfect owner. To overcome these issues, some remedies have been found and there also some disadvantages in the existing solutions and they are discussed below.

2 Review of Related Literature

Ghodke et al. [7] proposed an architecture named RTO cloud server, where all required documents of the vehicle such as vehicle pollution under control certificate, registration certificate, and insurance policy are scanned and stored. The proposed method will facilitate in reduction of a precious amount of time. The main concept is to provide a QR-code to maintain privacy. Kalra et al. [8] proposed a practical implementation idea of providing all services by biometric system of individuals. This will include complete database linkage between various governmental services like Electricity Bill Payment, Water Bill Payment, Bank Services, etc. Today's new technology of bio-metric has helps a lot by reducing the time, space and speed of human life by applying one small fingerprint.

Mehra et al. [9] implemented a trusted framework which is founded on blockchain along with other technologies, which would enable philanthropic donations to be as accountable and transparent as direct charity. Kumar et al. [10] implement a framework for academic and government which is discussed in brief for the adoption of FOSS in business enterprise, where its adoption is dynamic and accelerated. Kumar et al. [11] proposed a framework and model using valid authentication process which is secure for the personal data. Availability of the documents verification also verified by the owner which is accessible for very less time. Ramya et al. [5] implemented blockchain multichain, the process is faster because proof-of-work is not required. Due to reduction in the forgery of documents in land registration system, use-case involves recording the documents into blockchain and verifying it with digital locker.

Table 1 represents comparisons between some of the best hashing algorithms like SHA-1, Bcrypt, Scrypt, and Argon2id. The parameters considered for comparison are salt (byte) (an arbitrary value added to key), key length, capability on brute force attack, and memory access. By observing Table 1, we have found that the performance of parameters are nearly same except Argon2id. Argon2id has an uniqueness, that is, to hash a password the user can decide the hashing time and memory required dynamically [12, 13].

Table 1 Comparison of hashing algorithms

Hashing algorithms	Salt (bytes)	Key length (bytes)	Can oppose Brute force attack	Data independency and memory access
SHA-1 [14, 15]	Nill	128	No	No
Bcrypt [16, 17]	16	128	Slow it	No
Scrypt [18]	16	128	Slow it	No
Argon2id [19, 20]	16	128	Slow it	Yes

3 Proposed Framework for Enhanced Blockchain-Based Security for DigiLocker App

This section presents the proposed framework for enhanced blockchain-based security for DigiLocker Application. The proposed framework is mainly divided into two modules such as: (i) Document Upload and (ii) Document Verification. The framework is shown in Fig. 2.

Working of the Proposed Framework

- **Document Upload** This module is responsible for uploading the correct document in the DigiLocker. While a user wants to upload their scanned document in the DigiLocker system, first the document will be verified by the certifying/issuing authority. After verification of the uploaded document's originality and its owner, it will tagged with a unique id by using users Aadhaar number and one-time password(OTP) [21]. Next, the administrator will forward these documents for digital signature of the certified authority and also attach the expiry date (if any) to some or all documents. Then, the documents will be stored using encryption technique using Argon2id in the blockchain security environment of the system.
  Once we connect our total verification system with blockchain technology, the system will automatically in secure hand. Blockchain is a chain of blocks that contain raw information, and is a distributed ledger. After applying the blockchain mechanism, all raw information are stored in a hash format, which are not reversible, only authentic person can access the information and the valid user who are under contract. After that, a key will be generated and that key is encrypted by key stretching technique Argon2 [22].

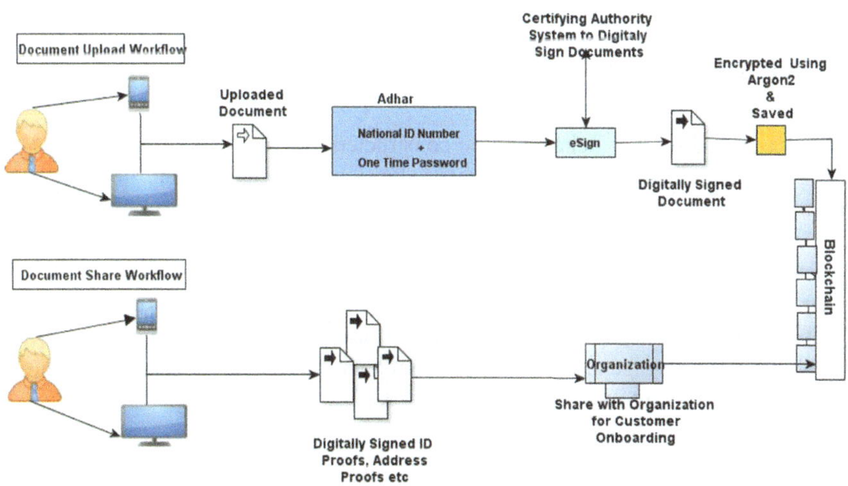

Fig. 2 Enhanced blockchain-based security for DigiLocker App

Argon2id is a key derivation function and designed for password hashing, with this control the GPU cracking attacks. After encrypted the documents the data are in distributed network of blockchained and in hash mode only the valid owner can see, access, extract data. It accesses the memory array in a password-independent order and password-established order, which reduces the possibility of time-memory exchange-off attacks. After applying Argon2, the first key is fed into the technique that outputs an enhanced key. It must be of sufficient length to make it infeasible to interrupt through brute force attack. The normal set of rules used ought to be at ease in the feeling that there needs to be no identified manner of taking a shortcut that would make it viable to calculate the enhanced key with less processor work than with the aid of the use of the key stretching algorithm itself. Argon2id is so secured for all user and create trust for system and society.

- **Document Verification** The verification system can work by sharing the documents to the valid authority or by showing the documents to them digitally also. The document can be verified by the legal authority who are under the platform of blockchain smart-contract rule.
  Using Ethereum concept of blockchain [23], the encryption keys are managed plus accessible only to the end user on their device. The user can likewise share the reports safely with different clients that they pick. An Ethereum smart-contract using SHA 256 encryption encourages this exchange and makes an unquestionable review trail of the equivalent which is more secure than current cloud-based file storage, and sharing with Aadhaar-based authentication mechanism that Government currently provides [23].
  The document can verified by using the exact hash key generated by the legal authority, which provides more authenticate to the system. Similarly, the documents can be shared to other authorized organizations also, they can take the profit of blockchain mechanism and get validated and authenticated user originality.

4 Benefits of Proposed Architecture

A decentralized digital locker may be constructed in a server-less environment by using Ethereum blockchain technology. It allows users to upload and store personal documents, that are encrypted and stored in a decentralized manner. Using Ethereum concept of blockchain, the encryption keys are maintained and accessible to only the registered user on their device. The user can also share the documents securely with other users that they choose. An Ethereum smart-contract facilitates this transfer and creates a verifiable audit trail of the same which is more secure than the current cloud-based file storage and sharing with Aadhaar-based authentication mechanism that can be provided to the environment.

5 Conclusion

In this paper, we have proposed a framework for securing individual information using secured verification method of blockchain-based validation technique. In the literature review section, it is observed that some security issues and memory access problems were raised in the encryption techniques SHA-1, Bcrypt, Scrypt. Considering this concern we have applied the Argon2id encryption method with blockchain technique to verify and validate the documents. Our proposed framework will ensure better security by maintaining secure parameters, data independency, memory accessing with slow hashing mechanisms according to the users requirement.

References

1. The "DigiLocker": A Digital India Initiative. https://digitallocker.gov.in
2. Vohra, S.M.: Digital India and its impact on Indian economy. SEMCOM Manag. Technol. Rev. (2018)
3. Bandgar, V.B.: Government initiatives for digital India. J. Adv. Libr. Sci. **6**(1), 328–335 (2019)
4. Rathore, S., Panwar, A.: Digital-locker services in India: an assessment of user adoption and challenges. In: Leveraging Digital Innovation for Governance, Public Administration, and Citizen Services: Emerging Research and Opportunities, pp. 101–131. IGI Global (2020)
5. Ramya, U.M., Sindhuja, P., Atsaya, R.A., Dharani, B.B., Golla, S.M.V.: Reducing forgery in land registry system using blockchain technology. In: International Conference on Advanced Informatics for Computing Research, pp. 725–734. Springer, Singapore (2018)
6. Pal, S.K.: Changing technological trends for E-governance. In: E-governance in India, pp. 79–105. Palgrave Macmillan, Singapore (2019)
7. Ghodke, A.V., Dagade, R.V.: Electronic secure vehicle verification system using advanced digilocker system. In: 2018 3rd International Conference for Convergence in Technology (I2CT), pp. 1–4. IEEE (2018)
8. Kalra, R., Verma, R., Nasa, C.: Aadhaar Thumb: one Platform to all Services (2018)
9. Mehra, A., Lokam, S., Jain, A., Sivathanu, M., Singanamalla, S., ONeill, J.: Vishrambh: trusted philanthropy with end-to-end transparency. In: HCI for Blockchain: a CHI 2018 Workshop on Studying, Critiquing, Designing and Envisioning Distributed Ledger Technologies, Montreal, QC, Canada (2018)
10. Kumar, R., Kumar, S., Tiwari, S.K.: Adoption of free and open source software in India. Int. J. Appl. Eng. Res. **13**(16), 12725–12731 (2018)
11. Kumar, V., Chaturvedi, A., Dave, M.: A solution to secure personal data when Aadhaar is linked with DigiLocker. Int. J. Comput. Netw. Inf. Secur. **10**(5), 37–44 (2018)
12. Valois, M., Lacharme, P., Le Bars, J.M.: Performance of password guessing enumerators under cracking conditions. In: IFIP International Conference on ICT Systems Security and Privacy Protection, pp. 67–80. Springer, Cham (2019)
13. Nielson, S.J., Monson, C.K.: More symmetric crypto: authenticated encryption and kerberos. In: Practical Cryptography in Python, pp. 249–292. Apress, Berkeley, CA (2019)
14. Biryukov, A., Dinu, D., Khovratovich, D.: Argon2: new generation of memory-hard functions for password hashing and other applications. In: 2016 IEEE European Symposium on Security and Privacy (EuroS and P), pp. 292–302. IEEE (2016)
15. Aggarwal, A., Chaphekar, P., Mandrekar, R.: Cryptanalysis of bcrypt and SHA-512 using distributed processing over the cloud. Int. J. Comput. Appl. **128**(16) (2015)
16. Malvoni, K., Knezovic, J.: Are your passwords safe: energy efficient bcrypt cracking with low-cost parallel hardware. In: 8th USENIX Workshop on Offensive Technologies (2014)

17. Wiemer, F., Zimmermann, R.: High-speed implementation of bcrypt password search using special-purpose hardware. In: 2014 International Conference on ReConFigurable Computing and FPGAs (ReConFig14), pp. 1–6. IEEE (2014)
18. Choe, J., Moreshet, T., Bahar, R.I., Herlihy, M.: Attacking memory-hard scrypt with near-data-processing (2019)
19. Josefsson, S.: The memory-hard Argon2 password hash function. memory (2015)
20. Maliberan, E.V.: Modified SHA1: a hashing solution to secure web applications through login authentication. Int. J. Commun. Netw. Inf. Secur. **11**(1), 36–41 (2019)
21. Kumar, V., Chaturvedi, A., Dave, M.: A solution to secure personal data when Aadhaar is linked with DigiLocker. Int. J. Comput. Netw. Inf. Secur. **10**(5), 37–44 (2018)
22. Wetzels, J.: Open sesame: The password hashing competition and Argon2 (2016). arXiv:1602.03097
23. VS, R., Raj, V.C., Eswaran, S., RU, S.: Optimization of digitalized document verification using e-governance service delivery platform (e-SDP). Int. J. Appl. Eng. Res. **11**(4), 2531–2539 (2016)

Maize Leaf Disease Detection and Classification Using Machine Learning Algorithms

Kshyanaprava Panda Panigrahi, Himansu Das, Abhaya Kumar Sahoo and Suresh Chandra Moharana

Abstract Plant diseases are the major cause of low agricultural productivity. Mostly the farmers encounter difficulties in controlling and detecting the plant diseases. Thus, early detection of these diseases will be beneficial for farmers to avoid further losses. This paper focuses on supervised machine learning techniques such as Naive Bayes (NB), Decision Tree (DT), K-Nearest Neighbor (KNN), Support Vector Machine (SVM), and Random Forest (RF) for maize plant disease detection with the help of the images of the plant. The aforesaid classification techniques are analyzed and compared in order to select the best suitable model with the highest accuracy for plant disease prediction. The RF algorithm results with the highest accuracy of 79.23% as compared to the rest of the classification techniques. All the aforesaid trained models will be used by the farmers for the early detection and classification of the new image diseases as a preventive measure.

Keywords Classification · Machine learning · Maize leaf disease prediction · Naive Bayes · KNN · Decision tree · Support vector machine · Random forest

1 Introduction

Agriculture is not only the main support and the major sector of the Indian economy but also an important source of income for Indian people. If agriculture goes wrong, nothing else will have a chance to go right in the country. The demand for food is growing exponentially as the production of agronomy is far too low. To overcome this, farmers, scientists, researchers, analysts, specialists, and government try to place further effort and strategies to increase agricultural production to accommodate the

K. P. Panigrahi · H. Das (✉) · A. K. Sahoo · S. C. Moharana
School of Computer Engineering, KIIT Deemed to be University, Bhubaneswar, Odisha, India

K. P. Panigrahi
e-mail: 1850006@kiit.ac.in

S. C. Moharana
e-mail: scmoharanafcs@kiit.ac.in

H. Das et al. (eds.), *Progress in Computing, Analytics and Networking*,
Advances in Intelligent Systems and Computing 1119,
https://doi.org/10.1007/978-981-15-2414-1_66

needs. There are enormous advancements in the agriculture sector with the help of technology. There are various factors such as global climate change, and plant diseases for which farmers are facing problems. Several reasons for the loss of crop production that leads to the suicidal cases of the farmers. The consumption of time, cost and accuracy for the assessment of the quality of the crops through visual inspection is a challenging task. To overcome this problem researchers had come with several solutions through the development of new technologies such as object detection, and image processing for the quality assessments. In this paper, image processing technology is used for the detection and classification [1–3] of diseases in crops. This image processing technique requires high-resolution images for detection and classification of diseases which was difficult to capture. Due to this reason, it is also a tedious task to predict diseases effectively and most accurately. This paper intends to develop a model that accurately detect and classify the diseases of leaves at the early stage of diseases by using machine learning algorithms [4] and take necessary steps to prevent from such leaf diseases. This paper uses the various supervised machine learning classification techniques such as NB [5], KNN [6, 7], DT [8], SVM [9], and RF [10, 11] for disease detection and classification from plant leafs and also a comparison is made among the several classification techniques. It also provides different techniques which will provide the most accurate result as compared to other techniques. These classification techniques are successfully applied in several applications [12] such as biomedical signal processing [13] and healthcare [14, 15].

In this paper, the maize (Zea mays) plant has been considered for the analysis of various diseases of leaves. In India, the maize is the most important food crops along with rice and wheat. It provides a major source of carbohydrate for human beings. Apart from this, it is planted for cooking oil, animal food, flour, and also used as the raw materials for making furfural. Maize is cultivated in several states of India such as Maharashtra, Andhra Pradesh, Tamil Nadu, Bihar, Karnataka, Madhya Pradesh, Uttar Pradesh, Gujarat, and Rajasthan. Generally, maize is harvested in the autumn month and is sown at the beginning of the monsoon season.

The major causes of maize diseases can be due to the biotic (bacteria, fungi, nematodes, and viruses) and abiotic (due to the effects of nutrient deficiency, humidity, and temperature). The pathogens of maize mostly attack leaves, fruits, and stalk. Here, some of the major diseases are (a) Cercospora leaf spot or gray leaf spot presented in Fig. 1. It is a small necrotic spot on leaves that expand in the rectangular lesions as the lesion matures. They get tanned and finally become gray. Managements are hybrid seed use with resistance to the disease and the use of foliar fungicides. (b) Common Rust presented in Fig. 2 is the presence of brown pustules. It is available both the upper and the lower surfaces of the leaves when pustules the rupture and the release powdery red spores. When there is a severe infection in the pustules that may appear on tassels and ears and leaves that begins to yellow color. Managements are the use of resistant hybrids and the use of Fungicide sprays such as folicur, oxychloride, AMISTAR copper, and Bravo (c) Northern Leaf Blight shown in Fig. 3 is the elliptical gray–green lesions on leaves, in the beginning, can be noticed. As

Fig. 1 Cercospora leaf spot

Fig. 2 Common rust

Fig. 3 Northern leaf blight

the pathogens process, these lesions become pale gray color to tan color. Managements are the follow-up crop rotation with non-host crop. The remainder section of the paper is organized as follows: Sect. 2 describes the related work of maize disease classification techniques, Sect. 3 presented a detailed explanation about several machine learning techniques, Sect. 4 presents the detailed procedure of the proposed method for maize leaf disease classification, Sect. 5 presents the comparative result analysis of all the classification techniques and finally Sect. 6 concludes the paper with the future scope of the work.

2 Related Work

Maize commonly known as corn is one of the key crops which is very versatile under several climatic situations. The disease of the maize can be on several parts such as stem, leaf or panicle. In this section, we have considered only the diseases that are related to the leaves. Ishaket et al. [16] presented the artificial neural network (ANN) model for classification of Phyllanthus Elegant Wall leaf diseases into two classes such as healthy or unhealthy. They have transformed the color structure of herb plant images by using image processing techniques. The images are classified based on the color and area of the leaf. Padol et al. [17] described the used linear SVM for classification of the leaf diseases. The preprocessing techniques are applied to the input images of the grapes and the disease regions. Those are detected with the help of the clustering algorithms from which the color and texture information are extracted. Mohanty et al. [18] used convolution neural network (CNN) for leaf disease detection. They have used large dataset images that are consist of healthy plant leaves and also affected in disease. They have experimented with three versions of datasets such as colored, grayscaled, and segmented. This CNN model is able to identify the 26 diseases with 14 crop species easily. Similarly, Dandawate et al. [19] detected the diseases of soybean leaf using SVM. This algorithm used the scale-invariant feature transform technique that automatically detected the plant diseases based on its shape. This assists the farmer with minimal efforts over the internet. Singh et al. [20] used a genetic algorithm for automatic leaf disease classification. The input image is preprocessed and segmented using a genetic algorithm to classify the diseases. The authors observed that the optimal result was obtained with a less computational cost. The author has recommended the use of fuzzy logic, ANN, and hybridization of several algorithms for the improvement of the recognition rate. Patil et al. [21] extracted the features of the tomato leaf. The leaf image segregated into red, green, and blue components. These features are used for the classification of diseases. Ghadge et al. [22] focused on the assistance of the farmers in suitable crop production based on soil quality. Machine learning algorithms are used for the prediction of the crops. The presence of nutrients in the soil is analyzed and predicted the production of the crops in a particular location. Hong et al. [23] proposed a model for the development of precision in the agriculture field. The prediction of soil moisture was developed to predict the moisture based on environmental situations. This prediction results in more accuracy for a long period of time. Dahikar et al. [24] used ANN for crop prediction based on parameters like nitrogen, potassium, temperature, PH, and rainfall. This suggests an appropriate fertilizer.

3 Machine Learning Algorithms for Classification

Classification is a supervised learning approach that maps the data into certain class labels in the twofold process: (1) in the learning phase (training step), the classification model is modeled that presents the predefined set of classes. Here, the classification algorithms are modeled by the classifier using the learning algorithm of the data with respect to their specific class labels and (2) the trained model that is created in the first step is used for the classification of the data. The test data are used to compute the performance of the trained model by considering several parameters such as accuracy, precision, recall, and f-measure. The detailed description of different types of classification models such as NB, K-NN, DT, SVM, and RF are used for analysis is described below.

3.1 Naive Bayes (NB)

It is a variant of a probabilistic classifier based on the principle of Bayes classifier. It assumes the presence of the prior probabilities of the patterns are know and the posterior probabilities are assigns to the class labels. With this hypothesis, the posterior probability calculates the maximum likelihood values of the data that are belongs to a particular class label. It is computed by using the product of the conditional probability of each feature by using Baye's theorem. Though this hypothesis typically does not hold in real-life environment it is pretty successful in numerous classification tasks.

3.2 The K-Nearest Neighbors (KNN)

It is a supervised machine learning, nonparametric algorithm typically used in pattern recognition. It is based on the principle of nearest neighbor rule used in the classification process for the machine learning tasks. In this technique, the classifier is used to train the pattern to classify the test pattern based on the similarity between the test pattern with every training pattern. The outcome of the k-NN classifier is a class membership value that it belongs to. It is classified based on the plurality vote of the neighbors with the object being assigned to the most commonly used class labels among its k-nearest neighbors. It behaves as a type of instance-based learning in which the operations are locally approximated and all the computations have differed till the end of the classification process.

3.3 Decision Tree (DT)

It is a supervised classification and regression algorithm in supervised learning that builds the classifiers by dividing the data into several smaller groups (tree structure) based on which division constructs the higher disproportion. The Gini index or entropy is one of the commonly used attribute selection measures that are typically used as disparity measures. The advantage of this algorithm is the interpretation of results could be easy for human beings. A DT can produce very less amount of training error if the tree could have learned without any constraint of the depth of the tree. There are different variants of decision trees such as ID3, C4.5, CART are most frequently used in different applications of data mining and machine learning.

3.4 Support Vector Machine (SVM)

It is a supervised machine learning classifier defined by the separating hyperplane. This algorithm finds an optimal hyperplane that maximizes the margin between the data points of both the classes in high dimensional space. The property of SVM is called kernel tricks that are useful for nonlinear classification. It is very expected to obtain more distinguishable features in the high dimensional feature space. It can be completed by transforming the features using several general functions such as linear, polynomial and radial basis function. The transformation of features could considerably increase the dimensions of feature space. Hence, it increases the training time of the classification process. It could transform the features into the higher proportions by computing the dot products without transforming the feature set.

3.5 Random Forest (RF)

It is an ensemble of randomized decision tree classifiers learning methods. It is operated by constructing multiple decision trees at the training time. The class labels of the testing dataset are measured based on the voting of each classification tree. The outcome of the classifier depends on the class labels that have the maximum voting by the classification trees. This algorithm uses bagging and randomness of features during building of each individual tree and tries to create an uncorrelated forest of trees that will predict the performance more accurately than that of the individual tree.

4 Proposed Technique for Disease Classification

This section describes the working principle and the detailed architecture of the proposed technique for the classification of maize diseases in Fig. 4. The proposed method has several components such as image acquisition, image preprocessing, image segmentation, feature extraction, classification, and performance evaluation. The detailed explanation of each process are as follows.

4.1 Image Acquisition

The image dataset, particularly for maize disease pictures, is available at the plant village website. Maize plants are the subsets that have the total number of 3823 images and four class labels of diseases such as common rust, gray leaf spot, northern leaf blight and healthy having 1192 images, 513 images, 956 images, and 1162 images respectively. These labeled images are considered for the training and testing of the disease classification.

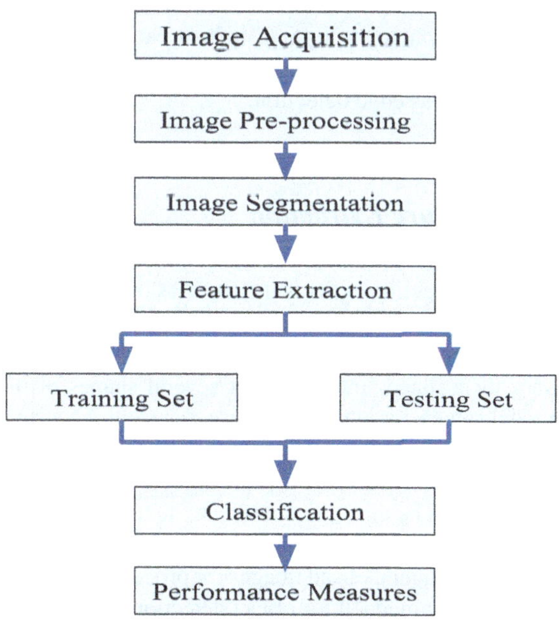

Fig. 4 Block diagram of the classification process

4.2 Image Preprocessing

The image preprocessing is necessary for the realization of the superior results in consequence steps due to the presence of dewdrops, dust, insect excrements on the plants. These effects are considered as the noise of the maize image. To overcome these problems the input RGB photo is transformed into a grayscale image to provide accurate results. In this case, the size of the pictures is very large for which the reduction in the image size is necessary. This image reduction is also useful to reduce memory size.

4.3 Image Segmentation

Image segmentation plays a crucial role in plant disease detection and classification. It simply divides the image into several objects or regions. It analyzes the image data to extract useful information for further processing. This image segmentation can be carried out in two ways based on similarities and discontinuities. In similarities, the images are partitioned based on some specific predefined criteria. Therefore, the label edge detection method is used in image segmentation and also it calculates the gradient of photograph intensities at each pixel within the image. But in discontinuities, the images are partitioned based on the sudden changes in the intensity of values such as edge detection.

4.4 Feature Extraction

Feature extraction extracts the features of the objects that are present in the images. These extracted features are used to illustrate an entity. These features extracted and categorized into three categories such as shape, color, and texture. The diseases may vary their shapes into different several shapes of the image due to diseases. The model can easily identify the diseases from the shape of the features. These shapes of the features vary in their axis, areas, and angles. The second parameter, i.e., color is an important feature of these three features. It differentiates the diseases from each other. The third parameter, i.e., texture describes how the patterns of the color are sprinkled in the images. RGB feature extraction extracts the color information from the frequently used images for processing and identification of patterns. RGB is highly recommended for object detection in the image. It has the significant change in color that easily identifies the images in the leaves. The value of RGB color can determine all probable colors that can be made from the three colored lights such as red, green, and blue. The standard value of RGB varies from 1 to 255 and the tasks are normalized in the range of 0–1. This experiment considers the grayscale pixel values as features for analysis.

5 Experimental Result Analysis

This section investigates the performance of different classification techniques such as NB, KNN, DT, SVM, and RF on maize disease detection dataset and found that the RF classification technique is superior to the rest of the classification techniques. This maize datasets are split into training data (90%) and testing data (10%). Maize plant disease dataset contains a total of 3.823 images and with four class labels. The details information about the class label information about the maize disease dataset is as follows: gray leaf spot, common rust, northern leaf blight, and healthy are 513, 1192, 985, and 1162, respectively. The implementation of these classifiers is made by using Python 3.3 running on Windows 7 operating system. An x64 based processor with a speed of 3.20 GHz, RAM of size 8.00 GB and system type of 64-bit operating system are used. Python software along with python machine learning library and package of Pandas is used. In this experiment, the size of the image is reduced from the original size to 100 × 100. The reason is that all the images have a unique size as the data set images were having different sizes. The labeled grayscale images are converted into a cv2 format image processing library in python. The formatted images are feed into a pickle so that this transformation can be performed again and again. After this step, this pickle file can be called into any of the classification algorithms. Then the codes of the particular models are implemented and the data is trained on the model and this model predicts the disease of an image accordingly. The classification accuracy is presented in Table 1 and Fig. 5. The other performance measures such as precession, recall, and F-measure of all the classification algorithms for maize disease datasets are presented in Fig. 6. We tend to classify whether the leaves are diseased or healthy and if it is affected with disease what type of disease

Table 1 Comparison between classification techniques

Classifiers	Accuracy (%)
SVM	77.56
NB	77.46
KNN	76.16
DT	74.35
RF	79.23

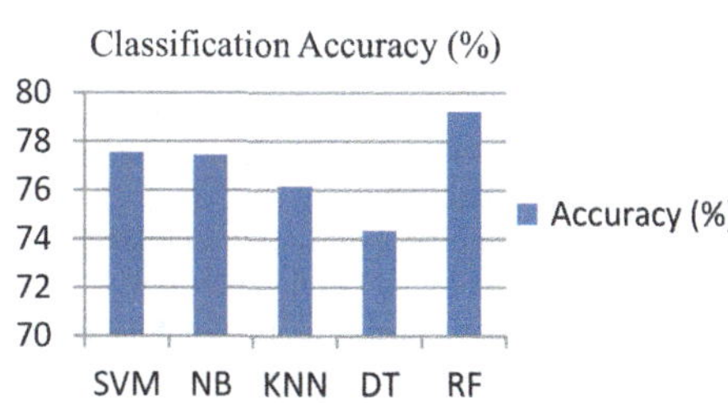

Fig. 5 Classification accuracy of classifiers

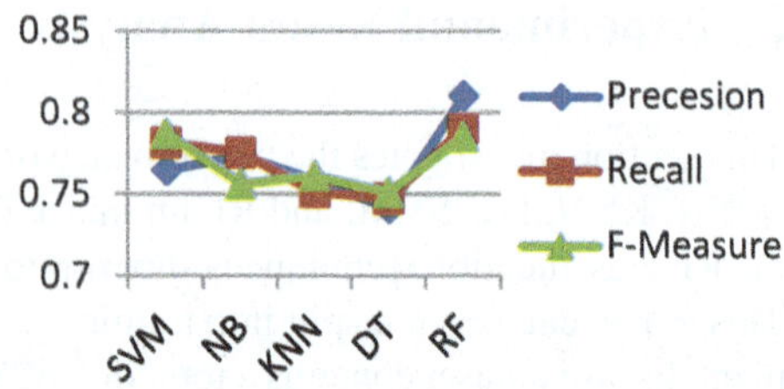

Fig. 6 Precision, recall, and F-measure of classifiers

it comes under so that the farmers will take the right action on specific diseases in early.

6 Conclusion

In this work, the supervised machine learning techniques namely NB, KNN, DT, SVM, and RF are used to detect the various maize leaf diseases. The proposed methodology has been applied by using labeled image data to train the classification model. It is observed that the highest accuracy recorded in the RF classifier among the rest of the classification models for disease detection in testing image data. The farmers can take necessary actions based on detected diseases at the earliest to avoid maize diseases. But there are some pitfalls are associated with each and every model in the classification process that may not be applicable for all the datasets. In the future, these models can be implemented by using several high-dimensional data sets with several other classification methods.

References

1. Das, H., Naik, B., Behera, H.S.: Classification of diabetes mellitus disease (DMD): a data mining (DM) approach. In: Progress in Computing, Analytics and Networking, pp. 539–549. Springer, Singapore (2018)
2. Sahani, R., Rout, C., Badajena, J.C., Jena, A.K., Das, H.: Classification of intrusion detection using data mining techniques. In: Progress in Computing, Analytics and Networking, pp. 753–764. Springer, Singapore (2018)
3. Das, H., Jena, A. K., Nayak, J., Naik, B., Behera, H.S.: A novel PSO based back propagation learning-MLP (PSO-BP-MLP) for classification. In: Computational Intelligence in Data Mining, vol. 2, pp. 461–471. Springer, New Delhi (2015)
4. Murty, M.N., Devi, V.S.: Pattern Recognition: an Algorithmic Approach. Springer Science & Business Media (2011)
5. Rish, I.: An empirical study of the naive Bayes classifier. In: IJCAI 2001 Workshop on Empirical Methods in Artificial Intelligence, vol. 3, no. 22, pp. 41–46. IBM, New York (2001)
6. Fix, E., Hodges Jr, J.L.: Discriminatory Analysis-Nonparametric Discrimination: consistency Properties. California Univ Berkeley (1951)
7. Cover, T., Hart, P.: Nearest neighbor pattern classification. IEEE Trans. Inf. Theory **13**(1), 21–27 (1967)

8. Quinlan, J.R.: Induction of decision trees. Mach. Learn. **1**(1), 81–106 (1986)
9. Cortes, C., Vapnik, V.: Support-vector networks. Mach. Learn. **20**(3), 273–297 (1995)
10. Ho, T.K.: Random decision forests. In: Proceedings of the Third International Conference on Document Analysis and Recognition, vol. 1, pp. 278–282. IEEE (1995)
11. Barandiaran, I.: The random subspace method for constructing decision forests. IEEE Trans. Pattern Anal. Mach. Intell. **20**(8) (1998)
12. Pattnaik, P.K., Rautaray, S.S., Das, H., Nayak, J.: Progress in computing, analytics and networking. In: Proceedings of ICCAN, p. 710 (2017)
13. Pradhan, C., Das, H., Naik, B., Dey, N.: Handbook of Research on Information Security in Biomedical Signal Processing, pp. 1–414. IGI Global, Hershey, PA (2018)
14. Sahoo, A.K., Mallik, S., Pradhan, C., Mishra, B.S.P., Barik, R.K., Das, H.: Intelligence-based health recommendation system using big data analytics. In: Big Data Analytics for Intelligent Healthcare Management, pp. 227–246. Academic Press (2019)
15. Dey, N., Das, H., Naik, B., Behera, H.S. (eds.).: Big Data Analytics for Intelligent Healthcare Management. Academic Press (2019)
16. Ishak, S., Rahiman, M.H.F., Kanafiah, S.N.A.M., Saad, H.: Leaf disease classification using artificial neural network. J. Teknologi, **77**(17) (2015)
17. Padol, P.B., Yadav, A.A.: SVM classifier based grape leaf disease detection. In: 2016 Conference on Advances in Signal Processing (CASP), pp. 175–179. IEEE (2016)
18. Mohanty, S.P., Hughes, D.P., Salathé, M.: Using deep learning for image-based plant disease detection. Front. Plant Sci. **7**, 1419 (2016)
19. Dandawate, Y., Kokare, R.: An automated approach for classification of plant diseases towards development of futuristic decision support system in Indian perspective. In: 2015 International Conference on Advances in Computing, Communications and Informatics (ICACCI), pp. 794–799. IEEE (2015)
20. Singh, V., Misra, A.K.: Detection of unhealthy region of plant leaves using image processing and genetic algorithm. In: 2015 International Conference on Advances in Computer Engineering and Applications, pp. 1028–1032. IEEE (2015)
21. Patil, J.K., Kumar, R.: Color feature extraction of tomato leaf diseases. Int. J. Eng. Trends Technol. **2**(2), 72–74 (2011)
22. Ghadge, R., Kulkarni, J., More, P., Nene, S., Priya, R.L.: Prediction of crop yield using machine learning. Int. Res. J. Eng. Technol. (IRJET), **5** (2018)
23. Hong, Z., Kalbarczyk, Z., Iyer, R.K.: A data-driven approach to soil moisture collection and prediction. In: 2016 IEEE International Conference on Smart Computing (SMARTCOMP), pp. 1–6. IEEE (2016)
24. Dahikar, S.S., Rode, S.V., Deshmukh, P.: An artificial neural network approach for agricultural crop yield prediction based on various parameters. Int. J. Adv. Res. Electron. Commun. Eng. (IJARECE) **4**(1) (2015)

The manufacturer's authorised representative in the EU is Springer Nature Customer Service Centre GmbH, Europaplatz 3, 69115 Heidelberg, Germany. If you have any concerns regarding our products, please contact ProductSafety@springernature.com

Printed and bound by CPI Group (UK) Ltd, Croydon, CR0 4YY
30/09/2026
02225509-0017